MATHEMATICS

A Cultural Approach

This book is in the
ADDISON-WESLEY SERIES IN MATHEMATICS

MATHEMATICS
A Cultural Approach

by

MORRIS KLINE
New York University

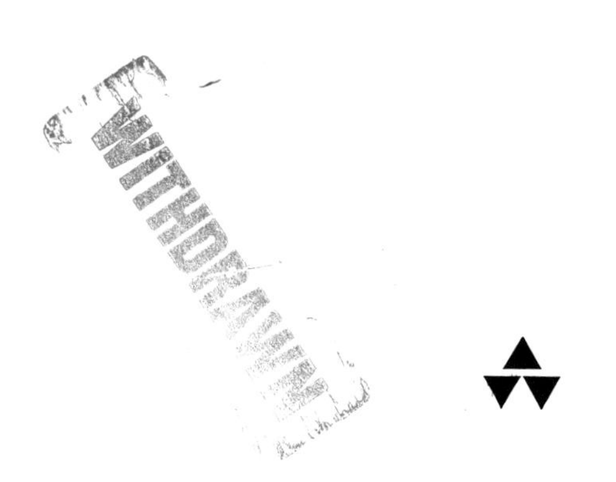

ADDISON-WESLEY PUBLISHING COMPANY, INC.
READING, MASSACHUSETTS · PALO ALTO · LONDON

Printed in the United States of America

Library of Congress Catalog Card No. 61-10970

Second Printing, June 1963

PREFACE

This text attempts to show what mathematics is, how mathematics has developed from man's efforts to understand and master nature, what the mathematical approach to real problems can accomplish, and the extent to which mathematics has molded our civilization and our culture. Though the essence of the material presented is mathematics, through motivations, applications, and implications the subject is shown to be intimately related to physical science, philosophy, logic, religion, literature, the social sciences, music, painting, and other arts.

The book is primarily intended for one-year terminal courses addressed to liberal arts students. It should serve in the training of teachers of elementary mathematics and of secondary-school teachers in nonmathematical subjects. It may also be helpful in giving high-school students some perspective on mathematics and in illustrating the subject's relevance to other fields. Indeed even the prospective specialist would do well to learn something *about* his subject before deciding to devote his life to it.

The content of this book ranges over elementary mathematics with the emphasis on ideas. No techniques are taught for the sake of techniques because technique *per se* is worthless knowledge and is not utilized later in life by the nonprofessional. The proofs of trigonometric identities are not important, but the fact that trigonometry has given man his understanding of the heavens is significant. Let us cease teaching scales to students who do not intend to play mathematical sonatas. In general, mathematical literacy, by which I mean understanding, is worth far more than technical proficiency. Hence there are very few exercises calling for drill, and the selection of exercises is, in general, based on the principle that a few, well chosen and thoroughly explored, are worth more than a thousandfold hastily done and poorly understood. To avoid mechanical imitation of stock procedures, any examples which may serve as a guide to the exercises are made an integral part of the text in the hope that students will be encouraged to read it, a practice by no means common among students of mathematics.

The material can be handled in various ways without loss of continuity. The first two chapters are straightforward reading matter, which, if desired, can be left to the student. The technical material can be started at several places. For students who have had recent training in arithmetic, algebra, and plane geometry, a course could begin with Chapter 7 or the latter part of Chapter 6. The chapters on philosophy, religion, literature, and the deductive approach to the social sciences can be left for student reading if time does not permit their inclusion or if the teacher believes that the material is best treated as reading assignments. Many of the mathematical chapters, such as the ones on projective geometry, non-Euclidean geometry, arithmetics and their algebras, and the chapters on trigonometric functions are logically independent. The material in the two chapters on the calculus is not used in the subsequent text.

Some teachers may be concerned that to teach the chapter on philosophy, for example, they should be experts in philosophy. But no mathematics teacher should be ashamed to admit the contrary. We are all laymen outside the field of our own specialty and we may frankly admit to our students that all we can do and all that the text is trying to do is to give some indication of the cultural influences of mathematics.

The present book borrows many themes from my previous writings, *Mathematics in Western Culture* and *Mathematics in the Physical World.* This is, of course, natural since both volumes stress basic values of mathematics. However, *Mathematics: A Cultural Approach* was specially written for text purposes and with the needs of young students in mind.

A few words may be in order to explain the philosophy underlying the approach used in this book. Its central tenet is that knowledge is not additive but an organic whole and that mathematics is an inseparable part of that whole. In Alfred North Whitehead's words, "There is only one subject matter for education, and that is Life in all its manifestations." The plan then is to approach knowledge through mathematics.

This objective is in direct opposition to the present practice not only of presenting mathematics as an abstract science isolated from all other branches of knowledge, but of teaching distinct subjects—algebra, geometry, trigonometry, and so forth—in the futile hope that students will see the interrelationships of the various branches and the importance of mathematics for other domains. Whitehead points out, "we offer children—Algebra, from which nothing follows; Geometry, from which nothing follows; Science, from which nothing follows, ..." To present branches of a subject in isolation from one another and from other domains of thought is to give students incomplete pieces of a most complicated jigsaw puzzle and to expect them to put the pieces together.

It is true that for various pedagogical reasons we must break down knowledge into separately taught subjects, but this compartmentalization should be compensated for as much as possible. We must overcome the disconnectedness which kills the vitality of our curricula. To help students see the interrelationships of the various branches of knowledge is the greater wisdom.

The thesis that we should teach mathematics as an approach to man's physical, social, and artistic worlds has a converse which is equally valid. The full significance of mathematics can be taught and appreciated only in terms of its intimate relationships to other branches of our culture. To separate mathematics from other human interests and endeavors is to present a hollow shell and leads to a perversion of the subject. To the extent that mathematics is isolated from other fields it loses importance. Abstractions taught independently of the totality from which they are abstracted are useless and meaningless. Of course, the ideal of a deductive structure is to be found in mathematics and is presented here. But this concept is one of method rather than content. It is also true that there is beauty in mathematics proper, but this is hardly to be found in most of the material that one must teach at the elementary level.

It may be helpful to discuss briefly the presentation of the mathematics itself. Physical problems often precede the introduction of a mathematical theme. The objective is not merely to provide motivation but also to illustrate in advance the very meaning of the mathematics to be discussed. From these physical problems one discovers what the mathematical procedure is intended to accomplish and one is then in a far better position to develop the relevant mathematical concepts. Thus the formation of concepts, the determination of the goals of the mathematical activity, and often even the method of proof are derived from physical and intuitive settings. Very little attempt has been made to provide formal, complete definitions or to supply what mathematicians would regard as rigorous proofs. The final formulation of a mathematical definition is normally of little meaning to anybody but the well-trained mathematician, and proofs rigorous from the standpoint of the professional mathematician go so far beyond the needs recognized by young people as to bewilder them and cause them to lose sight of the essential problem. (I say nothing about the fact that there is no absolute rigor.) Moreover the logic of discovery is far more exciting than the logic of the discovered.

Mathematicians seem to have forgotten that for over two thousand years Euclid's geometry was a model of rigor. Not even the greatest mathematicians observed its deficiencies. Similarly the foundation of the real number system was not created until the latter part of the nineteenth century. Neither Euler nor Gauss could have defined a real number, and it is unlikely that they would have enjoyed the gory details. But both managed to understand mathematics and to make a "fair" number of contributions to the subject. Rigorous proof is not nearly so important as proving the worth of what we are teaching; and most teachers, instead of being concerned about their failure to be sufficiently rigorous, should really be concerned about their failure to provide a truly intuitive approach. To paraphrase Pascal, rigorous proof is the slow and tortuous method by which those who do not know the truth discover it. The general principle, then, is that the rigor should be suited to the mathematical age of the student and not to the age of mathematics.

I have avoided symbolism wherever it is not a real help or a necessity. Symbolism burdens the memory and is often a bar to understanding. It has unfortunately become a fetish in mathematics, and mathematicians now erroneously believe that an idea expressed in symbols is somehow more valuable and more lofty than the same thought expressed in words. Nothing is more horrible in current texts than the misapplication of the symbolism of symbolic logic.

The arguments given for the material, approach, and goals of a course based on the present book are not intended to imply that there is one best course for all students. There is no doubt that some students will respond to mathematical challenges and not be too concerned with ultimate significance. Others will work for grades and be content if they earn them, no matter what they learn. But several years of experience with this very material and many years of experience with special courses for liberal arts students have convinced me that most students will respond to a cultural approach because it shows that mathematics has a point, whereas any appeal to intellectual discipline—a moot argument at best—or to the beauty of mathematics, or to the pleasures of mental activity is met with condescending scorn. For most students mathematics is not self-justifying. The mathematician must not presume that what interests him necessarily interests others. It is my conviction that it is the failure of the students to see the full significance of mathematics that has caused them to dislike it, do poorly in it, deprecate its value, and shrink from further involvement. On the other hand, if we do succeed in interesting students in our subject, we may get them to appreciate its values as a discipline, an art, and an engaging intellectual activity.

The present material is somewhat more difficult for teacher and student. It is easier for a teacher to present techniques and for a student to imitate them. The teacher must decide whether he wishes to go through perfunctory presentations of hackneyed and, for most students, irrelevant routines, or work hard with some worth-while material.

I wish to express my obligation to members of the Addison-Wesley staff for their receptiveness to the idea of a broad cultural text and to my wife for a vast amount of help in the editorial process.

M. K.

CONTENTS

"... I consider that without understanding as much of the abstruser part of geometry, as Archimedes or Apollonius, one may understand enough to be assisted by it in the contemplation of nature; and that one needs not know the profoundest mysteries of it to be able to discern its usefulness. ... I have often wished that I had employed about the speculative part of geometry, and the cultivation of the specious [symbolic] algebra I had been taught very young, a good part of that time and industry that I spent about surveying and fortification. ..."

ROBERT BOYLE

CHAPTER 1

WHY MATHEMATICS?

In mathematics I can report no deficience, except it be that men do not sufficiently understand the excellent use of the Pure Mathematics. . . .
Francis Bacon

One can wisely doubt whether the study of mathematics is worth while and can find good authority to support him. As far back as about the year 400 A.D., St. Augustine, Bishop of Hippo in Africa and one of the great fathers of Christianity, had this to say: "The good Christian should beware of mathematicians and all those who make empty prophecies. The danger already exists that the mathematicians have made a covenant with the devil to darken the spirit and to confine man in the bonds of Hell." Perhaps St. Augustine, with prophetic insight into the conflicts which were to arise later between the mathematically minded scientists of recent centuries and religious leaders, was seeking to discourage the further development of the subject. At any rate there is no question as to his attitude.

At about the same time that St. Augustine lived, the Roman jurists under the Code of Mathematicians and Evil-Doers ruled that "to learn the art of geometry and to take part in public exercises, an art as damnable as mathematics, are forbidden."

Even the distinguished seventeenth-century contributor to mathematics, Blaise Pascal, decided after studying mankind that the pure sciences were not suited to it. In a letter to Fermat written on August 10, 1660, Pascal says: "To speak freely of mathematics, I find it the highest exercise of the spirit; but at the same time I know that it is so useless that I make little distinction between a man who is only a mathematician and a common artisan. Also, I call it the most beautiful profession in the world; but it is only a profession; and I have often said that it is good to make the attempt [to study mathematics], but not to use our forces: so that I would not take two steps for mathematics, and I am confident that you are strongly of my opinion." Pascal's famous injunction was, "Humble thyself, impotent reason."

The philosopher Arthur Schopenhauer, who despised mathematics, said many nasty things about the subject, among others that the lowest activity of the spirit is arithmetic, as is shown by the fact that it can be performed by a machine. Many other great men, for example, the poet Johann Wolfgang Goethe and the historian Edward Gibbon, have felt likewise and have not hesitated to express themselves. And so the student who dislikes the subject can claim to be in good, if not living, company.

In view of the support he can muster from authorities, the student may well inquire why he is asked to learn mathematics. Is it because Plato, some 2300 years ago, advocated mathematics to train the mind for philosophy? Is it because the Church in medieval times taught mathematics as a preparation for theological reasoning? Or is it because the commercial, industrial, and scientific life of the Western world needs mathematics so much? Perhaps the subject got into the curriculum by mistake, and no one has taken the trouble to throw it out. Certainly the student is justified in asking his teacher the very question which Mephistopheles put to Faust:

> Is it right, I ask, is it even prudence,
> To bore thyself and bore the students?

Perhaps we should begin our answers to these questions by pointing out that the men we cited as disliking or disapproving of mathematics were really exceptional. In the great periods of culture which preceded the present one, almost all educated people valued mathematics. The Greeks, who created the modern concept of mathematics, spoke unequivocally for its importance. During the Middle Ages and in the Renaissance, mathematics was never challenged as one of the most important studies. The seventeenth century was aglow not only with mathematical activity but with popular interest in the subject. We have the instance of Samuel Pepys, so much attracted by the rapidly expanding influence of mathematics that at the age of thirty he could no longer tolerate his own ignorance and begged to learn the subject. He began, incidentally, with the multiplication table, which he subsequently taught to his wife. In 1681 Pepys was elected president of the Royal Society, a post later held by Isaac Newton.

In perusing eighteenth-century literature, one is struck by the fact that the journals which were on the level of our *Harper's* and the *Atlantic Monthly* contained mathematical articles side by side with literary articles. The educated man and woman of the eighteenth century knew the mathematics of their day, felt obliged to be *au courant* with all important scientific developments, and read articles on them much as modern man reads articles on politics. These people were as much at home with Newton's mathematics and physics as with Pope's poetry.

The vastly increased importance of mathematics in our time makes it all the more imperative that the modern person know something of the nature and role of mathematics. It is true that the role of mathematics in our civilization is not always obvious, and the deeper and more complex modern applications are not readily comprehended even by specialists. But the essential nature and accomplishments of the subject can still be understood.

Perhaps we can see more easily why one should study mathematics if we take a moment to consider what mathematics is. Unfortunately the answer cannot be given in a single sentence or a single chapter. The subject has many facets or, some might say, is Hydra-headed. One can look at mathematics as a language, as a particular kind of logical structure, as a body of knowledge about number and space, as a series of methods for deriving conclusions, as the essence of our knowledge of the physical world, or merely as an amusing intellectual activity. Each of these features would in itself be difficult to describe accurately in a brief space.

Because it is impossible to give a concise and readily understandable definition of mathematics, some writers have suggested, rather evasively, that mathematics is what mathematicians do. But mathematicians are human beings, and most of the things they do are uninteresting and some, embarrassing to relate. The only merit in this proposed definition of mathematics is that it points up the fact that mathematics is a human creation.

A variation on the above definition which promises more help in understanding the nature, content, and values of mathematics, is that mathematics is what *mathematics does*. If we examine mathematics from the standpoint of what it is intended to and does accomplish, we shall undoubtedly gain a truer and clearer picture of the subject.

Mathematics is concerned primarily with what can be accomplished by reasoning. And here we face the first hurdle. Why should one reason? It is not a natural activity for the human animal. It is clear that one does not need reasoning to learn how to eat or to discover what foods maintain life. Man knew how to feed, clothe, and house himself millenniums before mathematics existed. Getting along with the opposite sex is an art rather than a science mastered by reasoning. One can engage in a multitude of occupations and even climb high in the business and industrial world without much use of reasoning and certainly without mathematics. One's social position is hardly elevated by a display of his knowledge of trigonometry. In fact, civilizations in which reasoning and mathematics played no role have endured and even flourished. If one were willing to reason, he could readily supply evidence to prove that reasoning is a dispensable activity.

Those who are opposed to reasoning will readily point out other methods of obtaining knowledge. Most people are in fact convinced that their senses are really more than adequate. The very common assertion "seeing is believing" expresses the common reliance upon the senses. But everyone should recognize that the senses are limited and often fallible and, even where accurate, must be interpreted. Let us consider, as an example, the sense of sight. How big is the sun? Our eyes tell us that it is about as large as a rubber ball. This then is what we should believe. On the other hand, we do not see the air around us, nor for that matter can we feel, touch, smell, or taste it. Hence we should not believe in the existence of air.

To consider a somewhat more complicated situation, suppose a teacher should hold up a fountain pen and ask, What is it? A student coming from some primitive society might call it a shiny stick, and indeed this is what the eyes see. Those who call it a fountain pen are really calling upon education and experience stored in their minds. Likewise, when we look at a tall building from a distance, it is experience which tells us that the building is tall. Hence the old saying that "we are prone to see what lies behind our eyes, rather than what appears before them."

Every day we see the sun where it is not. For about five minutes before what we call sunset, the sun is actually below the geometrical horizon and

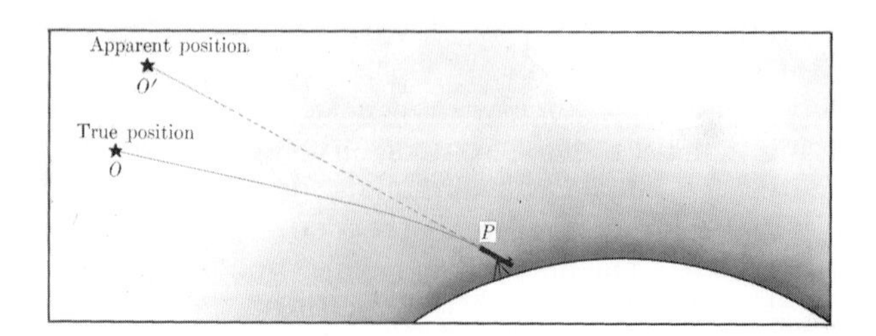

FIG. 1–1. Deviation of a ray by the earth's atmosphere.

should therefore be invisible. But the rays of light from the sun curve toward us as they travel in the *earth's* atmosphere, and the observer at P (Fig. 1–1) not only "sees" the sun but thinks the light is coming from the direction $O'P$. Hence he believes the sun is in that direction.

The senses are obviously helpless in obtaining some kinds of knowledge, such as the distance to the sun, the size of the earth, the speed of a bullet (unless one wishes to feel its velocity), the temperature of the sun, the prediction of eclipses, and dozens of other facts.

If the senses are inadequate, what about experimentation or, in simple cases, measurement? One can and in fact does learn a great deal by such means. But suppose one wants to find a very simple quantity, the area of a rectangle. To obtain it by measurement, one could lay off unit squares to cover the area and then count the number of squares. It is at least a little simpler to measure the lengths of the sides and then use a formula obtained by reasoning, namely, that the area is the product of length and width. In the only slightly more complicated problem of determining how high a projectile will go, we should certainly not consider traveling with the projectile.

As to experimentation, let us consider a relatively simple problem of modern technology. One wishes to build a bridge across a river. How long and how thick should the many beams be? What shape should the bridge take? If it is to be supported by cables, how long and how thick should these be? Of course one could arbitrarily choose a number of lengths and thicknesses for the beams and cables and build the bridge. In this event, it would only be fair that the experimenter be the first to cross this bridge.

It may be clear from this brief discussion that the senses, measurement, and experimentation, to consider three alternative ways of acquiring knowledge, are by no means adequate in a variety of situations. Reasoning is essential. The lawyer, the doctor, the scientist, and the engineer employ reasoning daily to derive knowledge that would otherwise not be obtainable or perhaps obtainable only at great expense and effort. Mathematics more than any other human endeavor relies upon reasoning to produce knowledge.

One may be willing to accept the fact that mathematical reasoning is an effective procedure. But just what does mathematics seek to accomplish with its reasoning? The primary objective of all mathematical work is to help man study nature, and in this endeavor mathematics cooperates with science. It may seem, then, that mathematics is merely a useful tool and that the real pursuit is science. We shall not attempt at this stage to separate the roles of mathematics and science and to evaluate the relative merits of their contributions. We shall simply state that their methods are different and that mathematics is at least an equal partner with science.

We shall see later how observations of nature are framed in statements called axioms. Mathematics then discloses by reasoning secrets which nature may never have intended to reveal. The determination of the pattern of motion of celestial bodies, the discovery and control of radio waves, the understanding of molecular, atomic, and nuclear structures, and the creation of artificial satellites are a few basically mathematical achievements. Mathematical formulation of physical data and mathematical methods of deriving new conclusions are today the substratum in all investigations of nature.

The fact that mathematics is of central importance in the study of nature reveals almost immediately several values of this subject. The first is the practical value. The construction of bridges and skyscrapers, the harnessing of the power of water, coal, electricity, and the atom, the effective employment of light, sound, and radio in illumination, communication, navigation, and even entertainment, and the advantageous employment of chemical knowledge in the design of materials, in the production of useful forms of oil, and in medicine are but a few of the practical achievements already attained. And the future promises to dwarf the past.

However, material progress is not the most compelling reason for the study of nature, nor have practical results usually come about from investigations so directed. In fact, to overemphasize practical values is to lose sight of the greater significance of human thought. The deeper reason for the study of nature is to try to understand the ways of nature, that is, to satisfy sheer intellectual curiosity. Indeed, to ask disinterested questions about nature is one of the distinguishing marks of mankind. In all civilizations some people at least have tried to answer such questions as: How did the universe come about? How old is the universe and the earth in particular? How large are the sun and the earth? Is man an accident or part of a larger design? Will the solar system continue to function or will the earth some day fall into the sun? What is light? Of course, not all people are interested in such questions. Food, shelter, sex, and television are enough to keep many happy. But others, aware of the pervasive natural mysteries, are more strongly obsessed to resolve them than any business man is to acquire wealth and power.

Beyond improvement in the material life of man and beyond satisfaction of intellectual curiosity, the study of nature offers intangible values of another sort, especially the abolition of fear and terror and their replacement by a deep, quiet satisfaction in the ways of nature. To the uneducated and to those uninitiated in the world of science, many manifestations of nature have appeared to be agents of destruction sent by angry gods. Some of the beliefs in ancient and even medieval Europe may be of special interest in view of what happened later. The sun was the center of all life. As winter neared and the days became shorter, the people believed that a battle between the gods of light and darkness was taking place. Thus the god Wodan was supposed to be riding through heaven on a white horse followed by demons, all of whom sought every opportunity to harm people. When, however, the days began to lengthen and the sun began to show itself higher in the sky each day, the people believed that the gods of light had won. They ceased all work and celebrated this victory. Sacrifices were offered to the benign gods. Symbols of fertility such as fruit and nuts, whose growth is, of course, aided by the sun, were placed on the altars. To symbolize further the desire for light and the joy in light, a huge log was placed in the fire to burn for twelve days, and candles were lit to heighten the brightness.

The beliefs and superstitions which have been attached to events we take in stride are incredible to modern man. An eclipse of the sun, a threat to the continuance of the light and heat which causes crops to grow, meant that the heavenly body was being swallowed up by a dragon. Many Hindu people believe today that a demon residing in the sky attacks the sun once in a while and that this is what causes the eclipse. Of course, when prayers, sacrifices, and ceremonies were followed by the victory of the sun or moon, it was clear that these rituals were the effective agent and so had to be pursued on every such occasion. In addition, special magic potions drunk during eclipses insured health, happiness, and wisdom.

To primitive peoples of the past, thunder, lightning, and storms were punishments visited by the gods on people who had apparently sinned in some way. The stories in the Old Testament of the flood and of the destruction of Sodom and Gomorrah by fire and brimstone are examples of such acts of wrath by the God of the Hebrews. Hence there was continual concern and even dread about what the gods might have in mind for helpless humans. The only recourse was to propitiate the divine powers, so that they would bring good fortune instead of evil.

Fears, dread, and superstitions have been eliminated, at least in our Western civilization, by just those intellectually curious people who have studied nature's mighty displays. Those "seemingly unprofitable amusements of speculative brains" have freed us from serfdom, given us undreamed of powers, and, in fact, have replaced negative doctrines by positive mathematical laws which reveal a remarkable order and uniformity in nature. Man has emerged as the proud possessor of knowledge which has enabled him to view nature calmly and objectively. An eclipse of the sun occurring on schedule is no longer an occasion for trembling but for quiet satisfaction that we know nature's ways. We breathe freely, knowing that nature will not be willful or capricious.

Indeed, man has been remarkably successful in his study of nature. History is said to repeat itself, but, in general, the circumstances of the supposed repetition are not the same as those of the earlier occurrence. As a consequence, the history of man has not been too effective a guide for the future. Nature is kinder. When nature repeats herself, and she does so constantly, the repetitions are exact facsimiles of previous events, and therefore man can anticipate nature's behavior and be prepared for what will take place. We have learned to recognize the patterns of nature and we can speak today of the uniformity of nature and delight in the regularity of her behavior.

The successes of mathematics in the study of inanimate nature have inspired in recent times the mathematical study of human nature. Mathematics has not only contributed to the very practical institutions such as banking, insurance, pension systems, and the like, but it has also supplied some substance, spirit, and methodology to the infant sciences of economics, politics, and sociology. Number, quantitative studies, and precise reasoning have replaced vague, subjective, and ineffectual speculations and have already given evidence of greater values to come.

As man turns to thoughts about himself and his fellow man, other questions occur to him which are as fundamental as any he can ask. Why is man born? What purposes does he serve or should he serve? What future awaits him? The knowledge acquired about our physical universe has profound implications for the origin and role of man. Moreover, as mathematics and science have amassed increasing knowledge and power, they have gradually encompassed the biological and psychological sciences, which in turn have shed further light on man's physical and mental life. Thus it has come about that mathematics and science have profoundly affected philosophy and religion.

Perhaps the most profound questions in the realm of philosophy are, What is truth and how does man acquire it? Though we have no final answer to these questions, the contribution of mathematics toward this end is paramount. For two millenniums mathematics was the prime example of truths man had unearthed. Hence all investigations of the problem of acquiring truths necessarily reckoned with mathematics. Though some startling developments in the nineteenth century altered completely our understanding of the nature of mathematics, the effectiveness of the subject, especially in representing and analyzing natural phenomena, has still kept mathematics the focal point of all investigations into the nature of knowledge. Not the least significant aspect of this value of mathematics has been the insight it has given us into the ways and powers of the human mind. Mathematics is the supreme and most remarkable example of the mind's power to cope with problems, and as such it is worthy of study.

Among the values which mathematics offers are its services to the arts. Most people are inclined to believe that the arts are independent of mathematics, but we shall see that mathematics has fashioned major styles of painting, architecture, and literature; and the service mathematics renders to music has not only enabled man to understand it, but has spread its enjoyment to all corners of our globe.

Practical, scientific, philosophical, and artistic problems have caused men to investigate mathematics. But there is one other motive which is as strong as any of these—the search for beauty. Mathematics is an art, and as such affords the pleasures which all the arts afford. This last statement may come as a shock to people who are used to the conventional concept of the true arts and mentally contrast these with mathematics to the detriment of the latter. But the average person has not really thought through what the arts really are and what they offer. All that many people actually see in painting, for example, are familiar scenes and perhaps bright colors. These qualities, however, are not the ones which make painting an art. The real values must be learned, and a genuine appreciation of art calls for much study.

So it is with mathematics. The subject offers concepts, structure, and methods of proof which embody the keenest and often liveliest of ideas, and results which are striking and gratifying to those prepared to appreciate them. Bertrand Russell, one of our great contemporary philosophers and mathematicians, has elaborated on this value of mathematics: "Mathematics, rightly viewed, possesses not only true but supreme beauty—a beauty cold and austere, like that of sculpture, without appeal to any part of our weaker nature, without the gorgeous trappings of painting or music, yet sublimely pure, and capable of a stern perfection such as only the greatest art can show. The true spirit of delight, the exaltation, the sense of being more than man, which is the touchstone of the highest excellence, is to be found in mathematics as fully as in poetry." Some of the greatest contributions to the subject matter of mathematics, portions of Euclidean geometry, projective geometry, and the theory of numbers, were motivated by, or developed to satisfy, aesthetic interests.

The artistic values of mathematics are to be found not only in mathematics proper but also in the mathematical structure of scientific theories. Just as the plastic arts shed new and uncommon light on men and familiar scenes, so do the mathematical theories of science shed new light on nature and give man an insight into the structure and functioning of nature which affords pleasures to scientists and students of nature. These ordered patterns which mathematics imposes on nature or reveals in nature and which describe and unite a multiplicity of seemingly diverse and disordered natural phenomena, such as the motions of the sun, moon, and planets, provide all the satisfactions of an art and rejoice the understanding as much as does the comprehension of an elaborate musical composition. Indeed, the search for rational accounts of natural phenomena and the basis for selection among alternative scientific theories can be understood only in terms of aesthetic principles which guide and spur on our scientists.

In addition to beauty of content, mathematics affords an outlet for all the artistic faculties of man. In anticipating the results later accepted on the basis of logical proof, mathematicians employ a highly developed intuition and imagination. Kepler, Newton, and Einstein were men of wonderful imaginative powers which enabled them not only to break away from age-long and rigid traditions but to set up new and revolutionary concepts. This indispensable use of intuition and imagination in the creation of proofs and conclusions has a high aesthetic value for both the creator and the student. Of course, as in any other art, struggles, frustrations, and heartbreak attend the creative process.

Nevertheless, we shall not insist on the aesthetic values of mathematics. It may be fairer to rest on the position that just as there are tone-deaf and color-blind people, so may there be some who temperamentally are intolerant of cold argumentation and the seemingly overfine distinctions of mathematics.

To many people, mathematics offers intellectual challenges, and it is well known that such challenges do engross humans. Games such as bridge, crossword puzzles, and magic squares are popular. Perhaps the best evidence is the attraction of puzzles such as the following: A wolf, a goat, and cabbage are to be transported across a river by a man in a boat which can hold only one of these in addition to the man. How can he take them across so that the wolf does not eat the goat or the goat the cabbage? Two husbands and two wives have to cross a river in a boat which can hold only two people. How can they cross so that no woman is in the company of a man unless her husband is also present? Such puzzles go back to Greek and Roman times. The mathematician Tartaglia, who lived in the sixteenth century, tells us that they were after-dinner amusements.

People do respond to intellectual challenges, and once one gets a slight start in mathematics, he encounters these in abundance. In view of the additional values to be derived from the subject, one would expect people to spend time on mathematical problems as opposed to the more superficial, and in some instances cheap, games which lack depth, beauty, and importance. The tantalizing and compelling pursuit of mathematical problems offers mental absorption, peace of mind amid endless challenges, repose in activity, battle without conflict, and the beauty which the ageless mountains present to senses tried by the kaleidoscopic rush of events. The appeal offered by the detachment and objectivity of mathematical reasoning is superbly described by Bertrand Russell. "Remote from human passions, remote even from the pitiful facts of nature, the generations have gradually created an ordered cosmos, where pure thought can dwell as in its natural home and where one, at least, of our nobler impulses can escape from the dreary exile of the actual world." The creation and contemplation of mathematics offer such values.

Despite all these arguments for the study of mathematics, the reader may have justifiable doubts. The idea that thinking about numbers and figures leads to deep and powerful conclusions which influence almost all other branches of thought may seem incredible. The study of numbers and geometrical figures may not seem a sufficiently attractive and promising enterprise. Not even the founders of mathematics envisioned the potentialities of the subject.

So we start with some doubts about the worth of our enterprise. We could encourage the reader with the hackneyed maxim, nothing ventured, nothing gained. We could call to his attention the daily testimony to the power of mathematics offered by almost every newspaper and journal. But such appeals are hardly inspiring. Let us proceed on the very weak basis that perhaps those more experienced in what the world has to offer may also have the wisdom to recommend worth-while studies.

Hence, despite St. Augustine, the reader is invited to tempt hell and damnation by engaging in a study of the subject. Certainly he can be assured that the subject is within his grasp and that no special gifts or qualities of mind are needed to learn mathematics. It is even debatable whether the creation of mathematics requires special talents as does the creation of music or great paintings, but certainly the appreciation of what others have done does not demand a "mathematical mind" any more than the appreciation of art requires an "artistic mind." Moreover, since we shall not draw upon any previously acquired knowledge, even this potential source of trouble will not arise.

Let us review our objectives. We should like to understand what mathematics is, how it functions, what it accomplishes for the world, and what it has to offer in itself. We hope to see that mathematics has content which serves the physical and social scientist, the philosopher, logician, and the artist; content which influences the doctrines of the statesman and the theologian; content which satisfies the curiosity of the man who surveys the heavens and the man who muses on the sweetness of musical sounds; and content which has undeniably, if sometimes imperceptibly, shaped the course of modern history. In brief, we shall try to see that mathematics is an integral part of the modern world, one of the strongest forces shaping its thoughts and actions, and a body of living thought inseparably connected with, dependent upon, and in turn valuable to all other branches of our culture. Perhaps we shall also see how by suffusing and influencing all thought it has set the intellectual temper of our times.

Recommended Reading

RUSSELL, BERTRAND: "The Study of Mathematics," an essay in the collection entitled *Mysticism and Logic,* Longmans, Green and Co., New York, 1925.

WHITEHEAD, ALFRED NORTH: "The Mathematical Curriculum," an essay in the collection entitled *The Aims of Education,* The New American Library, New York, 1949.

WHITEHEAD, ALFRED NORTH: *Science and the Modern World,* Chaps. 2 and 3, Cambridge University Press, Cambridge, 1926.

CHAPTER 2

THE COURSE OF MATHEMATICS: AN HISTORICAL ORIENTATION

An educated mind is, as it were, composed of all the minds of preceding ages.

Le Bovier de Fontenelle

2–1 Introduction. Our first objective will be to gain some historical perspective on the subject of mathematics. Although the logical development of mathematics is not markedly different from the historical, there are nevertheless many features of mathematics which are revealed by a glimpse of its history rather than by an examination of concepts, theorems, and proofs. Thus we may learn what the subject now comprises, how the various branches arose, and how the character of the mathematical contributions made by various civilizations was conditioned by these civilizations. This historical survey may also help us to gain some provisional understanding of the nature, extent, and uses of mathematics. Finally, a preview may help us to keep our bearings. In studying a vast subject, one is always faced with the danger of getting lost in details. This is especially true in mathematics, where one must often spend hours and even days in seeking to understand some new concepts or proofs.

2–2 Mathematics in early civilizations. Aside possibly from astronomy, mathematics is the oldest and most continuously pursued branch of human thought. Moreover, unlike science, philosophy, and social thought, very little of the mathematics that has ever been created has been discarded. Mathematics is also a cumulative development, that is, newer creations are built logically upon older ones, so that one must usually understand older results to master newer ones. These facts recommend that we go back to the very origins of mathematics.

As we examine the early civilizations, one remarkable fact emerges immediately. Though there have been hundreds of civilizations, many with great art, literature, philosophy, religion, and social institutions, very few possessed any mathematics worth talking about. Most of these civilizations hardly got past the stage of being able to count to five or ten.

In some of these early civilizations a few steps in mathematics were taken. In prehistoric times, which means roughly before 4000 B.C., several civilizations at least learned to think about numbers as abstract concepts. That is, they recognized that three sheep and three arrows have something in common, a quantity called three, which can be thought about independently of any physical objects. Each of us in his own schooling goes through this same process of divorcing numbers from physical objects. The appreciation of "number" as an abstract idea is a great, and perhaps the first, step in the founding of mathematics.

Another step was the introduction of arithmetical operations. It is quite an idea to add the numbers representing two collections of objects in order to arrive at the total instead of counting the objects in the combined collections. Similar remarks apply to subtraction, multiplication, and division. The early methods of carrying out these operations were crude and complicated compared with ours, but the ideas and the applications were there.

Only a few ancient civilizations, Egypt, Babylonia, India, and China, possessed what may be called the rudiments of mathematics. The history of mathematics, and indeed the history of Western civilization, begins with what occurred in the first two of these civilizations. The role of India will emerge later, whereas that of China may be ignored because it was not extensive and moreover had no influence on the subsequent development of mathematics.

Our knowledge of the Egyptian and Babylonian civilizations goes back to about 4000 B.C. The Egyptians occupied approximately the same region that now constitutes modern Egypt and had a continuous, stable civilization from ancient times until about 300 B.C. The term "Babylonian" includes a succession of civilizations which occupied the region of modern Iraq. Both of these peoples possessed whole numbers and fractions, a fair amount of arithmetic, some algebra, and a number of simple rules for finding the areas and volumes of geometrical figures. These rules were but the incidental accumulations of experience, much as people learned through experience what foods to eat. Many of the rules were in fact incorrect but good enough for the simple applications made then. For example, the Egyptian rule for finding the area of a circle amounts to using 3.16 times the square of the radius; that is, their value of π was 3.16. This value, though not accurate, was even better than the several values the Babylonians used, one of these being 3, the value found in the Bible.

What did these early civilizations do with their mathematics? If we may judge from problems found in ancient Egyptian papyri and in the clay tablets of the Babylonians, both civilizations used arithmetic and algebra largely in commerce and state administration, to calculate simple and compound interest on loans and mortgages, to apportion profits of business to the owners, to buy and sell merchandise, to fix taxes, and to calculate how many bushels of grain would make a quantity of beer of a specified alcoholic content. Geometrical rules were applied to calculate the areas of fields, the estimated yield of pieces of land, the volumes of structures, and the quantity of bricks or stones needed to erect a temple or pyramid. The ancient Greek historian Herodotus says that because the annual overflow of the Nile wiped out the boundaries of the farmers' lands, geometry was needed to redetermine the boundaries. In fact, Herodotus speaks of geometry as the gift of the Nile. This bit of history is a partial truth. The redetermination of boundaries was undoubtedly an application, but geometry existed in Egypt long before the date of 1400 B.C. mentioned by Herodotus for its origin. Herodotus would have been more accurate to say that Egypt is a gift of the Nile, for it is true today as it was then that the only fertile land in Egypt is that along the Nile; and this because the river deposits good soil on the land as it overflows.

Applications of geometry, simple and crude as they were, did play a large role in Egypt and Babylonia. Both peoples were great builders. The Egyptian temples, such as those at Karnak and Luxor, and the pyramids still appear to be admirable engineering achievements even in this age of skyscrapers. The Babylonian temples, called ziggurats, also were remarkable pyramidal structures. The Babylonians were, moreover, highly skilled irrigation engineers, who built a system of canals to feed their hot dry lands from the Tigris and Euphrates rivers.

Perhaps a word of caution is necessary with respect to the pyramids. Because these are impressive structures, some writers on Egyptian civilization have jumped to the conclusion that the mathematics used in the building of pyramids must also have been impressive. These writers point out that the horizontal dimensions of any one pyramid are exactly of the same length, the sloping sides all make the same angle with the ground, and the right angles are right. However, not mathematics but care and patience were required to obtain such results. A cabinetmaker need not be a mathematician.

Mathematics in Egypt and Babylonia was also applied to astronomy. Of course, astronomy was pursued in these ancient civilizations for calendar reckoning and, to some extent, for navigation. The motions of the heavenly bodies give us our fundamental standard of time, and their positions at given times enable ships to determine their location and caravans to find their bearings in the deserts. Calendar reckoning is not only a common daily and commercial need, but it fixes religious holidays and planting times. In Egypt it was also needed to predict the flood of the Nile, so that farmers could move property and cattle away beforehand.

It is worthy of note that by observing the motion of the sun, the Egyptians managed to ascertain that the year contains 365 days. There is a conjecture that the priests of Egypt knew that $365\frac{1}{4}$ was a more accurate figure but kept the knowledge secret. The Egyptian calendar was taken over much later by the Romans and then passed on to Europe. The Babylonians, by contrast, developed a lunar calendar. Since the duration of the month as measured from new moon to new moon varies from 29 to 30 days, the twelve-month year adopted by the Babylonians did not coincide with the year of the seasons. Hence the Babylonians added extra months, up to a total of seven, in every 19-year cycle. This scheme was also adopted by the Hebrews.

Astronomy served not only the purposes just described, but from ancient times until recently it also served astrology. In ancient Babylonia and Egypt the belief was widespread that the moon, the planets, and the stars directly influenced and even controlled affairs of the state. This doctrine was gradually extended and later included the belief that the health and welfare of the individual were also subject to the will of the heavenly bodies. Hence it seemed reasonable that by studying the motions and relative positions of these bodies man could determine their influences and even predict his future.

When one compares Egyptian and Babylonian accomplishments in mathematics with those of earlier and contemporary civilizations, one can indeed find reason to praise their achievements. But judged by other standards, Egyptian and Babylonian contributions to mathematics were almost insignificant, although these same civilizations reached relatively high levels in religion, art, architecture, metallurgy, chemistry, and astronomy. Compared with the accomplishments of their immediate successors, the Greeks, the mathematics of the Egyptians and Babylonians is the scrawling of children just learning how to write as opposed to great literature. They barely recognized mathematics as a distinct subject. It was a tool in agriculture, commerce, and engineering, no more important than the other tools they used to build pyramids and ziggurats. Over a period of 4000 years hardly any progress was made in the subject. Moreover, the very essence of mathematics, namely, reasoning to establish the validity of methods and results, was not even envisioned. Experience recommended their procedures and rules, and with this support they were content. Egyptian and Babylonian mathematics is best described as empirical and hardly deserves the appellation mathematics in view of what, since Greek times, we regard as the chief features of the subject. Some flesh and bones of concrete mathematics were there, but the spirit of mathematics was lacking.

The lack of interest in theoretical or systematic knowledge is evident in all activities of these two civilizations. The Egyptians and Babylonians must have noted the paths of the stars, planets, and moon for thousands of years. Their calendars, as well as tables which are extant, testify to the scope and accuracy of these observations. But no Egyptian or Babylonian strove, so far as we know, to encompass all these observations in one major plan or theory of heavenly motions. Nor does one find any other scientific theory or connected body of knowledge.

2–3 The classical Greek period. We have seen so far that mathematics, initiated in prehistoric times, struggled for existence for thousands of years. It finally obtained a firm grip on life in the highly congenial atmosphere of Greece. This land was invaded about 1000 B.C. by people whose origins are not known. By about 600 B.C. these people occupied not only Greece proper but many cities in Asia Minor on the Mediterranean coast, islands such as Crete, Rhodes, and Samos, and cities in southern Italy and Sicily. Though all of these areas bred famous men, the chief cultural center during the classical period, which lasted from about 600 B.C. to 300 B.C., was Athens.

Greek culture was not entirely indigenous. The Greeks themselves acknowledge their indebtedness to the Babylonians and especially to the Egyptians. Many Greeks traveled in Egypt and in Asia Minor. Some went there to study. Nevertheless, what the Greeks created differs as much from what they took over from the Egyptians and Babylonians as gold differs from tin. Plato was too modest in his description of the Greek contribution when he said, "Whatever we Greeks receive we improve and perfect." The Greeks not only made finished products out of the raw materials imported from Egypt and Babylonia, but they created totally new branches of culture. Philosophy, pure and applied sciences, political thought and institutions, historical writings, almost all our literary forms (except fictional prose), and new ideals such as the freedom of the individual are wholly Greek contributions.

The supreme contribution of the Greeks was to call attention to, employ, and emphasize the power of human reason. This recognition of the power of reasoning is the greatest single discovery made by man. Moreover, the Greeks recognized that reason was the distinctive faculty which humans possessed. Aristotle says, "Now what is characteristic of any nature is that which is best for it and gives most joy. Such to man is the life according to reason, since it is that which makes him man."

It was by the application of reasoning to mathematics that the Greeks completely altered the nature of the subject. In fact, mathematics as we understand the term today is entirely a Greek gift, though in this case we need not heed Virgil's injunction to fear such benefactions. But how did the Greeks plan to employ reason in mathematics? Whereas the Egyptians and Babylonians were content to pick up scraps of useful information through experience or trial and error, the Greeks abandoned empiricism and undertook a systematic, rational attack on the whole subject. First of all, the Greeks saw clearly that numbers and geometric forms occur everywhere in the heavens and on earth. Hence they decided to concentrate on these important concepts. Moreover, they were explicit about their intention to treat general abstract concepts rather than particular physical realizations. Thus they would consider the ideal circle rather than the boundary of a field or the shape of a wheel. They then observed that certain facts about these concepts are both obvious and basic. It was evident that equal numbers added to or subtracted from equal numbers give equal numbers. It was equally evident that two right angles are necessarily equal and that a circle can be drawn when center and radius are given. Hence they selected some of these obvious facts as a starting point and called them *axioms*. Their next idea was to apply reasoning, with these facts as premises, and to use only the most reliable methods of reasoning man possesses. If the reasoning were successful, it would produce new knowledge. Also, since they were to reason about general concepts, their conclusions would apply to all objects of which the concepts were representative. Thus if they could prove that the area of a circle is π times the square of the radius, this fact would apply to the area of a circular field, the floor area of a circular temple, and the cross section of a circular tree trunk. Such reasoning about general concepts might not only produce knowledge of hundreds of physical situations in one proof, but there was always the chance that reasoning would produce knowledge which experience might never suggest. All these advantages the

Greeks expected to derive from reasoning about general concepts on the basis of evident reliable facts. A neat plan, indeed!

It is perhaps already clear that the Greeks possessed a mentality totally different from that of the Egyptians and Babylonians. They reveal this also in the plans they had for the use of mathematics. The application of arithmetic and algebra to the computation of interest, taxes, or commercial transactions, and of geometry to the computation of the volumes of granaries was as far from their minds as the most distant star. As a matter of fact, their thoughts were on the distant stars. The Greeks found mathematics valuable in many respects, as we shall learn later, but they saw its main value in the aid it rendered to the study of nature; and of all the phenomena of nature, the heavenly bodies attracted them most. Thus, though the Greeks also studied light, sound, and the motions of bodies on the earth, astronomy was their chief scientific interest.

Just what did the Greeks seek in probing nature? They sought no material gain and no power over nature; they sought merely to satisfy their minds. Because they enjoyed reasoning and because nature presented the most imposing challenge to their understanding, the Greeks undertook the purely intellectual study of nature. Thus the Greeks are the founders of science in the true sense.

The Greek conception of nature was perhaps even bolder than their conception of mathematics. Whereas earlier and later civilizations viewed nature as capricious, arbitrary, and terrifying, and succumbed to the belief that magic and rituals would propitiate mysterious and feared forces, the Greeks dared to look nature in the face. They dared to affirm that nature was rationally and indeed mathematically designed, and that man's reason, chiefly through the aid of mathematics, would fathom that design. The Greek mind rejected traditional doctrines, supernatural causes, superstitions, dogma, authority, and other such trammels on thought and undertook to throw the light of reason on the processes of nature. In seeking to banish the mystery and seeming arbitrariness of nature and in abolishing belief in dreaded forces, the Greeks were the pioneers.

For reasons which will become clearer in a later chapter, the Greeks favored geometry. By 300 B.C., Thales, Pythagoras and his followers, Plato's disciples, notably Eudoxus, and hundreds of other famous men had built up an enormous logical structure, most of which Euclid embodied in his *Elements*. This is, of course, the geometry we still study in high school. Though they made some contributions to the study of the properties of numbers and to the solution of equations, almost all of their work was in geometric form, and so there was no improvement over the Egyptians and Babylonians in the representation of, and calculation with, numbers or in the symbolism and techniques of algebra. For these contributions the world had to wait many more centuries. But the vast development in geometry exerted an enormous influence in succeeding civilizations and supplied the inspiration for mathematical activity in civilizations which might otherwise never have acquired even the very concept of mathematics.

The Greek accomplishments in mathematics had, in addition, the broader significance of supplying the first impressive evidence of the power of human reason to deduce new truths. In every culture influenced by the Greeks, this example inspired men to apply reason to philosophy, economics, political theory, art, and religion. Even today Euclid is the prime example of the power and accomplishments of reason. Hundreds of generations since Euclid's days have learned from his geometry what reasoning is and what it can accomplish. Modern man as well as the ancient Greeks learned from the Euclidean document how exact reasoning should proceed, how to acquire facility in it, and how to distinguish correct from false reasoning. Although many people depreciate this value of mathematics, it is interesting nevertheless that when these people seek to offer an excellent example of reasoning, they inevitably turn to mathematics.

This brief discussion of Euclidean geometry may show that the subject is far from being a relic of the dead past. It remains important as a stepping stone in mathematics proper and as a paradigm of reasoning. With their gift of reason and with their explicit example of the power of reason, the Greeks founded Western civilization.

2–4 The Alexandrian Greek period. The intellectual life of Greece was altered considerably when Alexander the Great conquered Greece, Egypt, and the Near East. Alexander decided to build a new capital for his vast empire and founded the city in Egypt named after him. The center of the new Greek world became Alexandria instead of Athens. Moreover, Alexander made deliberate efforts to fuse Greek and Near Eastern cultures. Consequently, the civilization centered at Alexandria, though predominantly Greek, was strongly influenced by Egyptian and Babylonian contributions. This Alexandrian Greek civilization lasted from about 300 B.C. to 600 A.D.

The mixture of the theoretical interests of the Greeks and the practical outlook of the Babylonians and Egyptians is clearly evident in the mathematical and scientific work of the Alexandrian Greeks. The purely geometric investigations of the classical Greeks were continued, and two of the most famous Greek mathematicians, Apollonius and Archimedes, pursued their studies during the Alexandrian period. In fact, Euclid also lived in Alexandria, but his writings reflect the achievements of the classical period. For practical applications, which usually require quantitative results, the Alexandrians revived the crude arithmetic and algebra of Egypt and Babylonia and used these empirically founded tools and procedures, along with results derived from exact geometrical studies. There was some progress in algebra, but what was newly created by men such as Nichomachus and Diophantus was still short of even the elementary methods we learn in high school.

The attempt to be quantitative, coupled with the classical Greek love for the mathematical study of nature, stimulated two of the most famous astronomers of all time, Hipparchus and Ptolemy, to calculate the sizes and distances of the heavenly bodies and to build a sound and, for those times, accurate astronomical theory, which is still known as Ptolemaic theory. Hipparchus and Ptolemy also created the chief tool they needed for this purpose, the mathematical subject known as trigonometry.

During the centuries in which the Alexandrian civilization flourished, the Romans grew strong, and by the end of the third century B.C. they were a world power. After conquering Italy, the Romans conquered the Greek mainland and a number of Greek cities scattered about the Mediterranean area. Among these was the famous city of Syracuse in Sicily, where Archimedes spent most of his life, and where he was killed at the age of 77 by a Roman soldier. According to the account given by the noted historian Plutarch, the soldier shouted to Archimedes to surrender, but the latter was so absorbed in studying a mathematical problem that he did not hear the order, whereupon the soldier killed him.

The contrast between Greek and Roman cultures is striking. The Romans have also bequeathed gifts to Western civilization, but in the fields of mathematics and science their influence was negative rather than positive. The Romans were a practical people and even boasted of their practicality. They sought wealth and world power and were willing to undertake great engineering enterprises, such as the building of roads and viaducts, which might help them to expand, control, and administer their empire, but they would spend no time or effort on theoretical studies which might further these activities. As the great philosopher Alfred North Whitehead remarked, "No Roman ever lost his life because he was absorbed in the contemplation of a mathematical diagram."

Indirectly as well as directly, the Romans brought about the destruction of the Greek civilization at Alexandria, directly by conquering Egypt and indirectly by seeking to suppress Christianity. The adherents to this new religious movement, though persecuted cruelly by the Romans, increased in number while the Roman Empire grew weaker. In 313 A.D. Rome legalized Christianity and, under the Emperor Theodosius (379–395), adopted it as the official religion of the empire. Yet even before this time, and certainly after it, the Christians began to attack the cultures and civilizations which had opposed them. By pillage and the burning of books, they destroyed all they could reach of ancient learning. Naturally the Greek culture suffered, and many works wiped out in these holocausts are now lost to us forever.

The final destruction of Alexandria in 640 A.D. was the deed of the Arabs. The books of the Greeks were closed, never to be reopened in this region.

2–5 The Hindus and Arabs. The Arabs, who suddenly appeared on the scene of history in the role of destroyers, had been a nomadic people. They were unified under the leadership of the prophet Mohammed and began an attempt to convert the world to Mohammedanism, using the sword as their most decisive argument. They conquered all the land around the Mediterranean Sea. In the Near East they took over Persia and penetrated as far as India. In southern Europe they occupied Spain, southern France, where they were stopped by Charles Martel, southern Italy and Sicily. Only the

Byzantine or Eastern Roman Empire was not subdued and remained an isolated center of Greek and Roman learning. In rather surprisingly quick time as the history of nations goes, the Arabs settled down and built a civilization and culture which maintained a high level from about 800 to 1200 A.D. Their chief centers were Bagdad in what is now Iraq, and Cordova in Spain. Realizing that the Greeks had created wonderful works in many fields, the Arabs proceeded to gather up and study what they could still find in the lands they controlled. They translated the works of Aristotle, Euclid, Apollonius, Archimedes, and Ptolemy into Arabic. In fact, Ptolemy's chief work, whose title in Greek meant "Mathematical Collection," was called the *Almagest* (The Greatest Work) by the Arabs and is still known by this name. Incidentally, other Arabic words which are now common mathematical terms are algebra, taken from the title of a book written by Al-Khowarizmi, a ninth-century Arabian mathematician, and algorithm, now meaning a process of calculation, which is a corruption of the man's name.

Though they showed keen interest in mathematics, optics, astronomy, and medicine, the Arabs contributed little that was original. It is also peculiar that, although they had at least some of the Greek works and could therefore see what mathematics meant, their own contributions, largely in arithmetic and algebra, followed the empirical, concrete approach of the Egyptians and Babylonians. They could on the one hand appreciate and critically review the precise, exact, and abstract mathematics of the Greeks while, on the other, offer methods of solving equations which, though they worked, had no reasoning to support them. During all the centuries in which Greek works were in their possession, the Arabs manfully resisted the lures of exact reasoning in their own contributions.

We are indebted to the Arabs not only for their resuscitation of the Greek works but for picking up some simple but useful ideas from India, their neighbor on the East. The Indians, too, had built up some elementary mathematics comparable in extent and spirit with the Egyptian and Babylonian developments. However, after about 200 A.D., mathematical activity in India became more appreciable, probably as a result of contacts with the Alexandrian Greek civilization. The Hindus made a few contributions of their own, such as the use of special number symbols from 1 to 9, the introduction of 0, and the use of positional notation with base ten, that is, our modern method of writing numbers. They also created negative numbers. These ideas were taken over by the Arabs and incorporated in their mathematical works.

Because of internal dissension the Arab Empire split into two independent parts. The Crusades launched by the Europeans and the inroads made by the Turks further weakened the Arabs, and their empire and culture disintegrated.

2–6 Early and medieval Europe. Thus far Europe proper has played no role in the history of mathematics. The reason is simple. The Germanic tribes who occupied central Europe and the Gauls of western Europe were barbarians. Among primitive civilizations, theirs were primitive indeed. They had no learning, no art, no science, not even a system of writing.

The barbarians were gradually civilized. While the Romans were still successful in holding the regions now called France, England, southern Germany, and the Balkans, the barbarians were in contact with, and to some extent influenced by, the Romans. When the Roman Empire collapsed, the Church, already a strong organization, took on the task of civilizing and converting the barbarians. Since the Church did not favor Greek learning and since at any rate the illiterate Europeans had first to learn reading and writing, one is not surprised to find that mathematics and science were practically unknown in Europe until about 1100 A.D.

2–7 The Renaissance. Insofar as the history of mathematics is concerned, the Arabs served as the agents of destiny. Trade with the Arabs and such invasions of the Arab lands as the Crusades acquainted the Europeans, who hitherto possessed only fragments of the Greek works, with the vast stores of Greek learning possessed by the Arabs. The ideas in these works excited the Europeans, and scholars set about acquiring them and translating them into Latin. Through another accident of history another group of Greek works came to Europe. We have already noted that the Eastern Roman or Byzantine Empire had survived the Germanic and the Arab aggrandizements. But in the fifteenth century the Turks captured the Eastern Roman Empire, and Greek scholars carrying precious manuscripts fled the region and went to Europe.

We shall leave for a later chapter a fuller account of how the European world was aroused by the renaissance of the novel and weighty Greek ideas, and of the challenge these ideas posed to the European beliefs and way of life.* From the Greeks the Europeans acquired arithmetic, a crude algebra, the vast development of Euclidean geometry, and the trigonometry of Hipparchus and Ptolemy. Of course, Greek science and philosophy also became known in Europe.

The first major European development in mathematics occurred in the work of the artists. Imbued with the Greek doctrines that man must study himself and the real world, the artists began to paint reality as they actually perceived it instead of interpreting religious themes in symbolic styles. They applied Euclidean geometry to create a new system of perspective which permitted them to paint realistically. Specifically, the artists created a new style of painting which enabled them to present on canvas, scenes making the same impression on the eye as the actual scenes themselves. From the work of the artists, the mathematicians derived ideas and problems that led to a new branch of mathematics, projective geometry.

Stimulated by Greek astronomical ideas, supplied with data and the astronomical theory of Hipparchus and Ptolemy, and steeped in the Greek doctrine that the world is mathematically designed, Nicolaus Copernicus

---

* See Chapter 9.

sought to show that God had done a better job than Hipparchus and Ptolemy had described. The result of Copernicus' thinking was a new system of astronomy in which the sun was immobile and the planets revolved around the sun. This heliocentric theory was considerably improved by Kepler. Its effects on religion, philosophy, science, and on man's estimations of his own importance were profound. The heliocentric theory also raised scientific and mathematical problems which were a direct incentive to new mathematical developments.

Just how much mathematical activity the revival of Greek works might have stimulated cannot be determined, for simultaneously with the translation and absorption of these works, a number of other revolutionary developments altered the social, economic, religious, and intellectual life of Europe. The introduction of gunpowder was followed by the use of muskets and later cannons. These inventions revolutionized methods of warfare and gave the newly emerging social class of free common men an important role in that domain. The compass became known to the Europeans and made possible long-range navigation, which the merchants sponsored for the purpose of finding new sources of raw materials and better trade routes. One result was the discovery of America and the consequent influx of new ideas into Europe. The invention of printing and of paper made of rags afforded books in large quantities and at cheap prices, so that learning spread far more than it ever had in any earlier civilizations. The Protestant Revolution stirred debate and doubts concerning doctrines that had been unchallenged for 1500 years. The rise of a merchant class and of free men engaged in labor in their own behalf stimulated an interest in materials, methods of production, and new commodities. All of these needs and influences challenged the Europeans to build a new culture.

2-8 Developments from 1550 to 1800. Since many of the problems raised by the motion of cannon balls, navigation, and industry called for quantitative knowledge, arithmetic and algebra became centers of attention. A remarkable improvement in these mathematical fields followed. This is the period in which algebra was built as a branch of mathematics and in which much of the algebra we learn in high school was created. Almost all the great mathematicians of the sixteenth and seventeenth centuries, Cardan, Tartaglia, Vieta, Descartes, Fermat, and Newton, men we shall get to know better later, contributed to the subject. In particular, the use of letters to represent a class of numbers, a device which gives algebra its generality and power, was introduced by Vieta. In this same period, logarithms were created to facilitate the calculations of astronomers. The history of arithmetic and algebra illustrates one of the striking and curious features of the history of mathematics. Ideas that seem remarkably simple once explained were thousands of years in the making.

The next development of consequence, coordinate geometry, came from two men, both interested in method. One was René Descartes. Descartes is perhaps even more famous as a philosopher than as a mathematician, though he was one of the major contributors to our subject. As a youth Descartes was already troubled by the intellectual turmoil of his age. He found no certainty in any of the knowledge taught him, and he therefore concentrated for years on finding the method by which man can arrive at truths. He found the clue to this method in mathematics, and with it constructed the first great modern philosophical system. Because the scientific problems of his time involved work with curves, the paths of ships at sea, of the planets, of objects in motion near the earth, of light, and of projectiles, Descartes sought a better method of proving theorems about curves. He found the answer in the use of algebra. Pierre de Fermat's interest in method was confined to mathematics proper, but he too appreciated the need for more effective ways of working with curves and also arrived at the idea of applying algebra. In this development of coordinate geometry we have one of the remarkable examples of how the times influence the direction of men's thoughts.

We have already noted that a new society was developing in Europe. Among its features were expanded commerce, manufacturing, mining, large-scale agriculture, and a new social class—free men working as laborers or as independent artisans. These activities and interests created problems of materials, methods of production, quality of the product, and utilization of devices to replace or increase the effectiveness of manpower. The people involved, like the artists, had become aware of Greek mathematics and science and sensed that it could be helpful. And so they too sought to employ this knowledge in their own behalf. Thereby arose a new motive for the study of mathematics and science. Whereas the Greeks had been content to study nature merely to satisfy their own curiosity and to organize their conclusions in patterns pleasing to the mind, the new goal, effectively proclaimed by Descartes and Francis Bacon, was to make nature serve man. Hence mathematicians and scientists turned earnestly to an enlarged program in which both understanding and mastery of nature were to be sought.

However, Bacon had cautioned that nature can be commanded only when one learns to obey her. One must have facts of nature on which to base reasoning about nature. Hence mathematicians and scientists sought to acquire facts from the experience of artists, technicians, artisans, and engineers. The alliance of mathematics and experience was gradually transformed into an alliance of mathematics and experimentation, and a new method for the pursuit of the truths of nature, first clearly perceived and formulated by Galileo Galilei (1564–1642) and Newton, was gradually evolved. Its content, perhaps oversimply stated, is that experience and experiment were to supply basic mathematical principles and mathematics was to be applied to these principles to deduce new truths, just as new truths are deduced from the axioms of geometry.

The most pressing scientific problem of the seventeenth century was the study of motion. On the practical side, investigations of the motion of projectiles, of the motion of the moon and planets to aid navigation, and of the motion of light to improve the design of the newly discovered telescope and microscope, were the primary interests. On the theoretical side, the new heliocentric astronomy invalidated the older, Aristotelian laws of motion and called for totally new principles. It was one thing to explain why a ball fell to earth on the assumption that the earth was immobile and the center of the universe, and another to explain this phenomenon in the light of the fact that the earth was rotating and revolving around the sun. A new science of motion was created by Galileo and Newton, and in the process two brand-new developments were added to mathematics. The first of these was the notion of a function, a relationship between variables best expressed for most purposes as a formula. The second, which rests on the notion of a function but represents the greatest advance in method and content since Euclid's days, was the calculus. The subject matter of mathematics and the power of mathematics expanded so greatly that at the end of the seventeenth century Leibniz could say, "Taking mathematics from the beginning of the world to the time when Newton lived, what he had done was much the better half."

With the aid of the calculus Newton was able to organize all data on earthly and heavenly motions into one system of mathematical mechanics which encompassed the motion of a ball falling to earth and the motion of the planets and stars. This great creation produced universal laws which not only united heaven and earth but revealed a design in the universe far more impressive than man had ever conceived. Galileo's and Newton's plan of applying mathematics to sound physical principles not only succeeded in one major area but gave promise, in a rapidly accelerating scientific movement, of embracing all other physical phenomena.

We learn in history that the end of the seventeenth century and the eighteenth century were marked by a new intellectual attitude briefly described as the Age of Reason. We are rarely told that this age was inspired by the successes which mathematics, to be sure in conjunction with science, had achieved in organizing man's knowledge. Infused with the conviction that reason, personified by mathematics, would not only conquer the physical world but could solve all of man's problems and should therefore be employed in every intellectual and artistic enterprise, the great minds of the age undertook a sweeping reorganization of philosophy, religion, ethics, literature, and aesthetics. The beginnings of new sciences such as psychology, economics, and politics were made during these rational investigations. Our principal intellectual doctrines and outlook were fashioned then, and we still live in the shadow of the Age of Reason.

While these major branches of our culture were being transformed, eighteenth-century scientists continued to win victories over nature. The calculus was soon extended to a new branch of mathematics called differential equations, and this new tool enabled scientists to tackle more complex problems in astronomy, in the study of the action of forces causing motions, in sound, especially musical sounds, in light, in heat (especially as applied to the development of the steam engine), in the strength of materials, and in the flow of liquids and gases. Other branches which can be merely mentioned, such as infinite series, the calculus of variations, and differential geometry, added to the extent and power of mathematics. The great names of the Bernoullis, Euler, Lagrange, Laplace, d'Alembert, and Fourier, belong to this period.

2–9 Developments from 1800 to the present. During the nineteenth century, developments in mathematics came at an ever increasing rate. Algebra, geometry and analysis, the last comprising those subjects which stem from calculus, all acquired new branches. The great mathematicians of the century were so numerous that it is impractical to list them. We shall encounter some of the greatest of these, Karl Friedrich Gauss and Bernhard Riemann, in our work. We might mention also Henri Poincaré and David Hilbert, whose work extended into the twentieth century.

Undoubtedly the primary cause of this expansion in mathematics was the expansion in science. The progress made in the seventeenth and eighteenth centuries had sufficiently illustrated the effectiveness of science in penetrating the mysteries of the physical world and in giving man control over nature, to cause an all the more vigorous pursuit of science in the nineteenth century. In that century also, science became far more intimately linked with engineering and technology than ever before. Mathematicians, working closely with the scientists as they had since the seventeenth century, were presented with thousands of significant physical problems and responded to these challenges.

Perhaps the major scientific development of the century, which is typical in its stimulation of mathematical activity, was the study of electricity and magnetism. While still in its infancy this science yielded the electric motor, the electric generator, and telegraphy. Basic physical principles were soon expressed mathematically, and it became possible to apply mathematical techniques to these principles, to deduce new information just as Galileo and Newton had done with the principles of motion. In the course of such mathematical investigations, James Clerk Maxwell discovered electromagnetic waves of which the best known representatives are radio waves. A new world of phenomena was thus uncovered, all embraced in one mathematical system. Practical applications, with radio and television as most familiar examples, soon followed.

Remarkable and revolutionary developments of another kind also took place in the nineteenth century, and these resulted from a re-examination of elementary mathematics. The most profound in its intellectual significance was the creation of non-Euclidean geometry by Gauss. His discovery had both tantalizing and disturbing implications: tantalizing in that this new field contained entirely new geometries based on axioms which differ from Euclid's, and disturbing in that it shattered man's firmest conviction, namely that mathematics is a body of truths. With the truth of mathematics undermined, realms of philosophy, science, and even some religious beliefs went up in smoke. So shocking were the implications that even mathematicians refused to take non-Euclidean geometry seriously until the theory of relativity forced them to face the full significance of the creation.

For reasons which we trust will become clearer further on, the devastation caused by non-Euclidean geometry did not shatter mathematics but released it from bondage to the physical world. The lesson learned from the history of non-Euclidean geometry was that though mathematicians may start with axioms that seem to have little to do with the observable behavior of nature, the axioms and theorems may nevertheless prove applicable. Hence mathematicians felt freer to give reign to their imaginations and to consider abstract concepts such as complex numbers, tensors, matrices, and n-dimensional spaces, for which applications were not at hand or in prospect. This development was followed by an even greater advance in mathematics and, surprisingly, an increasing use of mathematics in the sciences.

Even before the nineteenth century, the rationalistic spirit engendered by the success of mathematics in the study of nature penetrated to the social scientists. They began to emulate the physical scientists, that is, to search for the basic truths in their fields and to attempt reorganization of their subjects on the mathematical pattern. But these attempts to deduce the laws of man and society and to erect sciences of biology, economics, and politics did not succeed, although they did have some indirect beneficial effects.

The failure to penetrate social and biological problems by the deductive method, that is, the method of reasoning from axioms, caused social scientists to take over and develop further the mathematical theories of statistics and probability, which had already been initiated by mathematicians for various purposes ranging from problems of gambling to the theory of heat and astronomy. These techniques have been remarkably successful and have given some scientific methodology to what were largely speculative domains.

This brief sketch of the mathematics which will fall within our purview may make it clear that mathematics is not a closed book written in Greek times. It is rather a living plant that has flourished and languished with the rise and fall of civilizations. Since about 1600 it has been a continuing development which has become steadily vaster, richer, and more profound. The character of mathematics has been aptly, if somewhat floridly, described by the nineteenth-century English mathematician James Joseph Sylvester. "Mathematics is not a book confined within a cover and bound between brazen clasps, whose contents it needs only patience to ransack; it is not a mine, whose treasures may take long to reduce into possession, but which fill only a limited number of veins and lodes; it is not a soil, whose fertility can be exhausted by the yield of successive harvests; it is not a continent or an ocean, whose area can be mapped out and its contour defined; it is as limitless as the space which it finds too narrow for its aspirations; its possibilities are as infinite as the worlds which are forever crowding in and multiplying upon the astronomer's gaze; it is as incapable of being restricted within assigned boundaries or being reduced to definitions of permanent validity as the consciousness, the life, which seems to slumber in each monad, in every atom of matter, in each leaf and bud and cell and is forever ready to burst forth into new forms of vegetable and animal existence."

Our sketch of the development of mathematics has attempted to indicate the major eras and civilizations in which the subject has flourished, the variety of interests which induced people to pursue mathematics, and the branches of mathematics that have been created. Of course, we intend to investigate more carefully and more fully what these creations are and what values they have furnished to mankind. One fact of history may be noted by way of summary here. Mathematics as a body of reasoning from axioms stems from one source, the classical Greeks. All other civilizations which have pursued or are pursuing mathematics acquired this concept of mathematics from the Greeks. The Arab and Western European were the next civilizations to take over and expand on the Greek foundation. Today countries such as the United States, Russia, China, India, and Japan are also active. Though the last three of these did possess some native mathematics, it was limited and empirical as in Babylonia and Egypt. Modern mathematical activity in these five countries and wherever else it is now taking hold was inspired by Western European thought and actually learned by men who studied in Europe and returned to build centers of teaching in their own countries.

2–10 The human aspect of mathematics. One final point about mathematics is implicit in what we have said. We have spoken of problems which gave rise to mathematics, of cultures which emphasized some directions of thinking as opposed to others, and of branches of mathematics, as though all these forces and activities were as impersonal as the force of gravitation. But ideas and thinking are conveyed by people. Mathematics is a human creation. Although most Greeks did believe that mathematics existed independently of human beings as the planets and mountains seem to, and that all that human beings do is discover more and more of the structure, the prevalent belief today is that mathematics is entirely a human product. The concepts, the axioms, and the theorems established are all created by human beings in man's attempt to understand his environment, to give play to his artistic instincts, and to engage in absorbing intellectual activity.

The lives and activities of the men themselves are also fascinating. While mathematicians produce formulas, no formula produces mathematicians. They have come from all levels of society. The special talent, if there is such, which makes mathematicians has been found in Casanovas and ascetics, among business men and philosophers, among atheists and the profoundly religious, among the retiring and the worldly. Some, like Blaise Pascal and Gauss, were precocious; Évariste Galois was dead at 21, and Niels Hendrik Abel at 27. Others, like Karl Weierstrass and Henri Poincaré, matured more normally and were productive throughout their lives. Many were modest; others extremely egotistical and vain beyond toleration. One finds scoundrels, such as Cardan, and models of rectitude. Some were generous in their recognition of other great minds; others were resentful and jealous and even stole ideas to boost their own reputations. Disputes about priority of discovery abound.

The point in learning about these human variations, aside from satisfying our instinct to pry into other people's lives, is that it explains to a large extent why the progress of the highly rational subject of mathematics has been highly irrational. Of course, the major historical forces, which we sketched above, limit the actions and influence the outlook of individuals, but we also find in the history of mathematics all the vagaries which we have learned to associate with human beings. Leading mathematicians have failed to recognize bright ideas suggested by younger men, and the authors died neglected. Big men and little men made unsuccessful attempts to solve problems which their successors solved with ease. On the other hand, some supposed proofs offered even by masters were later found to be false. Generations and even ages failed to note new ideas, despite the fact that all that was needed was not a technical achievement but merely a point of view. The examples of the blindness of human beings to ideas which later seem simple and obvious furnish fascinating insight into the working of the human mind.

Recognition of the human element in mathematics explains in large measure the differences in the mathematics produced by different civilizations and the sudden spurts made in new directions by virtue of insights supplied by genius. Though no subject has profited as much as mathematics has by the cumulative effect of thousands of workers and results, in no subject is the role of great minds more readily discernible.

Exercises

1. Name a few civilizations which contributed to mathematics.
2. What basis did the Egyptians and Babylonians have for believing in their mathematical methods and formulas?
3. Compare Greek and pre-Greek understanding of the concepts of mathematics.
4. What was the Greek plan for establishing mathematical conclusions?
5. What was the chief contribution of the Arabs to the development of mathematics?
6. In what sense is mathematics a creation of the Greeks rather than of the Egyptians and Babylonians?
7. Criticize the statement "Mathematics was created by the Greeks and very little was added since their time."

Topics for Further Investigation

To write on the following topics use the books listed under Recommended Reading.

1. The mathematical contributions of the Egyptians or Babylonians.
2. The mathematical contributions of the Greeks.

Recommended Reading

Bell, Eric T.: *Men of Mathematics,* Simon and Schuster, New York, 1937.

Childe, V. Gordon: *Man Makes Himself,* The New American Library, New York, 1951.

Eves, Howard: *An Introduction to the History of Mathematics,* Rinehart and Co., New York, 1953.

Neugebauer, Otto: *The Exact Sciences in Antiquity,* Princeton University Press, Princeton, 1952.

Smith, David Eugene: *History of Mathematics,* Vol. I, Dover Publications, Inc., New York, 1958.

Struik, Dirk J.: *A Concise History of Mathematics,* Dover Publications, Inc., New York, 1948.

CHAPTER 3

THE WAYS OF MATHEMATICS

Geometry will draw the soul toward truth and create the spirit of philosophy.

Plato

3–1 Introduction. Mathematics has its own ways of establishing knowledge, and the understanding of mathematics is considerably promoted if one learns first just what those ways are. In this chapter we shall study the concepts which mathematics treats; the method, called deductive proof, by which mathematics establishes its conclusions; and the principles or axioms on which mathematics rests. Study of the contents and logical structure of mathematics leaves untouched the subject of how the mathematician knows what conclusions to establish and how to prove them. We shall therefore present a brief and preliminary discussion of the creation of mathematics. This topic will recur as we examine the subject matter itself in subsequent chapters.

Since mathematics, as we conceive the subject today, was fashioned by the Greeks, we shall also attempt to see what features of Greek thought and culture caused these people to remodel what the Egyptians and Babylonians had pursued for several thousand years.

3–2 The concepts of mathematics. The first major step which the Greeks made was to insist that mathematics must deal with abstract concepts. Let us see just what this means. When we first learn about numbers we are taught to think about collections of particular objects such as two apples, three men, and so on. Gradually and rather subconsciously we begin to think about the numbers 2, 3, and other whole numbers without having to associate them with physical objects. We soon reach the more advanced stage of adding, subtracting, and performing other operations with numbers without having to handle collections of objects in order to understand these operations or to see that the results agree with experience. Thus we soon become convinced that 4 times 5 must be 20, whether these numbers represent quantities of apples, horses, or even purely imaginary objects. By this time we are really dealing with concepts or ideas, for the whole numbers do not exist in nature. Any whole number is rather an abstraction of a property which is common to many different collections or sets of objects.

The whole numbers then are ideas, and the same is true of fractions such as $2/3$, $5/7$, and so on. In the latter case, too, the formulation of the physical relationship of a part of an object to the whole, whether it refers to pies, bushels of wheat, or to a smaller monetary value in relation to a larger one, again leads to an abstraction. Mathematicians formulate operations with fractions, that is, combining parts of an object, taking one part away from the other, or taking a part of a part, in such a way that the result of any operation on abstract fractions agrees with the corresponding physical occurrence. Thus the mathematical process of, say adding $2/3$ and $4/5$, which yields $22/15$, expresses the addition of $2/3$ of a pie and $4/5$ of a pie, and the result tells us how many parts of a pie one would actually have.

Whole numbers, fractions, and the various operations with whole numbers and fractions are abstractions. Although this fact is rather easy to understand, we tend to lose sight of it and cause ourselves unnecessary confusion. Let us consider an example. A man goes into a shoe store and buys 3 pairs of shoes at 10 dollars per pair. The storekeeper reasons that 3 pairs times 10 dollars is 30 dollars and asks for 30 dollars in return for the 3 pairs of shoes. If this reasoning is correct, then it is equally correct for the customer to argue that 3 pairs times 10 dollars is 30 pairs of shoes and to walk out with 30 pairs of shoes without handing the storekeeper one cent. The customer may end up in jail, but he may console himself while he languishes there that his reasoning is as sound as the storekeeper's.

The source of the difficulty is, of course, that one cannot multiply shoes by dollars. One can multiply the number 3 by the number 10 and obtain the number 30. The practical and no doubt obligatory physical interpretation of the answer in the above situation is that one must pay 30 dollars rather than walk out with 30 pairs of shoes. We see, therefore, that one must distinguish between the purely mathematical operation of multiplying 3 by 10 and the physical objects with which these numbers may be associated.

The same point is involved in a slightly different situation. Mathematically $2/1$ is equal to $4/2$. But the corresponding physical fact may not be true. One may be willing to accept 4 half-pies instead of 2 whole pies, but no woman would accept 4 half-dresses in place of 2 dresses or 4 half-shoes in place of 1 pair of whole shoes.

The Egyptians and Babylonians did reach the stage of working with pure numbers dissociated from physical objects. But like young children of our civilization, they hardly recognized that they were dealing with abstract entities. By contrast, the Greeks not only recognized numbers as ideas but emphasized that this is the way we must regard them. The Greek philosopher Plato, who lived from 428 to 348 B.C. and whose ideas are representative of the classical Greek period, says in his famous work, the *Republic,* "We must endeavor that those who are to be the principal men of our State go and learn arithmetic, not as amateurs, but they must carry on the study until they see the nature of numbers with the mind only; . . . arithmetic has a very great and elevating effect, compelling the soul to reason about abstract number, and rebelling against the introduction of visible or tangible objects into the argument."

The Greeks not only emphasized the distinction between pure numbers and the physical applications of such numbers, but they preferred the former to the latter. The study of the properties of pure numbers, which they called *arithmetica,* was esteemed as a worthy activity of the mind, whereas the use of numbers in practical applications, *logistica,* was deprecated as a mere skill.

Geometrical thinking prior to the classical Greek period was even less advanced than thinking about numbers. To the Egyptians and Babylonians the words "straight line" meant no more than a stretched rope or a line traced in sand, and a rectangle was a piece of land of a particular shape. The Greeks began the practice of treating point, line, triangle, and other geometrical notions as concepts. They did of course appreciate that these mental notions are suggested by physical objects, but they stressed that the concepts differ from the physical examples as sharply as the concept of time differs from the passage of the sun across the sky. The stretched string is a physical object illustrating the concept of line, but the mathematical line has no thickness, no color, no molecular structure, and no tension.

The Greeks were explicit in asserting that geometry deals with abstractions. Speaking of mathematicians, Plato says, "And do you not know also that although they make use of the visible forms and reason about them, they are thinking not of these, but of the ideals which they resemble; not of the figures which they draw, but of the absolute square and the absolute diameter . . . they are really seeking to behold the things themselves, which can be seen only with the eye of the mind?"

On the basis of elementary abstractions, mathematics creates others which are even more remote from anything real. Negative numbers, equations involving unknowns, formulas, and other concepts we shall encounter are abstractions built upon abstractions. Fortunately, every abstraction is ultimately derived from, and therefore understandable in terms of, intuitively meaningful objects or phenomena. The mind does play its part in the creation of mathematical concepts, but the mind does not function independently of the outside world. Indeed the mathematician who treats concepts that have no physically real or intuitive origins is almost surely talking nonsense. The intimate connection between mathematics and objects and events in the physical world is reassuring, for it means that we can not only hope to understand the mathematics proper, but also expect physically meaningful and valuable conclusions.

The use of abstractions is not peculiar to mathematics. The concepts of force, mass, and energy, which are studied in physics, are abstractions from real phenomena. The concept of wealth, an abstraction from material possessions such as land, buildings, and jewelry, is studied in economics. The concepts of liberty, justice, and democracy are familiar in political science. Indeed, with respect to the use of abstract concepts, the distinction between mathematics on the one hand and the physical and social sciences on the other is not a sharp one. In fact, the influence of mathematics and mathematical ways of thinking on the physical sciences especially has led to ever increasing use of abstract concepts including some, as we shall see, which may have no direct real counterpart at all, any more than a mathematical formula has a direct real counterpart.

The very fact that other studies also engage in abstractions raises an important question. Mathematics is confined to some abstractions, numbers and geometrical forms, and to concepts built upon these basic ones. Abstractions such as mass, force, and energy belong to physics, and still other abstractions belong to other subjects. Why doesn't mathematics also treat forces, wealth, and justice? Certainly these concepts are also worthy of study. Did the mathematicians make an agreement with physicists, economists, and others to divide the concepts among themselves? The restriction of mathematics to numbers and geometrical forms is partly an historical accident and partly a deliberate decision made by the Greeks. Numbers and geometrical forms had already been introduced by the Egyptians and Babylonians, and their utility in daily life was established. Since the Greeks learned the rudiments of mathematics from these civilizations, the sheer weight of tradition might have caused them to continue the practice of regarding mathematics as the study of numbers and geometrical figures. But people as original and bold in thought as the Greeks would not have been bound merely by tradition, had they not found in numbers and geometrical forms sharp and clear notions which appealed to their delight in the processes of exact thinking. However, an even more compelling reason was their belief that numerical and geometrical properties and relationships were basic, that they underlay the phenomena of the physical world and the design of the entire universe. Hence to understand the world one should seek this mathematical essence. The brilliance and depth of their conception of the universe will be revealed more and more as we proceed.

When one compares the pre-Greek and Greek understanding of the concepts of mathematics and notes the sharp transition from the concrete to the abstract, another question presents itself. The Greeks eliminated the physical substance and retained only the idea. Why did they do it? Surely it is more difficult to think about abstractions than about concrete things. Also it would seem that an attempt to study nature by concentrating on just a few aspects of physical objects rather than on the objects themselves would fall far short of effectiveness.

Insofar as the emphasis on abstractions is concerned, the Greeks saw at once what any thinking people would see sooner or later. One advantage of treating abstractions is the gain in generality. When a child learns that $5 + 5 = 10$, he acquires in one swoop a fact which applies to hundreds of situations. Likewise a theorem proved about the abstract triangle applies to a triangular piece of land, a musical percussion instrument, and a triangle determined by three heavenly bodies at any instant of time. It has been said that the process of abstraction amounts to giving the same name to different things, but this very recognition that different objects possess the common property named in the abstraction carries with it the implication that anything true of the abstraction will apply to the several objects. Part of the secret of the power of mathematics is that it deals with abstractions.

Another advantage of abstraction was also clear to the Greeks. Abstracting from a physical situation just those properties which are to be studied frees the mind from burdensome and irrelevant details and enables one to concentrate on the features of interest. When one wishes to determine the area of a piece of land, only shape and size are relevant, and it is desirable to think only about these and not about the fertility of the soil.

The emphasis on mathematical abstractions by the classical Greeks was part and parcel of their outlook on the entire universe. They were concerned with truths, and leading philosophical schools, notably the Pythagoreans and the Platonists, maintained that truths could be established only about abstractions. Let us follow their argument. The physical world presents various objects to the senses. But the impressions received by the senses are inexact, transitory, and constantly changing; indeed, senses may be even deceived, as by mirages. However, truth, by its very meaning, must consist of permanent, unchanging, definite entities and relationships. Fortunately, the intelligence of man excited to reflection by the impressions of sensible objects may rise to higher conceptions of the realities faintly exhibited to the senses, and so man may rise to the contemplation of ideas. These are eternal realities and the true goal of thought, whereas mere "things are the shadows of ideas thrown on the screen of experience."

Thus Plato would say that there is nothing real in a horse, a house, or a beautiful woman. The reality is in the universal type or idea of a horse, a home, or a woman. The ideas, among which Plato emphasized Beauty, Justice, Intelligence, Goodness, Perfection, and the State, are independent of the superficial appearances of things, of the flux of life, and of the biases and warped desires of man; they are in fact constant and invariable, and knowledge concerning them is firm and indestructible. Real and eternal knowledge concerns these ideas, rather than sensuous objects. This distinction between the intelligible world and the world revealed by the senses is all-important in Plato.

To put Plato's doctrine in everyday language, fundamental knowledge does not concern itself with what John ate, Mary heard, or William felt. Knowledge must rise above individuals and particular objects and tell us about broad classes of objects and about man as a whole. True knowledge must therefore of necessity concern abstractions. Plato admits that physical or sensible objects suggest the ideas just as diagrams of geometry suggest abstract geometrical concepts. Hence there is a point to studying physical objects, but one must not lose himself in trivial and confusing minutiae.

The abstractions of mathematics possessed a special importance for the Greeks. The philosophers pointed out that, to pass from a knowledge of the world of matter to the world of ideas, man must train his mind to grasp the ideas. These highest realities blind the person who is not prepared to contemplate them. He is, to use Plato's famous simile, like one who lives continuously in the deep shadows of a cave and is suddenly brought out into the sunlight. The study of mathematics helps make the transition from darkness to light. Mathematics is in fact ideally suited to prepare the mind for higher forms of thought because on the one hand it pertains to the world of visible things and on the other hand it deals with abstract concepts.

FIG. 3–1. Polyclitus: *Spear-bearer* (*Daryphorus*). National Museum, Naples.

Hence through the study of mathematics man learns to pass from concrete figures to abstract forms; moreover, this study purifies the mind by drawing it away from the contemplation of the sensible and perishable and leading it to the eternal ideas. These latter abstractions are on the same mental level as the concepts of mathematics. Thus, Socrates says, "The understanding of mathematics is necessary for a sound grasp of ethics."

To sum up Plato's position we may say that while a little knowledge of geometry and calculation suffices for practical needs, the higher and more advanced portions tend to lift the mind above mundane considerations and enable it to apprehend the final aim of philosophy, the idea of the Good. Mathematics, then, is the best preparation for philosophy. For this reason Plato recommended that the future rulers, who were to be philosopher-kings,

FIG. 3–2. *Bust of Cæsar.* Vatican.

be trained for ten years, from age 20 to 30, in the study of the exact sciences, arithmetic, plane geometry, solid geometry, astronomy, and harmonics (music). The oft-repeated inscription over the doors of Plato's Academy, stating that no one ignorant of mathematics should enter, fully expresses the importance he attached to the subject, although modern critics of Plato read into these words his admission that one would not be able to learn it after entering. This value of mathematical training led one historian to remark, "Mathematics considered as a science owes its origins to the idealistic needs of the Greek philosophers, and not as fable has it, to the practical demands of Egyptian economics."

The preference of the Greeks for abstractions is equally evident in the art of the great sculptors, Polyclitus, Praxiteles, and Phidias. One has only to glance at the face in Fig. 3–1 to observe that Greek sculpture of the classical period dwelt not on particular men and women but on types, ideal types. Idealization extended to standardization of the ratios of the parts of the body to each other. Polyclitus believed, in fact, that there were ideal numerical ratios which fix the proportions of the human body. Perfect art must follow these ideal proportions. He wrote a book, *The Canon,* on the subject and constructed the "Spear-bearer" to illustrate the thesis. These abstract types contrast sharply with what is found in numerous busts and statues of private individuals and military and political leaders made by Romans (Fig. 3–2).

FIG. 3–3. *Parthenon,* Athens, Greece.

Greek architecture also reveals the emphasis on ideal forms. The simple and austere buildings were always rectangular in shape; even the ratios of the dimensions employed were fixed. The Parthenon at Athens (Fig. 3–3) is an example of the style and proportions found in almost all Greek temples.

EXERCISES

1. Suppose 5 trucks pass by with 4 men in each. To answer the question of how many men there are in all the trucks, a person reasons that 4 men times 5 trucks is 20 men. On the other hand, if there are 4 men each owning 5 trucks, the total number of trucks is 20 trucks. Hence 4 men times 5 trucks yields 20 trucks. How do you know that the answer is 20 men in one case and 20 trucks in the other?

2. If the product of 25¢ and 25¢ is obtained by multiplying 0.25 by 0.25 the result is 0.0625 or $6\frac{1}{4}$¢. Does it pay to multiply money?

3. Can you suggest some abstract political or ethical concepts?

4. Suppose 30 books are to be distributed among 5 people. Since 30 books divided by 5 people yields 6 books, each person gets 6 books. Criticize the reasoning.

5. A store advertises that it will give a credit of $1 for each purchase amounting to $1. A man who spends $6 reasons that he should receive a credit of $6 times $1, or $6. But $6 is 600¢ and $1 is 100¢. Hence 600¢ times 100¢ is 60,000¢, or $600. It

would seem that it is more profitable to operate with the almost worthless cent than with dollars! What is wrong?

6. What does the statement that mathematics deals with abstractions mean?

7. Why did the Greeks make mathematics abstract?

3–3 Idealization. The geometrical notions of mathematics are abstract in the sense that shapes are mental concepts which actual physical objects merely approximate. The sides of a rectangular piece of land may not be exactly straight nor would each angle be exactly 90°. Hence, in adopting such abstract concepts, mathematics does idealize. But in studying the physical world, mathematics also idealizes in another sense which is equally important. Very often mathematicians undertake to study an object which is not a sphere and yet choose to regard it as such. For example, the earth is not a sphere but a spheroid, that is, a sphere flattened at the top and bottom. Yet in many physical problems which are treated mathematically the earth is represented as a perfect sphere. In problems of astronomy a large mass such as the earth or the sun is often regarded as concentrated at one point.

In making such idealizations, the mathematician deliberately distorts or approximates at least some features of the physical situation. Why does he do it? The reason usually is that he simplifies the problem and yet is quite sure that he has not introduced any gross errors. If one is to investigate, for example, the motion of a shell which travels ten miles, the difference between the assumed spherical shape of the earth and the true spheroidal shape does not matter. In fact, in the study of a motion which takes place over a limited region, say one mile, it may be sufficient to treat the earth as a flat surface. On the other hand, if one were to draw a very accurate map of the earth, he would take into account that the shape is spheroidal. As another example, to find the distance to the moon, it is good enough to assume that the moon is a point in space. However, to find the size of the moon, it is clearly pointless to regard the moon as a point.

The question does arise, how does the mathematician know when idealization is justified? There is no simple answer to this question. If he has to solve a series of like problems, he may solve one using the correct figure, and another, using a simplified figure, and compare results. If the difference does not matter for his purposes, he may then retain the simpler figure for the remaining problems. Sometimes he can estimate the error introduced by using the simpler figure and may find that this error is too small to matter. Or the mathematician may make the idealization and use the result because it is the best he can do. Then he must accept experience as his guide in deciding whether the result is good enough.

To idealize by deliberately introducing a simplification is to lie a little, but the lie is a white one. Using idealizations to study the physical world does impose a limitation on what mathematics accomplishes, but we shall find that even where idealizations are employed, the knowledge gained is of immense value.

Exercises

1. Distinguish between abstraction and idealization.
2. Is it correct to assume that the lines of sight to the sun from two places A and B on the earth's surface are parallel?
3. Suppose you wished to measure the height of a flagpole. Would it be wise to regard the flagpole as a line segment?

3-4 Methods of reasoning. There are many ways, more or less reliable, of obtaining knowledge. One can resort to authority as one often does in obtaining historical knowledge. One may accept revelation as many religious people do. And one may rely upon experience. The foods we eat are chosen on the basis of experience. No one determined in advance by careful chemical analysis that bread is a healthful food.

We may pass over with a mere mention such sources of knowledge as authority and revelation, for these sources cannot be helpful in building mathematics or in acquiring knowledge of the physical world. It is true that in the medieval period of Western European culture men did contend that all desirable knowledge of nature was revealed in the Bible. However in no significant period of scientific thought has this view played any role. Experience, on the other hand, is a useful source of knowledge. But there are difficulties in employing this method. We should not wish to build a fifty-story building in order to decide whether a steel beam of specified dimensions is strong enough to be used in the foundation. Moreover, even if one should happen to choose workable dimensions, the choice may be wasteful of materials. Of course, experience is of no use in determining the size of the earth or the distance to the moon.

Closely related to experience is the method of experiment which amounts to setting up and going through a series of purposive, systematic experiences. It is true that experimentation fundamentally is experience, but it is usually accompanied by careful planning which eliminates extraneous factors, and the experience is repeated enough times to yield reliable information. However, experimentation is subject to much the same limitations as experience.

Are authority, revelation, experience, and even experimentation the only methods of obtaining knowledge? The answer is no. The major method is reasoning, and within the domain of reasoning there are several forms. One can reason by analogy. A boy who is considering a college career may note that his friend went to college and handled it successfully. He argues that since he is very much like his friend in physical and mental qualities, he too should succeed in college work. The method of reasoning just illustrated is to find a similar situation or circumstance and to argue that what was true for the similar case should be true of the one in question. Of course, one must be able to find a similar situation and one must take the chance that the differences do not matter.

Another common method of reasoning is induction. People use this method of reasoning every day. Because a person may have had unfortunate experiences in dealing with a few department stores, he concludes that all department stores are bad to deal with. Or, for example, experimentation would show that iron, copper, brass, oil, and other substances expand when heated, and one consequently concludes that *all* substances expand when heated. Inductive reasoning is in fact the common method used in experimentation. An experiment is generally performed many times, and if the same result is obtained each time, the experimenter concludes that the result will always follow. The essence of induction is that one observes repeated occurrences of the *same phenomenon* and concludes that the phenomenon will always occur. Conclusions obtained by induction seem well warranted by the evidence, especially when the number of instances observed is large. Thus the sun is observed so often to rise in the morning that one is sure it has risen even on those mornings when it is hidden by clouds.

There is still a third method of reasoning, called deduction. Let us consider some examples. If we accept as basic facts that honest people return found money and that John is honest, we may conclude unquestionably that John will return money that he finds. Likewise, if we start with the facts that no mathematician is a fool and that John is a mathematician, then we may conclude with certainty that John is not a fool. In deductive reasoning we start with certain statements, called premises, and assert a conclusion which is a *necessary* or *inescapable* consequence of the premises.

All three methods of reasoning, analogy, induction, and deduction, and other methods we could describe, are commonly employed. There is one essential difference, however, between deduction on the one hand and all other methods of reasoning on the other. Whereas the conclusion drawn by analogy or induction has only a probability of being correct, the conclusion drawn by deduction necessarily holds. Thus one might argue that because lions are similar to cows and cows eat grass, lions also eat grass. This argument by analogy leads to a false conclusion. The same is true for induction: although experiment may indeed show that two dozen different substances expand when heated, it does not necessarily follow that all substances do. Thus water, for example, when heated from 0° to 4° centigrade* does not expand; it contracts.

Since deductive reasoning has the outstanding advantage of yielding an indubitable conclusion, it would seem obvious that one should always use this method in preference to the others. But the situation is not that simple. For one thing analogy and induction are often easier to employ. In the case of analogy, a similar situation may be readily available. In the case of induction, experience often supplies the facts with no effort at all. The fact that the sun rises every morning is noticed by all of us almost automatically. Furthermore, deductive reasoning calls for premises which it may be impossible to obtain despite all efforts. Fortunately we can use deductive reasoning in a variety of situations. For example, we can use it to find the distance to the moon. In this instance, both analogy and induction are

---

* In scientific texts, "celsius" is considered to be the more precise term.

powerless, whereas, as we shall see later, deduction will obtain the result quickly. It is also apparent that where deduction can replace induction based on expensive experimentation, deduction is preferred.

Because we shall be concerned primarily with deductive reasoning, let us become a little more familiar with it. We have given several examples of deductive reasoning and have asserted that the conclusions are inescapable consequences of the premises. Let us consider, however, the following example. We shall accept as premises that all good cars are expensive and that the Locomobile is expensive. We might conclude that the Locomobile is a good car. This example purports to be deductive reasoning but, unfortunately, the reasoning is not correct. All good cars may be expensive, but this statement is not the same as the assertion that all expensive cars are good. The particular conclusion, then, is not a necessary consequence of the premises, or one might say that deduction has been attempted, but the argument is not valid.

There are many occasions when a deductive argument is advanced, and we have to decide whether the argument is valid or sound. A good way of picturing the argument which helps us to determine its soundness is called the circle test. Suppose that the deductive argument starts with the premises that

all parallelograms are quadrilaterals

and

figure $ABCD$ is a quadrilateral,

and ends with the conclusion that

figure $ABCD$ is a parallelogram.

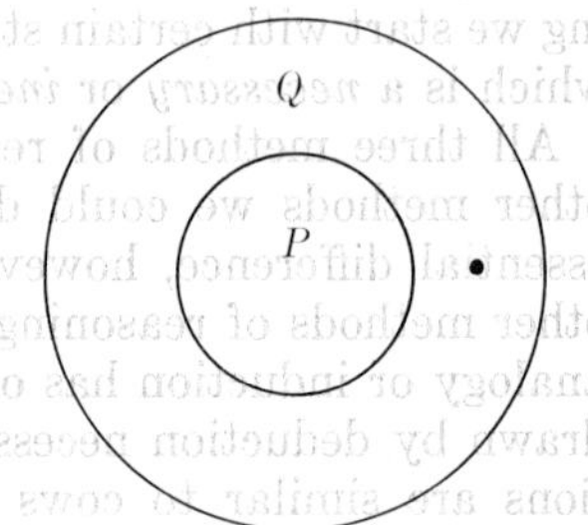

FIGURE 3–4

To test this argument we begin with the first premise and note that it deals with quadrilaterals and parallelograms. We represent all possible quadrilaterals as points in a circle Q (Fig. 3–4). Since the first premise asserts that all parallelograms are quadrilaterals, a circle P representing all parallelograms must fall *within* the circle Q. The second premise of the argument asserts that figure $ABCD$ is a quadrilatral. Hence the point representing figure $ABCD$ must fall within the circle Q which includes all quadrilaterals. However, the given facts do not oblige us to think of figure $ABCD$ as falling within the circle P. Hence we are not forced to conclude that figure $ABCD$ is a parallelogram; in other words, it is a quadrilateral but not necessarily a parallelogram. Since the conclusion of a valid deductive argument must *necessarily* follow from the premises and the above conclusion does not, the deductive argument is invalid or unsound.

We shall encounter numerous instances of deductive reasoning in our work. Although in most cases common experience will enable us to be sure that the reasoning is or is not valid, one can, when necessary, employ the circle test. The subject of deductive reasoning is customarily studied in logic, a discipline which treats more thoroughly the valid forms of reasoning. However, we shall not need to depend upon formal training in logic. In fact, mathematics itself is the superb field for the exercise of reason and the best experience in logic.

Exercises

1. A coin is tossed ten times and each time it falls heads. What conclusion does inductive reasoning warrant?
2. Characterize deductive reasoning.
3. What superior features does deductive reasoning possess compared with induction and analogy?
4. Can you prove deductively that George Washington was the best president of the United States?
5. Can one always apply deductive reasoning to prove a desired statement?
6. Can you prove deductively that monogamy is the best system of marriage?
7. Are the following purportedly deductive arguments valid?
   (a) All good cars are expensive. A Daffy is an expensive car. Therefore a Daffy is a good car.
   (b) All New Yorkers are good citizens. All good citizens give to charity. Therefore all New Yorkers give to charity.
   (c) All college students are clever. All young boys are clever. Therefore all young boys are college students.
   (d) The same premises as in (c), but the conclusion: All college students are young boys.
   (e) It rains every Monday and it is raining today; hence today must be Monday.
   (f) No decent people curse; Americans are decent; therefore Americans do not curse.
   (g) No decent people curse; Americans curse; therefore some Americans are not decent.
   (h) No decent people curse; some Americans are not decent; therefore some Americans curse.
   (i) No undergraduates have a bachelor-of-arts degree; no freshmen have a bachelor-of-arts degree. Therefore all freshmen are undergraduates.
8. If someone gave you a valid deductive argument but the conclusion was not true, where would you look for the difficulty?
9. Distinguish between the validity of a deductive argument and the truth of the conclusion.

3-5 Mathematical proof. We have seen so far in our discussion of reasoning that there are several methods of reasoning and that all are useful. These methods can be applied to mathematical problems. Let us suppose that one wished to determine the sum of the angles of a triangle. He could draw on paper many different triangles or construct some out of wood or metal and measure the angles. In each case he would find that the sum is as close to 180° as the eye and hand can determine. By inductive reasoning he could conclude that the sum of the angles in every triangle is 180°. As a matter

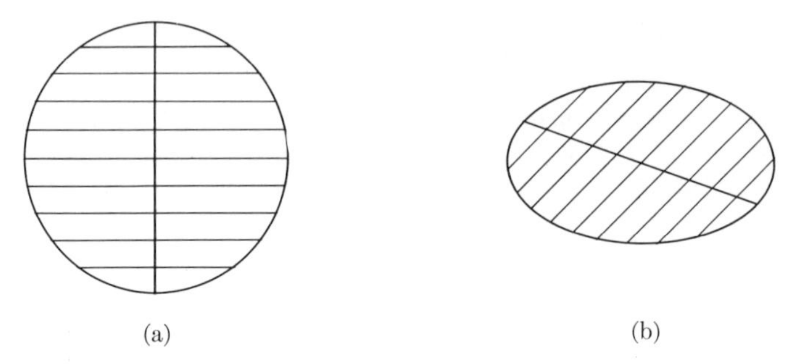

FIG. 3–5. The mid-points of parallel chords lie on a straight line.

of fact, the Babylonians and Egyptians did in effect use inductive reasoning to establish their mathematical results. They must have determined by measurement that the area of a triangle is one-half the base times the altitude and, having used this formula repeatedly and having obtained reliable results, they concluded that the formula is correct.

To see that reasoning by analogy can be used in mathematics, let us note first that the centers of a set of parallel chords of a circle lie on a straight line (Fig. 3–5a). In fact this line is a diameter of the circle. Now an ellipse (Fig. 3–5b) is very much like a circle. Hence one might conclude that the centers of a set of parallel chords of an ellipse also lie on a straight line.

Deduction is certainly applicable in mathematics. The proofs which one learns in Euclidean geometry are deductive. As another illustration we might consider the following algebraic argument. Suppose one wishes to solve the equation $x - 3 = 7$. One knows that equals added to equals give equals. If we added 3 to both sides of the preceding equation, we would be adding equals to an equality. Hence the addition of 3 to both sides is justified. When this is done, the result is $x = 10$, and the equation is solved.

Thus all three methods are applicable. There is a lot to be said for the use of induction and analogy. The inductive argument for the sum of the angles of a triangle can be carried out in a matter of minutes. The argument by analogy given above is also readily made. On the other hand, finding deductive proofs for these same conclusions might take weeks or might never be accomplished by the average person. As a matter of fact, we shall soon encounter some examples of conjectures for which the inductive evidence is overwhelming but for which no deductive proof has been thus far obtained even by the best mathematicians.

Despite the usefulness and advantages of induction and analogy, mathematics does not rely upon these methods to establish its conclusions. *All mathematical proofs must be deductive.* Each proof is a chain of deductive arguments, each of which has its premises and conclusion.

Before examining the reasons for this restriction to deductive proof, we might contrast the method of mathematics with those of the physical and social sciences. The scientist feels free to draw conclusions by any method of reasoning and, for that matter, on the basis of observation, experimentation, and experience. He may reason by analogy as, for example, when he reasons about sound waves by observing water waves or when he reasons about a possible cure for a disease affecting human beings by testing the cure on animals. In fact reasoning by analogy is a powerful method in science. The scientist may also reason inductively: if he observes many times that hydrogen and oxygen combine to form water, he will conclude that this combination will always form water. At some stages of his work the scientist may also reason deductively and, in fact, even employ the concepts and methods of mathematics proper.

To contrast further the method of mathematics with that of the scientist—and perhaps to illustrate just how stubborn the mathematician can be—we might consider a rather famous example. Mathematicians are concerned with whole numbers, or integers, and among these they distinguish the prime numbers. A prime is a number which has no integral divisors other than itself and 1. Thus 11 is a prime number, whereas 12 is not because it is divisible by 2 for example. Now by actual trial one finds that each of the first few even numbers can be expressed as the sum of two prime numbers. For example, $2 = 1 + 1$; $4 = 2 + 2$; $6 = 3 + 3$; $8 = 3 + 5$; $10 = 3 + 7$; . . . . If one investigates larger and larger even numbers, one finds without exception that every even number can be expressed as the sum of two primes. Hence by *inductive* reasoning one could conclude that *every* even number is the sum of two prime numbers.

But the mathematician does not accept this conclusion as a theorem of mathematics because it has not been proved deductively from acceptable premises. The conjecture that every even number is the sum of two primes, known as Goldbach's hypothesis because it was first suggested by the eighteenth-century mathematician Christian Goldbach, is an unsolved problem of mathematics. The mathematician will insist on a deductive proof even if it takes thousands of years, as it literally has in some instances, to find one. However a scientist would not hesitate to use this inductively well supported conclusion.

Of course, the scientist should not be surprised to find that some of his conclusions are false because, as we have seen, induction and analogy do not lead to sure conclusions. But it does seem as though the scientist's procedure is wiser since he can take advantage of any method of reasoning which will help him advance his knowledge. The mathematician by comparison appears to be narrow-minded or shortsighted. He achieves a reputation for certainty, but at the price of limiting his results to those which can be established deductively. How wise the mathematician may be in his insistence on deductive proof we shall learn as we proceed.

The decision to confine mathematical proof to deductive reasoning was made by the Greeks of the classical period. And they not only rejected all other methods of proof in mathematics, but they also discarded all the knowledge which the Egyptians and Babylonians had acquired over a period of four thousand years because it had only an empirical justification. Why did the Greeks do it?

The intellectuals of the classical Greek period were largely absorbed in philosophy and these same men, because they possessed intellectual interests, were the very ones who developed mathematics as a system of thought. The Ionians, the Pythagoreans, the Sophists, the Platonists, and the Aristotelians were the leading philosophers who gave mathematics its definitive form. The credit for initiating this step probably belongs to one school of Greek philosopher-mathematicians, known as the Ionian school. However, if credit can be assigned to any one person, it belongs to Thales, who lived about 600 B.C. Though a native of Miletus, a Greek city in Asia Minor, Thales spent many years in Egypt as a merchant. There he learned what the Egyptians had to offer in mathematics and science, but apparently he was not satisfied, for he would accept no results that could not be established by deductive reasoning from clearly acceptable axioms. In his wisdom Thales perceived what we shall perceive as we follow the story of mathematics, that the obvious is far more suspect than the abstruse.

Thales himself supplied the proof of many geometrical theorems. He also acquired great fame as an astronomer and is believed to have predicted an eclipse of the sun in 585 B.C. A philosopher-astronomer-mathematician might readily be accused of being an impractical stargazer, but Aristotle tells us otherwise. In a year when olives promised to be plentiful, Thales shrewdly cornered all the oil presses to be found in Miletus and in Chios. When the olives were ripe for pressing, Thales was in a position to rent out the presses at his own price. Thales might perhaps have lived in history as a leading businessman, but he is far better known as the father of Greek philosophy and mathematics. From his time onward, deductive proof became the standard in mathematics.

It is to be expected that philosophers would favor deductive reasoning. Whereas scientists select particular phenomena for observation and experimentation and then draw conclusions by induction or analogy, philosophers are concerned with broad knowledge about man and the physical world. To establish universal truths, such as that man is basically good, that the world is designed, or that man's life has purpose, deductive reasoning from acceptable principles is far more feasible than induction or analogy. As Plato put it in his *Republic,* "If persons cannot give or receive a reason, they cannot attain that knowledge which, as we have said, man ought to have."

There is another reason that philosophers favor deductive reasoning. These men seek truths, the eternal verities. We have seen that of all the methods of reasoning only deductive reasoning grants sure and exact conclusions. Hence this is the method which philosophers would almost necessarily adopt. Not only do induction and analogy fail to yield absolutely unquestionable conclusions, but many Greek philosophers would not have accepted as facts the data with which these methods operate, because these are acquired by the senses. Plato stressed the unreliability of sensory perceptions. Empirical knowledge, as Plato put it, yields opinion only.

The Greek preference for deduction had a sociological basis. Contrary to our own society wherein bankers and industrialists are respected most, in classical Greek society, the philosophers, mathematicians, and artists were the leading citizens. The upper class regarded earning a living as an unfortunate necessity. Work robbed man of time and energy for intellectual activities, the duties of citizenship and discussion. These Greeks did not hesitate to express their disdain for work and business. The Pythagoreans, who, as we shall see, delighted in the properties of numbers and applied numbers to the study of nature, derided the use of numbers in commerce. They boasted that they sought knowledge rather than wealth. Plato, too, maintained that knowledge rather than trade was the goal in studying arithmetic. Freemen, he declared, who allowed themselves to become preoccupied with business should be punished, and a civilization which is concerned mainly with the material wants of man is no more than a "city of happy pigs." Xenophon, the famous Greek general and historian, says, "What are called the mechanical arts carry a social stigma and are rightly dishonored in our cities." Aristotle wanted an ideal society in which citizens would not have to practice any mechanical arts. Among the Boeotians, one of the independent tribes of ancient Greece, those who defiled themselves with commerce were by law excluded from state positions for ten years.

Who did the daily work of providing food, shelter, clothing, and the other necessities of life? Slaves and free men ineligible for citizenship ran the businesses and the households, did unskilled and technical work, managed the industries, and carried on the professions such as medicine. They produced even the articles of refinement and luxury.

In view of this attitude of the Greek upper class towards commerce and trade, it is not hard to understand the classical Greek's preference for deduction. People who do not "live" in the workaday world can learn little from experience, and people who will not observe and use their hands to experiment will not have the facts on which to base reasoning by analogy or induction. In fact the institution of slavery in classical Greek society fostered a divorce of theory from practice and favored the development of speculative and deductive science and mathematics at the expense of experimentation and practical applications.

One may also assert of the Greeks that their preference for deductive reasoning was part and parcel of their search for beauty. In all their activities beauty was the ultimate objective. The understanding of nature was pursued because it pleased their minds. The order which they began to uncover showed them that nature properly understood was an artistic work which could arouse emotions as deeply satisfying as literature or poetry. Aristotle says, "Then will nature's purpose and her deep-seated laws be revealed in all things, all tending in her multitudinous work to one or another form of the beautiful." And in the field of ethics Socrates asks, "Do you think that the good is one thing and the beautiful another?" There is no doubt that the beauty of sharp, clean, convincing proofs satisfied the aesthetic sense of the Greeks. Such proofs bring home a truth with the force of great art and often contain an element of ingenuity which produces surprise and gratification. Mathematics to the Greeks was an art of the mind.

Over and above the various cultural forces which inclined the Greeks toward deduction were a farsightedness and a wisdom which mark true genius. The Greeks were the first to recognize the power of reason. The mind was a faculty not only additional to the senses but more powerful than the senses. The mind can survey all the whole numbers, but the senses are limited to perceiving only a few at a time. The mind can encompass the earth and the heavens; the sense of sight is confined to a small angle of vision. Indeed the mind can predict future events which the senses of contemporaries will not live to perceive. This mental faculty could be exploited. The Greeks saw clearly that if man could obtain some truths, he could establish others entirely by reasoning, and these new truths, together with the original ones, enabled man to establish still other truths. Indeed the possibilities would multiply at an enormous rate. Here was a means of acquiring knowledge which had been either overlooked or neglected.

This was indeed the plan which the Greeks projected for mathematics. By starting with some truths about numbers and geometrical figures they could deduce others. A chain of deductions might lead to a significant new fact which would be labeled a theorem to call attention to its importance. Each theorem added to the stock of truths which could serve as premises for new deductive arguments, and so one could build an immense body of knowledge about the basic concepts.

Although the Greeks may have been guilty of overemphasizing the power of the mind unaided by experience and observation to obtain truths, there is no doubt that in insisting on deductive proof as the sole method, they rose above the practical level of carpenters, surveyors, farmers, and navigators. At the same time they elevated the subject of mathematics to a system of thought. Moreover the preference for reason which they exhibited gave this faculty the high prestige which it now enjoys and permitted it to exercise its true powers. When we have surveyed some of the creations of the mind which succeeding civilizations building on the Greek plan contributed, we shall appreciate the true depth of the Greek vision.

Exercises

1. Compare Greek and pre-Greek standards of proof in mathematics. Reread the relevant parts of Chapter 2.

2. Distinguish science and mathematics with respect to ways of establishing conclusions.

3. Explain the statement that the Greeks converted mathematics from an empirical science to a deductive system.

4. Are the following deductive arguments valid?

(a) All even numbers are divisible by 4. Ten is an even number. Hence 10 is divisible by 4. (b) Equals divided by equals give equals. Dividing both sides of $3x = 6$ by 3 is dividing equals by equals. Hence $x = 2$.

5. Does it follow from the fact that the square of any odd number is odd that the square of any even number is even?

6. Criticize the argument: The square of every even number is even because $2^2 = 4$, $4^2 = 16$, $6^2 = 36$, and it is obvious that the square of any larger even number also is even.

7. If we accept the premises that the square of any odd number is odd and the square of any even number is even, does it follow deductively that if the square of a number is even, the number must be even?

8. Why did the Greeks insist on deductive proof in mathematics?

9. Let us take for granted that if a triangle has two equal sides, the opposite angles are equal and that we have a triangle in which all three sides are equal. Prove deductively that all three angles are equal in the triangle under consideration. You may also use the premise that things equal to the same thing are equal to one another.

10. How did the Greeks propose to obtain new truths from known ones?

3–6 Axioms and definitions. From our discussion of deductive reasoning we know that to apply such reasoning we must have premises. Hence the question arises, What premises does the mathematician use? Since the mathematician reasons about numbers and geometrical figures, he must of course have facts about these concepts. These cannot be obtained deductively because then there would have to be prior premises, and if one continues this process backward, there would be no starting point. The Greeks readily found premises. It seemed indisputable, for example, that two points determine one and only one line and that equals added to equals give equals.

To the Greeks the premises on which mathematics was to be built were self-evident truths, and they called these premises axioms. Socrates and Plato believed, as did many later philosophers, that these truths were already in our minds at birth and that we had but to recall them. And since the Greeks believed that axioms were truths and since deductive reasoning yielded unquestionable conclusions, they also believed that theorems were truths. This view is no longer held, and we shall see later in this book why mathematicians abandoned it. We now know that axioms are suggested by experience and observation. Naturally, to be as certain as we can of these axioms we select those facts which seem clearest and most reliable in our experience. But we must recognize that there is no guarantee that we have selected truths about the world. Some mathematicians prefer to use the word assumptions instead of axioms to emphasize this point.

The mathematician also takes care to state his axioms at the outset and to be sure as he performs his reasoning that no assumptions or facts are used which were not so stated. There is an interesting story told by former President Charles W. Eliot of Harvard which illustrates the likelihood of introducing unwarranted premises. He entered a crowded restaurant and handed his hat to the doorman. When he came out, the doorman at once picked Eliot's hat out of hundreds on the racks and gave it to him. He was amazed that the doorman could remember so well and asked him, "How did you know that was my hat?" "I didn't," replied the doorman. "Why, then, did you hand it to me?" The doorman's reply was, "Because you handed it to me, sir."

Undoubtedly no harm would have been done if the doorman had assumed that the hat he returned to President Eliot belonged to the man. But the mathematician interested in obtaining conclusions about the physical world might be wasting his time if he unwittingly introduced an assumption that he had no right to make.

There is one other element in the logical structure of mathematics about which we shall say a few words now and return to in a later chapter (Chapter 26). Like other studies mathematics uses definitions. Whenever we have occasion to use a concept whose description requires a lengthy statement, we introduce a single word or phrase to replace that lengthy statement. For example, we may wish to talk about the figure which consists of three distinct points which do not lie on the same straight line and of the line segments joining these points. It is convenient to introduce the word triangle to represent this long description. Likewise the word circle represents the set of all points which are at a fixed distance from a definite point. The definite point is called the center, and the fixed distance is called the radius. Definitions promote brevity.

Exercises

1. What belief did the Greeks hold about the axioms of mathematics?
2. Summarize the changes which the Greeks made in the nature of mathematics.
3. Is it fair to say that mathematics is the child of philosophy?

3-7 The creation of mathematics. Because mathematical proof is strictly deductive and merely reasonable or appealing arguments may not be used to establish a conclusion, mathematics has been described as a deductive science, or as the science which derives necessary conclusions, that is, conclusions which necessarily or inevitably follow from the axioms. This description of mathematics is incomplete. Mathematicians must also discover what to prove and how to go about establishing proofs. These processes are also part of mathematics and they are *not* deductive.

How does the mathematician discover what to prove and the deductive arguments that lead to the conclusions? The most fertile source of mathematical ideas is nature herself. Mathematics is devoted to the study of the physical world, and simple experience or the more careful scrutiny of nature suggests idea after idea. Let us consider here a few simple examples. Once mathematicians had decided to devote themselves to geometric forms, it was only natural that such questions should arise as, what are the area, perimeter, and sum of the angles of common figures? Moreover, it is even possible to see how the precise statement of the theorem to be proved would follow from direct experience with physical objects. The mathematician might measure the sum of the angles of various triangles and find that these measurements all yield results close to 180°. Hence the suggestion that the sum of the angles in every triangle is 180° occurs as a possible theorem. To decide the question, which has more area, a polygon or a circle having the same perimeter, one might cut out cardboard figures and weigh them. The relative weights would suggest the statement of the theorem to be proved.

After some theorems have been suggested by direct physical problems, others are readily conceived by generalizing or varying the conditions. Thus knowing the problem of determining the sum of the angles of a triangle, one might ask, What is the sum of the angles of a quadrilateral, a pentagon, and so forth? That is, once the mathematician begins an investigation which is suggested by a physical problem, he can easily find new problems which go beyond the original one.

In the domains of arithmetic and algebra direct calculation with numbers, which is analogous to measurement in geometry, will suggest possible theorems. Anyone who has played with integers, for example, has doubtless observed the following facts:

$$1 = 1,$$

$$1 + 3 = 4 = 2^2,$$

$$1 + 3 + 5 = 9 = 3^2,$$

$$1 + 3 + 5 + 7 = 16 = 4^2,$$

$$\dots \dots \dots$$

We note that each number on the right is the square of the number of odd numbers appearing on the left; thus in the fourth line, there are four numbers on the left side, and the right side is 4^2. The general result which these calculations suggest is that if the first n odd numbers were on the left side, then the sum would be n^2. Of course, this possible theorem is not proved by the above calculations. Nor could it ever be proved by such calculations, for no mortal man could make the infinite set of computations required to establish the conclusion for *every* n. The calculations do, however, give the mathematician something to work on.

These simple illustrations of how observation, measurement, and calculation suggest possible theorems are not too striking or very profound. We shall see in the course of later work how physical problems suggest more significant mathematical theorems. However, experience, measurement, calculation, and generalization do not include the most fertile source of possible theorems. And it is especially true in seeking methods of proof that more than routine techniques must be utilized. In both endeavors the most important source is the creative act of the human mind.

Let us consider the matter of proof. Suppose one has discovered by measurements that the sum of the angles of various triangles is 180°. One must now prove this result deductively. No obvious method will do the job. Some new idea is required, and the reader who remembers his elementary geometry will recall that the proof is usually made by drawing a line

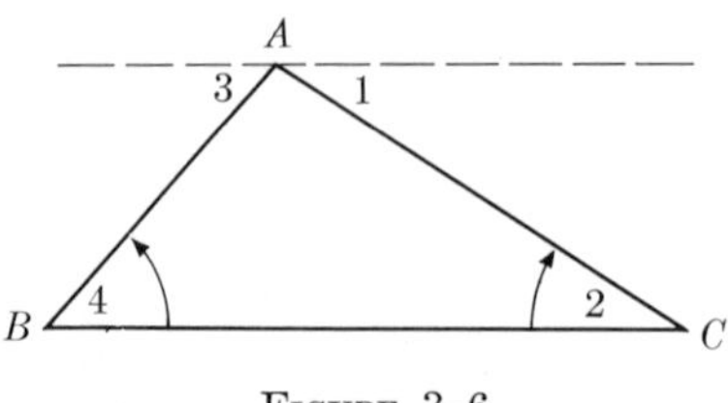

FIGURE 3–6

through one vertex (A in Fig. 3–6) and parallel to the opposite side. It then turns out as a consequence of a previously established theorem on parallel lines that the angles 1 and 2 are equal, as are the angles 3 and 4. However the angles 1, 3, and the angle A of the triangle itself do add up to 180°, and so the same is true for the angles of the triangle. This method of proof is not routine. The idea of drawing the line through A must be supplied by the mind. Some methods of proof seem so devious and artificial that they have provoked critical comments. The philosopher Arthur Schopenhauer called Euclid's proof of the Pythagorean theorem "a mouse-trap proof" and "a proof walking on stilts, nay, a mean, underhand proof."

The above example has been offered to emphasize the fact that ingenious mathematical work must be done in finding methods of proofs even after the question of what to prove is disposed of. In the search for a method of proof, as in finding what to prove, the mathematician must use audacious imagination, insight, and creative ability. His mind must see possible lines of attack where others would not. In the domains of algebra, calculus, and advanced analysis especially, the first-rate mathematician depends upon the kind of inspiration that we usually associate with the creation of music, literature, or art. The composer feels that he has a theme which when properly developed will produce true music. Experience and a knowledge of music aid him in arriving at this conviction. Similarly, the mathematician surmises that he has a conclusion which will follow from the axioms of mathematics. Experience and knowledge may guide his thoughts into the proper channels. Modifications of one sort or another may be required before a correct proof and a satisfactory statement of the new theorem are achieved. But essentially both mathematician and composer are moved by an afflatus which enables them to see the final edifice before a single stone is laid.

We do not know just what mental processes may lead to correct insight any more than we know how it was possible for Keats to write fine poetry or why Rembrandt was able to turn out fine paintings. One might say of mathematical creation what P. W. Bridgman, the noted physicist, has said of scientific method, that it consists of "doing one's damnedest with one's mind, no holds barred." There is no logic or infallible guide which tells the mind how to think. The very fact that many great mathematicians have tackled a problem and failed and that another comes along and solves it shows that the mind has something to contribute.

The preceding discussion of the creation of mathematics should correct several mistaken popular impressions. When creating a mathematical proof, the mind does not see the cold, ordered arguments which one reads in texts, but rather it perceives an idea or a scheme which when properly formulated constitutes the deductive proof. The formal proof, so to speak, merely sanctions the conquest already made by the intuition. Secondly, the deductive proof is not the preferable form by which to grasp the idea or method employed. In fact the deductive argument often conceals the idea because the logical form is not perspicuous to the intuition. At the very least the details of the arguments obscure the main threads. The value of the deductive organization of the proof is that it enables the creator and the reader to test the arguments by the standards of exact reasoning. Thirdly, there is the prevalent but mistaken notion that scientists and mathematicians must keep their minds open and unbiased in pursuing an investigation. They are not supposed to prejudge the conclusion. Actually the mathematician must first decide *what* to prove, and this conclusion not only does but must precede the search for the proof, or else he would not know where to head. This is not to say that the mathematician may not sometimes make a false conjecture. If he does, his search for a proof will fail or in the course of the search he will realize that he cannot prove what he is after, and he will correct his conjecture. But in any case he knows what he is trying to prove.

Exercises

1. Consider the parallelogram $ABCD$ (Fig. 3–7). By definition, the opposite sides are parallel. Now introduce the diagonal BD. Does observation suggest a possible theorem relating the triangles ABD and BDC?

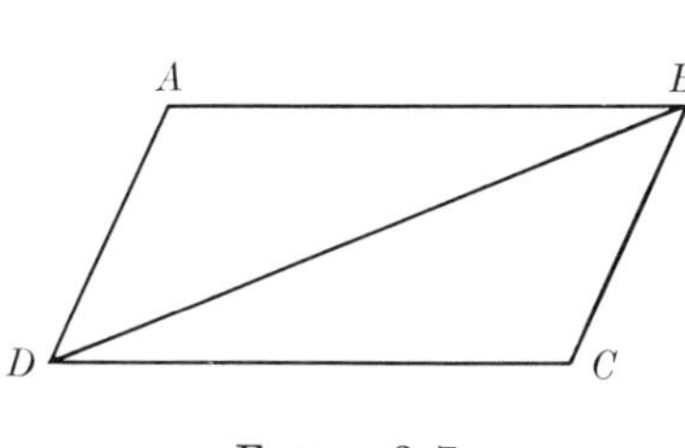

Figure 3–7

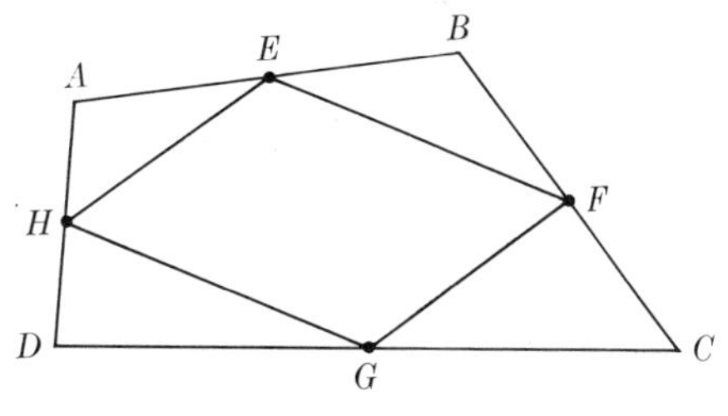

Figure 3–8

2. Consider any quadrilateral $ABCD$ (Fig. 3–8) and the figure formed by joining the mid-points E, F, G, H of the sides of the quadrilateral. Does observation or intuition suggest any significant fact about the quadrilateral $EFGH$?

3. The formula $n^2 - n + 41$ is supposed to yield primes for various values of n. Thus when $n = 1$, $1^2 - 1 + 41 = 41$, and this is a prime. When $n = 2$, $2^2 - 2 + 41 = 43$, and this is a prime. Test the formula for $n = 3$ and $n = 4$. Are the resulting values of the formula primes? Have you proved, then, that the formula always yields primes?

4. Can you specify conditions under which two quadrilaterals will be congruent, that is, have the same size and shape?

5. The following lines show some calculations with the sum of the cubes of whole numbers:

$$(1)^3 = 1,$$

$$(1)^3 + (2)^3 = 1 + 8 = 9 = 3^2 = (1 + 2)^2,$$

$$(1)^3 + (2)^3 + (3)^3 = 1 + 8 + 27 = 36 = 6^2 = (1 + 2 + 3)^2$$

$$(1)^3 + (2)^3 + (3)^3 + (4)^3 = 1 + 8 + 27 + 64 = 100 = 10^2 = (1 + 2 + 3 + 4)^2.$$

What generalization do these few calculations suggest?

3–8 Summary. Let us review the major changes in the nature of mathematics which the Greeks instituted. They decided, first of all, to study the abstract concepts of number and geometrical forms because they saw that these abstract concepts underlie a great variety of phenomena and indeed the design of the universe itself. Because they were interested in truths, the Greeks selected the deductive method of establishing conclusions as opposed to other methods of reasoning. To ensure truths there must be sound premises, and so the Greeks adopted axioms about numbers and geometrical figures which they believed were self-evident truths. These features now delimit the scope and method of proof of mathematics and are the essential features of the subject.

We have also noted that the creation of mathematics calls for intuition, imagination, and other processes which are not deductive. Moreover, the creative process is not direct and systematic as are the proofs. There are other distinctive features of mathematics, such as symbolism, which we shall discuss in a later chapter. Further, we shall find that mathematicians were forced in modern times to alter their understanding of the nature of mathematics.

Recommended Reading

Bell, E. T.: *The Development of Mathematics,* 2nd ed., Chaps. 2 and 3, McGraw Hill Book Co., New York, 1945.

Clagett, Marshall: *Greek Science in Antiquity,* Chap. 2, Abelard-Schuman, Inc., New York, 1955.

Cohen, Morris R. and E. Nagel: *An Introduction to Logic and Scientific Method,* Chaps. 1 to 5, Harcourt Brace and Co., New York, 1934.

Coolidge, J. L.: *The Mathematics of Great Amateurs,* Chap. 1, Oxford University Press, London, 1949.

Hamilton, Edith: *The Greek Way to Western Civilization,* Chaps. 1 to 3, The New American Library, New York, 1948.

Jeans, Sir James: *The Growth of Physical Science,* 2nd ed., Chap. 2, Cambridge University Press, Cambridge, 1951.

Taylor, Henry Osborn: *Ancient Ideals,* 2nd ed., Vol. I, Chaps. 7 to 13, The Macmillan Co., New York, 1913.

Wedberg, Anders: *Plato's Philosophy of Mathematics,* Almqvist and Wiksell, Stockholm, 1955 (for students of philosophy).

CHAPTER 4

NUMBER: THE FUNDAMENTAL CONCEPT

A marvelous neutrality have these things Mathematical, *and also a strange participation between things supernatural, immortal, intellectual, simple, and indivisible, and things natural, mortal, sensible, compounded and divisible.*

John Dee (1527–1608)

4–1 Introduction. Just as we are inclined to accept the sun, moon, and stars as our birthright and do not appreciate the grandeur, the mystery, and the knowledge which can be gleaned from the contemplation of the heavens, so are we inclined to accept our number system. There is, however, this difference. Many of us would not claim the latter and would gladly sell it for a mess of pottage. Because we are forced to learn about numbers and operations with numbers while we are still too young to appreciate them—a preparation for life which hardly excites our interest in the future—we grow up believing that numbers are drab and uninteresting. But the number system warrants attention not only as the basis of mathematics, but because it contains weighty and beautiful ideas which lend themselves to powerful applications.

Among past civilizations, the Greeks best appreciated the wonder and power of the concept of number. They were, of course, a people with great intellectual perception, but perhaps because they viewed numbers abstractly, they saw more clearly their true nature. The very fact that one can abstract from many diverse collections of objects a property such as "fiveness" struck the Greeks as a marvelous discovery. If one may use the ridiculous to accentuate the sublime, one may say that the Greek delight in numbers was the rational counterpart of the hysteria which many young and old Americans experience when they encounter numbers in the form of baseball scores and batting averages.

4–2 Whole numbers and fractions. The first Greeks who, to our knowledge, expressed their satisfaction with numbers and propounded a philosophy based on numbers which is extremely alive and vital today were the Pythagoreans. This group was founded in the middle of the sixth century B.C. by Pythagoras. We know rather little that is certain about this man. However, it seems very likely that he was born in 569 B.C. in a Greek settlement on the island of Samos in the Aegean Sea. Like many other Greeks he traveled to Egypt and to the Near East to learn what these older civilizations had to offer, and then settled in Croton, another Greek city in southern Italy. Pythagoras and his followers were among the early founders of the great Greek civilization, and so it is not surprising to find that the rational attitude which characterizes the Greeks was still surrounded in his times with mystical and religious doctrines prevalent in Egypt and its eastern neighbors. In fact the Pythagoreans were a religious sect as well as students of philosophy and nature.

Membership in the group was restricted, and the members were pledged to secrecy. Among their religious doctrines was the belief that the soul was tainted by the body. To purify the soul they maintained celibacy; their religious practices were also supposed to be efficacious in purifying the soul. At death the soul was reincarnated in another human or an animal. Like most mystics they observed certain taboos. They would not touch a white cock, walk on the highways, use iron to stir a fire, or leave the marks of ashes on a pot.

The secrecy of the group, its aloofness, and an attempted interference in the political affairs of Croton finally aroused the people of this city to drive out the Pythagoreans. We do not know for certain what happened to Pythagoras. One story has it that he fled to Metapontum, another Greek city in southern Italy, and was murdered there. However, the Pythagoreans continued to be influential in Greek intellectual life. One of their notable members was the philosopher Plato.

The Pythagoreans were impressed with numbers and, because they were mystics, attached to the whole numbers meanings and significances which we now regard as childish. Thus, they considered the number "one" as the essence or very nature of reason, for reason could produce only one consistent body of doctrines. The number "two" was identified with opinion, clearly because the very meaning of opinion implies the possibility of an opposing opinion, and thus of at least two. "Four" was identified with justice because it is the first number which is the product of equals. Of course, one can also be thought of as 1 times 1, but to the Pythagoreans one was not a number in the full sense because it did not represent quantity. The Pythagoreans represented numbers as dots in sand or with pebbles, and for each number the dots or pebbles had a special arrangement. Thus the number "four" was pictured as four dots suggesting a square, and so the square and justice were also linked. Foursquare and square shooter still mean a person who acts justly. "Five" signified marriage because it was the union of the first masculine number, three, and the first feminine number, two. (Odd numbers were masculine and even numbers feminine.) The number "seven" represented health and "eight," friendship or love.

We shall not pursue all the ideas which the Pythagoreans developed about numbers. What is significant about their work is that they were the first to study properties of whole numbers. As we shall see in a later chapter, they also possessed the vision of deep mystics and saw that numbers could be used to represent and even embody the essence of natural phenomena.

The speculations and results obtained by the Pythagoreans about whole numbers and ratios of whole numbers, or fractions as we prefer to call them, were the beginning of a long and involved development of arithmetic as a science as opposed to arithmetic as a tool for daily applications. During the

2500 years since the Pythagoreans first called attention to the importance of numbers, man has not only learned to better appreciate the idea but has invented excellent methods of writing quantity and of performing the four operations of arithmetic, i.e., ambition, distraction, uglification, and derision, as Lewis Carroll called them. While these methods of writing and operating with numbers are largely familiar, there are a few facts which are worthy of comment.

One of the most important members of our present number system is the mathematical representation of no quantity, that is, zero. We are accustomed to this number and yet usually fail to appreciate two facts about it. The first is that this member of our number system came rather late. The idea of using zero was conceived by the Hindus and, like other of their ideas, reached Europe through the Arabs. It had not occurred to earlier civilizations, even to the Greeks, that it would be useful to have a number which represents the absence of any objects. Connected with this late appearance of the number is the second significant fact, namely, that zero must be distinguished from nothing. Undoubtedly it was the inability of earlier peoples to perceive this distinction which accounts for their failure to introduce the zero. That zero must be distinguished from nothing is easily seen from several examples. A student's grade in a course he never took is no grade or nothing. He may, however, have the grade of zero in a course he has taken. If a person has no account in a bank, his balance is nothing. If he has a bank account, he may very well have a balance of zero.

Because zero is a number, we may operate with it; for example, we may add zero to another number. Thus $5 + 0 = 5$. By contrast $5 +$ nothing is meaningless or nothing. The only restriction on zero as a number is that one cannot divide by zero. Division by zero does, so to speak, produce nothing. Because so many false steps in mathematics result from division by zero, it is well to understand clearly why we cannot do this. The answer to a problem of division, say $6/2$, is some number which when multiplied by the divisor yields the dividend. In our example, 3 is the answer because $3 \cdot 2 = 6$. Hence the answer to $5/0$ should be a number which when multiplied by 0 gives 5. However, any number multiplied by 0 gives 0 and not 5. Thus, there is no answer to the problem of $5/0$. In the case of $0/0$ the answer should be some number which when multiplied by 0 yields the dividend 0. However, any number may then serve as a quotient because any number multiplied by 0 gives 0. But mathematics cannot tolerate such an ambiguous situation. If $0/0$ arises and any number may serve as an answer, we do not know what number to take and hence are not aided. It is as if we asked a person for directions to some place and he replied, Take any direction.

With the availability of zero, mathematicians were finally able to develop our present method of writing whole numbers. First of all we count in units and represent large quantities in tens, tens of tens, tens of tens of tens, etc. Thus we represent two hundred and fifty-two by 252. The left-hand 2 means, of course, two tens of tens; the 5 means 5 times 10; and the right-hand 2 means 2 units. The concept of zero makes such a system of writing quantities practical since it enables us to distinguish 22 and 202. Because ten plays such a fundamental role, our number system is called the *decimal system*, and ten is called the *base*. The use of ten resulted most likely from the fact that man counted on his fingers and, when he had used the fingers on his two hands, considered the number arrived at as a larger unit.

Because the position of an integer determines the quantity it represents, the principle involved is called *positional notation*. The decimal system of positional notation is due to the Hindus; however, the same scheme was used two millenniums earlier by the Babylonians, but with base 60 and in more limited form since they did not have zero.

The operations of arithmetic, addition, subtraction, multiplication, and division, are of course familiar to us, but it is perhaps not recognized that these operations are quite sophisticated and remarkably efficient. They date back to Greek times and gradually evolved, as improvements in the methods of writing numbers and the concept of zero were introduced. The Europeans picked up the methods from the Arabs. Previously the Europeans had used the Roman system of writing numbers, and the operations were based on that system. Partly because these latter methods were relatively cumbersome and partly because education was limited to a few people, those who acquired the art of calculation were regarded as skilled mathematicians. In fact the processes defied the average man so much that it seemed to him that those possessed of the ability must have magical powers. Good calculators were called practitioners of the "Black Art."

To appreciate the efficiency of our present methods we would have to learn the older ones and even acquire some facility in them, to make the comparison a fair one. But we cannot spare the time and effort. Perhaps the one point we should emphasize is how much our methods of arithmetic depend upon positional notation. This can be seen even in a simple problem of addition. To add 387 and 359 say, the written work is

$$\begin{array}{r} 387 \\ \underline{359} \\ 746 \end{array}$$

However, in performing this work, we think as follows. We add the units 7 and 9, the "tens" quantities 8 and 5, and the "hundreds" 3 and 3, separately. When we add the 7 and 9, we obtain 16. We recognize that 16 is $1 \cdot 10 + 6$, and so we add the $1 \cdot 10$ to the $13 \cdot 10$ already obtained from the 8 and 5. We say that we "carry" the $1 \cdot 10$ over, and instead of $13 \cdot 10$ we obtain $14 \cdot 10$. However, $14 \cdot 10$ is $(10 + 4) \cdot 10$ or $1 \cdot 10^2 + 4 \cdot 10$. Thus we write 4 in the tens' column, add the $1 \cdot 10^2$ to the $6 \cdot 10^2$ already obtained from the 3 and 3, and arrive at $7 \cdot 10^2$. All these steps are usually executed rather mechanically by writing the appropriate numbers in the units,' tens,' and hundreds' places and by using the process called carrying. Were we to analyze the processes of subtraction, multiplication, and division, we would again see how the steps which we learn mechanically in elementary school are just the skeletal processes of thinking suited to positional notation in base ten.

A word about fractions may also be in order. The natural method of writing fractions, for example, $2/3$ or $7/5$, to express parts of a whole presents no difficulties of comprehension. However the operations with fractions do seem to be somewhat arbitrary and mysterious. To add $2/3$ and $7/5$, say, we go through the following process:

$$\frac{2}{3} + \frac{7}{5} = \frac{10}{15} + \frac{21}{15} = \frac{31}{15}.$$

What we have done is to express each fraction in an equivalent form such that the denominators are now alike, and then add the numerators. We are not required by law to add fractions in this manner. It would, of course, be much simpler if we agreed to add fractions by adding the numerators and adding the denominators so that

$$\frac{2}{3} + \frac{7}{5} = \frac{9}{8}.$$

As a matter of fact, when we multiply two fractions, we do multiply the numerators and multiply the denominators so that it does seem as though the mathematicians prefer to be unnecessarily complicated about the addition of fractions.

The explanation of this seeming mathematical idiosyncrasy is simple: the operations with fractions are formulated to fit experience. When one has $2/3$ of a pie and $7/5$ of a pie, he has in all not $9/8$ but $31/15$ of a pie. In other words, if mathematical concepts and operations are to fit experience, the nature of the operations is forced upon us. In the case of multiplication of fractions it is correct that multiplying the numerators and multiplying the denominators will yield the fraction which represents the physical result. Thus to find $2/3$ of $7/5$, say, it is intuitively helpful to think of $2/3$ as $2 \cdot 1/3$. That $1/3$ of $7/5$ is $1/3$ of $21/15$, or $7/15$, is clear, and $2(1/3 \cdot 21/15)$ is then $14/15$. This result is obtained at once by multiplying the numerators and multiplying the denominators of the original fractions.

In the case of division of fractions, the most helpful way of seeing that the operation does what it should is to recognize first of all what the operation is supposed to produce. When we seek the solution to $(7/5)/(2/3)$, we wish to find some number which when multiplied by $2/3$ gives $7/5$. The number which answers this question is $(7/5) \cdot (3/2)$ or $21/10$, for if we multiply the result $21/10$ by the divisor, that is $(21/10) \cdot (2/3)$, we obtain the dividend $7/5$. The solution, $(7/5) \cdot (3/2)$, is obtained from the given quantity, namely $(7/5)/(2/3)$, by inverting the denominator $2/3$ and multiplying. Hence our definition of division of fractions is designed to answer the very question which the division poses.

Fractions, like the whole numbers, can be written in positional notation. Thus

$$\frac{1}{4} = \frac{25}{100} = \frac{20}{100} + \frac{5}{100} = \frac{2}{10} + \frac{5}{100}.$$

If we now agree to suppress the powers of 10, that is 10, 100, and higher powers where they occur, then we can write $\frac{1}{4} = 0.25$. The decimal point reminds us that the first number is really $\frac{2}{10}$, the second $\frac{5}{100}$, and so forth. The Babylonians had employed positional notation for fractions, but they used base 60 rather than base 10, just as they had for whole numbers. The decimal base for fractions was introduced by sixteenth-century European algebraists. Of course, the operations with fractions can also be carried out in decimal form.

The disappointing feature of the decimal representation of fractions is that some simple fractions cannot be represented as decimals with a finite number of digits. Thus when we seek to express $\frac{1}{3}$ as a decimal, we find that neither 0.3, nor 0.33, nor 0.333, and so on, suffices. All one can say in this and similar cases is that by carrying more and more decimal digits one comes closer and closer to the fraction, but no finite number of digits will ever be the exact answer. This fact is expressed by the notation

$$\frac{1}{3} = 0.333\ldots,$$

where the dots indicate that we must keep on adding threes to approach the fraction $\frac{1}{3}$ more and more closely.

From the standpoint of applications the fact that some fractions cannot be expressed as decimals with a finite number of digits does not matter because we can always carry enough digits to obtain an answer as accurate as the application requires.

Exercises

1. What is the principle of positional notation?
2. Why is the number zero almost indispensable in the system of positional notation?
3. What is the meaning of the statement that zero is a number?
4. What two methods are there of representing fractions?
5. What principle determines the definitions of the operations with fractions?

4–3 Irrational numbers. The Pythagoreans, as we noted earlier, were the first to appreciate the very concept of number, and sought to employ numbers to describe the basic phenomena of the physical and social worlds. Numbers to the Pythagoreans were also interesting in and for themselves. Thus they liked square numbers, that is, numbers such as 4, 9, 16, 25, 36, and so on, and observed that the sums of certain pairs of square numbers, or perfect squares, are also square numbers. For example, $9 + 16 = 25$; $25 + 144 = 169$; and $36 + 64 = 100$. These relationships can also be written as $3^2 + 4^2 = 5^2$; $5^2 + 12^2 = 13^2$; and $6^2 + 8^2 = 10^2$. The three numbers whose squares furnish such equalities are today called Pythagorean triples. Thus 3,4,5 constitute a Pythagorean triple because $3^2 + 4^2 = 5^2$.

The Pythagoreans liked these triples so much because, among other features, they have an interesting geometrical interpretation. If the two

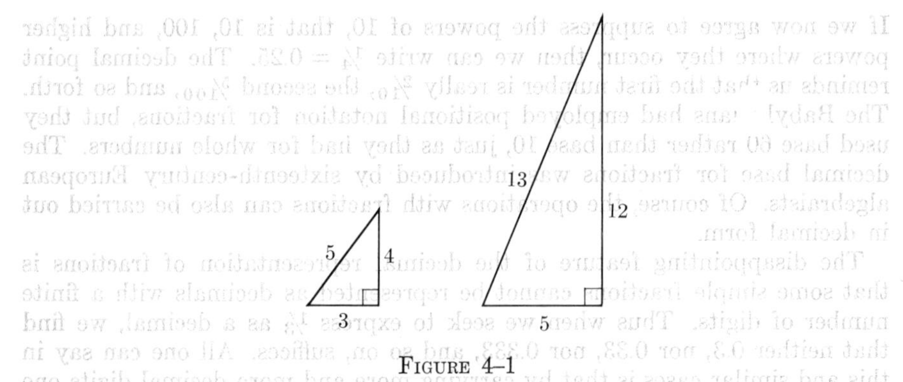

FIGURE 4–1

smaller numbers are the lengths of the sides or arms of a right triangle, then the third one is the length of the hypotenuse (Fig. 4–1). Just how the Pythagoreans knew this geometrical fact is not clear, but assert it they did. They also claimed that in *any* right triangle, the square of the length of one arm added to the square of the length of the other gives the square of the length of the hypotenuse. This more general assertion is still called the Pythagorean theorem and a proof of it, such as we learn in high-school geometry, was given about 200 years later by Euclid. Pythagoras is said to have been so overjoyed with this theorem that he sacrificed an ox to celebrate its discovery.

This theorem proved to be the undoing of a central doctrine in the Pythagorean philosophy and caused woe and misery to many mathematicians. But before we pursue this story, we should look into a few simple properties of the whole numbers which are embodied in the following exercises.

EXERCISES

1. Prove that the square of any even number is an even number. [*Suggestion:* By definition every even number contains 2 as a factor.]

2. Prove that the square of any odd number is an odd number. [*Suggestion:* Every odd number ends in 1, 3, 5, 7, or 9.]

3. Let a stand for a whole number. Prove that if a^2 is even, then a is even. [*Suggestion:* Use the result in Exercise 2.]

4. Establish the truth or falsity of the assertion that the sum of any two square numbers is a square number.

There are tragedies in mathematics also, and one of these struck the very group of mathematicians who deserved a better fate. The Pythagoreans had constructed, at least to their own satisfaction, a philosophy which asserted that all natural phenomena and all social and ethical concepts were in essence just whole numbers or relationships among whole numbers.* But one day it occurred to a member of the group to examine the seemingly simplest case of the Pythagorean theorem. Suppose each arm of a right triangle (Fig. 4–2) is 1 unit in length; how long, he asked, is the hypotenuse? The Pythagorean theorem says that the square of (the length of) the hypotenuse equals the sum of the squares of the arms. Hence if we call c the unknown length of the hypotenuse, then the theorem says that

$$c^2 = 1^2 + 1^2$$

or

$$c^2 = 2.$$

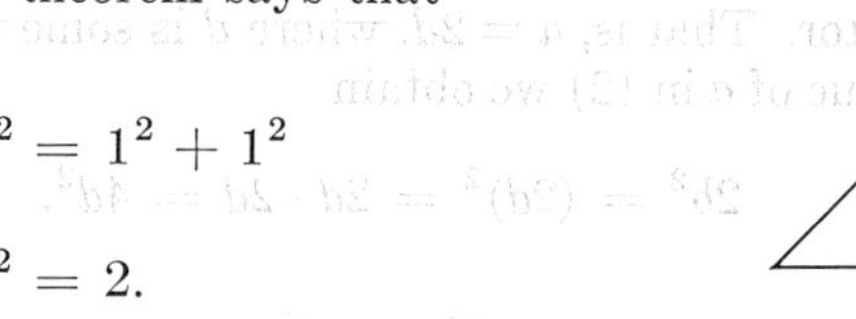

FIGURE 4–2

Now 2 is not a square number, that is, a perfect square, and so c is not a whole number. But it certainly seemed reasonable to this Pythagorean that c should be a fraction; that is, there should be a fraction whose square is 2. Even the simple fraction 7/5 comes close to being the correct value because $(7/5)^2 = 49/25$, and this is almost 2. However, simple trial does not easily yield a fraction whose square is 2. Hence this Pythagorean became worried, and he decided to investigate the question of whether there is a fraction whose square is 2. We shall examine his reasoning which, as far as we know, is the same as that given in Euclid's famous work on geometry, the *Elements*.

The number c which we seek to determine is one whose square is 2. Let us denote it by $\sqrt{2}$. All we mean by this symbol is that it represents a number whose square is 2. And now let us suppose that $\sqrt{2}$ is a fraction a/b, where a and b are whole numbers. Moreover, to make matters simpler, let us suppose that any factors common to a and b are cancelled. Thus if a/b were 8/6, for example, we would cancel the common factor 2 and write it as 4/3. Hence we have assumed so far that

$$\sqrt{2} = \frac{a}{b} \tag{1}$$

and that a and b have no common factors.

If equation (1) is correct, then by squaring both sides, a step which utilizes the axiom that equals multiplied by equals give equals (because we multiply the left side by $\sqrt{2}$ and the right side by a/b), we obtain

$$2 = \frac{a^2}{b^2}.$$

* See Chapter 8.

Again by employing the axiom that equals multiplied by equals yield equals, we may multiply both sides of this last equation by b^2 and write

$$2b^2 = a^2. \tag{2}$$

The left side of this equation is an even number because it contains 2 as a factor. Hence the right side must also be an even number. But if a^2 is even, then, according to Exercise 3 above, a must be even. If a is even, it must contain 2 as a factor. That is, $a = 2d$, where d is some whole number. If we substitute this value of a in (2) we obtain

$$2b^2 = (2d)^2 = 2d \cdot 2d = 4d^2. \tag{3}$$

Since then

$$2b^2 = 4d^2,$$

we may divide both sides of this equation by 2 and obtain

$$b^2 = 2d^2. \tag{4}$$

We now see that b^2 is an even number and so, by again appealing to the result in Exercise 3, we find that b is an even number.

What we have shown in the above argument is that if $\sqrt{2} = a/b$, then a and b must be even numbers. But at the very outset we had cancelled any common factors in a and b; yet we find that a and b still contain 2 as a common factor. This result contradicts the fact that a and b have no common factors.

Why do we arrive at a contradiction? Since our reasoning is correct, the only possibility is that the assumption that $\sqrt{2}$ equals a fraction is not correct. In other words, $\sqrt{2}$ cannot be a ratio of two whole numbers.

This proof is so neat that one can almost believe the legend that Pythagoras sacrificed an ox in honor of its creation. But there are at least two reasons for discrediting this tale. The first is that if all the legends telling of Pythagoras sacrificing an ox were true, he could not have had time for mathematics. The second reason is that the above proof was not a triumph for the Pythagoreans but a disaster. The symbol $\sqrt{2}$ is a number because it represents the length of a line, namely the hypotenuse of the triangle in Fig. 4–2. But this number is not a whole number or a fraction. The Pythagoreans had, however, developed an embracing philosophy which asserted that everything in the universe reduced to whole numbers. Clearly, then, this philosophy was inadequate. Indeed the existence of numbers such as $\sqrt{2}$ was such a serious threat to the Pythagorean philosophy that another legend, more credible, states that the Pythagoreans, who were at sea when the above discovery was made, threw overboard the member who made it, and pledged to keep the discovery secret.

But secrets will out, and later Greeks not only learned that $\sqrt{2}$ is neither a whole number nor a fraction, but they discovered that there is an indefinitely large collection of other numbers which are not whole numbers or fractions. Thus $\sqrt{3}$, $\sqrt{5}$, $\sqrt{7}$, and, more generally, the square root of any number which is not a perfect square, the cube root of any number which is not a perfect cube, and so on, are numbers which are not whole numbers or fractions. The number π, which is the ratio of the circumference of any circle to its diameter, is also neither a whole number nor a fraction. All these new numbers are called irrational numbers, the word "irrational" now meaning that these numbers cannot be expressed as ratios of whole numbers, although in Pythagorean times it meant unmentionable or unknowable.

If these irrational numbers are really so common and represent lengths of sides of triangles and circumferences of circles in terms of the diameters, why weren't they encountered before? Didn't the Babylonians and Egyptians run across them? They did. But since they were concerned only with having numbers serve their practical purposes, they used convenient approximations. Thus, when they encountered a length such as $\sqrt{2}$, they were content to use a value such as 1.4 or 1.41. For π, as we noted in an earlier chapter, they used values even as crude as 3. Not only did these peoples use such approximations, but they never realized that the most complicated fraction or decimal could never represent an irrational number exactly. The Egyptians and Babylonians treated irrational numbers and their mathematics in general rather lightheartedly. We may hail their blithe spirits, but mathematicians they never were.

The Greeks, as we know, were of a different intellectual breed and could not be content with approximations, but they also exhibited a weakness. Although they recognized that quantities exist which are neither whole numbers nor fractions, they were so convinced that the concept of number could not comprise anything else than whole numbers or fractions that they did not accept irrationals as numbers. Instead they thought of such quantities only as geometrical lengths or areas. Thus the Greeks never did develop an *arithmetic* of irrational numbers. In their astronomical work, for example, they used only whole numbers and fractions. The difficulty which the Greeks experienced also baffled all mathematicians up to modern times. The greatest mathematicians refused to accept irrationals as numbers and followed the Greek procedure of thinking about such quantities as lengths or areas. All these people wished that the Pythagoreans had thrown the irrational number overboard rather than the man who discovered them.

But the needs of society often oblige even mathematicians to face unpleasantnesses. In the seventeenth century, science began to develop at an amazing rate, and science needs quantitative results. It may be nice to know that $\sqrt{2}$ is a certain length and that $\sqrt{2} \cdot \sqrt{3}$ is an area, but this knowledge does not suffice when one needs numerical results. And so finally mathematicians had to accept the fact that if they were to treat numerically all the quantities that arise in scientific work, they must handle irrational numbers as numbers. The mathematicians' refusal over centuries to grant irrationals the status of numbers illustrates one of the surprising features of the history of mathematics. New ideas are often as unacceptable in this field as they are in politics, religion, and economics.

The situation, then, which must be faced squarely is that there are other numbers besides whole numbers and fractions. It is, of course, quite understandable that whole numbers and fractions should have been created and used first, for these numbers arise in the simplest physical situations man encounters. The irrational numbers on the other hand are not commonly encountered. Only the application of a theorem such as the Pythagorean theorem brings them to our attention, and even then one must go through a proof such as that examined above, to see that they are not whole numbers or fractions. But the fact that irrational numbers are late-comers does not mean that they are less acceptable or less genuine numbers. Just as we gradually add to our knowledge of the varieties of human beings and animals which exist in our physical world, so must we broaden our knowledge of the varieties of numbers and with true liberality accept these strangers on the same basis as the already familiar numbers.

However, if we are to use irrational numbers, we must know how to operate with them, that is, how to add, subtract, multiply, and divide them. We have already noted with whole numbers and fractions that if we wish the operations to fit experience, we must formulate the operations accordingly. So it is with the irrational numbers. We could define addition, multiplication, and the other operations as we please. But if we wish these operations to represent physical situations, we must define them properly. However, there is no real difficulty here. Since irrational numbers are quantities, as are whole numbers and fractions, we may use the latter as a guide to the proper operations with irrational numbers.

Let us consider a few examples which will be sufficient to indicate the general principles. Should we say that

$$\sqrt{2} + \sqrt{3} = \sqrt{5}?$$

To answer this question let us consider the analogous question: May we say that

$$\sqrt{4} + \sqrt{9} = \sqrt{13}?$$

It is clear in the latter case that $2 + 3$ does not equal $\sqrt{13}$, for $\sqrt{13}$ is certainly less than 4. Hence we should *not* add the radicands, that is, the 2 and the 3, in the preceding equation. One might then ask, How much is $\sqrt{2} + \sqrt{3}$? Since both summands are numbers, the sum is also a number, but it cannot be written more compactly than $\sqrt{2} + \sqrt{3}$. This inability to combine the summands is not something new or troublesome. When we add 2 and $\frac{1}{2}$, for example, the answers continues to be $2 + \frac{1}{2}$. We usually omit the plus sign and write $2\frac{1}{2}$, but the summands are really not combined.

Let us consider next whether

$$\sqrt{2} \cdot \sqrt{3} = \sqrt{6}.$$

Here too we shall see what the analogous operation with whole numbers suggests. Is it true that

$$\sqrt{4} \cdot \sqrt{9} = \sqrt{36}?$$

The answer is clearly yes, and so we shall agree that to multiply square roots we shall multiply the radicands. That is,

$$\sqrt{2} \cdot \sqrt{3} = \sqrt{6}.$$

The definitions of the operations of subtraction and division are also readily determined. Thus $\sqrt{3} - \sqrt{2}$ yields a definite number, but the difference cannot be written any more compactly than $\sqrt{3} - \sqrt{2}$.

For division, say $\sqrt{3}/\sqrt{2}$, the procedure, as in the case of multiplication, is suggested by observing that

$$\frac{\sqrt{9}}{\sqrt{4}} = \sqrt{\frac{9}{4}},$$

for this equation simply says that $3/2 = 3/2$. Hence we shall agree that

$$\frac{\sqrt{3}}{\sqrt{2}} = \sqrt{\frac{3}{2}}.$$

The general principle which these examples illustrate is that operations with irrational numbers are defined so as to agree with the same operations on whole numbers when the latter are expressed as roots. We could state our definitions in general form, but there is no need to do so.

In applications we often approximate irrational numbers by fractions or decimals because actual physical constructions cannot be built exactly anyway. Thus if we had to construct a length which strictly should be $\sqrt{2}$, we would approximate $\sqrt{2}$. Since $(1.4)^2 = 1.96$ and 1.96 is nearly 2, we could approximate $\sqrt{2}$ by 1.4. If we desired a more accurate approximation, we might determine to the nearest hundredth the number whose square approximates 2. Thus, since $(1.41)^2 = 1.988$ and $(1.42)^2 = 2.016$, we see that 1.41 is a good two-decimal approximation of $\sqrt{2}$. We could, of course, improve still more on the accuracy of the approximation. We should, however, realize that no matter how many decimal places we employed, we would never obtain a number which is exactly $\sqrt{2}$ because any decimal with a finite number of digits or a whole number plus such a decimal is just another way of writing a fraction, whereas $\sqrt{2}$, as the above proof showed, can never equal a quotient of two whole numbers.

The fact that we often approximate an irrational number when we wish to construct something raises a question which merits an answer. The question is, Why don't we approximate irrational numbers wherever they arise and forget about operations with irrationals as such? For example, to calculate $\sqrt{2} \cdot \sqrt{3}$, we could approximate $\sqrt{2}$ by, say 1.41, approximate $\sqrt{3}$ by 1.73, and then multiply 1.41 by 1.73. The answer is 2.44, and since $(2.44)^2$ is 5.95, we see that we have a good approximation to $\sqrt{6}$. If we wanted a more accurate answer, we could approximate $\sqrt{2}$ and $\sqrt{3}$ more closely and then multiply. One reason we do not approximate in mathematics proper is that mathematics is an exact science. It insists on reasoning as rigorous as human beings can perform. We pay a price for this rigor by expending more thought and effort, but we shall see that mathematics has made its contributions just because it insists on exactness.

There is also a practical advantage in working with irrational numbers as such. Let us suppose that some problem required us to calculate $(\sqrt{3})^4$, that is, $\sqrt{3} \cdot \sqrt{3} \cdot \sqrt{3} \cdot \sqrt{3}$. The person who insists on approximating would now approximate $\sqrt{3}$ to some number of decimal places, for example, 1.732, and then calculate $(1.732)^4$. While the practical person takes an hour to calculate and check his arithmetic, the mathematician would see at once that $\sqrt{3} \cdot \sqrt{3} \cdot \sqrt{3} \cdot \sqrt{3} = (\sqrt{3} \cdot \sqrt{3})(\sqrt{3} \cdot \sqrt{3}) = 3 \cdot 3 = 9$, and could spend the rest of the hour in refreshing sleep. Moreover, the mathematician's answer is exact, whereas the practical man's answer is not accurate even to the four decimal places with which he started, because the product of two approximate numbers is less accurate than either factor. To achieve an answer accurate to four decimal places, the practical man would have to use an approximation of $\sqrt{3}$ containing seven decimal places and then multiply.

The irrational number is the first of many sophisticated ideas which the mathematician has introduced to think about and cope with the real world. The mathematician creates these concepts, devises ways of working with them which fit real situations, and then uses his abstractions to think about the phenomena to which the ideas apply.

Exercises

1. Express the answers to the following problems as compactly as you can:

(a) $\sqrt{3} + \sqrt{5}$ (b) $\sqrt[3]{2} + \sqrt[3]{7}$ (c) $\sqrt[3]{2} + \sqrt{7}$ (d) $\sqrt{7} + \sqrt{7}$

(e) $\sqrt{3} \cdot \sqrt{7}$ (f) $\sqrt[3]{2} \cdot \sqrt[3]{5}$ (g) $\sqrt[3]{2} \cdot \sqrt[3]{4}$ (h) $\sqrt{12} \cdot \sqrt{3}$

(i) $\dfrac{\sqrt{5}}{\sqrt{2}}$ (j) $\dfrac{\sqrt{8}}{\sqrt{2}}$ (k) $\dfrac{\sqrt[3]{10}}{\sqrt[3]{2}}$

2. Simplify the following:

(a) $\sqrt{50}$ (b) $\sqrt{200}$ (c) $\sqrt{75}$ [*Suggestion*: $\sqrt{50} = \sqrt{25 \cdot 2} = \sqrt{25} \cdot \sqrt{2}$.]

3. Criticize the following argument: No irrational number can be expressed as a decimal with a finite number of decimal places. The number $\frac{1}{3}$ cannot be expressed as a decimal with a finite number of decimal places. Hence $\frac{1}{3}$ is an irrational number.

4–4 Negative numbers. One more addition to the number system which has considerably extended the power of mathematics comes from far-off India. Numbers are commonly used to represent an amount of money, in particular the amount of money which a person owns. Perhaps because the Hindus were in debt more often than not, it occurred to them that it would also be useful to have numbers which represent the amount of money one owes. They therefore invented what are now called negative numbers, while the previously known numbers are called positive numbers. Thus numbers which we denote by -3, $-5/2$, and $-\sqrt{2}$ came into existence. Where necessary to distinguish clearly positive from negative numbers or to emphasize what is positive as opposed to what is negative, one writes $+3$ or $+5/2$ instead of 3 or $5/2$.

It is not necessary, incidentally, to use such symbols as -3 to represent the negative counterpart to 3. Modern banks and large commercial corporations, which deal with negative numbers continually, often write these in red ink, whereas positive numbers are written in black ink. However, we shall find that placing the minus sign in front of a number to indicate a negative number is a convenience.

The use of positive and negative numbers is not limited to the representation of assets and debts. One represents temperatures below 0° as negative temperatures, while temperatures above 0° are positive. Likewise heights above and below sea level can be represented by positive and negative numbers, respectively. It is sometimes convenient to represent time after and before a specified event by positive and negative numbers. For example, using the birth of Christ as the event, the year 50 B.C. can just as well be described as the year -50.

To derive more use from the concept of negative numbers it must be possible to operate with them just as we operate with positive numbers. The operations with negative numbers and with negative and positive numbers together are easy to understand if one keeps in mind the physical significance of these operations. For example, suppose a man has assets of 3 dollars and debts of 8 dollars. What is his net wealth? Clearly the man is 5 dollars in debt. The same calculation is represented in terms of positive and negative numbers by stating that assets of 8 dollars must be taken from 3 dollars, that is, $3 - 8$, or that a debt of 8 dollars must be added to assets of 3 dollars, that is, $+3 + (-8)$. The answer is obtained by subtracting the smaller numerical value (that is, the smaller number without regard to sign) from the larger numerical value and giving the answer the sign attached to the larger numerical value. That is, we subtract 3 from 8 and call the answer negative because the larger numerical value, namely 8, has the minus sign attached to it.

Since negative numbers represent debts and subtraction usually has the physical meaning of "taking away" or "removing," then the subtraction of a negative number means the removal of a debt. Thus, if a person has assets of, say 3 dollars, but this figure already takes into account a debt of 8 dollars, the removal or cancellation of the debt leaves the person with assets of 11 dollars. Mathematically we say $+3 - (-8) = +11$. In words, to subtract a negative number we add the corresponding positive number.

Suppose a man goes into debt at the rate of 5 dollars per day. Then in 3 days after a given date he will be 15 dollars in debt. If we denote a debt of 5 dollars as -5, then going into debt at the rate of 5 dollars per day for 3 days can be stated mathematically as $3 \cdot (-5) = -15$. That is, the multiplication of a positive and a negative number yields a negative number whose numerical value is the product of the two given numerical values.

In the very same situation in which a man goes into debt at the rate of 5 dollars per day, his assets three days *before* a given date are 15 dollars more than they are at the given date. If we represent time before the given or zero date by -3 and the loss per day as -5, then his relative financial position 3 days *ago* can be expressed as $-3 \cdot (-5) = +15$; that is, to consider his assets three days ago, we would multiply the debt per day by -3, whereas to calculate the financial status three days in the future, we multiply by $+3$. Hence the result is $+15$ in the former case compared to -15 in the latter.

There is one more definition concerning negative numbers which is readily seen to be sensible. For the positive numbers and zero we say for obvious reasons that 3 is greater than 2, that 2 is greater than $\frac{1}{2}$, and that any positive number is greater than zero. The negative numbers are said to be less than the positive numbers and zero. Moreover, we say that -5 is less than -3, or that -3 is greater than -5. If one thinks of these various numbers as representing people's wealth, then the agreement concerning their order fits our usual understanding of relative wealth. A person whose financial status is -3 is wealthier than one whose status is -5; one is better off to be 3 dollars than 5 dollars in debt. Incidentally, the symbol $>$ is used to denote "greater than" as in $5 > 3$, and the symbol $<$ denotes "less than" as in $-5 < -3$.

The relative position of the various positive and negative numbers and zero is readily remembered if one visualizes these numbers as points on a line as shown in Fig. 4–3. The figure is really not different from that obtained by moving a thermometer scale into a horizontal position.

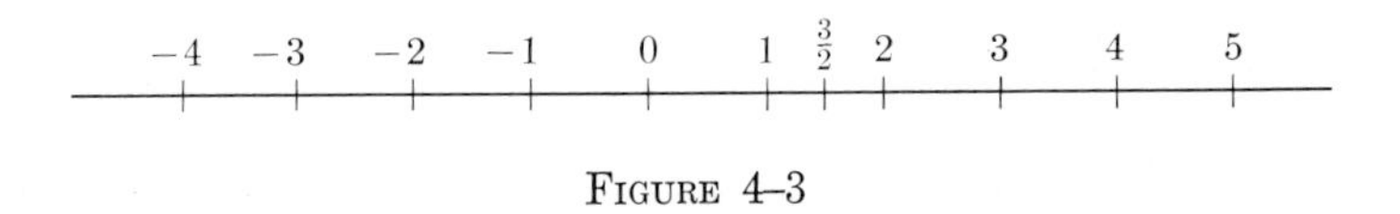

FIGURE 4–3

The above situations, which illustrate how the definitions of the operations with positive and negative numbers were suggested, are of course by no means the only ones in which positive and negative numbers are employed. Indeed the usefulness of negative numbers would hardly be great were this the case. However, these simple financial transactions show not only how mathematicians arrived at the definitions, but that there is no more mystery about negative numbers than about positive ones. The definitions represent in abstract form what takes place physically, and, as with all numbers, we can think in terms of the abstractions to arrive at a knowledge of physical happenings.

It may be of some comfort to the reader to know that the concept of negative numbers, like the concept of irrational numbers, was resisted by mathematicians for several hundred years. The history of mathematics illustrates the rather significant observation that it is more difficult to get a truth accepted than to discover it. The mathematicians to whom "number" meant whole numbers and fractions found it hard to accept negative numbers as true numbers. They, too, failed to realize for centuries that mathematical concepts are man-made abstractions which can be introduced at will if they can serve useful purposes.

Exercises

1. Suppose a man has \$3 and incurs a debt of \$5. What is his net worth?
2. Suppose a man owes \$5 and then incurs a new debt of \$8. Use negative numbers to calculate his financial condition.
3. Suppose a man owes \$5 and earns \$8. Use positive and negative numbers to calculate his net worth.
4. Suppose a man owes \$13, and a debt of \$8 is cancelled. Use negative numbers to calculate his net worth.
5. A man loses money in business at the rate of \$100 per week. Let us denote this change in his assets by -100 and let us denote time in the future by positive numbers and time in the past by negative numbers. How much will the man lose in 5 weeks? How much more was the man worth 5 weeks ago?

4–5 The axioms concerning numbers. In the preceding chapter we said that mathematics proceeds by deductive reasoning from explicitly stated axioms. Yet thus far in this chapter we have said nothing about axioms. The reason is simply that the axioms concerning numbers are such obvious properties that we use them automatically without realizing that we are doing so.

This situation may perhaps be better understood by means of an analogy. Whenever a child at play throws a ball up into the air, he expects that the ball will come down. He is really assuming that all balls thrown up will come down. Of course, this assumption is well founded in experience; nevertheless, the child's expectation that the ball will come down is a deduction from the assumption just stated and the additional premise that he is throwing a ball up into the air. Recognition of the fact that he has made an assumption makes clear the reasoning, conscious or unconscious, behind the act.

To understand the deductive process in the mathematics of numbers, as well as in geometry, we must recognize the existence and use of the axioms. We do not hesitate to say that $275 + 384 = 384 + 275$. Surely we did not add 384 objects to 275, count the total, then add 275 objects to 384, count that total, and check that the two totals agree. Rather, whenever in our experience we combined two groups of objects, we found that we obtained the same total collection regardless of whether we put the first group with the second or the second with the first. Of course, our evidence to the effect that the order of addition is immaterial is limited to a small number of cases, whereas in practice we use this fact with all numbers. Hence, we are really making an assumption, namely, that for *any* two numbers a and b, integral, fractional, irrational, and negative, the order of addition will not affect the result. Thus our assumption also includes the affirmation that $\sqrt{3} + \sqrt{5} = \sqrt{5} + \sqrt{3}$. It is important for another reason to recognize that this assumption is being made. Numbers are not apples or cows. They are abstractions from physical situations. Mathematics works with these abstractions in order to deduce information about physical situations. However, if the axioms are not well chosen, the deductions will not apply. Hence it is well to note what assumptions are being employed and to ascertain that they are well founded in experience.

Let us, therefore, note the axioms which we have been using and will continue to use. The first axiom is the one discussed in the preceding paragraph:

AXIOM 1. For any two numbers a and b, $a + b = b + a$.

The axiom is called the *commutative axiom of addition* because it says that we can commute or interchange the order of the two numbers to be added. We note that subtraction is not commutative, that is, $3 - 5$ does not equal $5 - 3$.

If we had to calculate $3 + 4 + 5$, we could first add 4 to 3 and then add 5 to this result, or we could add 5 to 4, and then add this result to 3. Of course, the result is the same in the two cases, and this is exactly what our second axiom says.

AXIOM 2. For any numbers a, b, and c, $(a + b) + c = a + (b + c)$.

This axiom is called the *associative axiom of addition* because we can associate the three numbers in two different ways in performing the addition.

The two axioms we have just discussed have their analogues for the operation of multiplication.

AXIOM 3. For any two numbers a and b, $a \cdot b = b \cdot a$.

This axiom is called the *commutative axiom of multiplication*. Incidentally, the dot which is used to denote multiplication is omitted if there is no danger of misunderstanding. Thus, we could as well write $ab = ba$. The axiom is clearly a property of numbers; yet we sometimes fail to recognize that it is applicable. Many a student hesitates to write $5 \cdot a$ instead of $a \cdot 5$. But the commutative axiom says that the two expressions are equal. We might note in this context that the operation of division is not commutative, for $4 \div 2$ does not equal $2 \div 4$.

AXIOM 4. For any three numbers a, b, and c, $(ab)c = a(bc)$.

This axiom is called the *associative axiom of multiplication*. Thus $(3 \cdot 4)5 = 3(4 \cdot 5)$.

The next axiom is not quite so obvious. It says, for example, that $3 \cdot 6 + 3 \cdot 5 = 3(6 + 5)$. In the present instance we can perform the calculation to see that the left and right sides are equal, but this is really not necessary. Suppose we had 157 cows in one herd and 379 in another, and each herd increased sevenfold. The total number of cattle is then $7 \cdot 157 + 7 \cdot 379$. But if the original two herds were one herd with $157 + 379$ cows, and this single herd increased sevenfold, we would have $7(157 + 379)$ cows. It is physically clear that we have the same number of cows now as before, that is, that $7 \cdot 157 + 7 \cdot 379 = 7(157 + 379)$. Stated in general terms, the axiom is:

AXIOM 5. For any three numbers a, b, and c, $ab + ac = a(b + c)$.

This axiom, called the *distributive axiom*, is very useful. For example, to calculate $571 \cdot 36 + 571 \cdot 64$ we can apply the axiom to state that this quantity is $571(36 + 64)$ or $571 \cdot 100$ or 57,100. We say often that we have *factored* the quantity 571 out of the sum $571 \cdot 36 + 571 \cdot 64$.

We should note that from

$$ab + ac = a(b + c)$$

we can also state that

$$ba + ca = (b + c)a,$$

because in each term of the first equation we can apply the commutative axiom of multiplication to change the order of the factors.

We often use the second form of the distributive axiom. Thus, suppose a is some number and we wish to calculate $5a + 7a$. We can replace this sum by $(5 + 7)a$, and obtain $12a$.

The distributive axiom is also applicable in the following situation. Suppose we have to calculate

$$\frac{296 + 148}{296}.$$

One might be tempted to cancel the two numbers 296. But this is incorrect. The given fraction means

$$\tfrac{1}{296}(296 + 148),$$

and the distributive axiom tells us that we may write instead

$$\tfrac{1}{296} \cdot 296 + \tfrac{1}{296} \cdot 148, \quad \text{or} \quad 1 + \tfrac{1}{2}.$$

In addition to the above axioms, we have the following evident properties of numbers:

AXIOM 6. Quantities equal to the same quantity are equal to each other.

AXIOM 7. If equal quantities are added to, subtracted from, multiplied with, or divided into equal quantities, the results are equal. However, division by zero is not permitted.

On the whole, the above axioms are obvious, and we really apply them almost automatically. Hence we shall not usually call attention to the fact that we are using them unless a particular step is in doubt. Thus we shall say that $2(a + b)$ is $2a + 2b$, without noting that we have applied the distributive axiom. However, we should not lose sight of the fact that the mathematics built upon the number system is a deductive system. This point needs emphasis because we begin to learn arithmetic at an early age by rote and thereafter we tend to operate with numbers mechanically without perceiving that we are constantly using axioms of numbers.

Exercises

1. Do you believe that $256(437 + 729) = 256 \cdot 437 + 256 \cdot 729$? Why?
2. Is it correct to assert that $a(b - c) = ab - ac$? [*Suggestion:* $b - c = b + (-c)$.]
3. Perform the operations called for in the following examples:

(a) $3a + 9a$	(b) $a \cdot 3 + a \cdot 9$	(c) $a(5 + \sqrt{2})$	(d) $7a - 9a$
(e) $3(2a + 4b)$	(f) $(4a + 5b)7$	(g) $a(a + b)$	(h) $a(a - b)$
(i) $2(8a)$	(j) $a(ab)$		

4. Carry out the multiplication: $(a + 3)(a + 2)$. [*Suggestion:* Regard $(a + 3)$ as a single quantity and apply the distributive axiom.]
5. Calculate $(n + 1)(n + 1)$.
6. If $3x = 6$, is $x = 2$? Why?
7. If $3x + 2 = 7$, is $3x = 5$? Why?
8. Is the equality $x^2 + xy = x(x + y)$ correct?
9. Is it correct to assert that $a + (bc) = (a + b)(a + c)$?

4–6 Applications of the number system. Something of the power, methodology, and subtlety of mathematical reasoning can already be seen in the applications which have been made of the several types of number. Indeed, we shall see that these resulted in significant scientific discoveries.

Let us begin with some rather simple matters. Suppose a man drives a car for one mile at 60 miles per hour and for another mile at 120 miles per hour. What is his average speed? We tend to answer this question by applying the common procedure for finding an average. Thus, if a man buys one pair of shoes for $5 and another for $10, the average price is $5 + $10 divided by 2, or $7.50. Hence it would seem as though the average speed in the above problem should be 60 + 120 divided by 2, or 90 miles per hour. However, this answer is not correct. The number 90 is an average in the arithmetic sense, but it is not the average we seek. The average speed should be that speed which would enable the man to drive the two miles *in the same time* as it took him to drive that distance at the two different speeds. Now, it took the man 1 minute to drive the first mile and it took him $\frac{1}{2}$ minute to drive the second mile. Hence it took him $1\frac{1}{2}$ minutes to drive 2 miles. We now ask, What average speed maintained for $1\frac{1}{2}$ minutes would cover 2 miles? Since the average speed multiplied by the total time should give the total distance, the average speed is the total distance divided by the total time, that is,

$$\text{average speed} = \frac{2}{3/2} = \frac{4}{3}.$$

The average speed is then $4/3$ miles per minute or 80 miles per hour.

The point of this example, not a momentous one to be sure, is merely that the unthinking, blind application of arithmetic does not produce the correct result. The notion of average speed serves a physical purpose, and unless we are clear about what average speed is supposed to mean, we shall not profit by the use of arithmetic.

Exercises

1. A man can row a boat in still water at 6 mi/hr. He plans to row upstream for 12 mi and then back in a river whose current flows at 2 mi/hr. Thus his speed upstream is 4 mi/hr and his speed downstream is 8 mi/hr. He reasons that his average speed is 6 mi/hr and that the entire trip of 24 mi should therefore take 4 hr. Is this reasoning correct?

2. Suppose that a merchant sells apples at a price of 2 for 5¢ and oranges at 3 for 5¢. To make his arithmetic simpler he decides to sell any 5 pieces of fruit for 10¢ or at the average price of 2¢ per piece. Thus, if he sells 2 apples and 3 oranges, he sells 5 pieces of fruit at 2¢ each and receives the same 10¢ as if he had sold them at the original separate prices. Is the merchant's average price correct? [*Suggestion:* Consider what results if he sells 12 apples and 12 oranges.]

3. Suppose the merchant wishes to sell a apples and b oranges to some customer at the prices given in Exercise 2. What should the average price be?

4. Given the data of Exercise 2, is there an average price which would be correct no matter how many apples and how many oranges are sold?

5. One man can dig a certain ditch in 2 days and another can dig the same ditch in 3 days. What is their average rate of ditch-digging per day?

Let us consider next an application of simple arithmetic to genetics. Suppose we have before us 2 red aces and 2 red kings from the usual deck of 52 cards. How many different pairs consisting of one ace and one king can be put together? Since each ace can be paired with either of 2 kings, there are 2 different pairs for any one ace. Since we have 2 aces, there are $2 \cdot 2$ or 4 different pairs.

Now let us suppose that we have 2 red aces, 2 red kings, and 2 red queens. How many different sets consisting of one ace, one king, and one queen can we form with the given cards? We saw above that there are 4 different pairs of aces and kings. With each of these 4 pairs we can place 2 different queens. Hence there are $4 \cdot 2$ or 8 different sets of 3 cards. We note that $4 \cdot 2 = 2 \cdot 2 \cdot 2 = 2^3$.

If we have 2 red aces, 2 red kings, 2 red queens, and 2 red jacks, the number of different sets, each consisting of one ace, one king, one queen, and one jack, can also be readily calculated. Each of the 8 choices of ace, king, and queen can be paired with each of the 2 jacks. Hence there are $8 \cdot 2$ or 16 choices in all. Now, $8 \cdot 2 = 4 \cdot 2 \cdot 2 = 2 \cdot 2 \cdot 2 \cdot 2 = 2^4$.

Clearly, if we had 10 different pairs of cards and had to make all possible choices of 10 cards, one from each pair, the number of all possible sets of 10 cards would be

$$2 \cdot 2 \cdot 2 \cdot 2 \cdot 2 \cdot 2 \cdot 2 \cdot 2 \cdot 2 \cdot 2 = 2^{10} = 1024.$$

This simple reasoning about cards has an important application to genetics. The reproductive cells (as well as ordinary cells) of the human male contain 24 pairs of chromosomes. When a sperm cell is formed from the reproductive cell, it contains 24 chromosomes, each coming from one of the 24 pairs. Hence a sperm cell can be formed in 2^{24} possible combinations. The reproductive cells of the human female also contain 24 pairs of chromosomes. An ovum formed from the female reproductive cell contains 24 chromosomes, each coming from one of the 24 pairs of the reproductive cell. Hence there are 2^{24} possible ways in which an ovum can be formed. In conception, any one sperm joins, or fertilizes, any one ovum. Since there are 2^{24} possible sperms and 2^{24} possible ova, the number of possible chromosome combinations for the fertilized ovum is then

$$2^{24} \cdot 2^{24} = 16{,}777{,}216 \cdot 16{,}777{,}216 = 281{,}474{,}976{,}710{,}656.$$

This is the number of possible variations in the genetic make-up of any one child a man and wife may have. Actually, the number of variations is somewhat larger. Each chromosome contains genes, and these determine the hereditary qualities. Biologists have found that any two paired chromosomes in a reproductive cell may exchange some genes, and this exchange gives rise to new varieties of sperm cells and ova.

Exercises

1. The usual deck of 52 cards contains 4 different aces and 4 different kings. How many different pairs of cards, each pair consisting of one ace and one king, can be formed from the aces and kings?

2. A manufacturer offers his automobile in 3 different colors, with or without a heater, and with or without a radio. How many different choices can a purchaser make?

3. A girl has 3 hats, 2 dresses, and 2 pairs of shoes. How many different costumes does she have?

4. There are six numbers on a die (singular of dice). How many different pairs of numbers can show up on a throw of a pair of dice? The two dice are to be marked so that a throw of a 2 on one die (say A) and of a 5 on the other (say B) can be distinguished from the reverse arrangement (5 on A and 2 on B).

We have already discussed the fact that our method of writing quantities uses the idea of positional notation in base ten (see Section 4–2). However, some civilizations used other numbers as a base. For example, the Babylonians, for reasons that are obscure, selected 60. This system was taken over by the Greek astronomers and was used in Europe for many mathematical and all astronomical calculations as late as the seventeenth century. It still survives in our practice of dividing hours and angles into 60 minutes and 60 seconds. In adopting ten as a base, Europe followed the practice of the Hindus. Let us challenge history and see whether we can derive some advantage from a change to a new base.

We shall choose base six. The quantities from zero to five would be designated by the symbols 0, 1, 2, 3, 4, 5, as in base ten. The first essential difference comes up when we wish to denote six objects. Since six is to be the base, we would no longer use the special symbol 6, but place the 1 in a new position to denote 1 times the base, just as in base ten the 1 in 10 denotes one times the base, or the quantity ten. Hence, to write six in base six, we would write 10, but now the symbols 10 mean 1 times six plus 0. Thus the symbols 10 can denote two different quantities, depending upon the base employed. Seven in base six would be written 11, because in base six these symbols mean 1 times six + 1, just as 11 in base ten means 1 times ten + 1. Again the symbols 11 represent different quantities, depending upon the base implied. As another example, to denote twenty-two in base six we write 34, because these symbols now mean 3 times six + 4.

In base ten, to write numbers larger than ninety-nine, we use a third position, the hundreds' place, to indicate tens of tens. Similarly in base six, when we reach numbers larger than thirty-five, we use a third position to denote sixes of sixes. Thus thirty-eight would be written in base six as 102, wherein the one means 1 times six times six, the 0 means 0 times six, and the 2 denotes just 2 units. To express very large numbers we would use four-place numbers, five-place numbers, and so forth.

We can perform the usual arithmetic operations in base six. However, we would have to learn new addition and multiplication tables. For example, in base ten we write $5 + 3 = 8$, whereas in base six, eight must be written 12. Hence our addition table would have to state that $3 + 5 = 12$. Likewise, our new multiplication table would have to list $3 \cdot 5 = 23$, because fifteen is $2 \cdot 6 + 3$ or, in base six, 23. When we learned to use base ten, we had to memorize the result of adding each number from 0 to 9 to every number from 0 to 9, and the result of multiplying each number from 0 to 9 with every number from 0 to 9. For base six we would have to learn to add (and multiply) only numbers from 0 to 5 to (or with) the numbers of this set. Thus our addition and multiplication tables would be shorter, and we would learn arithmetic sooner as youngsters. We might even pass the hurdle of arithmetic so easily that we might get to like mathematics. The only disadvantage of base six would be that to represent large quantities we would have to use more digits. For example, the quantity fifty-four, written as 54 in base ten, must be written as 130 in base six, because fifty-four equals $1 \cdot 6^2 + 3 \cdot 6 + 0$.

There are people who campaign for the adoption of base twelve, because it offers special advantages. For one thing, more fractions can be written as finite decimals in base twelve. Thus $\frac{1}{3}$ must be written as the unending decimal 0.333 . . . in base ten, but can be written as 0.4 in base twelve, since in this base 0.4 means $\frac{4}{12}$. Also, since the English system of denoting length calls for 12 inches in one foot, we could, for example, express 3 feet and 6 inches as 36 inches in base twelve, whereas to express this number of inches in base ten, we must first calculate $3 \cdot 12 + 6$ and then write 42 in base ten. To a limited extent we could use base twelve in our method of recording time. In the United States the day has two sets of twelve hours, and in base twelve the hours of the day would run from 0 to 20. Whereas determining what 7 hours after 6 o'clock will be requires at present some computation, under the addition table for base twelve we would state at once that $7 + 6 = 11$. However, base ten is now so widely used that a change to another base for ordinary daily use or commerce is hardly likely.

In the subject of bases we have an idea that was pursued for centuries, largely as an interesting and amusing speculation, but which suddenly became highly important in science and even in the commercial world. For several centuries mathematicians worked on the design of machines which would perform arithmetical computations quickly and thus remove a good deal of the drudgery of arithmetic. Although some types of computing machines were invented and used, mathematicians saw their golden opportunity in the electronic devices developed by modern radio engineers. The key is the radio vacuum tube, which can be made to pass current by applying a voltage to it or can be kept inactive. Two maneuvers are thus possible. In base two all numbers require only two symbols, the 0 and the 1. A typical number would be 1011, which means $1 \cdot 2^3 + 0 \cdot 2^2 + 1 \cdot 2 + 1$. This number can be recorded by the machine by employing four tubes, one for the units' place, another for multiples of 2, a third for multiples of 2^2, and a fourth for multiples of 2^3. To record 1011, the first, second, and fourth tubes can be activated, and the third, which records the third place in the number, kept inactive. The currents passed by the tubes which "fire" are recorded by the machine in special circuits. Another number can then be fed into the machine. Let us suppose that it is to be added to the first number. The result of having two 1's in the same place means, in base 2, that the sum is to be 0 and a 1 carried over to the next place. This operation is readily performed by the circuits. While this description of an electronic computer certainly doesn't begin to present the ingenious ideas which engineers and mathematicians have incorporated, we may perhaps see that the workings of a vacuum tube are ideally suited to operations in base two.

To take advantage of the fact that computers can perform calculations in base two, the numbers to be worked on are converted beforehand from base ten to base two and then fed to the machine along with other instructions. The machine then operates in this latter base. The result is, of course, reconverted to base ten.

Because computers work with microsecond speed, they are exceedingly valuable in any commercial or industrial organization which must process a lot of numerical data. Calculations in banks, insurance companies, and in industry, which used to require an immense amount of human labor, are now performed by machines. Computers are the first in a new series of machines which keep track of great quantities of data, select information from millions of cards on which data are recorded, plan factory operations, direct machinery, and may soon provide translations of foreign-language publications.

Electronic computing machines are an enormous boon to science and mathematics also. The arithmetic required to extract concrete information from mathematical formulas is often so lengthy that it would take years to perform these calculations. Computing machines do such work in hours. Moreover, mathematicians no longer hesitate to work on problems which will lead to extensive computations, because they now know that their work will not be in vain.

Computing machines may help us to learn more about the action of the human brain. According to biologists, the nerve cells in the nerve chains and the cells in our brains respond to electrical impulses much as a vacuum tube does. Just as a tube will "fire" when it receives electrical current beyond a certain minimum value and remain inactive otherwise, so do the nerve cells in the nerve chains transmit an electrical impulse to whatever organ they may lead when this impulse exceeds a threshhold value; otherwise they are inactive. Computing machines also have a memory; that is, partial results of calculations are stored automatically in a special device, called the memory. When these results are needed, the memory device releases them. Thus the result of an addition process might be stored in the memory device until the result of some multiplication process is obtained and then, if so instructed, the machine will add these two results. The machine's memory device, then, functions somewhat like the human memory. Hence, in two respects at least, electronic computing machines simulate the actions of human nerves and memory. Though machines are, in speed, accuracy, and endurance, superior to the human brain, one should not infer, as many popular writers are now suggesting, that machines will ultimately replace brains. Machines do not think. They perform the calculations which they are directed to perform by people who have the brains to know what calculations are wanted. Nevertheless, we undoubtedly have in the machine a useful model for the study of some functions of the human brain and nerves.

Exercises

1. Construct an addition table for base six.
2. Construct a multiplication table for base six.
3. Construct an addition and multiplication table for base two.

4. The following numbers are in base ten:

   9, 10, 12, 36, 48, 100.

   Write the respective quantities in base six.

5. The following numbers are in base six:

   5, 10, 12, 20, 100.

   Write the respective quantities in base ten.

6. Write the fraction $\frac{1}{2}$ as a decimal in base six.

7. The quantity 0.2 is in base six. Write the corresponding quantity in base ten.

8. The number 101 is written in some unknown base and equals ten. What is the base?

9. Find the least number of weights needed to weigh, to the nearest pound, objects weighing from 0 to 63 lb. The scale to be used contains two pans and the weights are to be put in one pan. [*Suggestion:* Consider the problem of representing all numbers from 0 to 63 in base 2.]

Some of the most remarkable uses of numbers, which have led to profound discoveries, are found in the study of the structure of matter. During the early and middle part of the nineteenth century, certain basic experimental facts about the varieties of matter found in nature led, after some inevitable fumbling on the part of John Dalton, Amadeo Avogadro, and Stanislav Cannizzaro, to the theory that all matter is made up of atoms. Thus hydrogen, oxygen, chlorine, copper, aluminum, gold, silver, and all other varieties of matter are composed of atoms. Experimental techniques were developed to measure the relative weights of the atoms of different elements. The convenient unit of weight chosen was $1/16$ the weight of the oxygen atom, so that the weight of the atom of oxygen is 16. The weight of the hydrogen atom then proved to be 1.0080, that of copper 63.54, that of gold 197.0, and so on. By this time, too, a number of chemical properties of these various elements, such as their melting temperatures, boiling temperatures, and their ability to combine with other elements to form compounds, had been determined.

The question which had begun to stir the chemists was whether there existed any law or principle which utilized the atomic weights of these elements. The crowning discovery was made in 1869 by Dimitri Ivanovich Mendeléev (1834–1907). He found, as he began to arrange the elements in order of increasing atomic weight, that every eighth element, starting from a given one, had chemical properties similar to the first one. Thus, the gases fluorine and chlorine, the latter the eighth element starting from fluorine, both combine readily with metals. However, as he continued to place each of the 63 different elements known in his time in the eighth position after the one with similar chemical properties, he saw that he had to leave blank spaces. Mendeléev was so much impressed with the periodicity of chemical properties that he did not hesitate to leave the blank spaces and to affirm that there must be elements to fill these spaces. Since each of these missing elements should have chemical properties similar to those of the element found in the eighth position preceding or succeeding the missing element, he could even predict some of the properties of the unknown elements. Mendeléev described the properties of three of the missing elements, and his immediate successors discovered them. These are now called scandium, gallium, and germanium. Still later, others were found. The interesting fact about Mendeléev's work from the mathematical standpoint is that he had no physical explanation of why elements eight positions removed from one another should have similar chemical properties. He knew only that the number eight was the key to the arrangement, and he followed this mathematical guide faithfully. Long after Mendeléev's time, other elements, for example helium, were found, which do not fit into this arrangement, but his periodic table is still the basic one which all students of chemistry learn today.

Simple arithmetic continued to play a leading role in subsequent developments of atomic theory. The continuing study of the atomic weights of various elements and their chemical properties showed that elements formerly regarded as pure were really not so. Thus, there are two kinds of hydrogen. These have similar chemical properties, but different atomic weights; in fact, one is twice as heavy as the other. Since both were previously called hydrogen, and since they do, in any case, have similar chemical properties, these two forms of hydrogen are called isotopes of hydrogen. Likewise, there is not one substance, oxygen, but there are three, of atomic weights 16, 17, and 18. Uranium, a very important element today, has two isotopes of atomic weights 238 and 235.

The startling fact which emerged from the discovery of isotopes is that when all isotopes are distinguished and the relative weights of the distinct elements determined, the weight of any one element is within 1% of a whole number. Such a fact can hardly be accidental. The explanation would seem to be that all these elements are really multiples of a single element, namely the lighter isotope of hydrogen, which has the least weight of all elements. In other words, the various elements which previously appeared to be entirely different substances, now were seen to be just smaller or larger collections of the same element, but arranged in special ways peculiar to the substance. (Strictly speaking, the fundamental building block is not the lighter isotope of hydrogen, but what is now called the proton. The lighter hydrogen isotope also has an electron whose weight is insignificant by comparison.)

If all of the different elements are really just aggregates of the lighter hydrogen atom, it should be possible to remove some atoms and convert one substance into another. Thus, we should be able to convert mercury, which is the next heavier element after gold, into gold. And we can. What the medieval alchemists hoped to do on mystical and superficial grounds, we can now do on the basis of far better scientific knowledge. Unfortunately, the cost of converting mercury into gold is too great to make it worth while. But we do have uses for the transmutation of elements which are, in our age, more valued, and which we shall describe in a moment.

A scientist who has a theory cannot afford to overlook even one detail, trivial as it may seem, which does not square with his theory. If all elements are merely combinations of the lighter hydrogen atom, then their weights should be exact multiples of the weight of this atom instead of having values within 1% of such weights. (The electrons in the atoms do not account for the difference.) This discrepancy must be explained. The lightest isotope of oxygen had, somewhat arbitrarily, been given weight 16. With this rather arbitrary standard, the lighter hydrogen atom has weight 1.008 rather than exactly 1. But helium, which consists of 4 hydrogen atoms, proves to have weight 4.0028. However, if it consists of 4 hydrogen atoms, its atomic weight should be 4 times 1.008, or 4.032. The difference, $4.032 - 4.0028$, or about 0.03, is the discrepancy which must be accounted for. Now it so happens that Einstein, working in an entirely different field, the theory of relativity, had already shown that mass can be converted into energy. Energy can take different forms. It can be the heat created by burning coal or wood, or it can be radiation such as comes to us from the sun. At the moment the precise form of it does not matter. What does matter is the thought which occurred to scientists that perhaps, when 4 hydrogen atoms are fused to form helium, the missing 0.03 of matter is converted into energy in the process. Hence the fusion of elements should release energy. And experiments showed that this is indeed what happens. The energy which is released is called the binding energy, and it is this energy which is released when a hydrogen bomb is exploded.

In this brief account of the role of arithmetic in chemistry and atomic theory, we have said almost nothing about the great thinking and brilliant experiments which physicists and chemists contributed. Our interest has been to show how the use of simple numbers supplies scientists with a powerful tool. Of course, the mathematics of numbers remains to be developed, and we shall learn how much more can be accomplished with slightly more advanced tools. But we can already see something of what the Pythagoreans envisioned when they spoke of numbers as the essence of reality.

Topics for Further Investigation

1. The Egyptian method of writing whole numbers and fractions.
2. The Babylonian method of writing whole numbers and fractions.
3. The Roman method of writing whole numbers and fractions.
4. The fundamental arithmetical laws of atomic theory. (Use the references to Holton and Roller and to Bonner and Phillips).
5. Pythagorean number theory.

Recommended Reading

Ball, W. W. Rouse: *A Short Account of the History of Mathematics,* Chaps. 1 and 2, Dover Publications, Inc., New York, 1960.

Bonner, F. T. and M. Phillips: *Principles of Physical Science,* Chap. 7, Addison-Wesley Publishing Co., Inc., Reading, Mass., 1957.

COLERUS, EGMONT: *From Simple Numbers to the Calculus,* Chaps. 1 to 8, Wm. Heineman Ltd., London, 1954.

DANTZIG, TOBIAS: *Number, the Language of Science,* 4th ed., Chaps. 1 to 6, The Macmillan Co., New York, 1954 (also in a paperback edition).

EVES, HOWARD: *An Introduction to the History of Mathematics,* pp. 53–60, Rinehart and Co., Inc., New York, 1953.

GAMOW, GEORGE: *One Two Three . . . Infinity,* Chap. 9, The New American Library, New York, 1953.

HOLTON, G. and D. H. D. ROLLER: *Foundations of Modern Physical Science,* Chaps. 22 and 23, Addison-Wesley Publishing Co., Inc., Reading, Mass., 1958.

JONES, BURTON W.: *Elementary Concepts of Mathematics,* Chaps. 2 and 3, The Macmillan Co., New York, 1947.

MILLER, DENNING: *Popular Mathematics,* Chap. 1, Coward-McCann, Inc., New York, 1942.

SMITH, DAVID EUGENE: *History of Mathematics,* Vol. I, pp. 1–75, Vol. II, Chaps. 1 to 4, Dover Publications, Inc., New York, 1953.

CHAPTER 5

ALGEBRA, THE HIGHER ARITHMETIC

Algebra is the intellectual instrument which has been created for rendering clear the quantitative aspect of the world.

Alfred North Whitehead

5–1 Introduction. Mathematics is concerned with reasoning about certain special concepts, the concepts of number and the concepts of geometry. Reasoning about numbers—if one is to go beyond the simplest procedures of arithmetic—requires the mastery of two facilities, vocabulary and technique, or one might say, vocabulary and grammar. In addition, the entire language of mathematics is characterized by the extensive use of symbolism. In fact, it is the use of symbols and of reasoning in terms of symbols which is generally regarded as marking the transition from arithmetic to algebra, though there is no sharp dividing line.

The task of learning the vocabulary and techniques of algebra may be compared with that which faces the prospective musician. He must learn to read music and he must develop the technique for playing an instrument. Since our goal in mathematics is far more the acquisition of an understanding than the attainment of professional competence, the problem of learning the vocabulary and techniques will hardly be a severe one.

5–2 The language of algebra. The nature and use of the language of algebra are readily illustrated, although the illustration is at the moment a trivial one. Most readers have encountered parlor number games, of which the following is an especially simple example. The leader of the game says to any member of the group: Take a number; add 10; multiply by 3; subtract 30; and give me your answer. And now, says the leader, I shall tell you the number you chose originally. To the amazement of the audience he does so immediately. The secret of his method is absurdly simple. Suppose the subject chooses the number a. Then adding 10 yields $a + 10$. Multiplication by 3 means $3(a + 10)$. By the distributive axiom, this quantity is $3a + 30$. Subtraction of 30 yields $3a$. The leader has only to divide by 3 the number given to him to tell the subject what his original choice was. If the leader wishes to be especially impressive, he can ask the subject to perform many more computations that will yield a simple and known multiple of the original number, and he can give the original number just as readily. By representing in the language of algebra the operations which he asks the subject to perform and by noting what the operations amount to, the leader can easily see how the final result is related to the original number chosen.

The language of algebra involves more than the use of a letter to represent a number or a class of numbers. The expression $3(a + 10)$ contains, in addition to the usual plus sign of arithmetic, the parentheses which denote that the 3 multiplies the entire quantity $a + 10$. The notation b^2 is a shorthand expression for $b \cdot b$ and is read b-square. The word *square* enters here because b^2 is the area of a square whose side is b. Likewise, the notation b^3 means $b \cdot b \cdot b$ and is read b-cube. The word *cube* is suggested by the fact that b^3 is the volume of a cube whose side is b. The expression $(a + b)^2$ means that the entire quantity $a + b$ is to be multiplied by itself. An expression such as $3ab^2$ means 3 times some quantity a and that product multiplied by the quantity b^2. In addition, the notation uses the convention that numbers and letters following one another with no symbol in between any two are to be multiplied together. Another important convention stipulates that if a letter is repeated in an expression, it stands for the same number throughout. For example, in $a^2 + ab$ the value of a must be the same in both terms. Thus algebra uses many symbols and conventions to represent quantities and operations with quantities.

Why do mathematicians bother with such special symbols and conventions? Why must they place hurdles in the way of would-be students of their subject? The answer is not that mathematicians are trying to introduce hurdles; nor are they seeking to impress people by making their subject look awesome. Rather the symbolism of algebra and the symbolism of mathematics in general are an unfortunate necessity. The most weighty reason is comprehensibility. Symbolism enables the mathematician to write lengthy expressions in a compact form so that the eye can see quickly and the mind can retain what is being said. To describe in words even the simple expression $3ab^2 + abc$ would require the phrase, "The product of 3 times a number multiplied by a second number which is multiplied into itself added to the product of the first number, the second number, and still a third number." It is unfortunate that our eyes and minds are limited. The long and complicated sentences that would be required if ordinary language were used could not be remembered and, in fact, can be so involved as to be incomprehensible.

In addition to comprehensibility, there is the advantage of brevity. The expression in ordinary language of what is covered in typical texts on mathematics would require tomes of two to ten or fifteen times the customary size of such books.

Still another advantage is clarity. Ordinarily English or, for that matter, any other language is ambiguous. The statement, "I read the newspaper," can mean that one reads newspapers regularly, once in a while, or often, or that one has read the newspaper, presumably the paper of the day. One must judge by the context just what this sentence means. Such ambiguity is intolerable in exact reasoning. By using symbols for specific ideas mathematics avoids ambiguity or, to put the matter positively, each symbol has its own precise meaning, and so the resulting expressions are clear.

Symbolism is one of the sources of the remarkable power of algebra. Suppose that one wished to discuss equations of the form $2x + 3 = 0$, $3x + 7 = 0$, $4x - 9 = 0$, and the like. The particular numbers which appear in these equations do not happen to be important in the discussion; in fact, one wishes to include all equations in which the product of some number and x is added to some other number. The way to represent all possible equations of this form is

$$ax + b = 0. \tag{1}$$

Here a stands for any number, and so does b. These numbers are known, but their precise value is not stated. The letter x stands for some unknown number. By reasoning about the general form (1) the mathematician covers the millions of separate cases which arise when a and b have specific values. Thus, by means of symbolism, algebra can handle a whole class of problems in one bit of reasoning.

Of course, it is unfortunate that one must learn the elements of a new language to master some mathematics. But one could with much justice complain that the French people insist on their language, the Germans on theirs, and so on. Obviously English is the best language, and the French and Germans are exhibiting provincialism by insisting on holding on to their respective languages. The language of mathematics has the additional merit of being universal.

There are justifiable criticisms of the symbolism of algebra, although they are hardly major ones. Mathematicians are greatly concerned about the accuracy of their reasoning, but pay little attention to the aesthetics or appropriateness of their symbolism. Very few symbols suggest their meaning. The signs $+$, $-$, $=$, $\sqrt{\ }$ are easy to write, but they are historical accidents. No mathematician has bothered to replace these by at least prettier ones, perhaps $\diamond$ for plus. The seventeenth-century mathematician Gottfried Wilhelm Leibniz, who did spend days on the choice of symbols in an effort to make them suggestive, was an exception. There are even inconsistencies in symbolism which, once recognized, fortunately do not impair the clarity. For example, when two letters, such as ab, are written together with no symbol between them, then it is understood that multiplication is meant. However two numbers, such as $3\frac{1}{2}$, with no symbol between them, mean $3 + \frac{1}{2}$.

Symbolism entered algebra rather late. The Egyptians, Babylonians, Greeks, Hindus, and Arabs knew and applied a great deal of the algebra which we learn in high school. But they wrote out their work in words. Their algebraic style is in fact called rhetorical algebra because, except for a few symbols, they used ordinary rhetoric. It is significant that symbolism entered mathematics in the sixteenth and seventeenth centuries when pressure to improve the efficiency of mathematics was applied by science. The idea of using symbols was no longer new, but mathematicians were undoubtedly stimulated to extend the application of symbolism and to adopt it readily.

Exercises

1. Why does mathematics use symbols?
2. Criticize the statement that all men are created equal.
3. In the following symbolic expressions the letters stand for numbers. Write out in words what the expressions state.

(a) $a + b$ (b) $a(a + b)$ (c) $a(a^2 + ab)$ (d) $3x^2y$

(e) $(x + y)(x - y)$ (f) $\dfrac{x + 3}{7}$ (g) $\frac{1}{7}(x + 3)$

4. Does $(x + 3)/7 = (1/7)(x + 3)$? [*Suggestion:* What do these symbolic expressions say in words?]
5. Write in symbols: (a) three times a number plus four; (b) three times the square of a number plus four.

5–3 Algebraic transformations. Symbolism is a means to an end. The function of algebra is not to display symbols but to convert or transform expressions from one form to another which may be more useful for the problem in hand.

Let us consider an example. Suppose that in the course of some mathematical work we encounter the expression

$$(x + 4)(x + 3). \tag{2}$$

The letter x in this expression may stand for some number whose value we do or do not happen to know, or it may stand for any one of some class of numbers. What matters is that x stands for a number. If x is a number, then $x + 4$ is a number. We may now apply the distributive axiom, which states that for any numbers a, b, and c,

$$a(b + c) = ab + ac. \tag{3}$$

If we compare (2) and (3) we see that (2) has the form of (3) if we think of $x + 4$ in (2) as the a of (3). Then by applying the distributive axiom to (2) we may assert that

$$(x + 4)(x + 3) = (x + 4)x + (x + 4)3. \tag{4}$$

We also know that there is another form of the distributive axiom, namely

$$(b + c)a = ba + ca.$$

If we apply this axiom to each of the terms on the right side of (4), we see that

$$(x + 4)x = x^2 + 4x \qquad \text{and} \qquad (x + 4)3 = 3x + 12.$$

If we substitute these last two results on the right side of (4), we have

$$(x + 4)(x + 3) = x^2 + 4x + 3x + 12 = x^2 + (4 + 3)x + 12 \qquad (5)$$

or

$$(x + 4)(x + 3) = x^2 + 7x + 12. \qquad (6)$$

Before discussing what this example illustrates, let us note that we do not usually carry out the multiplication of $(x + 4)$ by $(x + 3)$ in this long and rather cumbersome fashion. Instead we write

$$\begin{array}{r} x + 4 \\ x + 3 \\ \hline 3x + 12 \\ x^2 + 4x \quad\quad \\ \hline x^2 + 7x + 12. \end{array}$$

The partial product $3x + 12$ results from multiplying $x + 4$ by 3, and the partial product $x^2 + 4x$ results from multiplying $x + 4$ by x. The two partial products are then added. This manner of carrying out the multiplication is faster, but fails to indicate explicitly that we have used the distributive axiom several times.

The main point of the above example is that we have transformed the expression $(x + 4)(x + 3)$ into the expression $x^2 + 7x + 12$. We do not maintain that the latter expression is more attractive than the former, but it may be more useful in a particular mathematical application. On the other hand, we might, in some situation, find ourselves with the expression $x^2 + 7x + 12$ and, by recognizing that it is equal to $(x + 4)(x + 3)$, be able to make progress toward some significant conclusion. In this latter transformation we say that we have *factored* $x^2 + 7x + 12$ into $(x + 4)(x + 3)$. Which of the two forms is more useful depends upon the application in hand. At the moment we should merely see that algebra is concerned with the technique of such transformations, and that a skilled mathematician should be able to perform them rapidly. Since we shall not become too involved in complicated technical processes, we shall not spend much time in developing skills.

The problem of factoring to which we referred in the preceding paragraph does arise reasonably often. For example, one usually starts with an expression such as $x^2 + 6x + 8$ and seeks to transform it into a product of factors of the form $(x + a)(x + b)$. The original expression is said to be of second degree because it contains x^2 but no higher *power* of x. The factors are first-degree expressions because each contains x but no higher power of x. The problem is to find the correct values of a and b so that the product $(x + a)(x + b)$ will equal the original expression. We know from our work on multiplication [see equation (5)] that

$$x^2 + (a + b)x + ab = (x + a)(x + b).$$

Hence to factor the second-degree expression, we should look for two numbers a and b whose sum is the coefficient of x and whose product is the constant. Thus to factor $x^2 + 6x + 8$, we look for two numbers whose sum is 6 and whose product is 8. By mere trial of the possible factors of 8 we see that $a = 4$ and $b = 2$ will meet the requirements, that is:

$$x^2 + 6x + 8 = (x + 4)(x + 2).$$

Exercises

1. Transform to an equal expression:

(a) $3x \cdot 5x$ (b) $(x + 4)(x + 5)$ (c) $(3x + 4)(x + 5)$

(d) $(x - 3)(x + 3)$ (e) $(x + \frac{5}{2})(x + \frac{5}{2})$ (f) $(x + \frac{5}{2})(x - \frac{5}{2})$

2. Factor the following expressions. Experiment with numbers to find the correct factors.

(a) $x^2 + 9x + 20$ (b) $x^2 + 5x + 6$ (c) $x^2 - 5x + 6$

(d) $x^2 - 9$ (e) $x^2 - 16$ (f) $x^2 + 7x - 18$

3. Prove that $x(x^2 + 7x) = x^3 + 7x^2$.

4. Can you think of a way of testing or verifying (not proving) that

$$x^2 + 5xy + 6y^2 = (x + 3y)\ (x + 2y)$$

for all values of x and y?

5. Write out in words the equivalent of $(x - 3)(x + 3) = x^2 - 9$.

6. A high school girl had to simplify $(a^2 - b^2)/(a - b)$. She reasoned that a^2 divided by a gives a. Minus divided by minus gives plus. And b^2 divided by b gives b. Hence the answer is $a + b$. Is the answer correct? Is the argument correct?

7. There is a well known "proof" that $2 = 1$. The proof runs as follows. Suppose a and b are two numbers such that

$$a = b.$$

We may multiply both sides of this equation by a and obtain

$$a^2 = ab.$$

Now we may subtract b^2 from both sides and obtain

$$a^2 - b^2 = ab - b^2.$$

By factoring we may replace the left and right sides of this equation by

$$(a - b)(a + b) = b(a - b).$$

Division of both sides of this equation by $a - b$ yields

$$a + b = b.$$

Since $a = b$, we may as well write

$$2b = b.$$

But now we can divide both sides of this last equation by b, and there results

$$2 = 1.$$

Find the flaw in this proof.

5–4 Equations involving unknowns. The study of algebraic transformations as such is not very interesting. It is much like the grammar of a language. The significant uses of these transformations occur in larger investigations which we shall undertake later. However, a direct use of the processes of algebra does arise in the problem of finding unknown quantities, a problem not without some interest in itself and one which also arises in the course of broader investigations.

A somewhat practical, though by no means vital, example is the following. The radiator of a car contains 10 gallons of liquid 20 per cent of which is alcohol. The owner wishes to draw off a quantity of liquid and replace it by pure alcohol so that the resulting mixture contains 50 per cent alcohol. How many gallons of liquid should he draw off?

Now the very practical person who refuses to use mathematics can handle this situation very readily. He can draw off 5 gallons of the mixture and replace it by 5 gallons of alcohol. Then the mixture will certainly contain at least 50 per cent alcohol because even the remaining 5 gallons contain some alcohol. However, if the final mixture need contain only 50 per cent alcohol, then the practical person has wasted alcohol and therefore money. If he draws off 4 gallons, 6 gallons will be left, and since 20 per cent of this is alcohol, the alcoholic content is $1\frac{1}{5}$ gallons. If he now adds 4 gallons of alcohol he will have $5\frac{1}{5}$ gallons of alcohol, or more than 50 per cent, in the 10 gallons. On the other hand, if he draws off only 3 gallons, 20 per cent of the remaining 7 gallons is $\frac{7}{5}$ gallons of alcohol, and the addition of 3 more will yield $4\frac{2}{5}$ gallons of alcohol out of 10, or less that 50 per cent. The correct answer lies somewhere between 3 and 4, but where? Instead of continuing to guess let's use a little algebra.

Let x be the number of gallons of the mixture to be drawn off and to be replaced by an equal amount of pure alcohol. Then the number of gallons remaining of the original mixture is $10 - x$. Of this 20 per cent, or $\frac{1}{5}$, is alcohol, so that of the $10 - x$ gallons, $(\frac{1}{5})(10 - x)$ is alcohol. After the x gallons are replaced with pure alcohol, the amount of alcohol in the tank will be $(\frac{1}{5})(10 - x) + x$. We should like to fix x so that the amount of alcohol should be 50 per cent of 10 gallons, or 5 gallons. Hence we seek the value of x which satisfies the equation

$$\frac{1}{5}(10 - x) + x = 5. \tag{7}$$

Now we can apply the distributive axiom to start off our transformations and write

$$\tfrac{1}{5} \cdot 10 - \tfrac{1}{5}x + x = 5. \tag{8}$$

The terms $-\frac{1}{5}x + x$ or $x - \frac{1}{5}x$ amount to $\frac{4}{5}x$. Hence (8) is equivalent to

$$2 + \tfrac{4}{5}x = 5. \tag{9}$$

If we now subtract 2 from both sides of this equation, the result will still be an equality because equals subtracted from equals give equals. Then

$$\tfrac{4}{5}x = 3.$$

We now multiply both sides of this equation by 5/4, and since equals multiplied by equals give equals, we have

$$x = 3 \cdot \tfrac{5}{4}. \tag{10}$$

Hence the answer is that the owner of the car should draw off $3\frac{3}{4}$ gallons of the original liquid. We knew before we applied algebra that the answer lies between 3 and 4, and we now know exactly where.

The more significant point made by this example, however, is that we started with equation (7) which expresses the condition to be satisfied by the unknown quantity x and that, by executing a series of almost mechanical steps justified by axioms about numbers, we arrived at a new equation, (10), which tells us what we wish to know. In other words, we performed a series of transformations which carried us from one equation to another and we profited thereby. The answer is not sensational, but we see how the manipulation of symbols gives us new information.

There is another point which the above example illustrates, at least in a minor way. Once we formulate equation (7) we forget all about the physical situation and concentrate solely on the equation. Nothing that is not relevant to the problem, i.e., to the problem of determining the number x, interferes with our thinking. Ernst Mach, a famous scientist of the late nineteenth century, said that mathematics is characterized by "a total disburdening of the mind," and we can now see what he meant. The make of the car, the shape of the radiator, the fact that the owner may be concerned with protecting the liquid in the radiator from freezing, and any other facts which have nothing to do with determining x can be forgotten. We disburden our minds of everything but the quantitative facts expressed in equation (7), and proceed to handle quantitative relationships only.

Equation (7) is rather simple. It is called a *linear* or *first-degree equation* because the unknown x occurs to the first power only. Let us consider a second example which will again illustrate the transformation value of algebra, but which also has other interesting features. Suppose that one ship is at A (Fig. 5–1) and another is at B, exactly 10 miles north of A. The ship at B is steaming east at the rate of 2 miles per hour. The ship at A is capable of traveling at a speed of 5 miles per hour and wishes to intercept the other ship. To set his course properly the captain of the ship at A must know where the two will meet.

Let us suppose that C is the point where they will meet. If the captain can determine the distance BC, he will head along the hypotenuse of a right triangle whose arms are AB and BC. Let us therefore denote the distance BC by x. Now that we seem to have labeled all relevant quantities, we encounter the first puzzling aspect of this problem, namely, that we do not have any equation to find x. Without this, of course, we can only sit and do some wishful thinking. Yet we do have enough information to set up such an equation.

What we have overlooked is a physical fact which is implied by the given information: The *time that the ship at B will take to travel to C must be the same as the time it will take the ship at A to reach C.* Since the ship at B travels at 2 miles per hour, it will take $x/2$ hours to reach C. To calculate the time required by the ship at A to reach C, we need the distance AC. We do not know AC, but we can at least express its value by means of the Pythagorean theorem of geometry. This theorem says in the present instance that

$$AC^2 = 100 + x^2.$$

Then

$$AC = \sqrt{100 + x^2}.$$

The time required for the ship at A to travel the distance AC at 5 miles per hour is

$$\frac{\sqrt{100 + x^2}}{5}.$$

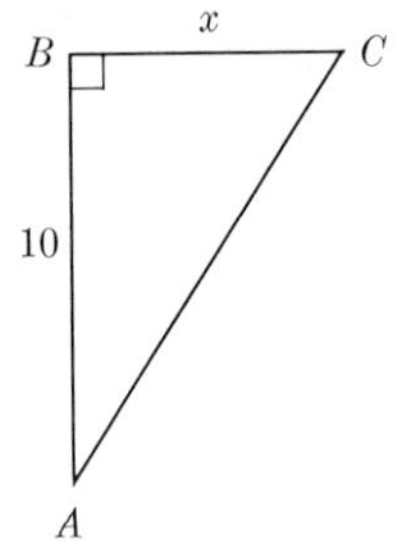

FIGURE 5–1

We next equate the time required for the ship at B to travel the distance BC and the time required for the ship at A to travel the distance AC. This equation is

$$\frac{x}{2} = \frac{\sqrt{100 + x^2}}{5}. \tag{11}$$

We now have an equation to work with. Let us see whether we can transform it so that it will yield a value for x. Since the square root is annoying, let us square both sides, i.e., multiply the left side by itself and the right side by itself. Since the left side equals the right side we are in effect multiplying equals by equals, and so the step is justified. Squaring both sides, we obtain

$$\frac{x^2}{4} = \frac{100 + x^2}{25}. \tag{12}$$

Since fractions are also annoying, let us multiply both sides by 100. We choose 100 because both 25 and 4 divide evenly into 100. Thus

$$100 \cdot \frac{x^2}{4} = 100 \cdot \frac{100 + x^2}{25}.$$

We may apply our operations with fractions to write

$$25x^2 = 4(100 + x^2).$$

Application of the distributive axiom yields

$$25x^2 = 400 + 4x^2. \tag{13}$$

Now we subtract $4x^2$ from both sides, and because equals subtracted from equals yield equals, we obtain

$$21x^2 = 400. \tag{14}$$

Division of both sides by 21, which is a division of equals by equals, yields

$$x^2 = \frac{400}{21}. \tag{15}$$

Now we ask ourselves what number squared yields 400/21. Certainly $\sqrt{400/21}$ is one possibility. But a negative number squared or multiplied by itself is also positive. Hence there are two possible answers:

$$x = \sqrt{400/21} \qquad \text{and} \qquad x = -\sqrt{400/21}. \tag{16}$$

Let us accept both of these for the moment and dispose first of a purely arithmetical question. How much is $\sqrt{400/21}$? Well, we can divide 21 into 400 and obtain 19.05 to two decimal places. We must now find $\sqrt{19.05}$. There is an arithmetic process for finding the square root of a number, but for our purposes it will be sufficient to estimate the answer. Clearly 4 is too small and 5 is too large. By sheer trial we find that $(4.3)^2 = 18.49$ and $(4.4)^2 = 19.36$. Hence the correct value lies between 4.3 and 4.4. If we wished to have a more accurate answer, we could now try 4.31, 4.32, and so on, until we found a result which came as close to 19.05 as possible, and so obtain an answer to the nearest hundredths' place. We shall accept 4.4 as good enough for our purposes and thus we may say that

$$x = 4.4 \qquad \text{and} \qquad x = -4.4. \tag{17}$$

And now we have more than we want; we have two answers, whereas we sought only one. Of course, we wish to use the positive answer because the x we seek stands for a length which is positive. This is the value which has the proper physical meaning in our problem. But the question, How did the negative value of x get into the picture, remains open. The answer involves a rather important point about the nature of mathematics and its relation to the physical world. The mathematician starts with concepts and axioms which express some idealized facts about the world, and proceeds to apply these concepts and axioms to solve physical problems. In the present case the methods used lead to two solutions. Hence the methods may involve new elements which are not present in the physical world, even though the intent was to stay close to it. Thus, squaring both sides of equation (11), a justifiable mathematical step, introduced a new solution, for, if our original equation had been

$$\frac{x}{2} = -\frac{\sqrt{100 + x^2}}{25}, \tag{18}$$

we would have obtained the same equation, (12) and everything we did thereafter would have applied to (18) as well as (11). Hence, in this case, we can see specifically where mathematics departs from the physical situation.

The main point to be noted is then that, although mathematical concepts and operations are formulated to represent aspects of the physical world, mathematics is not to be identified with the physical world. However, it tells us a good deal about that world if we are careful to apply it and interpret it properly. We shall find that this point, which eluded the best thinkers until the late nineteenth century, will acquire increasing importance as we proceed.

There is another valuable lesson to be learned from the solution of the problem we have just examined. When we arrived at step (13), we combined terms in x^2 and then proceeded to find x. The subsequent work led to a fair amount of arithmetic. An engineer working with the same problem and perhaps satisfied with an approximate answer might argue that the term $4x^2$ is small compared with the term $25x^2$ and so disregard it. Instead of our next equation, (14), his new equation would then read

$$25x^2 = 400,$$

and by dividing both sides of this equation by 25, he would obtain

$$x^2 = 16.$$

It now follows that

$$x = 4 \quad \text{and} \quad x = -4.$$

Thus 4 is an approximate answer. Engineers often are satisfied with such approximations because, in constructing actual objects of wood and steel, they cannot meet a specified value exactly. Not only can't one measure exactly, but tools and machines also introduce errors. By neglecting $4x^2$ in (13) the engineer gained the advantage of finding the approximate answer much more readily than we were able to determine the correct answer even to one decimal place only.

In the present problem the saving is trivial, but approximation may make a lot of difference in more difficult problems. Whereas the mathematician, who seeks exact answers, will work months and years on a problem, the engineer will often settle for an approximate answer and obtain it far more easily. The point we are making is not that the engineer is smarter. To get on with his job the engineer must arrive at an answer quickly, whereas the mathematician's job is to obtain a correct answer, no matter how long it takes. Both are true to the objectives and spirit of their own work. Moreover, in making approximations, the engineer raises a question which he may not be able to answer. How good is his approximation? After all, while physical constructions and measurements are not exact, beams must fit. Hence the engineer should really ascertain that the approximation is good enough for his purposes. If he can tolerate an error of only 0.1 of an inch, he must make sure that his approximations do not introduce a larger error.

In really difficult problems the engineer will make approximations and, usually with the aid of a mathematician, determine the error introduced. If he cannot do so, he will often overdesign; that is, if the approximate result shows that a beam supporting a building need be only 1 inch thick, he may make it two inches thick and thereby hope that he has more than allowed for the error. Is he certain even with this precaution that his beam will hold up? No. Big bridges have collapsed because such calculations and additional precautionary measures were not enough. A recent example was the Tacoma bridge in the State of Washington. The bridge did not withstand the force of the wind and collapsed.

Exercises

1. The speed of sound in an iron rod is 16,850 ft/sec, and the speed in air is 1100 ft/sec. If a sound originating at one end of the rod is heard one second sooner through the rod than through the air, how long is the rod?

2. An airplane which can fly at a speed of 200 mi/hr in still air flies a distance of 800 mi with the wind in the same time as it flies 640 mi against the wind. What is the speed of the wind? [*Suggestion:* If x is the speed of the wind, then the speed of the plane when flying with the wind is $200+x$; the speed of the plane when flying against the wind is $200-x$.]

3. A bridge AB is 1 mi (5280 ft) long in winter and expands 2 ft in the summer. For simplicity suppose that the shape in summer is the triangle ACB shown in Fig. 5-2. How far does the center of the bridge drop in summer, that is, how long

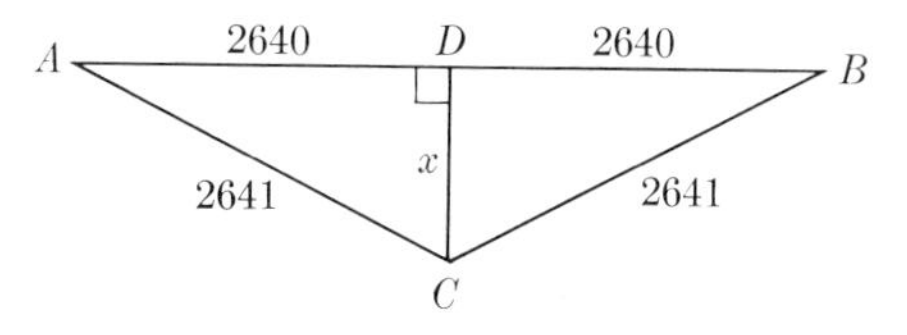

Figure 5-2

is CD? Before calculating the answer, estimate it. To calculate, use the Pythagorean theorem and estimate the square root to the nearest foot.

4. The population of town A is 10,000 and is increasing by 600 each year. The population of town B is 20,000 and is increasing by 400 each year. After how many years will the two towns have the same population?

5. A rope hanging from the top of a flagstaff is 2 ft longer than the staff. When pulled out taut, it reaches a point on the ground 18 ft from the foot of the staff. How high is the staff?

6. A publisher finds that the cost of preparing a book for printing and of making the plates is $5000. Each set of 1000 printed copies costs $1000. He can sell the books at $5 per copy. How many copies must he sell to at least recover his costs?

7. We may certainly say that

$$\tfrac{1}{4}\text{ dollar} = 25\text{ cents}.$$

We take the square root of both sides and obtain

$$\tfrac{1}{2}\text{ dollar} = 5\text{ cents}.$$

What is wrong?

8. A glass which is half full certainly contains as much liquid as a glass which is half empty. Then

$$\tfrac{1}{2}\text{ full} = \tfrac{1}{2}\text{ empty}.$$

If we multiply both sides by 2 we obtain

$$1\text{ full} = 1\text{ empty},$$

or a full glass contains as much as an empty glass. What is wrong?

5–5 The general second-degree equation. Our discussion of the solution of equations in the preceding section dealt with two types of equations, first-degree equations illustrated by equation (7) and second-degree equations illustrated by equation (14). No difficulties can arise in the process of solving first-degree equations, i.e., equations which, by proper algebraic operations, can be expressed in the form

$$ax + b = 0, \tag{19}$$

where a and b are definite numbers and x is the unknown. Equation (19) can readily be solved for x.

The case of second-degree equations is not so simple. We were fortunate that equation (14) led to (15), and that by taking the square root of both sides we obtained the two solutions, or roots as they are called. However we might have to solve an equation such as

$$x^2 - 6x + 8 = 0. \tag{20}$$

This equation is more complicated than (14) because (20) also contains the first-degree term in x.

In solving equation (20), we still do not encounter much trouble. We know from our work on transforming algebraic expressions that the left-hand side of (20) can be factored; that is, the equation can be written as

$$(x - 2)(x - 4) = 0. \tag{21}$$

We now see that when $x = 2$, the left side is zero because

$$(2 - 2)(2 - 4) = 0.$$

When $x = 4$, the left side is again zero because

$$(4 - 2)(4 - 4) = 0.$$

Hence the solutions or roots are

$$x = 2 \qquad \text{and} \qquad x = 4.$$

Now suppose we had to solve the second-degree equation

$$x^2 + 10x + 8 = 0. \tag{22}$$

This time it is not possible to find simple factors of the left side. Equations such as (22) do arise in real problems. Hence the mathematician considers the question, Is there a method which will solve such second-degree equations? Naturally he studies those he can solve to see whether they furnish any clue to such a method.

Examination of equation (20) reveals an interesting fact. The roots are 2 and 4. The sum of these two numbers is 6, and the coefficient, or multiplier, of x is -6. The product of 2 and 4 is 8, and 8 is the constant term, that is, the term free of x. These facts might be a coincidence, and so the mathematician would investigate whether they hold for other simple equations. Consider the very simple equation:

$$x^2 - 4 = 0. \tag{23}$$

Here the roots are $+2$ and -2. Their sum is 0, and we note that the term in x is missing, which means it is $0 \cdot x$. The product of the roots is -4, precisely the constant term in (23). Presumably we have some facts about the roots, but how can we use them?

Equations of the form (23) are easy to solve, since one only has to take a square root. Perhaps the method we should seek is one which reduces all equations of the type (20) to the type (23). But how do we do this? The sum of the roots in (23) is zero. The sum of the roots in (20) is 6, and this is the negative of the coefficient of x. If we added to each root of (20) one-half the coefficient of x, that is, -3, the sum of the roots would be zero. What this suggests, then, is to *form a new equation whose roots are the roots of the old one, each increased* by one-half the coefficient of x.* Since the coefficient of x is -6, we let

$$y = x + (-3) = x - 3$$

or

$$x = y + 3. \tag{24}$$

* We use the term "increased" here, even though in the example we add a negative quantity to each root and really decrease the value of the roots.

If we substitute this value of x in (20), we obtain

$$(y+3)^2 - 6(y+3) + 8 = 0.$$

We now calculate the square in the first term, carry out the multiplication in the second term, and find that

$$y^2 + 6y + 9 - 6y - 18 + 8 = 0$$

or

$$y^2 - 1 = 0.$$

Then

$$y^2 = 1$$

and

$$y = 1 \quad \text{and} \quad y = -1.$$

But from (24) we see that

$$x = 1 + 3 \quad \text{and} \quad x = -1 + 3$$

or

$$x = 4 \quad \text{and} \quad x = 2.$$

Thus we obtain *without factoring* the very same roots of equation (20) that we found previously by factorization.

Now let us reconsider equation (22), namely

$$x^2 + 10x + 8 = 0. \tag{22}$$

Since the roots cannot be obtained by any apparent method of factoring, let us see whether the idea just tried works here also; that is, let us form a new equation whose roots are the roots of (22) increased by one-half the coefficient of x. The roots of (22) are represented by x. Then we shall form a new equation whose roots y are:

$$y = x + \tfrac{10}{2} = x + 5. \tag{25}$$

From (25) we have

$$x = y - 5. \tag{26}$$

We substitute this value of x in (22) and obtain

$$(y-5)^2 + 10(y-5) + 8 = 0.$$

We perform the indicated multiplications and obtain

$$y^2 - 10y + 25 + 10y - 50 + 8 = 0.$$

By combining terms we find that

$$y^2 - 17 = 0$$

or

$$y^2 = 17.$$

Then

$$y = \sqrt{17} \qquad \text{and} \qquad y = -\sqrt{17}.$$

We now use (26) to state that

$$x = \sqrt{17} - 5 \qquad \text{and} \qquad x = -\sqrt{17} - 5. \tag{27}$$

We have found the two roots of (22) without factoring.

Exercises

1. Find the roots of the following equations by factoring the left-hand side:

(a) $x^2 - 8x + 12 = 0$ (b) $x^2 + 7x - 18 = 0$

2. Find the roots of each of the equations in Exercise 1 by forming a new equation whose roots are "larger" than those of the original equation by one-half the coefficient of x.

3. Solve the following equations by the method of forming a new equation whose roots are "larger" than those of the original equation by one-half the coefficient of x.

(a) $x^2 + 12x + 9 = 0$ (b) $x^2 - 12x + 9 = 0$

The method of solving second-degree equations by forming a new equation seems to work, but we have no proof that it will always work. To secure a general proof we shall use one of the basic devices of algebra; that is, instead of working with particular equations, we shall consider the general second-degree equation

$$x^2 + px + q = 0. \tag{28}$$

Here p and q are letters, each of which can stand for *any* given real number. The use of the letters p and q must be distinguished from the use of x to stand for the specific unknown roots of the equation. Now we follow the method employed to solve equations (20) and (22); that is, we form a new equation whose roots are the roots of (28), each increased by one-half the coefficient of x. This means that we introduce the expression

$$y = x + \frac{p}{2}.$$

Then

$$x = y - \frac{p}{2}. \tag{29}$$

We substitute this value of x in (28) and obtain

$$\left(y - \frac{p}{2}\right)^2 + p\left(y - \frac{p}{2}\right) + q = 0.$$

By squaring the first term and multiplying through by p in the second one, we obtain

$$y^2 - py + \frac{p^2}{4} + py - \frac{p^2}{2} + q = 0.$$

The terms involving py cancel. Moreover, $p^2/4 - p^2/2 = -p^2/4$. Hence

$$y^2 - \frac{p^2}{4} + q = 0.$$

By adding $p^2/4$ to both sides and subtracting q from both sides, we obtain

$$y^2 = \frac{p^2}{4} - q.$$

Hence

$$y = \sqrt{\frac{p^2}{4} - q} \quad \text{and} \quad y = -\sqrt{\frac{p^2}{4} - q}.$$

In this general case, we cannot determine the numerical value of the square root, but we can leave the result in this form. We now see from equation (29) that

$$x = \sqrt{\frac{p^2}{4} - q} - \frac{p}{2} \quad \text{and} \quad x = -\sqrt{\frac{p^2}{4} - q} - \frac{p}{2}. \tag{30}$$

This result is remarkable.* We have shown that the roots of any equation of the form (28) (that is, no matter what p and q are) are given by the expressions (30).

We really have accomplished more than we sought to accomplish. We sought a method of solving an equation such as (22). We not only have found such a method, but, since the result (30) holds for *any* such equation, we do not have to go through the entire process each time; we proceed by simply substituting the proper value of p and q in (30). Thus if we compare equations (22) and (28), we see that the p in (22) is 10 and q is 8. Hence

* In many books a method is given for solving the general second-degree equation $ax^2 + bx + c = 0$. If we divide this equation by a, we obtain $x^2 + (b/a)x + (c/a) = 0$. This equation is now of the same form as (28), where $p = b/a$ and $q = c/a$. If we enter these values of p and q in (30), we get the roots

$$x = -\frac{b}{2a} + \frac{\sqrt{b^2 - 4ac}}{2a} \quad \text{and} \quad x = -\frac{b}{2a} - \frac{\sqrt{b^2 - 4ac}}{2a}.$$

let us substitute 10 for p and 8 for q in (30). We find

$$x = \sqrt{\frac{100}{4} - 8} - \frac{10}{2} \quad \text{and} \quad x = -\sqrt{\frac{100}{4} - 8} - \frac{10}{2},$$

or

$$x = \sqrt{17} - 5 \quad \text{and} \quad x = -\sqrt{17} - 5. \tag{31}$$

This is exactly the result obtained in (27).

By working with the general form $x^2 + px + q = 0$ instead of equations with specific numbers as coefficients, we have shown how to solve *any* second-degree equation. This general result could never be derived from equations with numerical coefficients because there are infinitely many such equations, and one could not investigate them all. Thus the use of letters to represent any one of a class of numbers gives mathematics a power and generality which achieves what could not be accomplished in many lifetimes of effort with particular equations. Of course, to people who do not care to solve one quadratic the ability to solve all is no boon. But even these people have benefited indirectly. The preceding theory illustrates how the mathematician, when called upon to solve the same type of problem repeatedly, seeks a general method which will handle all of them.

The use of letters such as p and q, which has made an enormous difference in the effectiveness of mathematics, seems like a small idea once understood, and yet it is a rather recent development. From the time of the Babylonians and Egyptians to about 1550, all the equations solved had numerical coefficients. Although many algebraists realized that the method they used for one set of numerical coefficients would work for any other, they had no general proof. The idea of employing general coefficients in algebraic equations, an idea which, as we shall see, was taken over into other domains of mathematics, is due to François Vieta (1540–1603), a great French mathematician. The remarkable fact about Vieta is that he was a lawyer who worked for the kings of France. Mathematics was just a hobby to him, but one at which he "worked" extensively. Vieta was fully conscious of what he had done by introducing literal coefficients. He said that he was introducing a new kind of algebra which he called *logistica speciosa,* that is calculation with whole species, as opposed to the numerical work of his predecessors which he called *logistica numerosa.*

We could consider other examples of how the processes of algebra permit us to solve equations involving unknowns, but we shall not devote more time to the subject. What is important is the recognition that by means of algebra we can extract information from some given facts. It is also important to see how readily and mechanically the processes of solving equations yield the desired information. In fact, one of the curious things about mathematics that clearly emerges even from our brief work in algebra is that mathematics which is concerned with reasoning nevertheless creates processes which can be applied almost mechanically, that is, without reasoning. The thinking is, so to speak, mechanized and this mechanization enables us to solve complicated problems in no time. We think up processes so that we don't have to think.

It may be necessary to caution the reader again that while the techniques of transformations are necessary to perform useful and interesting mathematical work, they are not the substance of mathematics. If all that one learns in mathematics is the ability to execute these techniques, however quickly and accurately, he will not see the real purpose, nature, and accomplishments of mathematics. To a large extent, techniques are a necessary evil, like practicing scales on a piano, in order to be able to play large and beautiful compositions. Naturally those who wish to be professional mathematicians must learn as many of these techniques as possible.

Exercises

1. Solve by means of (30) the following equations:

(a) $x^2 - 8x + 10 = 0$ (b) $x^2 + 8x + 10 = 0$

(c) $x^2 - 6x - 9 = 0$ (d) $2x^2 + 8x + 6 = 0$

(e) $x^2 - 8x + 16 = 0$

5–6 The history of equations of higher degree. The search for generality in mathematics began in the sixteenth century. One type of generality became possible when Vieta showed how to treat a whole class of equations by means of literal coefficients. Another direction which the search for generality took was the investigation of equations of degree higher than the second.

The first of the notable mathematicians to pursue the mathematics of equations of higher degree and certainly the greatest combination of mathematician and rascal is Jerome Cardan. He was born in Pavia, Italy, in 1501 to somewhat disreputable parents, although his father was a lawyer, doctor, and minor mathematician. Cardan had no upbringing worth speaking about and was sickly during the first half of his life. Despite these handicaps, he studied medicine and became so celebrated a physician that he was invited to treat prominent people in many countries of Europe. At various times he was professor of medicine, and he also lectured on mathematics at several Italian universities.

He was aggressive, high-tempered, disagreeable, and even vindictive, as if anxious to make the world suffer for his early deprivations. Because illnesses continued to harass him and prevented him from enjoying life, he gambled daily for many years. This experience undoubtedly helped him to write a now famous book, *On Games of Chance,* which treats the probabilities in gambling. He even gives advice on how to cheat, which was also gleaned from experience.

A product of his age in many respects, Cardan collected and published prolifically legends, false philosophical and astrological doctrines, folk cures, methods of communion with spirits, and superstitions. Apparently he himself believed in spirits and in astrology. He cast horoscopes, many of which proved to be false. Toward the end of his life he was imprisoned for casting the horoscope of Christ, but was soon pardoned, pensioned by the Pope, and lived peacefully until his death in 1576. In his *Book of My Life,* an autobiography, he says that despite his years of trouble he has to be grateful, for he had acquired a grandson, wealth, fame, learning, friends, belief in God, and he still had fifteen teeth.

Part of Cardan's rascality concerns our present subject. The mathematicians of the sixteenth century had undertaken to solve higher-degree equations, for example, equations of the third degree such as

$$x^3 - 6x = 8.$$

Among them was another famous man, Nicolò of Brescia, better known as Tartaglia (1499–1557), whom we shall meet occasionally in other contexts. Tartaglia had discovered a method for solving third-degree equations, and Cardan wished to publish this method in a book he was writing on algebra, which later appeared under the title *Ars Magna,* the first major book on algebra in modern times. After refusing to divulge the method, Tartaglia finally acquiesced, but asked Cardan to keep it secret. However, Cardan wished his book to be as important as possible and so published the method, though acknowledging that it was Tartaglia's. From this book, which appeared in 1545, the mathematical world learned how to solve third-degree equations. In this same book Cardan also published a method of solving fourth-degree equations discovered by one of his own pupils, Lodovico Ferrari (1522–1565). Although general coefficients were not in use as yet, it was clear that all third- and fourth-degree equations could be solved. In other words, the solutions could be expressed in terms of the coefficients by means of the ordinary operations of algebra, i.e., addition, subtraction, multiplication, division, and roots (though not necessarily square roots), in just about the manner in which (30) expresses the solutions of a second-degree equation in terms of the coefficients p and q.

And now the mathematicians' interest in generality took over. Since the *general* equations of the first, second, third, and fourth degree could be solved, what about fifth-, sixth- and higher-degree equations? It seemed certain that these equations could also be solved. For three hundred years many mathematicians worked on this basic problem and made almost no progress. And then a young Norwegian mathematician, Niels Henrik Abel (1802–1829), showed at the age of 22 that fifth-degree equations could not be solved by the processes of algebra. Another youth, Évariste Galois (1811–1832), who failed twice to pass the entrance examinations for the École Polytechnique and spent just one year at the École Normale, demonstrated that all general equations of degree higher than the fourth cannot be solved by means of the operations of algebra. In a letter he wrote the night before he was killed in a duel, Galois explained his ideas and showed how a new and general theory of the solution of equations could be developed. Galois' ideas gave algebra a totally new turn. Instead of being a tool, a series of techniques for the transformation of expressions into more useful ones, it became a beautiful body of knowledge which can be of interest in itself. Unfortunately we cannot undertake to study Galois' ideas, or the Galois theory as it is called, because there are more basic things to be learned first.

This brief account of the search for generality in the solution of equations has been given here because it illustrates many important features of mathematics. One is the persistence, stubborness if you will, of mathematicians over hundreds of years. Another is the experience that the search for generality leads to new and important developments, even though at the outset the generality is sought for its own sake. Today, the solution of higher-degree equations is a most practical matter, and we owe to Galois the most revealing insight into this subject. We also find in this history of the theory of equations a major example of how mathematicians find problems on which to work, problems of significance drawn from other problems which have humble and practical origins such as simple equations involving unknowns.

Topics for Further Investigation

1. The rise of symbolism in algebra.
2. The history of the solution of equations.

Recommended Reading

BALL, W. W. ROUSE: *A Short Account of the History of Mathematics,* pp. 201–243, Dover Publications, Inc., New York, 1960.

COLERUS, EGMONT: *From Simple Numbers to the Calculus,* Chaps. 9 to 13, Wm. Heinemann Ltd., London, 1954.

MILLER, DENNING: *Popular Mathematics,* pp. 138–215, Coward-McCann, Inc., New York, 1942.

ORE, OYSTEIN: *Cardano, The Gambling Scholar,* Chaps. 1 to 5, Princeton University Press, Princeton, 1953.

SAWYER, W. W.: *A Mathematician's Delight,* Chap. 7, Penguin Books Ltd., Harmondsworth, England, 1943.

SMITH, DAVID E.: *History of Mathematics,* Vol. II, pp. 378–470, Dover Publications, Inc., New York, 1958.

CHAPTER 6

THE NATURE AND USES OF EUCLIDEAN GEOMETRY

Circles to square and cubes to double
Would give a man excessive trouble.
Matthew Prior

6–1 The beginnings of geometry. Just as the study of numbers and its extensions to algebra arose out of the very practical problems of keeping track of property, trading, taxation, and the like, so did the study of geometry develop from the desire to measure the area of pieces of land (or geodesy in general), to determine the volumes of granaries, and to calculate the dimensions and amount of material needed for various structures.

The physical origin of the basic figures of geometry is evident. Not only the common figures of geometry but the simple relationships, such as perpendicularity, parallelism, congruence, and similarity, derive from ordinary experiences. A tree grows perpendicular to the ground, and the walls of a house are deliberately set upright so that there will be no tendency to fall. The banks of a river are parallel. A builder constructing a row of houses according to the same plan wishes them to have the same size and shape, that is, to be congruent. A workman or machine producing many pieces of a particular item makes them congruent. Models of real objects are often similar to the object represented, especially if the model is to be used as a guide to the construction of the object.

The science of geometry, indeed, the science of mathematics, was founded by the Greeks of the classical period. We have already described the major steps: the recognition that there are abstract concepts or ideas such as point, line, triangle, and the like, which are distinct from physical objects, the adoption of axioms which contained the surest knowledge about these abstractions man can obtain, and the decision to prove deductively any other facts about these concepts. The Greeks converted the disconnected, empirical, limited geometrical facts of the Egyptians and Babylonians into a vast, systematic, and thoroughly deductive structure.

Although the Greeks also studied the properties of numbers, they favored geometry. The reasons are pertinent. First of all, the Greeks liked exact thinking, and found that this faculty was more readily applied to geometry. Possible theorems are rather easily gleaned from the visualization of geometrical configurations. The neat correspondence between deductively established conclusions and intuitive understanding further increases this appeal of geometry. That one can draw pictures to represent what one is thinking about in geometry has its drawbacks. One is prone to confuse the abstract concept with the picture and to accept unconsciously properties of the picture. Of course, the idea of a triangle must be distinguished from the triangle drawn in chalk or pencil, and no properties of the picture may be used unless they are contained in the axioms or in some previously proved theorem. The Greeks were careful to make this distinction.

Secondly, the Greek philosophers who founded mathematics were intrigued with the design and structure of the universe, and they studied the heavens, certainly the most impressive spectacle in nature, to fathom the design. The shapes and paths of the heavenly bodies and the over-all plan of the solar system were of interest. On the other hand, they hardly saw any value in the ability to describe the exact locations of the moon, sun, and planets and to predict their precise locations at a given time, information of importance in calendar reckoning and in navigation.

Thirdly, since commerce and daily business were handled in large part by slaves, and were in any case in low regard, the study of numbers, which served such purposes, was subordinated. Why worry about the uses of numbers for measurement and trade if one does not measure or trade? One does not need the dimensions of even one rectangle to speculate about the properties of all rectangles.

The Greek philosophers emphasized an aspect of reality which is today, at least in scientific circles, neglected. To the Greeks of the classical period the reality of the universe consisted of the forms which matter possessed. Matter as such was formless and therefore meaningless. But an object in the shape of a triangle was significant by the very fact that it was triangular.

Finally, there were purely mathematical grounds for the Greek emphasis on geometry. The Greeks were the first to recognize that quantities such as $\sqrt{2}$, $\sqrt{3}$, $\sqrt[3]{2}$, etc., are neither whole numbers nor fractions, but they failed to recognize that these were new types of numbers, and that one could reason with them. To handle all types of quantities, they conceived the idea of treating them as line segments. As line segments, the hypotenuse of a right triangle (Fig. 4–2) and the arms have the same character, despite the fact that if the arms are each 1 unit long, the hypotenuse has the irrational length $\sqrt{2}$. To execute their plan of treating all quantities geometrically, the Greeks converted the algebraic processes developed in Egypt and Babylonia into geometrical ones. We could illustrate how the Greeks solved equations geometrically, but their methods are no longer favored. For science and engineering, the knowledge that a certain line segment solves an equation is not nearly so useful as a numerical answer which can be calculated to as many decimal places as needed. But the classical Greeks, who regarded exact reasoning as paramount in importance and who deprecated practical applications, found the solution of their difficulty in geometry and were content with this solution. Geometry remained the basis for all exact mathematical reasoning until the seventeenth century, when the needs of science forced the shift to number and algebra and the ultimate recognition that these could be built up as logically as geometry. In the intervening centuries arithmetic and algebra were regarded as practical disciplines.

Of course, the Greek conversion of exact mathematics to geometry was, from our present viewpoint, a backward step. Not only are the geometrical methods of performing algebraic processes insufficient for science, engineering, commerce, and industry, but they are by comparison clumsy and lengthy. Moreover, because Greek geometry was so complete and so admirable, mathematicians following in the Greeks' footsteps continued to think that exact mathematics must be geometrical. As a consequence, the development of algebra was unnecessarily delayed.

6–2 The content of Euclidean geometry. The major book on geometry of the classical Greek era is Euclid's *Elements,* a work on plane and solid geometry. Written about 300 B.C., it contains the best results produced by dozens of fine mathematicians during the period from 600 to 300 B.C. The work of Thales, the Pythagoreans, Hippias, Hippocrates, Eudoxus, members of Plato's Academy, and many others furnished the material which Euclid organized. His text was not the first to be written, but unfortunately we do not have copies of the earlier ones. It is quite certain that the particular axioms one finds in the *Elements,* the arrangement of the theorems, and many of the proofs are all due to Euclid. The geometry texts used in high schools today in essence reproduce Euclid's work, although these contemporary versions usually contain only a small part of the 467 theorems and many corollaries found in the *Elements.* Euclid's version is so marvelously knit together that most readers are amazed to see so many profound theorems deduced from the few self-evident axioms.

Though the reader may already be familiar with the basic theorems of Euclidean geometry, we shall take a few moments to review some features of the subject and the nature of the accomplishment. We might note first the structure of Euclid's *Elements.* He begins with some definitions of the basic concepts: point, line, circle, triangle, quadrilateral, and the like. Although modern mathematicians would make some critical comments about these definitions, we shall not discuss them at present.*

Euclid then states ten axioms on which all subsequent reasoning is based. We shall note these merely to see that they do indeed describe apparently unquestionable properties of geometric figures. The first five axioms are:

AXIOM 1. Two points determine a unique straight line.

AXIOM 2. A straight line extends indefinitely far in either direction.

AXIOM 3. A circle may be drawn with any given center and any given radius.

AXIOM 4. All right angles are equal.

AXIOM 5. Given a line l (Fig. 6–1) and a point P not on that line, there exists in the plane of P and l one and only one line m, which does not meet the given line l.

In a separate *definition* Euclid defines parallel lines to be any two lines in the same plane which do not meet, that is, do not have any points in common. Thus, Axiom 5 asserts the existence of parallel lines.

---

* See Chaper 26.

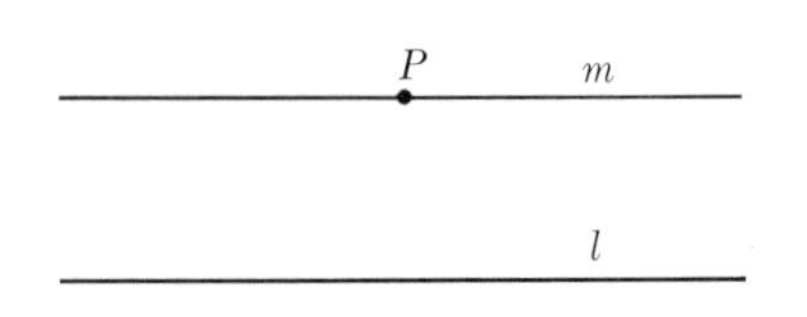

FIG. 6–1. The parallel axiom.

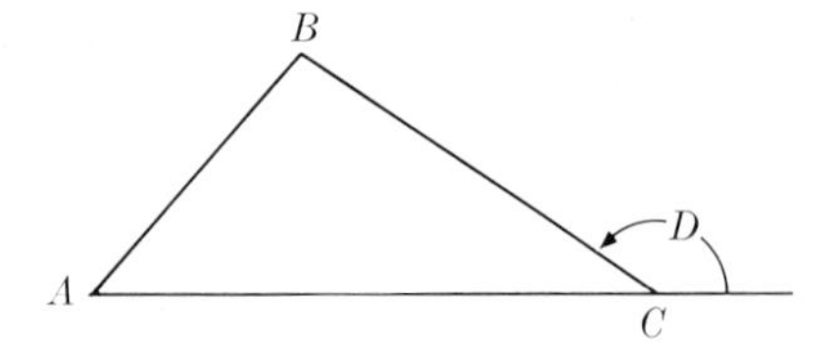

FIG. 6–2. An exterior angle of a triangle is greater than either remote interior angle.

The remaining five axioms are:

AXIOM 6. Things equal to the same thing are equal to each other.
AXIOM 7. If equals be added to equals, the sums are equal.
AXIOM 8. If equals be subtracted from equals, the remainders are equal.
AXIOM 9. Figures which can be made to coincide are equal (congruent).
AXIOM 10. The whole is greater than any part.

The formulations of these axioms are not quite the same as those prescribed by Euclid. Axiom 5 is, in fact, different from Euclid's; but is stated here in the form which is most likely to be familiar to the reader. The differences between Euclid's versions and those introduced by later mathematicians are not important for our present purposes, and so we shall not take time now to note them. (See Chapter 26.)

After stating his axioms, Euclid proceeded to prove theorems. Many of these theorems are indeed simple to prove and obviously true of the geometrical figures involved. But Euclid's purpose in proving them was to play safe. As we shall see in later chapters, many a conclusion seems obvious but is false. Of course, the major proofs are those which establish conclusions that are not at all obvious and, in some cases, even come as a surprise.

Partly to refresh our memories about some theorems of Euclidean geometry and partly to note once again the deductive procedure of mathematics, let us review one or two proofs. A basic theorem of Euclidean geometry asserts the following:

THEOREM I. An exterior angle of a triangle is greater than either remote interior angle of the triangle.

Before proving this theorem, let us be clear about what it says. Angle D, in Fig. 6–2, is called an exterior angle of triangle ABC because it is outside the triangle and is formed by one side, BC, and an extension of another side, AC. With respect to angle D, angles A and B are remote interior angles of triangle ABC, whereas angle C is an adjacent interior angle. Hence we have to prove that angle D is larger than angle A and larger than angle B. Let us prove that angle D is larger than angle B.

The problem before us is a tantalizing one because, while it does seem visually obvious that angle D is greater than angle B, there is no apparent method of proof. An idea is needed, and this is supplied by Euclid. He tells us to bisect side BC (Fig. 6–3), to join the mid-point E of BC to A, and to

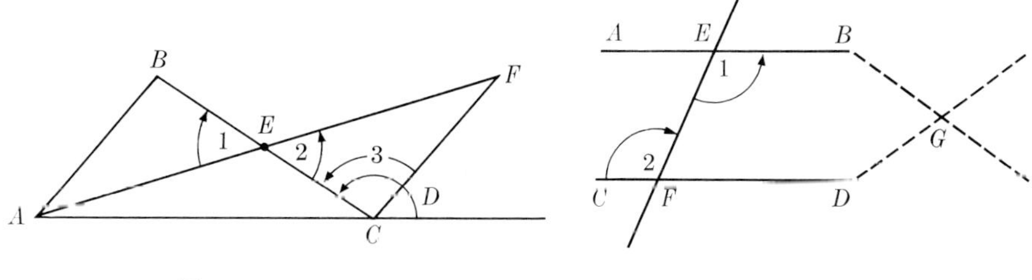

FIGURE 6–3

FIGURE 6–4

extend AE to the point F, so that $AE = EF$. He then proves that triangle AEB is congruent to triangle CEF, that is, that the sides and angles of one triangle are equal, respectively, to the sides and angles of the other. This congruence is easy to prove. Euclid had previously proved that vertical angles are equal, and we see from Fig. 6–3 that angles 1 and 2 are vertical angles. Further, the fact that E is the mid-point of BC means that $BE = EC$. Moreover, we constructed AE to equal EF. Hence, in the two triangles in question, two sides and the included angle of one triangle are equal to two sides and the included angle of the other. But Euclid had previously proved that two triangles are congruent if merely two sides and the included angle of one are equal to two sides and the included angle of the other. Since these facts are true of our triangles, the two triangles must be congruent.

Because triangles AEB and CEF are congruent, angle B of the first triangle equals angle 3 of the second one. We know that angle 3 is the angle to choose in the second triangle as the angle which corresponds to B, because angle B is opposite AE, and angle 3 is opposite the corresponding equal side EF. The proof is practically finished. Angle D is larger than angle 3 because the whole, angle D in our case, is greater than the part, angle 3. Hence angle D is also greater than the equal angle B because angle B has the same size as angle 3.

We have now proved a major theorem, and we should see that a series of simple deductive arguments leads to an indubitable result.

And now let us prove another, equally important theorem which will exhibit one or two other features of Euclid's work:

THEOREM II. If two lines are cut by a transversal so as to make alternate interior angles equal, then the lines are parallel.

Again let us see what the theorem means before we consider its proof. In Fig. 6–4, AB and CD are two lines cut by the transversal EF. The angles 1 and 2 are called alternate interior angles, and we are told that they are equal. The theorem asserts that, under this condition, AB must be parallel to CD. As in the case of the preceding theorem, the assertion is seemingly correct, and yet the method of proof is by no means apparent.

Here Euclid uses what is usually called the indirect method of proof; that is, he supposes that AB is not parallel to CD. Two lines that are not parallel must, by definition, meet somewhere. Thus AB and CD meet, let us say, in the point G. But now EG, GF, and FE form a triangle. Angle 2 is an exterior angle of this triangle and angle 1 is a remote interior angle. Since we have the theorem that in *any* triangle an exterior angle is greater than either remote interior angle, it follows that angle 2 must be greater than angle 1. But, in the above figure, we were given as fact that angle 2 equals angle 1. We have arrived at a contradiction which, if we did not make any mistakes in reasoning, has only one explanation: somewhere we introduced a false premise. We find that the only questionable fact is the assumption that AB is not parallel to CD. But there are only two possibilities, namely, that AB is parallel to CD or that it is not parallel to CD. Since the latter supposition led to a contradiction, it must be that AB is parallel to CD. Thus the theorem is proved.

We should be sure to note that the indirect method of proof is a deductive argument. The essence of the argument is that if AB is not parallel to CD, then angle 2 must be greater than angle 1. But angle 2 is not greater than angle 1. Hence it is not true that AB is not parallel to CD. But AB is or is not parallel to CD. If nonparallelism is not true then parallelism must hold.

Though we shall use a few other theorems of Euclidean geometry in subsequent work, we shall not present their proofs. We are now reasonably familiar with the nature of proof in geometry, and so we shall merely state the theorems when we wish to use them.

Perhaps one other point about the contents of the *Elements* warrants attention. A superficial survey of the many different theorems may leave one with the impression that the Greek geometers proved what they could and produced merely a mélange. But there are broad themes in Euclidean geometry, and these are pursued systematically. The first major theme is the study of conditions under which geometric figures must be congruent. This is a highly practical subject. Suppose, for example, that a surveyor has two triangular pieces of land and wishes to show that they are equal or congruent. Must he measure all the sides and all the angles of the first piece and show that they are of the same size, respectively, as the sides and angles of the second piece? Not at all! There are several Euclidean theorems which can aid the surveyor. If he can show, for example, that two angles and the included side of the first triangle equal, respectively, the two angles and the included side of the second one, then Euclid's theorem tells him that the triangular pieces of land must be equal.

A second major theme in Euclid's work is the similarity of figures, that is, figures with the same shape. We have already mentioned that models of houses, ships, and other large structures are often built to assist in planning. One may wish to know what conditions will guarantee the similarity of the model and the actual structure. Let us suppose that the model or some part of it is triangular in shape. One of Euclid's theorems tells us that if the corresponding sides of two triangles have the same ratio, then the two triangles will be similar. Thus, if the model is constructed so that each side of the model is 1/100 of the corresponding side of the actual structure, we know that the model will be similar to the structure. This similarity is useful because, by definition, two triangles are similar if the

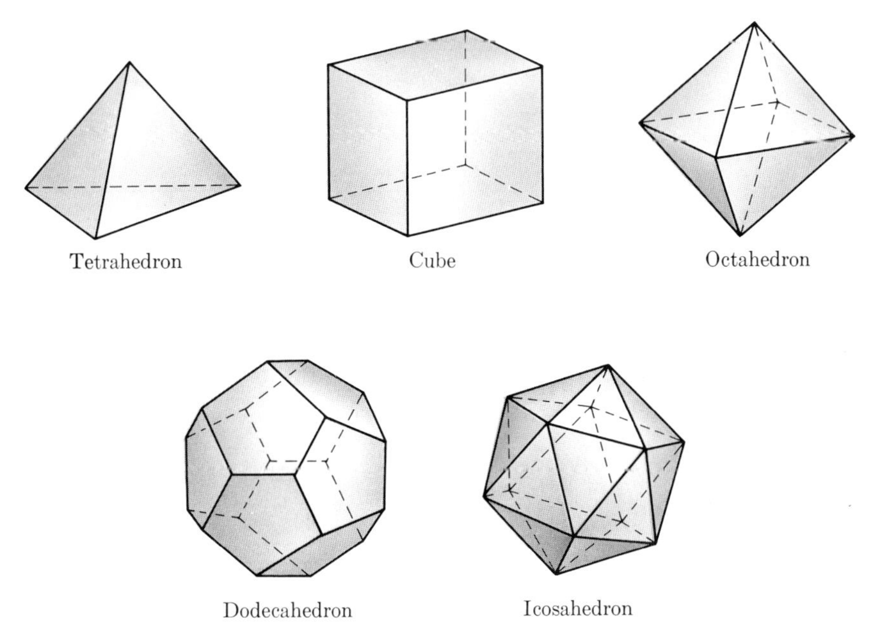

FIG. 6-5. The five regular polyhedra.

angles of one equal the corresponding angles of the other. Hence, an engineer can measure the angles of the model and know precisely what the angles of the actual structure will be.

Suppose that two figures are neither congruent nor similar. Could they have some other significant property in common? One answer, clearly, is area. And so Euclid considers conditions under which two figures may have the same area, or, in Euclid's language, be equivalent.

There are many other themes in Euclid, such as interesting properties of circles, quadrilaterals, and regular polygons. He also considers the common solid figures such as pyramids, prisms, spheres, cylinders, and cones. Finally, Euclid devotes considerable space to a class of figures which all Greeks favored, the regular polyhedra (Fig. 6-5).

EXERCISES

1. What essential fact distinguishes axioms from theorems?

2. Why were the Greeks willing to accept the statements 1 through 10 above as axioms?

3. Use the indirect method of proof to show that if two angles of a triangle are equal, then the opposite sides are equal. [*Suggestion:* Suppose that angle A (Fig. 6-6) equals angle C, but that BC is greater than BA. Lay off $BC' = BA$ and draw AC'. Use the theorem that the base angles of an isosceles triangle are equal and Theorem I above.]

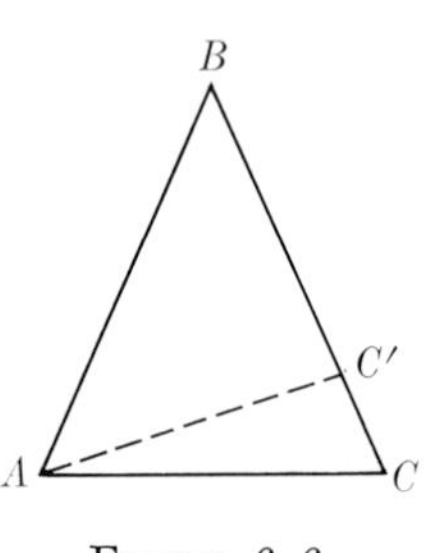

FIGURE 6–6

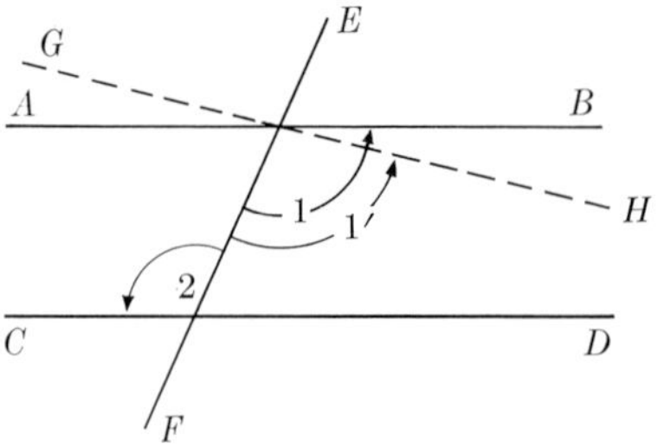

FIGURE 6–7

4. Use the indirect method of proof to show that if two lines are parallel, alternate interior angles must be equal. [*Suggestion:* Suppose angle 1 in Fig. 6–7 is greater than angle 2. Then draw GH so that angle 1′ equals angle 2. Now use Theorem II and Axiom 5.]

5. In Section 3–6, we have briefly outlined the proof that the sum of the angles of a triangle is 180°. Write out the full proof.

6. Under what conditions would two parallelograms be congruent?

7. What conditions would ensure the similarity of two rectangles?

8. A right triangle has an arm 1 mi long and a hypotenuse 1 mi plus 1 ft long. How long is the other arm? Before you apply mathematics, use your imagination to estimate the answer. To work out the problem, use the Pythagorean theorem which says that the square of the hypotenuse equals the sum of the squares of the arms.

9. A farmer is offered two triangular pieces of land. The dimensions are 25, 30, and 40 ft and 75, 90, and 120 ft, respectively. Since the dimensions of the second one are 3 times the dimensions of the first, the two triangles are similar. The price of the larger piece is 5 times the price of the smaller one. Use intuition, measurement, or mathematical proof to decide which is the better buy in the sense of price per square foot.

10. Suppose a roadway is to be built around the earth and each point on the surface of the roadway is to be 1 ft above the surface of the earth (Fig. 6–8). Given that the radius of the earth is 4000 mi or 21,120,000 ft, estimate by how much the length of the roadway would exceed the circumference of the earth. Then use the fact that the circumference of a circle is 2π times the radius and calculate how much longer the roadway would be.

11. Criticize the statement: Euclid assumes that two parallel lines do not meet.

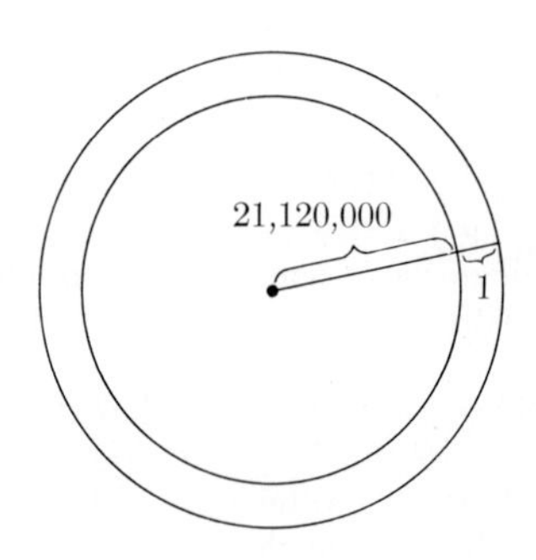

FIGURE 6–8

6–3 Some mundane uses of Euclidean geometry. The creation of Euclidean geometry was motivated by the desire to learn the properties of figures in the world about us. Let us see now whether the knowledge can be applied to the world to good advantage.

Suppose a farmer has 100 feet of fencing at his disposal and he wishes to enclose a rectangular piece of land. Since the perimeter will be 100 feet, the farmer can enclose a piece of land 10 feet by 40 feet, 15 feet by 35 feet, 20 feet by 30 feet, or of still other dimensions, all of which yield a perimeter of 100 feet. The farmer plans to garden in the enclosed plot and therefore wishes the enclosed area to be as large as possible. He notes that the dimensions 10 by 40 would yield an area of 400 square feet; the dimensions 15 by 35 enclose 525 square feet; and the dimensions 20 by 30 enclose 600 square feet. Evidently the area can vary considerably despite the fact that the perimeter in each case is 100 feet. The question then arises, What dimensions would yield the maximum area?

Our first task in seeking to answer this question is to make some reasonable conjecture about these dimensions. We might then be able to prove that the conjecture is correct. Since in the present instance it is easy to play with the numbers involved, let us make a little table of dimensions (always yielding a perimeter of 100 feet) and the corresponding area.

Dimensions, in feet	*Area, in square feet*
1 by 49	49
5 by 45	225
10 by 40	400
15 by 35	525
20 by 30	600

Study of the table suggests that the more nearly equal the dimensions are, the larger is the area. Hence one might readily conjecture that if the dimensions were equal, that is, if the rectangle were a square, the area would be a maximum.

We can see at once that the dimensions 25 by 25 give an area of 625 square feet, and this area is larger than any of the areas in the table. So far our conjecture is confirmed. However, we could not be sure that some other dimensions, perhaps $24\frac{1}{2}$ by $25\frac{1}{2}$, would not do even better. Moreover, even if we could be certain that the square furnishes the largest area among all rectangles with a perimeter of 100 feet, the question would arise whether the square would continue to be the answer for some other perimeter. Hence, let us see whether we can prove the general theorem that *of all rectangles with the same perimeter, the square has maximum area.*

Figure 6–9 shows the rectangle $ABCD$. Since this rectangle is not a square, let us erect on the longer side a square which has the same perimeter. Thus, the square $EFGD$ has the same perimeter as $ABCD$. We now denote equal segments by the same letters. The perimeter of the rectangle is then $2x + 2u + 2y$, and the perimeter of the square is $2x + 2v + 2y$. Since the

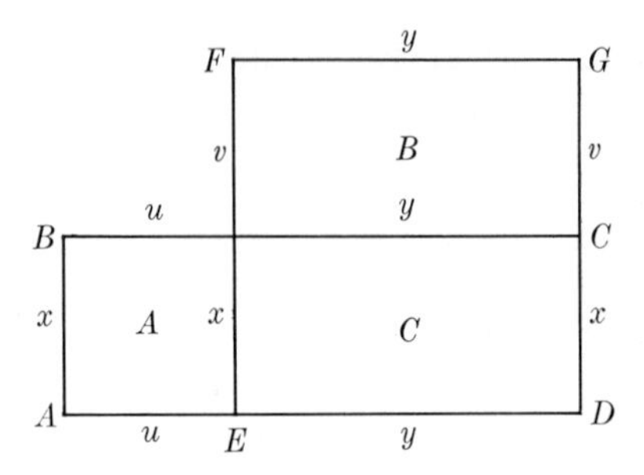

FIG. 6–9. Of all rectangles with the same perimeter the square has the greatest area.

two figures have the same perimeter, we have

$$2x + 2v + 2y = 2x + 2u + 2y.$$

If we subtract $2x$ and $2y$ from both sides of this equation and then divide both sides by 2, we obtain

$$v = u. \tag{1}$$

Moreover, because the square has equal sides,

$$y = x + v. \tag{2}$$

If we now multiply the left side of equation (2) by the left side of equation (1), and do the same for the right sides, the results must be equal. Hence,

$$yv = u(x + u),$$

or, by the distributive axiom,

$$yv = ux + u^2.$$

Since $yv = ux$ plus an additional area, it must be that yv is greater than ux. Now yv is area B in the figure, and ux is area A. Thus B is greater than A, and so $B + C$ is greater than $A + C$. But $B + C$ is the area of the square, and $A + C$ is the area of the rectangle. Hence the square has more area than the rectangle.

We have proved that a square has more area than a rectangle of the same perimeter, *no matter what this perimeter may be.* A little thinking proves in a few minutes what may have taken man hundreds of years to learn through trial and error.

The result is far more useful than may appear at first sight. Suppose a house is to be built. The major consideration is to have as much floor area or living space as possible. Now the perimeter of the floor determines the number of feet of wall that will be needed and hence the cost of the walls. To obtain the maximum floor area for a given cost of walls, the shape of the floor should be square.

A farmer who seeks the rectangle of maximum area with given perimeter might, after finding the answer to his question, turn to gardening, but a mathematician who obtains such a neat result would not stop there. He might ask next, Suppose we were free to utilize any quadrilateral rather than just rectangles, which one of all quadrilaterals with the same perimeter has maximum area? The answer happens to be a square, though we shall not prove it. The mathematician might then consider the question, Which pentagon of all pentagons with the same perimeter has maximum area? One can show that the answer is the regular pentagon, that is, the pentagon whose sides are all equal and whose angles are all equal. Now the square also has equal sides and equal angles. Hence it would seem that if one compares all polygons of same perimeter and same number of sides, then the one with equal sides and equal angles, i.e. the regular polygon, should have maximum area. This general result can also be proved.

But now an obvious question comes to the fore. The square has maximum area among all quadrilaterals of the same perimeter. The regular pentagon has maximum area among all pentagons of the same perimeter. Suppose that we compared the regular pentagon with the square of the same perimeter. Which would have more area? The answer, perhaps surprising, is the regular pentagon. And now the conjecture seems reasonable that of two regular polygons with the same perimeter, the one with more sides will have more area. This is so. Where does this result lead? One can form regular polygons of more and more sides, which all have the same perimeter. As the number of sides increases, the area increases. But as the number of sides increases, the regular polygon approaches the circle in shape. Hence the circle should have more area than any regular polygon of the same perimeter. And since the regular polygon has more area than an arbitrary polygon, *the circle has more area than any polygon with the same perimeter.* This result is a famous theorem.

Now the sphere, among surfaces, is the analogue of the circle among curves. Hence, a reasonable conjecture would be that the spherical surface bounds more volume than any other surface with the same area. This conjecture can be proved. Nature obeys this mathematical theorem. For example, if one blows up a rubber balloon, the balloon assumes a spherical shape. The reason is that the rubber must enclose the volume of air blown into the balloon and the rubber must be stretched. But rubber contracts as much as possible. For a spherical shape, less surface area is required to contain a given volume of gas than for any other shape. Hence, with the spherical shape, the rubber is stretched as little as possible.

The problem of bounding the greatest possible area with a perimeter of given length has a variation whose solution shows how ingenious mathematical reasoning can be. Suppose that a person has a fixed amount of fencing at his disposal and wishes to enclose as much area as possible along a river front in such a way that no fencing is required along the shore itself. The question now is, What should the shape of the boundary curve be? According to a legend, which may or may not have a factual basis, this prob-

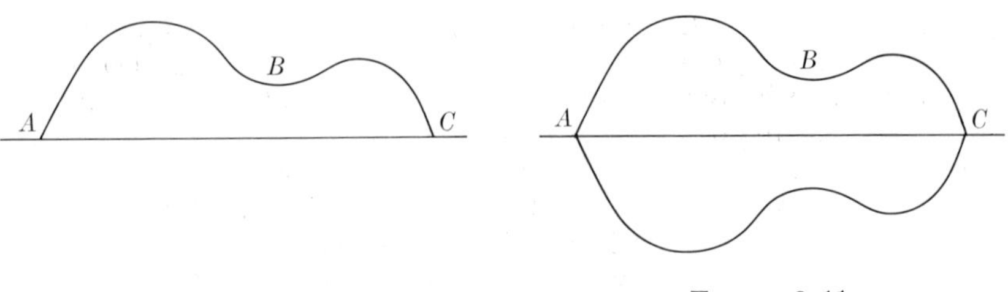

FIGURE 6–10

FIGURE 6–11

lem was solved thousands of years ago by Dido, the founder of the city of Carthage on the Mediterranean coast of Africa. Dido, the daughter of the king of the Phoenician city of Tyre, ran away from home. She took a fancy to this land on the Mediterranean, and made an agreement to pay a definite sum of money for as much land as "could be encompassed by a bull's hide." Dido thereupon took a bull's hide, cut it up into thin long strips, tied the strips together, and used this length to "encompass land." She chose an area along the shore, because she was smart enough to realize that no hide would be needed along the shore. But there still remained the question of what shape to use for the boundary formed by the hide, that is, for ABC of Fig. 6–10. Dido decided that the most favorable shape was a semicircle, enclosed that shape, and built a city there.

A sequel to this story, which has nothing to do with the mathematics of Dido's problem, is not without relevance to the history of mathematics. Shortly after she founded Carthage, Aeneas, a refugee from Troy, intent on getting to Italy to found his own city, was blown ashore along with his compatriots. Dido took a fancy to Aeneas also, and did her best to persuade him to remain at Carthage, but despite the best of hospitality, Aeneas could not be diverted from his plan, and soon sailed away. Rejected and scorned, Dido was so despondent that she threw herself on a blazing pyre just as Aeneas sailed out of the harbor. And so an ungrateful and unreceptive man with a rigid mind caused the loss of a potential mathematician. This was the first blow to mathematics which the Romans dealt.

Dido's fate was a tragic end to a brilliant beginning, for her solution to the geometrical problem described above was correct. The answer is a semicircle. We do not know how Dido found the answer, but it can be obtained very neatly. The way to prove it is by complicating the problem. Suppose that, instead of bounding an area on one side of the seashore, which we idealize as the line AC (Fig. 6–11), we try to solve the problem of enclosing an area on both sides of AC with *double* the length of hide Dido had for one side, i.e., now we seek to solve the problem by determining the maximum area which can be completely enclosed by a perimeter of given length. The answer to this problem is a circle. If, therefore, we choose a semicircle for arc ABC, it will contain maximum area on one side of the shore. For if there were a more favorable shape than the semicircle, the mirror image in AC of that shape would, together with the original, do

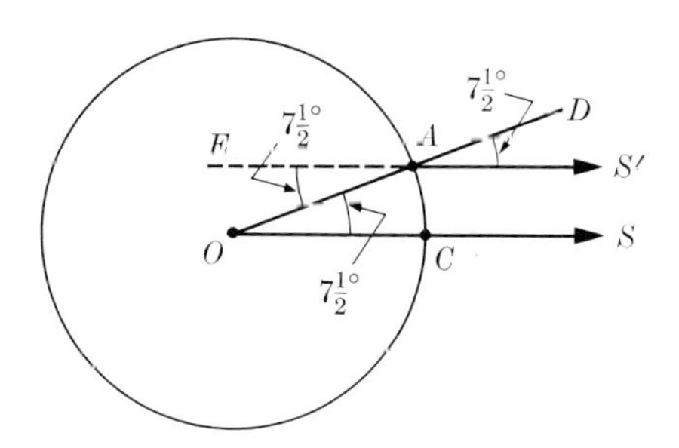

FIG. 6-12. Eratosthenes' method of deducing the circumference of the earth.

better than the circle and yet have the same perimeter as the circle. But this is impossible.

Our last few pages have dealt with problems which grew out of determining the rectangle of maximum area with given perimeter. We can see from the lines of thought pursued how the mathematician can raise one question after another on this same theme of figures with maximum area and given perimeter and will find the answers to these questions. Moreover, many of these answers prove to be applicable to physical problems.

The first reasonably accurate calculation of the size of the earth was made by a simple application of Euclidean geometry. One of the most learned men of the Alexandrian Greek world, Eratosthenes (275–194 B.C.), a geographer, mathematician, poet, historian, and astronomer, used the following plan. At the summer solstice, the sun shone directly down into a well at Syene (C in Fig. 6-12). As Eratosthenes well appreciated, this meant that the sun was directly overhead. At the same time, at the city of Alexandria, 500 miles north of Syene, the direction of the sun was AS', whereas the overhead direction was OAD. Now the sun is so far away that the lines AS' and CS could be taken to be parallel. Eratosthenes measured the angle DAS' and found it to be $7\frac{1}{2}°$. But this angle equals the vertical angle OAE, and the latter and angle AOC are alternate interior angles of parallel lines. Hence angle AOC is also $7\frac{1}{2}°$, or $7\frac{1}{2}/360$, or $1/48$ of the entire angle at O. Then arc AC is $1/48$ of the entire circumference. Since AC is 500 miles, the entire circumference is $48 \cdot 500$ or 24,000 miles.

Strabo, a Greek geographer who lived in the first century B.C., tells us that after Eratosthenes obtained this result, he realized that one might sail from Greece past Spain across the Atlantic Ocean to India. This is, of course, what Columbus attempted. Fortunately or unfortunately, the geographers who lived after Eratosthenes, notably Poseidonius (first century B.C.) and Ptolemy (second century A.D.), gave other results which were interpreted by Columbus (because of some uncertainty about the units of distance used by these early scientists) to mean that the circumference of the earth is 17,000 miles. Had he known the correct value, he might never have undertaken to sail to India because the greater distance might have daunted him.

Exercises

1. Suppose that DF (Fig. 6–13) is the course of a railroad, and A and B are two towns. It is desired to build a station somewhere on DF so that the station will be equally distant from A and B. Where should the station be built? One draws the line AB and, at its mid-point, erects the perpendicular CE. The point E on DF is equidistant from A and B. Prove this statement.

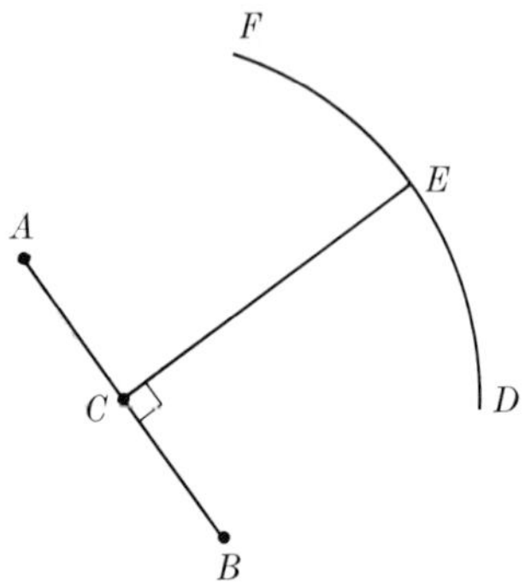

Figure 6–13

2. A pinhole camera is a practical device if a long exposure time is possible. In fact, one of the best pictures of the scene following the explosion of the first atomic bomb was made with a pinhole camera. The principle involves similar triangles. The object AB being photographed (Fig. 6–14) appears on the film inside the box as $A'B'$. If one draws OD perpendicular to AB, the extension of OD to D' will be perpendicular to $A'B'$. Then triangles OAD and $OA'D'$ are similar. Now suppose the sun, whose radius is AD, is photographed. We know that OD is 93,000,000 mi. Suppose that OD', the width of the box, is 1 ft. The length $A'B'$ is readily measured and is found to be 0.009 ft. What is the radius of the sun?

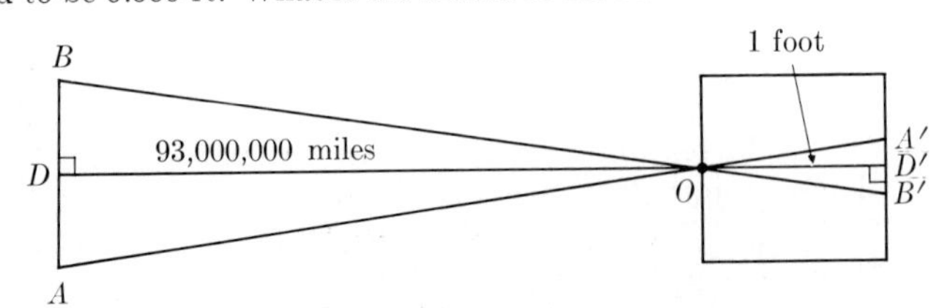

Figure 6–14

3. A farmer has 400 yd of fencing and wishes to enclose a rectangle of maximum area. What dimensions should he choose?

4. A farmer has p yd of fencing and wishes to enclose a rectangle of maximum area. What dimensions should he choose?

5. A farmer plans to enclose a rectangular piece of land alongside a lake; no fencing is required along the shoreline AD (Fig. 6–15). He has 100 ft of fence and wishes the area of the rectangle to be as large as possible. What dimensions should he choose?

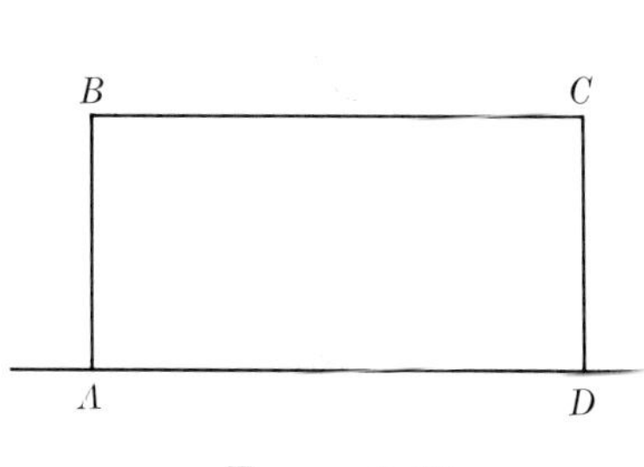

Figure 6–15

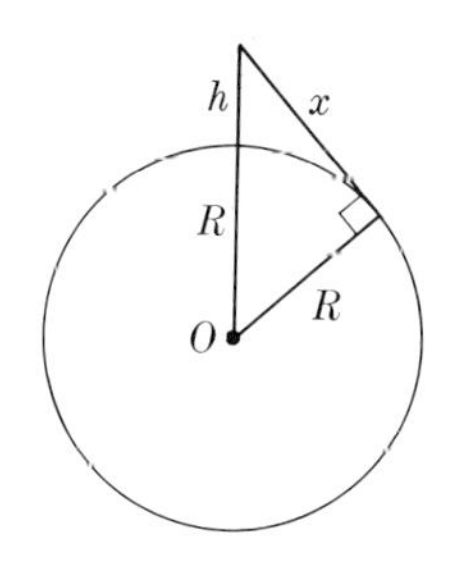

Figure 6–16

6. Of any two numbers whose sum is 12, the product is greatest for 6 and 6; that is, $6 \cdot 6$ is greater than $5 \cdot 7$, $4 \cdot 8$, $3\frac{1}{2} \cdot 8\frac{1}{2}$, and so forth. Can you explain why this is so? [*Suggestion:* Think in geometrical terms.]

7. Suppose h is the known height of a mountain, and R is the radius of the earth (Fig. 6–16). How far is it from the top of the mountain to the horizon; that is, how long is x? [*Suggestion:* Use the fact that the line of sight from the top of the mountain to the horizon is tangent to the circle shown, and that a radius of a circle drawn to the point of tangency is perpendicular to the tangent.]

8. Having obtained the exact answer to Problem 7, can you suggest a good approximate answer which would suffice for many applications and yet make calculation easier?

9. A boy stands on a cliff $\frac{1}{2}$ mi above the sea. How far away is the horizon?

10. Knowing that of all rectangles with the same perimeter, the square has maximum area, prove that of all rectangles with the same area, the square has the least perimeter. [*Suggestion:* Use the indirect method of proof. Suppose, then, that the square has more perimeter than the rectangle of the same area and consider the square which has the same perimeter as the rectangle.]

6–4 Euclidean geometry and the study of light. Light is certainly a pervasive phenomenon. Man and the physical world are subject daily to the light of the sun, and the process of vision of course is dependent upon light. Hence it is to be expected that the Greeks, the first great students of nature, would investigate this phenomenon. Plato and Aristotle had much to say on the nature of light, and the Greek mathematicians also tackled the subject. It has continued to be a primary concern of mathematicians and physicists right down to the present day. Despite man's continuous experience with light, the nature of this occurrence is still largely a mystery. Through mathematics and through Euclidean geometry in particular, man obtained his first grip on the subject. Two books by Euclid were the beginning of the mathematical attack.

In ordinary air, light is observed to travel along straight lines. This preference of light for the simplest and shortest path is in itself of significance. But Euclid proceeded beyond this point to study the behavior of light under reflection in a mirror, and discovered a now famous mathematical law of light.

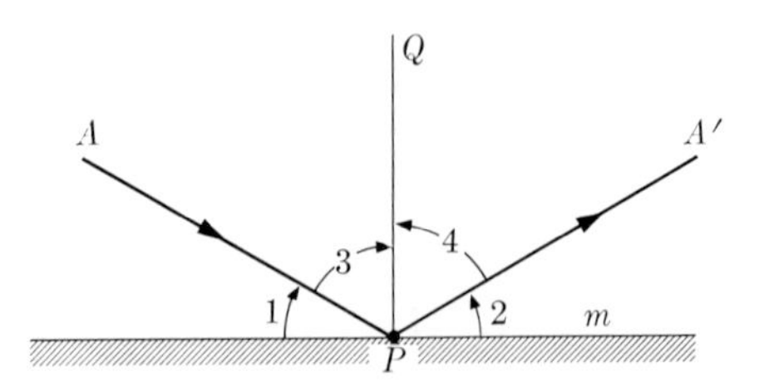

FIG. 6–17. The law of reflection of light.

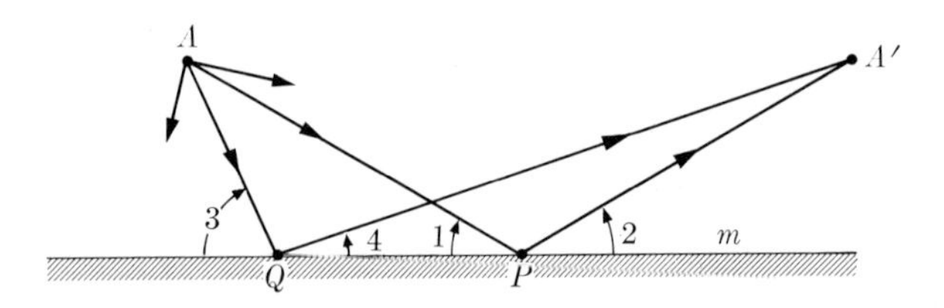

FIGURE 6–18

Suppose light issuing from A (Fig. 6–17) takes the path AP to the point P on the mirror m. As we all know, the light is reflected and takes a new direction, PA'. The significant fact about this reflection, which was pointed out by Euclid, is that the reflected ray, i.e., the line PA' along which the reflected light travels, always takes a direction such that angle 1 equals angle 2. Angle 1 is called the angle of incidence and angle 2 the angle of reflection.* It is, of course, very obliging of light to follow such a simple mathematical law. As a consequence, we are able to prove other facts rather readily.

Assume there is a source of light at A (Fig. 6–18), and rays of light spread out in all directions from A. Many of these will strike the mirror. But through a definite point A' only one of these rays will pass, namely the ray PA', for which angle 1 equals angle 2. To prove that only one ray from A will pass through A', let us suppose that another ray, AQ, is also reflected to A'. Now angle 2 is an exterior angle of triangle $A'QP$. Hence

$$\measuredangle 2 > \measuredangle 4.$$

Angle 3 is an exterior angle of triangle AQP, and so

$$\measuredangle 3 > \measuredangle 1.$$

Since angle 1 equals angle 2, we see from the two preceding inequalities that

$$\measuredangle 3 > \measuredangle 4.$$

* It is more common to introduce the perpendicular PQ to the mirror and to call angle 3 the angle of incidence and angle 4 the angle of reflection. However, if angle 1 equals angle 2, then angle 3 equals angle 4.

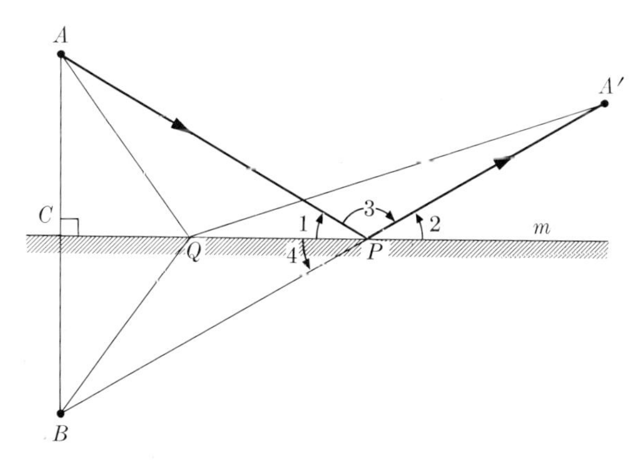

FIG. 6–19. The shortest path from A to A' is the one for which $\measuredangle 1 = \measuredangle 2$.

Then QA' cannot be the reflected ray corresponding to the incident ray AQ because the reflected ray must make an angle with the mirror which equals angle 3.

The more interesting point, which was first observed and proved by the Greek mathematician and engineer Heron (first century A.D.), is that the unique ray from A (Fig. 6–19) which does reach A' after reflection in the mirror travels the shortest possible path in going from A to the mirror and then to A'. In other words, $AP + PA'$ is less than $AQ + QA'$, where Q is any point on the mirror other than P, the point at which the angle of incidence equals the angle of reflection.

How can we prove this theorem? Nature not only sets problems for us, but often solves them too, if we are but keen enough in our observations. If a person at A' sees in the mirror the reflection of an object at A, he must be looking in the direction $A'P$ and actually sees the image of A at B. Hence, perhaps we should bring B into our thinking. Closer observation shows that the mirror image of an object is on the perpendicular from A to the mirror and, moreover, seems to be as far behind the mirror as the object is in front. That is, AB seems to be perpendicular to the mirror and AC seems to equal CB.

Let us use this suggestion. We construct the perpendicular from A to the mirror, thus obtaining AC, and extend AC by its own length to B. Now it is not hard to see that triangles ACQ and BCQ are congruent because QC is common to both triangles, the angles at C are right angles, and $AC = CB$. Hence, $AQ = BQ$, because they are corresponding parts of congruent triangles. Likewise triangles ACP and BCP are congruent and $AP = BP$. We wish to prove that

$$(AP + PA') < (AQ + QA').$$

But now, since $AP = BP$ and $AQ = BQ$, it will be enough to prove that

$$(BP + PA') < (BQ + QA)'. \tag{3}$$

Well, we have exchanged one difficulty for another, but perhaps this second one is easier to overcome. Physically one looks directly along $A'P$ and sees B. If we could prove that BPA' is a straight line, then, of course, the inequality (3) would be proved because BQ and QA' are the other two sides of triangle $A'BQ$, and the sum of these two sides must be greater than the third side. Our goal, then, is to prove that $A'PB$ is a straight line.

We know that

$$\measuredangle 1 + \measuredangle 3 + \measuredangle 2 = 180° \tag{4}$$

because m is a straight line. But angle 1 equals angle 4 because triangle PCA and PCB are congruent. Also, according to the law of reflection, angle 2 equals angle 1. If, therefore, in (4) we replace angle 1 by angle 4 and angle 2 by angle 1, we have

$$\measuredangle 4 + \measuredangle 3 + \measuredangle 1 = 180°. \tag{5}$$

Hence, $A'PB$ is a straight line and the inequality (3) is proved. Then the light ray, in going from A to m to A', really travels the shortest path.

This behavior of light rays is striking. It seems to show that nature is interested in accomplishing its ends by the most efficient means. We shall find this theme to be a recurring one, and it will be seen to have broad applicability.

We have proved a theorem about light rays, but we have also proved somewhat more. As far as the mathematics is concerned, the lines AP and PA' are any lines which make equal angles with m, and the fact that they are light rays plays no role. What we have proved, then, is a theorem of geometry, namely: *Of all the broken line paths from a point A to a point on a line and then to a point A' on the same side as A, the shortest path is the one fixed by the point P on m for which AP and $A'P$ make equal angles with m.* This theorem has applications in quite different domains (see the exercises). It is worth noting how the study of light gives rise to purely mathematical theorems. The converse of this theorem is, incidentally, equally true and is presented in the exercises.

To carry the story of light just a little farther, let us suppose now that we have two mirrors m and n which form a right angle at a corner (Fig. 6–20), and that *any* light ray AP enters the corner and is reflected by mirror m at P. The ray proceeds to mirror n and is reflected there at Q to emerge as the ray QA'. Is there anything of interest about the entire process of two successive reflections? There is, and one might discover it by experimenting with flashlight beams and mirrors or one might use a little mathematics. Since m and n are straight lines, we can say at once that

$$\measuredangle 1 + \measuredangle 5 + \measuredangle 2 + \measuredangle 3 + \measuredangle 6 + \measuredangle 4 = 360°. \tag{6}$$

But angle 2 and angle 3 are acute angles of a right triangle, and so their sum is 90°. And since angle 1 equals angle 2 and angle 4 equals angle 3,

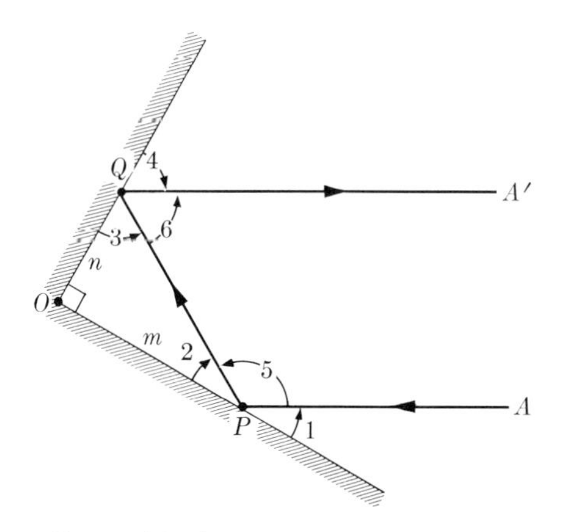

FIG. 6-20. Any ray reflected by two mirrors forming a right angle emerges parallel to the direction of incidence.

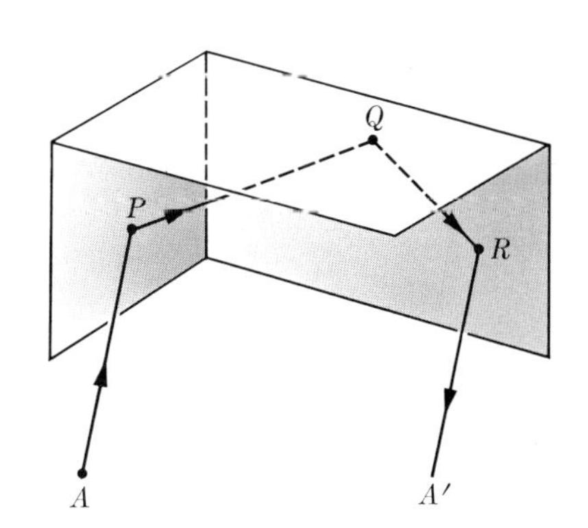

FIG. 6-21. A light ray reflected from all three faces emerges parallel to the incident ray.

the sum of angle 1 and angle 4 is 90°. If we subtract these four angles from (6), we obtain

$$\measuredangle 5 + \measuredangle 6 = 180°.$$

But now we can apply a theorem of Euclidean geometry which we shall state because we did not review it above. *If the interior angles on one side of a transversal* (PQ in Fig. 6-20) *to two lines add up to* 180°, *then the lines are parallel.* What this theorem asserts in our case is that PA and QA' are parallel. Hence the ray which emerges is parallel to the incident ray. Thus, any light ray entering the corner is reflected parallel to itself.

This result seems to be of academic interest only, but a slight extension of it, which we shall merely describe, has a very useful application in our time. Suppose three mirrors form a corner where they meet one another at right angles, as do the two walls and ceiling of a room. If a light ray AP (Fig. 6-21) should strike one mirror, be reflected to Q at a second mirror, and then to R at the third mirror, it will emerge parallel to the incident ray. That is, RA' is parallel to AP. In practical applications, a corner reflector, which consists of four such corners back to back, is suspended from a balloon, and the balloon, after being filled with helium, rises. A radio set on the ground sends out a radio beam composed of many rays which behave much like light rays (see Chapter 25). The radio rays enter one of the corners and, because they are reflected in the direction from which they came, are reflected back to the radio set on the ground. Since all of the rays entering the corner return to the source, a strong signal is received at that point. Hence the radio set is able to determine the direction of the balloon by the fact that it receives a strong reflected signal when its beam is pointed in the direction of the four corners attached to the balloon. Now, the balloon drifts with wind. By following the balloon and noting its direction at one-minute intervals, and by obtaining the distance to the balloon (radar sets provide such information), one can calculate the velocity with which the balloon drifts and therefore the wind velocity at the level of the balloon. This information is very useful in weather forecasting, and such observations and calculations are now made every day all over the world.

EXERCISES

1. Where is the mirror image of a point A which is in front of a plane mirror?

2. Suppose that m (Fig. 6–22) is the shore of a river and a pier is to be built somewhere along m so that merchandise can be trucked from the pier to two inland towns, A and A'. Where should the pier be built so that the total trucking distance from the pier to A and from the pier to A' is a minimum?

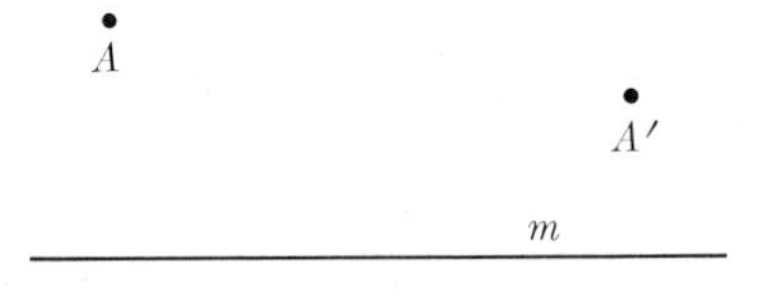

FIGURE 6–22

3. A billiard player wishes to hit the ball at A (Fig. 6–23) in such a way that it will strike side m of the table and then hit the ball at A'. Now billiard balls behave like light rays, that is, the angle of reflection equals the angle of incidence. At what point on m should the billiard player aim?

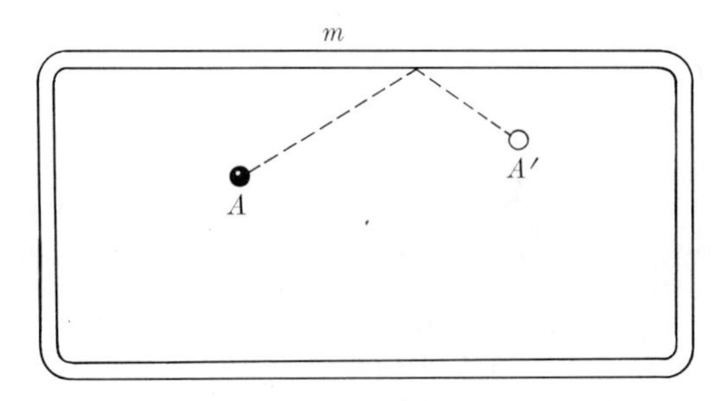

FIGURE 6–23

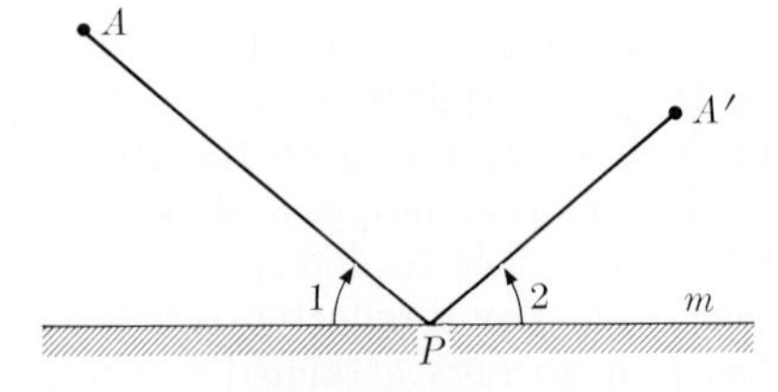

FIGURE 6–24

4. A billiard ball starting from a point A on the table (Fig. 6–23) strikes two successive sides and then travels along the table. What can you say about the final in relation to the original direction of travel?

5. In the text we proved that if angle 1 equals angle 2 (Fig. 6–24), than $AP + PA'$ is the shortest path from A to any point on the mirror to A'. Prove the converse, namely, that if $AP + PA'$ is the shortest path, then $\measuredangle 1$ must equal $\measuredangle 2$. [*Suggestion:* Use the indirect method of proof. If $\measuredangle 1$ does not equal $\measuredangle 2$, then one can find another point, P', on m for which the angles made by AP' and $A'P'$ with m are equal.]

6–5 Conic sections. The *Elements* of Euclid dealt with plane figures which can be built up with line segments and circles, with the corresponding solid figures which can be built up with pieces of a plane, such as prisms and the regular polyhedra, and with the sphere. But the classical Greeks also studied another class of curves which they called conic sections because they were originally obtained by slicing a cone with a plane. The resulting curves, the parabola, ellipse, and hyperbola, were treated by Euclid in a separate book. Unfortunately, no copies of this book have survived. But a little after Euclid's time another famous Greek geometer, Apollonius, wrote a book entitled *Conic Sections*, which is known to us and which is about as exhaustive in its treatment of these curves as the *Elements* are about figures formed by lines and circles.

Conic sections were introduced, as already noted, by cutting a conical surface with a plane. However, the curves themselves can be considered apart from the surface on which they lie. For example, the circle is also one of the conic sections. Yet we know that the circle can be defined as the set of all points which are at a fixed distance from a given point, and this definition does not involve the cone at all. Indeed, insofar as properties and applications of these curves are concerned, it is far more convenient to disregard the conical surface and concentrate on the curves themselves.

Let us consider, therefore, the direct definitions of conic sections. To define the parabola, we start with a fixed point F and a fixed line d (Fig. 6–25). We then consider the set of all points, each of which is equally

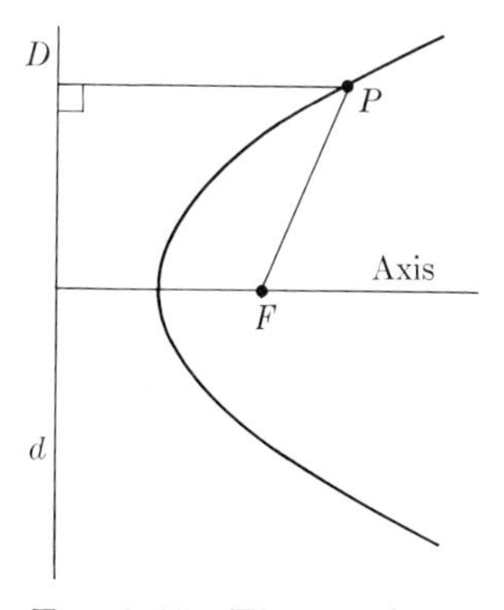

FIG. 6–25. The parabola.

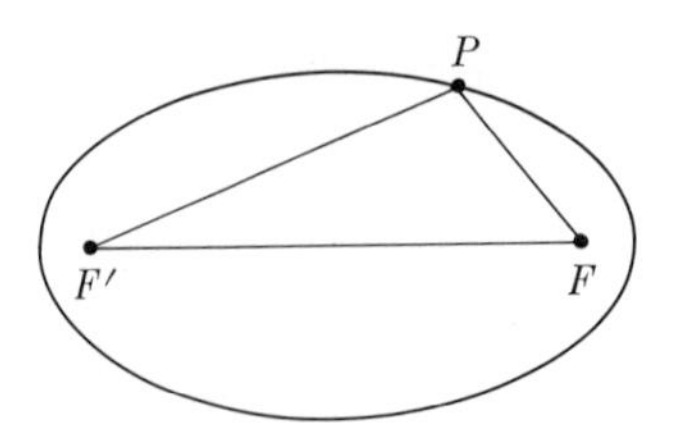

FIG. 6–26. The ellipse.

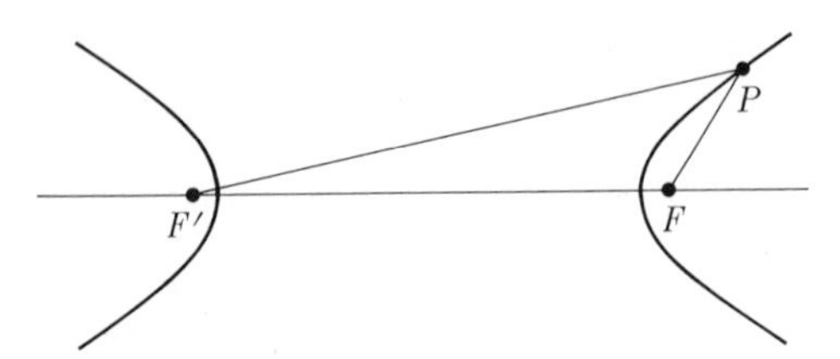

FIG. 6–27. The hyperbola.

distant from F and d. Thus the point P in Fig. 6–25 is such that $PF = PD$. The collection of all points, each of which is equidistant from F and d, fills out a curve called the parabola. The point F is called the *focus* of the parabola, and the line d is called the *directrix.*

Each choice of a point F and line d determines a parabola. Hence there are infinitely many different parabolas. The general shape of all such curves is, however, about the same. Each is symmetric about the line which passes through F and is perpendicular to d. This line is called the *axis* of the parabola. Each parabola passes between its focus and directrix and opens out as it extends farther and farther from the directrix.

The direct definition of the ellipse is also simple. We start with two fixed points F and F' (Fig. 6–26) and consider any constant quantity greater than the distance F to F'. If, for example, the distance from F to F' is 6, we may choose 10 as the constant quantity. One then determines all points for each of which the distance from F and the distance from F' add up to 10. This collection of points is called an ellipse. Thus, if P is a point for which $PF + PF'$ equals 10, then P lies on the ellipse determined by F, F', and the quantity 10. The points F and F' are called the *foci* of the ellipse.

By changing the distance FF' or the quantity 10, one obtains another ellipse. Some ellipses are long and narrow; others are almost circular. All are symmetric about the line FF' and about the line perpendicular to and midway between F and F'.

The direct definition of the hyperbola also calls for choosing two fixed points F and F', called foci, and a constant quantity which, however, must be less than the distance from F to F'. If FF' is 6, then the constant quantity can, for example, be 4. We now consider any point P for which the difference $PF' - PF$ equals 4. All such points lie on the right-hand portion of Fig. 6–27, whereas the points for which $PF - PF' = 4$ lie on the left-hand portion of the figure. The two portions together are the hyperbola; each portion is a branch of the hyperbola.

As for the ellipse, each choice of the distance FF' and the constant quantity determines a hyperbola. Here, too, the curve is symmetric about the line FF' and about a line perpendicular to and midway between F and F'. One branch opens to the right and the other to the left.

We shall not prove that the curves we have defined by means of focus and directrix or by means of foci and constant quantities are the same as those obtainable by slicing a conical surface. In our future work we shall use the direct definitions.

Exercises

1. Since the circle is also a conic section, it should be included among one of the three types—parabola, ellipse, and hyperbola. From the shapes of these curves it would appear that the circle falls among the ellipses. Can you see how the circle may arise as a special kind of ellipse?

2. Suppose that we have an ellipse for which $F'F$ is 6 and the constant quantity is 10. If the point P of the ellipse lies on the line $F'F$ to the right of F, how much is PF?

3. For the ellipse, why must the constant quantity be chosen greater than the distance $F'F$?

4. Given a parabola for which the distance from focus to directrix is 10, how far from the focus is that point on the parabola which lies on the axis?

6-6 The conic sections and light. Next to straight line and circle, conic sections are the most valuable curves mathematics has to offer for the study of the physical world. We shall examine here the uses of the parabola in the control of light.

Let P be any point on the parabola (Fig. 6-28). By the tangent to the parabola at P we mean the line through P which meets the parabola in just that one point and lies entirely outside the curve. From the standpoint of the control of light, the curve possesses a most pertinent property. If P is any point on the curve and F is the focus, then the line FP and PV, the line through P parallel to the axis, where V is any point on this parallel, make equal angles with the tangent at P. That is, angle 1 equals angle 2.

Before proving the geometrical property just stated, let us see why it is significant. If a light ray issues from some source of light at F and strikes a parabolic mirror at P, it will be reflected in accordance with the law that

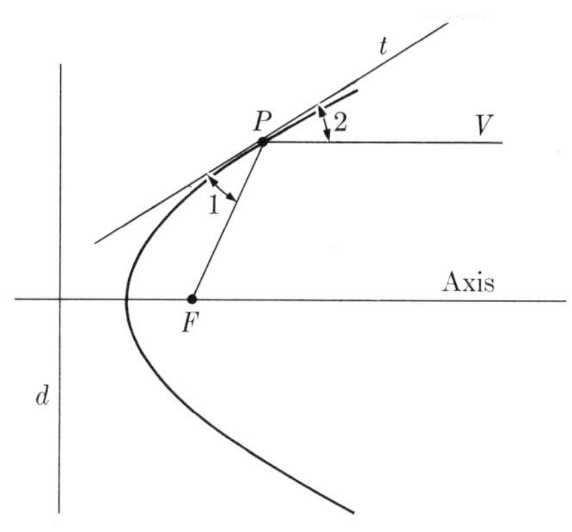

Fig. 6-28. The reflecting property of the parabola.

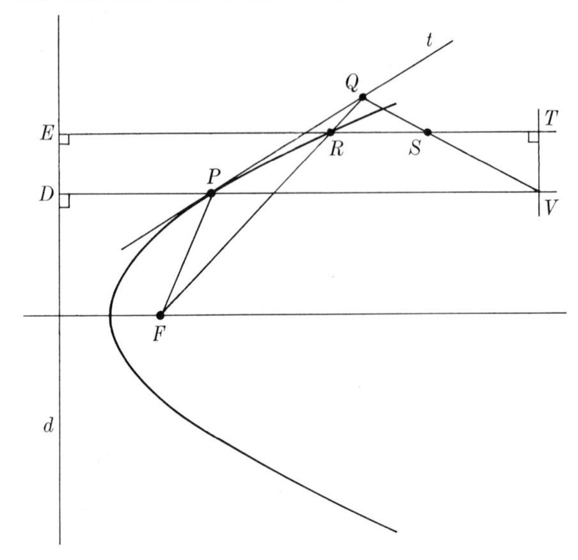

FIGURE 6–29

the angle of incidence equals the angle of reflection. The curve acts at P as though it had the direction of the tangent. Then angle 1 is the angle of incidence. Because angle 1 equals angle 2, the reflected ray will be PV. Hence the reflected ray will travel out parallel to the axis of the parabola. Now, P is any point on the parabola. Hence any ray leaving F and striking the parabola will, after reflection, travel out parallel to the axis of the parabola, and the reflected light will form a powerful beam in one direction. We thus obtain a concentration of light.

Let us now prove that PF and PV make equal angles with the tangent at P. We denote the tangent by t (Fig. 6–29). Further let V be any definite point on PV. To prove the equality of the angles, we shall prove that $FP + PV$ is shorter than $FQ + QV$, where Q is *any* other point on the tangent t. Then it follows that FP and VP make equal angles with t. (See Exercise 5 of Section 6–4.)

We label the point R where FQ cuts the parabola and draw RT parallel to the axis. The point T is directly above V; that is, TV is perpendicular to the axis. We next obtain a useful fact about $FQ + QV$. We see directly that

$$FQ + QV = FR + RQ + QS + SV. \tag{1}$$

But, since the sum of two sides of a triangle is greater than the third side, we have

$$RQ + QS > RS;$$

and since the hypotenuse of a right triangle is greater than either arm,

$$SV > ST.$$

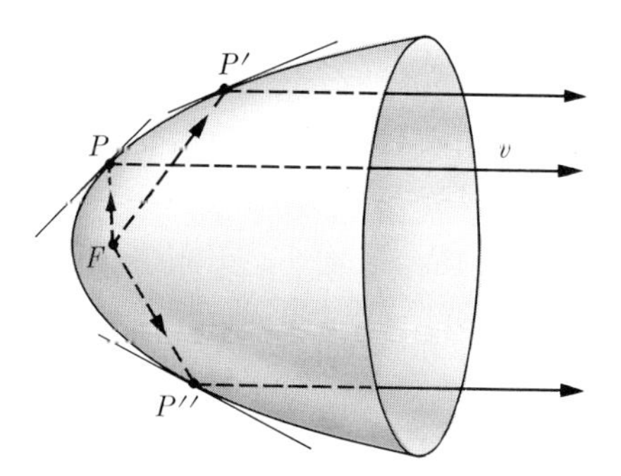

FIG. 6–30. The paraboloidal mirror.

Then, by substituting the smaller quantities in the right side of (1), we obtain

$$FQ + QV > FR + RS + ST = FR + RT. \tag{2}$$

Now let us turn to $FP + PV$. Since P is a point on the parabola, its distance from focus and directrix must be equal. Thus,

$$FP + PV = DP + PV. \tag{3}$$

Likewise because R also is a point on the parabola,

$$FR + RT = ER + RT. \tag{4}$$

But

$$DP + PV = ER + RT \tag{5}$$

because both sides of the equation denote perpendicular distances between the parallel lines DE and VT. Since (5) shows that the right sides of (3) and (4) are equal, the left sides must also be equal. Then

$$FP + PV = FR + RT. \tag{6}$$

We have but to compare (6) and (2) to see that

$$FQ + QV > FP + PV. \tag{7}$$

We now recall the point already made, namely, that if P is the point on the line t for which $FP + PV$ is less than for any other point Q, then FP and VP make equal angles with t. Consequently, any light ray issuing from F will be reflected by a parabolic mirror in the direction parallel to the axis.

The parabolic mirror's power to concentrate light in one direction is very useful. The commonest application is found in automobile headlights. In each headlight there is a small bulb. Surrounding this bulb is a surface (Fig. 6–30), called a paraboloid, which is formed by rotating a parabola about its axis. (The surface is, of course, silvered so that it will reflect.)

Light issuing in millions of directions from the bulb, which is placed at the focus of the paraboloidal mirror, strikes the mirror, is reflected along the axis of the paraboloid, and illuminates strongly whatever lies in that direction. The effectiveness of this arrangement may be judged from the fact that the light thrown forward by bulb and mirror is about 6000 times as intense as that thrown in the same direction by the bulb alone. The reflecting property of the paraboloidal mirror is utilized in many devices, for example, searchlights and flashlights.

The reflecting property of a paraboloidal mirror can be used in reverse. If a beam of parallel light rays enters such a mirror while traveling parallel to the axis, each ray will be reflected by some point on the surface in accordance with the law of reflection. But since FP (Fig. 6–30) and VP make equal angles with the tangent, the reflected ray will travel along PF and all reflected rays will arrive at the focus F. Hence there will be a great concentration of light at F.

This concentration of light is used effectively in telescopes. The light emitted by stars is so faint that it is necessary to collect as much as possible in order to obtain a clear image. The axis of the telescope is therefore directed toward the star, and, because this source is so far away, the rays enter the telescope practically parallel to the axis, travel down the telescope to a paraboloidal mirror at the back, and are reflected to the focus of the mirror.

We shall find in a later chapter that radio waves behave very much like light rays. Hence paraboloidal reflectors made of metal are used to concentrate radio waves issuing from a small source into a powerful beam. Conversely, a paraboloidal antenna can pick up faint radio signals and produce a relatively strong signal at the focus. Since radio is used today for hundreds of purposes, the paraboloidal radio antenna is a very common instrument.

We see from this brief account that the conic sections are immensely valuable. Some of the most momentous applications have yet to be described and will be taken up in later chapters.

How did the Greeks come to study these curves? As far as we know, the conic sections were discovered in attempts to solve the famous construction problems of Euclidean geometry, i.e., to trisect any angle, to construct a square equal in area to a given circle, and to construct the side of a cube whose volume is twice that of a cube of given side. The constructions were to be performed subject to the restriction that only a straight edge (not a ruler) and a compass be used. Having obtained the curves, the Greeks continued to work on them, partly because they were interested in geometrical forms and partly because they discovered the uses of these curves in the control of light. Apollonius himself wrote a book entitled *On Burning Glasses,* whose subject was the parabola as a means of concentrating light and heat, and there is a story that Archimedes constructed a huge parabola which focused the sun's rays on the Roman ships besieging his city of Syracuse, and thus set them on fire.

We see in the history of conic sections one more example of how mathematicians, pursuing a subject far beyond the immediate problems which give rise to it, come to make important contributions to science.

Exercises

1. Let Q be any point outside of an ellipse (Fig. 6–31). Prove that $F_2Q + F_1Q$ is greater than a, where a is the sum of the distances of any point on the ellipse from the foci. [*Suggestion:* Introduce the point P where F_2Q cuts the ellipse.]

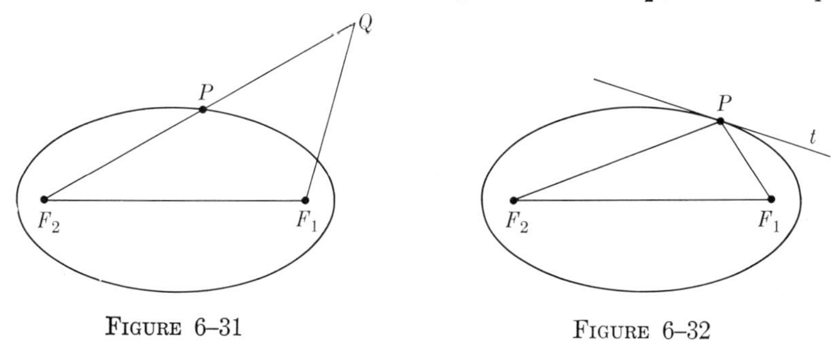

Figure 6–31

Figure 6–32

2. Let t be the tangent at any point P of an ellipse (Fig. 6–32). Let F_2 and F_1 be the foci. Prove that F_2P and F_1P make equal angles with t. [*Suggestion:* Use the result of Exercise 1 and Exercise 5 of Section 6–4.]
3. In view of the result of Exercise 2, what do you expect to happen to the light rays issuing from a source placed at the focus F_2 of the ellipse?
4. When the distance between the two foci F_2 and F_1 of an ellipse approaches 0, the ellipse approaches a circle in shape What do the lengths F_2P and F_1P become when F_2 and F_1 coincide? What theorem about circles follows as a special case of the result in Exercise 2?

6–7 The cultural influence of Euclidean geometry. If the development of mathematics had ceased with the creation of Euclidean geometry, the contribution of the subject to the molding of Western civilization would still have been enormous, for Euclidean geometry was and still is an overwhelming demonstration of the power and effectiveness of our reasoning faculty. The Greeks loved to reason and applied it to philosophy, political theory, and literary criticism. But philosophy breaks down into philosophies whose relative merits become the object of much dispute between the adherents of one school and those of another. Plato's *Republic* may indeed be the perfect answer to the quest for a satisfactory political system, but we must still be convinced of this fact. And literary criticism certainly does not lead to universally accepted standards and the creation of universally acclaimed literature. In Euclidean geometry, however, the Greeks showed how reasoning which is based on just ten facts, the axioms, could produce thousands of new conclusions, mostly unforeseen, and each as indubitably true of the physical world as the original axioms. New, unquestionable, thoroughly reliable, and usable knowledge was obtained, knowledge which obviated the need for experience or which could not be obtained in any other way.

The Greek, therefore, demonstrated the power of a faculty which had not been put to use in other civilizations, much as if they had suddenly shown the world the existence of a sixth sense which no one had previously recognized. Clearly, then, the way to build sound systems of thought in any field was to start with truths, apply deductive reasoning carefully and exclusively to these basic truths, and thus obtain an unquestionable body of conclusions and new knowledge.

The Greeks themselves recognized this broader significance of Euclidean geometry, and Aristotle stressed that the Euclidean procedure must be the aim and goal of all sciences. Each science must start with fundamental principles relevant to its field and proceed by deductive demonstrations of new truths. This ideal was taken over by theologians, philosophers, political theorists, and the physical scientists. We shall see later on how widely and how deeply it influenced subsequent thought.

By teaching mankind the principles of correct reasoning, Euclidean geometry has influenced thought even in fields where extensive deductive systems could not be or have not thus far been erected. Stated otherwise, Euclidean geometry is the father of the science of logic. We pointed out in Chapter 3 that certain ways of combining statements lead to unquestionable conclusions, provided the original premises are unquestionable. These ways are called principles or methods of deductive reasoning. Where did we get these principles? The answer is that the Greeks learned to recognize them in their work on Euclidean geometry and then appreciated that these principles apply to all concepts and relationships. If one argues from the premises that all bankers are wealthy and some bankers are intelligent to the conclusion that some intelligent men are wealthy, he is using a principle of valid reasoning discovered in the work on Euclidean geometry. The indirect method of proof which we applied earlier in this chapter owes its recognition to the same source. Toward the end of the classical Greek period, Aristotle formulated the valid principles of reasoning and created the science of logic. In particular, he called attention to some basic laws of logic, such as the principle of contradiction, which says that no proposition can be both true and false, and the principle of the excluded middle, which states that any proposition must be either true or false.

It is because Euclidean geometry applies these principles of reasoning so clearly and so repeatedly that this subject is often taught as an approach to reasoning. The Greeks themselves stressed the value of mathematics as a preparation for the study of philosophy. Whether this is the best way of learning the general lessons of reasoning may perhaps be disputable, but there is no doubt that historically this is the way in which Western man learned to reason. And it is pertinent that even current texts on logic use mathematical examples quite freely because these illustrate the principles clearly, unobscured by irrelevant implications or by vagueness in the concepts and relations employed.

The most portentous fact about Euclidean geometry is that it inspired a large-scale mathematical investigation of nature. From the outset the geometrical studies were an investigation of nature. But as the Greeks proved more and deeper theorems and these theorems continued to agree perfectly with observations and measurements, the Greeks became convinced that through mathematics they were learning some of the secrets of the design of this world. It became clear that mathematics was the instrument for this investigation, and the results fostered the expectation that the further application of mathematics would reveal more and more of that design. Just how far the Greeks were emboldened to carry this venture will be apparent in the next two chapters. From the Greeks the Western world learned that mathematics was the extraordinarily powerful instrument with which to explore nature.

6-8 Euclidean geometry within the Greek cultural world. Throughout the four thousand years during which mathematics was in the hands of the Egyptians and Babylonians, the subject stagnated. Then a new people, just a few hundred thousand in number, suddenly emerged, took over a few empirical rules and transformed them into a science. Within a few hundred years these people created thousands of theorems pregnant with meaning about the physical world. In the history of intellectual developments, the Greek transformation of mathematics is a miracle. But the miracle was not Greek mathematics so much as the Greek mind, a mind which expressed itself also in philosophy, logic, political thought, literature, and art. Since these creations of the Greeks are part of our own cultural world, it is desirable to understand them. A most illuminating approach can be made through Euclidean geometry. Indeed the chief characteristics of that geometry are also the main features of Greek cultural contributions in other areas.

Euclidean geometry is simple, not in the sense that the proofs of the theorems are simple, but in that it is deliberately confined to simple structures, i.e., to figures that can be built up from straight line and circle. This is obviously so in the plane geometry of Euclid's *Elements,* but is also true of solid figures and conic sections. Among the solid figures, the sphere is no more than the figure generated by rotating a circle about its axis. The cone is generated by rotating the right triangle ABC (Fig. 6-33) about the

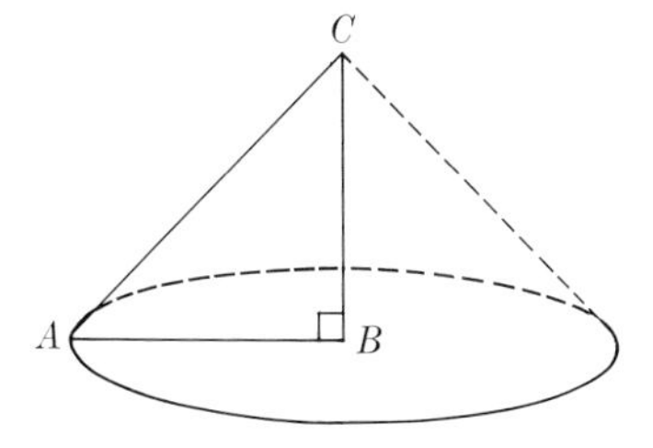

FIG. 6-33. The cone generated by rotating a right triangle around one side.

FIG. 6–34. A Gothic cathedral.

vertical arm BC. A plane is generated by moving a straight line parallel to itself. And the conic sections are, as we know, the curves which result when a plane intersects a conical surface. It would have been easy to introduce some mechanical devices other than straight edge and compass, which are, of course, the physical counterparts of straight line and circle, and generate more complicated curves, as was in fact done in the Alexandrian period and in modern times. These curves could then have been studied mathematically. But the classical Greeks would have none of this. The restriction to straight line and circle kept geometry simple and aesthetically appealing.

This quality of simplicity is found in Greek architecture, sculpture, and literature. One has only to compare the plain, restrained Greek temples (Fig. 3–3 in Chapter 3) with the Gothic cathedrals (Fig. 6–34) erected in medieval times to see this quality. The Greek temple is certainly simple in shape and almost bare compared to the multistructured cathedrals often adorned with hundreds of figures. In Greek sculpture the naked figure is all that one usually finds. By contrast, the majority of Roman sculptures emphasizes special dress, military ornaments, shield and sword, medals, and often surrounding, subsidiary figures. Greek literature, too, relies upon simple, clear, restrained expression. Greek art is plain art, and yet achieves beauty.

Euclidean geometry is limited in several senses. It is limited, as we have already noted, to the figures derivable from straight edge and compass, and to theorems derivable from a fixed set of axioms stated at the outset. It is limited also in the sense that the consideration of figures is always restricted to finite portions. The circle is necessarily so restricted and, for that reason among others, was a favorite among the Greeks. But the straight line was also "confined." Whereas in our discussion of geometry we have followed the modern practice of thinking about the straight line as something extending indefinitely far in either direction, Euclid would not have it so. In fact, to him the straight line is always bounded at each end and is what we call a segment. In his axiom corresponding to our second axiom above, he says that a straight line can be extended *as far as necessary* in either direction, the words implying a finite terminus.

The Greek temple, too, is a limited structure which can be seen and encompassed almost in one glance; it is entirely within the range of the onlooker's eye. This character of the architecture is again best appreciated by comparing it with Gothic cathedrals. The latter are complicated and vast. One must examine them section by section. Even in any one part the eye seems to lose sight of the building proper and to find itself staring outward and upward into space. The nave presents what seems to be an unending series of arches, numerous aisles and chapels, which branch off, and impressive altars visible from a distance. Finiteness is vanquished in the almost invisible, awe-inspiring, dim interiors. The imagination is stirred to contemplate the infinite.

Greek science and Greek philosophy not only preferred the limited, but almost feared the infinite or unlimited. The simplest type of motion for the Greeks was not, as it is for us, motion along a straight line, because there is no natural end to such a motion. Circular motion, on the other hand, can be grasped as a whole and was considered the basic type. In philosophy, the Greeks avoided concepts such as infinite space or the infinite divisibility of a straight line. Aristotle says the infinite is imperfect, formless, and therefore unthinkable, although he, of course, did think about it on occasions. The Pythagoreans associated limited with good and infinite with evil. Typical is the remark of Sophocles, "Nothing that is vast enters into the life of mortals without a curse."

The static character of Greek geometry is another clue to Greek culture. The Greeks did not study changing figures or how one figure can be continuously altered to assume the shape of another, as an ellipse, for example, can be altered to become a circle. The figures were fixed and when compared, as, for example, similar figures were compared, no thought of how one may become the other by expansion or contraction arose. One gets this same feeling of fixedness in surveying the Greek temple. It seems rooted and solidly founded where it stands, whereas the tall spires of the Gothic cathedrals suggest aspiration and even motion to heaven. Greek drama is also static. There is little or no action. A plot does not unfold with new developments from act to act. Rather the essential facts are presented at the outset, and the plot concerns itself with the solution of the problem presented. The subsequent developments are minor. In political thought, Plato, Aristotle, and other Athenian philosophers admired the fixed and rigid institutions of Sparta and compared them most favorably with the more fluid conditions in Athens.

The inevitableness of the theorems which follow from the axioms of geometry also had its counterpart in various phases of Greek culture. The Greeks believed in fate or destiny. People's lives are predetermined, and life is essentially a witnessing of what fate has in store. The role of fate is a principle theme in Greek drama. Events occur as they do because they must. The web of fate clearly grips all characters, and they act out of compulsion. There is no conception of individual will.

The Greeks converted mathematics from a series of rules for concrete tasks of daily life to the study of ideal forms. Here, too, they expressed their cultural outlook. These forms and the necessary relations among them were universal and therefore considered worthy of study. In fact, Plato thought the physical world far less important than the eternal, ideal world which the mind could grasp through suggestions derived from the physical world. In their art the Greeks were after the essence of man, the eternal qualities of people and objects rather than of particular men (Fig. 3-1).

The sculpture of the classical age chose the most perfect form in nature, the human form, and refined and idealized it. Faces and postures of statues of this period show no emotion or concern. Indeed, the absence of facial expressions would lead one to believe that Greek gods and Greek people neither thought, nor laughed, nor worried. Their demeanor is calm and serene even in sculpture depicting dramatic actions. Particular emotions are, after all, a matter of the moment, whereas the Greek sculptors were depicting the permanent in the nature of man. By contrast, Roman sculpture honored generals and leaders, and Roman architecture concentrated on roads, baths, and viaducts, needs of life on earth. Greek political thought emphasized the perfect state, as Plato did in his *Republic,* and Aristotle in his *Politics;* the improvement of the lot of the average man in Athens was not even considered. The Romans, who achieved a reputation for the administration of a far-flung empire and for the creation of a code of law, wrote no works on the ideal state. For the Greeks, what is universal and everlasting is valuable.

Perhaps the paramount quality in all Greek works is the appreciation of beauty. That they found this beauty in bounded, static, universal types is less significant than that they sought beauty, a beauty which appealed to the mind even more than to the eye or ear. Greek mathematics is an expression of the search for beauty, the elegance of deductive structure, the beauty of reasoning, the clarity of concepts, and the perfection of forms. Reason and beauty are not separable in Greek culture. And mathematics, as the art of the mind, united both. In fact, the Greeks valued geometry because it taught men to contemplate the abstract, the ideal, and the beautiful, as much as they valued learning about the wonders of nature. That the Greeks sought and created beauty in their philosophy, literature, and arts, needs no substantiation even to modern man. Greek philosophy and literature are still read, not as historical curiosities of the past, but because they possess beauty. Greek dramas are still performed. That we can enjoy these works despite the passage of 2000 years is a testimony to their quality and to the universal problems and values which they incorporate. The classical Greek age has been unsurpassed in the beauty, the ideality, and the universality of its mathematics, art, literature, and philosophy.

Recommended Reading

BALL, W. W. R.: *A Short Account of the History of Mathematics,* pp. 13–63, Dover Publications, Inc., New York, 1960.

BOYS, C. VERNON: *Soap Bubbles,* Dover Publications, Inc., New York, 1959.

COURANT, R. and H. ROBBINS: *What is Mathematics?,* pp. 329–338, pp. 346–361, Oxford University Press, New York, 1941.

EVES, HOWARD: *An Introduction to the History of Mathematics,* pp. 51–82, pp. 110–122, Rinehart and Co., New York, 1953.

KLINE, M.: *Mathematics and the Physical World,* Chaps. 6 and 17, T. Y. Crowell Co., New York, 1959.

SAWYER, W. W.: *Mathematician's Delight,* Chaps. 2 and 3, Penguin Books, Harmondsworth, England, 1943.

SMITH, DAVID EUGENE: *History of Mathematics,* Vol. I., Chap. 3, Vol. II, Chap. 5, Dover Publications, Inc., New York, 1958.

TAYLOR, LLOYD WM.: *Physics, The Pioneer Science,* Chaps. 29–32, Dover Publications, Inc., New York, 1959.

CHAPTER 7

CHARTING THE EARTH AND THE HEAVENS

Thrice happy souls! to whom 'twas given to rise
To truths like these, and scale the spangled skies!
Far distant stars to clearest view they brought,
And girdled ether with their chains of thought.
So heaven is reached:—not as old they tried
By mountains piled on mountains in their pride.

Ovid

7–1 The Alexandrian world. The course of mathematics is very much dependent upon the caprices of man. What more the classical Greeks might have produced had they been able to continue their way of life uninterruptedly, we shall never know. In 352 B.C. Philip II of Macedonia, a province to the north of Athens and outside the pale of Greek culture, started out to conquer the world. He defeated Athens in 338 B.C. In 336 B.C. Alexander the Great, Philip's son, took over the Macedonian armies, completed the conquest of Greece, conquered Egypt, and penetrated Asia as far east as India, and Africa as far south as the cataracts of the Nile. For a new capital he chose a site in Egypt which was central in his empire. Too big a man to be hampered by modesty, he called the capital Alexandria. Alexander drew up plans for the city and for populating it, and the work was begun. Alexandria did become the center of the Hellenistic world, and even 700 years later was still called the noblest of all cities.

Alexander was the most cosmopolitan of men and sought to break down barriers of race and creed. Hence he encouraged and invited Greeks, Egyptians, Jews, Romans, Ethiopians, Arabs, Indians, Persians, and Negroes to settle in the city. At that time the Persian culture was flourishing, and so Alexander made special efforts to fuse Greek and Persian ways of life. He himself married Statira, daughter of Darius, in 325 B.C. and compelled 100 of his generals and 10,000 of his soldiers to marry Persians. After his death written orders were found to transport large groups of Asians to Europe and vice versa.

Alexander died while still engaged in reconstructing the world, and his empire split into three parts. Of these Egypt proved to be the most significant from the standpoint of mathematical progress. Alexander had indeed chosen a good site for his capital. Located at the junction of Asia, Africa, and Europe, it became the center of trade, which brought wealth to the city. The successors of Alexander who ruled Egypt and who adopted the title of Ptolemy were wise men. They appreciated the cultural greatness of classical Greece and decided to make Alexandria a great cultural center. Under their direction part of the wealth was used to beautify the city with splendid buildings, baths, parks, theaters, temples, libraries, and a national archive. They also erected a famous building devoted to the Muses of literature, art, and science, called the Museum, and adjacent to it, an enormous library to house manuscripts. At its height this library was said to contain 750,000 works, an enormous number in view of the fact that in those days "books" were written and reproduced by hand. The Ptolemies invited scholars from all over the world to work there and supported them. Euclid and Eratosthenes, to speak for the moment of men we have already met, lived and worked at this center; Apollonius was educated there; and we shall meet other luminaries shortly. These men, coming from all over the world, brought knowledge of their lands, people, animals, and vegetation to Alexandria, and this in itself helped to make Alexandria cosmopolitan.

The scholars set to work in the fields of mathematics, science, philosophy, philology, astronomy, history, geography, medicine, jurisprudence, natural history, poetry, and literary criticism. Fortunately, Egyptian papyrus, cheaper than parchment, was available for books, and so many more works could not only be written but copied. Alexandria became in fact the center of the book-copying trade of the ancient world. The scholars undertook not only to create and write, but they sent expeditions all over the world to gather knowledge. At Alexandria they built a huge zoological garden and a botanical garden to house the species of animals and plants brought back by these expeditions.

Alexander had planned to fuse cultures in his new empire, and at Alexandria his goal was realized. The culture which developed there was indeed different from that of classical Greece, for reasons that are of interest because they account for the kind of mathematics the Alexandrians produced. First of all, the rather sharp segregation between free men and slaves which existed in Athens was destroyed. The scholars came from all parts of the world and from all economic levels and took a natural interest in the scientific, commercial, and technical problems of commerce, industry, engineering, and navigation. Although Athens also was primarily a sea power and lived on trade, Alexandrian commerce and navigation were far more widespread. Hence there developed an intense interest in astronomy and in geography, i.e., in the subjects enabling man to tell time, navigate over land and sea, build roads, and determine boundaries of the empire. Free men engaged in commerce are naturally more concerned with materials, methods of production, and new ventures. Finally, though the nucleus of the scholars gathered at Alexandria was Greek, it was exposed to the influence of the practical Egyptians to whom mathematics, to the extent that it was used in ancient Egypt, was a tool for engineering, commerce, and state administration.

The results of the new outlook and interests are readily detected. First of all, there was a sharp increase in mechanical devices, which of course aid men in their work. Even training schools to educate young people in mechanics were established. Pulleys, wedges, tackles, geared devices, and a mileage-measuring instrument such as is found in the modern automobile were invented. Archimedes, the greatest intellect of the Alexandrian world, constructed a planetarium which reproduced the motions of the heavenly bodies and designed a pump for raising water from a river to land. He used pulleys to launch a heavy galley for King Hiero of Syracuse. Instruments to improve astronomical measurements were also invented.

Another science whose beginnings may be found in Alexandria is the study of gases. The Alexandrians, notably Heron (about first century A.D.), a famous mathematician and engineer, learned that the steam created by heating water seeks to expand and that compressed air can also exert force. Heron is responsible for many inventions which used these forces. Temple doors opened automatically when a coin was deposited. Inside the temple another coin inserted in a machine blessed the donor by automatically sprinkling holy water upon him. Fires lit under the altar created steam, and the mystified and awe-struck audience observed gods who raised their hands to bless the worshippers, gods shedding tears, and statues pouring out libations. Doves rose and descended under the unobservable action of steam. Guns similar to the toy bee-bee gun were operated by compressed air. Steam power was used to drive automobiles in the annual religious parade along the streets of Alexandria.

The Alexandrians also studied water power and applied it. They invented improved water clocks (used in the courts to limit the time allowed to lawyers), fountains in which figures moved under water pressure, pumps to bring water from wells and cisterns, musical organs worked by water pressure, and a water-spraying device operating on exactly the same principle as that applied in contemporary lawn-sprinklers.

The study of sound and light was intensified. We have already mentioned Euclid's and Heron's studies on the reflection of light by mirrors. Books on optics were written not only by Euclid and Apollonius but also by Heron, the astronomer Ptolemy (whom we shall discuss shortly), and others. Indeed the Alexandrians were the first to concern themselves with a second basic phenomenon of light, refraction, which we shall encounter in this chapter.

Chemical and medical skills, if not a science of chemistry, show a marked advance in Alexandria. The Egyptians had previously acquired some knowledge in these areas, as we know from their ability to embalm. However, metallurgical studies, including the first text on the subject, and the investigation of chemicals, including poisons and their uses, were essentially new developments. Dissection of bodies, forbidden in classical Greece, was permitted, and the Alexandrian world produced the beginnings of anatomy and the most famous doctor of the ancient world, Galen.

Where was mathematics in this scheme of things? The Greeks brought to Alexandria a fully formed, mature, and philosophically oriented mathematics which had little bearing on practical problems. Although the great Alexandrian mathematicians continued to display the Greek genius for theory and abstraction, they combined with that an interest in the world about them and in practical problems. To the classical Greek concern with qualitative properties such as congruence, the Alexandrians added a new theme, *quantitative* results which are useful in a variety of ways.

To illustrate the combination of old and new we might note that while Euclid chronologically belongs to the Alexandrian period, his mathematical work is in essence a recapitulation of the work done in the classical period. Thus Euclid tells us, for example, that the ratio of the area of any circle to the square of its radius is the same for all circles. In symbols, if A is the area of any circle of radius r, then

$$\frac{A}{r^2} = k,$$

where k is the same number for *all* circles. But now suppose that we wish to find the area of a particular circle. Does Euclid's theorem help us? Not directly. We know from the preceding equation that for any circle

$$A = kr^2,$$

where k is a contant. But how much is k? This quantity which we usually denote by π is an irrational number. It is not readily computed and, because it is irrational, can be expressed as a decimal only approximately. One of Archimedes' great achievements, which also illustrates the interest in quantitative knowledge, is his determination that π lies between $3\frac{1}{7}$ and $3\frac{10}{71}$. The achievement is all the more remarkable because neither the classical nor the Alexandrian Greeks had an efficient system for writing and operating with numbers.

As a matter of fact, Archimedes (287–212 B.C.) is the man whose work best illustrates the character of Alexandrian Greek mathematics. He derived many formulas for the areas and volumes of geometric figures, and his results, as opposed to those of Euclid and Apollonius, made actual computations possible. At the same time, Archimedes also pursued the classical Greek interest in proof and in beautiful mathematical results. In this area, he was proudest of his proof that the ratio of the volume of a sphere inscribed in a cylinder (Fig. 7–1) is to the volume of the cylinder as 2 is to 3.

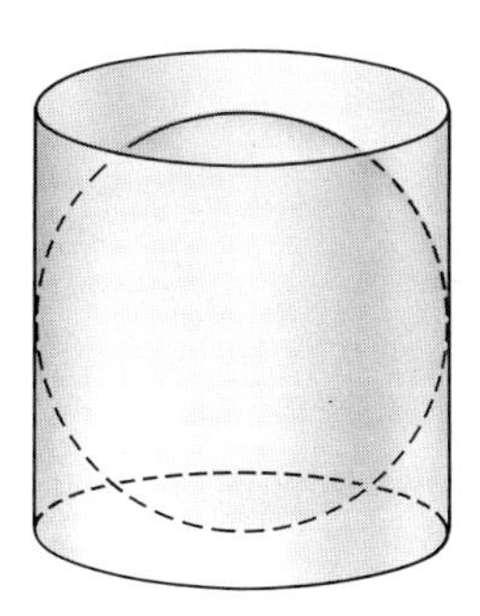

FIG. 7–1. The volume of a sphere inscribed in a cylinder is two-thirds the volume of the cylinder.

He also proved that the same ratio holds for the areas of the sphere and the cylinder. Archimedes was so pleased with this result that he asked that it be inscribed on his tombstone. After Archimedes was killed by a Roman soldier during the Roman conquest of Syracuse, the Romans built an elaborate tomb on which they inscribed this theorem. It was this inscription which enabled Cicero to recognize the tomb on a visit to Syracuse two hundred years later.

Even in his physical studies Archimedes displayed this combination of theoretical and practical interests. He took up the subject of the lever, a device which had been used in Egypt and Babylonia for thousands of years. Like a true Greek, he produced a scientific work, *On the Lever,* along the lines of Euclidean geometry; that is, he started from axioms and proved theorems about the lever. He did the same with the subjects of floating bodies and centers of gravity of various surfaces and volumes. To these achievements must be added his inventions, of which we have already spoken.

The work of some other giants of the Alexandrian civilization also illustrates the combination of theoretical and practical interests. Eratosthenes (273 B.C.–192 B.C.), director of the library at Alexandria, was distinguished in mathematics, poetry, philology, philosophy, and history. He was the first outstanding mathematical geographer and geodesist. The calculation of the circumference of the earth, which we studied in the preceding chapter, is one of his great achievements. He collected and integrated all available geographical knowledge, introduced methods of surveying, made maps, and compiled all of this information in his *Geographica.*

Eratosthenes was also an astronomer. He constructed some new instruments, made many astronomical measurements, and, among other applications, used his astronomical knowledge to improve the calendar. As a result of his work, an old Greek calendar based on a year of 12 months each containing 30 days was replaced by the Egyptian year of 365 days, to which Eratosthenes added an extra day every fourth year. This calendar was adopted by the Romans when Julius Caesar called in Sosigenes, an Alexandrian, to reform the calendar. Julius contributed his name. The Julian calendar was taken over by the Western world with the slight modification that we omit the leap year in three out of every four centuries.

The work in geography and astronomy, continued by such famous men as Strabo (*ca.* 63 B.C.–*ca.* 15 B.C.), Poseidonius (first century B.C.), and many others, was crowned by the achievements of two of the greatest men of the Alexandrian world, Hipparchus and Ptolemy. Hipparchus (second century B.C.), about whom we know rather little, lived at Rhodes, but was in close touch with the developments in Alexandria. After criticizing Eratosthenes' *Geographica,* he refined the method of locating places on the earth by systematically employing latitude and longitude. He improved astronomical instruments, "measured" irregularities in the moon's motion, catalogued about 1000 stars, and estimated the length of the solar year as 365 days, 5 hours, and 55 minutes, i.e., he overestimated by about $6\frac{1}{2}$ minutes. One of his notable astronomical discoveries was the precession of the equinoxes, a slow change in the time of occurrence of the spring and fall equinoxes. Hipparchus is the creator of the most famous and most useful astronomical theory of antiquity, about which we shall learn more later.

The work of Hipparchus is known to us largely through the writings of the mathematician, astronomer, geographer, and cartographer Claudius Ptolemy. Ptolemy, who is believed to be Egyptian—he was no relation to the Greek rulers of Egypt—lived from about 100 to 178 A.D. One of his influential achievements was his *Guide to Geography*, or *Geographica*, the most comprehensive work of antiquity on this subject. This book, which contains the latitude and longitude of 8000 places, almost every place on the earth known then, estimates of the size and extent of the habitable world, and methods of map making, summarized the geographical knowledge of the ancient world and became the standard atlas for over a thousand years. Better known is Ptolemy's great work on astronomy, the *Mathematical Syntaxis* or *The Mathematical Collection,* which the Arabs called *Al Megiste* (an Arabic and Greek combination meaning "the greatest")—later Anglicized as *The Almagest.* This book contains the full development of Hipparchus' and Ptolemy's astronomical theory, generally known as Ptolemaic theory, which dominated astronomy until about 1600 A.D. when it was superseded by the work of Copernicus and Kepler.

7–2 Basic concepts of trigonometry. The theoretical sciences of geography and astronomy require their own mathematical tool, trigonometry. Hipparchus and Ptolemy created this branch of mathematics whose first presentation is found in Ptolemy's *Almagest.* With this simple branch of mathematics it is possible to calculate the sizes and distances of the heavenly bodies as easily as one calculates the area of a rectangle. In presenting the trigonometry of Hipparchus and Ptolemy, we shall not use their notation and proofs; however, the modern approach is not essentially different.

Let us consider the two right triangles shown in Fig. 7–2 and let us suppose that angle A equals angle A'. Since all right angles are equal, angle C equals angle C'. One of the key theorems in Euclidean geometry states that the sum of the angles in any triangle is 180°. Since all three angles in each

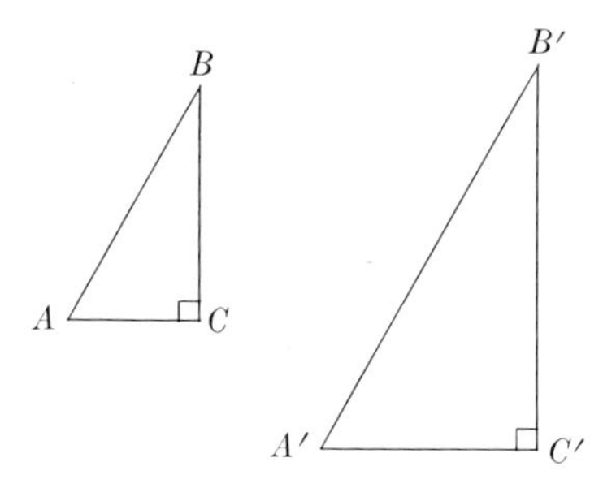

FIG. 7–2. Similar right triangles.

triangle add up to the same amount and two angles of one are equal to two of the other, the third angles must be equal; i.e., angle B equals angle B'.

Now another theorem of Euclidean geometry states that if two triangles are similar, the ratio of any two sides in one equals the ratio of the corresponding sides in the other. Thus, for example,

$$\frac{BC}{AB} = \frac{B'C'}{A'B'}.$$

Let us note here that triangle $A'B'C'$ is *any* other right triangle which has an acute angle, A', equal to angle A. Hence for any such triangle the ratio $B'C'/A'B'$ must equal BC/AB. Therefore, if we could compute this ratio—it is a number of course—for any one right triangle containing a given angle A, we would know it for all right triangles having an acute angle equal to A.

Before we pursue this idea, let us observe that what we said about the ratio BC/AB applies to any other ratio of two sides of triangle ABC. Of the many ratios we can form three are especially useful and are given names. These ratios are:

$$\text{sine } A = \frac{\text{side opposite angle } A}{\text{hypotenuse}} = \frac{BC}{AB},$$

$$\text{cosine } A = \frac{\text{side adjacent to angle } A}{\text{hypotenuse}} = \frac{AC}{AB},$$

$$\text{tangent } A = \frac{\text{side opposite angle } A}{\text{side adjacent to angle } A} = \frac{BC}{AC}.$$

The angle A is written alongside the name of each ratio. This practice is necessary not only because the very use of such words as opposite and adjacent depends upon which angle of the triangle we are talking about, but also because the values of the ratios depend upon the size of the angle. It is very common to abbreviate these names as sin, cos, and tan, respectively.

Since we intend to employ these ratios, our first task should be to see whether we can compute them for angles of various sizes. First of all let us get some general notion of how these ratios vary with the angle. Let us consider sin A as an example. We have already pointed out that the values of these ratios for a given angle A are the same in any right triangle containing A. To study the variation of sin A as A changes, we can then take right triangles whose hypotenuse is 1. We know from the very definition of sin A that it is the ratio of the side opposite angle A to the hypotenuse. Since sin A equals BC/AB and $AB = 1$, then sin $A = BC$. When A is small (Fig. 7–3), BC or sin A is small. We should expect, then, that for an angle close to 0°, the sine of that angle should be close to 0. On the other hand,

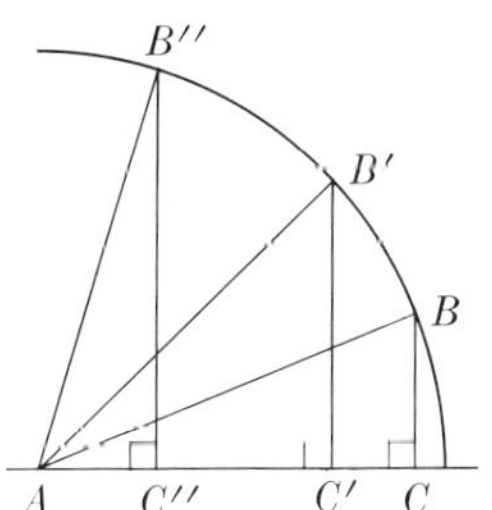

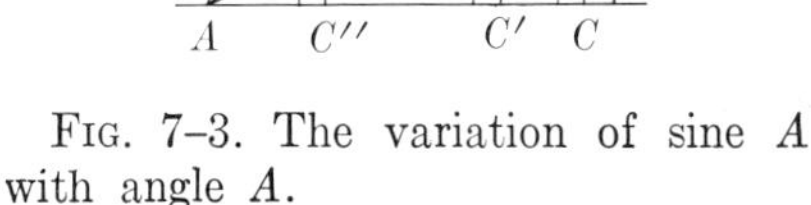
Fig. 7–3. The variation of sine A with angle A.

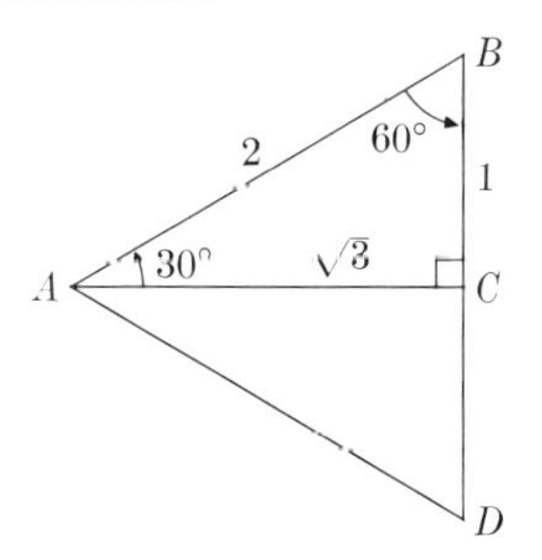

Figure 7–4

as Fig. 7–3 shows, when angle A increases and the hypotenuse is kept one unit in length, the opposite side must increase; hence the sine ratio must increase. When angle A is very close to 90°, as in triangle $AC''B''$, the side $B''C''$ is almost as large as AB''; hence sin A must be close to 1. When angle A is 90°, it can no longer be an acute angle of a right triangle, but because sin A approaches 1 as A approaches 90°, it is agreed that in this special case we shall take sin A to be 1. Likewise, we take sin 0° to be 0. The general point of this discussion is that sin A varies from 0 to 1 as A varies from 0° to 90°.

Next let us take a particular angle and let us see whether we can calculate the three ratios. We shall choose 30°. Consider the equilateral triangle ABD (Fig. 7–4). We know that in such a triangle each angle is 60°. If we now draw the angle bisector AC, then angle BAC is 30°. Moreover, triangle ACB is a right triangle because triangles ACB and ACD are congruent, and hence the two angles at C must be equal. Since the sum of these two angles is 180°, each must be 90°. Triangle ACB is, then, a right triangle containing an acute angle of 30°.

Now it does not matter how long we take AB to be, for we saw earlier that we may compute the ratios in any right triangle containing the given acute angle. Let us therefore choose a convenient number, say 2, for the length of AB. Since ABD is equilateral, side $BD = 2$. But because triangles ACB and ACD are congruent, $CB = CD$. Hence $CB = 1$. We now find the length of AC. The Pythagorean theorem says that

$$(AC)^2 + (CB)^2 = (AB)^2;$$

therefore

$$(AC)^2 = (AB)^2 - (CB)^2.$$

Since $AB = 2$ and $CB = 1$,

$$(AC)^2 = 4 - 1,$$

or

$$AC = \sqrt{3}.$$

We can now use the definitions of sine, cosine, and tangent to state at once that

$$\sin 30^\circ = \frac{\text{side opposite}}{\text{hypotenuse}} = \frac{1}{2};$$

$$\cos 30^\circ = \frac{\text{side adjacent}}{\text{hypotenuse}} = \frac{\sqrt{3}}{2};$$

$$\tan 30^\circ = \frac{\text{side opposite}}{\text{side adjacent}} = \frac{1}{\sqrt{3}}.$$

As a dividend for our patience we get more information from the above reasoning than we sought. Let us note that angle B, which is 60°, is also an acute angle in a right triangle, and we know the lengths of the sides. Hence, since $\sin B$ is the side opposite angle B divided by the hypotenuse, we have

$$\sin 60^\circ = \frac{\sqrt{3}}{2}.$$

Similarly, by applying the definitions of cosine and tangent we obtain

$$\cos 60^\circ = \tfrac{1}{2}, \qquad \tan 60^\circ = \sqrt{3}.$$

We must admit that in undertaking to find the ratios belonging to 30° we selected a simple case. For most angles the ratios are not so easily found, and a good deal of geometry must be applied. The process of determining the ratios for angles from 0° to 90° is not particularly fascinating. Fortunately these values were obtained by Hipparchus and Ptolemy and compiled in a table to be found in Ptolemy's *Almagest*. (These tables were checked and extended by many later mathematicians.) Hence let us take over their results which appear in the "Table of Trigonometric Ratios" (Table I of the appendix).

The table gives the sine, cosine, and tangent values for each angle from 0° to 90°. For angles from 0° to 45° we use the left-hand column and the headings across the top of the page. For example, alongside of 30° and under tangent we find 0.5774. This number is the approximate decimal value of $1/\sqrt{3}$. To find the sine, cosine, or tangent of an angle from 45° to 90° we use the right-hand column and the column designations at the *bottom* of the page. For example, to find sin 60° we look for 60° in the right-hand column and above the word sine we find 0.8660. This number is the approximate decimal value of $\sqrt{3}/2$.

Our table does not give the ratios for angles which contain minutes and seconds as well as degrees. There are tables which do so, but we shall not bother with them because the idea is the same. Where we need the value of a ratio for an angle not in the table, it will be supplied in the text proper.

Let us note that we can use these tables in reverse. If, for example, we are given $\tan A = 1.7321$, we can look down one tangent column and up the other until we come to 1.7321. We can then look to the left (or to the right, depending upon where we locate this number), and find the angle which has the given tangent value. In the present case we must choose the angle at the right, namely 60°. If the table does not contain the exact value given, it will suffice, for our purposes, to choose the one nearest to it.

Exercises

1. Use the isosceles right triangle shown in Fig. 7–5 to compute sin 45°, cos 45°, and tan 45°.

2. Use the Table of Trigonometric Ratios to find

(a) sin 20°	(b) sin 70°	(c) cos 35°
(d) cos 55°	(e) tan 15°	(f) tan 80°

3. Use Fig. 7–3 in the text to determine the range of cosine values as angle A varies from 0° to 90°.

4. Use Fig. 7–3 in the text to determine the range of tangent values as angle A varies from 0° to 90°.

5. Show that when A and B are the two acute angles of a right triangle, then $\sin A = \cos B$ and $\cos B = \sin A$.

6. Prove that $\cos(90° - A) = \sin A$ and that $\sin(90° - A) = \cos A$.

7. Show that $\sin^2 A + \cos^2 A = 1$ for any acute angle A. Here the notation $\sin^2 A$ means $(\sin A)(\sin A)$ or the square of $\sin A$. Can the result be used to compute trigonometric ratios? If so, how?

8. State the definitions of sine, cosine, and tangent of angle D in terms of the sides DE, EF, and FD of the triangle shown in Fig. 7–6.

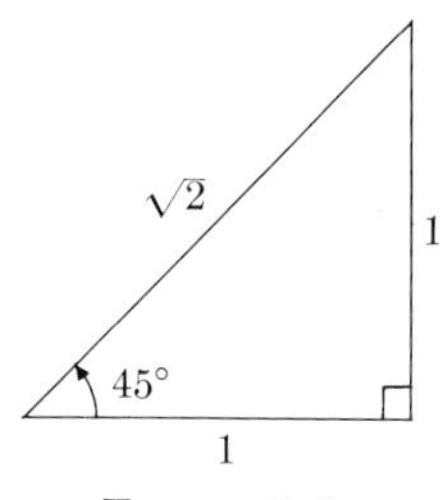

FIGURE 7–5

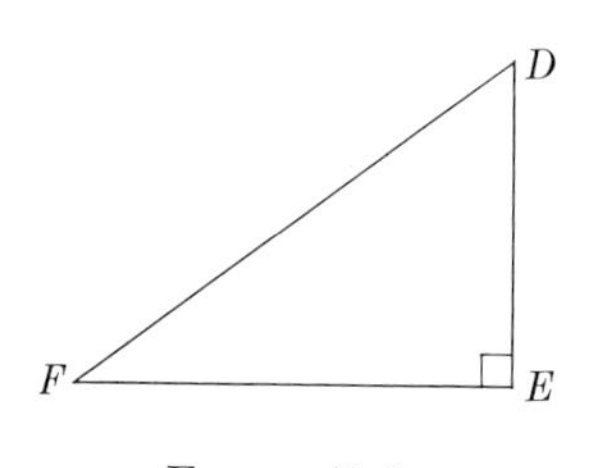

FIGURE 7–6

7–3 Some mundane uses of trigonometric ratios. Before we venture onto vast stretches of the earth's surface or into the heavens, let us see what we can do with trigonometric ratios in rather simple, homely situations. Suppose we had to find the height of the cliff BC in Fig. 7–7. Of course, we could climb the cliff, let a rope down from point B until it just reaches C, pull up the rope, and measure the length which stretched from B to C. There is, however, an easier method which is especially recommended to people who do not like heights.

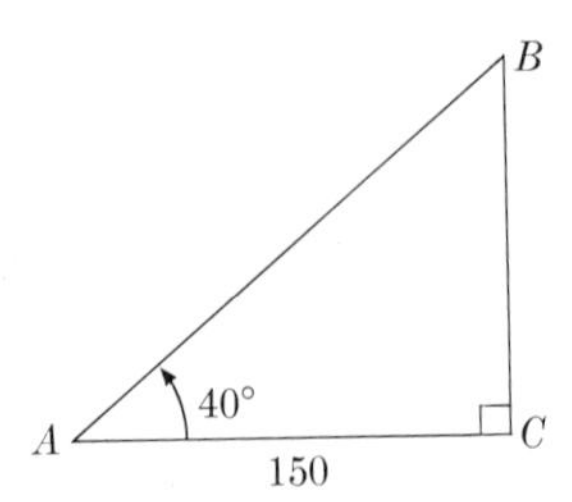

FIG. 7–7. Determining the height of a cliff.

Instead of climbing the cliff, one can walk along the ground from C to any convenient point A. The distance from C to A is then measured; let us suppose it proves to be 150 feet. At A, a person measures the angle between the horizontal AC and the line of sight from A to B. A surveyor would use a transit for this purpose, but there are simpler devices, called protractors, which one can carry in his pocket. Suppose that angle A turns out to be 40°. We are interested in side BC and we know side AC. The fact that these two sides are the side opposite angle A and the side adjacent to angle A suggests that we use the tangent ratio and write

$$\tan 40^\circ = \frac{BC}{150}.$$

This equation involves numbers, and we can therefore apply the axiom that equals multiplied by equals give equals, to justify multiplying both sides by 150. We obtain

$$150\,(\tan 40^\circ) = BC.$$

Now tan 40° can be found in the table which Hipparchus and Ptolemy so considerately prepared, and proves to be 0.8391. Hence

$$BC = 150(0.8391) = 126.$$

The answer, then, is 126 feet. We ignore the decimals because the given information is presumably accurate only to the nearest foot.

EXERCISES

1. To measure the width BC of a canyon (Fig. 7–8), a surveyor at C walks along the edge (preferably alongside the edge) to some convenient point A. He then measures AC and the angle A. Suppose AC is 300 ft and angle A is 56°. How large is BC?

2. At some point on the ground, located at a distance from the Empire State Building in New York City, an observer finds that the angle between the horizontal and the line of sight to the top is 5° (Fig. 7–9). The building is 1248 ft high. How far away is the observer?

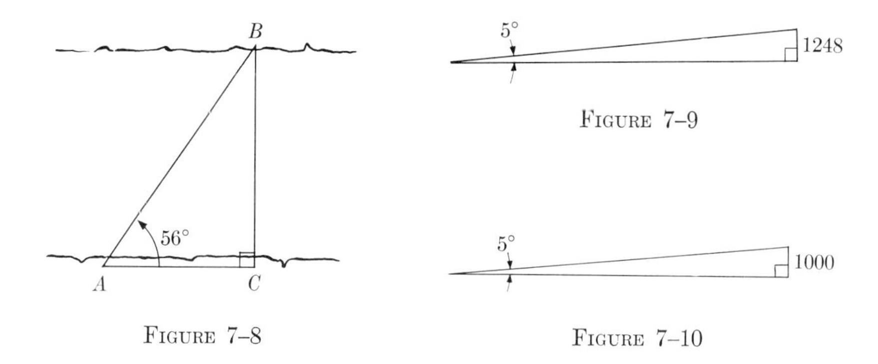

FIGURE 7-8

FIGURE 7-9

FIGURE 7-10

3. A railroad line is being planned which must rise to 1000 ft (Fig. 7-10) at a "grade" of 5°. How long must the line be?

4. A lighthouse beacon is 400 ft above sea level (Fig. 7-11), and the sea around it is obstructed by rocks extending as far as 300 ft from the base of the lighthouse. A sailor on a ship's deck 20 ft above sea level measures the angle between his horizontal and the line of sight to the top of the beacon and finds it to be 50°. Is his ship clear of the rocks?

5. The Alexandrian Greek mathematician and engineer Heron showed how one could dig a tunnel under a mountain by working from both ends simultaneously and have the borings meet. He chose a convenient point A on one side, a convenient point B on the other, and finally point C for which angle ACB is 90° (Fig. 7-12). He next measured AC and BC and found their lengths to be 100 ft and 75 ft, respectively. Now, said Heron, it is possible to calculate angles A and B. He then instructed the workers at A to follow a line which made the calculated angle with AC, and gave analogous directives to the workers at B. How did he calculate angle A and angle B?

6. Trigonometric ratios can be used to compute the radius of the earth, whence, of course, the circumference can be determined by plane geometry. The method is an alternative to Eratosthenes' procedure. From a point A which is 3 mi above the

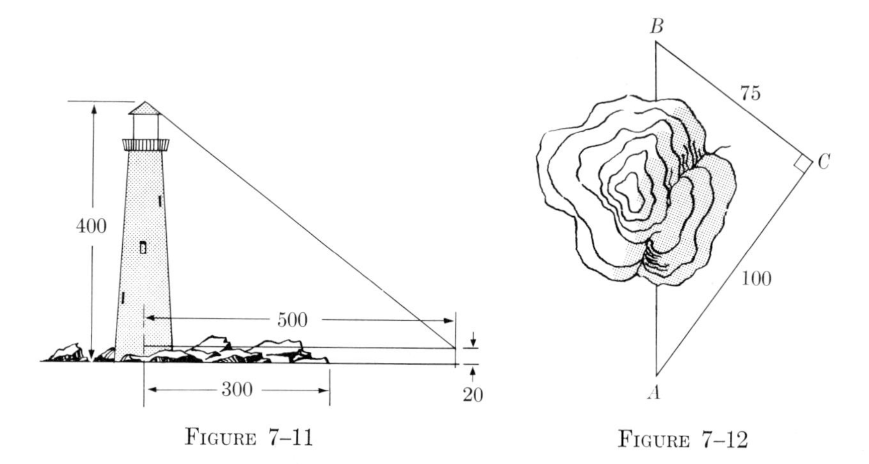

FIGURE 7-11

FIGURE 7-12

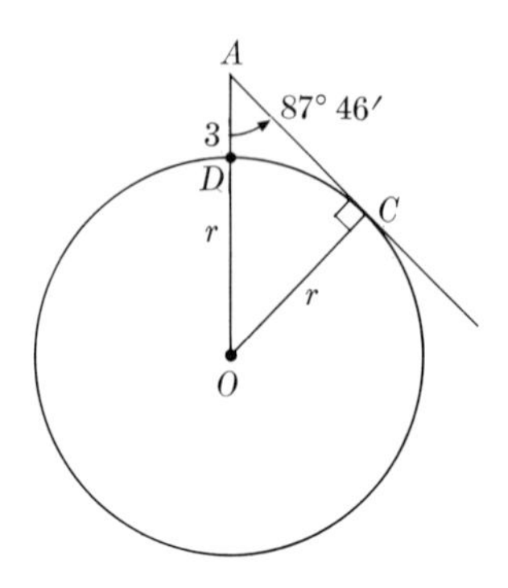

FIGURE 7–13

surface of the earth (A can be the top of a mountain or an airplane), an observer looks to the horizon. His line of sight, AC in Fig. 7–13, is just tangent to the earth's surface. According to a theorem of Euclidean geometry, the radius OC of the earth is perpendicular to the tangent at C. Hence triangle ACO is a right triangle. Suppose that the size of angle A is 87° 46′. Let us denote the length of OC by r. Then OD is also r. We can now say that

$$\sin 87°46' = \frac{r}{r+3}.$$

Given that sin 87° 46′ is 0.99924, calculate r.

7–4 Charting the earth. We have already related that geography was one of the major interests of the Alexandrians. Here Hipparchus and Ptolemy, helped by the trigonometry they had created, made great strides. Let us see how they determined the locations of important places and how they calculated the distances between such places.

Hipparchus proceeded by employing systematically an idea already advanced prior to his time, namely the scheme of latitude and longitude. The earth is, of course, a sphere. Let us consider circles with center O, which is the center of the earth, each going through the North and South Poles, N and S in Fig. 7–14. Thus NWS is one half of such a circle; the other half runs in back of our figure and is therefore invisible. Likewise NVS is one half of another such circle. Obviously we can think of such a circle through N, S, and any other point on the earth's surface. Each half circle from N to S is called a longitude line or a meridian of longitude.

To distinguish among these many lines, we introduce another circle, $XWVU$, which is perpendicular to the longitude lines and halfway between the two poles. This circle is called the equator. Now one of the longitude lines, say NWS, is chosen as the starting line, so to speak. (Today this line goes through the city of Greenwich, England.) We consider next any other line, such as NVS in our figure. The *angle* VOW formed at the earth's center, O, by the lines VO and OW is called the longitude of any point on NVS. Thus longitude is an angle. To distinguish the meridians of longitude

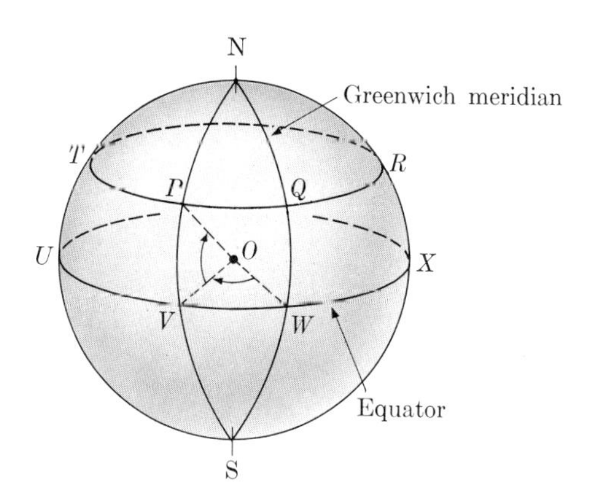

FIG. 7–14. Latitude and longitude.

on the left of *NWS* from those on the right, we use the term "west longitude" to designate the angles determined by the former, and apply the term "east longitude" to those formed by the latter.

Thus any point on the earth's surface has a definite longitude. However all points on the half circle *NVS* have the same longitude. How shall we distinguish any one of these points from the others? The answer is: by introducing horizontal circles going around the earth. The equator is one such circle, and the circle $TPQR$ of our figure is another. Clearly we can introduce many such circles lying in planes parallel to the equator. These circles are called circles of latitude. Again we have the problem of distinguishing among these circles. This is solved by introducing angles formed at the center of the earth, such as POV of Fig. 7–14, where P is any point on a circle of latitude, O is the center of the earth, and V is on the equator and on the meridian of longitude through P. The angle POV is called the latitude of P. If P is north of the equator, it is said to have north latitude; if it is south of the equator, it is called south latitude. Thus points on the same meridian of longitude are distinguished by their differing latitudes.

The point P is a typical point on the earth's surface, and its position is now described by its latitude and longitude. For example, it might have 30° north latitude and 50° west longitude. In this case, angle POV is 30°, and angle VOW is 50°. Any point north or south of P, that is, on the same meridian, will have the same longitude as P but a different latitude. Any point east or west of P, that is, on the same circle of latitude, will have the same latitude as P but a different longitude.

We have described what is meant by the latitude and longitude of any point on the earth's surface, but how do we determine the latitude and longitude for any given point P? (After all, we cannot penetrate to the center of the earth to measure the angles POV and VOW.) There are numerous methods available. We shall describe a simple one just to see that the latitude and longitude of places on the earth can be determined. Sup-

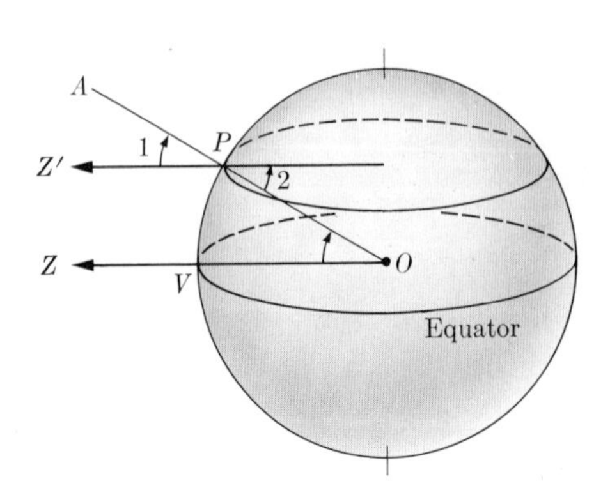

FIG. 7–15. The determination of latitude at a point on the earth's surface.

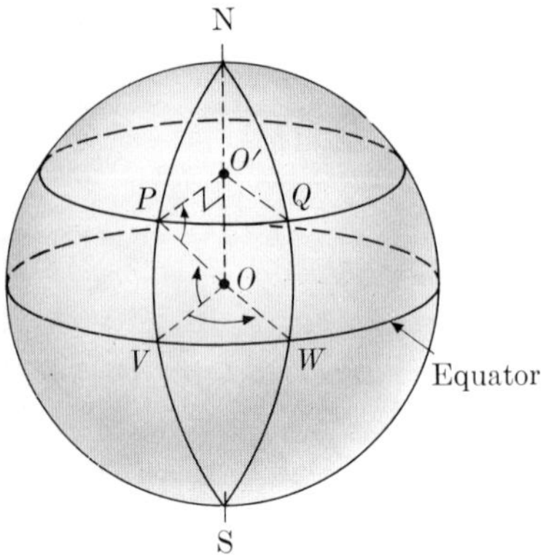

FIGURE 7–16

pose we seek the latitude of some point P (Fig. 7–15). On the day of the spring equinox, that is about March 21, the sun is in the plane of the equator and, at noon on that day, it is also in the plane of the meridian of longitude. For a person at P, the overhead direction is PA, and the direction to the sun is PZ'. Now the sun is so far away that PZ' and VZ can be taken to be parallel lines. Then angle 2 equals the angle of latitude POV because they are alternate interior angles of parallel lines. But angle 1 equals angle 2 because they are vertical angles. Hence angle 1 equals the latitude of P. But angle 1 can be measured. It is the angle between the direction of the sun and the overhead direction at P. Thus the latitude of P can be determined.

There are other methods of measuring latitude as well as methods for finding the longitude of places on the earth. It is of interest that the methods of measuring latitude are more readily applied. The problem of determining longitude accurately aboard a ship at sea was not resolved until the middle of the eighteenth century. We shall have more to say about this later.

We may suppose, then, that the latitude and longitude of places on the earth can be determined. Can we now determine how far apart two places are? We can and we shall illustrate the process. Suppose P (Fig. 7–16) is New York City, which has a north latitude of 41° and a west longitude of 74°. Hence angle POV is 41°, and angle VOW is 74°. Let us answer first the question, How far north of the equator is New York City? This question is easy to answer. The distance we seek is the arc PV. But POV is 41° and arc PV is the arc of a circle whose radius is the radius of the earth. Hence arc PV is that part of the circumference of the earth which 41° is of 360°; that is, if we take the circumference of the earth to be 25,000 miles, then

$$PV = \tfrac{41}{360} \cdot 25{,}000 = 2847.$$

Thus New York City is 2847 miles north of the equator.

Now let us calculate how far west New York City is of the point Q which has the same latitude and has longitude 0°. This point Q is actually the location of Morella, Spain, a small town about 200 miles east of Madrid. Since the longitude of New York City is 74°, angle VOW is 74°. But the distance we seek is not arc VW but arc PQ. Now arc PQ is on the circle of latitude through P. This circle has its center at O' on the straight line through NS, and its radius, $O'P$, is not the radius of the earth. If we could calculate $O'P$, then we could calculate the circumference of the circle of latitude, and since angle $PO'Q$ is also 74°, we could calculate arc PQ.

Our problem then reduces to finding $O'P$. We can find it. The radius $O'P$ is a side of the triangle $OO'P$. Moreover, OO' is perpendicular to $O'P$. Hence we have a right triangle. Since $O'P$ and OV are parallel, angle $O'PO$ equals the latitude of P because $O'PO$ and POV are alternate interior angles of parallel lines. Hence in triangle $O'PO$,

$$\cos 41° = \frac{O'P}{OP}$$

or

$$O'P = OP \cos 41°.$$

Now OP is the radius of the earth, or 4000 miles. From our table we find that cos 41° is 0.7547. Hence

$$O'P = 4000 \cdot 0.7547 = 3019.$$

We may now calculate arc PQ. This arc is 74/360 of the circumference of the circle whose radius is 3019 miles. Hence

$$PQ = \tfrac{74}{360} \cdot 2\pi \cdot 3019.$$

Using the approximate value of 3.14 for π, we find that

$$PQ = 3897.$$

Thus New York City is 3897 miles west of Morella, Spain.

We have computed the distance between two points on the same meridian of longitude and the distance between two points on the same circle of latitude. We could investigate how to calculate the distance between two points on the earth's surface which have neither the same longitude nor the same latitude. However, we have seen enough of the method to comprehend how the trigonometric ratios can be used. Only one point may be worthy of note here. Suppose that P and Q (Fig. 7–17) are two points on the surface of the earth, and we now consider the question, What is the distance between them? We cannot mean the straight-line distance between P and Q because this does not lie on the earth's surface. The distance along the surface of the earth from P to Q must then be an arc of a curve. Which curve shall we choose? If we choose a circle *whose center O is the center of the earth* and

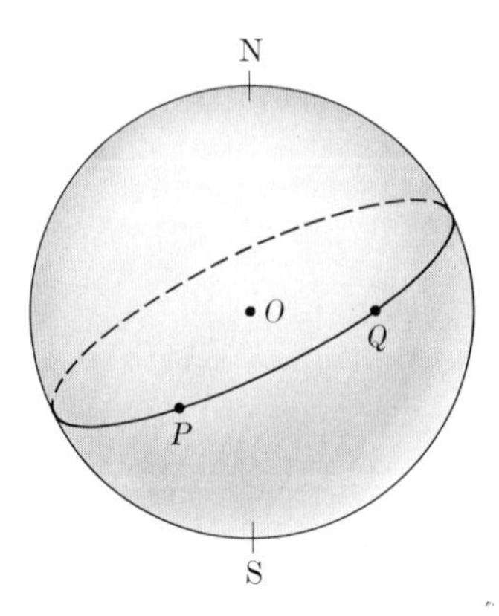

Fig. 7–17. A great circle on the earth's surface.

which passes through P and Q, then we shall have what is called a *great circle*. The shorter of the two arcs from P to Q along this great circle is the shortest distance from P to Q along the surface of the sphere. This theorem of spherical geometry, which we shall not prove, is noteworthy, because it tells us what route ships and planes should take if they are to save time and expense.

Let us consider this theorem in connection with travel by the shortest route between two points such as New York City and Morella, Spain, which are on the same circle of latitude. Although in this case, one wishes to reach a point due east or west (depending on the direction of the trip) of a given point, the circle of latitude is *not* the shortest route because it is not a great circle. We saw in fact that the circle of latitude has O' as its center (Fig. 7–16), whereas the center of the earth is O.

Determining the latitude and longitude of places on the earth and their distances apart is valuable not only for navigation but for map making. Both Hipparchus and Ptolemy made maps of the ancient world. Although we shall not describe their mathematical methods, we would like to call attention to the problem of making a map. A map is supposed to be a reproduction on flat paper of the relative locations of places on the earth. Now the earth is a sphere, and the one deficiency of this most prized figure of the Greeks is that one cannot take a sphere, cut it open, and lay it flat without creasing, folding, stretching, or tearing the material. One can see this readily if he peels an orange and then tries to flatten out the skin.

Since it is not possible to flatten a sphere without distorting it, any attempt to reproduce on flat paper the relationships that exist on the sphere must involve a distortion of areas, or the relative directions of one place from another, or distances. Hipparchus and Ptolemy therefore invented several methods of map making each of which has features useful for one or more purposes. Thus some methods preserve area, others direction, and still others project great circles into straight lines so that the shortest distance on the sphere between two points is represented by the shortest distance on the map. No map can be a true representation in all respects.

Exercises

1. To determine the latitude of a point P (Fig. 7–18) on the surface of the earth, an observer at P measures the angle between the horizontal at P and the direction of the North Star. He finds this angle to be 30°. What is the latitude of P?

2. As one travels north along a meridian from the South Pole to the North Pole, how does his latitude change?

3. Suppose that one travels west from some point on the 0° meridian. How does his longitude change?

4. Of the two circles of latitude, 30° north and 40° north, which has the larger radius?

5. If a man changes his latitude by 2° in traveling along a meridian (Fig. 7–19), how far does he travel?

6. Suppose a man travels due west along the 41° circle of latitude and changes his longitude by 5° (Fig. 7–20). How far does he travel?

7. In one day (24 hours) the earth rotates through 360°. Hence a person has in effect moved around in a complete circle. How far has a person traveled who is at 41° latitude?

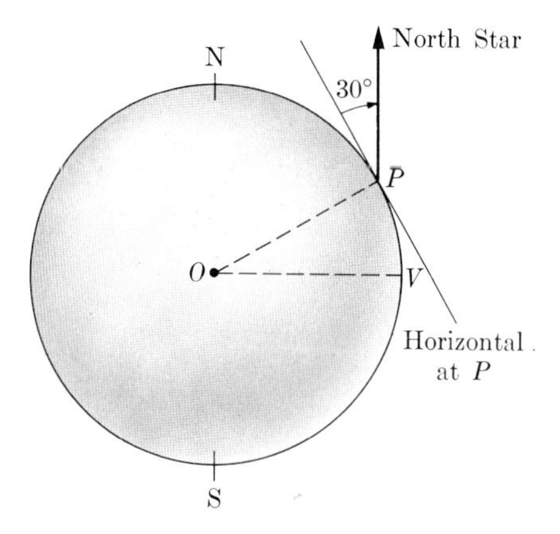

Figure 7–18

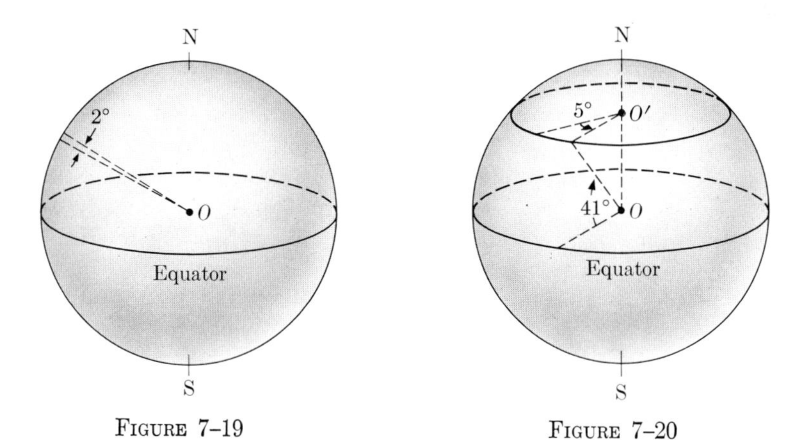

Figure 7–19

Figure 7–20

7–5 Charting the heavens. From the determination of the latitude and longitude of places on the earth's surface and of distances between places, Hipparchus and Ptolemy proceeded to the far more ambitious problem of calculating the sizes and distances of the heavenly bodies. The classical Greeks had indeed speculated about these sizes and distances, but since they relied far more upon aesthetically pleasing principles than upon keen observation, measurement of angles, and numerical calculation, their conclusions were often absurd.

The Alexandrian Greeks made the decisive step in quantitative astronomy. They were, as we have noted, more disposed to measure. Moreover many of them, including Hipparchus himself, had improved the astronomical instruments and the sundials and water clocks which helped to fix more accurately the time at which observations were made. Hipparchus and Ptolemy also had at their disposal in Alexandria a wealth of astronomical data which the Egyptians, Babylonians, and Alexandrians had compiled over many centuries. Let us see how these men "triangulated" the heavens. We shall not reproduce their exact procedures but merely show the essential principles.

We shall consider first how one can find the distance to the moon. Suppose that P and Q (Fig. 7–21) are two points on the earth's equator which are chosen to satisfy the following conditions: The moon is to be directly overhead at P; that is, the moon, M, regarded as a point, is to be on the line from the center of the earth, E, through P. The moon is in this position at certain times each month. The point Q is chosen such that the moon is just visible from it. This means that the moon is clearly visible from points closer to P but not visible from points farther away from P and along the equator. Another way of saying the same thing is that the line MQ is tangent to the equator at Q. Let us draw the line EQ. The angle EQM is a right angle because the radius of a circle drawn to the point of contact of a tangent is perpendicular to the tangent.

We now have a right triangle. Moreover, EQ is the radius of the earth and this is known. The angle at E is the difference in longitude between the points P and Q, and since the longitudes of places on the earth are known, so is angle E. A modern value for it is $89° \ 4'$, a value far more ac-

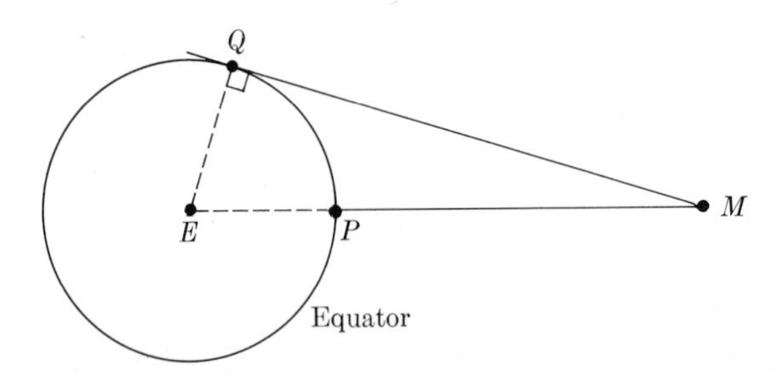

FIG. 7–21. Finding the distance to the moon.

curate than Hipparchus or Ptolemy could have obtained with their instruments. The calculation of EM is now child's play, for

$$\cos E = \frac{EQ}{EM}.$$

The value of $\cos E$ or $\cos 89° 4'$, taken from a larger trigonometric table than ours, is 0.0163. Moreover EQ is 4000 miles. Then

$$0.0163 = \frac{4000}{EM}.$$

If we multiply both sides of this equation by EM and then divide by 0.0163, we obtain

$$EM = \frac{4000}{0.0163} = 245{,}000.$$

Our data yield $EM = 245{,}000$ miles, and if we now subtract EP, the radius of the earth, we find that PM, the distance from the surface of the earth to the moon, is 241,000 miles. Hipparchus arrived at the figure of about 280,000 miles because his angular measure of E was not so accurate.*

Precisely the same method can be used to find the distance to the sun. The point M (Fig. 7-21) would now represent the sun. However, because the distances PM and QM are much larger in the case of the sun, angle E is larger and very close to 90°. Moreover, the angle must be measured very accurately because a small error in the angle will cause a large error in the value of PM. For this reason the result of Hipparchus and Ptolemy, of the order of millions of miles, was, as they realized, not very accurate (see Exercise 1).

Let us now find the radius of the moon. Whereas in the preceding calculation we regarded the moon as a point, this idealization will obviously not do in finding the radius. Instead let us regard the moon as a small sphere with center M and radius MR (Fig. 7-22). At a point E on the earth's surface, one measures the angle between the line EM which has the direction

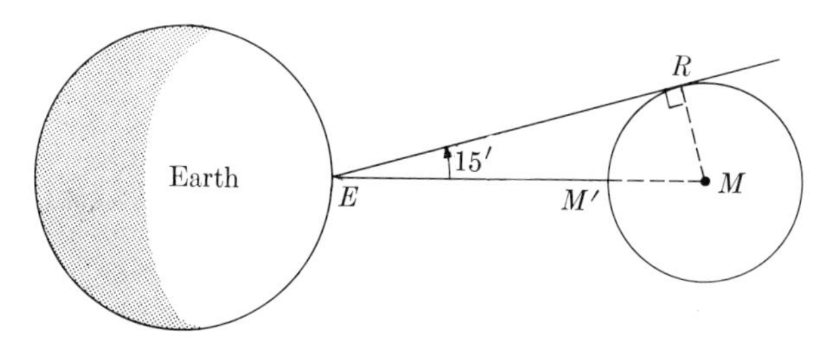

FIG. 7-22. Determining the radius of the moon.

* The distance of the earth from the moon varies over the year. The above value is about an average value.

from E to the center of the moon, and the line ER which is tangent to the moon's surface. This angle proves to be 15′. We know the distance from earth to moon, at least when the moon is idealized as a point. Let us use this distance, even though it is not exactly EM in our figure. We shall see that the error introduced is minor. Hence EM for us is 241,000 miles. We shall use again the Euclidean theorem that a radius of a circle drawn to the point of tangency of a tangent is perpendicular to the tangent. For our figure this theorem says that MR is perpendicular to ER. Then in the right triangle EMR we have

$$\sin E = \frac{MR}{EM}.$$

Now angle $E = 15'$ and $\sin 15'$, taken from a table giving sine values for angles in minutes, is 0.0044. Moreover EM is 241,000. Hence

$$0.0044 = \frac{MR}{241{,}000}.$$

Then

$$MR = 241{,}000 \cdot 0.0044 = 1060.$$

Thus the radius of the moon is 1060 miles. We can see now that the error introduced by using 241,000 miles as the distance to the center of the moon cannot be great because the radius of the moon is only 1060 miles. The distance of 241,000 miles is really the distance EM' since, in determining the distance to the moon, we could observe only the surface. (For a more accurate calculation of the moon's radius see Exercise 3.) It is of interest that Hipparchus obtained the result of 1,333 miles; his measurement of angle E was not as accurate as the modern one.

The method just used to find the radius of the moon can also be applied to find the radius of the sun. The point M in Fig. 7–22 becomes the center of the sun, and the distance EM becomes the distance to the sun (see Exercise 2). Angle E is about the same for this case as for the moon, as one might expect from the fact that when the moon is between the earth and the sun, the moon just about eclipses the sun.

We can find the distances to the moon and sun and the radii of the moon and sun by making measurements on the surface of the earth. But now suppose that we wish to calculate the distance from Venus to the sun. If we were to use the preceding methods we should have to make measurments on the surface of Venus. Of course, we all expect to be able to make the trip to Venus shortly, and can then make the measurements. In the meantime, to satisfy our curiosity, we shall employ a somewhat less direct method.

Let us regard all three bodies, the earth, the sun, and Venus, as points and let us suppose that the paths of earth and Venus are circular. At any time the three bodies are the vertices of triangle ESV' in Fig. 7–23(a). From the earth we can observe the size of angle E, which, of course, changes as the earth and Venus move around the sun.

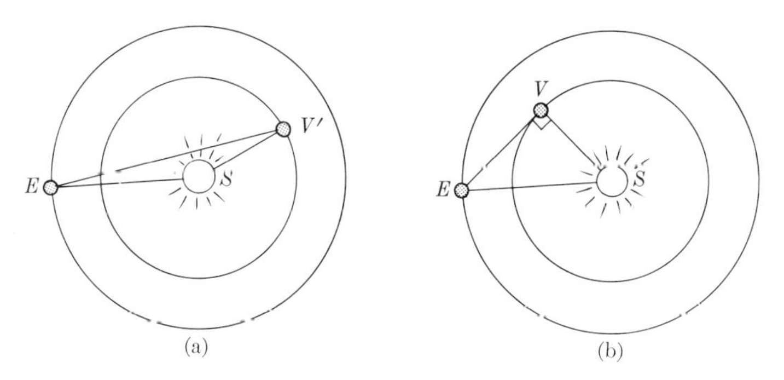

FIG. 7-23. Determining the distance of Venus from the sun.

A neat fact, which emerges from a study of Fig. 7-23(b), is that when angle E is a maximum, then the line from earth to Venus is tangent to the path of Venus around the sun. For when the angle at E is a maximum, the line from the earth to Venus is farthest from ES and still meets the circle on which Venus travels. But such a line must be tangent to the circle. A tangent to a circle is perpendicular to the radius drawn to the point of contact. Hence the radius SV (Fig. 7-23b) is perpendicular to EV. What we should do, then, is measure the angle E at various times of the year and find out when it is largest. At this time EV is perpendicular to SV.

Measurements show that the largest value of angle E is 47°. If in Fig. 7-23(b), we use 47° for E and the fact that angle V must then be a right angle, we have

$$\sin 47^\circ = \frac{SV}{ES}.$$

From our tables we find that $\sin 47^\circ = 0.7314$. The distance ES is 93,000,000 miles. Hence

$$0.7314 = \frac{SV}{93{,}000{,}000}.$$

Then

$$SV = 93{,}000{,}000 \cdot 0.7314 = 68{,}000{,}000.$$

Thus the distance from Venus to the sun is 68,000,000 miles.

We can begin to see from these examples how Hipparchus and Ptolemy gave mankind its first reasonable values for the dimensions of our solar system. The figures they produced were staggering to the Greeks because these people believed that our solar system and universe were far smaller.

The crowning achievement of Hipparchus and Ptolemy was the creation of a new astronomical theory which described the paths of the heavenly bodies and enabled man to predict their position. We shall consider their theory in the next chapter.

Exercises

1. Let us use the method given in the text to find the distance to the sun. We know that in Fig. 7–24, QE is the radius of the earth, or 4000 mi. The angle at E is the difference in longitude between P and Q and in our case is $89° 59' 51''$. Given that $\cos E = 0.000043$, find ES.

2. Let us apply the method of the text to find the radius of the sun (Fig. 7–25). The distance to the sun, ES, is 93,000,000 mi. Angle E is measured and found to be about $16'$. Given that $\sin 16' = 0.0046$, find the radius SR.

3. In the text (Fig. 7–22) we found the radius of the moon without considering the distance $M'M$. By a slight bit of extra work we can take this radius into account. Let us denote $M'M$, which equals RM, by r. Then, since EM' is 241,000 miles and angle E equals $15'$, we have

$$\sin 15' = \frac{r}{241{,}000 + r}.$$

Use the value of $\sin 15'$ given in the text to find r.

4. Use the method in the text to find the distance from Mercury to the sun. The relevant angle E in this case is $23°$.

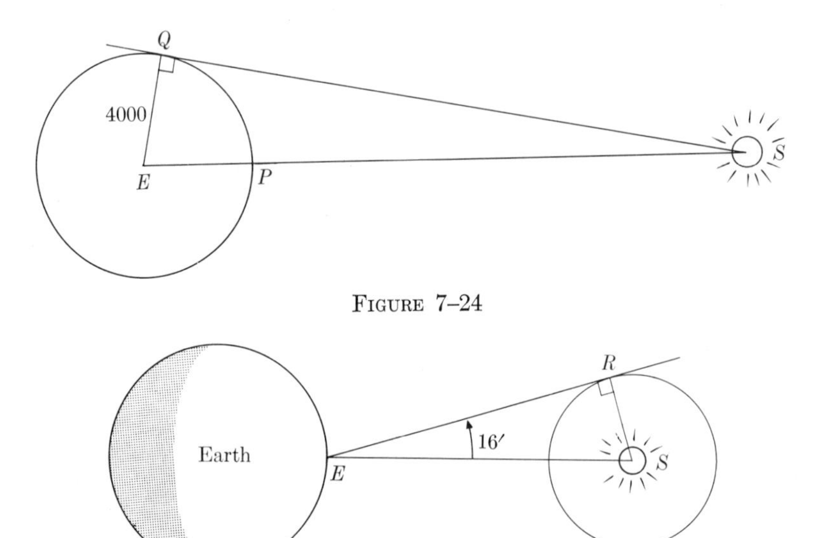

Figure 7–24

Figure 7–25

7–6 Further progress in the study of light. We saw in the preceding chapter that Euclid had already formulated one basic law for the behavior of light, namely the law of reflection. The Alexandrians undertook to study a second basic phenomenon of light, namely the change in the direction of light as it passes from one medium to another.

We often recognize that something strange does happen when light goes from air to water, say, because a straight rod when partially immersed in

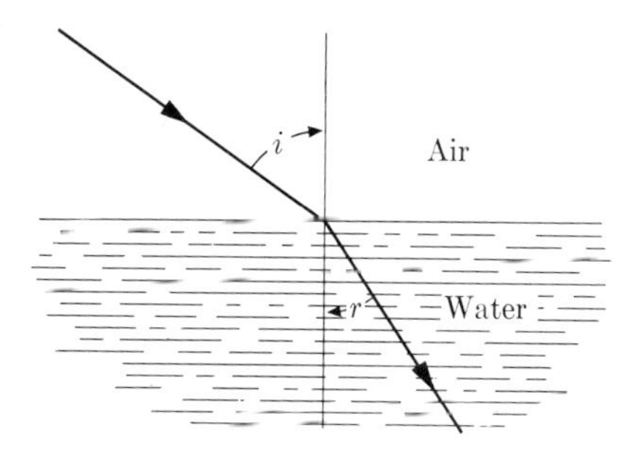

FIG. 7-26. Refraction of light.

water seems to bend sharply at the water level. Also if one should shine a flashlight beam into water, he would observe the sudden change in the direction of the beam as it enters the water. This bending of light is called refraction. The Alexandrians sought to determine the extent of this change in direction. Specifically, if i is the angle (Fig. 7-26) which the direction of the incident light ray makes with the perpendicular to the surface which separates the two media, say air and water, and if r is the angle which the refracted light makes with this same perpendicular, then what the Alexandrians sought is the relationship between angle i and angle r. But the Alexandrians and Ptolemy, in particular, who worked very hard on this problem, were baffled. They did observe that as i increased, r increased, but the increase did not occur in any simple manner. Moreover, the r which corresponds to a given i is not the same for any two different media. Thus, if the first medium should be air, then for the same i, the value of r for glass would be different from that for water.

Ptolemy did not succeed in arriving at the correct law, but he developed the mathematical tool which finally enabled the Dutchman Willebrord Snell and the Frenchman René Descartes to discover and express it. It was found in the seventeenth century that light travels with a finite velocity and that this velocity is different in different media. Let us suppose that we have two media bordering each other as shown in Fig. 7-26, and let v_1 be the velocity of light in the upper medium and v_2 the velocity in the lower medium. Then Snell and Descartes demonstrated by arguments we shall not reproduce that

$$\frac{\sin i}{\sin r} = \frac{v_1}{v_2}. \tag{1}$$

Thus it is the sine ratio which proves to be the key to this phenomenon of light. We see here, as we shall see many times later, how—as more mathematical ideas are placed at our disposal—we can take hold of more natural phenomena.

To familiarize ourselves with the law of refraction, we shall consider a concrete example. Let us suppose that the two media in question are air and water. The ratio of v_1 to v_2 in this case is 4 to 3. If we are then given a

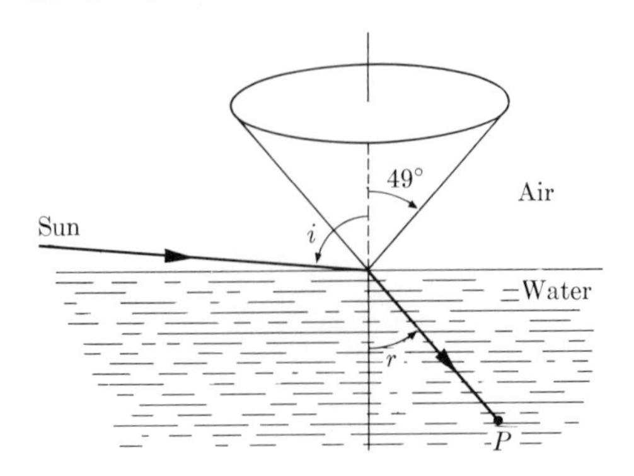

FIG. 7–27. The fish-eye view of the world.

value of i, say 30°, we can find r. Since $\sin 30° = \frac{1}{2}$, it follows from (1) that

$$\frac{1/2}{\sin r} = \frac{4}{3}.$$

Therefore

$$\frac{1}{2} = \frac{4}{3} \sin r.$$

If we multiply both sides of this equation by $\frac{3}{4}$, we obtain

$$\sin r = \frac{3}{8} = 0.3750.$$

We have now but to find the angle whose sine value is 0.3750. From our table we see that $r = 22°$ to the nearest degree. Hence, as the light enters the water, the angle between its direction and the perpendicular will change from 30° to 22°.

We now know how light refracts. Can we put this knowledge to use? Let us assume that the sun is close to the horizon. Of course, light from the sun streams out in all directions, but some rays will travel horizontally. Suppose that the surface over which the light travels is the surface of a large body of water (Fig. 7–27). Then some rays will enter the water at a very large angle of incidence i, in fact, so close to 90° that we shall consider angle i to be 90°. The question we shall discuss is, How large is angle r in this case? To obtain an answer, we follow the procedure described in the preceding paragraph. This time, angle i is 90°, and sin 90° is 1. If we substitute this value in formula (1), we have

$$\frac{1}{\sin r} = \frac{4}{3},$$

or

$$\sin r = \frac{3}{4}.$$

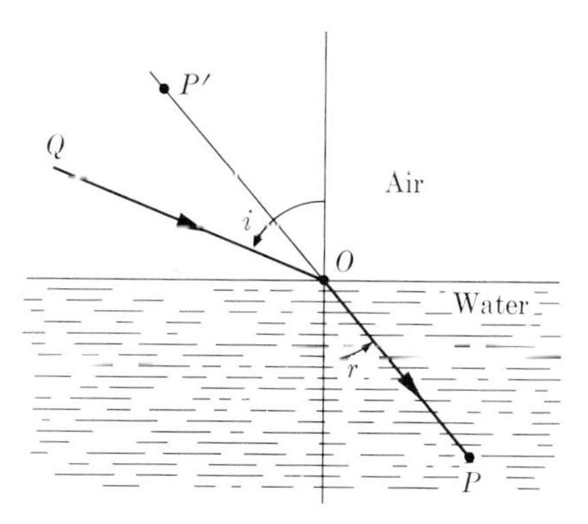

FIGURE 7–28

Reference to the table shows that $r = 49°$. Here the change in direction is considerable. Before we draw any further conclusions, let us note that 90° is the largest angle of incidence possible. If angle i should be less than 90°, angle r will be less than 49°. Hence any light entering the water will take a direction which makes an angle between 0° and 49° with the perpendicular.

Now suppose that a person located at the point P (Fig. 7–27) in the water sees the light ray OP coming toward him. Since the light travels along the direction OP, he will conclude that the sun is located above the water in the direction PO. Moreover, if light from any source enters the water at an angle of incidence less than 90°, the angle of refraction will be even less than 49°. In such cases a person in the water observing this light will conclude that the light comes from a source whose angle of incidence is less than 49°. The point of this discussion is that a person at P will believe that all objects in the air are situated within a 49°-angle from the perpendicular because the light from them will seem to come from a direction within this range. The region in the air extending in all directions within 49° of the perpendicular at O is the interior of a cone. Hence to a person in the water all objects seem to lie within this cone. We have of course presumed that the person in the water does not know the refractive effect of light or that he at least does not know how much light is bent under refraction. It is fairly certain that fish do not know mathematics, and so the inference that all objects above water must be within the 49°-cone around the perpendicular is called the fisheye view of the world.

To be a little clearer about the possible error into which one may be led, suppose that light comes to a person in a submarine at P (Fig. 7–28) and that the direction of the light is OP. The object which emits the light will then appear to lie along the line PO. If, to hit the object, he shoots a bullet in the direction PO, the bullet will enter the air and follow the direction POP'. But the object at which he believes to be shooting is located along the direction OQ.

Let us now reverse the roles of air and water and let us suppose that the light originates in the water. Assume, in fact, that a beam of light is shot

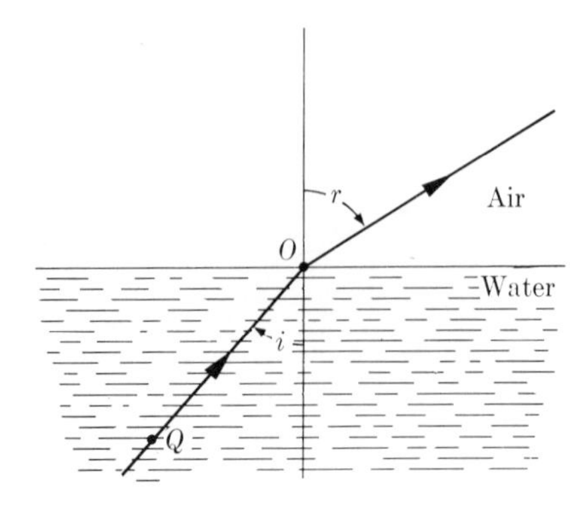

FIG. 7–29. Refraction from water into air.

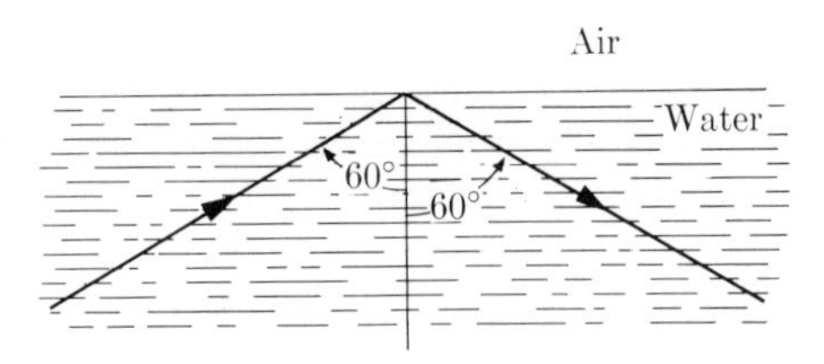

FIG. 7–30. Total reflection.

in the direction of QO of Fig. 7–29. The angle of incidence is now the angle marked i in the figure. The angle of refraction is the angle r. Since the ratio of the two velocities of light, that is v_1/v_2, is now $3/4$, the law of refraction (1) becomes

$$\frac{\sin i}{\sin r} = \frac{3}{4}$$

or

$$\sin r = \frac{4}{3} \sin i. \tag{2}$$

We see that $\sin r$ is greater than $\sin i$ and that therefore r must be greater than i. This is as it should be because light in going from water to air will bend *away* from the perpendicular.

Let us now suppose that angle i is greater than 49°, say 60°, and let us seek to determine the corresponding angle of refraction, r. Then, since $\sin 60° = 0.8660$, we find by (2) that

$$\sin r = \tfrac{4}{3}(0.8660) = 1.155.$$

We see then that $\sin r$ is greater than 1. Unfortunately, there is no angle whose sine value is greater than 1, and so there is no angle of refraction. Thus mathematics predicts that the light cannot leave the water. Is this the case

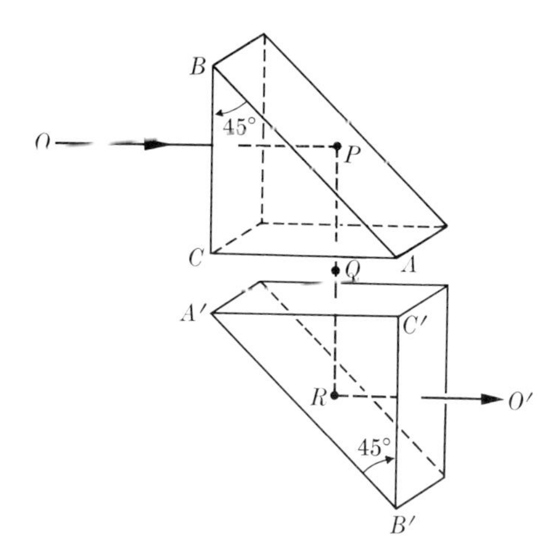

FIG. 7-31. Parallel displacement of light by means of total reflection in two prisms.

physically? Well, no decently behaving light ray would wish to disobey the mathematical law of refraction. And none does. The light remains in the water. But what does it do? The answer is that the light is *reflected* from the boundary between air and water. Since the light must return to the water, it may as well do what it has already learned how to do under the process of reflection, namely, be reflected at an angle equal to the angle of incidence which in the present case is 60° (Fig. 7-30).

Thus we have discovered that if light seeks to pass from one medium to a second one in which the velocity is greater, then for all angles larger than a certain angle (49° for water and air), the light is not refracted but reflected. This particular angle, the largest for which refraction is still possible, is called the *critical angle*, and the phenomenon that for all greater angles of incidence the light is reflected is called *total reflection*.

This phenomenon is indeed a surprising one. It means that a surface, such as the surface of the water in the above example, serves as a mirror for some angles of incidence. Now mirrors are very useful devices. Generally they are made by silvering the back of a glass plate. Would there be any use for the phenomenon of total reflection in view of the fact that here too we encounter a reflecting surface? As a matter of fact, the phenomenon is put to use in a number of familiar instruments.

Let us consider the following situation (Fig. 7-31), where ACB and $A'C'B'$ are two prism-shaped pieces of glass with the faces AC and $A'C'$ parallel to each other. Both prisms are shaped as isosceles right triangles. Suppose that OP is a light ray which first strikes the face BC perpendicularly. Here the angle of incidence is 0°. Hence the angle of refraction is also 0°, and the light therefore goes through unchanged in direction. The light ray strikes the face BA at an angle of 45°. Now if the prisms are made of flint glass, the critical angle is 37°. Thus the light ray strikes the face BA at an angle of incidence greater than the critical angle. In accordance with the phenomenon of total reflection, the light is reflected at an angle of 45° to AB and follows the direction PQ. The light strikes the faces AC and $A'C'$ at an angle of incidence of 0° and so goes right through unchanged. It then strikes the face $A'B'$ at an angle of 45°. Since this angle of incidence also is greater than the critical angle, the light is again totally reflected at an angle of 45° with $B'A'$ and takes the direction RO'. Thus the final ray, RO', has the same direction as the original ray, OP, but is displaced by the distance PR.

We might well ask, Does this combination of prisms have any practical value? One application is the periscope. The two prisms are at opposite ends of a long vertical tube. Now OP is the light received above water and RO' is the light received below. One could very well use two silvered mirrors at BA and $A'B'$ and obtain the same result. But silvered mirrors tarnish with age and lose their effectiveness. Moreover, well-made glass prisms reflect almost all the light that falls on a face such as BA, whereas a silvered mirror reflects only about 70% of the incident light; the rest is absorbed or scattered in all directions. Hence the prism not only outlasts the silvered mirror but is much more efficient.

Another application of the above combination of two prisms is made in binoculars. The two tubes which first receive the light are deliberately placed rather far apart so that the field of vision is large. But the eye pieces of the binoculars cannot be farther apart than the distance between a person's eyes. In each half of a binocular, the incident light is displaced as OP is displaced to RO'. Then the two incoming rays, one in each of the main tubes, can be far apart, whereas the two emerging rays are no farther apart than the eyes of a person.

Another use of total reflection is made in cutting diamonds. The critical angle for this substance is 24°. Hence light, having entered the piece of crystal, will be totally reflected if it strikes any face at an angle of incidence greater than 24°. The diamond is cut in such a way that the light received will strike many faces at angles greater than the critical angle and so be shuttled back and forth inside the diamond many times. The result is that the light leaves the diamond in all directions and the eye gets the effect of brilliance in the diamond itself. A similar process takes place in leaded glass, and hence this type is preferred for windows which are intended to give an effect of brilliance.

Total reflection is but one phenomenon of the refractive effect of light. The most common use of the refractive effect of light is in lenses. If light streams out from an object at P (Fig. 7–32) in all directions, some of the rays will strike the lens at points such as Q_1, Q_2, and Q_3. There their directions will change because they are entering glass. Thus the rays PQ_1, PQ_2, and PQ_3 may take the directions Q_1R_1, Q_2R_2, and Q_3R_3, respectively. At the right-hand surface of the glass, the rays re-enter the air and, since the medium in which the light is traveling changes, the light rays bend again. By properly shaping the lens surfaces, that is $Q_1Q_2Q_3$ on the left and $R_1R_2R_3$ on the right, the light from P may be made to concentrate at S. All

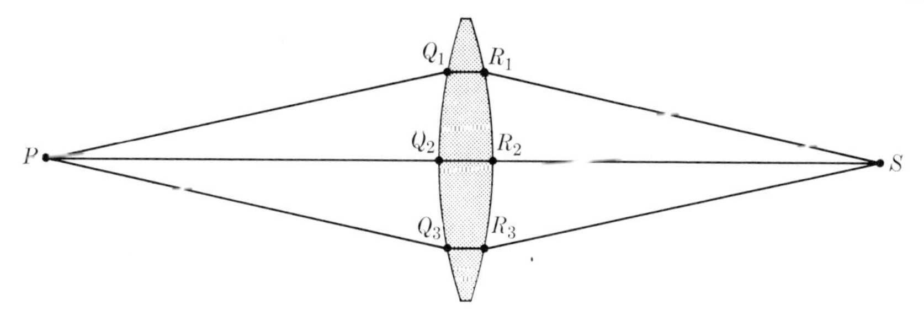

FIG. 7-32. Refraction by a lens.

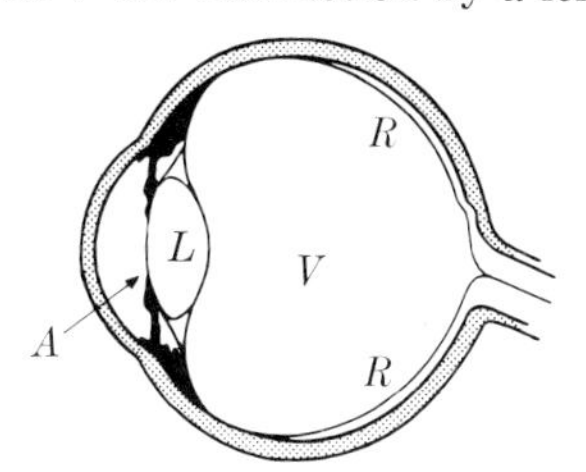

FIG. 7-33. The eye.

optical instruments, such as telescopes, microscopes, binoculars, and cameras, contain lenses of this kind.

The eye itself is a complicated refracting device. When light enters the eye (Fig. 7-33), it passes through a liquid (denoted by A in the figure), called the aqueous humor, then through the lens, L, which is made of a fibrous jelly, and finally it enters another liquid, V, called the vitreous humor. Although all three media have some refractive effect upon the light, most of the refraction occurs when it encounters the aqueous humor. To be perceived, light rays that enter the eye must strike the retina, R, in the rear. The eye has a ciliary muscle which changes the shape of the lens and therefore the direction of the light rays passing through the eye so that the rays are directed toward the retina. Eyes which for one reason or another cannot direct the rays to the retina must be aided by additional lenses in eyeglasses. Clearly the science of medicine profits immensely from the mathematical and physical knowledge acquired about the action of the eye.

In the camera, the lens or lenses are fixed in shape. The film acts as does the retina in the eye. Since the shapes of the lenses are fixed, the distances of the lenses from the film can be varied to enable the refracted light to reach the proper places on the film.

We have been discussing the law of refraction and some of the remarkable effects which take place at a sharp boundary between the two media. But the refractive effect of light is equally striking and important when there is a gradual change in the nature of the medium through which the light passes. Let us consider the passage of light through air, which is not a uniform medium. Generally it is more dense near the ground and thinner at higher altitudes. Hence, when light comes to a person at P (Fig. 1–1) from the sun at O, the light ray follows a curved path as it travels through the earth's atmosphere because it is continually refracted. For the observer at P the direction of the incoming light is $O'P$, and hence he thinks that the source lies along the direction PO'. This is the reason that we are often deceived about the true position of the sun (see Chapter 1).

The refractive effect of light is, as we can see, a peculiar phenomenon. Why does light behave this way? We do not understand what light is and so cannot analyze the substance itself to learn why it refracts, but we have another kind of explanation which sheds light on nature's operations. The clue lies in the law of refraction. We note that refraction depends upon the velocity of light in the medium. The seventeenth-century mathematician Pierre de Fermat, whom we shall meet again, pondered on this fact and, after analyzing the law of refraction, found an important principle. Suppose that light travels from the point P in air (Fig. 7–34) to the point Q in water and bends at O in accordance with the law of refraction. Were the light to follow the straight-line path from P to Q instead of the broken-line path POQ, it would travel a shorter distance. Let us note, however, that the distance $O'Q$ in *water* would be longer than OQ. Because the velocity in water is smaller than in air, the light might lose more time in traveling the path $O'Q$ instead of OQ than it might save by traveling the shorter distance PO' instead of PO in air. By a mathematical argument Fermat showed that light takes the path which requires *least time.*

But is this fact true for other phenomena of light? When light travels from one point to another in a uniform medium, it takes the straight-line path. It would seem as though in this case light chooses the criterion of shortest path and not that of least time. But in a uniform medium the velocity of light is constant, and so the shortest path requires the least time. Let us consider next what happens when light goes from a point P to a mirror and then to a point Q. We proved in Chapter 6 that light takes the shortest path. But here too the light travels in one medium and, because the

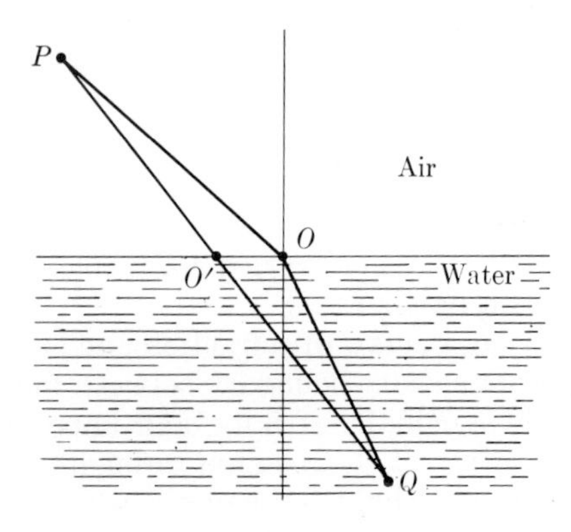

FIG. 7–34. Light takes the path requiring least time.

medium is uniform, the velocity is constant; hence the shortest path again means least time. It would appear from Fermat's analysis that nature is wise. It knows mathematics and employs it in the interest of economy.

We have gotten a little ahead of our story by presenting the mathematical law of refraction and Fermat's analysis of the deeper implications of this law. The Alexandrian Greeks had grappled with the phenomenon of refraction and, as we noted earlier, supplied the key in the concept of the trigonometric ratios, but did not attain the law itself or see its meaning in terms of least time. But by providing this tool and by charting the earth and heavens, the Alexandrians extended enormously man's mathematical understanding of the physical world. The power of mathematics to describe and analyze nature's ways was advanced well beyond the stage at which Euclid and Apollonius had left it. The crowning achievement of the Alexandrians is yet to be related.

Exercises

1. Given that the ratio of the velocity of light in air to that in water is 4 to 3 and that the angle of incidence of a light ray originating in the air and striking the surface of the water is 45°. What is the angle of refraction?

2. Suppose a light ray traveling in glass strikes the boundary of the glass and seeks to enter the air beyond the boundary. The velocity of light in the glass is two thirds of its velocity in air. What angles of incidence can the light ray have and still penetrate into the air?

3. Prove that a light ray passing through a plate of glass (Fig. 7–35) emerges parallel to its original direction but is somewhat displaced.

4. Suppose that one measures the angle of incidence, i, and the angle of refraction, r, for a light ray passing from air into a plate of glass, and assume that angle i proves to be 50° and angle r, 45°. The velocity of light in air is 186,000 mi/sec. What is the velocity of light in the glass?

5. What is the mathematical theme of this chapter?

6. Is it correct to say that the trigonometry of the Alexandrian Greeks is an extension of Euclidean geometry?

7. Contrast the classical and the Alexandrian Greek activities in mathematics.

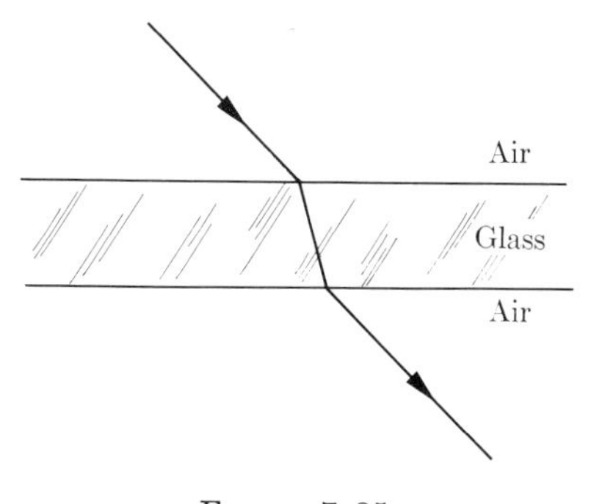

Figure 7–35

Topics for Further Investigation

1. The mathematics of lenses. Use the references to Taylor, or to Sears and Zemansky, or look up any elementary physics book.
2. The mathematics of map making. Use the references to Brown, Raisz, Deetz, or Chamberlin.
3. The history of mathematics during the Alexandrian period. Use the references to Smith and Ball.

Recommended Reading

Ball, W. W. Rouse: *A Short Account of the History of Mathematics,* 4th ed., Chaps. 4 and 5, Dover Publications, Inc., New York, 1960.

Brown, Lloyd A.: *The Story of Maps,* Little Brown and Co., Boston, 1944.

Chamberlin, Wellman: *The Round Earth on Flat Paper,* National Geographic Society, Washington, D.C., 1947.

Deetz, Charles H. and Oscar S. Adams: *Elements of Map Projection,* pp. 1–52. U.S. Department of Commerce, Special Publication No. 68, 1938.

Miller, Denning: *Popular Mathematics,* pp. 295–351, Coward-McCann Inc., New York, 1942.

Parsons, Edward A.: *The Alexandrian Library,* The Elsevier Press, Amsterdam, 1952.

Raisz, E.: *General Cartography,* McGraw-Hill Book Co., New York, 1948.

Sawyer, W. W.: *Mathematician's Delight,* Chap. 13, Penguin Books, Harmondsworth, England, 1943.

Sears, Francis W. and Mark Zemansky: *University Physics,* 3rd ed., Chaps. 39–43, Addison-Wesley Publishing Co., Inc., Reading, Mass., 1959.

Smith, David E.: *History of Mathematics,* Vol. I, Chap. 4, Dover Publications, Inc., New York, 1958.

Taylor, Lloyd W.: *Physics, The Pioneer Science,* pp. 442–470, Dover Publications, Inc., New York, 1959.

CHAPTER 8

THE MATHEMATICAL ORDER OF NATURE

Great men! elevated above the common standard of human nature, by discovering the laws which celestial occurrences obey, and by freeing the wretched mind of man from the fears which the eclipses inspired.

Pliny

8–1 The Greek concept of nature. The Greeks, as we now know, molded the nature of mathematics, constructed Euclidean geometry and trigonometry, and applied their theoretical results to objects in space, to the behavior of light, to mapping the earth, and to determining the sizes and distances of heavenly bodies. But these extensive and magnificent achievements within mathematics proper and in its applications do not exhibit the full greatness of the Greek genius, and are indeed dwarfed by the Greeks' grand conception of the universe itself.

Possessed with insatiable curiosity and courage, they asked and answered the questions which occur to many, are tackled by few, and are resolved only by individuals of the highest intellectual caliber. Is there any plan underlying the workings of the entire universe? Are planets, men, animals, plants, light, and sound merely physical accidents or are they part of a grand plan? Because they were dreamers enough to arrive at new points of view, the Greeks fashioned a conception of the universe which has dominated all subsequent Western thought. They affirmed that nature is rationally and indeed mathematically designed. All phenomena apparent to the senses, from the motions of planets in the heavens to the stirrings of leaves on a tree, can be fitted into a precise, coherent, intelligible pattern. The Greeks were the first people with the audacity to conceive of such law and order in the welter of phenomena and the first with the genius to uncover a pattern to which nature conforms. They dared to ask for and they found a design underlying the greatest spectacle man beholds, the motion of the brilliant sun, the changing shapes of the many-hued moon, the piercing shafts of the planets, the broad panorama of lights from the canopy of stars, and the seemingly miraculous eclipses of the sun and moon.

8–2 Pre-Greek and early Greek views of nature. To appreciate the originality and boldness of the steps which the Greeks took in this direction, one must compare their attitude with what preceded. To all pre-Greek civilizations and later ones which lay beyond the Greek pale, nature appeared arbitrary, capricious, mysterious, and even terrifying. The ancient Egyptians and Babylonians did note the periodic motions of the sun and moon. But the motions of the planets made no sense at all. These bodies moved with varying speeds at different times of the year; at times they stood still;

and often they reversed their courses. They appeared and disappeared. The few regularities which were observed in these motions were beclouded by the many irregularities.

If these two ancient peoples had any expectation at all that the universe would continue to function in the future as it had in the past, it was because they believed that sun, moon, and planets were gods who would most likely behave in a gentlemanly and beneficent manner. In the complex actions of nature, they saw no glimpse of plan, order, or law. They scarcely dreamed of design and certainly conceived no embracing theories.

Lacking any clear conception of the motions of heavenly bodies, these societies formed what are to us ludicrous ideas about their structure and function. A typical ancient Babylonian view regarded the earth as a mountain sprung up from the depths of the water and still surrounded by water. Heaven was a solid vault surrounding the earth and resting on the water. The gods resided above this vault of heaven. Each day the sun which, like the moon and stars, was a living deity, came out through a door in the dwelling of the gods, shone throughout the day, and then retired through another door at night. The earth itself was hollow and divided into zones, one below the other. In one of these zones lay the dead. Some of these old Babylonian views appear in the Hebrew and Christian Bibles. For example, the account of creation in Genesis, which defies critical interpretation, says: "Let there be a firmament in the midst of the waters and let it divide the waters from the waters." The Biblical account then affirms that the sun and moon are in this firmament.

Even the Greeks of about 1000 B.C. accepted such fanciful accounts of the universe, accounts which are found in Homer and Hesiod. There were many gods, each of whom played some role in the creation and maintenance of the universe. Indeed the names Jupiter, Saturn, Venus, Mercury, and Mars are merely the Roman names for the Greek gods, and the Greek names, such as Aphrodite for Venus and Hermes for Mercury, were replacements for Babylonian names. These gods not only determined but even intervened in the affairs of man.

The belief common to Babylonians, Egyptians, and early Greeks that the heavenly bodies were gods who controlled events on the earth was reinforced by a number of observations. That the crops grew under the influence of the sun and that animals mated and rivers overflowed at special times of the year readily identified with particular positions of the sun and planets seemed to show clearly the superhuman influence of the heavenly bodies. It was the formalization and codification of these observances which led to the science of astrology. To know the future one had to know the courses of planets and stars. This pseudoscience was widely accepted long after the belief that heavenly bodies were gods disappeared, and it had, in fact, a strong hold on medieval Europe.

Rather suddenly, or so at least our knowledge of history indicates, rational accounts of the structure of the universe and of the motions of heavenly bodies appeared in the Greek city of Miletus located in Ionia, a region of

Asia Minor. There is the theory that the Miletans, far from home and therefore free of the tyranny of beliefs which a society imposes on its members and yet repelled by the strange doctrines they encountered among the peoples of the Near East, were propelled into thinking for themselves. Certainly from 600 B.C. onward rational views dominate the picture. These Greeks and their successors were the first to reveal the passionate desire for knowledge, the love of reason, and the conviction that nature not only is rational but that an examination of nature's ways would reveal the order inherent in the physical world. The new thesis is proclaimed by the Ionian Anaxagoras: "Reason rules the world."

The early rational theories are crude from a modern standpoint, but the new outlook is evident. Thus Thales (*ca.* 640–*ca.* 562 B.C.), who must have observed that water nourishes plants and human beings and that heat is usually accompanied by moisture, argued that everything is ultimately water and that mist and earth are forms of water. The universe he believed to be a mass of water containing a bubble. At the bottom of this bubble is a flat disc, the earth, which floats on the water. The rains come from the top of the bubble on which the heavenly bodies float, each composed of water in an incandescent state. Thales also observed that eclipses of the sun occur because the moon comes between the earth and the sun and that the moon receives its light from the sun. What is especially striking about Thales' constructions is that he does not maintain that they describe literally what exists. He believes they are of value because they provide a rational pattern.

Another philosopher, Anaxagoras (500–428 B.C.), taught a new theory of creation, according to which the heavenly bodies were hurled out into space by rotation of the original cosmos. As to the individual bodies, he maintained that the sun is a mass of red hot iron and that the moon is made of the same material as the earth. Some Greek writers credit Anaxagoras with being the first to point out that the moon does not give off its own light, but merely reflects the light of the sun. On this basis, Anaxagoras could show that the phases of the moon, that is, the variation in the portion that is lit up during one month, are due to the changing position of the moon relative to the earth and sun. He also could proclaim that an eclipse of the moon results merely from its passing through the earth's shadow and that an eclipse of the sun occurs when the moon passes between earth and sun. He believed that there were other planetary systems and people on them. That such doctrines were radical may be inferred from the fact that Anaxagoras was accused of, and temporarily imprisoned for, "frittering away the deities into unnecessary causes and blind forces."

Dozens of such theories were constructed by other Greeks of the sixth and fifth centuries B.C. To us these theories seem childish and naive and, indeed, compared to modern scientific theories, they are. These men, eager to build an entire world picture in one swoop, relied too much upon brilliant intuitions. But the supreme merit of their thinking is not so much what they said as their approach. They evidently made some observations and then proceeded to give material and objective explanations which appealed to reason instead of resorting to pure fancy or to gods, animals, spirits, and other mythical agents invoked by their predecessors.

8–3 The Pythagorean and Platonic views of nature. The decisive step leading to the construction of precise and verifiable scientific theories in place of vague and largely speculative accounts was the involvement of mathematics. This step was made by the Pythagoreans. We have already noted the prepossession of these people with the concept of number, though admixed with mystical and religious doctrines. In their philosophy of nature the Pythagoreans began with the principle that number is the essence of all substance. Unlimited space furnishes the material for particular forms of matter. But to the Pythagoreans any form was a pattern of discrete points arranged, as small pebbles might be, to build up the form. Hence the forms reduced to numbers. Since number is the essence of any object, the explanation of natural phenomena could be achieved only through number.

Among numbers the Pythagoreans regarded ten as ideal. They concluded that there must be ten moving bodies in the heavens. The Greeks knew and recognized as moving bodies Mercury, Venus, Mars, Jupiter, Saturn, the sun, moon, and the sphere of stars. The Pythagoreans declared that the earth also moves around a fixed central fire which one does not see because the inhabited side of the earth faces away from this fire. Since only nine moving bodies were thus accounted for, the Pythagoreans asserted that there must be a tenth one, called Antikhthon, or counter-earth, which also revolved around the central fire. However, this counter-earth was always on the far side of the central fire, and so one did not see it either. The earth revolved around the central fire once in 24 hours, the moon once in $29\frac{1}{2}$ days, and the sun once a year. The light of the sun was a reflection of light from the central fire. The Pythagoreans, it will be noted, believed the earth to be in motion, and were the first to declare that the earth and other heavenly bodies were spheres and moved in perfect circles. It is true that aesthetic preferences rather than scientific arguments led them to these conclusions, but they were not daunted by the boldness of their assertions.

The Pythagoreans are noted for having observed numerical relationships in musical sounds. The sounds of two equally taut, plucked strings harmonize only when their lengths have special ratios. One such harmonious combination arises when the relative lengths are 2 to 1, and in this case the musical interval between the two sounds is called an octave. Another is the ratio 3 to 2, and the corresponding musical interval is called the fifth. Still another is the ratio 4 to 3, and the musical interval is called the fourth. These intervals were pleasing, said the Pythagoreans, because the numbers 1, 2, 3, and 4 were involved and because these added up to 10. The Pythagoreans constructed musical scales, each a succession of notes separated by musical intervals which would permit the various chords composed of these notes to be harmonious sounds.

To the Pythagoreans the concept of harmony implied far more than agreeable combinations of sounds. It meant the existence of pleasing number relationships in any natural phenomena. Thus they believed that the moving heavenly bodies gave off sounds, and that the pitches of these sounds increased with the speed of the motion. They also believed that the planets farther from the central fire moved more rapidly, and so the sounds of the farther planets were higher pitched than those emitted by the nearer planets. However, all these sounds harmonized because the distances and speeds possessed the proper number relationships; that is, because the number relationships were pleasing, the "music of the spheres" was agreeable.

The natural philosophy of the Pythagoreans is hardly very substantial. Aesthetic principles commingled with an obsession to find number relationships certainly led to assertions transcending observational evidence. Nor did the Pythagoreans develop any one branch of physical science very far. One can justifiably call their theories superficial. But whether by a lucky stroke or by intuitive genius the Pythagoreans did hit upon two doctrines which later proved to be all important. The first is that nature is built in accordance with mathematical principles, and the second that number relationships reveal the order in nature. They underlie and unify the seeming diversity exhibited by nature. The Pythagoreans said in fact that numbers and number relationships are the essence of nature. This statement will assume deeper meaning when we get to modern times.

Perhaps because mathematics developed considerably in the intervening century, the principle that nature is mathematically designed emerged more sharply and was applied more substantially in Plato's time. Plato was indeed a Pythagorean but a master in his own right who influenced Greek thought in a most important century, the fourth century B.C. He was the founder of an academy in Athens, a university which attracted the leading thinkers of his day and which, in fact, endured for nine hundred years.

Plato's own doctrines were extreme. Reality to him was not to be found in the physical world but in a system of ideas and in an ideal plan of the universe which God himself had created and contemplates. The visible and sensible world is just a vague, dim, and imperfect realization of these ideas. Moreover, the ideas were perfect and eternal, whereas the physical world is imperfect and decays. One might say that, unlike the Pythagoreans, Plato did not wish to comprehend the physical world through mathematics but aimed at understanding the mathematical plan itself which observation of the physical world suggested very imperfectly.

There is a passage in Plato's dialogue, the *Timaeus,* which illustrates very clearly Plato's natural philosophy. He had brought his pupil and hearer, Glaucon, to the point of comprehending something of the nature of science, and Glaucon is now desirous to show that he has profited by Plato's teaching. The subject of astronomy arises and Glaucon then says, "Astronomy is one of the sciences which you require, because it makes men's minds look upwards, and study things above." "Well," says Socrates (who speaks for Plato), "perhaps any one can see it—except me—I cannot see it . . . Your notice of the 'study of things above' is certainly a very magnificent one. You seem to think that if a man bends his head back and looks at the ceiling he 'looks upwards' with his mind as well as his eyes. You may be right and I may be wrong: but I have no notion of any science which makes the mind look upwards, except a science which is about the permanent and invisible . . . If a man merely looks up and stares at sensible objects, his mind does not look upwards, even if he were to pursue his studies swimming on his back in the sea."

Plato then describes the real science of astronomy. The visible figures in the heavens are far inferior to the true objects, namely those objects that are to be apprehended by reason and mental conceptions, not by vision. The varied configurations which the sky presents to the eye are to be used only as diagrams to assist in the study of higher truths. We must treat astronomy, like geometry, as a series of problems suggested by visible things. True astronomy deals with the laws of motion of true stars in a mathematical heaven of which the visible heaven is but an imperfect expression. True astronomy must leave the actual heavens alone. It is clear, incidentally, that Plato, like the classical Greeks in general, was indifferent to the practical problems of navigation, calendar reckoning, and the measurement of time.

Although the planets, at least as seen from the earth, do not appear to follow any regular course (the word "planet" means in fact "wanderer," and the planets were referred to as the vagabonds of the sky), Plato was sure—because "God eternally geometrizes"—that there was a mathematical pattern underlying and governing the motions of all heavenly bodies. Plato's own attempts to find such a plan were crude, largely because he would not devote himself to a careful study of the actual motions. But he did pose to his colleagues and students the problem of devising a mathematical scheme that would call for regular motions and yet account for the irregular motions we see, the problem he described as "saving the appearances."

8–4 Greek astronomical theories. One of Plato's pupils, Eudoxus (408–355 B.C.), who later became one of the most famous of Greek mathematicians, did take on this problem and, by creating the first major astronomical theory known to history, made one of the great and ingenious contributions to the demonstration of the mathematical design of nature. We shall not present the details of his theory. It was constructed before Hipparchus and Ptolemy calculated the sizes and distances of the heavenly bodies, and so Eudoxus did not have the data on which to build an accurate system. The defects in the theory were soon recognized.

The problem of finding the design of planetary motions continued to engage the minds of the Greeks, possibly because they were not distracted by the "heavenly" stars of stage, screen, and radio with whom many modern minds seem to be preoccupied. One of the solutions advanced but rejected is worthy of mention. Aristarchus, who lived about 270 B.C. and who had made many estimates of the sizes and distances of heavenly bodies, though with methods cruder than those developed later by Hipparchus and Ptolemy, proposed the theory that the planets move in circles about the sun.

Aristarchus, to our knowledge, did not attempt to show that such a theory would fit the data known to his time. But the theory was not acceptable to his contemporaries and successors because it was totally at variance with Greek conceptions of the universe and Greek physics. For one thing, the Greeks already knew that simple circular motion would not do because the distance of the earth from the sun was known not to be constant. One piece of evidence was that the apparent diameter of the sun varied with the seasons. Another objection to Aristarchus' plan arose from the knowledge that the earth consisted of heavy matter; it was inconceivable that such a heavy body could be in motion. The planets, on the other hand, were supposed to be made of some light substance and so their motion was feasible. This distinction between the physical constitution of the earth and that of the planets was almost universally accepted up to the seventeenth century. Moreover, if the earth were in motion, why did objects on the earth not fall behind? Greek physics had no answer to this argument.

The supreme achievement of all Greek efforts aimed at exhibiting the mathematical design of the universe is the astronomical theory of Hipparchus and Ptolemy. These two men, as we noted in the preceding chapter, had created the mathematical method that enabled them to determine the sizes and distances of the sun, moon, and several planets, the method which, as Ptolemy put it, gave them the tool needed to base astronomy "on the incontrovertible ways of arithmetic and geometry." They also had older Egyptian and Babylonian observations at their disposal as well as innumerable others made by Hipparchus himself at Rhodes and by the observatory in Alexandria. They tackled the plan of organizing all this knowledge into one comprehensive scheme.

In the astronomy of Hipparchus and Ptolemy, which we now refer to as the Ptolemaic theory, the earth is the center of the universe and stationary. To account for the motion of a planet P (Fig. 8–1), these men assumed that P moves at a constant speed along a circle whose center is Q. At the same time that P moves around Q, Q is supposed to be moving in a circle and at a constant speed around the earth, E. The circle on which P moves is called an epicycle, and the circle on which Q moves is called the deferent. Hipparchus and Ptolemy could, of course, choose the radii of the two circles and the speeds at which P and Q move on their respective circles so that the motion of P agreed with the observed positions of the particular planet. For each planet the choice of radii and speeds was different.

Actually the above scheme did not give these men enough latitude; i.e., they found it impossible to fit the motions of all the planets to a system of two circles. In some cases they used a combination of three circles (Fig. 8–2). Thus the planet P moved on the circle with center Q; during this motion, Q moved on the circle with center R, and R moved in a circle about the earth, E. Again the radii of the circles and the speeds of the moving points were at the disposal of Hipparchus and Ptolemy.

Their astronomical system contained also some minor devices which enabled them to fit a system of such circles to the motion of any one heavenly

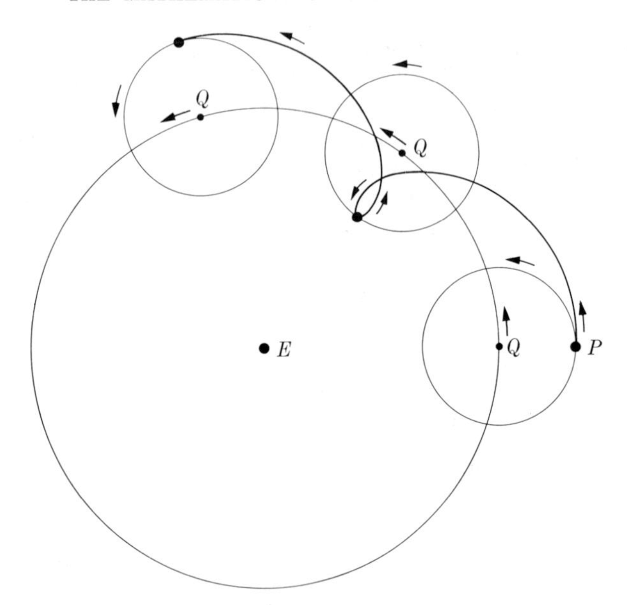

FIG. 8–1. A planet moves on its epicycle, which in turn moves around the deferent.

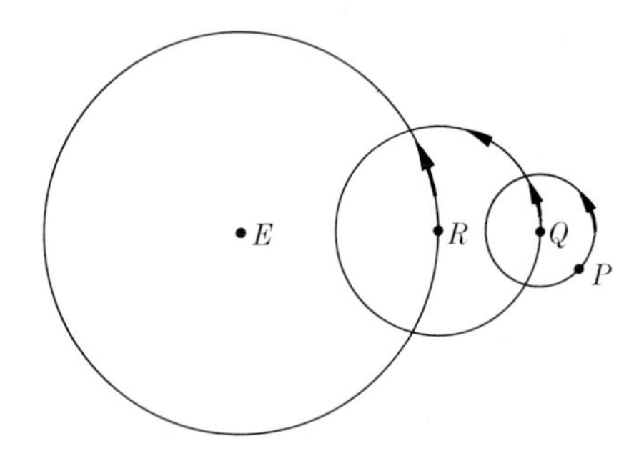

FIG. 8–2. Two epicycles and a deferent.

body, but the essential principle is the use of deferent and epicycle. It should be noted that the motion of a planet as viewed from the earth is actually quite complicated and yet, by the above scheme, is readily understood in terms of a combination of circular motions. This theory accounted for planetary motions within the accuracy of observations attained in Alexandrian times. From the time of Hipparchus an eclipse of the moon could be predicted to within an hour or two. Predictions of the sun's motion were not so precise, but we must recall here a point made in the preceding chapter, namely, that calculations of the sun's distances at various times were not exact because the requisite angles were too small to be measured accurately.

The scheme we have just described is contained in Ptolemy's *Almagest*, the book mentioned earlier (Chapter 7). This theory was quantitatively so precise that it was accepted as the true design of the heavens until the work of Copernicus and Kepler displaced it. It is significant, however, that Ptolemy at least laid no claims to truth. He had constructed a mathematical scheme which accounted for the motions of the celestial bodies, a theory which worked, but he did not profess that God had so designed the universe. Unfortunately people's confidence in the *truth* of a doctrine increases with the length of time it holds sway, and since Ptolemaic theory was accepted for about 1500 years, people came to regard it as an absolute and unchallengeable truth. No other product of the entire Greek era rivals the *Almagest* in the profound influence it exerted on conceptions of the universe and none, except Euclid's *Elements*, achieved such unquestioned authority.

The theory of Hipparchus and Ptolemy is the final Greek answer to Plato's problem of rationalizing the appearances in the heavens and is the first really great scientific synthesis. Whereas the Greeks of the classical period were convinced on philosophical and intuitive grounds that nature was rationally designed, Ptolemaic theory provided overwhelming, concrete evidence.

8–5 The evidence for the mathematical design of nature. Let us look back for a moment to see the total evidence which the Greeks could muster for their momentous doctrine that nature is mathematically designed. The astronomical theory of Hipparchus and Ptolemy was certainly the most impressive evidence not only because it dealt with the grandest natural spectacle but because it showed design in a maze of phenomena whose outward appearances scarcely suggested design. To this achievement we must add Euclidean geometry. We have already pointed out the larger significance of this body of knowledge; it demonstrated that the shapes and sizes of earthly figures conform to a reasoned system of doctrines. One might very well prove on the basis of self-evident axioms and of reasoning that satisfies the mind that the sum of the angles of a triangle is 180°. But when one constructs triangle after triangle for various purposes and finds in every case that the sum is indeed 180°, one cannot escape the implication that this and the other theorems of Euclidean geometry express essential principles of nature. Moreover, because these principles are all part of one reasoned body of knowledge, it seems clear that nature is designed in accordance with a reasoned plan.

In the domains of light and sound (music), the progress made by the Greeks was not nearly so impressive, but they had produced the law of reflection and they did know and use the properties of curved mirrors to concentrate light. Plato and Aristotle offered theories of vision, but these were nonmathematical and hardly backed by any experimental verification. The Greeks were sure that further investigation would reveal additional laws, and almost every Greek mathematician worked on light. Many, among them Euclid, Archimedes, Apollonius, Heron, and Ptolemy, wrote mathematical books on the subject. The development of a mathematical theory of musical sounds was initiated by the Pythagoreans and, as in the case of light, pursued by many later Greeks.

The Greeks also applied mathematics to various other classes of natural phenomena and found the mathematical laws applicable. Archimedes wrote a still famous book on the mathematical laws of the lever. Another of his works investigated the weight and stability of various shapes placed in water. It was primarily motivated by the experience that a ship whose shape is not well chosen may readily overturn in water. Still another study dealt with the centers of gravity of various shapes, an important bit of knowledge if bodies are to be balanced or remain upright.

Phenomena of motion were also studied by the Greeks. Here too they adopted what seemed to be self-evident principles and made deductions which fitted their limited experience. In the Aristotelian theory of matter all objects were composed of lightness, heaviness, wetness, and dryness. Those in which lightness dominated (for example, fire) always sought to rise. Those in which heaviness dominated (for example, metals) sought to fall. Every object had a natural place and, when not hindered, sought it. Thus the natural place of light objects was a region near the moon, whereas heavy objects tended to congregate at the center of the universe which was, of course, the center of the earth. Force is required to set an object in motion, and a measure of this force was the product of the weight and the velocity given to the body. Also, a force must constantly be applied to keep a body in motion or else the motion would cease. Forces are transmitted by material agents. Thus one body must strike another to transmit motion to the latter.* The Greeks made progress in other scientific fields such as geography and geodesy which we discussed somewhat in the preceding chapter.

In all of the fields discussed above mathematics was, at the very least, considerably involved. In fact, in the classical period mathematics *meant* arithmetic, geometry, astronomy, and music and, by the end of the Alexandrian period, it had come to mean, in addition, mechanics (motion, the lever, the hydrostatics of Archimedes), optics, geodesy, and logistics (practical arithmetic).

From these scientific investigations one major fact stood forth: the universe *is* mathematically designed. Mathematics is immanent in nature, it is the truth about its structure, or, as Plato would have it, the reality of the physical world. Moreover, human reason could penetrate the divine plan and reveal the mathematical structure of nature. Almost all of the mathematical and scientific research which has taken place since Greek times has been inspired by the conviction that there is law and order in the universe, and that mathematics is the key to this order.

The Greek miracle has not been rivaled, not even by our modern civilization. A relative handful of people produced in a few hundred years supreme works not only in mathematics and science but in literature, art, music,

---

* See also Section 15–3.

logic, and in many branches of philosophy. Yet their achievements have been subject to some criticisms which are worthy of note because they bear on what happened subsequently. The Greeks insisted on exact thinking and would not accept and work with ideas they could not formulate to their satisfaction. On this account they rejected the irrational numbers, turned to geometry to express all kinds of quantities*, and consequently failed to develop algebra which is a basic tool in scientific work today. Experience teaches us that it is usually better to work with an idea that seems to have some intuitive soundness and trust that continued efforts will ultimately clarify it. The idea that must wait to see light until it is perfectly fashioned will die a-borning.

The Greeks have also been accused of failing to uncover a greater number of physical principles because of their refusal to observe and experiment. They would not use their hands and preferred to accept aesthetically pleasing doctrines which appealed to the mind rather than check these against experience. Because the philosophical geometers did not "live" in the physical world, they failed to assist engineering, navigation, and the trades. This second charge is more applicable to the classical Greeks. As we have seen, the Alexandrians (and Aristotle in the classical period) did observe, did apply their mathematics, and did seek to aid other human activities. Indeed the alliance between theory and practice had begun to reach a point in Alexandria which is decidedly modern. Alexandria at its height was so close to the threshold of the modern era that if the 1500 years from the time of Ptolemy to 1600 A.D. were wiped from the books of history, the transition from Alexandrian to Western European civilization would seem natural.

Closely connected with this second criticism is a third. The Greeks were unwilling to observe and experiment, but they also had a positive conviction which in their eyes made these means of obtaining knowledge unnecessary. They believed that truths come from the mind. These truths were either innate and merely recalled or suggested by experience and immediately recognized by the mind to be self-evident. Thus the principles which were at the basis of their geometry and others that served similarly for the sciences were vouched for by the mind. In fact, Plato based an argument for the immortality of the soul on this fact. Since the axioms and theorems of geometry were to him innate truths and thus independent of experience, man possessed truths which were independent of the body. These truths were in our souls; hence the soul must be an entity independent of the body and its existence need not terminate with the body. The belief that truths can be found by exploring the mind instead of the physical world persisted until modern times. We shall learn later just what happened to it.

These criticisms of the Greek approach cannot be taken too seriously, for the implication that the Greeks might have accomplished still more had they not been hampered by the attitudes and beliefs just described is hardly

* See Section 6–1.

justified. It was their belief in the power of the mind which caused them to apply the reasoning faculty in the first place and to bring into being the very concept of mathematics. The progress they did make is astounding, and the failure to go further was due not nearly so much to any inherent limitations as to historical events over which the Greeks had little control.

8–6 The destruction of the Greek world. It is accurate to say of the Greeks that God proposed them but man disposed of them. During the first few hundred years in which the culture at Alexandria was still flourishing, the Romans had begun to extend their empire and conquered the Greek cities in Italy and Sicily. Then the Romans secured a hold on Egypt itself. During the reign of Cleopatra, the last of the Ptolemy dynasty, the Roman military force besieged Egypt. The Romans were forced to withdraw and, to protect their withdrawal, set fire to the Egyptian fleet riding at anchor in the harbor of Alexandria. The fire swept inland and destroyed the great library in the city. It is estimated that half a million manuscripts, the treasure of almost three centuries of book collecting, were wiped out. Upon the death of Cleopatra in 31 B.C. the Romans secured control of Egypt and from that time on proved to be a destructive influence on Alexandrian culture.

Although a few great Romans admired the Greek works, on the whole the Romans had only contempt for theoretical knowledge. With no spirit of apology Cicero says, "The Greeks held the geometer in the highest honor, accordingly, nothing made more brilliant progress among them than mathematics. But we have established as the limit of this art its usefulness in measuring and counting." The Romans were equally contemptuous of philosophy. Cicero's criticism, "There is nothing so absurd that it has not already been discussed by the philosophers," is representative of the Roman attitude. Theoretical mathematics was regarded as an exotic study, and the few who wished to pursue it went to Alexandria. It is a striking fact that the Roman and Greek civilizations existed over the same centuries, but in all that time there was not one Roman mathematician.

The Romans devoted themselves to useful constructions, which they regarded as a holy undertaking. The high priest of the pre-Christian Roman religion was called *Pontifex Maximus,* the greatest bridge builder, a title incidentally taken over by the Popes since the fifth century A.D. From ancient times on, the buildings of Rome, the walls, the bridges, temples, churches, water viaducts, and fountains were consecrated structures.

Indirectly the Romans were responsible for another sequence of events which proved destructive to Greek learning. The Romans drew wealth from the lands they conquered at the expense, of course, of the inhabitants. And the lot of the people in Egypt and the Near East became an unhappy one. The new religious movement, Christianity, which stressed brotherhood, ethics, and rewards in an afterlife, appealed to the oppressed and attracted many followers.

Though there was some opposition to pagan culture because the Christians, who originally were a Jewish sect, opposed polytheism, the practice of slavery, and the oppression of one people by another, many pagan doctrines and holidays were at first adopted to make the transition from the pagan to the Christian religion more attractive. But the Roman persecution of the Christians embittered them toward all pagan learning. The Christians grew stronger, and in 313 A.D. the Romans were forced by political circumstances to legalize Christianity. A little later, under Theodosius (379–95), Christianity became the official religion of the empire, and pagan religions were proscribed. As the Christians grew more powerful, they condemned and attacked pagan learning. The fathers of the Church ridiculed astronomy and physical science and advanced the Bible as the source of all knowledge. The doctrine of a spherical earth became a popular target. As the Christians put it, it was absurd to believe that people could live on the other side with their feet above their heads, that there could be places where rain and snow fell upward, and that seas, towns, and mountains could hang downward. Eusebius (*ca.* 340 A.D.), bishop of Caesarea and advisor to Constantine, said he had only contempt for scientific activity. It was far more important to direct the soul to higher truths. Soon Christians were forbidden to contaminate themselves with pagan learning. Even the pagan virtues were described as at best splendid vices. The Christians burned the Greek works; the Greek scholars in the museum at Alexandria were driven from the city; and the famous Temple of Serapis was destroyed, although there is some question whether the books housed in it were burned. In 529 A.D. the Eastern Roman emperor Justinian closed all Greek schools of philosophy at Athens, including Plato's Academy.

The "thinkers" who remained at Alexandria corrupted rather than improved knowledge. They of course neglected the sciences and mathematics, exhausted themselves in metaphysical disputes, attempted to explore the invisible world, sought to free the soul from the body, and even claimed to have methods of discoursing with demons and spirits. They tried to reconcile Plato and Aristotle on subjects of which both philosophers were ignorant and ended by converting philosophy into magic.

The destruction of what remained at Alexandria, Christian and pagan, was completed by the Mohammedans. The Arabs had been inspired by Mohammed to adopt a new religion. Mohammed died in 632 A.D., but his successors undertook to convert the world by the sword. They conquered Alexandria in 646 and burned the Museum. They argued that if the books there contained anything contrary to the teachings of Mohammed, they were wrong, and if in agreement, superfluous. With this stroke the dusk settled on Alexandria.

Although the Museum was destroyed and the scholars dispersed, Greek learning did ultimately become an integral part of European civilization and culture. Just how the Greek creations found a new home in Western Europe through one of the quirks of history we shall relate in a later chapter.

Exercises

1. What essential differences can you find between the pre-Greek and the Ptolemaic view of the heavens?
2. What is the Pythagorean doctrine concerning the essence of reality?
3. What is the meaning of the statement that Ptolemaic theory is a geocentric theory?
4. Describe the basic idea in Ptolemaic theory.
5. Suppose a planet moves on an epicycle at twice the speed with which the center of the epicycle moves on the deferent. Suppose, further, that the radius of the deferent is three times the radius of the epicycle. Sketch the path of the planet around the earth.
6. What is meant by the rationality of nature?
7. How does Ptolemaic theory support the belief in the mathematical design of nature?
8. How does Euclidean geometry tend to establish the mathematical design of nature?

Topics for Further Investigation

1. The accomplishments of Greek physical science.
2. The astronomical theory of Eudoxus.
3. The astronomical theory of Aristarchus.
4. The astronomical theory of Ptolemy.

Recommended Reading

Clagett, Marshall: *Greek Science in Antiquity,* Abelard-Schuman, Inc., New York, 1955.

Dampier-Whetham, Wm. C. D.: *A History of Science,* Chap. 1, Cambridge University Press, Cambridge, 1929.

Dreyer, J. L. E.: *A History of Astronomy,* 2nd ed., Chaps. 1 through 9, Dover Publications, Inc., New York, 1953.

Farrington, Benjamin: *Greek Science,* 2 vols., Penguin Books, Harmondsworth, England, 1944 and 1949.

Jeans, Sir James: *The Growth of Physical Science,* 2nd ed., Chaps. 1 through 3, Cambridge University Press, Cambridge, 1951.

Jeans, Sir James: *Science and Music,* pp. 160–190, Cambridge University Press, Cambridge, 1947.

Kuhn, Thomas S.: *The Copernican Revolution,* Chaps. 1 through 3, Harvard University Press, Cambridge, 1957.

Sambursky, S.: *The Physical World of the Greeks,* Routledge and Kegan Paul, London, 1956.

Sarton, George: *A History of Science,* Vols. I and II, Harvard University Press, Cambridge, 1952 and 1959.

Singer, Charles: *A Short History of Science,* Chaps. 1 through 4, Oxford University Press, London, 1953.

CHAPTER 9

THE AWAKENING OF EUROPE

Solicit not thy thoughts with matters hid,
Leave them to God, Him serve and fear.
. . . . be lowly wise;
Think only what concerns thee and thy being.

John Milton

9–1 The medieval civilization of Europe. It is perhaps a comfort after reading about the destruction of the Greek civilization to turn to a new one—the civilization of western Europe. We know that Europe did acquire the Greek creations and built upon them a vast, scientifically oriented civilization. How did this come about? To answer this question and to understand the special nature of subsequent developments in Europe, we must note a few historical facts.

The Germanic tribes, who have occupied western and central Europe as far back as history goes and who are the forefathers of most Americans, were barbarians. We know very little about their early history because they had no writing and hence no records were kept. Facts about these tribes were first acquired when the Romans invaded territories north of the Alps. From Roman historians, notably Tacitus (first century A.D.), we know that the Germanic tribes possessed a very primitive civilization. Tacitus describes them as honest, hospitable, hard-drinking, hating peace, and proud of the loyalty of their wives. Their dwellings were huts of timber and straw located in woods and surrounded by crude fortifications. Animal skins and coarse linens served for clothes, while herds of cattle, hunting, and the cultivation of grain crops provided food. Industry was unknown; just enough iron was mined to provide crude weapons. Trade was effected through barter and supplemented by plundering other tribes and more civilized regions. There were no arts, no science, and no learning. The chief activities were eating, sleeping, carousing, and fighting other tribes. Since such activities are also characteristic of peoples we call civilized, we may say that to that extent the Germanic tribes were civilized.

Some of the tribes were ruled by kings and others by valorous leaders. The political organization, such as it was, was strengthened by religious bonds. All these tribes worshipped the sun, the moon, fire, the earth, and special deities who governed the daily affairs of life. They believed in divination and in human sacrifices to the gods of whom they had many. Since the number of myths in any civilization seems to be roughly proportional to the number of gods, they also had many myths. Of course, like all other civilizations, the Germanic tribes called their myths higher truths. A life spent in battle and a glorious death were considered the best preparation for the afterlife promised by the priests.

Although the Romans won many battles with the Germanic tribes, these peoples were never really subdued and continued to raid and pillage outlying regions of the Roman Empire. As the Empire grew weaker for a variety of reasons which we cannot survey here, the barbarians finally conquered it, and barbarians became kings of Rome and what was left of the Empire. Only a small region around Constantinople, which we call the Eastern Roman or Byzantine Empire, managed to remain independent and isolated. The Eastern Roman Empire, incidentally, also withstood the Mohammedans, who in the seventh century conquered Egypt, the Near East, and the lands bordering the Mediterranean Sea.

By the time that the Roman Empire collapsed in the fifth century A.D., the Catholic Church had become a strong organization with good leadership. It gradually converted the heathens to Christianity, established schools in Europe, and taught reading, writing, and ethics. Moreover, it perpetuated and imposed the legal and political organization of Rome. The Christian influence was certainly beneficial in that it produced a more stable state of affairs and even induced the barbarians to remain at peace for longer periods of time, a restraining influence which the barbarians did not resent because they soon learned that civilization had its advantages. With a little thought they found that peaceful interludes permitted them to develop methods of mass destruction and so do as much killing at intervals as previously in constant warfare.

Cities and small states governed by powerful leaders were established in Europe. Trade between cities developed, producing the wealth necessary to support scholarship. But study was almost entirely confined to understanding the word of God as fostered, expounded, and dictated by the Fathers of the Church. Those Greek works which had survived destruction by Romans, Christians, and Mohammedans lay almost unnoticed in neglected public buildings, in private libraries, or in the isolated, beleaguered Eastern Roman Empire. In fact up to about 1200 A.D., the Church battled against the reading of Greek works. Greek and Latin masterpieces contained a mythology which had to be blotted out. The morality they described was opposed to Christian ethics. The Greek and Roman emphasis on life in this world was regarded as at best misguided. Of what use were philosophy, literature, science, health, and a physical life compared to salvation of the soul? Why read Greek poets when one should ponder the precepts of the Gospels? Why endeavor to make life on earth agreeable and comfortable when it is but an insignificant prelude to eternal life elsewhere? Why ask or seek to answer questions about natural phenomena when the nature of God and the relation of the human soul to God was yet to be explored and understood? Greek and Roman learning was condemned as impious and heretical. All intellectual and even artistic interests were absorbed in theological questions.

What little knowledge of nature was deemed necessary in the life prescribed by the Church was derivable, so the Christian leaders said, from the Bible. St. Augustine (354–430), a man learned in Greek and Christian thought, even declared that the authority of the Scriptures is greater than the capacity of the human mind. Unfortunately the Biblical statements about the nature and structure of the physical world are of Babylonian origin and hence decidedly inferior to the knowledge acquired by the Greeks.

Typical of medieval science was the famous *Topographia Christiana* written by Kosmas, a widely traveled merchant and later a churchman of unknown rank. Kosmas, who lived in sixth-century Alexandria, that is, before the Mohammedan conquest, was certain that the earth was at the bottom of the universe. He argued that it was too heavy to be at the center. Moreover, the earth must be flat. If it were spherical, how could it be swamped by the deluge in the days of Noah? Also, since the Bible tells us that man lives on the face of the earth, there can be no antipodes. In fact, if the earth were spherical, the sky would have to surround the earth, but the Bible says that the earth is firmly fixed on its foundations. To those pagans who asserted that there were people on the opposite side of the earth, Kosmas countered that these people could not be standing upright. All these facts and many more Kosmas derived from the Divine Scriptures which a lawful Christian could not doubt. Kosmas' work was widely read until the twelfth century.

Medieval natural science contained many other doctrines which are equally interesting. In contradiction to Kosmas, the most common belief placed the earth at the center of the universe because the earth contained man, who, as the most important creation of God, would naturally be at the center. However, Kosmas' doctrine that the earth must be flat was accepted and defended with the argument originally offered by Saint Augustine: if there were men on the opposite side, they could not attain salvation because they did not possess the Gospel. Such a situation was unthinkable. Hell was inside the earth and was shaped like a funnel, with sinners sitting in rows along the sloping wall. Satan lived at the bottom. The Garden of Eden was on the earth, but surrounded by a wall of fire and hence inaccessible.

The earth was believed to be inhabited by some remarkable, albeit monstrous, creatures. Most important were the devil and his demon assistants. Similar to man were the satyrs, who had crooked noses, horns on their foreheads, and goatlike feet. Satyrs came in many varieties; some were headless, others one-eyed, some one-footed, and some had enormous ears. There were centaurs, minotaurs and dog-headed men, giants, men who never ate or drank, men who never aged, and men without mouths. Equally marvelous were the varieties of animals "known" to exist, unicorns and salamanders on land and dragons in the seas.

Of course, some of the actual phenomena of nature were observed, and questions raised about them. The very few medieval intellectuals who pursued more sober lines of thought offered a kind of explanation which is satisfying to some minds. They believed that natural processes were mainly means to an end, i.e., they adopted what is called a teleological viewpoint. Thus rain existed to nourish the crops. Crops and animals existed to provide food for man. Sickness was a punishment from God. Plagues and earthquakes were expressions of God's anger. In general, all explanations focused on the phenomenon's value to, or effect on, man. Man was the center of the universe not merely geographically but also in terms of the ultimate purposes served by nature.

Although nature existed to serve man, man himself existed on this earth only to serve an apprenticeship during which he prepared his soul for a life in heaven with God—or elsewhere. Life on earth was but an unimportant prelude, to be endured but not enjoyed. To prepare for the afterlife man had to wrest his soul from a stubborn flesh which was guilty of original sin. Participation in the bounty of nature, food, clothing, and sex, tainted the soul and so had to be severely restricted. Medieval man, certain of his sins and doubtful of salvation, had to bend all his efforts to attain redemption. By earning divine grace man could escape from this foul earth to the divine empyrean.*

The common beliefs in the existence of strange creatures and in the trichotomy of earth, heaven, and hell, and the more esoteric rational, purposive explanation of nature which we have described were characteristic until about 1100 A.D. Some secular and ecclesiastical intellectuals did take an interest in the physical world, but their influence was negligible. On the whole, teachings deviating from ecclesiastical doctrines were declared heretical and were suppressed.

9–2 Mathematics in the medieval period. We see that a new civilization did arise in Europe, but from the standpoint of the perpetuation of mathematical learning or the creation of mathematics, it was totally ineffective. Although this civilization did spread ethical teachings, fostered Gothic architecture and great religious paintings, no scientific, technical, or mathematical concept gained any foothold. In none of the civilizations which have contributed to the modern age was mathematical learning reduced to so low a level.

Superficially mathematics did seem to play an important role. In the medieval schools the standard curriculum consisted of seven subjects, the quadrivium and the trivium. The quadrivium comprised arithmetic, the science of pure numbers; music as an application of numbers; geometry, or the study of magnitudes such as length, area, and volume at rest; and astronomy, the study of magnitudes in motion. But the scope of these studies was terribly limited. Even the first universities of Europe, which began to function about 1100 A.D., offered merely a minimum of arithmetic and geometry. Arithmetic consisted of simple calculations mingled with complex superstitions. Geometry was confined to the first part of Euclid, far less than we learn in high-school courses today. The most advanced point reached in some of these institutions of learning was the very elementary theorem that the base angles of an isosceles triangle are equal. This theorem was called *"pons asinorum,"* the bridge of asses. Various

---

* A term of medieval cosmology referring to the "highest heaven" or paradise.

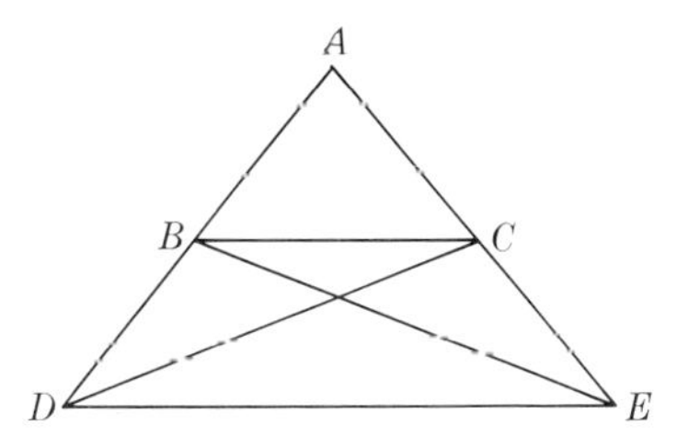

FIG. 9-1. The figure used by Euclid to prove that the base angles of an isosceles triangle are equal.

reasons are given for the choice of this name. One ascribes its origin to the fact that the figure which Euclid used to establish his proof (Fig. 9-1) had the shape of a simple bridge between two banks sloping down so steeply that only an ass could mount it. However, the more commonly given reason is that the proof of even this simple theorem could not be mastered by all students, and hence the poor ones could proceed no further in geometry. They were like asses who balk at crossing a bridge. Of course the retort of the "fools" was that only an ass would want to master the theorem.

The little mathematics kept alive in the schools served various purposes in the medieval period. Some astronomy was pursued to keep the calendar. Here a minimum of arithmetic and geometry sufficed for the accuracy needed, just as it did in ancient Egypt and Babylonia. This work was usually performed by monks because the clergy was the most learned class.

Astronomy and therefore elementary mathematics played a larger role in medieval life in that they provided the factual information needed for astrology, which was regarded as a science. The history of astrology, as noted in Chapter 2, goes back to Egyptian and Babylonian times. Early in the Christian era it was in disfavor with the Roman emperors and was banned. At that time, the word "mathematics" as distinguished from "geometry" meant largely astrology, and professors of astrology were called *mathematicii*. Hence one finds laws such as the Roman Code of Mathematics and Evil Deeds which damned and forbade the art of mathematics. Later Roman and Christian emperors, while banishing astrologers from their kingdoms, employed them in their own courts as advisors. If there was anything in this business of foretelling the future, the rulers were not going to overlook it, but neither were they going to let anyone else get hold of this knowledge.

The key doctrine of astrology is that the stars and the planets control events on earth, including the lives and fortunes of human beings. The positions of the stars and planets are the main factors in determining what occurs. Hence by predicting their positions one could predict the future. Earthquakes, revolutions, floods, success in business ventures, and individual health were foretold from the positions of the stars and planets relative to the signs of the zodiac.

In medieval times the "science" of astrology was universally accepted. Men such as Roger Bacon, who in the thirteenth century was already a clear and outspoken champion of the experimental method in science, subscribed to astrology. One of the founders of modern astronomy, Johannes Kepler, cast horoscopes for many years and, in his youth at least, apparently believed in them. It was not until the seventeenth century that France and England passed laws forbidding the insertion of astrological predictions in almanacs. However, horoscopes can be bought even today at magazine stands.

The "science" of astrology was the basis for medieval medicine. Since stars and planets controlled the body, knowledge of their positions and movements could be used by physicians to tell people what to expect or how to guide themselves in accordance with the will of the heavenly bodies. Physicians told people when to shave, bathe, purge themselves, or be bled. They kept records of the constellations present at the birth, marriage, sicknesses, and death of people and used these to correlate heavenly and earthly events.

Perhaps one more medieval use of mathematics is worthy of mention. Plato's belief that the study of mathematics trains the mind for philosophy was taken over by the Church which, however, substituted theology for philosophy. Clearly the interest here was not in mathematics as such but as a preparation for grasping the subtle reasoning which the Church employed to build and strengthen the foundations of religious doctrines.

This brief sketch of medieval civilization, rather one-sided because we have been largely concerned with its relation to mathematics, may give some idea of what was indigenous to Europe, or at least what Europe, building on a meagre legacy from Rome, produced under the leadership of the Church. Until 1100 A.D. the medieval period did not produce any significant culture in intellectual spheres. Its intellectual state is characterized by indistinctness of ideas, mysticism, dogmatism, and a reliance upon scriptural authorities, who were constantly consulted, analyzed, and commented on. The mystical leanings caused people to elevate vague ideas into realities and even accept them as religious truths. Such theoretical science as existed was stationary. The Christian religion emphasized its own ideas, duties, way of life, and values so much that it made inquiries about nature inspired by curiosity seem frivolous and unworthy of man. To satisfy those few who possessed intellectual interests, the Christian leaders extended their doctrines to cover the functioning of the entire universe. Theology embraced all knowledge, and Fathers of the Church authored universal systems. But they did not conceive or search for principles other than those contained in Christian theology.

The medieval attempt to organize man and nature under one comprehensive scheme was not too successful because of the lack of sheer facts. The paucity of knowledge was concealed by richness of elaboration, fine and overdrawn distinctions, subtleties of reasoning, and varieties of argumentation. The emphasis of the age, as a result of the dominance of a religious leadership, was on the life of the soul. The spiritual world was as much and more of a reality than the physical world.

9-3 First revival of the Greek works. Whether or not the civilization of medieval Europe might in due time have given rise to mathematical activity will never be known. Any survey of the attitudes of the learned class or of the masses indicates that it is hardly likely. But dramatic changes, largely initiated by non-European forces, drastically altered the Christian world. The earliest influence tending to transform thought and life in medieval Europe may be credited to the Arabs. While the Church was slowly civilizing the European barbarians and establishing the Christian way of life, the Arabs, perhaps more ruthless in proselytizing and certainly more dynamic and aggressive, succeeded in establishing their own civilization and culture in southern Europe, North Africa, and the Near East. Though fanatic in the advancement of their own religion, once their empire was stabilized, the Arabs displayed great tolerance toward alien ideas and learning, readily absorbed the mathematics and science of the Greeks and Hindus, and built cultural centers in Spain and the Near East. They translated the Greek works into Arabic and added commentaries and contributions of their own to mathematics, astronomy, medicine, optics, meteorology, and science in general.

By about 1100 A.D. Europeans were trading freely with Arabs. The Crusades, which attempted to wrest Palestine from the Arabs, brought further contacts between Christians and Moslems. Through these channels the Europeans became aware of the Greek works and Arab additions. They were so fascinated by this material that they aroused themselves to acquire it. Wealthy merchants, princes, and popes sent agents to the Arab centers to purchase manuscripts. Many Europeans went to live in Spain and learned Arabic in order to read the works and translate them into Latin. Others were assisted by Jewish and Arab scholars in making the translations. Plato, Aristotle, Euclid, Ptolemy, and the Greek literary works were avidly grasped.

The translations of the twelfth century had some effect on the intellectual life of Europe. Criticisms of medieval thought, of Aristotle's philosophy, which had become firmly imbedded in Christian theology, and of Aristotle's physics were made in the universities of Chartres and Oxford by men who had already begun to think for themselves, largely under the inspiration of the newly available Greek works. The most outstanding figure, far ahead of his time, was Roger Bacon (1214–1292).

Bacon studied at the universities of Paris and Oxford. He knew many languages including Arabic, and his knowledge of science covered the ancient works and current discoveries. His enormous learning is revealed in his writings on mathematics, mechanics, optics, astronomy, geography, chronology, chemistry, music, medicine, grammar, logic, metaphysics, ethics, and theology. He was somewhat bound to the civilization of his era, and so did not fail to include magic and astrology and to stress that theology is the end of all learning.

The highly exceptional feature of Bacon's thought was that he saw the need for, and values to be derived from, experimentation. He says, "Argument concludes a question but it does not make us feel certain, or acquiesce in the contemplation of a truth, except the truth also be found to be so by experience." He realized that experimentation can lead to new ideas and can be used to test results obtained by theory. Long before the Renaissance this *Doctor Mirabilis* called attention to the importance of studying nature and to the significant role of mathematics in this study. Not only did he understand how new ideas and inventions are acquired, but he also had some grasp of the fruitful method of scientific inquiry which was developed four centuries later. He foresaw and had confidence in the kind of civilization which progress in science could make possible.

Bacon had the courage to urge resistance to the dogmas so forcibly imposed upon the people of Europe and, in particular, he opposed Aristotle's teachings. He says, "If I had power over the works of Aristotle, I would have them all burnt; for it is only a loss of time to study in them, and a cause of error, and a multiplication of ignorance beyond expression."

Criticism of Aristotle, one of the leading Greeks, seems hard to understand at a time when Greek works were the very ones exciting the admiration of Europe. But the Greek science of the classical age, which Aristotle of course presented, was necessarily only a beginning and erroneous in many respects. In his effort to be complete and systematic Aristotle offered concepts, theories, and explanations which had very little basis in reality or were not fruitful (see Chapter 14). Unfortunately the Church, having used some of Aristotle's doctrines at the outset and already engaged in the process of incorporating even more in the scholastic reconstruction of theology, regarded all of Aristotle as sacrosanct. To undertake any reconstruction of thought or to gain a hearing for any new principles of science, it was necessary to abolish the complete and uncritical acceptance of Aristotle.

Bacon ended up in prison as did many other intellectual leaders who had begun to assert the independence of human reason and the importance of observation and experimentation.

The influence of the revived Greek knowledge is exemplified by the work of another man who was closer to mathematics. Leonardo of Pisa, also called Fibonacci, was the first to make Greek mathematical knowledge somewhat more widely available. Leonardo, the son of a business man, traveled extensively in Egypt, Syria, Greece, and Sicily and thus became aware of the Greek works. In 1202 he published *Liber abaci* (Book of the Abacus), which contains the knowledge the Arabs possessed in arithmetic and algebra. In his *Practica geometriae* he presented some of the Arab material on geometry and trigonometry. The first of these books was for centuries a standard source of mathematical knowledge.

The initial burst of new intellectual life in Europe was short-lived. Of the works being made available the medievalists selected Aristotle for admiration and study. They were impressed by the vast store of opinions and facts which he offered, his acute distinctions, his cogent reasons and argumentation, his comprehensiveness and logical arrangement of knowledge. Also Aristotle's teleology fitted the idea of God as final cause. Even

Aristotle's doctrine that the heavens were composed of an ethereal substance which was perfect and indestructible was convenient because it provided the proper dwelling for God. Finally, because Christian doctrines had already been founded in part upon Aristotle's philosophy, his ideas were more congenial. Plato's works, by contrast, less systematic, less comprehensive, less positive, and more subtle, were hard to grasp. Moreover, since he left open many problems of philosophy, his ideas were less satisfying. In Aristotle's works a vast new body of information became available which, because it was from a source already accepted in Catholic theology, could not be denied, but had to be reconciled with existing doctrines. A group of philosophers, known as the Scholastics, of whom Thomas Aquinas is the most noted, undertook this task. Like all Greek writings, Aristotle's stressed a reasoned approach to his conclusions, whereas the Church had relied upon faith. Hence Aquinas' problem was also to reconcile reason and faith.

We cannot digress here even to the extent of sketching Aquinas' monumental reconstruction of Catholic thought. It must suffice to say that his *Summa Theologiae,* one of the landmarks in philosophy, did afford a comprehensive, thorough, and systematic reform which had pervasive and enduring effects. Aquinas used strict deductive reasoning to erect a system of theology which accounted for all phenomena the average man was likely to encounter and answered all questions he was likely to raise about the universe and man's place and purpose in it. Of course, Aquinas based his system of thought on principles drawn from the Bible, the writings of the Church Fathers, and Aristotle. Explanations and answers to common questions were given in terms of forms, qualities, causes, origins, and ends of things. He stressed the importance of understanding God and striving for salvation. Faith also played a role and, in a sense, was even superior to reason since it was called in as a basis for accepting doctrines inaccessible to reason. It was, so to speak, a higher reason or extension of reason to spheres where reason was inadequate.

The predominant effect of Aquinas' work, precisely because it was so comprehensive and rationally presented, was to stifle thought. There was no need any more to ponder problems of man, nature, and God. The answers were at hand and had the backing of an organization that could readily silence those who were not satisfied.

Other factors too delayed changes in Europe. The revived Greek knowledge reached only a few scholars. Manuscripts were expensive, and hence could not easily be obtained by many who had the desire to read them. The few universities already in existence were controlled by the Church, and professors were not free to teach what they thought correct or best. Moreover, Europe in the period from 1100 to 1600 was broken up into a number of independent dukedoms, principalities, more or less democratic or oligarchic city states, and the Papal states. The wars among all these political units were continuous and absorbed the energies of the people. The Crusades, beginning about 1100, wasted a fantastic number of lives. Finally, the

Black Death in the second half of the fourteenth century took about one third of the population of Europe and set back the entire civilization.

9–4 Revolutionary influence in Europe. Radical changes did ultimately alter the life of Europe, but not until the fifteenth century did these changes begin to be appreciable. It was Italy which led the way. Although the history of the Italian states in this century was marred by incessant intrigues, mass murders, and destructions wrought by local wars and by the intervention of France, Spain, and Germany, the very fluid nature of political conditions and the existence of some democratic governments were favorable to the rise of the individual. Wars against the Papacy itself, which was a leading political and military power in that era, not only freed people from political control but encouraged intellectual opposition as well. By virtue of its geographical position Italy acquired great wealth during the latter part of the Middle Ages, and the princes used some of this to support scholars.

In the fifteenth century Italy made new contacts with the Greek heritage. The Pope and local princes had ambassadors in Constantinople, the capital of the Eastern Roman empire, which still possessed the largest collection of ancient manuscripts. Moreover, ambassadors of this Empire came to Italy several times in the first half of the fifteenth century, largely to seek help against the Turks. The Italians learned about the Greek works and, like the Europeans of three centuries earlier, sought eagerly to possess them. In addition, some Greek scholars discouraged by the poverty in Eastern Europe and Alexandria migrated to Italy. When the Turks finally captured Constantinople in 1453, a flood of these men bringing their manuscripts with them came to Italy.

A dynamic force in the European acquisition of Greek works was Cosimo de' Medici (1389–1464), a wealthy banker and leading citizen of Florence. He hired a regular staff of agents to search for the manuscripts in Arab centers and in Constantinople, had them brought to Florence, and employed copyists to make copies available. In 1444 Cosimo established the Medici Library, the first public library in Europe. His grandson, Lorenzo the Magnificent (1449–1492), supported the search for manuscripts and their study even more than Cosimo. A little later Pope Sixtus IV founded the Vatican library, still world-famous, and Federigo Montefeltro, Duke of Urbino, also found one, The Platonic Academy in Florence, another institution sponsored by Cosimo for the study of rediscovered Greek manuscripts, became the home of the richest intellectual life of the fifteenth century. Greek was taught there so that Europeans could learn to read the Greek works. Lorenzo's son, Pope Leo X (reigned 1513–1521), who himself knew the Greek and Latin classics, invited writers and artists to Rome, sponsored the search for ancient manuscripts, and founded the University of Rome and a printing press for Greek works.

By financing the geographical explorations of the fifteenth and sixteenth centuries, which were intended to discover new trade routes, the merchants affected the life of Europe. The discovery of America and of a route to China around Africa resulted in acquainting Europe with strange lands, beliefs, customs, religions, and ethical doctrines. Catholics met Mohammedans, Chinese, and the American Indian. To the broadening influence of trade itself was added knowledge which conflicted sharply with the doctrines and way of life hitherto accepted in Europe. Questioning of the accepted doctrines and values and ultimately revolt from them were bound to ensue.

The merchant class and the large classes of artisans and free laborers introduced new interests. Employers and employees sought material gain, and so looked for commodities, machinery, and natural phenomena which might be employed to advantage. The rulers of the Italian cities and states also spurred on these interests. They coveted power and magnificence and, to acquire the necessary wealth, favored trade, industries, and inventions. The cities competed to surpass one another in skills, devices, and quality of merchandise. These groups, though selfishly motivated, were nevertheless effective in orienting the civilization toward the physical world and in fostering the accumulation of empirical knowledge.

To appreciate the importance of the new interest in the physical world, one must note the variety of industries, arts, and commercial activities which flourished by the fifteenth century. Mining, the production of metals, glass, pottery, and furniture, the weaving and dyeing of cloth, the conversion of leather to useful forms, the brewing of alcoholic liquids, printing, map making, and the manufacture of agricultural implements, armaments, cannons and muskets, all required work with materials. Enormous and expensive building projects, notably churches, palaces, fortifications, and public works, posed physical problems and also involved materials. The flourishing activity in painting and sculpture utilized paints, metals, glass, ceramics, and stone. Navigators, lens designers, druggists, and doctors were of necessity absorbed in physical problems. The attempts to use water power, wind power, and air pressure to supplant human labor became a most serious field of endeavor. Thus nature became the concern of business men, artists, engineers, mechanics, and technicians. In fact, the Renaissance is the beginning of the interest in mechanization, materials, and inventions which is characteristic of modern industry.

The Protestant Revolution, or the Reformation as it is called, also upset the old culture in Europe. We are not concerned here with justification of the break from the Church. But Luther fanned the fire of discontent which had spread throughout Europe. Not only did the Protestants differ in their interpretation of religious doctrines, but to attract adherents they often favored ideas and practices which the Church opposed. Thus, because the Church opposed interest on loans, the Protestants favored it. Also Protestantism professed that private judgement, rather than the decrees of the Pope, must be the basis for belief. Disputes about the nature of the sacrament, the validity of the control of the Church by Rome, and the meaning of passages of the Scriptures raised doubts in many people, who were thus emboldened to question all religious teachings and to turn to other sources of knowledge, notably the physical world itself.

Several discoveries and inventions of the late medieval period had effects far greater than one might at first expect. In the twelfth century the Europeans learned from the Chinese about the compass. The introduction of the compass was important because it was an immense aid to navigators on long sea voyages. The explorers who dared the Atlantic might not have been willing to do so without it. The phenomenon was so striking that, even in the unscientific thirteenth century, Picard Petrus Peregrinus was moved to carry out a series of experiments on the subject, one of the few notable experimental studies in the medieval period.

The introduction of gunpowder in the thirteenth century produced as its most obvious effects changes in methods of warfare and the design of fortifications. It also introduced a new physical problem, the motion of projectiles. An indirect result was the granting of more power to the common man because with a musket he could be effective in warfare. Previously only those who could afford expensive armor, that is, the wealthy nobles, could wield military power.

The invention of printing (about 1450) was immensely important in helping to spread Greek knowledge across Europe. In fact, we have already mentioned that undoubtedly one reason for the slow spread of the first revival of Greek works was that prior to the invention of printing, only the wealthy and the scholars they supported could get hold of manuscripts. Another invention, paper made of cotton and later of rags which replaced costly parchment, also helped to make books plentiful and cheap. Many editions and translations of Greek works were printed in the century following these inventions. They helped to bridge the gulf between the learned and the untutored just at the time when great numbers were seeking to obtain knowledge. Only those who possessed the manuscripts with their beautiful calligraphy did not welcome the crude and relatively ugly books produced by printing.

Advances in the subject of optics had a vast effect on future scientific activity. The first was the discovery made in the thirteenth century that lenses can be used to magnify objects and thus aid in the examination of materials and natural phenomena. Late medieval scientists, notably Roger Bacon, Robert Grosseteste, Witelo, Peckham, and Theodoric of Freiburg, studied the refraction of light by lenses, determined the focal lengths of some lenses, tried combinations of lenses, produced colors from sunlight by passing it through a crystal, and obtained a better understanding of the rainbow. Lens grinders began to produce spectacles. Early in the seventeenth century, two of them discovered that a pair of lenses held at some distance from each other could be used to make distant objects seem close. Thus the telescope became available and was immediately applied to astronomy with results we shall describe later. At about the same time, it was found that a combination of lenses would do even better than a single lens to magnify nearby objects, and the microscope was invented. The investigation of the biological world and the revelation of hitherto unsuspected small-scale phenomena soon followed.

9-5 Intellectual revolt. It was to be expected that the insular world of medieval Europe accustomed for centuries to one rigid, dogmatic system of thought would be shocked and aroused by the series of events we have just described. The European world was in revolt. As John Donne put it, "All in pieces, all coherence gone." Europe revolted against scholastic tyranny of thought, obsolete authority, and restrictions on the physical life. It revolted against the Scriptures as the source of all knowledge and the authority for all assertions. It revolted against enforced conformity to the established canons of conduct.

The extent of the break from medieval ethics is exemplified by the writings of the erudite Lorenzo Valla (1406–57), man of letters, tutor, and private secretary to Alfonso of Aragon, King of Naples. In his bold dialogue *On Pleasure* Valla contrasts the life of the spirit with the life of the flesh and unhesitatingly chooses the latter. He denies all values in ascetism and holiness and upholds the physical appetites. Chastity was to Lorenzo a violation of nature's laws and an intolerable torment. Pleasure is the highest good and joys could be derived from physical activities. Man should investigate the laws of his own being rather than God's and obey them, confident that nature can do no wrong. In fact, whatever violates nature's laws must be wrong. Thus Valla proclaims the new outlook. The reader may be familiar with the more indirect expression of the same theme in the writings of Boccaccio, who celebrates sensuous beauty and natural passions.

A leading figure in the revolt from the old modes of thought is Leonardo da Vinci (1452–1519). Because he saw how most scholars accepted as authoritative all that they read, he distrusted the men who took their learning only from books and professed their knowledge so dogmatically. He describes them as puffed up and pompous, strutting about, and adorned only by the labors of others whom they merely repeated. These were only the reciters and trumpeters of other people's learning. Leonardo determined to learn for himself and made exhaustive studies of plants, animals, the human body, light, the principles of mechanical devices, rocks, the flight of birds, and hundreds of other subjects. Although he is most often remembered as one of the great masters of painting, he also was a psychologist, linguist, botanist, zoologist, anatomist, geologist, musician, sculptor, architect, and engineer.

Cardinal Nicholas of Cusa (1401–64) is another outstanding figure of the Renaissance, who in his famous *De Docta Ignorantia* attacked the scholars and spoke of the learned as possessing only instructed ignorance. He too rejected the Aristotelian doctrines because they were inadequate and because they were accepted so gullibly.

Many others, Cardinal Bernadinus Telesius (1508–1588), the monk Thomas Campanella (1568–1639), the priest Giordano Bruno (1548–1600), the humanist professor Peter Ramus (1515–1572), and the Lutheran reformer Philip Melanchthon (1497–1560), to mention some of the names, attacked Aristotle and realized that new bases for scientific activities were needed. But these men were not too clear as yet about the new principles and new knowledge which should replace the rejected doctrines. They did, however, seek the freedom to speculate, the refutation of error, and the promotion of independent thought.

Many scholars turned to exhaustive studies of the Greek authors, to translations, and to compilations. They gave to these works the same infinitely detailed and critical attention that they and others had formerly given to biblical documents.

The writings of Luca Pacioli (1445–1514) show this tendency. He was a monk, who in 1499 published *Summa de Arithmetica, Geometrica, Proportione, et Proportionalita*. It is sad to note that his work contained almost nothing that could not already be found in Leonardo of Pisa's books, which shows how little medieval Europe progressed in mathematics over 300 years. But as a full, almost encyclopedic, account of the knowledge available to Europe by 1500 it was enormously helpful.

More interesting as a transitional figure is Jerome Cardan whom we met in Chapter 5. He wrote a great number of works which exhibit a critical attitude only in the sense that he traced the origins of stories, miracles, and "facts" to the authorities. However, he accepted freely any number of medieval superstitions, legends, accounts of supernatural events, pseudosciences, and even magical medical treatments. He believed in the significance of dreams, ghosts, portents, palmistry, and astrology which to him were sciences. He also wrote volumes on moral aphorisms and on the varieties of beings and bodies which fill the universe. Among these were spirits which took the form of sylphs, salamanders, gnomes, and ondines. Communion with these spirits was the highest aim in life.

Cardan's writings in the above fields were compilations; much of the material, incidentally, he stole from Leonardo da Vinci, who was a friend of Cardan's father. In his mathematical and scientific work, however, he shows the new influences. His still famous *Ars Magna* (1545), which contains a full account of the algebraic methods known to the Arabs, also contains results due to himself and his contemporaries. He is the first European mathematician of consequence. Some indication of what was new in his work was given in Chapter 5.

Pacioli and Cardan are mathematical figures in the movement commonly known as humanism. The humanists, and we speak now of those active in all fields, have been criticized because they idolized the past too much and looked backward rather than forward. They slavishly accepted the Greek works and pored over them, even undertaking extensive philological studies to determine the meanings of dubious words. To their credit may be noted that they prepared the atmosphere for the revival of reason, spread the Greek ideas through Europe, sought to restrict religion and theology to their own spheres, secularized education, and stressed the individual, experience, and the natural world.

Exercises

1. What uses did medieval civilization make of mathematics?

2. Which doctrines of medieval Christianity were favorable and which were unfavorable to the study of mathematics?

3. The following problem was supposed to be a difficult one in Leonardo of Pisa's *Liber Abaci*. If A gets from B 7 denare, then A's sum is 5 times B's money; if B gets from A 5 denare, then B's sum is 7 times A's holdings. How much does each have to begin with?

4. The following problem in Leonardo of Pisa's book goes back in essential content to Egyptian times. Seven old women go to Rome; each woman has 7 mules; each mule carries 7 sacks; each sack contains 7 loaves; with each loaf are 7 knives; each knife is enclosed in 7 sheaths. What is the total number of objects and people?

5. In medieval times a collection entitled *Problems for Quickening the Mind* offered the following already ancient teasers:

(a) A dog chasing a rabbit which has a start of 150 ft jumps 8 ft every time the rabbit jumps 7. In how many leaps does the dog overtake the rabbit?

(b) One pipe can fill a cistern in 5 hr and another in 7 hr. How long will it take both pipes operating simultaneously?

(c) A wolf, a goat, and a cabbage are to be rowed across a river in a boat holding only one of these three objects besides the oarsman. How should he carry them across so that the goat should not eat the cabbage or the wolf devour the goat?

6. Another hoary teaser is the following: A man goes to a tub of water with two jars, one holding 3 pt and the other 5 pt. How can he bring back exactly 4 pt?

7. What mathematical ideas did the Europeans contribute up to the year 1500?

8. Can you reconstruct Euclid's proof of the theorem that the base angles of an isosceles triangle are equal (Fig. 9–1)? [*Suggestion:* By hypothesis, sides AB and AC of triangle ABC are equal. Prolong AB to D and AC to E so that $BD = CE$. Now use congruent triangles.]

9–6 New doctrines of the Renaissance. The period devoted to the collection and study of the classics was followed by one in which intellectuals groped for positive doctrines and methods to replace the by now discredited medieval culture. We cannot trace in detail the oscillations of thought, the mixture of medieval fantasy and rational speculations, the commingling of fine observations with outmoded principles, all of which one finds especially in the sixteenth century. Many European thinkers finally broke away from the authority of Aristotle and the Church Fathers and from the endless rationalizing on the basis of dogmatic principles which were vague in meaning and unrelated to experience, and chose human inquiry rather than divine authority.

It was from the Greek works that the leaders in this intellectual revitalization of Europe derived the principles of a new approach to man and the universe. They learned that man could enjoy a physical life and find pleasure in food, sports, and the development of his own body. Beauty was not a snare, and pleasure not a sin. Man, the unworthy creature, who had been commanded to regard himself as a sinner, to spend his life in abstinence, penance and abjectness, and to prepare for death, the only real event of life, could find dignity in his own being, and demand a full life on this earth as his birthright. In place of sin, death, and judgment, men should seek beauty, pleasure, and joy. The Renaissance world began to see man as the goal of God rather than God as the goal of man.

The human spirit was emancipated and inspired to refashion its ideals of existence. Perhaps the most important decision was to turn to nature herself as the source of knowledge. "Back to nature" became the new cry. The call to nature was ardently sounded by Thomas Campanella, philosopher and champion of a new movement:

> The world's the book where the eternal sense
> Wrote his own thoughts; the living temple where,
> Painting his very self, with figures fair
> He filled the whole immense circumference.
> Here then should each man read, and gazing find
> Both how to live and govern, and beware
> Of godlessness; and, seeing God all-where,
> Be bold to grasp the universal mind.

Europeans turned to nature's laws instead of divine pronouncements gleaned from the Scriptures, to the universe of God instead of God. Man himself was included in the study of nature.

Leonardo is a representative figure in this shift to nature as the prime focus. He almost boasts that he is not a man of letters and that he chose to learn from experience. His observations and inventions recorded in his notebooks give evidence of his extensive and detailed physical studies. He says, "If you do not rest on the good foundation of nature, you will labor with little honor and less profit." Sciences which arise in thought and end in thought do not give truths because no experience enters into these purely mental reflections, and without experience no thing is sure.

A new school of biologists arose, of whom Andreas Vesalius (1514–64) was the leader. His *On the Structure of the Human Body* (1543) may be regarded as the beginning of modern anatomy. Although this work is based on Galen, he corrected many of Galen's errors and added new observations. Vesalius asserted that the true Bible is the human body, and he dissected corpses to learn the human structure. William Harvey (1578–1657), the famous seventeenth-century doctor, voices the spirit of Vesalius in the preface to his book *On the Movement of the Heart and the Blood:* "I profess to learn and to teach anatomy, not from books, but from dissections; not from the positions of philosophers, but from the fabric of nature." Harvey also followed Galen but, like Vesalius, added new material derived from his own observations and thought. Andrew Cesalpinus (1520–1603), the botanist, clearly advocated starting from observation and then proceeding through careful differentiation of the species observed to inductive truths.

We shall see in the next chapter how the artists, too, turned to the study of nature and to new goals in painting which obliged them to study anatomy, perspective, light, and mechanics. Regard for the primacy of observation forced Johannes Kepler to devise revolutionary doctrines in astronomy.

Indeed, experience became the source of all basic scientific laws and, in this respect, usurped the role of mind.

Of course, one must not expect sharp breaks in intellectual developments. Leonardo, Vesalius, Harvey, and others we have cited caught the spirit of the new outlook. But almost inevitably—in view of their early training and the influence of the world about them—they still had one foot in antiquity. Harvey, for example, revered Aristotle and Galen and thought that he was being true to them.

The second guiding principle adopted by the Europeans of the Renaissance was to let reason alone be the judge of what to accept. Revelation, faith, and authority were to be dropped as support for assertions about man and the universe, and reason was to be applied freely to all problems man sought to solve. Although the Church itself had used reason to erect its own theology, it had said that some matters were beyond reason. Moreover, the results obtained by reasoning were not put forth to be scrutinized rationally but rather to be accepted. In the Renaissance, mind replaced faith as the sovereign authority, and man was encouraged to apply it to the problems besetting his age.

The new impulse to study nature and the decision to apply reason instead of relying upon authority were forces which might in themselves have led to mathematical activity. But the Europeans also had the Greek works. From the Greeks the Europeans learned that nature is mathematically designed, and that this design is harmonious, aesthetically pleasing, and the inner truth about nature. Nature is not only rational, simple, and orderly but it acts in accordance with inexorable and immutable laws.

Almost from the beginning of the period in which Greek works began to be known in Europe, one finds leading thinkers impressed with the importance of the mathematical study of nature. Roger Bacon, whose pioneering thoughts we have already mentioned, believed that the laws of nature are but the laws of geometry. Mathematical truths are identical with things as they are in nature. Moreover mathematics is basic to the other sciences because it takes cognizance of quantity. Leonardo, too,—although his knowledge of Greek works was rather limited and his appreciation of what mathematical proof means almost nil—had caught the new spirit. He says that only by holding fast to mathematics can the mind safely penetrate to the essence of nature. "No human inquiry can be called true science unless it proceeds through mathematical demonstrations." He also says, "The man who discredits the supreme certainty of mathematics is feeding on confusion and can never silence the contradictions of sophistical sciences, which lead to eternal quackery." Leonardo was not a mathematician, and his understanding of the principles of mechanics, the study of bodies at rest and in motion, was intuitive and but a dim foreshadowing of the work of Galileo and Newton, but he had prophetic vision. He says in one of his notebooks, "Mechanics is the paradise of the mathematical sciences because in it we come to the fruits of mathematics." Leonardo does stress the role of theory in science and says, "Theory is the general; experiments are the soldiers." However he did not appreciate the precise role of theory or foresee what later became the true method of science. He, in fact, lacked methodology. Copernicus and Kepler, whom we shall study in more detail later, were also convinced that the world is mathematically and harmoniously designed, and this belief sustained them in their scientific endeavors.

Galileo speaks of mathematics as the language with which God wrote the great book—the universe—and unless one knows this language, it is impossible to comprehend a single word. René Descartes, father of coordinate geometry, was convinced that nature is but a vast geometrical system. He says that he "neither admits nor hopes for any principles in Physics other than those which are in Geometry or in abstract Mathematics, because thus all the phenomena of nature are explained, and some demonstrations of them can be given." Certainly by 1600 the conviction that mathematics is the key to nature's behavior had taken firm hold and stimulated the great scientific work which was to follow.

To the intellectuals of the Renaissance mathematics appealed for still another reason. The Renaissance, as we have seen, was a period in which medieval civilization and culture were discredited and new influences, information, and revolutionary movements were sweeping Europe. These men sought new and sound bases for the erection of knowledge, and mathematics offered such a foundation. Mathematics remained the one accepted body of truths amid crumbling philosophical systems, disputed theological beliefs, and changing ethical values. Mathematical knowledge was certain knowledge and offered a secure foothold in a morass. The search for truth was redirected toward mathematics.

9–7 The religious motivation in the study of nature. The decisions to study nature, to apply reason, and to seek the mathematical design of nature led to a revival of mathematical activity and to the emergence of great mathematicians. But the thinking of these men took a turn which is of interest because it shows one of the strong motivations for mathematical activity over a couple of centuries and because it played a role in the subsequent cultural history.

The mathematicians and scientists of the Renaissance were brought up in a religious world which stressed the universe as the handiwork of God. The scientists whom we shall meet shortly, Copernicus, Brahe, Kepler, Pascal, Galileo, Descartes, Newton, and Leibniz, accepted this doctrine. These men were in fact orthodox Christians. Copernicus was a member of the Church. Kepler studied for the ministry although he did not take orders. Newton was deeply religious and, when late in life he felt too exhausted to pursue creative scientific work, turned to religious studies. Some of the mathematicians were even given to excesses of religious zeal. John Napier (1550–1617), one of the inventors of logarithms, not only took an active part in the religious controversies of his day—as did Blaise Pascal—but regarded as his main task in life to show that the Pope was Antichrist. In 1593 he published an immensely popular defense of Protestantism in which he "proved" this assertion. He also "proved" that God would put an end to the world in 1700 and, of course, destroy the Antichrist. To help defeat the Spanish Catholic enemies of England and Scotland he invented weapons.

However, in the sixteenth century the new goal in the intellectual world became to study nature through mathematics and indeed to uncover the mathematical design of nature. Now Catholic teachings had by no means included this last principle, which is Greek. How then was the attempt to understand God's universe to be reconciled with the search for the mathematical laws of nature? The answer was to add a new doctrine, namely, that God had designed the universe mathematically. Thus the Catholic doctrine postulating the supreme importance of seeking to understand God and his creations took the form of a search for God's mathematical design of nature. Indeed the work of the sixteenth, seventeenth, and even some eighteenth-century mathematicians was a religious quest, motivated by religious beliefs, and justified in their minds because their work served this larger purpose. The search for the mathematical laws of nature was an act of devotion. It was the study of the ways and nature of God which would reveal the glory and grandeur of his handiwork. The Renaissance scientist was a theologian studying nature instead of the Bible. Copernicus, Kepler, and Descartes speak repeatedly of the harmony which God imparted to the universe through his *mathematical* design. Mathematical knowledge, being in itself truth about the universe, is as sacrosanct as any line of the Scriptures and indeed superior because it is clear, undisputed knowledge. Galileo says, "Nor does God less admirably discover Himself to us in Nature's actions than in the Scripture's sacred dictions." Man could not hope to perceive the divine plan as clearly as God himself understood it, but man could with humility and modesty seek to at least approach the mind of God.

One can go further and assert that these men were sure of the existence of mathematical laws underlying natural phenomena and persisted in the search for them because they were convinced *a priori* that God had incorporated them into the construction of the universe. Each discovery of a law of nature was hailed as evidence testifying more to God's brilliance than to the ingenuity of the investigator. Kepler in particular wrote paeans to God on the occasion of each discovery. The beliefs and attitudes of the mathematicians and scientists exemplify the larger cultural phenomenon which swept Renaissance Europe. The Greek works impinged on a deeply devout Christian world, and the intellectual leaders born in one and attracted by the other fused the doctrines of both.

Exercises

1. In view of what we know about Greek and medieval attitudes toward the physical world and mathematical activities in these two cultures, would you draw any conclusion about the connection between interest in the physical world and the pursuit of mathematics?
2. What events and influences led to a revival of interest in mathematics?
3. How did Renaissance scientists and mathematicians reconcile the Greek doctrine that the world is mathematically designed and the Christian doctrine that the universe is the creation of God?

Topics for Further Investigation

1. The rise of algebra in the sixteenth century.
2. Hindu and Arab mathematics.
3. The life and work of Roger Bacon.
4. The life and work of Jerome Cardan.
5. The life and work of Leonardo da Vinci.

Recommended Reading

Ball, W. W. Rouse: *A Short Account of the History of Mathematics,* 4th ed., Chaps. 6 to 12, Dover Publications, Inc., New York, 1960.

Cardan, Jerome: *The Book of My Life,* E. P. Dutton and Co., New York, 1930.

Cajori, Florian: *A History of Mathematics,* 2nd ed., pp. 83–129, The Macmillan Co., New York, 1938.

Crombie, A. C.: *Augustine to Galileo,* Chaps. 1 to 5, Falcon Press, London, 1952. Also published in paperback under the title *Medieval and Early Modern Science,* 2 vols., Doubleday and Co. Anchor Books, New York, 1959.

Crombie, A. C.: *Robert Grosseteste and the Origins of Experimental Science,* Oxford University Press, London, 1953.

Dampier-Whetham, William C. D.: *A History of Science,* pp. 65–138, Cambridge University Press, London, 1929.

Da Vinci, Leonardo: *Philosophical Diary,* Philosophical Library, Inc., New York, 1959.

Easton, Stewart C.: *Roger Bacon and His Search for a Universal Science,* Columbia University Press, New York, 1952.

MacCurdy, Edward: *The Notebooks of Leonardo da Vinci,* George Braziller, New York, 1954.

Ore, Oystein: *Cardano, The Gambling Scholar,* Princeton University Press, Princeton, 1953.

Randall, John Herman Jr.: *The Making of the Modern Mind,* rev. ed., Chaps. 1 to 9, Houghton Mifflin Co., Boston, 1940.

Russell, Bertrand: *A History of Western Philosophy,* pp. 324–545, Simon and Schuster, New York, 1945.

Smith, David Eugene: *History of Mathematics,* Vol. 1, Chaps. 5 to 8, Dover Publications, Inc., New York, 1958.

Vallentin, Antonina: *Leonardo da Vinci,* The Viking Press, New York, 1938.

CHAPTER 10

MATHEMATICS AND PAINTING IN THE RENAISSANCE

Mighty is geometry; joined with art, resistless.

Euripides

10–1 Introduction. The new currents of thought in the European Renaissance, the search for new truths to replace the discredited ones, the turn to the study of nature to obtain reliable facts, and the revived Greek conviction that the essence of nature's behavior should be sought in mathematical laws, bore fruit first in the field of art rather than science. While philosophers and scientists sought to unearth basic facts which might somehow be incorporated into their yet to be formulated new scientific method, and while mathematicians were still digesting the Greek works and awaiting inspiration for new themes, the artists, particularly the painters, reacted far more quickly and revolutionized the art of painting.

That the painters turned to mathematics to formulate their new style of painting is a little surprising, but the phenomenon has an explanation. The painters of the fourteenth, fifteenth, and sixteenth centuries were the architects and engineers of their time. They were also the sculptors, inventors, goldsmiths, and stonecutters. They designed and built churches, hospitals, palaces, cloisters, bridges, dams, fortresses, canals, town walls, and weapons. Thus Leonardo da Vinci, in offering his services to Lodovico Sforza, ruler of Milan, promises to serve as engineer, constructor of military works, and designer of war machines, as well as architect, sculptor, and painter. The artist was even expected to predict the motion of cannon balls, a by no means simple problem for the mathematics of those times. In view of these manifold activities the painter necessarily had to be something of a scientist.

Further, the Renaissance painter, unlike the builder of Gothic cathedrals, was influenced by the current doctrines which proclaimed that he learn truths from nature and that the essence of natural phenomena is best expressed through mathematics. Again, in comparison with his predecessors, he had the advantage of gleaning some mathematical knowledge from the newly recovered Greek works that were exciting the Europeans. The Renaissance painters went so far in assimilating this knowledge and in applying mathematics to painting that they produced the first really new mathematics in Europe. In the fifteenth century they were the most accomplished and also the most original mathematicians.

10–2 Gropings toward a scientific system of perspective. Before we examine just how Renaissance painters employed mathematics and thereby revolutionized the art of painting, let us see what had been going on in this field. There are various schemes, or systems of perspective as they are called, for organizing the subject matter which the artist desires to put on plaster or canvas. The two major types are the conceptual and the optical. If a conceptual system is adopted, the objects in a painting are arranged in accordance with a principle which may have intellectual appeal, but which is unconcerned about the resemblance with the actual scene being painted. For example, ancient Egyptian painting and relief work were organized conceptually. The sizes of the people portrayed often reflected the order of the subjects' importance. If a king were involved, he would, as the most important person, be portrayed largest. His servants would be drawn much smaller. The same figure might be shown in part from a frontal view and in part from a side view. Thus a head would be painted in profile, but the eye was viewed frontally; the trunk was also full-faced, but the legs again were in profile. If a dog and a man were to be painted and if the dog were actually behind the man, he would be portrayed above the man; if in front, he would be portrayed below.

In contrast to a conceptual system, an optical scheme seeks to portray a scene so that the painting makes somewhat the same appeal to the eye as does the scene itself. The painter might rely upon his own devices and judgment to create this effect or he might use some formal system. From the little we know of Greek and Roman painting we can say that the intent was to employ an optical system.

Rather early in the medieval period painting became an extensive activity. Kings, princes, and church leaders commissioned works of art to enhance buildings. The system which the medieval painters used until about 1300 was conceptual. Their objective was to portray and embellish the central themes in the Christian drama. Since the intent was to stir up religious feelings rather than to present real scenes, people and objects were drawn in accordance with conventions which had acquired symbolic meaning. Thus people were placed in unnatural, stylized positions; the general impression was one of flatness; and the entire painting had a two-dimensional effect. The backgrounds were usually solid gold to suggest that the action or people existed in some supra-earthly region.

Examples of this style of representation are abundant. A classic example of the late medieval period is found in Simone Martini's (1285–1344) "Majesty" (Fig. 10–1). Clearly this is no real scene. The background is gold. Despite the assemblage the scene looks flat; the throne especially lacks depth. There is hardly the suggestion of a floor on which the figures stand, and these appear lifeless and unrelated to one another. Moreover, sizes are not important. This painting also illustrates another conceptual device used in medieval painting, known as terraced perspective. To show a group of people arranged in depth, those farther back are placed somewhat above those in front.

Toward the end of the thirteenth century, the painters began to be influenced by the Renaissance. Since the preoccupation with religious themes still existed and paintings, in fact, continued to be commissioned mainly by church officials, the same subjects appear but in more realistic settings. The

FIG. 10-1. Simone Martini: *Majesty*. Pallazzo Communale, Siena.

painters had turned to the observation of nature and saw a real world, physical beings, earth, sea, and air. Their paintings reveal this interest in natural scenes by reflecting their efforts to render space, depth, mass, volume, and other visual effects largely through the use of lines, surfaces, and other geometric forms. To achieve naturalism they also tried to render emotions and to depict drapery folding around parts of the body as drapery actually does. People began to look like real individuals instead of types. Mysticism gradually gave way to realism and art became more and more secular.

Cimabue (*ca.* 1300), Cavallini (*ca.* 1250–1330), Duccio (1255–1318), and Giotto (1266–1337) were the leaders of the new movement to inject realism into painting and to incorporate the beauty of nature. Giotto, in particular, is often called the father of modern painting. In the works of the men cited and in those of their immediate successors, we can readily observe the search for an optical system of perspective.

Duccio's "Last Supper" (Fig. 10–2) shows what could be a real scene and offers an ambitious attempt at depth. The receding wall and ceiling lines create this effect. Moreover, pairs of lines which are parallel in the actual scene and symmetrically placed with respect to the center are drawn so as to meet on a vertical line through the center of the painting. This scheme is called vertical perspective and was developed further by other painters.

FIG. 10–2. Duccio: *Last Supper*. Opera del Duomo, Siena.

In addition, the lines of the middle portion of the ceiling come together in one area.

On the whole the picture is not too successful. The table seems to slant toward the front. The objects on the table are too much in the foreground and appear to be on the point of sliding off or toppling over. The table and the room are not seen from the same point of view. The various parts lack proportion. The failure of the painting to depict depth properly causes one to look from side to side instead of into the painting. An interesting feature characteristic of the period is the setting in a partially boxed-in room. The artists were beginning to treat nature, but for the moment limited themselves to scenes which had both interior and exterior components. They were already looking into space and were about to venture into the wide world.

Giotto painted with the definite goal of reproducing visual perceptions and spatial relations, and his paintings tend to produce the effect of photographic copies. His figures possess mass, volume, and vitality, are grouped appealingly, and are interrelated. His "Birth and Naming of St. John the Baptist" (Fig. 10–3) is typical. The partially boxed-in interior is again evident as is the use of lines and surfaces. The side walls are drawn small or foreshortened to suggest depth. The ground plane is a clear surface.

FIG. 10–3. Giotto: *Birth and Naming of St. John the Baptist.* Church of Santa Croce, Florence.

Although Giotto's paintings are not visually correct and although he introduced no new principles, his results are far better than those of his predecessors. He chose homelike scenes, gave human feelings to his figures, and distributed them in space. He catches shades of emotions and expresses them through the features and postures of the bodies. There is no mysticism nor ecstatic piety; "real" angels, Christ, and disciples stand before us. He was aware of the progress he had made and he delighted in showing his skill.

Duccio and his contemporaries attempted outdoor scenes, but the results are not good. Duccio's "Entry into Jerusalem" (Fig. 10–4) is an ambitious attempt at both naturalism and depth. Figures and buildings are piled on top of each other, and the whole scene is terribly crowded. Yet there is movement into the picture, away from the observer.

A step forward in the achievement of realism was made by Ambrogio Lorenzetti (fourteenth century). His outdoor panoramas are the best of this period. However, from the standpoint of the significant development which was to follow, his "Presentation in the Temple" (Fig. 10–5) is more worthy of attention. There is a definite foreground or horizontal plane as opposed to the background or vertical plane. The lines on the floor clearly

FIG. 10–4. Duccio: *Entry into Jerusalem.* Opera del Duomo, Siena.

recede and meet in one point. Other pairs of receding parallel lines meet in respective points of a vertical line. Also significant is the gradual decrease (foreshortening) in the size of the floor blocks to suggest distance. But the floor and the rest of the painting are not unified.

These few samples of fourteenth-century Renaissance painting show the increasing efforts to achieve naturalism, real scenes, and three-dimensionality. The innovators were groping for an effective technique but did not succeed. Visualization and sheer artistic skill were not enough.

Fig. 10–5. Ambrogio Lorenzetti: *Presentation in the Temple.* Uffizi, Florence.

10–3 Realism leads to mathematics. There was rather little progress in the second half of the fourteenth century because the Black Death seriously disturbed the life of Europe and decimated the population. The fifteenth century witnessed, as we noted in the preceding chapter, a new flood of Greek works to Italy, a new series of translations, and enormous support for artists. The Greek ideals, spread by the academies and libraries we have mentioned, became better known and were discussed enthusiastically in

Italy. Secularization was hastened, and the artist acquired a heightened interest in humanity and in the study of nature, and a zeal for science.

To achieve an accurate delineation of actual objects and a system of painting which would yield sound portraiture, painters studied nudes, the body in various postures, anatomy, expression, light, and color. Madonna and Son were portrayed as human beings suffering human emotions, and Church history was enacted by real people. Religious themes became predominantly a conventional or habitual outlet for the depiction of the real world. Later, instead of humanizing religious themes, the artists turned to glorifying man and nature. The ascetic, mystical, and devotional attitudes were dropped entirely. Still later pagan subjects were adopted. The glory and gladness of nature, the delight in physical existence, the pleasures of earth, sea, and air are the new values. Painting became entirely secular.

In their striving for realism the artists went one step further and decided that their function was to imitate nature, to depict what they saw as realistically as they could. Nature was to be the authority for what appeared on canvas, and painting was to be the science of reproducing nature accurately. Ghiberti (1378–1455), who did two principal pairs of doors for the Baptistery in Florence, says, "In modeling these reliefs I strove to imitate nature to the utmost . . . I sought to understand how forms strike upon the eye, and how the theoretical part of sculptural and pictorial art should be managed. Working with the utmost care and diligence I introduced into some of my panels as many as a hundred figures; these I modeled upon different planes, so that those nearest to the eye might appear larger, and those more remote smaller in proportion." The objective of painting, says Leonardo da Vinci, is to reproduce nature and the merit of a painting lies in the exactness of the reproduction. Even a purely imagined scene must appear to the spectator as if it existed exactly as pictured. Painting was to be a veridical reproduction of reality.

But how was the reproduction to be achieved? Here, too, the Renaissance artist adopted a Greek ideal. By the fifteenth century he had become thoroughly familiar and imbued with the Greek doctrine that mathematics is the essence of the real world. Hence to penetrate to the real substance of the theme he sought to display on canvas, the Renaissance artist believed that he must reduce it to its mathematical content. To capture the essence of forms, the organization of objects in space, and the structure of space the artist decided that he must find the underlying mathematical laws. Just what this thought means is well illustrated by Leonardo da Vinci's "Study in Proportion" (Fig. 10–6). In this sketch he tries to fit the structure of the ideal man to the square and circle. To Leonardo the true design of an object such as the human body is revealed by the geometrical properties portrayed.

But realistic painting includes more than the mathematical properties of the objects being portrayed. The eye sees the painting, and this must create on the eye the same impression as the scene itself. Also, since vision and the light which carries the scene to the eye are involved, these too must be analyzed.

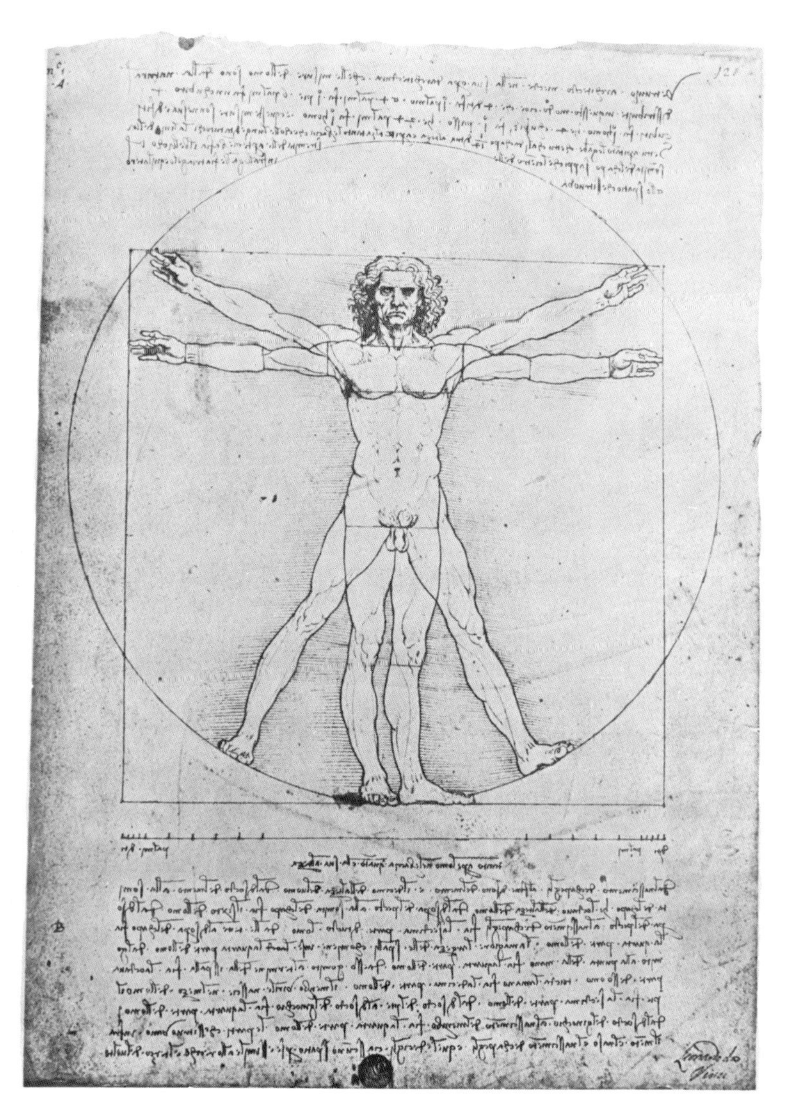

FIG. 10-6. Leonardo da Vinci: *Study in Proportion.*

It so happens that light played a most important role in Renaissance thought. Light was believed to be the fundamental reality. The origin of this belief is interesting. Saint Augustine and the Neo-Platonists had regarded light as the analogy of Divine Grace and as the symbol for the illumination of the human intellect by Divine Truth. These figurative thoughts were taken literally by the medievalists, who maintained that light penetrates everything and unites the world with God. Light was the cause of all phenomena and the substance of all bodies, the basic corporeal form. The planets, for example, exerted their influence on man and his world through light. Thus the study of light, that is optics, was basic in understanding the physical world, and the study of nature had to proceed through optics. In the fifteenth century optics was the most important branch of physics. Since light reaches us through the sense of vision, this sense was the most important one. Nature is what the eye sees. Leonardo says, "The eye, which is called the window of the soul, is the chief means whereby the understanding may most fully and abundantly appreciate the infinite works of nature; and the ear is the second, inasmuch as it acquires its importance from the fact that it hears the things which the eye has seen."

But the study of light also led to mathematics. From Greek times on, as we have already seen in earlier chapters, light had been shown to be subject to mathematical laws. Indeed, the few mathematical laws of light were about the only precise knowledge about the phenomenon which the Greek and Renaissance worlds possessed, because the nature of light itself was a mystery. And so, to study the impress of scene and painting on the eye, the artists were once again led to mathematics.

Thus, although the artists made extensive and intensive physical studies of light and shade, color, the chemistry of pigments, the laws of movement and balance, the eye, anatomy, and the effect of distance on sight, they were chiefly dominated by the new thought that mathematics must be used to achieve realism in painting and that geometry is the key to the solution of this problem. Thereupon they created and perfected a totally new mathematical system of perspective which enabled them to "place reality on their canvases."

10–4 The basic idea of mathematical perspective. The mathematical system of perspective which the Renaissance painters created and which is known as the system of focused perspective was founded about 1425 by the architect and sculptor Brunelleschi (1377–1446). His ideas were furthered and written down by the architect and painter Leone Battista Alberti (1404–1472). It is not Alberti's artistic work which entitles him to fame but his technical knowledge. He studied architecture, painting, perspective, and sculpture, wrote several books explaining theoretical matters to artists, and exercised enormous influence. In his *Della Pittura* (1435) Alberti says that learning is essential to the artist. The arts are learned by reason and method; they are mastered by practice. He says further that the first necessity of a painter is to know geometry and that painting by incorporating and revealing the mathematical structure of nature can even improve on nature.

The mathematical scheme was developed and perfected by Paolo Uccello (1397–1475), Piero della Francesca (1416–1492), and Leonardo da Vinci (1452–1519). The system these men and others created and which Leonardo called the rudder and guide rope of painting has been used since the Renaissance by all artists who seek exact depiction of reality, and is taught in art schools today.

In their study of light, vision, and the representation of objects on canvas, these artists discovered the following facts. Suppose that a person looks at a real scene from a fixed position. Of course, he sees with both eyes, but each eye sees the same scene from a slightly different position. Although in ordinary vision we need both sensations to give us some perception and measure of depth, this perception is really not very good. Experience teaches us how to interpret the combined sensations, as Leonardo points out in his *Treatise on Painting*. The Renaissance artists decided to concentrate on what one eye sees and to compensate for the deficiency by shading, shadows where pertinent, and by what is known as aerial perspective, that is the gradual diminution of the intensity of colors with distance.*

Let us imagine that lines of light are drawn from one eye to various points on the objects in the scene. This collection of lines is called a *projection*. Let us imagine next, as did Alberti, Leonardo, and the German artist Albrecht Dürer (1471–1528), that a glass screen is interposed between the eye and the scene itself. Thus when one looks out of a window at a scene outside, the window serves as the glass screen. The lines of the projection will pierce the glass screen, and we may imagine a dot placed on the screen where each line pierces it. The figure formed by these dots on the screen is called a *section*. The most important fact which the Renaissance artists discovered is that *this section makes the same impression on the eye as does the scene itself,* for all that the eye sees is light traveling along a straight line from each point on the object to the eye, and if the light emanates from points on the glass screen but travels along the very same lines, it should still create the same impression. Hence this section, which is two-dimensional, is what the artist must place on the canvas to create the correct impression on the eye. Dürer used the word "perspective" because the Latin verb from which it is derived means "to see through."

Before we investigate just how the painter is to put this section on canvas, let us study the idea of projection and section. Fortunately some woodcuts made by Dürer, who learned the mathematical system of perspective in Italy and then returned to Germany to teach it to his countrymen, are very helpful. The woodcuts are in Dürer's text *Underweysung der Messung mit dem Zyrkel und Rychtscheyed* (1525). The first of these, "The Designer of the Sitting Man" (Fig. 10–7), shows an artist looking through a glass screen; he holds his eye at a fixed position, and marks on the screen the point at which a line of light from his eye to some point on the man's body pierces the screen.

The second woodcut, "The Designer of the Lying Woman" (Fig 10–8), shows the artist again holding his eye at a fixed position and noting on paper the points where the lines of light from his eye to the woman pierce the screen. To facilitate the process of reproducing the correct location of the dots on the paper, he has divided screen and paper into little squares.

* The difference between a drawing made according to the laws of perspective and a three-dimensional picture is clear when one views a stereoscopic drawing with both eyes through colored glasses.

FIG. 10–7. Albrecht Dürer: *Designer of the Sitting Man.*

FIG. 10–8. Albrecht Dürer: *Designer of the Lying Woman.*

FIG. 10–9. Albrecht Dürer: *Designer of the Lute.*

The third woodcut, "The Designer of the Lute" (Fig. 10–9), delineates on the screen the section which the eye would see if it viewed the lute from the point on the wall where the rope is attached.

These woodcuts, then, illustrate what the artists meant by a section on a glass screen. Of course, a section depends upon the position of the glass screen as well as on the position of the observer. But this implies no more than that there can be many different paintings of the same scene. Thus, for example, two paintings can be the same except for size, and size is determined by the distance between glass screen and eye. Two paintings may differ in that one shows a frontal view and the other represents the same scene viewed somewhat from the side. The difference is due to a change in the observer's position.

10–5 Some mathematical theorems on perspective drawing. Let us accept then the principle that the canvas must contain the same section that a glass screen placed between the eye of the painter and the actual scene would contain. Since the artist cannot look through his canvas at the actual scene and may even be painting an imaginary scene, he must have theorems which tell him how to place his objects on the canvas so that the *painting will,* in effect, *contain the section made by a glass screen.*

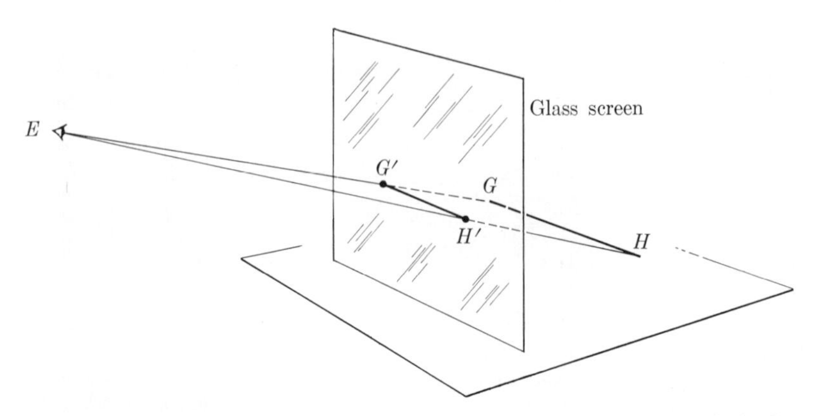

FIG. 10–10. The image of a line horizontal and parallel to the screen is horizontal.

Suppose then that the eye at E (Fig. 10–10) looks at the horizontal line GH and that GH is parallel to a vertical glass screen. The lines from E to the points of GH lie in one plane, namely the plane determined by the point E and the line GH, for a point and a line determine a plane. This plane will cut the screen in a line, $G'H'$, because two planes which meet at all meet in a line. It is apparent that the line $G'H'$ must also be horizontal, but we can prove this fact and so be certain. We can imagine a vertical plane through GH. Since GH is parallel to the screen and the latter is also vertical, the two planes must be parallel. The plane determined by E and GH cuts these parallel planes, and a plane which intersects two parallel planes intersects them in parallel lines. Hence $G'H'$ is parallel to GH, and since GH is horizontal, so is $G'H'$. But GH was any horizontal line parallel to the screen. *Hence the image on the screen of any horizontal line parallel to the screen or picture plane must be horizontal.* Hence, in a painting which is to contain what this glass screen contains, the line $G'H'$ must be drawn horizontally.

We can present practically the same argument to show that the image of any vertical line, which is automatically parallel to the vertical screen, must appear on the screen as a vertical line. *Thus all vertical lines must be drawn vertically.*

Now let us consider a somewhat more complicated situation. Suppose that AB and CD (Fig. 10–11) are two parallel, horizontal lines in an actual scene. Moreover, assume that these lines are *perpendicular* to the screen. The eye is at E. If we now imagine that lines go from E to each point of AB, these lines, that is the projection, will lie in one plane for the point E, and the line AB will determine this plane by virtue of the theorem of solid geometry already mentioned. Similarly, E and the line CD determine another plane. The screen cuts the two planes we have just described. The sections must lie on the screen, and our problem is to determine where they should lie.

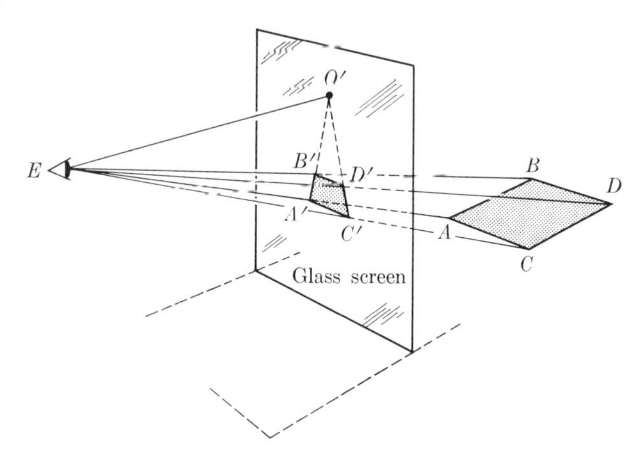

Fig. 10–11. The images of two horizontal, parallel lines which are perpendicular to the screen meet at a point on the screen.

Of course, the intersection of two planes is a line, and so the section corresponding to AB and that corresponding to CD will be lines, $A'B'$ and $C'D'$, respectively. Moreover, as the eye at E looks farther and farther out along the parallels AB and CD, the lines of sight will become more and more horizontal. As the eye follows AB and CD to infinity, so to speak, the lines from E tend to merge into one horizontal line which will be *parallel* to AB and CD. This line from E will pierce the screen at some point, say O', and this point corresponds to the imaginary point O where AB and CD seem to meet at infinity. Of course, AB and CD are parallel and do not meet, but it is convenient to think of them as meeting at a point at infinity. Indeed, the eye gets the impression that they do meet. Then the line EO' will be perpendicular to the screen because it is parallel to AB and CD and these two lines are perpendicular to the screen. The point O' corresponds to the imagined meeting point at infinity of AB and CD, but because this point does not actually exist, O' is called the *principal vanishing point.* It vanishes in the sense that it does not correspond to any actual point on AB or CD, whereas other points on $A'B'$ or $C'D'$ do correspond to actual points on AB or CD, respectively.

Now the lines AB and CD extend out to infinity to the hypothetical meeting point O; that is, ABO and CDO are lines in the real scene. The sections of these lines, $A'B'O'$ and $C'D'O'$, must therefore meet at O'. What we have shown then is that $A'B'$ and $C'D'$ must be placed on the screen so that they meet at O', and O' is the foot of the perpendicular extending from the eye to the screen. Let us now note that AB and CD are *any* horizontal lines perpendicular to the screen. *Hence all horizontal lines which are perpendicular to the screen must be drawn so as to go through* O', the principal vanishing point, which is the foot of the perpendicular from the eye to the screen.

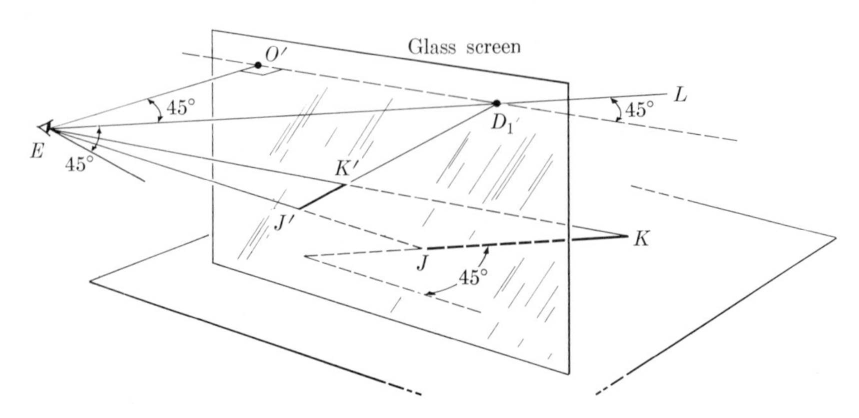

FIG. 10–12. The image of a horizontal line which makes a 45°-angle with the screen goes through a diagonal vanishing point.

We may draw another important conclusion from the preceding situation. The distances AC and BD are equal, for they are the distances between parallel lines. However, the corresponding images $A'C'$ and $B'D'$ are not equal because the lines $A'B'$ and $C'D'$ converge to O'. Moreover, $B'D'$ will be shorter than $A'C'$ because it is closer to O'. But $B'D'$ corresponds to the actual distance BD which is farther from the screen than AC is. Hence lengths which are farther from the screen must be drawn shorter than equal lengths closer to the screen. This fact is often described by the statement that, to obtain proper perspective in a painting, *lengths farther away from the observer must be foreshortened.*

We shall establish one more theorem about perspective drawing. Let us now suppose that JK (Fig. 10–12) is a horizontal line which makes an angle of 45° with the screen. Assume that the eye at E looks out along the line JK toward infinity. Then the line from the eye to the point at infinity on JK will be parallel to JK. Since JK is horizontal, the new line, EL in Fig. 10–12, will also be horizontal. It will pierce the screen at some point, say D_1, and will also make an angle of 45° with the screen. The triangle D_1EO' is a right triangle because EO' is perpendicular to the screen. In view of the acute angles of 45°, $O'D_1 = EO'$. Then the point D_1 is as far from O' as E is. The projection from E to the various points of JK cuts the screen in some line, $J'K'$, say. As the eye continues to follow JK toward infinity, the projection cuts the screen in points lying on an extension of $J'K'$, and we have already established that when the eye looks toward infinity on JK, the projection cuts the screen at D_1. Hence $J'K'$ must go through D_1. We now have another important result. The image of any horizontal line which makes an angle of 45° with the screen must go through the point D_1 which lies on the screen, on the same level as E, but is as far to the right of the principal vanishing point as E is from the principal vanishing point. The point D_1 is called a *diagonal vanishing point.*

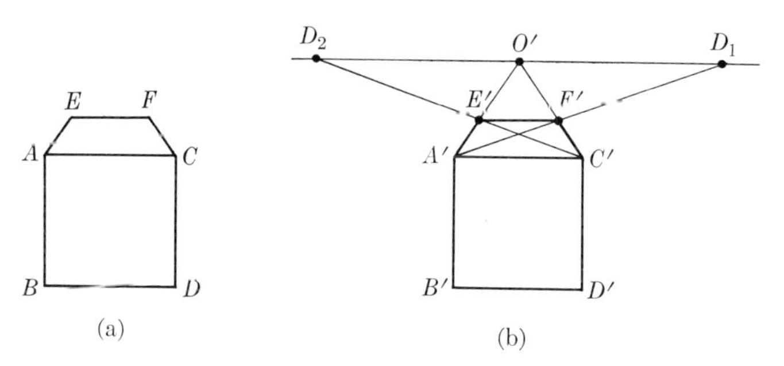

FIG. 10–13. (a) Actual cube. (b) Perspective construction of a cube.

Had we considered instead of JK lines which make an angle of 135° with the screen, we would have found that their images must go through a point D_2 which lies as far to the left of O' as E is from O'. The point D_2 is also called a diagonal vanishing point.

We see, then, that the points O', D_1, and D_2 correspond to points at infinity in the actual scene. As a matter of fact, all points on the horizontal line $D_2O'D_1$ correspond to points at infinity in the actual scene, and this line is called the *vanishing line*. It is the image of what one might call the horizon in the actual scene, that is, the points at infinity toward which the eye gazes when it looks in a horizontal direction.

These few theorems that we have proved will now be applied to draw a cube. Let us suppose that in drawing the cube, the observer is looking directly at the front face and that his eye is a little above the top face so that he can see this top face. The artist should now imagine that a vertical screen is placed parallel to the front face of the cube. Of the four front edges of the actual cube two, AB and CD of Fig. 10–13(a), are vertical, and so their images on the screen will be vertical. Hence they will be vertical on the actual drawing, which must contain what the screen contains. The distance between these two lines depends upon how large the picture is to be, and the size, in turn, depends upon where the screen is placed between eye and cube. This decision must be made by the artist. (In more detailed work than we shall pursue, he must choose a point E' above O' on the canvas such that $E'O'$ equals the distance of the eye from O'. The point E' enters into more advanced constructions which we shall not discuss.) Thus we obtain the lines $A'B'$ and $C'D'$ (Fig. 10–13b), although the positions of A', B', C', and D' remain to be determined. The two horizontal edges of the front face of the cube are parallel to the screen and so must be drawn as parallel, horizontal lines on the canvas. Since the front face is a square, the distances $A'B'$, $B'D'$, $D'C'$, and $C'A'$ must now be drawn in such a way that these sides are equal. Thus we get a square $A'B'D'C'$.

We must next tackle the top face of the cube since this is also visible. Two edges of the top face, AE and CF, are perpendicular to the screen and are horizontal lines in the actual cube. Such lines must be drawn to meet in the principal vanishing point. Thus the artist must choose a point on the canvas which, were he looking through a glass screen to the actual edges, would be the foot of the perpendicular from his eye to the screen. This point must be somewhere above the edge $A'C'$ and equidistant from A' and C'. Choosing for O' a point far above $A'C'$ will make it appear that the observer's eye is on a level high above the top face. Choosing O' close to $A'C'$ will produce the opposite effect. The lines $A'E'$ and $C'F'$ must then go through O'.

We have yet to determine where the back edge EF must appear on the drawing. If the artist is looking at the actual cube so that he faces $ABDC$ directly, then the line AF makes a 45°-angle with the screen. Hence the corresponding line $A'F'$ must go through the diagonal vanishing point to the right of O'. The painter must have in mind a constant distance from his eye to O' and, as one of our theorems showed, the distance $O'D_1$ must equal this constant distance. He must therefore fix D_1 and then draw $A'D_1$. Where this line intersects $C'O'$, the point F' must lie. Similarly, the actual line CE makes an angle of 135° with the screen. Hence the artist must now fix the point D_2 as far to the left of O' as D_1 is to the right and draw $C'D_2$. Where this line cuts $A'O'$, the point E' must lie. He can now draw $E'F'$. The line $E'F'$ will be shorter than $A'C'$ despite the fact that, on the actual cube, EF is as long as AC. The difference illustrates the principle of foreshortening.

The construction of the perspective picture of a cube is a very simple problem and hardly begins to illustrate the theorems one must know and apply to draw actual scenes realistically. The treatment of curves is especially difficult. For example, actual circles and spheres cannot, in general, be drawn as circles unless their centers happen to lie on the perpendicular from the eye to the screen. In all other cases, they must be drawn as ellipses or as arcs of parabolas or hyperbolas, depending upon their position relative to the observer. This fact becomes clear if one considers that the lines from the eye to each point on the edge of the circle or sphere, the projection, in other words, form a cone and that the section of this cone on the screen will be one of the conic sections discussed in Chapter 6. We shall not investigate the more complicated theorems because to do so would require a course in the subject and because the detailed theorems are of interest only for the specific purpose of learning to paint realistically. We may have seen enough of the basic principles to appreciate that the problem of painting realistically is handled by the application of a thoroughly mathematical system.

We know that the construction of a painting in accordance with the focused scheme presupposes a definite fixed position of the painter in relation to the scene. To view properly a painting so constructed, the observer should place himself in precisely the position the painter used in planning the painting. Otherwise the observer will get a distorted view. Strictly speaking, paintings in museums should be hung so that the observer can conveniently take that position.

FIG. 10-14. Uccello: *Rout of San Romano*. Uffizi, Florence.

10-6 Renaissance paintings employing mathematical perspective. Renaissance painters achieved their goal of devising a mathematical system which permitted the realistic representation of actual scenes and joyously hastened to employ it. Realistic paintings constructed in accordance with the focused scheme of perspective begin to appear about 1430. One of the major contributors to the mathematical science of perspective is Paolo Uccello. Giorgio Vasari, the famous painter and biographer of painters, tells us that Uccello would spend long nights in his study working out the vanishing points of his paintings and, when called to bed by his wife, replied, "How sweet a thing is this perspective." The nineteenth-century critic John Ruskin says, "He went off his head with love of perspective." Uccello deliberately undertook difficult problems in perspective drawing because he took pleasure in solving them and exploring the possibilities of the system. Painting was to Uccello an opportunity to solve problems in perspective and to display his mastery of the subject.

His "Rout of San Romano" (Fig. 10-14), one of the first paintings treating a secular outdoor theme, is an excellent example of his work. Uccello seeks a bold effect. He is particularly concerned with foreshortening. The fallen horses, pieces of armor, lances, and other objects in the foreground are strikingly and carefully foreshortened. These are difficult feats which he tackled deliberately. The figures in the distance lose size in direct proportion to their distances from the front, a principle not exactly in accord with the mathematical scheme described above, but this was Uccello's device for achieving reasonable foreshortening. Consequently his mastery of depth was not perfect. Nor did he color effectively or use atmospheric perspective, that is, decrease the intensity of color with distance. He was a

FIG. 10–15. Piero della Francesca: *The Flagellation*. Ducal Palace, Urbino.

FIG. 10–16. Piero della Francesca: *Architectural view of a city*. Kaiser Friedrich Museum, Berlin.

mathematician concerned with design and geometrical laws. Critics say he was too much distracted by perspective to apply his full powers to painting.

The artist who contributed key principles of mathematically determined perspective, including new methods of construction, and who was the best mathematician of his times is Piero della Francesca. This highly intellectual painter with a passion for geometry planned all his works mathematically to the last detail. Each scene to be painted was a mathematical problem. The placement of each figure was calculated to ensure its correctness in relation to other figures and to the painting as a whole. He loved geometrical forms so much that he used them for hats, parts of the body, and other details in his paintings. Piero practically identified painting and perspective. His *De prospettiva pingendi,* a treatise on painting and perspective in which he uses Euclid's deductive method, presents perspective as a science and provides sample constructions illustrating how perspective problems are to be handled. Though incidental to our purposes, it is worth noting that Piero painted the first Renaissance portraits of real people, the Duke and Duchess of Urbino, Federigo de Montefeltro and his wife Battista Sforza.

Like Alberti, Piero sought the harmony immanent in nature and believed that mathematics displayed it. In the last 20 years of his life he wrote three treatises to show how geometry might be employed to reduce the visible world to mathematical order.

There are numerous examples which illustrate Piero's excellent perspective. His "Flagellation" (Fig. 10-15) is one of the best. As in all of his paintings a geometric framework underlies the design. The principal vanishing point is chosen to be near the figure of Christ. This device of placing the principal vanishing point within the most important area in the painting is deliberate because the eye tends to focus on that vanishing point. All objects are carefully foreshortened; this is especially noticeable in the marble blocks on the floor and in the beams. The immense labor which went into the calculation of these sizes is indicated by a drawing in the book referred to above wherein he explains a similar construction.

Piero achieves unity of the various parts by means of the system of perspective. All parts are mathematically tied together to produce this synthesis. Indeed, it was somewhat because of this effect that the Renaissance painters valued the system and were excited about it. The examples shown here should be compared with the fourteenth-century works (Section 10-2), where unity is lacking. The entire layout of Piero's painting is so carefully planned that movement is sacrificed to the unity of design.

To illustrate the power of perspective Piero painted several scenes of cities. His "Architectural View of a City" (Fig. 10-16) gives a striking illusion of depth. These examples of Piero's paintings show his obsession for perspective and his great technique.

Leonardo da Vinci's work provides excellent examples of paintings embodying mathematical perspective. Leonardo prepared for painting by deep and extensive studies in anatomy, perspective, geometry, physics, and chemistry. In his *Treatise on Painting,* a scientific treatise on painting and

FIG. 10–17. Leonardo da Vinci: *Study for the Adoration of the Magi.* Uffizi, Florence.

FIG. 10–18. Leonardo da Vinci: *Adoration of the Magi.* Uffizi, Florence.

perspective, Leonardo gives his views. He opens with the statement, "Let no one who is not a mathematician read my works." Painting, he says, is a science which should be founded on the study of nature and, like all sciences, must also be based on mathematics. He scorns those who think they can ignore theory and by mere practice produce art: "Practice must be founded on sound theory." Painting, which he regarded as superior to architecture, music, and poetry, is a science because it deals with the geometry of surfaces.

The detailed mathematical studies which Leonardo undertook in preparation for his paintings are illustrated by one of several sketches he made for his "Adoration of the Magi" (Fig. 10–17). The painting itself, which was never completed, is shown in Fig. 10–18. His "Last Supper" is another excellent example of mathematical perspective, but is so well known that we shall reproduce instead "The Annunciation" (Fig. 10–19). Although the action takes place in the foreground and the chief figures are far apart, they and the distant scene in the rear are all brought together by the perspective structure.

A man who painted in various styles, but who was capable of excellent perspective, is Sandro Botticelli (1444–1510). In many paintings he sought and achieved realism, although in others, such as "The Birth of Venus," the treatment is purely allegorical. His "The Annunciation of the Virgin" (Fig. 10–20) presents the same theme as Leonardo does and shows his mastery of perspective which unifies the scene in the background with the principal figures.

Raphael (1483–1520) supplies many superb paintings which exhibit excellent perspective. In his "School of Athens" (Fig. 10–21) he boldly tackles an enormous scene encompassing a vast number of people within a magnificent architectural setting. The portrayal of depth, the harmonious organization, coherence, and exactness of proportions achieved despite the difficulty of the undertaking are extraordinary. This picture, especially, shows how perspective unifies a composition and ties figures at the sides to the central theme.

The history of this painting is of interest. Pope Julius II (1443–1513) was impressed with ancient learning and regarded Christianity as the climax of Jewish religious thought and Greek philosophy. He wished to have his idea embodied in paintings and commissioned both Michelangelo and Raphael to develop this theme. Michelangelo treated it in his frescoes on the ceiling of the Sistine Chapel, where he shows the human race led to Christ through a long line of Jewish prophets and pagan sibyls. Raphael executed the same theme in a somewhat different manner. In four frescoes which cover the walls of the Pope's principal official room, the Camera della Segnatura, he teaches that the human soul is to aspire to God through each of its faculties: reason, the artistic capacity, the sense of order and good government, and the religious spirit. "The School of Athens" glorifies reason and naturally exhibits the people who excelled in the intellectual sphere. Plato and Aristotle are the central figures. Plato points upward to the eternal

FIG. 10–19. Leonardo da Vinci: *Annunciation.* Uffizi, Florence.

FIG. 10–20. Botticelli: *Annunciation of the Virgin.* Uffizi, Florence.

FIG. 10–21. Raphael: *School of Athens.* Vatican.

FIG. 10–22. Raphael: *Fire in the Borgo.* Vatican.

FIG. 10–23. William Hogarth: *False Perspective.*

ideas and Aristotle down to the earth as the field of experience. At Plato's left is Socrates. In the left foreground Pythagoras writes in a book. The right foreground shows the bald-headed Euclid; Archimedes stoops to demonstrate a theorem; Ptolemy holds up a sphere. All the way to the right is Raphael himself.

Raphael offers so many examples of excellent perspective that it is difficult to limit oneself to one or two representative samples. His "The Fire in the Borgo" (Fig. 10–22) shows exquisite depth, perfect handling of figures in various positions, the proper foreshortening, and again the unification of a scene in which many actions take place.

Let us now test our understanding of the subject and see how many mistakes we can find in William Hogarth's "False Perspective" (Fig. 10–23), a drawing which he composed as a frontispiece to John Joshua Kirby's text on perspective.

10–7 Other values of mathematical perspective. We could offer countless examples illustrating the application of the mathematical system of perspective by the Renaissance masters.* All painters of this period in which western European art reached one of its pinnacles employed it and employed it well. Many, among them Uccello and Piero, were obsessed by it and painted scenes viewed from unusual positions just to solve the mathematical problems involved. The essential difference between the art of the Renaissance and that of the Middle Ages is the introduction of the third dimension, and Renaissance painting is characterized by the importance attached to realism, to the realistic rendering of space, distance, and forms, achieved by means of the mathematical system of perspective. Through it, the process of seeing was rationalized; the extended world was brought under control; and the rational interests of the painters were satisfied.

The development of perspective led not only to new ideas in mathematics, as we shall see shortly, but the technique had other uses of great value. It provided scientists with an effective weapon for studying nature accurately. Botanists, zoologists, and physiologists had been handicapped by their inability to convey to others an accurate description of the human body, animals, and plants. In fact, the needs of these scholars also prompted the study of perspective since they, too, were faced with the problem of representing a three-dimensional situation on flat paper. One of Dürer's woodcuts (Fig. 10–8) illustrates this relationship of physiology to perspective. Coming at a time when printing had been developed and the method of using wood blocks (a die for printing, cut in relief on wood) to reproduce drawings in books was discovered, the science of perspective was doubly useful. In 1543 Vesalius and John of Calcar produced the first fully illustrated anatomy. The use of photographs from about 1850 on has somewhat replaced the need for accurate drawing in these areas.

Another practical application, which results from the introduction of photography, lies in the field known as photogrammetry. The camera produces a picture exactly in accordance with the system of mathematical perspective. The lens takes the place of the eye and the film is the screen. However, in this case the screen lies behind the eye. The rays of light go from the scene through the lens to the film, and the film makes a section of these lines of light. In aerial photography various pictures are taken of a given scene during which the camera is held at different angles to the ground. The task of photogrammetry is the reconstruction of the actual scene from the pictures, a process which includes obtaining distances between objects,

* The same theme is treated in the author's *Mathematics in Western Culture*. Other examples can be found there.

heights of buildings or hills, and other information. The mathematics of perspective furnishes the answers to such questions.

In the history of culture the accomplishments of the Renaissance artists have broad significance. Their apparent goal was to gaze at nature and to depict what they saw on canvas, but their true, more profound objective was to uncover the very secrets of nature. The Renaissance artist was a scientist, and painting was a science not merely in the sense that it had a highly technical and even mathematical content, but because it was inspired by the ultimate goal of science, understanding nature. Art and science are never separated in the thinking and work of Ghiberti, Alberti, and Leonardo, for example. Leonardo's *Paragone, A Comparison of the Arts* (*Treatise on Painting*) contains a chapter on "Painting and Science" in which he asserts that painting seeks the truths of nature. The artist of that period regarded himself as the servant of science. These men who explored and represented nature with methods peculiar to their art were motivated precisely by the spirit and objectives of the scientists who studied astronomy, light, motion, and other phenomena. They were in fact the forerunners in spirit and goals of the great physical scientists of modern times and they revealed truth in a form which means more to many people than the deep and intricate analyses of modern mathematical physics. That mathematics proved to be the foundation of painting and thereby enabled painting to reveal the structure of nature was no more than fitting, for the Greeks had already shown that mathematics was the essence of design, and later scientists were to confirm this fact in ever more striking fashion.

The works of the Renaissance artists are hung in art museums. They could, with as much justification, be hung in science museums. The lover of Renaissance art is consciously or unconsciously a lover of science and mathematics.

Exercises

1. Relate the "back to nature" movement of the Renaissance to the development of a mathematical system of perspective.
2. Distinguish between conceptual and optical systems of perspective.
3. Which artists did most to create a mathematical system of perspective?
4. What is the principle of projection and section in the theory of perspective?
5. Draw the rear wall, and the visible portions of the side walls, ceiling, and floor of a room as seen by an observer in the room whose eye is looking directly at the rear wall.
6. Add to the drawing of the preceding exercise a square table, two of whose edges are parallel to the rear wall.
7. Draw a cube positioned in such a way that one edge is closest to you and that neighboring edges make angles of 45° and 135°, respectively, with the canvas. Go as far as you can with the theorems at your disposal.
8. State three theorems of the geometry of perspective drawing.
9. Find six errors of perspective in Hogarth's drawing.

Topics for Further Investigation

1. The influence of mathematics on Renaissance painting
2. Theories of the artists on human proportions. Use the reference to Panofsky's *Meaning in the Visual Arts.*
3. Vision and painting. Use the reference to Helmholtz.

Recommended Reading

Bunim, Miriam: *Space in Medieval Painting and the Forerunners of Perspective,* Columbia University Press, New York, 1940.

Blunt, Anthony: *Artistic Theory in Italy,* Oxford University Press, London, 1940.

Clark, Kenneth: *Piero della Francesca,* Oxford University Press, New York, 1951.

Cole, Rex V.: *Perspective,* Seeley, Service and Co., Ltd., London, 1927.

Coolidge, Julian L.: *Mathematics of Great Amateurs,* Chap. 3, Oxford University Press, London, 1949.

da Vinci, Leonardo: *Treatise on Painting,* Princeton University Press, Princeton, 1956.

Fry, Roger: *Vision and Design,* pp. 112–168, Penguin Books Ltd., Baltimore, 1937.

Helmholtz, Herman von: *Popular Lectures on Scientific Subjects,* Vol. II, Chap. 3 entitled "On the Relations of Optics to Painting," Longmans, Green and Co., London, 1881.

Ivins, Wm. M. Jr.: *Art and Geometry,* Harvard University Press, Cambridge, 1946.

Johnson, Martin: *Art and Scientific Thought, Part Four,* Faber and Faber, Ltd., London, 1944.

Kline, Morris: *Mathematics in Western Culture,* Chap. 10, Oxford University Press, New York, 1953.

Lawson, Philip J.: *Practical Perspective Drawing,* McGraw-Hill Book Co., Inc., New York, 1943.

Panofsky, Erwin: "Dürer as a Mathematician," pp. 603–621 of James R. Newman: *The World of Mathematics,* Simon and Schuster, New York, 1956.

Panofsky, Erwin: *Meaning in the Visual Arts,* Chap. 6, Doubleday Anchor Books, New York, 1955.

Pope-Hennessy, John: *The Complete Work of Paolo Uccello,* Phaidon Press, London, 1950.

Porter, A. T.: *The Principles of Perspective,* University of London Press Ltd., London, 1927.

Vasari, Giorgio: *Lives of the Most Famous Painters, Sculptors and Architects,* E. P. Dutton, New York, 1927, and many other editions.

CHAPTER 11

PROJECTIVE GEOMETRY

The gods did not reveal all things to men at the start; but as time goes on, by searching, they discover more and more.

Xenophanes

11–1 The problem suggested by projection and section. The origins of mathematical ideas are far more novel and surprising than is commonly believed. Practical and scientific problems no doubt most frequently suggest new areas of exploration. However, not only are there other sources, but these often give rise to major branches of mathematics, some of which become valuable tools in scientific and practical endeavors. The questions raised by the painters during their work on the mathematics of perspective caused them and, later, professional mathematicians to develop the subject known as projective geometry. This subject, the most original creation of the seventeenth century, is now one of the principal branches of mathematics.

Let us see just what problems led to this development. The basic mathematical concepts in the system of focused perspective are projection and section. A projection, we may recall, is a set of lines of light from the eye to the points of an object or scene; a section is the pattern formed by the intersection of these lines with a glass screen placed between the eye and the object viewed.

Let us consider, as an example, the section of the projection of a square. Suppose that the square is horizontal (Fig. 11–1) and is viewed from a point somewhere above the level of the plane in which it lies. Furthermore, let us suppose that a vertical glass screen is placed parallel to the front and

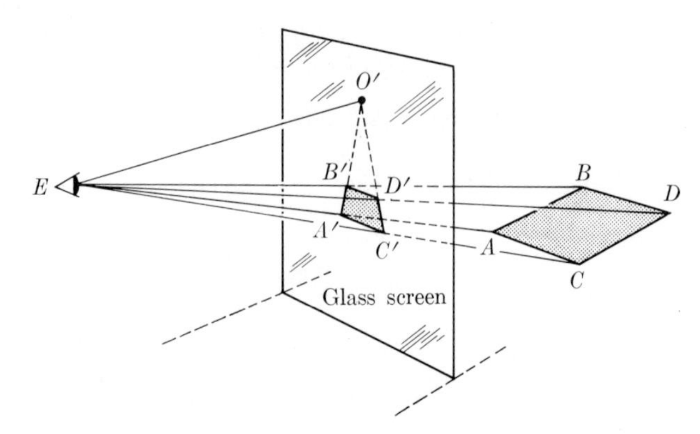

FIG. 11–1. The section of the projection of a horizontal square.

back edges of the square. We know from our work on perspective that the section on the screen of the two sides of the square which are perpendicular to the glass screen will consist of two line segments which tend to meet in the principal vanishing point. That is, the extensions of $A'B'$ and $C'D'$ meet at O'. Since the sides AC and BD of the square are parallel to the screen, they give rise to the parallel sections $A'C'$ and $B'D'$.

The section $A'B'D'C'$ is not a square because the sides $A'B'$ and $C'D'$ are not parallel. Nor is it a rectangle. The angles, then, of the section do not equal the corresponding angles of the original figure. The size of the section clearly depends upon where the glass screen is placed. Hence neither the lengths of the sides of the section nor the area equals the corresponding quantities in the original figure. In the language of Euclidean geometry, we may say that the section is neither congruent, similar, nor equivalent to the original figure. But the section does create the same impression on the eye as the original figure does. Hence it should possess some properties in common with the original. The question then becomes, What geometrical properties do the section and original figure have in common?

It is a natural step from this question to the next one. If two observers view the same scene from different positions, two different projections are formed (Fig. 11-2). If a section of each of these projections is made, then, in view of the fact that the sections are determined by and suggest the same scene, the sections should possess common geometrical properties. What are they?

These questions, which were originally raised by the artists, could occur to anybody who thinks about what he sees. Suppose that an observer looks

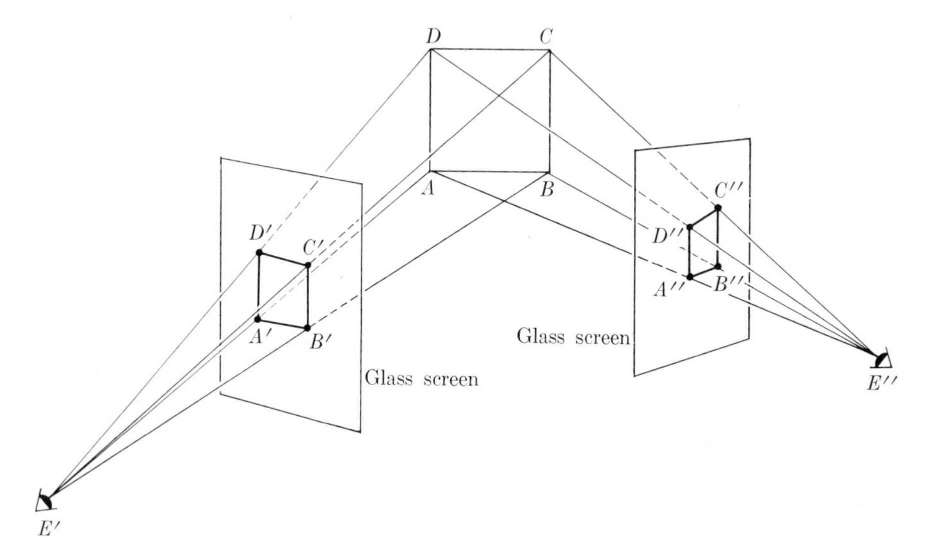

FIG. 11-2. Two different sections of two different projections of a square.

at a rectangular picture frame from various positions. The figure he actually perceives varies with his position. Thus he would be led to ask the same question that we raised above, What properties do these various shapes have in common? Again, as a man walks along a street and near a street lamp, his shadow changes in size and shape. The projection here consists of the lines of light from the street lamp, which takes the place of the eye, extending to the outlines of the man's body and then continuing to the ground. The section made by the plane of the ground is the shadow, although in this case the lamp and the section are on opposite sides of the actual object. The mathematical question raised by this example is, What properties do the various shadows have in common?

Another familiar, contemporary example illustrating projection and section is the photograph. In this case, the eye is the lens of the camera, and rays of light proceed from the scene through the lens to the film. The section is made by the film. We know that different sections can be obtained by placing the camera closer or farther from the scene being photographed or by tilting the camera. Yet all photographs of the same scene should contain some common geometrical properties.

The study of properties common to a figure and a section of a projection of that figure, or to two sections of the same projection, or to two sections of two different projections of the same figure has led to new concepts and theorems which today comprise a whole new branch of geometry, namely projective geometry. We shall attempt in this chapter to gain some understanding of the nature of this subject.

11–2 The work of Desargues. If we study any section of the projection of a figure and consider its relation to the original figure, then a few facts are readily noted. Mathematically the eye is a point, and this point and any line in the actual figure determine a plane, which is the projection determined by the line. The glass screen which cuts the projection is also a plane, and since two nonparallel planes meet in a line, the section is a line. Hence corresponding to a line in the actual scene there is a line in the section. We may therefore say that the property of linearity is common to an actual line and any section of a projection of that line. Similarly, it is easy to visualize that two intersecting lines of the actual figure will generate two intersecting lines in the section. This, then, is another minor mathematical property which is common to actual object and section. It follows that a triangle will give rise to a triangular section, although the shape of the triangle in the section will not necessarily be the same as that of the original triangle. Likewise a quadrilateral will correspond to a quadrilateral.

But the discovery of these few properties which a figure and its section or two sections of the same projection have in common hardly elates one nor does it give us any significant answers to the question of what properties figures and sections or various sections have in common. The first man to explore the problem and come up with nontrivial answers was Girard Desargues (1593–1662). Desargues was not a professional mathematician;

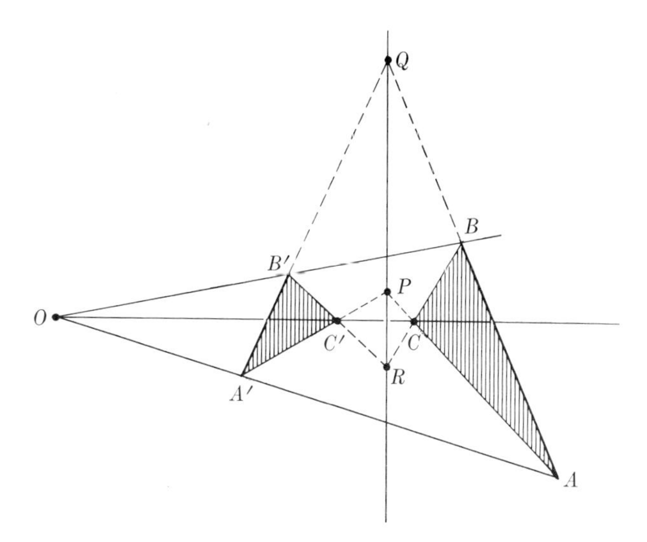

FIG. 11-3. Desargues' theorem.

he was a self-educated architect and engineer. His motive in tackling the subject was to help his colleagues. He believed that he could compile in compact form the many theorems on perspective that were useful to architects, engineers, painters, and stone-cutters. He even invented a special terminology which he thought would be more comprehensible to craftsmen and artists than the usual mathematical language. Of his motives Desargues writes: "I freely confess that I never had taste for study or research either in physics or geometry except in so far as they could serve as a means of arriving at some sort of knowledge of the proximate causes . . . for the good and convenience of life, in maintaining health, in the practice of some art, . . . having observed that a good part of the arts is based on geometry, among others the cutting of stone in architecture, that of sundials, that of perspective in particular." Desargues began by organizing numerous theorems and published one book on the construction of sundials and another on the application of his own geometric theories to stone-cutting and masonry. He lectured in Paris about 1626, and wrote a pamphlet on perspective ten years later. Desargues' chief contribution, a book on projective geometry, appeared in 1639.

The basic theorem of projective geometry, a theorem now fundamental in the entire field of mathematics, was formulated and proved by Desargues and is named after him. It illustrates how mathematicians responded to the questions raised by perspective.

Suppose the eye at point O (Fig. 11-3) looks at a triangle ABC. The lines from O to the various points on the sides of the triangle constitute, as we know, a projection. A section of this projection will then contain a triangle, $A'B'C'$, where A' corresponds to A, B' to B, and C' to C. Alternatively, we may regard both triangles as sections of a projection of a third triangle. The two triangles, ABC and $A'B'C'$, are said to be perspective from the point O. Desargues' theorem states an important geometrical fact which relates triangles ABC and $A'B'C'$. If we prolong the corresponding sides AC and $A'C'$, they will meet in a point P; the sides AB and $A'B'$ extended will meet in a point Q; and the extensions of BC and $B'C'$ will meet in a point R. Then, the theorem asserts, P, Q, and R will lie on a straight line. In more compact language the theorem says:

If two triangles in different planes are perspective from a point, the three pairs of corresponding sides meet in three points which lie on one straight line.

The proof of this theorem is simple. The lines AC and $A'C'$ lie in one plane because OAA' and OCC' are two intersecting lines and, as such, determine a plane. Then AC and $A'C'$ will meet in a point because any two lines in one plane meet in a point.* Let us denote this point by P. The same argument shows that AB and $A'B'$ meet in a point Q, and that BC and $B'C'$ meet in a point R.

We now wish to show that P, Q, and R lie on one line. But P, Q, and R lie in the plane of triangle ABC because P lies on AC, Q on AB, and R on BC. Likewise P, Q, and R lie in the plane of triangle $A'B'C'$. Now the points common to two planes must lie on one line, the line of intersection of the two planes. Hence, P, Q, and R lie on one line.

The reader may be troubled about the assertion in Desargues' theorem that each pair of corresponding sides of the two triangles must meet in a point. He may ask, If these sides happen to be parallel, does the theorem fail? Desargues had taken account of this possibility. We observed in the preceding chapter that a set of lines which are parallel in a particular scene being painted may have to be drawn on the canvas so as to meet in a point. In this case, there is a point in the section which does not correspond to any point in the scene itself. This breakdown of the correspondence between points in the scene and points in the section can be repaired by agreeing that any set of parallel lines is to be regarded as having a point in common. Where is this point? The answer is that it cannot be visualized, although the student is often advised to think of it as being at infinity, a bit of advice which essentially amounts to answering a question by not answering it. However, whether or not one can visualize the point common to parallel lines, a point distinct from the usual, finitely located points of the lines, it is convenient to say that they have a point in common. In addition, it is agreed that two or more parallel lines are to have just *one* point in common as do nonparallel lines. Hence we say of *any* two lines in projective geometry that they meet in one and only one point. This agreement is further recommended by the argument that projective geometry is concerned with problems which arise from the phenomenon of vision, and we never *see* parallel

* We shall neglect for the moment the special case of two parallel lines.

lines, as the familiar example of apparently converging railroad tracks illustrates.

One more agreement must be made about these new points introduced in projective geometry. We just agreed that any set of parallel lines has one point in common. Since there are many sets of parallel lines in one plane, each set having its own direction, there are many such new points in the plane of projective geometry. It is agreed that all these new points lie on one new line, sometimes called the line at infinity.

Let us now turn back to Desargues' theorem. If each of the three pairs of corresponding sides of the triangles presented in Desargues' theorem consisted of parallel lines, it would follow from our agreements that the two lines of each pair intersect in one point and that the three points of intersection lie on one line, the line at infinity. These conventions or agreements about points and a line at infinity obviate the necessity for making special statements when parallel lines happen to be involved in the theorems. This is as it should be, because the property of parallelism plays no role in projective geometry as opposed to Euclidean geometry. The reader who balks at accepting the agreements about parallel lines may nevertheless accept the theorems of projective geometry, with the mental reservation that they fail to state the truth when parallel lines are involved.

Desargues discovered another fundamental property which is common to a figure and to a section of a projection of that figure. Let us consider the figure consisting of the line l on which any four points are selected and designated A, B, C, and D (Fig. 11–4). We may form a projection of this figure from an arbitrary point O and cut this projection with the usual glass screen to obtain a section. The original figure and the point O determine a plane. Then the section consists of the line l' on which A' corresponds to A, B' to B, and so on. Since length in the figure differs from length in a section or, to use more technical phraseology, length is not invariant under projection and section, we should not expect that $A'B'$ would equal AB, or that any segment on l would equal the corresponding segment on l'. One might next consider the ratio CA/CB and venture that perhaps this ratio would equal the corresponding ratio $C'A'/C'B'$. This

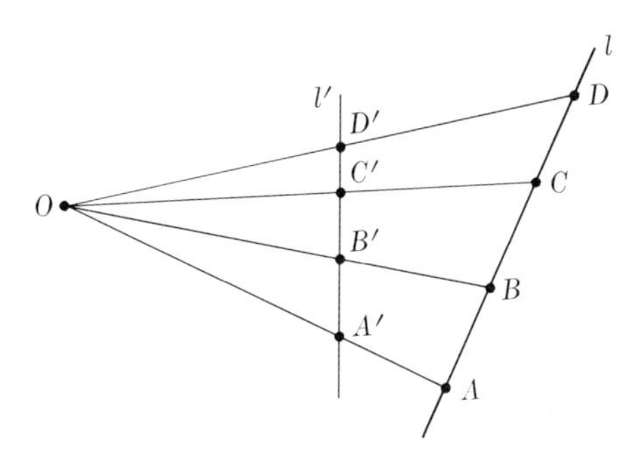

Fig. 11–4. A section of a projection of four points on a line.

conjecture is not correct. However, one can prove the surprising fact, namely that

$$\frac{CA/CB}{DA/DB} = \frac{C'A'/C'B'}{D'A'/D'B'}.$$

Thus this ratio of ratios, or cross ratio as it is called, is a projective invariant. This is a very surprising fact. It does not matter where the points A, B, C, and D lie on the line l or which points are labeled A, B, C, or D. The cross ratio of the lengths they determine and the cross ratio of the corresponding lengths in the section will be the same.

Incidentally the fact that the cross ratio of four points on a line is the same in the section as in the original figure permits us to check the correctness of a painting executed in accordance with the system of focused perspective. If four points A', B', C', D' in the painting correspond to four points A, B, C, D which lie on one line in the original scene, then the cross ratio of the first set must equal the cross ratio of the second set. This fact, however, is not too useful in constructing the painting itself.

Exercises

1. What fact or facts may you assert about: (a) a section of a projection of an equilateral triangle; (b) a section of a projection of a square?

2. Why should one expect that a figure and a section of a projection of that figure should possess some geometrical properties in common?

3. Figure 11–5 shows two triangles lying in the same plane, the plane of the paper. Moreover, these two triangles are perspective from the point O.

(a) What is the difference between this figure and the one considered in the text? (b) Verify by actually drawing lines that the pairs of corresponding sides of the two triangles in Fig. 11–5 meet in three points which lie on one straight line. (c) What generalization of Desargues' theorem is suggested by the result in (b)?

4. Given the points and lengths in Fig. 11–6, what is the cross ratio of the lengths determined by A, B, C, and D?

5. What is the meaning of the statement that a geometrical property is invariant under projection and section?

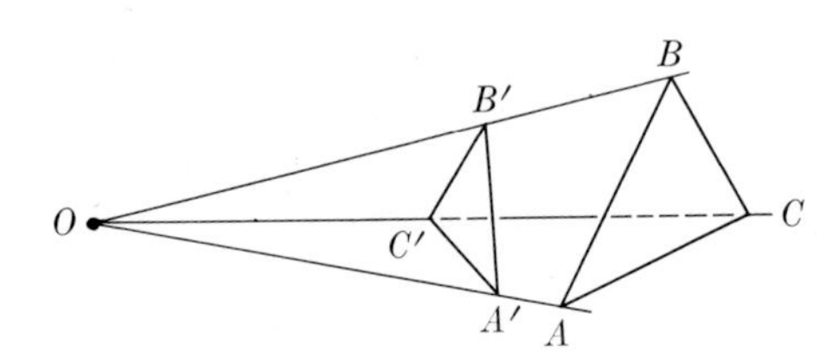

Figure 11–5

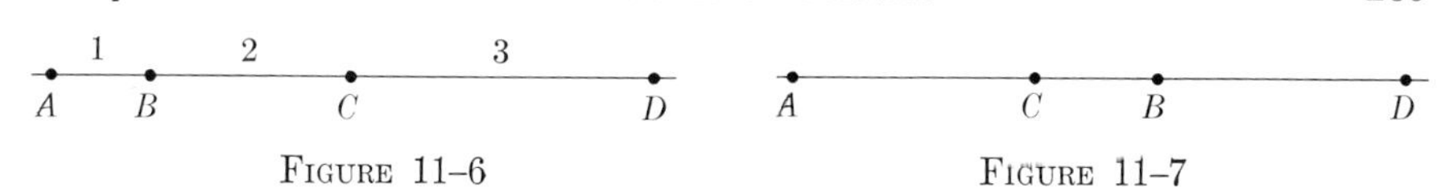

FIGURE 11-6 FIGURE 11-7

6. In projective geometry, lengths are sometimes regarded as directed. For the positions shown in Fig. 11-7, CA would be taken to be negative, whereas CB is positive because it is in the opposite direction. Then the cross ratio

$$\frac{CA/CB}{DA/DB}$$

in the present case is negative. When the cross ratio is -1, the four points A, B, C, D are said to form a *harmonic* set. Suppose now that D is moved indefinitely far to the right and C is moved so as to keep the cross ratio -1. Can you discover any special property of the point C in relation to A and B?

11-3 The work of Pascal. In the domain of projective geometry, Desargues' ideas were further advanced by Blaise Pascal (1623–1662). This man of many contradictory qualities, beset by deep emotional conflicts, is commended to us also by his superb original work in other branches of mathematics, physics, literature, and theology. His father, a judge and tax commissioner, recognized that his son was bright and guided his education. He decided that Blaise should not tackle mathematics until he was 16 years old, but somehow the boy got started on his own and learned a good deal quickly.

When Pascal was still a youth in Paris his father took him to the weekly sessions of a group of noted intellectuals, Roberval, Mersenne, Mydorge, and others. There Blaise met Desargues and as a result became interested in studying properties of geometric figures which remain invariant under projection and section. At the age of 16 he proved a famous theorem, still called Pascal's theorem, which we shall examine shortly. He then wrote the *Essay on Conic Sections,* which contains many original results. Mathematics became his great passion. To aid his father he conceived the idea of having a machine perform arithmetic operations, and he constructed the first successful computing machine. He also was one of the notable precursors of Newton and Leibniz in the creation of the calculus and, together with Pierre de Fermat, another great French mathematician whom we shall meet later, founded the theory of probability.*

Pascal made some famous physical experiments which confirmed the discovery by Evangelista Torricelli, a pupil of Galileo, that the air presses down upon us, or that the air has weight, and he also clarified and furthered the study of pressure in liquids, a study which, in technical parlance, is known as hydrostatics.

It was clear to Pascal that the data of experience are the starting points of knowledge and he respected and superbly exercised the power of reason.

---

* See Chapter 30.

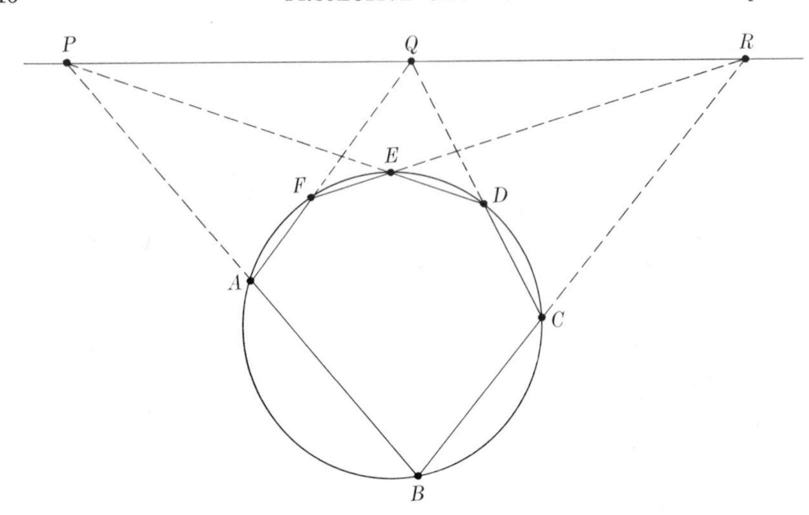

FIG. 11–8. Pascal's theorem.

His *Spirit of Geometry,* an essay on method and rules of thought, may well be classed with Descartes' *Discourse on Method,* another landmark on the role of reason. But as he grew older, he became more and more dissatisfied with the limited results attained by reason. About ten years before he died, he began to find emptiness in the knowledge of nature and acquired some distaste for it. "Don't overrate science," he cautioned. He became convinced that the truths of mathematics were not broad enough to encompass all of man's world. He would frequently say that all the sciences could not comfort one in days of affliction, but that the doctrines of Christian truth would comfort one at all times both in affliction and in one's ignorance of these sciences. Famous are his epigrams, "The heart has its reasons which the reason knows nothing of" and "Nothing that has to do with faith can be the concern of reason." More and more he turned to religion. He had been brought up as a Catholic, but he would not accept the strict dogmatic theology of the powerful Jesuits. He became a Jansenist and his *Provincial Letters,* one of his famous literary works, is filled with anti-Jesuitical polemics. In his *Pensées,* another literary classic, he penned many more of his thoughts on religion. The conflict between science and faith ended with a victory for religion. Ironically, Pascal, the defender of faith, helped immensely to found the ensuing Age of Reason.

Typical of theorems in projective geometry is the one conceived and proved by Pascal. This theorem, like Desargues', states the property of a geometrical figure which is common to the figure and to any section of any projection of that figure; that is, it states a property of a geometrical figure which is invariant under projection and section.

Pascal has this to say: Draw any six-sided polygon (hexagon) inscribed in a circle and letter the vertices A, B, C, D, E, F (Fig. 11–8). Prolong a pair of opposite sides, AB and DE say, until they meet in a point, P.

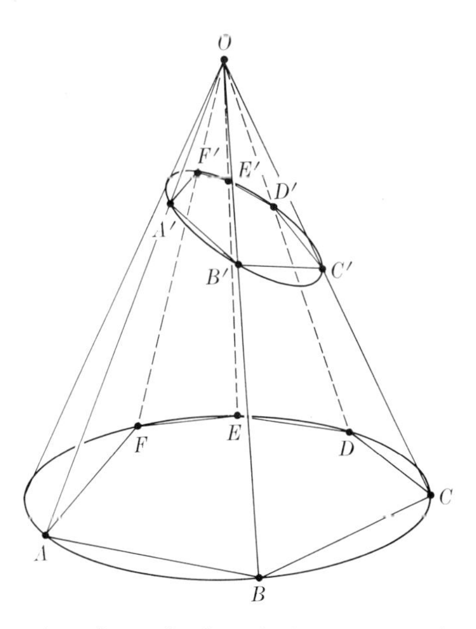

FIG. 11-9. A section of a projection of a hexagon inscribed in a circle.

Extend another pair of opposite sides, AF and CD say, until they meet in a point, Q. Finally, prolong the third pair until they meet in a point, R. Then, Pascal asserts, P, Q, and R will always lie on one straight line. The mathematician, with his usual passion for brevity, says:

If a hexagon is inscribed in a circle, the pairs of opposite sides intersect in three points which lie on one straight line.

We shall not give the proof of this theorem because it would require more time than we should devote to the subject. The statement, however, offers another illustration of the type of theorems investigated in projective geometry.

As stated, Pascal's theorem seems to have little to do with properties common to all sections of a projection. However, let us visualize (Fig. 11-9) a projection of the figure involved in Pascal's theorem and a section of this projection. The projection of the circle is a cone, and a section of this cone will not necessarily be a circle, but, as we know from the work of the Greeks, it may be an ellipse, a parabola, or a hyperbola, that is, a conic section. To each side of the hexagon inscribed in the circle there corresponds a side of the hexagon inscribed in this conic section, and to each intersection of lines in the original figure there belongs an intersection of the corresponding lines in the section. Finally, since the points P, Q, R lie on one line in the original figure, the corresponding points will be on one line in the section. Hence Pascal's theorem states a property of a circle which continues to hold in any section of any projection of that circle.

Exercises

1. Draw a circle, choose any six points on it as the vertices of a hexagon, find the points of intersection of the three pairs of *opposite* sides, and see whether they lie on a straight line.

2. Draw any two straight lines. Choose three points on one and label them A, B, C. Choose three points on the second line and label them A', B', C'. Find the point of intersection of AB' and $A'B$; do the same for AC' and $A'C$, and for BC' and $B'C$. Do you observe any interesting fact about these three points of intersection?

11–4 The principle of duality. It would be pleasant to relate that the innovations of Desargues and Pascal were immediately appreciated by their fellow mathematicians and that the potentialities in their methods and ideas were eagerly seized upon and further developed. Actually this pleasure is denied to us. Perhaps Desargues' novel terminology baffled his contemporaries just as many people today are baffled and repulsed by the language of mathematics. At any rate, except for Descartes, Pascal, and Fermat, Desargues' colleagues exhibited the usual reactions to radical ideas: they dismissed them, called Desargues crazy, and forgot projective geometry. Desargues himself became discouraged and returned to the practice of architecture and engineering. Every printed copy of Desargues' book, originally published in 1639, was lost. Pascal's work on conics and his other studies on projective geometry, though published in 1640, also remained unknown until almost 1800. Fortunately, a pupil of Desargues, Philippe de La Hire, made a manuscript copy of Desargues' book which was accidentally picked up in a bookshop by the nineteenth-century geometer Michel Chasles, and thus the world finally learned the full extent of Desargues' major work. Apart from some results which La Hire used and which were incorrectly credited to him for 150 years, Desargues' and Pascal's discoveries had to be remade one by one by the nineteenth-century geometers.

Another reason for the neglect of projective geometry during the seventeenth and eighteenth centuries was that analytic geometry, created by Desargues' contemporaries, Descartes and Fermat, and the calculus, developed chiefly by Newton and Leibniz during the latter half of the seventeenth century, proved to be so useful in the rapidly expanding branches of physical science that mathematicians concentrated on these subjects.

The study of projective geometry was revived through a series of accidents and events almost as striking as those which had first stimulated interest in this discipline. The problem of designing fortifications attracted the geometrical talents of Gaspard Monge (1746–1818), the inventor of descriptive geometry. It is relevant that this subject, though distinct from projective geometry, uses projection and section. Monge was a most inspiring teacher, and there gathered about him a host of bright pupils, among them Charles J. Brianchon (1785–1864), L. N. M. Carnot (1753–1823), and Jean Victor Poncelet (1788–1867). These men were so impressed by Monge's geometry that they sought to show that geometric methods could accom-

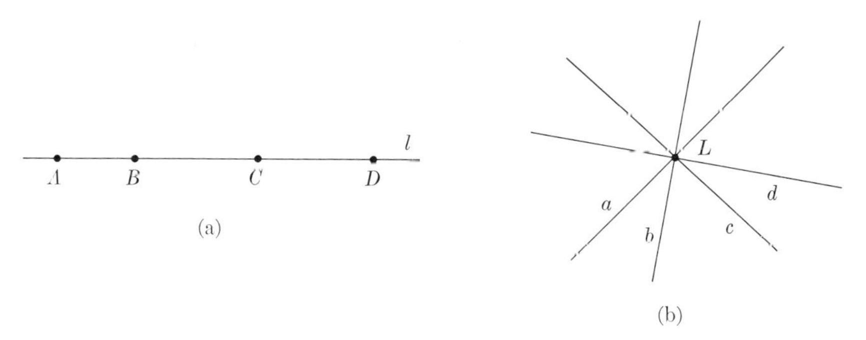

FIG. 11-10. (a) A set of points on a line; (b) a set of lines on a point.

plish as much and more than the algebraic or analytical methods of treating geometry introduced by Descartes. Carnot in particular wished "to free geometry from the hieroglyphics of analysis." As if to take revenge on Descartes whose creation had caused the abandonment of pure geometry, the early nineteenth-century geometers made it their objective to outdo Descartes.

The revival of projective geometry was launched dramatically by Poncelet. While serving as an officer in Napoleon's army, he was captured during the campaign in Russia and spent the year 1813-1814 in a Russian prison. There Poncelet reconstructed without the aid of any books all he had learned from Monge and Carnot and then proceeded to create new results in projective geometry. He was the first mathematician to appreciate fully that this subject was indeed a totally new branch of mathematics, and he consciously sought properties of geometrical figures which were common to all sections of any projection of a given figure. A group of French and, later, a group of German mathematicians continued Poncelet's work and developed intensively the subject of projective geometry.

The many accomplishments of this period were capped by the discovery of one of the most beautiful principles in all mathematics—the principle of duality. It is true in projective geometry, as in Euclidean geometry, that any two points determine one line or, as we now prefer to put it, *any two points lie on one line.* But in projective geometry it is also true that *any two lines* determine, or *lie on, one point.* (The reader who has refused to accept the convention that parallel lines in Euclid's sense are also to be regarded as having a point in common will have to forego the next few paragraphs and pay for his stubbornness.) It will be noted that the second italicized statement can be obtained from the first one by merely interchanging the words point and line. We say in projective geometry that we have dualized the original statement or that one is the dual of the other. If we are discussing a *set of points on a line* and interchange "point" and "line," we obtain the phrase a *set of lines on a point.* Figure 11-10 illustrates the two dual statements.

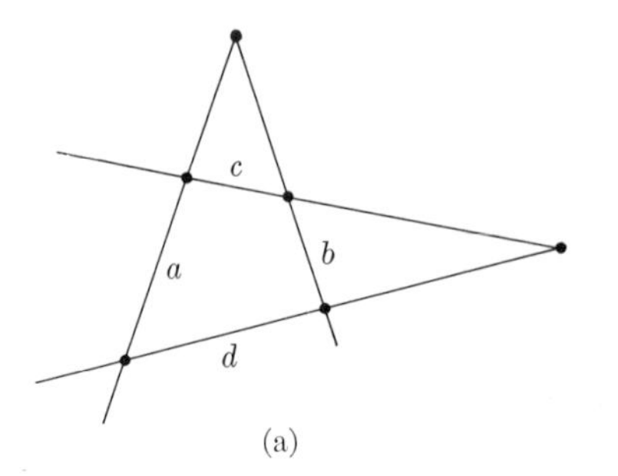

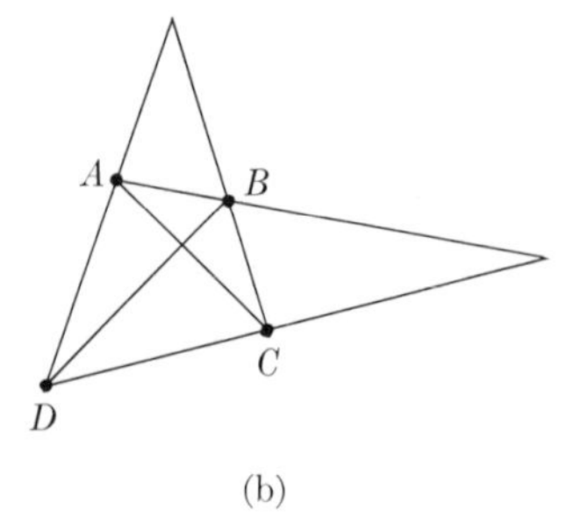

FIG. 11–11. (a) A quadrilateral; (b) a quadrangle.

A triangle consists of *three points not all on the same line and the lines joining them.* We could speak of *three lines not all on the same point and the points joining them.* We usually do not speak of a point as joining two lines; rather, we refer to such a point as the point of intersection of the lines. But the meaning is clear either way. The figure described by the rephrased or dualized statement is again a triangle. Because the dual figure of the triangle is a triangle, the triangle is called self-dual.

In projective geometry the quadrilateral is defined as a figure consisting of *four lines and the six points in which the lines join in pairs.* This use of the term "quadrilateral" differs slightly from the one common in Euclidean geometry. A picture of a quadrilateral is shown in Fig. 11–11(a). We can equally well speak of a figure consisting of *four points and the six lines which join the points in pairs* (Fig. 11–11b). This new figure is called a quadrangle. Hence quadrilateral and quadrangle are dual figures.

We seem to be able to take the statement describing any figure and by dualizing the statement obtain a new figure. Let us try next something more ambitious. We shall dualize Desargues' theorem. We shall consider the case where the two triangles and the point O from which the triangles are perspective all lie in one plane, and see what results when we interchange point and line. We shall use the fact already noted that the dual of a triangle is a triangle.

Desargues' Theorem

If we have two triangles
such that lines joining
corresponding vertices
lie on one point, O,
then corresponding sides
join in three points
which lie on one straight line.

Dual of Desargues' Theorem

If we have two triangles
such that points which join
corresponding sides
lie on one line, o,
then corresponding vertices
are joined by three lines
which lie on one point.

If we examine the new statement we see that it is actually the converse of Desargues' theorem; that is, the hypothesis and the conclusion in Desargues' theorem are now interchanged. Hence by interchanging point and line we have discovered a possible theorem. It would be too much to ask that the proof of the new statement should be obtainable from the proof of the original theorem by interchanging point and line. Although it is too much to ask, the gods have been generous beyond our merits, for the new proof can indeed be obtained in this way.

The principle of duality, as thus far described, tells us how to obtain a new statement or theorem from a given one involving points and lines. But projective geometry also deals with curves. How should one dualize statements describing curves? The clue lies in the fact that a curve is after all but a collection of *points* satisfying some condition. For example, the circle is the set of all points at a fixed distance from a given point. The principle of duality suggests, then, that the figure dual to a given curve might be a collection of *lines* satisfying the condition dual to the condition defining the given curve. (However the definition given for the circle is not in the form which can be dualized.) This collection of lines may also be called a curve, for a collection of lines suggests a curve as well as does a collection of points (Fig. 11–12) It is called a line curve.

For conic sections, the figure dual to a point conic, that is, a conic regarded as a collection of points, turns out to be the collection of tangents to that point conic. Thus if the conic section is a circle, the dual figure is the collection of tangents to that circle. This collections of tangents suggests the circle as well as does the usual collection of points, and we shall call this collection of tangents the line circle.

We have dualized statements about simple figures with suggestive results. Let us now see whether the application of the principle of duality to theorems on curves is equally productive. As a test we shall dualize Pascal's theorem.

Pascal's theorem	*Dual of Pascal's theorem*
If we take six points $A, B, C, D, E,$ and F on the point circle, then the lines which join A and B and D and E join in the point P; the lines which join B and C and E and F join in the point Q; the lines which join C and D and F and A join in the point R. The three points P, Q, and R lie on one line, l.	If we take six lines $a, b, c, d, e,$ and f on the line circle, then the points which join a and b and d and e are joined by the line p; the points which join b and c and e and f are joined by the line q; the points which join c and d and f and a are joined by the line r. The three lines p, q, and r lie on one point, L.

Figure 11–13 illustrates the content of the dual statement.

Geometrically the dual statement has the following meaning: Since the line circle is the collection of tangents to the point circle, the six lines on the line circle are any six tangents to the point circle, and these six tangents form a hexagon circumscribed about the point circle. Hence the dual

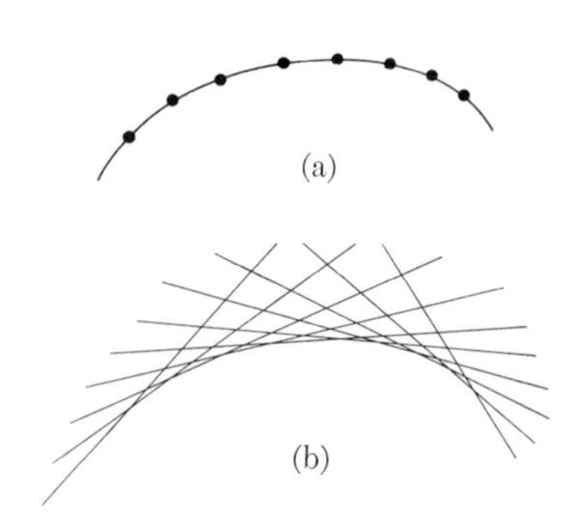

Fig. 11–12. (a) A point curve; (b) a line curve.

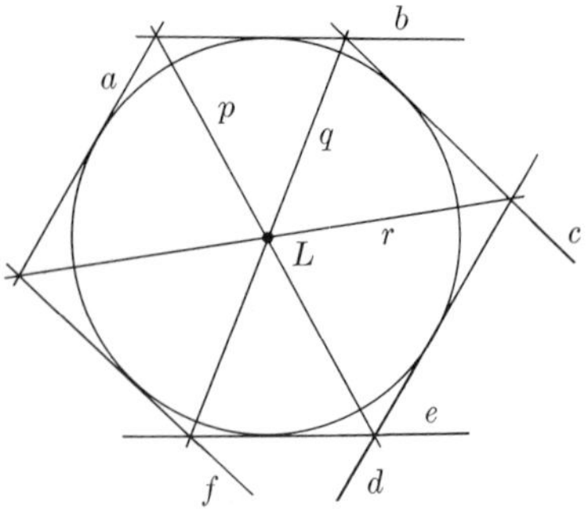

Fig. 11–13. Brianchon's theorem, the dual of Pascal's theorem.

statement tells us that if we circumscribe a hexagon about a point circle, the lines joining opposite vertices of the hexagon, lines p, q, and r in the dual statement, meet in one point. This dual statement is indeed a theorem of projective geometry. It is called Brianchon's theorem after the man who discovered it by applying the principle of duality to Pascal's theorem.

The *principle of duality* in projective geometry says that we can interchange point and line in a theorem about figures lying in one plane and obtain a meaningful statement. Moreover—although nothing said so far justifies the assertion—the new, or dual, statement will itself be a theorem; that is, it can be proved. However, it is possible to show by one proof that every rephrasing of a theorem of projective geometry in accordance with the principle of duality must lead to a theorem. The principle of duality is a remarkable property of projective geometry. It reveals the symmetry in the roles which point and line play in the structure of that geometry, and this symmetry in turn reveals that line and point are equally fundamental concepts.

The principle of duality also gives us some insight into the process of creating mathematics. Whereas the discovery of this principle as well as of theorems such as Desargues' and Pascal's calls for imagination and genius, the discovery of new theorems by means of the principle is an almost mechanical procedure.

Exercises

1. Given the figure consisting of four points no three of which are on the same line, what is the dual figure?
2. State the principle of duality.
3. Given the figure consisting of four points all on one line, what is the dual figure?
4. Given the figure consisting of three points on one line, a fourth point not on that line, and the lines joining any two of these points, what is the dual figure?
5. In what way is the principle of duality a means of discovering new theorems?

11–5 The relationship between projective and Euclidean geometries. Projective geometry offers many more exciting concepts than we can hope to survey. Let us see rather what the broader features of the subject are. The basic concept is projection and section and the main goal is to find properties of geometric figures which hold for any section of any projection of those figures. A careful examination of the properties which prove to be invariant under projection and section shows that these properties deal with the collinearity of points, that is, with points lying on the same line; with the concurrence of lines, that is with a set of lines meeting in one point; with cross ratio; and with the fundamental roles of point and line as exhibited by the principle of duality. On the other hand, Euclidean geometry, which, of course, was well known to the nineteenth-century projective geometers, deals, for example, with the equality of lengths, angles, and areas. A comparison of these two classes of properties suggests that the projective properties are simpler than those treated in Euclidean geometry. One might say that projective geometry deals with the very formation of the geometrical figures whose congruence, similarity, and equivalence (equal area) are discussed in Euclidean geometry.

With hindsight to aid us, we can see that there should be a geometry more fundamental than Euclidean geometry. Anybody first perceives the position in space of trees, houses, roads, and other objects and only then thinks of distances and sizes. When traveling, one must first choose a particular road to follow before being concerned with how far to move along the road. That is, position and relative position precede distance in importance both practically and logically.

Hence one might suspect that logically projective geometry is the more fundamental and encompassing subject and that Euclidean geometry is in some sense a specialization. This conjecture is correct. The clue to the relationship between the two geometries may be obtained by again examining projection and section. Consider a geometric figure, a rectangle say (Fig. 11–14). Let a projection of this figure be formed from an arbitrary point O, and then let a section of this projection be made by a plane *parallel*

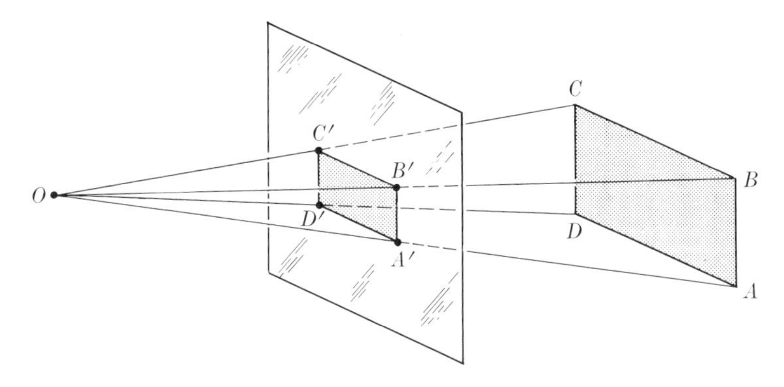

FIG. 11–14. Similar figures related by projection and section.

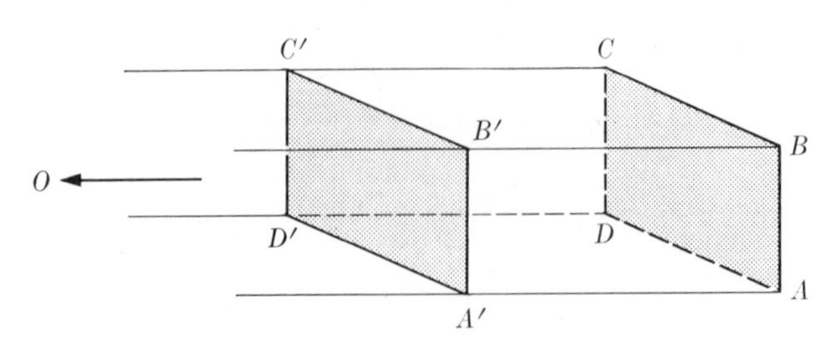

FIG. 11–15. Congruent figures related by projection and section.

to the plane of the rectangle. By applying some theorems of Euclidean geometry one can prove that the section will be a rectangle similar to the original one. Hence the relationship of similarity is a special type of projective relationship in which the plane of the section and the plane of the original figure are parallel.

If now the point O moves indefinitely far to the left, the lines of the projection come closer and closer to being parallel to each other. When the center of the projection is the "point at infinity"(!), these lines are parallel; then the section made by a plane parallel to the rectangle is a rectangle congruent to the original one (Fig. 11–15). This last type of projection, called parallel projection, thus yields congruent sections. In other words, from the standpoint of projective geometry, the relationships of congruence and similarity which are so intensively studied in Euclidean geometry can be studied through projection and section for special projections. We see therefore that Euclidean geometry is not only a logical subdivision of projective geometry, but we can now look upon it in a new light, namely, as a study of properties of geometric figures which are invariant under special projections.

Although projective geometry was initiated by Desargues for the very practical purpose of extending and systematizing theorems which might help the artists, it is not highly significant from the standpoint of applications to art or science. The subject has been developed and cultivated by mathematicians who sought and found pleasure in its ideas. The Renaissance artists and geometers opened up new themes which the Greek world did not grasp, the investigation of the properties of intersections of lines, cross ratio, duality, projection and section, and, above all, the theme of properties invariant under projection and section. Projective geometry is now a vast branch of mathematics because it does offer latitude to intuition, new methods of proof, elegant results, and aesthetically satisfying ideas. This subject born of art makes its primary contribution to mathematics as an art.

Exercises

1. What major mathematical problem was suggested by the artists' use of projection and section?
2. How would you distinguish projective geometry from Euclidean geometry with respect to the properties of geometrical figures?
3. Write a short essay on how the rise of realistic painting in the Renaissance stimulated a new mathematical development.

Recommended Reading

Bell, E. T.: *Men of Mathematics,* Chaps. 5 and 13, Simon and Schuster, New York, 1937.

Ivins, Wm. M. Jr.: *Art and Geometry,* Harvard University Press, Cambridge, 1946.

Kline, Morris: "Projective Geometry," an article in James R. Newman: *The World of Mathematics,* pp. 622–641, Simon and Schuster, New York, 1956.

Mortimer, Ernest: *Blaise Pascal: The Life and Work of a Realist,* Harper and Bros., New York, 1959.

Sawyer, W. W.: *Prelude to Mathematics,* Chap. 10, Pelican Books Ltd., England, 1955.

Young, Jacob W. A.: *Monographs on Topics of Modern Mathematics,* Chap. 2, Dover Publications, Inc., New York, 1955.

Young, John W.: *Projective Geometry,* The Open Court Publishing Co., Chicago, 1930.

CHAPTER 12

THE REVOLUTION IN ASTRONOMY

If there is anything that can bind the heavenly mind of man to this dreary exile of our earthly home and can reconcile us with our fate so that one can enjoy living—then it is verily the enjoyment of . . . the mathematical sciences and astronomy.

Johannes Kepler

12–1 Introduction. The first fruits of the Renaissance were produced by writers and artists. But the response of the scientists was not too far behind, and in our next few chapters we shall see how the problems of the new era became clear and what the mathematicians and scientists did to solve them.

The creation we are about to examine, known as the heliocentric theory, that is, an astronomical theory in which the sun rather than the earth is the center of the solar system, played a double role in the transition from medieval to modern culture. It was itself a consequence of the new influences and ideas flooding Europe. It in turn revolutionized European thought. Moreover, the heliocentric theory affords the most impressive illustration of the enormous influence mathematics has exerted on the formation of our modern culture. Mathematical arguments led, as we shall see, to a totally new conception of the universe and to the reopening of the most fundamental questions man can raise.

In the fifteenth century the only sound and useful astronomical theory was the geocentric system of Hipparchus and Ptolemy which we examined in Chapter 8. This was the theory accepted by professional astronomers and applied to calendar reckoning and navigation. It was, however, a rather sophisticated creation in that its strength lay entirely in the mathematical effectiveness of the scheme. The deferents and epicycles had no physical significance in themselves nor did the theory give any physical or intuitive reasons that the planets should move on epicycles attached to deferents. More acceptable therefore to the people at large was Aristotle's modification of Eudoxus' astronomical system. The details of this modification and further modifications made by medievalists do not matter here. The principal idea was that each planet was attached to a sphere and each such sphere was moved by a series of other spheres which transmitted their motion to one another. The prime mover of the several series of spheres was an outer one powered and regulated by intelligences and spirits controlled by God who dwelt in the empyrean, the outermost sphere.

12–2 The work of Copernicus. The author of the next great celestial drama, Nicolaus Copernicus, who lived about 1400 years after Ptolemy, was so much influenced by Greek thought that he could well have done his work in Greek times. But history delayed this development, and the intervening events lent momentous import to it when it finally took place. Copernicus was born in Poland in 1473 and, after studying mathematics and science at the University of Cracow, decided to go to Italy, the center of the revived Greek learning. At the University of Bologna, which he entered in 1497, he studied astronomy under the influential professor of mathematics and astronomy, Domenico Maria de Novara, a champion of Pythagorean doctrines, who also taught Copernicus how to make astronomical observations. For the next ten years he studied medicine and law and secured a doctor's degree in both fields. He also became learned in Greek and mathematics. In 1500 Copernicus was appointed a canon of the Cathedral of Frauenberg in East Prussia, but he did not assume his duties until 1512 when he had finished his studies in Italy. The job which entailed mainly the management of estates owned by the Cathedral left Copernicus with plenty of time to make astronomical observations and to think about the relevant theory.

After years of reflection and observation Copernicus finally evolved a new theory of planetary motions and in 1530 circulated a brief account of it among his friends. The theory impressed the distinguished scholar Cardinal Nicolaus von Schonberg, Archbishop of Capua, and Tiedeman Giese, Bishop of Culm, a devotee of the liberal arts. The reigning Pope, Clement VII, also approved of the work and requested publication. But Copernicus hesitated because he feared that the theory might offend other Church officials or a succeeding pope less favorable to the sciences. About ten years later, his friend George Joachim Rheticus, a professor at the University of Wittenberg, persuaded Copernicus to proceed with the publication, and Rheticus took over the work involved. The classic, *On the Revolutions of the Heavenly Spheres,* appeared in 1543. When the printed work reached Copernicus, he was already paralyzed by an apoplectic stroke. It is unlikely that he was able to read it, for he died soon afterwards.

As we have already noted, when Copernicus began to think about astronomy, the Ptolemaic theory was the only sound and effective system in existence. This theory had become somewhat more complicated during the intervening centuries in that more epicycles had been added to those introduced by Ptolemy, to make the theory fit the increased amount of observational data gathered largely by the Arabs. In Copernicus' time the theory required a total of 77 circles to describe the motion of the sun, moon, and the five planets known then.

Copernicus had studied the Greek works and had become convinced that the universe was mathematically and harmoniously designed. Harmony demanded a more pleasing theory than the complicated extensions of Ptolemaic theory, and Copernicus was disturbed by this discord. He read that some Greek authors, notably Aristarchus, had suggested the possibility that the sun might be stationary and that the earth revolved about the sun and rotated on its axis at the same time. Copernicus decided to explore

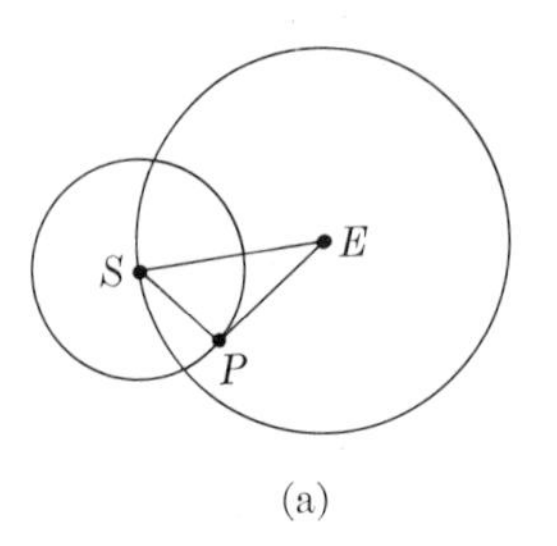

(a)

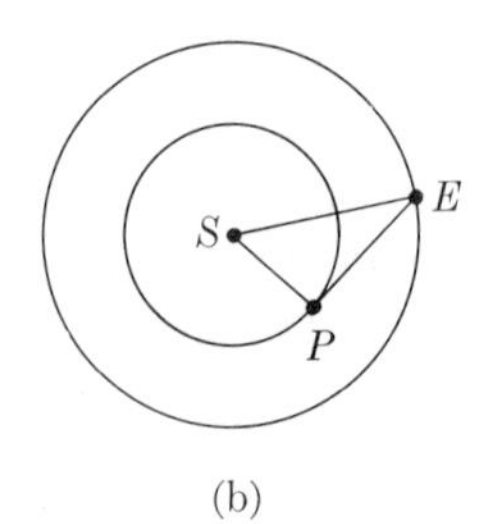

(b)

FIG. 12–1. (a) Geocentric and (b) heliocentric views of a planet's motion.

this possibility. He was in a sense overimpressed with Greek thought, for he, too, believed that the motions of heavenly bodies must be circular or, at worst, a combination of circular motions since circular motion was natural motion. Moreover, he also accepted the belief that each planet must move at a constant speed on its epicycle, and that the center of each epicycle must move at a constant speed on the circle which carried it. Such principles were axiomatic for him. Copernicus even adds an argument which shows the somewhat mystic character of sixteenth-century thinking. He says that a variable speed could be caused only by a variable power; but God, the cause of all motions, is constant.

The upshot of such reasoning was that Copernicus used the scheme of deferent and epicycles to describe the motions of the heavenly bodies, with, however, the all important difference that the sun was at the center of each deferent, while the earth itself became a planet moving about the sun and rotating on its axis. Nevertheless, he achieved considerable simplification.

To see the advantage which Copernicus gained by placing the sun at the center of each deferent, let us consider the motion of a planet under both points of view. In Fig. 12–1(a) the sun, S, moves on a circle around the fixed earth, E, and a planet, P, moves in a circle around the sun. The path of the sun is the deferent, and the circle on which P moves is the epicycle. Thus to describe the motion of the planet around the earth, a deferent and an epicycle are needed. In Fig. 12–1(b) the sun is the chief reference point, and the planet, P, moves on a circle around the sun. Of course, it is only the relative positions of the heavenly bodies that we can know. But the relative positions of earth, sun, and planet are the same in the two figures. Since the heliocentric view permits us to comprehend the motion of P in terms of one circle only whereas the geocentric view requires two, an appreciable simplification has been effected. The above figure illustrates only the principle Copernicus introduced. Actually several epicycles and a deferent would be needed to correctly describe the motion of P. Nonetheless the number is reduced under the heliocentric view and this reduction applies to each of the heavenly bodies. We can see therefore how the adoption of the heliocentric view enabled Copernicus to describe all motions with a total of 34 circles instead of the 77 required under the geocentric view.

The heliocentric theory just sketched did not work too well for all planets; that is, the predictions of the directions of some planets differed from the observations by angles of as much as 10′. Of course, Ptolemy too had encountered such differences, and he had introduced modifications of his basic scheme, the principal one being to have the earth near but not at the center of the deferent for some planets. Copernicus tried the same scheme by placing the sun near but not at the center of some of the deferents. The errors were not eliminated, but Copernicus was so sure that the heliocentric view must be correct that he never thought to abandon this concept. He could only hope that subsequent modifications or perhaps improved observations would produce better agreement between theory and observation.

12–3 The work of Kepler. The problem of improving astronomical theory was taken over by one of the most intriguing figures in the history of science, Johannes Kepler. He was born in 1571 in Weil, a city in the German duchy of Württemberg. His father was a soldier of fortune who finally settled down as a tavern keeper. While attending school, Johannes had to help in the tavern. He was later withdrawn from school and had to work as a laborer in the fields. In 1584, destined for the ministry, he was sent as a charity student to a Protestant seminary and subsequently to a college where he received a bachelor's degree. At the University of Tübingen, where he earned a master's degree in philosophy, he studied theology and mathematics. There he learned privately about the work of Copernicus from one of his professors, Michael Mästlin, who publicly had to teach Ptolemaic theory. Kepler became enthusiastic about the new theory, but the superiors of the Lutheran church did not share his enthusiasm. Since Kepler himself had begun to revolt from the narrowness of the current Lutheran thought, he abandoned the intention to enter the ministry and accepted the position of Professor of Mathematics and Morals at a "Gymnasium" in Graz, capital of Styria, Austria, where he lectured on mathematics, astronomy, rhetoric, and the Latin classics.

When the liberal Catholic ruler of Styria was succeeded by an intolerant one, Kepler, who could have stayed by converting to Catholicism, decided to leave. In 1600 he secured a position as assistant to the famous astronomer Tycho Brahe, who then worked at Prague and had been engaged in making extensive new observations, the first such major undertaking since Greek times. These observations, together with others which Kepler made himself, were most valuable to Kepler in his later work. In a life continually beset with hardships, he regarded access to these data as his one good fortune. When Brahe died in 1601, Kepler succeeded to the position of Imperial Mathematician to Emperor Rudolph II of Austria, King of Bohemia. After Kepler had served ten years, the Emperor found that he could no longer pay Kepler's salary, and Kepler had to look for a new position. But this was not easy. The Catholic states rejected him because he was a

Protestant, and the Protestants questioned his devoutness. Finally, in 1612 Kepler secured the position of Provincial Mathematician at Linz.

There Kepler decided to remarry—his first wife and one child had died during his stay in Prague—and he went about the task rationally. He made a list of eligible women, assigned positive numerical values to the virtues and negative numbers to the defects of each, and by adding obtained the algebraic value of each potential spouse.* It is, however, widely known that women will often let emotional reactions override rational considerations, and the highest-ranking candidate refused Kepler's offer. However, Kepler did find a willing partner.

A few years of peace were followed by new hardships. In 1615 his mother was accused of witchcraft, and Kepler had to defend her without daring to attack the belief in witches. Two more children died. The Protestants in Linz were unfriendly to Kepler. Nor was his salary adequate. In 1620 the Catholic Duke Maximilian of Bavaria conquered Linz, and Kepler's situation became intolerable. He searched for a new position while struggling to write, publish some of his books, and collect money due him. He became ill, and, while on a journey to Regensburg in 1630, caught pneumonia and died. This was the life of a man who created the key laws of astronomy and who made first-class contributions to mathematics, mechanics, optics, the theory of vision, observational astronomy, and to the methodology of science.

Kepler's scientific reasoning is fascinating. Like Copernicus, he was a mystic and, like Copernicus, he believed that the world was designed by God in accordance with some simple and beautiful mathematical plan. This belief dominated all his thinking. Kepler was also somewhat superstitious and, partly because he believed in astrology, at least as a youth, and partly because he was asked to make astrological predictions for his employers at Graz and at Prague, he spent many hours in that activity. To satisfy Emperor Rudolph he wrote a book entitled *On the More Certain Foundations of Astrology* (1602). Later in life he used to tell his clients in astrology, "What I say will or will not come to pass" and comforted himself over the loss of time devoted to the casting of horoscopes with the thought that just as nature had provided all animals with ways of acquiring food so God had provided astronomers with astrology. This daughter of astronomy, he said, nursed her mother. Kepler possessed a great imagination which led him into flights of fancy, and all of his writings are marked by ecstatic outbursts which occur in the midst of very sober and scientifically sound passages.

But Kepler also had qualities which we now associate with scientists. He could be coldly rational. His fertile imagination triggered the conception of new theoretical systems. But he knew that theories must fit observations and, in his later years, saw even more clearly that empirical data may indeed suggest the fundamental principles of science. Copernicus, too,

---

* This method of choosing a wife is not recommended. Mathematics, at its present stage of development, is not able to solve such problems.

wanted his theory to fit observational data; yet he held to the heliocentric view, although the differences between theoretical predictions and astronomical data were greater than might be accounted for by experimental errors alone. Kepler, on the other hand, sacrificed his most beloved theories when he saw that they did not fit observational data, and it was precisely this incredible persistence in refusing to tolerate discrepancies which any other scientist of his day would have disregarded that led him to espouse radical ideas. He also had the humility, patience, and energy to perform extraordinary labor which mark great men.

Kepler began his work in astronomy by searching for the neat mathematical principle which, he believed, God must have used to design the universe. Now the Greeks had found that there are just five regular polyhedra (Fig. 6–5). To Kepler, who was impressed by this fact, it therefore seemed inevitable that God would use these figures to design the universe and accordingly he says in the preface to his *Mystery of the Cosmos* (1596): "I undertake to prove that God, in creating the universe and regulating the order of the cosmos, had in view the five regular bodies of geometry as known since the days of Pythagoras and Plato, and that he has fixed according to those dimensions, the number of the heavens, their proportions, and the relations of their movements."

But how did God employ these five regular bodies? After much thought Kepler found what he believed to be God's plan. He had already become convinced that the sun must be the center of the solar system. Hence he supposed that Saturn, the farthest planet known in his day, moved on a sphere whose center was the sun. He next supposed that a cube was inscribed (Fig. 12–2) in this sphere. In this cube he imagined another sphere to be inscribed on which Jupiter was supposed to move. Jupiter's sphere in turn enclosed the tetrahedron, and in this regular polyhedron a third sphere was inscribed on which Mars moved. By using all five polyhedra he had room for six spheres, and this was the number of planets known at the time. The correspondence between the number of spheres permitted by his scheme and the number of planets seemed an overwhelming argument to Kepler.

Once the radius of the outermost sphere is fixed, the radii of the others are determined by the fact that the successive polyhedra and spheres are inscribed one within the other. Kepler hoped that if the radius of the largest sphere were chosen to equal the radius of the most distant planet, the radii of the other five spheres would automatically fit the distances of the other five planets from the sun. However, this was not the case; nor would simple motion on a sphere describe the path of any planet. Hence for many years Kepler sought a modification of this basic plan which would satisfy the data. It was not until he had Brahe's observations that he became convinced the plan would not work. Here he showed his regard for facts; although clearly enamored of his mathematical theory, he was willing to sacrifice it.

With Brahe's observations available, Kepler tried again, and this time his faith in the mathematical design of nature was rewarded. In his book *On*

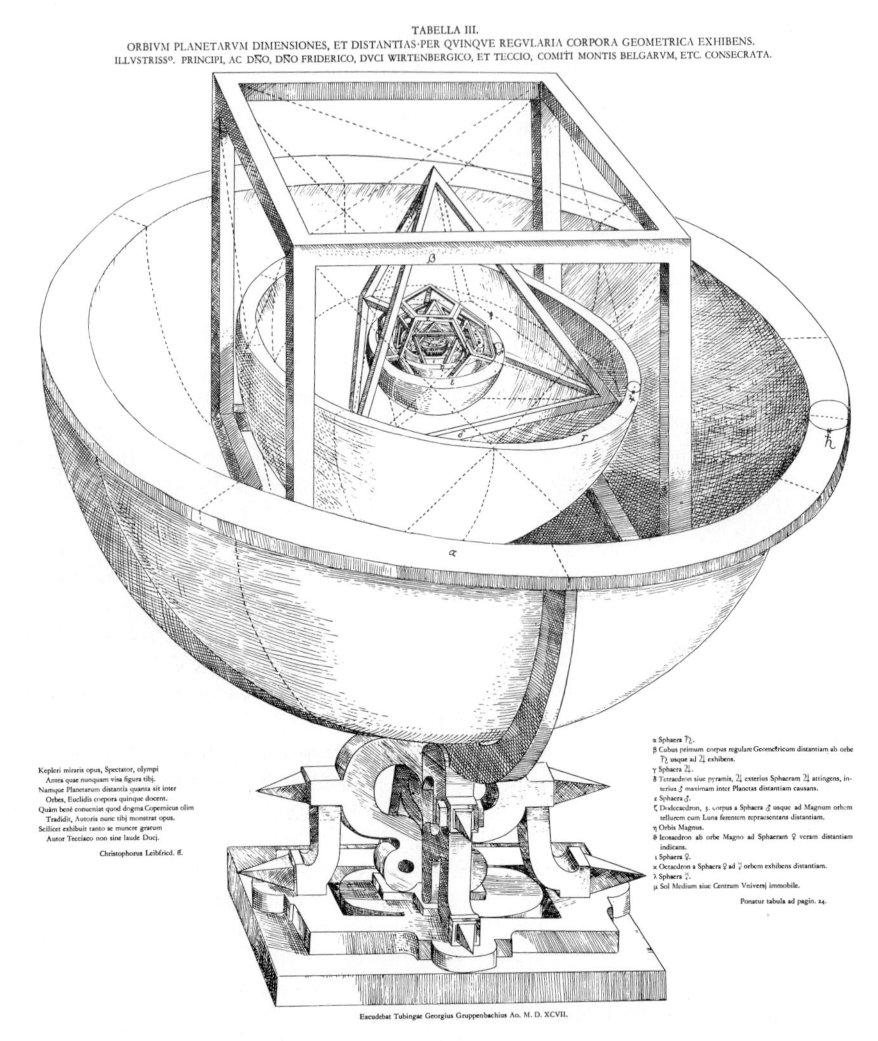

FIG. 12–2. Kepler's scheme for employing the five regular polyhedra.

the Motion of the Planet Mars, published in 1609, Kepler announced the first two of his three famous laws of planetary motion. The first of these is especially remarkable, for Kepler broke with the tradition held for 2000 years that circles or spheres must be used to describe heavenly motions. Instead of resorting to deferent and several epicycles which both Ptolemy and Copernicus had used to describe the motion of any one planet, Kepler found that a single ellipse would do. His first law states that each planet moves on an ellipse and that the sun is at one (common) focus of each of these elliptical paths (Fig. 12–3). The other focus of each ellipse is merely a mathematical point at which nothing physical exists.

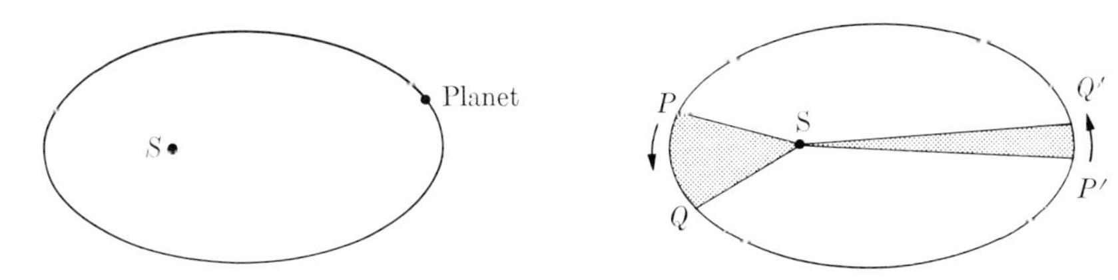

Fig. 12-3. Each planet moves in an ellipse about the sun.

Fig. 12-4. Kepler's law of equal areas.

Kepler's first law utilizes a geometrical figure which had been introduced and studied by the Greeks of the classical period. Had the ellipse and its properties not yet been known, Kepler himself, in his effort to find a curve which would fit the data, might perhaps have discovered it. But the concept of this curve is by no means an obvious one, and if Kepler had been faced with the double problem of abstracting the proper path from a multitude of data *and* conceiving the ellipse, he might possibly have ended up in an impasse. By working out the properties of this curve Euclid, Apollonius, and Archimedes determined the course of our civilization just as decisively as if they had stood at Kepler's side.

Kepler's first law is of immense value in comprehending readily the paths of the planets. But astronomy must go much further if it is to be interesting in itself and useful. It must tell us how to predict the positions of the planets. If one finds by observation that a planet is at a particular position, P say in Fig. 12-4, he might like to know when it might be at some other position, a solstice or an equinox, for example. What is needed is the velocity with which the planets move along their respective paths.

Here, too, Kepler made a radical step. Copernicus, as we noted earlier, and the Greeks had always used constant velocities. A planet moved along its epicycle so as to cover equal arcs in equal times, and the center of each epicycle also moved at a constant velocity on another epicycle or on a deferent. But Kepler's observations told him that a planet moving on its ellipse does not move at a constant speed. Kepler searched hard and long for the correct law of velocities and found it. What he discovered was that if a planet moves from P to Q (Fig. 12-4) in, say one month, then it will also move from P' to Q' in one month, provided that the *area* PSQ equals the *area* $P'SQ'$. Since P is nearer the sun than P' is, the arc PQ must be larger than the arc $P'Q'$ if the areas PSQ and $P'SQ'$ are equal. Hence the planets do not move at a constant velocity. In fact, they move faster when closer to the sun.

Kepler was overjoyed to discover this second law. Although it is not so simple to apply as a law of constant velocity, it nonetheless confirmed his fundamental belief that God had used mathematical principles to design the universe. God had chosen to be just a little more subtle, but a mathematical law clearly determined how fast the planets moved.

Another major problem remained open. What law described the distances of the planets from the sun? It was this question that he had sought to answer by using the five regular solids, but, as we have noted, the theory based on the existence of these solids did not fit the data. The problem was now complicated by the fact that a planet's distance from the sun was not constant but varied from a least to a greatest value (see Fig. 12–4). Hence Kepler searched for a new principle which would take this fact into account. Now he believed that nature was not only mathematically but harmoniously designed and he took this word "harmony" very literally. Thus he believed that there was a music of the spheres which produced a harmonious tonal effect, not one given off in actual sounds but discernible by some translation of the facts about planetary motions into musical notes. He followed this lead and after an amazing combination of mathematical and musical arguments, arrived at the law that if T is the period of revolution of any planet and D is its mean distance from the sun, then $T^2 = kD^3$, where k is a constant which is the same for all the planets. This statement is Kepler's third law of planetary motion and the one which he triumphantly announced in his book *The Harmony of the World* (1619).

The entire derivation is the strangest mixture of imagination, fancy, and scientific work. Pure fantasy, serious musical composition, mathematical interpretation of the music, and cold arithmetic follow one another in this long rendition of the extraordinary work he did over a period of years. Flushed with success, he opens the fifth and last book of his *Harmony* with the exultation: "What I prophesied two and twenty years ago as soon as I had discovered the five solids among the heavenly bodies; what I firmly believed before I had seen the *Harmonics* of Ptolemy; what I promised my friends in the title of this book which I named before I was sure of my discovery; what sixteen years ago I regarded as a thing to be sought; that for which I joined Tycho Brahe for which I settled in Prague; . . . . for which I devoted the best part of my life to astronomical contemplations, . . . at length I have brought to light, and have recognized its truth beyond my most sanguine expectations."

12–4 The objections to a heliocentric theory. The work of Copernicus and Kepler is by far the most dramatic, startling, and influential development in the formation of modern culture. The first surprising feature, and one which in itself makes their work astounding, is the sharp break from existing thought. Copernicus and Kepler were educated in a milieu which accepted the geocentric theory of Ptolemy as almost unquestionable truth. Moreover, both were scientifically cautious. Nor did either really have at his disposal any unusual observations which conflicted sharply with Ptolemy's theory. Copernicus, as a matter of fact, was not a great observer and did not seem to mind leaving his work somewhat at odds with observations. For this reason, Tycho Brahe rejected the Copernican theory, as did François Vieta. Kepler did have access to more numerous and more reliable data and showed greater tenacity in making the theory fit the data, but there was nothing in these observations which suggested that an entirely new theory must be introduced.

In their writings, Copernicus and Kepler give only minor reasons for being dissatisfied with Ptolemaic theory and minor arguments for adopting a new line of thought. The Greek astronomer Aristarchus had suggested that it might be simpler to think of planets moving around the sun, but Aristarchus did not work out such a theory. Moreover, this idea was discarded by Ptolemy and others because, as we shall see in a moment, it conflicted with a number of other Greek doctrines. Copernicus himself could not resolve these conflicts. He gives the rather feeble argument that his failure to see any other solution to the current arguments between the advocates of Aristotelian astronomy and the Ptolemaists compelled him to think along new lines. He further says that he did not like Ptolemy's device of placing the earth somewhat off center with respect to certain deferents, although he himself resorted to the same scheme in his efforts to improve agreement with observation. There was talk in Copernicus' time of an infinite universe, of other planetary systems like ours, and of life on other planets, but such speculations could hardly be taken seriously without some scientific evidence in their favor. Clearly not one of these arguments or difficulties suggests a total break with the past. Whereas almost every major creation of history has been preceded by ground-breaking efforts and slow positive advances which, in retrospect at least, make the decisive step appear to be natural, Copernicus in adopting the heliocentric hypothesis and Kepler in accepting it and in conceiving and formulating further tradition-shattering ideas made gigantic and almost superhuman innovations.

While casting aside the weight of centuries, Copernicus and Kepler could give only token rebuttals to the numerous scientific arguments against a moving earth which Ptolemy had advanced against Aristarchus. How could the heavy earth be put into motion? That the other planets were in motion even according to Ptolemaic theory was explained by the doctrine that these were made of special light matter and therefore easily moved. About the best answer Copernicus could give was that it was natural for a sphere to move. Another scientific argument against the earth's rotation proclaimed that rotation would cause objects to fly off into space just as an object on a rotating platform will fly off. Copernicus had no answer to this argument. To the further objection that a rotating earth should itself fly apart, Copernicus replied weakly that since the earth's motion was natural, it could not destroy the body. Then he countered by asking why the sky did not fall apart under the very rapid daily motion which the geocentric theory called for. Yet another objection declared that if the earth rotated from west to east, an object thrown up into the air should fall back to the west of its original position because the earth moved on while the object was in the air. If, moreover, the earth revolved about the sun, then, since the velocity of an object is proportional to its weight, or so at least Greek and Renaissance physics maintained, lighter objects on the earth should be left behind. Even the air should be left behind. To the last argument Copernicus replied that air is earthy and so moves in sympathy with the earth.

These scientific objections to a moving earth were weighty ones and could not be dismissed as the stubbornness of doubters who refused to see the truth. The substance of the matter is that a rotating and revolving earth did not fit in with the physical theory of motion due to Aristotle and common in Copernicus' time.

Another class of scientific arguments against a heliocentric theory came from astronomy proper. The most serious one stemmed from the fact that the heliocentric theory regarded the stars as fixed. In six months the earth changes its position in space by 186 million miles. Hence if one notes the direction of a particular star at one time and again six months later, he should observe a difference in direction. But this difference was not observed in Copernicus' and Kepler's time. Copernicus answered that the stars are so far away that the difference in direction was too small to be observed. However, his explanation did not satisfy the critics, who countered that if the stars were that distant, then they should not be clearly observable. In this instance, Copernicus' answer was correct. The change in direction over a six-month period for the nearest star is an angle of 0.76″, and this was first detected by the mathematician Bessel in 1838 who, of course, by that time had a good telescope at his disposal.

A further, powerful argument against a moving earth contended that we do not feel any motion despite the fact that the earth is presumably moving around the sun at 18 miles per second and that a person on the equator is rotating at the rate of about 0.3 miles per second. Our senses, on the contrary, tell us that the sun is moving in the sky. Of course, there are counterarguments which would mean more to us today because we have the experience of traveling at high speeds. A person traveling in an airplane at 400 miles per hour does not feel the motion. But to the people of Copernicus' time the argument that we do not feel ourselves moving at the very high speeds called for by the new astronomy was convincing.

Beyond the scientific difficulties and the argument based on sense impressions there was the weight of religious thought. The entire Christian theology was built on the notion that the universe revolved around man. Man was God's most important creation, and everything in the universe, including the sun, the moon, and the stars, was designed to serve man. It was very convenient for this theology that from Greek times on the prevailing astronomical doctrines had regarded the earth as the center of the universe, for where else should this most important creature man be but at the center. The geocentric view made it possible to suggest that heaven was indeed in the heavens, a definite sphere to which man could aspire, and hell was in the dark, forbidding interior of the earth which occasionaly erupted through volcanoes and showed that hell-fire and brimstone really existed. But the new theory put man on one of the many planets rushing around the sun with the all too obvious implication that there was nothing exceptional about this planet as opposed to the others. Indeed, the heliocentric theory forces one to think in terms of the entire planetary system and hence there is no reason to single out the planet earth or, in particular, to concentrate on the creature man. Where was heaven in this new doctrine? Beyond the farthest planet? Was it still reasonable to believe that hell was inside the earth? Copernicus and Kepler were themselves deeply religious. Yet they firmly and tenaciously espoused the new view despite the fact that it went against the entire theology which dominated their times.

12-5 The arguments for the heliocentric theory. In view of the numerous and sound arguments against the heliocentric theory and the challenge it posed to the prevailing religious thinking of the times, what made Copernicus and Kepler take up this long-discarded thought and pursue it so courageously? For what most other men would call a mess of pottage, they broke with established physics, philosophy, religion, and common sense.

Both loved astronomy for its beauty. Copernicus says, "For what could be more beautiful than the heavens which contain all beautiful things." In the preface to his *Mystery of the Cosmos,* Kepler says, "Happy the man who devotes himself to the study of the heavens; he learns to set less value on what the world admires the most; the works of God are for him above all else, and their study will furnish him with the purest of enjoyments." Both were convinced that the universe is mathematically designed and hence that a true pattern of the motions was inherent. Moreover, this design was instituted by God, and God would surely have used a simple and harmonious pattern. But Ptolemaic theory had become so encumbered in the sixteenth century that it was no longer simple or beautiful. Hence Copernicus and Kepler believed, when each found a more harmonious and simpler theory, that their work was indeed a description of the divine order of things.

There are many passages in Copernicus' *On the Revolutions of the Heavenly Spheres* and in Kepler's writings which bear unmistakable testimony to this motivation as the central force in the search for a new theory and to their conviction that they had found the right one when it proved to be simpler. Copernicus says of his theory: "We find therefore, under this orderly arrangement, a wonderful symmetry in the universe, and a definite relation of harmony in the motion and magnitude of the orbs, of a kind that it is not possible to obtain in any other way." Kepler remarks of his later work wherein he had already introduced the elliptical theory of motion, "I have attested it as true in my deepest soul and I contemplate its beauty with incredible and ravishing delight." And he concludes his *Harmony of the World* with a paean to God: "The wisdom of the Lord is infinite; so also are His glory and His power. Ye heavens, sing His praises! Sun, moon and planets glorify Him in your ineffable language. Celestial harmonies, all ye who comprehend His marvelous works, praise Him. And thou, my soul, praise thy Creator. It is by Him and in Him that all exists. That which we know best is comprised in Him, as well as in our vain science. To Him be praise, honor and glory throughout eternity." The work of Copernicus and Kepler is the work of men searching the universe for the harmony which their religious convictions assured them must exist and which must be describable mathematically and simply because God had so designed the universe. What distinguishes their religious convictions from those of their contemporaries is that they did not tie themselves to literal interpretations of the Holy Writings. They searched for the word of God in the heavens.

It is also true that Copernicus and Kepler were to a certain extent mystics and in this respect they were products of their times. Copernicus started with *a priori* beliefs on how nature *must have been* designed. He says that the universe must be spherical, divine, and endowed with the perfection of the Creator. He also advanced the idea that the sun is a living force which dominates the universe and supplies the life force. Copernicus speaks of the sun as regal and, since he held the Platonic belief that immobility is more noble than mobility, he preferred the sun to be stationary. He writes, "The earth conceives from the sun and the sun rules the family of stars." Again, "For who could in this most beautiful temple place this lamp in another or better place than that from which at the same time it can illuminate the whole." Kepler too, regarded the sun as the force which activates the universe. To account for the planets' motions around the sun, he proposed lines of forces stretching from the sun to each planet and forcing the planets to move as the sun rotated, just as the hub of a wheel transmits force to the rim through the spokes. Of course, such an idea could not be pursued too far because the "lines of force" to the several planets would interfere with one another. Later, when the phenomenon of magnetism became widely known through the experimental work of William Gilbert, Kepler tried to introduce the hypothesis that the sun exerted a magnetic force on the planets. However, he also spoke of a gravitational force between masses such as the earth and sun. Another illustration of mystic thinking is Kepler's attempt to derive his third law by arguing from the premise of a musical harmony. He believed that the sun had a soul which enjoyed the harmony he had brought to light.

Of course, what distinguishes the mysticism and pure fancy in the thinking of Copernicus and Kepler from characteristic medieval and Renaissance nonsense is their scientific principle that theories must fit facts. Facts were the final authority and in this respect Copernicus and Kepler are the first modern scientists, superior even to the Greeks, who were content with a mathematical theory which suited philosophical principles and rough observations. Francis Bacon had attacked Copernicus' work with the charge, "All these are the speculations of one, who cares not what fictions he introduces into nature, provided his calculations answer." To this Copernicus and Kepler would certainly reply that what truly matters *is* that the calculations fit. In Kepler's case especially it was his refusal to ignore even minor discrepancies between theory and observations that led him to elliptical paths as opposed to combinations of circles.

The core of the argument which Copernicus and Kepler presented for the heliocentric theory was its mathematical simplicity. Their philosophical and religious convictions assured them that the world is mathematically and simply designed; accordingly the fact that a heliocentric view was mathematically simpler than the geocentric one determined their position. The mathematical simplicity of the new view was, in fact, the sole argument they could advance. Only persons possessed of the unshakable conviction that mathematics is the essence of the design of the universe and that the omnipotent mathematician would necessarily prefer simplicity would dare to advance such a radical theory and would have had the courage to defend it against the opposition it was sure to and did encounter in those times.

The new theory appealed to astronomers, geographers, and navigators because it simplified their theoretical and arithmetical work. Hence many of these men adopted the new view just as a mathematical convenience, even though they were not convinced of its truth. While this argument emphasizing the applicability of the new theory carried little weight with Copernicus and Kepler, it nonetheless had the effect of making more and more people think in terms of a heliocentric view, and, since one tends to accept as true what is familiar, there is no doubt that these practical aspects did, in the long run, help to gain adherents for the theory.

Support for the new theory came from an unexpected development. Early in the seventeenth century the telescope was invented, and Galileo, upon hearing of this invention, built one himself. He then proceeded to make observations of the heavens which startled his age. He detected four moons of Jupiter (we now can observe twelve), and this discovery showed that a moving planet can have satellites. Hence it was likely that the earth, too, could be in motion and yet have a satellite, our moon. Galileo saw irregular surfaces and mountains on the moon, spots on the sun, and a bulge around the equator of Saturn (which we now call the rings of Saturn). Here was further evidence that the planets were like the earth and certainly not perfect bodies composed of some special ethereal substance, as Greek and medieval thinkers had believed. The Milky Way, which had hitherto appeared to be just a broad band of light, could be seen with the telescope to be composed of thousands of separate stars, each of which gave off light. Thus, there were other suns and presumably other planetary systems suspended in the heavens. Moreover, the heavens clearly contained more than seven moving bodies, a number which had been accepted as sacrosanct. Copernicus had predicted that if human sight could be enhanced, then man would be able to observe phases of Venus and Mercury, that is, to observe that more or less of each planet's hemisphere facing the earth is lit up by the sun, just as the naked eye can discern the phases of the moon. Galileo did discover the phases of Venus, but fearing attack announced the fact in an anagram: "The mother of the loves [Venus] imitates the form of Cynthia [the moon]." He also showed that Ptolemaic theory did not allow for the occurrence of these phases.

All of Galileo's observations were made with a telescope of such limited power that, as has been said, it is remarkable he could find Jupiter, let alone the moons of Jupiter. Many of his discoveries were in direct support of a heliocentric theory; others served primarily to challenge current beliefs and to at least prepare some minds for a more objective examination of the new theory. Galileo, himself, though he lectured on Ptolemaic theory until 1605, had been converted to Copernicanism by a work of Kepler. He acknowledges this in 1597 in a letter to Kepler. In 1611 he openly declared for

Copernicanism. His own observations convinced him that the Copernican system was correct, and in the classic *Dialogue on the Great World Systems* he defended it strongly. He says in this work, "I cannot find any bounds for my admiration that reason was able in Aristarchus and Copernicus to overwhelm their senses with such force as to make herself mistress of their beliefs." By the middle of the seventeenth century the scientific world was willing to proceed on a heliocentric basis.

12–6 The scientific import of the heliocentric theory. We have mentioned some of the practical advantages which the heliocentric theory offered, but its broader implications were of far greater significance. Neither Copernicus nor Kepler could explain why objects on the earth stayed with it during its rotation and revolution around the sun. There was also the question how the earth and the other "heavy" planets were kept in motion around the sun. Since motion is a common and fundamental phenomenon, it became clear that a new science of motion which would replace the no longer adequate Aristotelian science was needed. This problem dominated the seventeenth century, and we shall see shortly its importance for mathematics.

The work of Copernicus and Kepler thrust mathematics forward into the world of science in a manner that not only emphasized its role but paved the way for modern abstract theories. The Greeks were convinced that mathematics is the key to the understanding of nature. However, their mathematics described what the senses perceived and was in accord with sense perceptions. Copernicus and Kepler placed mathematics above the senses, so to speak. For the sake of a simpler mathematical theory, they asked the world to accept a moving earth, despite the fact that this motion is not perceptible. The senses are to be denied credence if a better mathematical theory can be obtained as a consequence. Of course, this statement applies only within proper limits. No scientific theory is acceptable if its predictions cannot be confirmed by observations or experiments, and to this extent at least the senses must rule. So, in an astronomical theory, the predictions of the locations of planets and eclipses, for example, must fit observations. But other sense impressions deemed unimportant for that science, such as the perception of the earth's motion, may be denied if a more satisfactory mathematical theory can thereby be designed.

The next step in this line of thinking is the construction of far-reaching and encompassing mathematical theories which may have no basic sensuous, intuitive content or physical meaning at all, provided the theory can predict key phenomena. We shall see shortly what this means in the new theory of motion created by the seventeenth century and we shall see it all the more clearly in the electromagnetic theory of Maxwell. These theories and others which we can merely mention but shall not be able to examine, such as the theory of relativity and atomic theory, are largely mathematical structures with only minor physical content. The senses are almost helpless in providing physical understanding, and indeed the kind of mathematical space employed in the theory of relativity, for example, cannot be visualized at all by the senses. The point of this discussion is that Copernicus and Kepler inaugurated a trend in scientific thinking in which mathematics secured ascendancy over the senses. Our difficulties in comprehending major modern scientific theories result precisely from the fact that these theories are not built up by physical and readily visualizable arguments but by vast mathematical abstractions which ignore and even defy the senses.

Prior to the era of Copernicus and Kepler, scientists and mathematicians had believed that nature is simply designed, but the belief in simplicity had not been stressed. It was regarded as more or less self-evident. Hence in any choice between theories it seemed clear that the simpler one was to be preferred. However, Copernicus and Kepler emphasized simplicity to the point of preferring a theory which was at odds with practically all other thinking of the times. Moreover they made it quite clear that *mathematical* simplicity was the criterion of truth. Thereafter this criterion became a fundamental one, and we shall see it in operation in the work of Galileo, Newton, and their successors.

One word of caution regarding the work of Copernicus and Kepler. These men believed that the heliocentric theory was true for the reasons already cited. This is not the view we hold today. If the criterion is truth, then heliocentric theory is not to be preferred to Ptolemaic theory. Scientific theories, we now believe, are the work of man. The mind supplies the patterns which organize observations. We may indeed prefer the heliocentric theory because it is simpler and agrees better with observations, but we do not regard it as the last word. Another theory, which still may not be the truth, may be conceived and produce even better results. As a matter of fact, one was—the theory of relativity. We shall not anticipate too much. The evolution of the concept of truth as it applies to mathematics and mathematical theories of science will be a continuing concern.

12-7 The cultural influences of the heliocentric theory. In Greek times, as we have often mentioned, the concept of a mathematically designed universe had impressed intellectuals. To medieval Europe it was, of course, unknown for the simple reason that the scholars of this period were not acquainted with the Greek works. Moreover in the early Renaissance the doctrine had lost a great deal of force because Ptolemaic theory, the most impressive evidence for it, had become complicated and could well have been regarded as a patchwork which served its purpose of describing and predicting heavenly events but hardly represented the inherent design of nature. By the time Kepler had completed his work, the simplicity had been restored and the accuracy of prediction increased. There seemed to be no question that the heavenly bodies followed the pattern described by Kepler's laws, and confidence in the mathematical design of nature was not only restored but uplifted. Nature followed a regular pattern and man could rely upon the predictability of nature's behavior.

This doctrine of the regularity and lawfulness of nature was to receive even more substantial support in the following centuries, and its effect on the lives and thoughts of modern man is incalculable. But its influence made itself felt even in the early seventeenth century. To appreciate what the doctrine implies one must compare it with the then current beliefs. Nature was regarded not only as the creation of God, but God was endowed with absolute power over the world and man. He could do with them as he willed. Of course, there was purpose in every divine act, and natural events were so interpreted. Sickness was an individual punishment inflicted by God. Plagues and earthquakes were summary punishments for collective misdeeds. The belief in magic and in superhuman beings who could make phenomena of nature serve their will was rife. The belief in witchcraft was so strong that when Kepler's own mother was accused of sorcery, he did not dare to attack that superstition. He could only argue that his mother was innocent. All such beliefs were vitiated by the doctrine of the lawfulness of nature. God might still be the designer but natural phenomena obeyed invariable laws and did not suddenly take place to mete out divine punishment or obey the commands of demons.

The phenomena of nature which were at the disposition of God had, of course, included the motions of the heavenly bodies and, in keeping with medieval thought, it followed that events in the heavens had special meaning and intent for the lives of men. Hence the science of astrology was enormously influential. As we noted in an earlier chapter, every court had an astrologer who served as special advisor to the king, prince, or duke, and whose prophecies based on coming heavenly events were accepted as the truth. The most intelligent rulers did not question astrological predictions.

But the doctrine that nature is regular and lawful negates the basis for all belief in astrology. If heavenly bodies follow invariable paths which will repeat themselves endlessly, how could there be any special significance in an astronomical conjunction of two heavenly bodies for an event about to take place on earth? Further, the purely mathematical content which is the substance of the new astronomy disregards man completely. Its strictly impersonal character denies by implication but unmistakably any relationship to the affairs and interests of man. Finally, and perhaps most importantly, the heliocentric theory places earth in the very same class as the other planets. If other heavenly bodies exert any special powers or influences, then so should the earth. But this planet clearly is composed of inert rock, clay, and water, and thus another belief was shown to be superstitious nonsense.

The most weighty and disturbing implications of the heliocentric theory involved man himself. We have already had occasion to mention that one of the objections to the heliocentric theory was that it displaced man from the center of the universe to a position of insignificance. Man was just a speck clinging precariously to one of many spinning balls which whirled around the sun. He was just a bit of matter in an infinite universe, a drop of water in an ocean, a mite in the great events of nature. Could one continue to believe that man was the most important creation of God, that all events in the universe were designed to serve man, and that man could attain a life in a heaven located presumably above the planets?

Giordano Bruno was the first to fully grasp the significance of the Copernican hypothesis. This globe, he perceived, was no longer the hub of the universe. Though deeply religious, he realized that the literal interpretation of many Christian doctrines became untenable under this new view. He declared the idea of a corporeal Deity sitting above the clouds and judging man to be grotesque. The concept of a God descending to earth as man and man ascending to God seemed to Bruno unintelligible, and the doctrine of fall and redemption of man no longer defensible. Bruno also had ideas about the physical universe which we are just beginning to take seriously today, such as that there must be an infinite number of worlds like our own. He proclaimed these conclusions and others boldly and vehemently.

Bruno's preachings brought home to man the new conception of the universe. The earth is not only not central but is not even distinguished among the planets. It is neither the smallest nor the largest, neither nearest nor farthest from the sun. Nor is our solar system unique. The sun, which even Kepler accepted as the center of the universe, is just an ordinary star. It is neither the smallest nor the largest star and neither the brightest nor the dimmest. The other stars are almost surely surrounded by planets, and it is highly probable that on these planets there are living beings perhaps superior to man.

In view of such possibilities man had to question his role in the universe. What is the purpose of his life? What goals should he set for himself? What values should he attach to life and to the various activities in which he can engage? Man seemed to be on his own and had to accept responsibility for and fashion his own destiny. And if the answers he previously accepted were no longer correct, what confidence could he continue to have in those sources of truth which he had accepted so trustingly? Evidently he had to re-examine every philosophical system based on the hypothesis that man is a special creation and that design and purpose determined his life and future. The doctrines man fashioned to fit the new conception of the universe will be taken up in the space of a few chapters. They amounted to the creation of a new culture.

The conflict between the heliocentric theory and current religious doctrines had other momentous consequences. Curiously, the work of Copernicus and Kepler might have stirred up very little opposition had it been performed one hundred years earlier. At that time many Church leaders had themselves said that Biblical statements did not need to be taken too literally. And early in the sixteenth century, Copernican theory was accepted by religious leaders as true in philosophy though false in theology; i.e., it was considered to be a reasonable doctrine for science but, of course, not the truth. Copernicus' book, however, was published in 1543, about 25 years after the Protestant Revolution, and both Catholics and Protestants had become alarmed by any movement which tended to undermine religious beliefs.

Leaders of both faiths joined in the attack on the heliocentric doctrine and on its sponsors. To defend their vested interests they fell back on literal interpretation of the Scriptures. Rather mild was the criticism made by one cardinal who suggested that Copernicus should be more concerned with how to go to heaven than with how the heavens go. Decidedly less tolerant was Martin Luther who called Copernicus an "upstart astrologer" and "a fool who wishes to reverse the entire science of astronomy. . . ." And resorting to the typical counterargument of his era, Luther added, "as the Holy Scripture shows, it was the sun and not the earth which Joshua ordered to stand still." Calvin, too, quoted the Scriptures: "Who will venture to place the authority of Copernicus above that of the Holy Spirit? Do not the Scriptures say that the sun runs from one end of the heavens to the other? That the foundations of the earth are fixed and cannot be moved?"

When mere verbal attacks failed to discourage leanings and convictions in favor of the heliocentric theory, the Church applied the power and threat of the Inquisition. In 1600 Giordano Bruno was burned for advocating the heliocentric theory and other doctrines opposed by the Church. We noted earlier that in 1611 Galileo openly endorsed the heliocentric theory. In 1616 he was called before the Roman Inquisition but was let off with a mere reproof. The Inquisition declared the Copernican doctrine to be heretical and in 1620 forbade all publications on the subject. Despite this prohibition, Pope Urban VIII gave Galileo permission to publish a book on the subject, provided that he would feature mathematical rather than doctrinal aspects. Thus in 1632 Galileo published his *Dialogue on the Great World Systems* and, to please the Church, stated in the preface that the heliocentric theory was only a product of the imagination. Galileo was again called by the Inquisition and compelled to declare, "The falsity of the Copernican system cannot be doubted. . . ." His book was put on the Index of Prohibited Books. As a matter of fact, the Church did not sanction any writings in favor of the heliocentric theory until 1822.

Some people took up the challenge posed by clerical opposition to freedom of thought, speech, and writing, and, by fighting for the right to advocate the heliocentric theory, advanced the cause of freedom for the human mind. Copernicus, Kepler, and Galileo were heroes in this battle. Copernicus quoted Ptolemy: "He who is to follow philosophy [science] must be a free man in mind." In a letter to Pope Paul III, he makes clear his willingness to attack narrow-minded theologians and scholars: "If perhaps there are babblers who, although completely ignorant of mathematics, nevertheless take it upon themselves to pass judgment on mathematical questions and, improperly distorting some passages of the Scriptures to their purpose, dare to find fault with my system and censure it, I disregard them even to the extent of despising their judgment as uninformed." It so happens that a Lutheran theologian, Andreas Osiander, who assisted in the publication of Copernicus' book, added an unsigned preface which stated that the new work was just a hypothesis which allows computation of the heavenly motions on geometrical principles. Whoever takes for truth what was designed for different purposes, he added, will emerge from the study of astronomy a greater fool than he was before he approached it. Of course, Osiander added this preface because he thought it would forfend suppression of the book. Copernicus, already old and ill when the manuscript was being prepared for publication, was not aware of the dissimulation; he would not have tolerated this surrender to authority.

Kepler, too, defended his right to advocate his ideas. He lost favor with Catholics and Protestants alike because he defended heliocentrism, and his stand cost him a number of jobs. However, he was willing to suffer poverty rather than yield the right to advocate what he believed. Galileo's contribution to freedom of thought is also unquestionable. That, as an infirm man of 69 years, he recanted under threat of torture by the Roman Inquisition can hardly detract from the independence and courage he showed until that time.

The heliocentric theory prevailed, and it prevailed because a few great minds were willing to fight for what they thought was the truth. Although other forces later also entered the fray and broke the power of religious leaders to dictate what men can believe, the heliocentric theory was one of the great issues to which many rallied in order to defend freedom of thought.

Exercises

1. What is a geocentric astronomical system? a heliocentric astronomical system?
2. Did Copernicus break completely with Greek astronomy?
3. Is the sun at the center of the Keplerian system?
4. What innovations did Kepler introduce into the Copernican system?
5. To reconstruct Kepler's improvement on Copernicus, suppose that a planet P moves once around its epicycle while the center of the epicycle moves once completely around the sun. What path does the planet seem to follow in relation to the sun? Would it be simpler to accept this single path as opposed to the combination of epicycle and deferent?
6. What scientific objections were there to an earth in motion?
7. Why did Copernicus and Kepler advocate the new heliocentric theory?
8. Why do you accept the heliocentric theory?
9. State Kepler's first law of planetary motion.
10. State Kepler's second law of planetary motion.
11. If we take the earth's average distance from the sun, 93,000,000 mi, as the *unit of distance* and the earth's period of revolution, 1 year, as the *unit of time*, then Kepler's third law says that $T^2 = D^3$. If the average distance of Neptune from the sun is 2,797,000,000 mi, how long does it take the planet to complete one revolution around the sun?
12. Utilize the history of the heliocentric theory to show how mathematics has influenced western European culture.

Recommended Reading

ARMITAGE, ANGUS: *Sun, Stand Thou Still,* Henry Schuman, New York, 1947. Also in paperback under the title *The World of Copernicus.*

ARMITAGE, ANGUS: *Copernicus,* W. W. Norton and Co., New York, 1938.

BAUMGARDT, CAROLA: *Johannes Kepler, Life and Letters,* Victor Gollancz Ltd., London, 1952.

BONNER, FRANCIS T. and MELBA PHILLIPS: *Principles of Physical Science,* Chap. 1, Addison-Wesley Publishing Co., Reading, Mass., 1957.

BURTT, E. A.: *The Metaphysical Foundations of Modern Physical Science,* 2nd ed., Chap. 2, Routledge and Kegan Paul, London, 1932.

BUTTERFIELD, HERBERT: *The Origins of Modern Science,* Chap. 2, The Macmillan Co., New York, 1951.

CASPAR, MAX: *Johannes Kepler,* Abelard-Schuman, New York, 1960.

COHEN, I. B.: *The Birth of a New Physics,* Doubleday and Co., Anchor Books, New York, 1960.

DE SANTILLANA, GIORGIO: *The Crime of Galileo,* University of Chicago Press, Chicago, 1955.

DRAKE, STILLMAN: *Discoveries and Opinions of Galileo,* Doubleday & Co., Anchor Books, New York, 1957.

DREYER, J. L. E.: *A History of Astronomy From Thales to Kepler,* Dover Publications, Inc., New York, 1953.

DREYER, J. L. E.: *Tycho Brahe, A Picture of Scientific Life and Work in the Sixteenth Century,* A. and C. Black, Edinburgh, 1890.

GADE, JOHN A.: *The Life and Times of Tycho Brahe,* Princeton University Press, Princeton, 1947.

GALILEI, GALILEO: *Dialogue on the Great World Systems,* The University of Chicago Press, Chicago, 1953. Other editions also exist of this work originally published in 1632.

HOLTON, GERALD and DUANE H. D. ROLLER: *Foundations of Modern Physical Science,* Chaps. 8 to 10, Addison-Wesley Publishing Co., Reading, Mass., 1958.

KOYRÉ, ALEXANDRE: *From the Closed World to the Infinite Universe,* Chaps. 1 to 4, The Johns Hopkins Press, Baltimore, 1957.

KUHN, THOMAS S.: *The Copernican Revolution,* Harvard University Press, Cambridge, 1957.

NICOLSON, MARJORIE: "Kepler, The Somnium and John Donne" in PHILIP P. WIENER and AARON NOLAND: *Roots of Scientific Thought,* Basic Books, Inc., New York, 1957.

SMITH, PRESERVED: *A History of Modern Culture,* Vol. I, Chap. 2, Henry Holt & Co., New York, 1930.

WOLF, ABRAHAM: *A History of Science, Technology, and Philosophy in the Sixteenth and Seventeenth Centuries,* 2nd ed., Chaps. 2, 3, and 6, George Allen and Unwin, Ltd., London, 1950.

CHAPTER 13

COORDINATE GEOMETRY

In order to seek truth it is necessary once in the course of our life to doubt as far as possible all things.

Descartes

13–1 Descartes and Fermat. Doubts as to the soundness of the knowledge and outlook possessed by medieval Europe had already been raised during the Renaissance. The revived Greek knowledge, great explorations, new inventions, the rise of an artisan class with problems of its own which could not be answered by purposive or teleological explanations of natural phenomena, the advocacy of experience as the source of all knowledge, and the new view of the universe presented by the heliocentric theory all tended to undermine the old foundations. No one appreciated more the need for the reconstruction of all knowledge than did René Descartes (1596–1650).

Born to moderately wealthy parents in La Haye, France, Descartes received an excellent formal and traditional education at the Jesuit College of La Flèche. But while still at school, he had already become critical of the truths which so many of his contemporaries and teachers professed so confidently and he began to question the kind of knowledge that was being imparted to him. Of the traditional studies, he said, eloquence has incomparable force and beauty, and poetry has its ravishing graces and delights. However these attainments he judged to be gifts of nature rather than the fruits of study. He respected theology because it pointed out the path to heaven and he, too, aspired to heaven, but "being given assuredly to understand that the way is not less open to the most ignorant than to the most learned, and that the revealed truths which lead to heaven are above our comprehension," he did not presume to subject these truths to his impotent reason. Philosophy, he agreed, "affords the means of discoursing with an appearance of truth on all matters, and commands the admiration of the more simple," but, though cultivated for ages by the most distinguished men, it had not produced doctrines which were beyond dispute. Law, medicine, and other professions secure riches and honor for their practitioners, but since these subjects borrow principles from philosophy, they could not be solid structures, and fortunately he was not obliged to pursue them to better his fortune. Logic he also deprecated because its syllogisms and the majority of its other rules are of use only in the communication of what one already knows or in speaking without judgment about things of which one is ignorant. It does not in itself proffer knowledge. Treatises on morals contain useful precepts and exhortations to be virtuous but no evidence that these are founded on truths.

Because he had a critical mind and because he lived at a time when the world outlook which had dominated Europe for a thousand years was being vigorously challenged, Descartes could not be satisfied with the tenets so forcibly and so dogmatically pronounced by his teachers and other leaders. He felt all the more justified in his doubts when he realized that he was in one of the most celebrated schools of Europe and that he was not an inferior student. At the end of his course of study he concluded that all his education had advanced him only to the point of discovering man's ignorance.

At the age of 20, after having graduated from the University of Poitiers, where he studied law, Descartes decided to learn some things that were not in books. He began by living a gay life in Paris, after which he retired for a period of reflection in a quiet corner of that city. To see the world he joined an army, participated in military campaigns, and traveled. Finally he decided to settle down.

Because Descartes thought he could more easily find peace and seclusion in the saner atmosphere of Holland, he secured in 1628 a house in Amsterdam. There he devoted himself over a period of twenty years to critical and profound thinking about the nature of truth, the existence of God, and the physical structure of the universe. There he created his best works. As he continued to write, he and his audience became more and more impressed with the greatness of his work. Lucid thoughts set forth in literary classics which revealed the clarity, precision, and effectiveness of the French language made Descartes famous and his philosophy popular.

His retirement from the world was broken by an invitation to serve as a tutor to Queen Christina of Sweden. Reluctant as he was to leave his comfortable home, he could not resist the attraction of royalty and so moved to Stockholm. The queen preferred to begin her day at 5 a.m. by studying in an icy library, and her tutor was obliged to meet her at that hour. This regimen was too much for the frail Descartes. His flesh was weak, and his spirit unwilling. He caught cold and died in the year 1650.

From the profound thinking and writings carried on during the years in Holland there emerged new foundations of knowledge. Descartes is the acknowledged father of modern mathematics and philosophy, and he founded a new cosmology which dominated the seventeenth century until it was ultimately displaced by the work of Galileo and Newton. Convinced that the knowledge he had acquired in school was either unreliable or worthless, Descartes swept away all opinions, prejudices, dogmas, pronouncements of authorities, and, so he believed, preconceived notions. He began reconstruction by seeking a new method of obtaining sure and reliable knowledge. The answer, he says, came to him in a dream while he was on one of his campaigns.

The "long chains of simple and easy reasonings by which geometers are accustomed to reach the conclusions of their most difficult demonstrations" led him to the conclusion that "all things to the knowledge of which man is competent are naturally connected in the same way." He decided, then, that a sound body of philosophy could be deduced only by the methods of the geometers, for only they had been able to reason clearly and unimpeachably and to arrive at universally accepted truths. Having concluded that mathematics "is a more powerful instrument of knowledge than any other that has been bequeathed to us by human agency," he sought to distill from a study of the subject some general principles which would provide a method of obtaining exact knowledge in all fields, a method which he called a "universal mathematic."

Following the pattern of mathematics which builds on axioms, he decided that he would accept nothing as true which was not so clear and distinct to his mind as to exclude all doubt. He would begin, in other words, with unquestionable, self-evident truths. The next principle of his method was to break down larger problems into smaller ones; he would proceed from the simple to the complex. Then he would write out the steps of his reasoning and review them so thoroughly that nothing would be inadvertently assumed or necessary arguments omitted. These four principles are the core of his method.

However, he first had to find those simple, clear, and distinct truths which would play the part in his philosophy that axioms play in mathematics proper. And here Descartes took a backward step. Whereas his age was turning to experience as the reliable source of knowledge, Descartes looked into his mind. After much critical reflection he decided that he was sure of the following truths: (a) I think, therefore I am. (b) Each phenomenon must have a cause. (c) An effect cannot be greater than the cause. (d) The mind has within it the ideas of perfection, space, time, and motion.

He then proceeded to reason on the basis of these axioms. The full story of his search for method and of the application of the method to actual problems of philosophy is presented in his famous *Discourse on Method* (1637). In later writings he continued to follow the procedure outlined in this work and thereby founded the first great modern system of philosophy. We shall have occasion later (Chapter 20) to discuss some of Descartes' philosophical doctrines. But what is relevant at the moment is that the truths of mathematics and mathematical method served as a beacon to a great thinker lost in the intellectual storms of the seventeenth century and enabled him to develop a philosophy which was more rational, less mystical, and less bound to theology than the systems produced by all his European predecessors.

To show what his new method could accomplish in fields outside of philosophy, Descartes applied it to geometry and published these results in his *Geometry*, an appendix to his *Discourse on Method.* But before we examine how Descartes revolutionized method in geometry, we must note another great seventeenth-century thinker, Pierre de Fermat, who, equally concerned with improving geometrical methods, independently arrived at the same broad idea.

In contrast to Descartes' adventurous, romantic, and purposive life Fermat's was highly conventional. He was born in 1601 to a French leather merchant. After studying law at Toulouse, he earned his living as a lawyer and served as King's councillor for the parliament of Toulouse, a position much like that of a modern district attorney. Fermat's home life was also quite ordinary. He married and brought up five children. His evenings he devoted to study. Whereas Descartes cared little for knowledge as such or for the beauty and harmony in mathematics and the arts but sought truths and useful knowledge, Fermat was faithful to the Greek ideals of speculative knowledge and intellectual pleasures. He was a student of Greek literature, wrote poetry, joined in the solution of the scientific problems of his day, and above all regaled himself in all branches of mathematics. Despite the brief amount of time he could devote to study, he made fundamental contributions to algebra, the calculus, the mathematical theory of probability, coordinate geometry, and the theory of numbers. Fermat's mathematical achievements entitle him to the honor of being considered one of the best mathematicians the world has had.

13–2 The need for new methods in geometry. Descartes and Fermat developed a new approach to geometry, specifically, an algebraic method of representing and analyzing curves. Why were the mathematicians of the seventeenth century so much concerned with ways of working with curves? The general reason is that the rise of science and the vast expansion of commercial and industrial activities had raised problems involving curves. Let us see what these problems were.

The heliocentric theory of Copernicus and Kepler won gradual acceptance during the seventeenth century, and scientists and mathematicians, at least, began to apply it intensively to the purely scientific problem of understanding the heavenly motions and to the more practical concerns, such as navigation, wherein astronomical knowledge was essential. Now the heliocentric theory called for the use of ellipses and, to some extent, of parabolas and hyperbolas. Many new facts about these curves were needed.

In the seventeenth century the idea that a ship's longitude could be determined most easily and accurately by means of a clock was actively pursued. The precise details of this method of determining longitude do not matter at the moment, but it is relevant that there were no clocks at that time which could be carried conveniently aboard ships. The adaptation of a spring and a pendulum to a clock was initiated by several scientists, notably Galileo Galilei, Robert Hooke, and Christian Huygens. The motion of the bob of a pendulum and of objects suspended from springs (see Chapter 23) is studied by the use of curves.

The very motion of ships at sea raised problems involving curves. Although the ideal path of a ship on a spherical surface is a great circle, the actual path cannot always be that since ships must obviously detour around land. Moreover, on maps the ideal and actual paths have to be represented by even more complicated curves whose shapes depend on the method of projection used to draw the flat map.

The increased interest in light raised numerous problems involving curves. When light travels long distances through the earth's atmosphere, it is gradually bent or refracted and hence follows a curved path (see

Chapter 1). Since observations of the positions of heavenly bodies may be in error because the path of light is curved, it is obviously necessary to know something about these curved paths to correct our observations. In 1604 Kepler wrote his *Optical Part of Astronomy* on this subject. A knowledge of curves was needed to design lenses used in telescopes and microscopes. Both of these instruments had been invented in the early part of the seventeenth century and attracted considerable attention. Lenses for spectacles had by this time been in use for 300 years, but improvement in design was a constant problem. Both Descartes and Fermat were very much interested in optics, and Descartes in particular did a great deal of work on the design of lenses. A good part of his findings are contained in the essay *Dioptrics* which, incidentally, appeared with *Geometry* as one of three appendices to his *Discourse on Method.* Fermat, too, was a notable contributor to optics; his *principle of least time,* which we described earlier (Chapter 7), still stands as a basic postulate in that subject.

Another class of problems calling for the study of curves was presented by the increasing use of cannons. The balls or shells shot from cannons are called projectiles, and a number of questions were raised concerning their motion. What paths or curves do projectiles follow? How do the paths depend upon the angle at which the cannon is inclined? What is the range or horizontal distance traveled by the projectile? And how is the path affected by the initial velocity imparted to the ball?

The motion of projectiles was but one of a wider class of problems involving motion. As we shall note more fully in a later chapter, the motion of objects on and near the surface of the earth became an active concern in the early seventeenth century because the heliocentric theory raised basic problems in this domain. Since all such motions take place along straight-line or curved paths, these more fundamental problems of motion also led to the study of curves.

Of course, this study was not new to mathematics. The Greeks had studied extensively the line, circle, and conic sections and had deduced hundreds of theorems about these curves. These contributions were known to seventeenth-century Europe. Why, then, did Fermat and Descartes decide that mathematics needed new methods of working with curves? The reason was stated by Descartes. He complained that Greek geometry was so much tied to figures "that it can exercise the understanding only on condition of greatly fatiguing the imagination." Descartes also deplored that the methods of Euclidean geometry were exceedingly diverse and specialized and did not allow for general applicability. Each theorem required a new type of proof, and much imagination, effort, and ingenuity had to be expended to find such proofs. The Greeks, with ample time at their disposal and no concern for immediate application, were not troubled by the lack of general procedures. However, these conditions no longer obtained in the seventeenth century. Moreover, new curves were needed for the applications which were of importance to the seventeenth century, and the Greek geometrical methods did not seem to be effective in these cases.

There is another limitation inherent in Greek geometry which the seventeenth century could no longer tolerate. One might indeed determine by some geometrical argument what type of curve a projectile shot from a cannon follows and one might prove some geometrical facts about this curve, but geometry could never answer such questions as how high the projectile would go or how far from the starting point it would land. The seventeenth century sought quantitative or numerical information because such data are paramount in practical applications.

It was clear to Descartes and Fermat that entirely new methods of working with curves were needed. Descartes, impatient with the methods of the Greeks and their disinterest in application, says, "I have resolved to quit only abstract geometry, that is to say, the consideration of questions which serve only to exercise the mind, and this, in order to study another kind of geometry, which has for its object the explanation of the phenomena of nature."

Both Descartes and Fermat, working independently of each other, saw clearly the potentialities of algebra for the representation and study of curves. This realization was not entirely a bolt from the blue. A great deal of progress had been made in algebra during the latter half of the sixteenth and the early part of the seventeenth century, much of it contributed by Descartes and Fermat themselves. Cardan, Tartaglia, Vieta, Descartes, and Fermat had extended the theory of the solution of equations (cf. Chapter 5), had introduced symbolism, and had established a number of algebraic theorems and methods. What impressed Descartes especially was that algebra enables man to reason efficiently. It mechanizes thought, and hence produces almost automatically results that may otherwise be difficult to establish. This value of algebra has already been pointed out earlier in this book, but historically it was Descartes who clearly perceived and called attention to this feature. Whereas geometry contained the truth about the universe, algebra offered the science of method. It is, incidentally, somewhat paradoxical that great thinkers should be enamored of ideas which mechanize thought. Of course, their goal is to get at more difficult problems, as indeed they do.

Exercises

1. How did mathematics help Descartes build his philosophy?
2. What were the four steps of mathematical method emphasized by Descartes?
3. Why did Descartes and Fermat seek a new method of deriving properties of curves?
4. What scientific problems of the seventeenth century required more knowledge about curves?

13–3 The concepts of equation and curve. We can understand more readily what Descartes and Fermat accomplished if we examine a somewhat modernized version of their idea. In his general study of methodology,

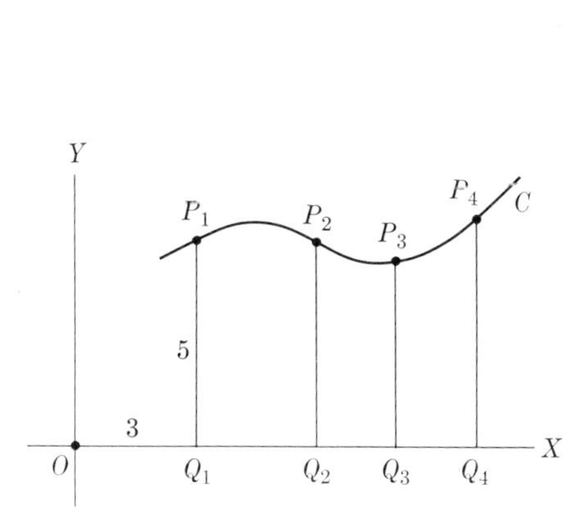

FIGURE 13-1

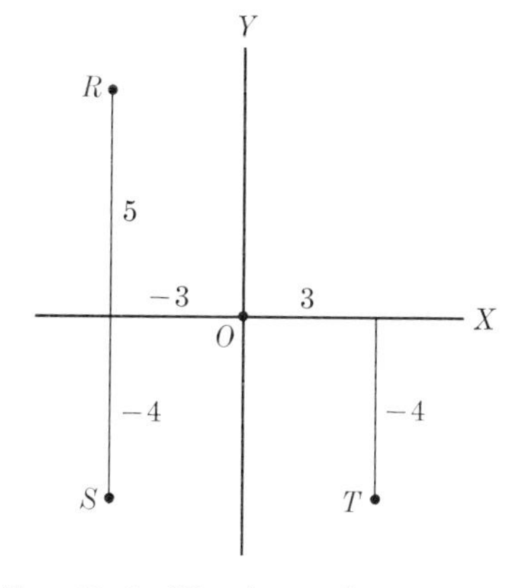

FIG. 13-2. Plotting points on a rectangular Cartesian coordinate system.

Descartes had decided to solve problems by proceeding from the simple to the complex. Now the simplest figure in geometry is the straight line, and so Descartes sought to analyze curves by working with straight lines. He observed, first of all, that if one introduces (Fig. 13-1) a horizontal line, OX, then the shape of a curve C could be studied by observing how vertical line segments such as Q_1P_1, Q_2P_2, Q_3P_3, . . . changed in length.

The next step was to express this information in arithmetical terms. The position of Q_1, for example, could be specified by stating its distance from a fixed point O, called the origin, on the horizontal line. The length Q_1P_1 could certainly be specified by a number. Thus the position of P_1 would be determined by two numbers, the length OQ_1 and the length Q_1P_1. The length OQ_1, which is 3 in Fig. 13-1, is called the *abscissa* of P_1, and the length P_1Q_1, which is 5 in Fig. 13-1, is called the *ordinate*. The line OX is called the X-axis, and the line OY, which is perpendicular to OX and which shows the direction in which Q_1P_1 is taken, is called the Y-axis.

Stated in more general terms, Descartes' and Fermat's first step was to describe the position of any point P on a curve by two numbers, an abscissa and an ordinate. The first expresses the distance or length from O along the X-axis to the point Q directly below P, and the second denotes the distance or length from Q to P along a line perpendicular to the X-axis or parallel to the Y-axis. The pair of numbers is called the coordinates of P and is written thus: $(3, 5)$.

To distinguish points which are reached by proceeding along the X-axis to the right of O from those which are reached by proceeding to the left of O, distances to the left are represented by negative numbers. Thus to arrive at the point R of Fig. 13-2 one proceeds 3 units to the left along the X-axis and 5 units upward in the direction of the Y-axis. The coordinates of R are therefore -3 and 5 and are represented as $(-3, 5)$. The distinction between upward and downward is also made by using positive and negative numbers, and as a consequence the coordinates of S in Fig. 13–2 are -3 and -4, and the coordinates of T are 3 and -4.

Thus far, then, Descartes and Fermat had a simple scheme for representing the position of any point in the plane by means of numbers, these numbers being distances from two arbitrarily chosen but fixed axes. To each point there corresponds a pair of numbers, and to each pair of numbers a unique point. The system using axes and coordinates to represent points is called a *rectangular Cartesian coordinate system.*

To represent a curve such as C of Fig. 13–1, one could list the coordinates of the many points on the curve, that is, the coordinates of $P_1, P_2, P_3, \ldots$ But such a representation would hardly be convenient, for each curve consists of an infinite number of points. Descartes and Fermat had a better idea. First of all they introduced the letters x and y to stand for the coordinates of *any* one of the points on the curve. When x and y have specific numerical values, for example, 2 and 3, respectively, then they refer, of course, to a definite point. Otherwise they represent an arbitrary point. The use of x and y here is analogous to using the words "man" and "woman" to represent any man or woman in the United States, whereas John and Mary would describe a particular couple.

Now, if one looks at the curve C of Fig. 13–3 and observes the abscissas and ordinates of the points on this curve, he notices that as the abscissas increase (as one looks from left to right), the corresponding ordinates at first increase and then decrease. The behavior of the ordinates changes as the abscissas change. Might it not be possible to describe the relationship between these abscissas and ordinates by specifying how large any ordinate is when the corresponding abscissa is named? This description should hold for all points on the curve and apply to that curve and no other. The answer is yes, and the general description turns out to be an algebraic equation involving x and y, the coordinates of an arbitrary point. This statement is a bit vague; let us see, therefore what it means in concrete examples.

Suppose, first, that the curve happens to be a straight line inclined 45° to the horizontal.* To describe the line algebraically, we introduce a horizontal line which passes through any point O of the given line (Fig. 13–4) and consider this horizontal line as the X-axis of our coordinate system. The Y-axis is then a vertical line through O. Consider *any* point P on the line. The coordinates of this general point P are the x and y shown in the figure. Now Euclidean geometry tells us that the triangle OQP is an isosceles right triangle. Hence $OQ = QP$ or $x = y$. Therefore the line OP appears to be characterized by the algebraic equation

$$y = x \tag{1}$$

because for *any* point P on the line, the coordinates are such that $y = x$.

* In coordinate geometry and in higher mathematics, in general, the word "curve" includes straight lines.

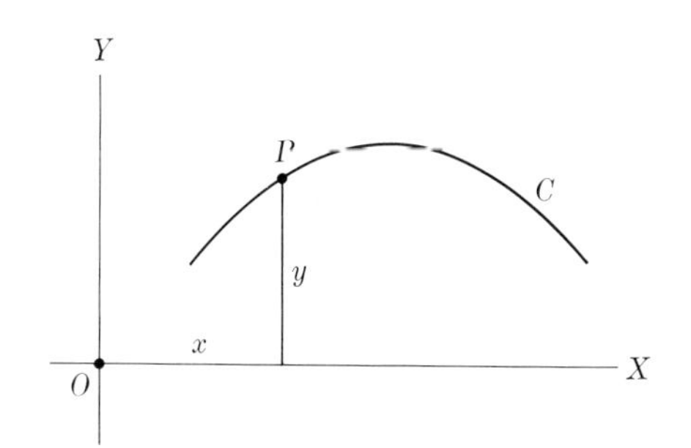

FIGURE 13-3

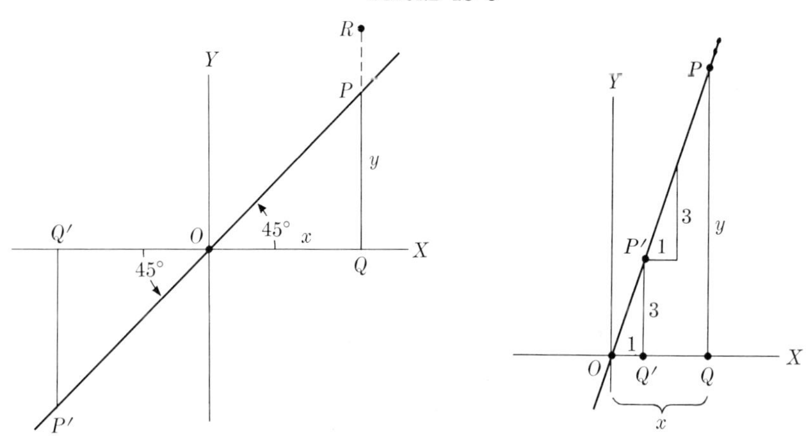

FIG. 13-4. A straight line on a rectangular Cartesian coordinate system.

FIG. 13-5. A straight line of slope 3 on a rectangular Cartesian coordinate system.

We should note that this equation also describes points, such as P', which lie to the left of the vertical axis. Thus, for example, the abscissa of P' may be -4; now angle $Q'OP'$ is also 45°, and hence the ordinate of P' must also be -4. Since abscissa and ordinate are equal, $x = y$ also holds for P'. Then the all-important fact about the line $P'P$ is that it may be described algebraically by the equation $y = x$. In other words, we may say that the coordinates of any point on the line satisfy the equation $y = x$. On the other hand, points not on that line, such as R, will have ordinates unequal to their abscissas because, while R has the same abscissa as P, the ordinate of R is larger than the ordinate of P, and hence for R, y does not equal x.

Let us consider a second example. One may describe the line of Fig. 13-4 by saying that it rises one unit for each unit of horizontal distance or, in the customary expression, that it has slope 1. We now consider a line which rises more steeply, for example, one which rises 3 units for each horizontal distance of 1 unit (Fig. 13-5). Again let P be *any* point on the line. The coordinates of this general point P are then (x, y). Then from the similar triangles $OQ'P'$ and OQP we may argue that

$$\frac{y}{x} = \frac{3}{1}.$$

Hence

$$y = 3x \tag{2}$$

is the equation of the line.

The nature of equations such as $y = x$ and $y = 3x$ requires some attention. In elementary algebra we also treat equations, for example, $x^2 - 5x + 6 = 0$ or $2x + 3 = 7$. In these latter equations, however, x represents some definite but unknown quantity and our aim is to find the value or values of x. On the other hand, when we represent a curve by an equation such as $y = 3x$, we are not seeking to determine unknowns. In fact, x and y are not unknown. They represent the coordinates of *any* point on the line. Thus $x = 3$ and $y = 9$ are one pair of values satisfying the equation; $x = 4$ and $y = 12$ constitute another such pair; there are millions of others. The end product of the process of finding the equation of a curve is then an equation involving x and y which states the relationship between x and y peculiar to all points on the curve. Of course, if one wishes to find the coordinates of a particular point on the curve, he can, provided he knows the abscissa of the point, substitute this number for x in the equation and now solve for the ordinate. Thus, if the given line has the equation $y = 3x$ and we wish to determine the ordinate of the point whose abscissa is $2\frac{1}{2}$, we substitute $2\frac{1}{2}$ for x and immediately find that the ordinate of this point is $7\frac{1}{2}$.

Let us consider next another example which will enlarge our understanding of these new equations. Suppose we are given two straight lines as shown in Fig. 13–6. The line OP is the one we have just discussed and its equation is $y = 3x$. The line $O'P'$ is supposed to be parallel to OP and 2 units above it; that is, PP' is 2. What is the equation of the line $O'P'$?

To answer this question, we again seek the relation between x and y which holds for any point on this line. Now P and P' have the same x-value or abscissa, namely OQ. But the ordinate of P' is larger than that of P by the amount PP'. The distance PP' is given to be 2. Since the two lines are parallel, the vertical distance between a point on OP and a point on $O'P'$ will always be 2. Hence, whereas the ordinate of each point on the line OP is always 3 times the abscissa, the ordinate of each point on $O'P'$ will be 3 times the abscissa plus 2. That is, the equation of $O'P'$ is

$$y = 3x + 2. \tag{3}$$

We should note that although the straight line $O'P'$ is identical except for position with the straight line OP, its equation is different. The difference results from the fact that OP passes through the origin O of the system of coordinates, whereas $O'P'$ does not. *Hence the very same curve may be represented by a different equation if its position with respect to the coordinate axes is changed.*

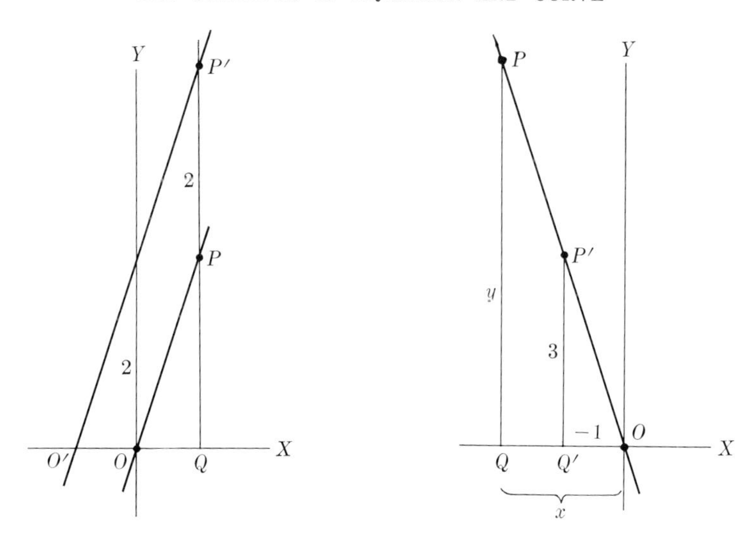

FIG. 13–6. Two parallel lines of slope 3 on a rectangular Cartesian coordinate system.

FIG. 13–7. A line with negative slope.

Why do we bother with different equations for the same curve? If a straight line having a slope of 3 can always be placed on a coordinate system so that its equation is $y = 3x$, why do we have to consider the more complicated form $y = 3x + 2$? The answer is that if one wishes to work with two such straight lines simultaneously and keep these lines in the same relative positions which they may happen to have in some physical application, one cannot assign to them an identical position on the set of axes.

While we are discussing equations of straight lines, we wish to note one more case which will be of interest later. Let us compare the line OP of Fig. 13–7 with the line OP of Fig. 13–5. The line in Fig. 13–7 may be said to "fall" as one views it from left to right, whereas that in Fig. 13–5 rises. Alternatively, we may say that the ordinates in Fig. 13–7 *decrease* as the abscissas increase. What is the equation of the line OP of Fig. 13–7? Suppose that the coordinates of some point P' on the line are $(-1, 3)$. The point P is an arbitrary point of the line OP, so let us denote its coordinates by x and y. Again we have the similar triangles OQP and $OQ'P'$. Insofar as mere lengths, without regard to sign, are concerned, we can say that

$$\frac{y}{x} = \frac{3}{1}.$$

However, we know that for the line OP, the ordinate of any point is always opposite in sign to the abscissa of that point. That is, when y is a positive number, x is a negative one and conversely. Hence, in the present case, not $y = 3x$, but

$$y = -3x \tag{4}$$

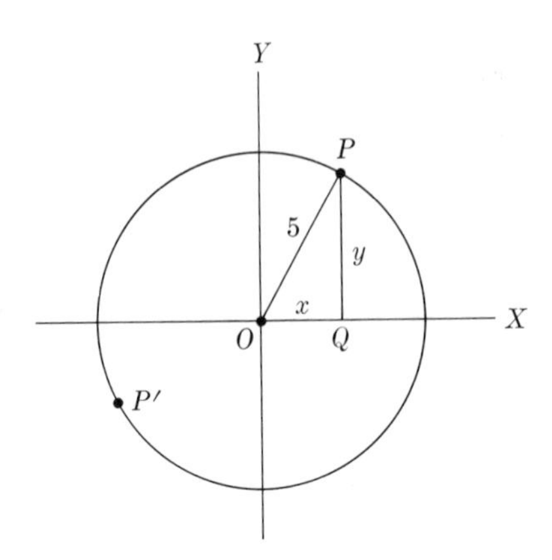

FIG. 13–8. A circle on a rectangular Cartesian coordinate system.

is the equation of the line OP. Thus the fact that OP falls to the right is reflected in the negative sign of the coefficient of x. To distinguish, with respect to slope, lines which fall to the right from those which rise to the right, we say that the former have negative slope. Thus the slope of OP is -3.

To illustrate Descartes' and Fermat's idea once more we shall seek the equation of a circle. Suppose that we choose our axes such that the origin is at the center of the circle (Fig. 13–8). Now the circle is defined to be the collection of all points which are at the same distance from one point, called the center. Let us assume that this distance is 5 units. (The quantity 5 is, of course, the length of the radius.) Since each point on the circle is described by a pair of coordinates, the problem of finding the algebraic equation of this circle can be solved by answering the following question. What property or relationship do the coordinates of points on the circle possess which distinguishes them from those of other points? Geometrically, the coordinates have the distinguishing property that the points on the circle are those and only those which are 5 units away from the center. If we consider a typical point P on the circle and let x and y represent its two coordinates, then we see that the lengths x, y, and 5 form a right triangle. According to the Pythagorean theorem of Euclidean geometry, the square of OP must equal the square of OQ plus the square of QP; hence

$$x^2 + y^2 = 5^2. \tag{5}$$

This same statement also holds for points on the circle such as P', for even though the coordinates of P' are negative, their squares are positive and so satisfy equation (5). Equation (5) is the algebraic representation of the circle. It says in words that the square of the abscissa of any point on the curve plus the square of the ordinate equals the square of the radius.

To decide algebraically whether any point belongs to the circle represented by equation (5), we have to test whether the coordinates of the point satisfy the equation. Thus the point whose abscissa is 3 and whose ordinate is 4 belongs to the circle under discussion because, substituting 3 for x and 4 for y in equation (5), we see that the resulting left side equals the right side, that is, $3^2 + 4^2 = 5^2$. As another example, let us consider the point whose abscissa is 2 and whose ordinate is $\sqrt{21}$. Again, substituting 2 for x and $\sqrt{21}$ for y in equation (5), we find that

$$2^2 + (\sqrt{21})^2 = 25,$$

because the square of $\sqrt{21}$ is 21. Thus the point whose coordinates are $(2, \sqrt{21})$ lies on the circle.

Exercises

1. What is meant by the coordinates of a point?
2. State in your own words what the equation of a curve is.
3. Find the equation of the straight line which (a) rises 2 units for each horizontal distance of one unit and passes through the origin of the coordinate system chosen; (b) makes an angle of 30° with the X-axis and passes through the origin; [*Suggestion:* $\tan 30° = 1/\sqrt{3}$.] (c) falls 4 units for each unit of horizontal distance traversed and passes through the origin; (d) passes through the origin and has slope 4; (e) passes through the origin and has slope -4.
4. Find the coordinates of one point which lies on the curve whose equation is

$$\text{(a)}\ x + 2y = 71; \quad \text{(b)}\ x^2 + y^2 = 36.$$

5. Does the point whose coordinates are $(-3, 5)$ lie on the curve whose equation is $x^2 + 2y^2 = 59$?
6. Determine whether the point whose coordinates are $(3, -2)$ lies on the curve whose equation is $x^2 + y^2 = 4x + 1$.
7. Describe the curve whose equation is

$$\text{(a)}\ y = 3x + 7; \quad \text{(b)}\ x^2 + y^2 = 49; \quad \text{(c)}\ x^2 + y^2 = 20;$$
$$\text{(d)}\ x + 2y = 6; \quad \text{(e)}\ y^2 = 20 - x^2.$$

8. Would latitude and longitude serve as a coordinate system for points on the surface of the earth?
9. Can a curve have more than one equation? If so, how is this possible?
10. Can you say anything about the slope of the line whose equation is $y = mx + 2$?
11. What are the coordinates of the point of intersection of the line $y = 3x + 7$ and the Y-axis?
12. What are the coordinates of the point of intersection of the line $y = 3x + b$ and the Y-axis?
13. What is the slope of the line whose equation is $y = mx + b$, and what are the coordinates of the point where the line cuts the Y-axis?
14. The equation of a circle with center at the origin and radius 1 is

$$x^2 + y^2 = 1. \tag{a}$$

The equation of a circle with center at the origin and radius 2 is

$$x^2 + y^2 = 4. \tag{b}$$

If we subtract equation (a) from equation (b), we obtain the result $0 = 3$. What is wrong?

13–4 The parabola. The curve most widely used, next to the straight line and circle, is the parabola. Let us see how this curve is represented algebraically. Recalling the definition of the parabola given in Chapter 6, we start with a fixed line d, called the directrix, and a fixed point F (Fig. 13–9), called the focus. We now consider all points each of which is equidistant from d and F. This set of points is called a parabola. Thus if P is a typical point on the parabola determined by d and F, then the distance from P to F must equal the distance from P to d, that is

$$PF = PD. \tag{6}$$

To obtain the equation of this curve, we first introduce a set of coordinate axes. We know from the discussion of the straight line that the same curve may have different equations, depending upon how one chooses the axes in relation to the curve. Mathematicians have learned by experience that a simple equation results if the axes are chosen in the following way (see Fig. 13–9). Let the Y-axis be the line through F and perpendicular to d. The X-axis, which is, of course, perpendicular to the Y-axis, is drawn halfway between F and d.

Since the point F and the line d are fixed, the distance from F to d, namely FQ, is fixed. Let us suppose that this distance is 6 units. Then the distance OF is 3 units, and the distance OQ is also 3 units because the X-axis is halfway between F and d. Then the coordinates of F are $(0, 3)$. We now wish to express equation (6) in algebraic terms. Let P be any point on the parabola. Then its coordinates are (x, y). We see that PF is the hypotenuse of a right triangle whose sides are x and $y - 3$. Hence by the Pythagorean theorem,

$$PF = \sqrt{x^2 + (y - 3)^2}.$$

The perpendicular distance from P to d is $y + 3$. Thus, in algebraic terms,

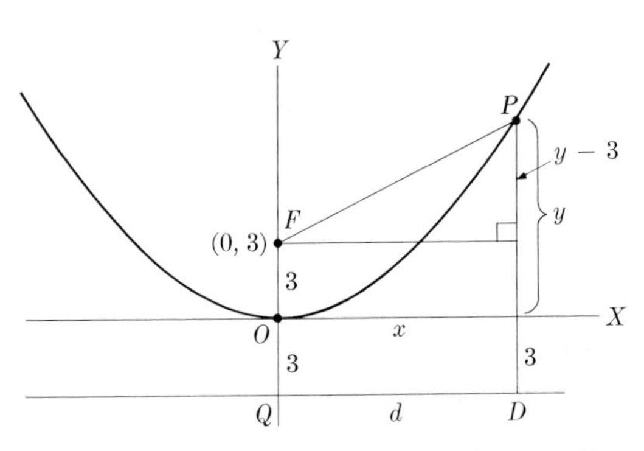

FIG. 13–9. A parabola on a rectangular Cartesian coordinate system.

equation (6) states that

$$\sqrt{x^2 + (y - 3)^2} = y + 3. \tag{7}$$

Equation (7) is the equation of the parabola.

Next we shall perform some algebraic manipulations to simplify the form of this equation. We begin by squaring both sides of the equation, an operation which amounts to multiplying equals by equals. Squaring the left side removes the radical, and squaring the right side yields $y^2 + 6y + 9$. Hence we now have

$$x^2 + (y - 3)^2 = y^2 + 6y + 9.$$

We now write out in full the square called for on the left side and obtain

$$x^2 + y^2 - 6y + 9 = y^2 + 6y + 9.$$

Subtracting y^2 and 9 from both sides yields

$$x^2 - 6y = 6y.$$

We add $6y$ to both sides and obtain $x^2 = 12y$ or, as we prefer to write it,

$$12y = x^2.$$

Dividing both sides by 12, we obtain

$$y = \frac{1}{12}\,x^2. \tag{8}$$

This equation is much simpler than (7) and yet expresses the same fact.

We might note incidentally that the number 12 in the denominator is twice the distance from F to d. Had we called this distance a, the resulting equation would have read

$$y = \frac{1}{2a}\,x^2. \tag{9}$$

In Fig. 13-9 we drew a curve which resembles a parabola, but we did not know at the time its position in relation to the axes we chose. To obtain a rough idea of this position, let us substitute into equation (8) any positive value of x, say 5. Then $y = 25/12$. This tells us that $(5, 25/12)$ is a point on the curve (Fig. 13-10). But if we substitute -5 for x, we obtain the same result for y. Thus $(-5, 25/12)$ is another point on the curve. These two points are symmetrically situated with respect to the Y-axis. Moreover, no matter which abscissa and ordinate we calculate, the negative of that abscissa will always yield the same ordinate because the abscissa has to be squared. This means that to each point on the curve to the right of the Y-axis there corresponds a point symmetrically situated to the left of the

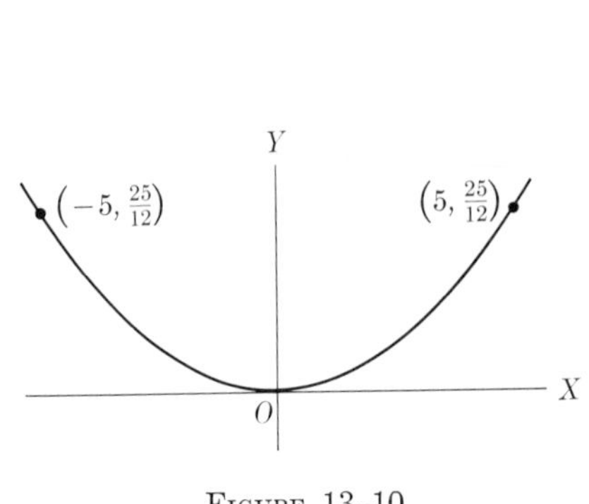

FIGURE 13–10

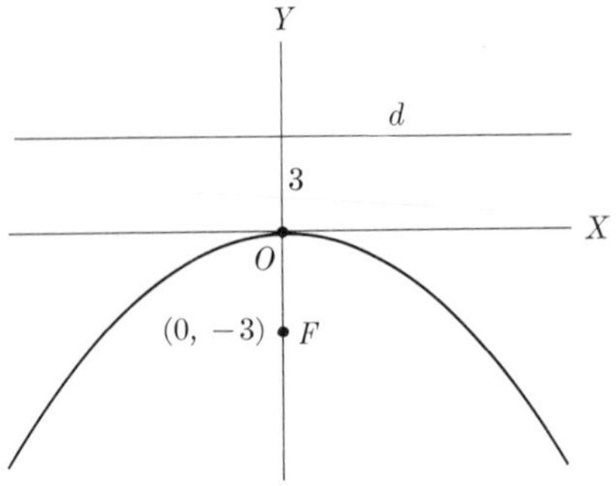

FIG. 13–11. A parabola with focus below the directrix.

Y-axis. Hence, once we have determined the shape of the curve to the right of the Y-axis, we automatically know what the curve looks like to the left.

If we are interested only in the shape of the curve, it is now sufficient to observe that as x increases from zero to any value, x^2 increases, and $x^2/12$ also increases. This means that the points on the curve move out and up from the origin. Of course, there are many curves that meet this description. To obtain a more precise picture of the curve we need to calculate a few sets of coordinates. Thus, when $x = 3$, $y = 9/12$, so that $(3, 3/4)$ are the coordinates of a point on the curve. We should note, too, that the curve lies entirely above the X-axis except, of course, for the point at the origin, because for every value of x, y is zero or positive.

In work involving parabolas and their equations it is sometimes convenient to consider one whose focus lies below the directrix. If we now choose the axes as shown in Fig. 13–11, what is the equation of the parabola? We could, of course, obtain the answer by going through steps analogous to those contained in equations (6) through (8). However, there is no need to do so since Figs. 13–9 and 13–11 differ only in the sign of the y-values: whereas the y-values in the former are positive, the y-values in the latter must be negative. Hence the equation of the parabola in Fig. 13–11 is

$$y = -\frac{1}{12}x^2. \tag{10}$$

Let us consider Fig. 13–11, but suppose now that distances *downward* on the Y-axis are chosen to be positive. How does this change affect the equation of the parabola? The suggested situation does not really differ from that shown in Fig. 13–9. If we imagine this whole figure rotated about the X-axis through 180°, that is through half a rotation, we obtain the situation we have just proposed. Since the position of the curve in relation to the X- and Y-axes is exactly the same in Fig. 13–11 as in Fig. 13–9, the parabola has the equation

$$y = \frac{1}{12}x^2. \tag{11}$$

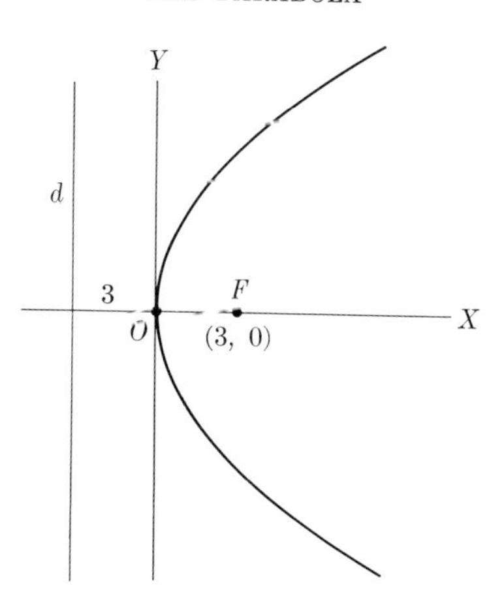

FIG. 13–12. A parabola with focus to the right of the directrix.

Let us consider one more variation. Suppose that directrix and focus happen to lie as in Fig. 13–12. We choose the X- and Y-axes as shown, with the Y-axis halfway between focus and directrix. The upward direction on the Y-axis is positive. What is the equation of the parabola determined by this choice of focus, directrix, and axes? To answer this question we have but to compare Figs. 13–9 and 13–12. The X-axis in Fig. 13–9 plays the role of the Y-axis in Fig. 13–12, and the Y-axis in Fig. 13–9 plays the role of the X-axis in Fig. 13–12; in other words, the roles of abscissa and ordinate are exchanged. Hence the equation of the parabola in Fig. 13–12 is

$$x = \frac{1}{12}\,y^2. \tag{12}$$

These last few equations illustrate some of the various forms which the equation of the parabola can take. Which of these one should use is a matter of convenience in application, as we shall see in later chapters.

EXERCISES

1. Determine the shapes of the following parabolas by choosing a number of values of x, calculating the corresponding values of y, and plotting the points whose coordinates are thereby determined:

(a) $y = 3x^2$, (d) $x = 2y^2$,
(b) $y = \frac{1}{10}x^2$, (e) $x = \frac{1}{2}y^2$,
(c) $y = -3x^2$, (f) $2x = y^2$.

2. Compare the curves of Exercises 1(a) and 1(b).

3. For the parabola shown in Fig. 13–13, the distance from focus to directrix is 6. What is the equation of the parabola?

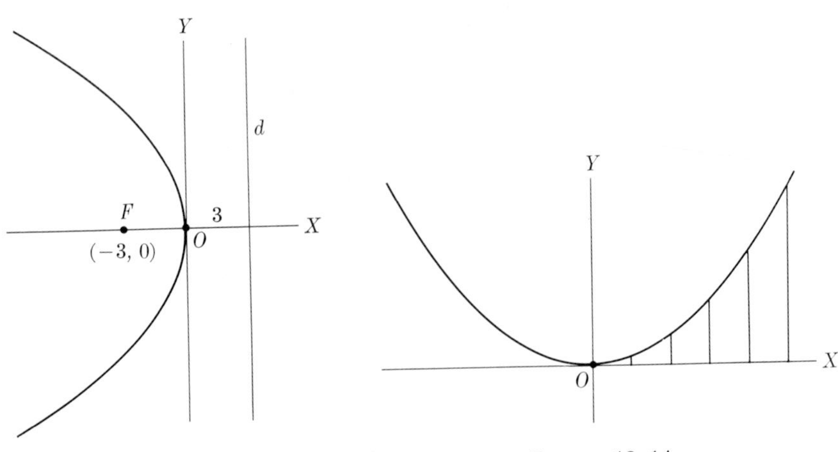

FIGURE 13–13

FIGURE 13–14

4. If the equation of a parabola is $y = \frac{1}{8}x^2$, what are the coordinates of the focus? Describe the position of the directrix.

5. The following exercises specify the position of focus and directrix of a parabola in relation to a coordinate system. Find the equation of the parabola.

(a) Focus $(0, 4)$; directrix parallel to, and 4 units below, the X-axis.

(b) Focus $(0, 6)$; directrix parallel to, and 6 units below, the X-axis.

(c) Focus $(0, -5)$; directrix parallel to, and 5 units above, the X-axis.

(d) Focus $(4, 0)$; directrix parallel to, and 4 units to the left of, the Y-axis.

6. Suppose that the designer of a bridge has decided to use the parabolic cable $y = x^2$ for the range $x = -5$ to $x = 5$ (Fig. 13–14). The roadbed of the bridge is to be the X-axis. Compute the lengths of straight wire needed to suspend the roadbed from the cable at $x = 1, 2, 3, 4,$ and 5.

7. Suppose that the designer of a bridge has decided to use the parabolic cable $y = x^2$ for the range $x = -5$ to $x = 5$ (Fig. 13–15). The roadbed is to have the shape $y = -\frac{1}{10}x^2$. Compute the length of straight wire needed to suspend the roadbed from the cable at $x = 1, 2, 3, 4,$ and 5.

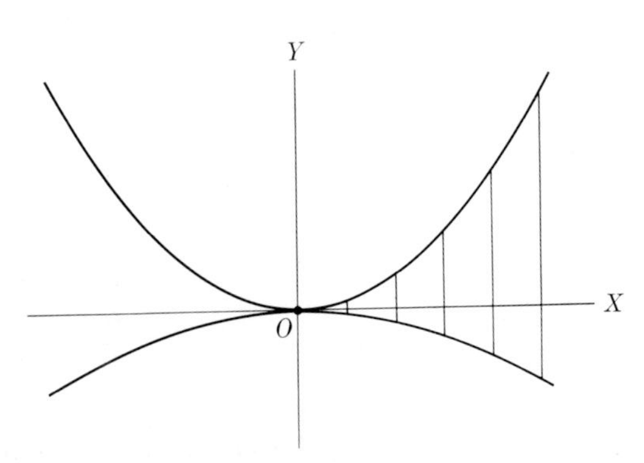

FIGURE 13–15

13-5 Finding a curve from its equation. The great merit of the idea conceived by Fermat and Descartes is that it permits us to represent a curve algebraically and, as we shall see later, learn much about the curve by working with the equation. But another value of their idea, hardly secondary in importance, is that any equation in x and y determines a curve. Hence by merely writing down any equation we please and by finding out what curve belongs to the equation we can discover many new curves. Although we shall not, at this time, make any sensational discoveries, let us see how the process of determining the curve of an equation can be carried out.

Suppose we consider the equation

$$y = x^2 - 6x \tag{13}$$

and try to determine the curve which has this equation. A direct method would be to calculate coordinates of points on the curve and then plot these points. Thus, when $x = 2$, $y = -8$, and so $(2, -8)$ are the coordinates of a point on the curve. By calculating many such sets of coordinates and by plotting the corresponding points one can determine the shape of the curve. However, one often learns more by applying a little algebra.

If equation (13) had been $y = x^2$, we would know at once by comparing with equation (9) that it is a parabola with distance from focus to directrix equal to $\frac{1}{2}$. Let us see whether a little algebraic juggling with equation (13) might bring it into a form which will permit us to identify the curve. By adding 9 to both sides of equation (13) we obtain

$$y + 9 = x^2 - 6x + 9.$$

Now our knowledge of algebra tells us that the right side is $(x - 3)^2$. Hence we have

$$y + 9 = (x - 3)^2. \tag{14}$$

Suppose we introduce new letters x' and y' such that

$$x' = x - 3 \qquad \text{and} \qquad y' = y + 9. \tag{15}$$

Then substitution in equation (14) yields

$$y' = x'^2. \tag{16}$$

The curve corresponding to this equation is the parabola shown in Fig. 13-16, where it is graphed with respect to the X'- and Y'-axes. Of course, we wish to find the curve corresponding to equation (13), not that described by (16). However, equations (15) provide the necessary connection. The equation $x' = x - 3$, or $x = 3 + x'$, tells us that the abscissas x of points belonging to (13) should be 3 more than the abscissas x' of points belonging to (16). How can we increase by 3 the abscissas of each point in Fig. 13-16?

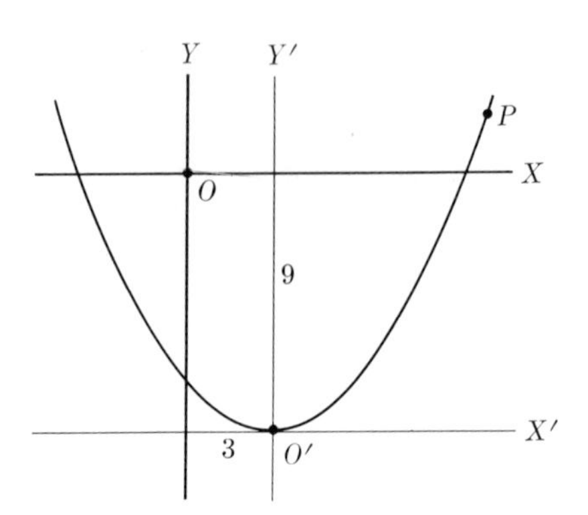

FIGURE 13–16

The answer is simple. We draw a new Y-axis 3 units to the left of the Y'-axis. Then the x-value of a typical point such as P is 3 units more than its x'-value. We now use the second equation in (15), that is, $y' = y + 9$, or $y = y' - 9$. This equation says that the y-values of points should be 9 units less than the y'-values. How can we reduce by 9 the ordinates of the curve in Fig. 13–16? We have already indicated the essential trick. We introduce a new X-axis 9 units above the X'-axis. Now the y-value of P is 9 units less than the y'-value. Hence with respect to the X- and Y-axes, the points of the parabola $y' = x'^2$ have the correct x- and y-values called for by equation (13). Since we did not change the curve in any way by introducing the X- and Y-axes but merely changed the axes, the curve of equation (13) is a parabola placed with respect to the X- and Y-axes as shown in Fig. 13–16.

We were a bit lucky in analyzing equation (13) because the introduction of new coordinates in equation (15) reduced equation (13) to equation (16) whose curve we already knew. If this change to a familiar form is not possible, then the initial equation may indeed represent some new curve, and by analyzing the equation we might get to know the properties of this new curve. The results of such studies would be a further addition to the stock of knowledge about equations and their corresponding curves, a type of knowledge which the professional mathematician builds up for his work just as a writer may build up a bigger and bigger vocabulary. Thus through the notion of equation and curve Fermat and Descartes opened up to mathematicians a vast variety of new curves.

EXERCISES

1. For equation (13) of the text calculate the coordinates of a number of points on the curve and plot the points. Then sketch in the curve. Does your graph look like the one in Fig. 13–16?
2. Determine the curve whose equation is $y = x^2 - 10x$.

3. Determine the curve whose equation is $y = -x^2 + 6x$. [*Suggestion:* Note that the given equation is the same as $-y = x^2 - 6x$ and use the results obtained for equation (13) of the text.]

4. Sketch the curve whose equation is $y = -x^2 + 6x$ by finding and plotting the coordinates of a number of points on the curve.

5. Knowing the curve which corresponds to $y = x^2 - 6x$, can you determine the curve which corresponds to $y = x^2 - 6x + 9$?

6. What does one mean by the statement that a curve can be associated with any equation in x and y?

7. Sketch the curve whose equation is

$$\text{(a) } y = x^3; \quad \text{(b) } y = x^3 + 9; \quad \text{(c) } y = \frac{1}{x}.$$

Does the sketch of part (c) suggest one of the conic sections?

13–6 The ellipse. Another very widely used curve is the ellipse. Let us review the definition given in Chapter 6. We start with two fixed points, F and F', called foci, and a constant quantity which is greater than the distance FF'. We now consider all points, the sum of whose distances from F and F' is the constant quantity. This set of points is called an ellipse. To be more concrete, suppose that the distance FF' is 6 and that the constant quantity is 10. If P is a point such that $PF + PF'$ is 10, then P is a point on the ellipse.

From the standpoint of coordinate geometry, the first thing of interest about the ellipse is its equation. Let us see whether we can find it. As in the case of the parabola, experience has taught mathematicians that the resulting equation will be simplest if the line FF' (Fig. 13–17) is chosen as the X-axis and the Y-axis is chosen to be the line perpendicular to the X-axis and halfway between F and F'. Let us consider the ellipse for which the length FF' is 6 units. Then the coordinates of F are $(3, 0)$ and those of F' are $(-3, 0)$. Now let P be any point on the curve and let us denote its

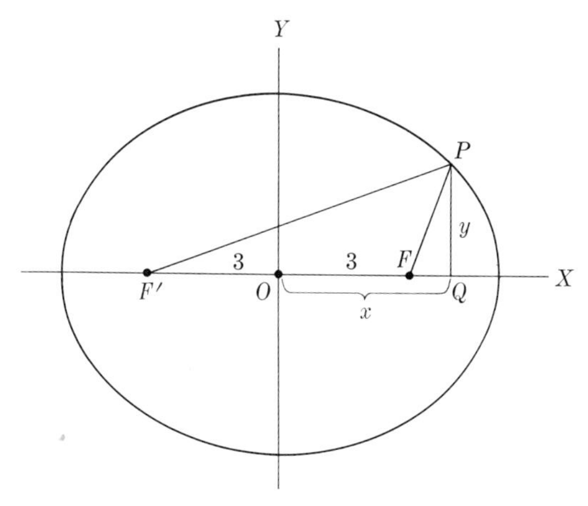

FIG. 13–17. An ellipse on a rectangular Cartesian coordinate system.

coordinates by (x, y). If the constant quantity which determines the ellipse is 10, then the condition which any point P on the ellipse satisfies is

$$PF + PF' = 10. \tag{17}$$

We wish to express this condition algebraically. The procedure is straightforward. The distance PF is the hypotenuse of the right triangle PQF whose arms are $x - 3$ and y. Hence $PF = \sqrt{(x-3)^2 + y^2}$. The distance PF' is the hypotenuse of the right triangle PQF' whose arms are $x + 3$ and y. Hence $PF' = \sqrt{(x+3)^2 + y^2}$. Thus equation (17) amounts to

$$\sqrt{(x-3)^2 + y^2} + \sqrt{(x+3)^2 + y^2} = 10. \tag{18}$$

We are now in the same position that we were in when we arrived at equation (7) for the parabola. We could maintain that (18) is the equation of the ellipse, for it is indeed the condition which the coordinates (x, y) of any point on the ellipse satisfy. However, as for the parabola, a little algebra applied to (18) will simplify the equation. We shall not carry out the algebraic steps explicitly because they are uninteresting, and it is not important for us to acquire great facility in algebra. The result is

$$16x^2 + 25y^2 = 400. \tag{19}$$

We know what an ellipse looks like. But we do not know how our ellipse lies in relation to the axes chosen in Fig. 13–17. An analysis of equation (19) will supply the answer. First of all let us note that if (a, b) should happen to be the coordinates of a point which satisfy equation (19), that is, if

$$16a^2 + 25b^2 = 400, \tag{20}$$

then the sets of coordinates $(-a, b)$, $(a, -b)$, and $(-a, -b)$ will also satisfy the equation, because the substitution of any one of these latter three pairs of coordinates will yield the same equation as (20). Figure 13–18 shows where the various points (a, b), $(-a, b)$, $(a, -b)$, and $(-a, -b)$ lie in relation to the axes. We see, for example, that (a, b) and $(-a, b)$ are

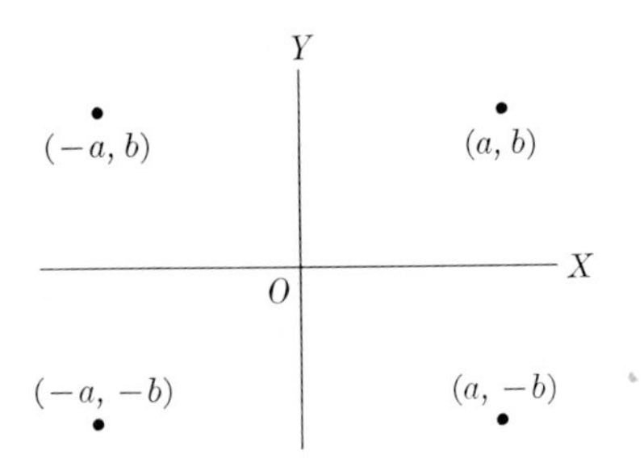

FIGURE 13–18

symmetrically placed with respect to the Y-axis. What we have learned so far is that if the ellipse contains a point (a, b) which lies in the first quadrant, it contains the point $(-a, b)$ which is symmetrically situated with respect to the Y-axis; it contains the point $(a, -b)$ which is symmetrically situated with respect to the X-axis; and it contains the point $(-a, -b)$ which is symmetric to $(-a, b)$ with respect to the X-axis. Hence, if we can determine which points lie on the ellipse in the first quadrant, we can, by symmetry, decide what the ellipse looks like in the other three quadrants.

The shape of the ellipse in the first quadrant is easily determined. We have but to calculate the coordinates of a number of points in the first quadrant and plot the points carefully with respect to the coordinate axes. By symmetry we obtain the shape of the curve in the other three quadrants. The final graph is that shown in Fig. 13–17.

We could investigate the equations of other curves and the curves corresponding to other equations. But what we have done should make the primary idea clear. To each curve there corresponds an equation which describes that curve. The equation depends upon how we choose the axes, but once this choice is made, the equation is unique. Conversely, given an equation involving x and y, we can find the curve which this equation describes, namely the collection of points whose coordinates satisfy the equation.

Exercises

1. For equation (19) of the text, calculate the coordinates of the points whose abscissas are 0, 1, 2, 3, 4, 5. Plot the points.
2. Sketch the ellipse whose equation is $9x^2 + 16y^2 = 144$.
3. Calculate the length of the X-axis which is contained within the ellipse represented by equation (19). Does this length have any relation to any of the quantities which determine the ellipse? This length is called the major axis of the ellipse.
4. Suppose that the constant quantity which defines an ellipse, 10 in the example discussed in the text, is retained, but the distance between the foci F and F' is 0. What changes must one make in equation (18)? Can you now simplify the equation and recognize the curve that it represents?
5. Kepler's first law of planetary motion says that the path of each planet is an ellipse with the sun at one focus. Let us suppose that equation (19) of the text is the equation of some planet's path and that the sun is at F. What is the planet's distance from the sun when it crosses the positive X-axis and what is the distance when it crosses the negative X-axis?

13–7 The equations of surfaces. The mathematician has but to get hold of an idea, and he will develop it for all that it is worth. It had already occurred to Descartes that the idea of equations for curves might be extended to finding equations for surfaces. This possibility was soon explored.

A curve can lie in one plane, but surfaces such as a sphere or the ellipsoid (of which the surface of the earth and the surface of a football are ex-

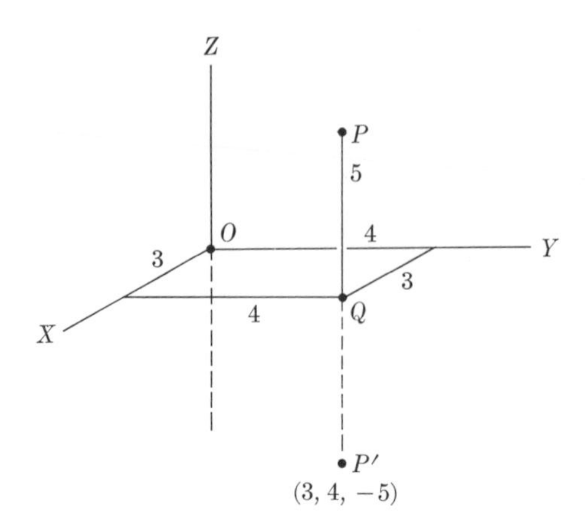

FIG. 13–19. A three-dimensional rectangular Cartesian coordinate system.

amples), do not lie in one plane. They exist in three-dimensional space. To pursue the idea of finding equations for surfaces, we must first introduce coordinates for points in space. This is readily done. One introduces three mutually perpendicular lines (Fig. 13–19) as axes instead of the two lines used for points in the plane. These are called the coordinate axes. The X- and Y-axes determine a plane called the XY-plane. Similarly, the X- and Z-axes determine the XZ-plane, and the Y- and Z-axes define the YZ-plane.

The location of a point P in space is described by three numbers. For example, the point P of Fig. 13–19 is described by (3, 4, 5). The number 5 describes the perpendicular distance of P above the XY-plane, while 3 and 4 are the x- and y-coordinates of Q, the foot of the perpendicular from P to this plane. Alternatively one can say that if one proceeds a distance of 3 units along the X-axis, then a distance of 4 units along a parallel to the Y-axis, and finally travels upward a distance of 5 units along the perpendicular to the XY-plane, he will arrive at the point P. To represent all points in space, we must, as in the two-dimensional system, use negative numbers also. Thus points below the XY-plane have negative third coordinates.

Let us now consider the problem of finding the equation of a surface. We shall use the sphere as an example. A surface, like a curve, has some defining property which states just which points belong to it. By definition, the sphere is the set of all points in space at a given distance, the radius, from a fixed point called the center. To be concrete let us suppose that all points of our sphere are 5 units from the center and that the sphere is located so that its center is at the origin of our three-dimensional coordinate system (Fig. 13–20). The general point on a surface is represented by three letters, x, y, z. Thus the coordinates of the general point P are (x, y, z). Let us now express algebraically the fact that the distance of any point (x, y, z) on the sphere is 5 units from the origin. The lengths x, y, and z are shown

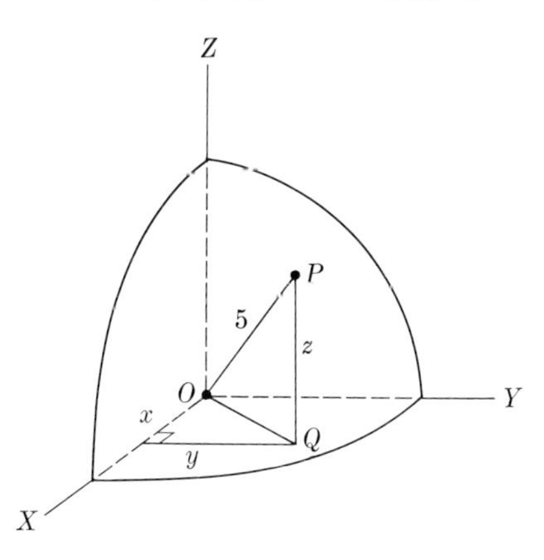

Fig. 13–20. A sphere on a three-dimensional rectangular Cartesian coordinate system.

in Fig. 13–20. Now x and y are the arms of a right triangle (lying in the XY-plane) whose hypotenuse is OQ. Then by the Pythagorean theorem,

$$x^2 + y^2 = OQ^2. \tag{21}$$

Further OQ and z are the arms of the right triangle OQP, whose hypotenuse is OP or 5 units. Hence

$$OQ^2 + z^2 = 25. \tag{22}$$

But OQ^2 has the value given by equation (21). If we substitute this value in equation (22), we obtain

$$x^2 + y^2 + z^2 = 25. \tag{23}$$

This is the equation of a sphere in the sense that the left side equals the right side when and only when the coordinates of a point on the sphere are substituted for x, y, and z.

We could readily obtain the equations of a few other surfaces, such as plane, paraboloid, and ellipsoid. But we shall not do so because we shall not utilize three-dimensional coordinate geometry and the procedure involved is a more or less apparent extension of a familiar concept.

Exercises

1. Plot the points whose coordinates are

$$\text{(a)}\ (1, 2, 3); \quad \text{(b)}\ (1, 2, -3); \quad \text{(c)}\ (-1, 2, 3).$$

2. What group of geometrical figures does an equation in x, y, and z represent?
3. Describe the surface whose equation is $x^2 + y^2 + z^2 = 49$.
4. We know that an equation such as $x + y = 5$ represents a straight line. What does the equation $x + y + z = 5$ represent?

13–8 Four-dimensional geometry. Our experience is limited to figures lying in a plane and in space. But our intellects are not. The idea of a four-dimensional world and of figures in it had tantalized mathematicians such as Pascal at a time when coordinate geometry was still being fashioned. During the next 200 years the subject was mentioned occasionally, but it was not taken seriously until some startling developments (which we shall consider in Chapter 26) caused a number of mathematicians, notably Bernhard Riemann, to investigate it. Four-dimensional geometry proved to be more than a speculation, for some of the deepest developments in modern science, notably the theory of relativity, use this concept. Let us see how coordinate geometry can be employed to portray a four-dimensional world.

We have seen that the equation of a circle of radius 5 in a two-dimensional coordinate system is

$$x^2 + y^2 = 25, \tag{24}$$

and we have just seen that the equation of a sphere of radius 5 in a three-dimensional coordinate system is

$$x^2 + y^2 + z^2 = 25. \tag{25}$$

Even idle speculation would suggest that we at least write down the equation

$$x^2 + y^2 + z^2 + w^2 = 25 \tag{26}$$

and consider what meaning it might have. It would seem reasonable to interpret x, y, z, and w as the coordinates of a point in four-dimensional space and, in analogy with equations (24) and (25), to interpret equation (26) as the equation of a hypersphere in four-dimensional space. This is exactly what mathematics does. However, mathematics does *not* suppose that there is anywhere a real four-dimensional space in which four mutually perpendicular axes can be set up. Nor do mathematicians claim that they, wise and farsighted as they think they are, can visualize figures in a four-dimensional space. It follows that no one else can either.

Four-dimensional geometry is entirely a creation of the mind; it is a geometry without pictures. One speaks of the coordinates (x, y, z, w) as representing a point, and one uses the term hypersphere as though it were a real geometrical figure corresponding to equation (26), but these geometrical terms are merely a convenience and a carry-over from two- and three-dimensional geometry. The words are suggestive but not descriptive of actual figures.

Suppose it is agreed then that four-dimensional geometry is indeed a mental creation. Is it of any value? There is excellent reason to study this "geometry", and there is excellent use for it. We can understand these facts better if we backtrack a bit. Consider the equation $x^2 + y^2 = 25$. It represents, as we know, a circle. But where is this curve that knows no end, this "arc unbroken", the cherished figure of the Greeks? Every geometrical fact we know or can establish about the geometrical circle has its algebraic equivalent and can be derived algebraically from the equation of the circle. Hence the equations of the circle and of other curves in plane, or two-dimensional, geometry are a complete substitute for their geometrical counterparts, and we could, if we wished to, eliminate geometry altogether. In four-dimensional "geometry" we have only the equations, but we talk about them as though they represented figures in a four-dimensional world. The properties of these figures are completely specified by the equations. What is lacking is the possibility of actually constructing such figures.

So far then we have tried to see what meaning this four-dimensional geometry has. Now we wish to know how it is used. One application is made in studying physical events wherein time plays a role. Consider for example, the motion of a planet. The location of a planet is described by three coordinates x, y, and z. But the instant at which the planet occupies that location is also important. An eclipse of the sun, for example, occurs because planet and sun are in certain positions at the same instant. Hence the full description of the position of a heavenly body requires four coordinates, the fourth being a value of time. The path of a celestial object is described by equations involving four letters, usually x, y, z, and t.

But is there any value to thinking geometrically about equations involving four letters? There is. Let us consider the usual sphere for a moment. Some curves on this sphere, a circle of latitude, for example, lie in one plane, and hence only two-dimensional geometry is needed to visualize some curves which lie on a three-dimensional sphere. Similarly, the path on which a planet moves in the four dimensions of space and time may be a curve which can exist in three-dimensional space. This curve may be part of a "geometrical structure" which lies in four-dimensional space, just as the circle of latitude is part of a sphere and yet the curve can be visualized. This visualization aids the understanding. This same visualization might be possible for the paths of other planets and one can consequently better understand their motions. Yet the proper interrelationship of these several paths can be represented only in four-dimensional space just as the relationship of the circles of latitude to one another can be represented only in three-dimensional space. We see therefore that it is helpful to think in terms of geometrical figures lying in a four-dimensional space.

This brief presentation of four-dimensional geometry may give some further indication of the direction in which scientific thought has been moving with the aid of mathematics. Copernicus asked the world to accept a theory of planetary motions which violated some sense impressions for the sake of a better mathematical account. The utilization of a four-dimensional geometry which has no sensuous or visual content means complete reliance upon the mind.

13-9 Summary. From the purely mathematical standpoint coordinate geometry offers a brand-new thought, the representation of geometrical figures by equations. It also offers, as Descartes and Fermat had expected, a new mathematical method of deriving properties of figures from equations. For example, the fact that a curve is symmetric with respect to some line is readily seen from the equation, as we observed in the case of parabola and ellipse.

But Descartes' and Fermat's union of algebra and geometry means far more than a new mathematical method of working with curves. The forms of all physical objects which are studied for any reason whatsoever are, at least when idealized, curves and surfaces. The fusilage of an airplane, the wings of an airplane, the hull of a boat, and the shape of a projectile are surfaces. The paths of all moving objects, a ball thrown by a child, an electron expelled from an atom, a ship on the ocean, a plane in the air, the planets in the heavens, and the tracks of light, are curves. These surfaces and curves can be represented by equations, and the shapes or motions studied by applying algebra to these equations. In other words, Descartes and Fermat made possible the algebraic representation and the study by algebraic means of the various objects and paths of interest to scientists. In addition, algebra supplies quantitative knowledge. This method of working with curves and surfaces is so basic in science that Descartes and Fermat may very well be called the founders of mathematical physics. Part of Descartes' greatness and perhaps the largest part of his contribution was his vision of what his method accomplished; he said he had "reduced physics to mathematics." The investigation of nature which Renaissance Europe had determined to undertake was enormously expedited as we shall soon see. The story of coordinate geometry illustrates how an interest in geometric method became immensely valuable for science and engineering.

Exercises

1. In what sense does a four-dimensional geometry exist?
2. What geometrical language would be appropriate to describe the figure whose equation is $x + y + z + w = 5$?
3. Did Descartes and Fermat introduce a new method for working with curves? If so, describe it.
4. Does coordinate geometry replace Euclidean geometry?

Topics for Further Investigation

1. The life and work of René Descartes.
2. The life and work of Pierre de Fermat.
3. Four-dimensional geometry.

Recommended Reading

Abbott, E. A.: *Flatland, A Romance of Many Dimensions,* Dover Publications, Inc., New York, 1952.

Descartes, Rene: *La Géométrie* (the original French and an English translation), Dover Publications, Inc., New York, 1954.

Descartes, René: *Discourse on Method,* Penguin Books Ltd., Harmondsworth, England, 1960 (also many other editions).

Haldane, Elizabeth S.: *Descartes, His Life and Times,* J. Murray, London, 1905.

Manning, H. A.: *The Fourth Dimension Simply Explained,* Peter Smith, New York, 1941.

Miller, Denning: *Popular Mathematics,* pp. 216–294, Coward-McCann, Inc., New York, 1942.

Sawyer, W. W.: *Mathematicians' Delight,* Chap. 9, Penguin Books Ltd., Harmondsworth, England, 1943.

Scott, J. F.: *The Scientific Work of René Descartes,* Taylor and Francis, Ltd., London, 1952.

Whitehead, Alfred N.: *An Introduction to Mathematics,* Chaps. 9 and 10, Henry Holt and Co., New York, 1939 (also in paperback).

CHAPTER 14

THE MATHEMATIZATION OF SCIENCE

Do not try to know everything if you wish to know anything.

Democritus of Abdera

14–1 Mathematics and modern science. The year 1600 can be considered as the beginning of a new era, an era which has witnessed such an enormous expansion of science that almost every historian recognizes the domination of science as the distinguishing feature of our civilization and culture. We have, in fact, all become familiar with the statement that we live in a scientific age. It is also true that since 1600 mathematics has grown immensely so that what the Greeks created, though extraordinary for a small nation, is insignificant quantitatively compared to the achievements attained in modern times. These two developments, the expansion of science and of mathematics, did not proceed independently. The remarkable progress which science has made in the last few hundred years is due primarily to a new conception of how science should proceed and in this revised methodology mathematics plays an essential role. Secondly, the advances made in the sciences brought forth new problems which, because mathematics is so instrumental in science, gave rise to fresh mathematical investigations. The resulting mathematical creations, in turn, gave new and often unforeseen power to science. And so the interaction has continued right down to the present day.

We intend in the present chapter to study the origins of modern science and the nature of scientific method because we wish to show the interplay of mathematics and scientific activity. As we follow in many of the succeeding chapters the developments in the interdependent scientific and mathematical disciplines, we shall see how the new developments provided motivation and suggestions for mathematical creations. And as we examine the application of mathematics to scientific problems, we shall learn about the most significant contributions which mathematics makes to the modern world. In this connection we might anticipate and correct a common misconception. Mathematics is often considered to be a handy tool of science, but we shall find that this is just a fraction of the full story. Insofar as one can describe the exact relationship in one short sentence, it would be much more to the point to say that mathematics is the basic instrument of science and the essence of its best theories.

14–2 The growth of modern science. Scientific activity is not new to the modern world. We have seen that the Greeks had created a comprehensive and entirely applicable astronomical system—the Ptolemaic theory. Their studies of light had also produced some results. They had investigated mechanical phenomena such as the lever, the use of air and steam pressure, geography, musical sounds, and motion, and had founded the science of medicine.

Renaissance Europe had acquired the Greek knowledge and the Greek interest in the physical world. Some new scientific activity in which mathematics would have played a role would surely have resulted, but the Greek influence alone would not have produced the remarkable activity which began in the seventeenth century. The physical investigations of the modern world surpass a thousandfold those of the Greeks because new forces were active.

A strong impulse came from the sharp and baffling problems which recent developments had thrust upon the Europeans. The most important of these was the class of problems involving motion. The heliocentric theory gradually won acceptance, but the questions it raised remained unanswered. Why do objects stay with a rotating and revolving earth? Why do objects fall toward the center of the earth? What keeps the earth and the other planets moving around the sun? The introduction of gunpowder, we recall, made possible the use of cannons and muskets which could send projectiles over great distances. This new phenomenon also raised a host of questions. What are the paths of such projectiles? How high and how far could they reach with a given initial velocity and direction of fire? How should they be aimed to hit a specified target or to obtain the greatest range, that is, land as far as possible from the starting point? The few principles of motion that Europe had acquired from the Greeks were invalidated by the heliocentric theory, and the study of motion, taken up by Galileo and Newton, dominated the seventeenth and eighteenth centuries.

The geographical explorations of the sixteenth century and the expanding commercial interests of Europe, which now traded with Africa, India, and China, raised a number of problems which bore most heavily on the science of astronomy. Astronomy is of fundamental importance in the determination of latitude and longitude. For ships at sea this application is crucial, for only by determining their latitude and longitude do they know where they are and whether they are following the correct course. Latitude is determined rather easily. A ship's officer has but to note the direction of the sun when it is highest in the sky on a given day (this happens at noon for the ship's location) and, knowing the day of the year, he can then determine the latitude. The determination of longitude is, however, a much more difficult matter. One method in use was based on the moon's position relative to stars and planets. This position could be correlated with position on the earth, provided one knew exactly where the moon should be at any given time. However, knowledge of the moon's path was not precise enough, and the problem of lunar motion engaged the best mathematicians of the seventeenth and eighteenth centuries. Another method of determining longitude depends upon the availability of a clock. A ship's captain could readily determine noon at his position. With a clock which recorded noon at a location whose longitude was known, the captain could tell the difference in time between the two locations. Thus difference in time can be translated readily into a difference in longitude. Unfortunately there were no clocks which operated satisfactorily on board a ship.

The problem of determining longitude was so important that Spain, Portugal, France, and Britain offered prizes for advances in the method. In the seventeenth century Britain established a Commission for the Discovery of Longitude which offered a handsome prize for a serviceable clock.

The work in optics, stimulated by the invention of the telescope and microscope, increased sharply in the seventeenth century. Because the telescope opened up vast spaces and the microscope revealed minute organisms swarming with activity in every square inch of space, scientists investigated lenses with the hope of improving them and even of uncovering other valuable phenomena of light. Men such as Descartes and Newton designed telescopes, and practically every scientist and mathematician of the seventeenth century worked in optics. The resulting progress dwarfed all that preceding centuries had achieved in the same area.

Many other basic scientific problems faced the European civilization of the seventeenth century, and we shall discuss these later. There was, however, another kind of motivation which accounts in large part for the remarkable increase in scientific activity. On the whole, science in Greek times was concerned with the understanding of nature, an understanding which satisfied intellectual curiosity. It was hardly applied to practical problems. Indeed the philosophically minded mathematicians and scientists looked down on problems of trade and the crafts. In Alexandria this cleavage was bridged somewhat, but even there very little use was made of scientific knowledge to improve methods of performing heavy labor or to produce new devices which might make life easier, safer, healthier, or fuller. The practical-minded Arabs first began to think about deriving material benefits from scientific knowledge. Unfortunately they believed in magic, and so their methods of attacking this problem did not succeed, although one must recognize that the alchemistic work of the Arabs and their European disciples was not dominated entirely by the search for gold but also by the desire to produce more useful metals from baser ones. A few men of the medieval period, notably Roger Bacon, did perceive more significant ways in which science could be used to help man, but the age was not oriented toward, or possessed of, the knowledge for concerted work in this direction.

In the sixteenth century many writers urged the scholars to learn and advance the practical arts. Rabelais, for example, suggests that even young princes should study how the objects used in ordinary life are made, and he describes how Gargantua and his tutor visited goldsmiths, jewelers, watchmakers, alchemists, coiners, and other craftsmen.

The new motivation for scientific work, mastery of nature for the welfare of man, became a strong force when such prominent and respected thinkers as Francis Bacon and René Descartes advocated this goal. Bacon criticizes the Greeks. He says that the interrogation of nature should be pursued not to delight scholars but to serve man. It is to relieve suffering, to better the mode of life, and to increase happiness. Let us put nature to use. Knowledge should bear fruit in works; science should be applied to industry. In Bacon's words, let us ascend to knowledge and descend to work. Man should reconstitute his knowledge to apply it to the relief of man's estate. "The true and lawful goal of science is to endow human life with new powers and inventions." In his *New Atlantis* he proposed a college of scientific investigators provided with vast space and equipment who were to work on ideas potentially applicable to improving the conditions of life. Bacon foresaw that science could provide man with "infinite commodities," endow "human life with inventions and riches," and minister to the conveniences and comfort of man.

Descartes, too, is explicit about employing science for practical ends. He says, "It is possible to attain knowledge which is very useful in life, and instead of that speculative philosophy which is taught in the schools, we may find a practical philosophy by means of which, knowing the force and action of fire, water, air, the stars, heavens and all other bodies that environ us, as distinctly as we know the different crafts of our artisans, we can in the same way employ them in all those uses to which they are adapted, and thus render ourselves the masters and possessors of nature."

The founder of modern chemistry, Robert Boyle, expressed the same thought: "The good of mankind may be much increased by the naturalist's insight into the trades." The mathematician and philosopher Leibniz, about whom we shall learn more later, proposed in 1669 the organization of a society devoted to making inventions in mechanics and discoveries in chemistry and physiology which would be useful to people. He, too, wanted to put knowledge to use. He called the universities monkish and said that they were absorbed in trifles. They possessed learning but no judgment. Instead he urged the pursuit of real knowledge, mathematics, physics, geography, chemistry, anatomy, botany, zoology, and history. To Leibniz the skills of the artisan and the practical man were more valuable than the learned subtleties of professional scholars.

We should realize that these men were proposing much more than the solution of existing practical problems. They were proposing profound and extensive investigations which might go far beyond current needs and lead to ideas and inventions whose nature could hardly be specified beforehand.

The potentialities in science gripped people and governments. The "Invisible College," an informal group of scientists who met regularly to exchange information on new discoveries, was chartered by the British government to form the Royal Society of London, and directed to cultivate "such knowledge as has a tendency to use." In 1666 Louis XIV founded the Academy of Sciences in Paris. An astronomical observatory was erected in Paris in 1667, and the Greenwich observatory of England was founded in 1675. Leibniz's project, the Berlin Academy of Sciences, finally came into existence in 1700.

Scientific activity in seventeenth century Europe expanded for still another reason. We have mentioned that the Renaissance witnessed the emergence of a new social class. Laborers, artists, technicians, engineers, and craftsmen, a group which made up the bulk of society, entered the economic, social, and political scene as free men instead of serfs. They envisioned and began to demand the pleasures and privileges previously enjoyed by the small group of wealthy men, nobility, and clergy. The commoners sought to improve their lives and, in seeking the knowledge which might help them to attain that goal, they gradually became aware of science. This change in the society of Europe might be compared roughly with the rise today of new nations in Africa and Asia which, having overthrown colonial masters, are now seeking to learn and employ modern methods of agriculture and production to improve the lives of their people.

Access to the intellectual life and scientific activity of Europe was not readily gained by members of the working class, who came from poor families and were unable to spend years in acquiring an education Books were not generally available; although printing was invented about 1450, book prices remained high for at least a century or more after that date. Moreover the books, manuscripts, and printed material which first spread Greek knowledge were in Latin, whereas the ordinary people spoke the languages of their native regions.

We cannot present here in detail the long history of the rise of the common man, his fight for education and for recognition of his social and economic aspirations, and his contributions to the artistic and intellectual progress of the Western world. We must content ourselves with mentioning that schools were formed, notably in Italy, to teach mathematical and scientific knowledge to the people at large. Some of the more fortunate of the lowly born, for example, Luca Pacioli and Albrecht Dürer, acquired learning and wrote books which sought primarily to popularize and spread the knowledge they had gained. Many of these books were deliberately written in the national languages to reach the people. Dürer wrote in German, Galileo in Italian, and Descartes in French for just this reason. Leibniz argued that the German language should replace Latin because German would be understood by the masses. He also condemned Latin for its alliance with the thinking of the past and for permitting men to use big and impressive words which masked their ignorance.

Progress was slow. As late as the eighteenth century Voltaire said, "Higher education is not for cobblers or kitchenmaids." The problem is by no means solved today, even in the United States, which has done more than any other country to spread education. Nevertheless the common man began to take his place among the learned and brought with him the problems and interests of the business man, the engineer, the technician, and the artisan. The number of potential scientists and the number of new problems which reached the scientific market place were immeasurably increased.

The challenge to master nature and the entry of the practical-minded common man broadened the character of science. Speculative thought became allied with practical interests. Theory descended to aid practice, and practitioners felt impelled to secure the aid of theory. Mathematics and science embraced applied science, technology, industry, the arts, crafts, agriculture, mining, and dozens of other fields of human activity. Whereas the technological advances of the late Middle Ages were products of mechanical ingenuity, sheer inventiveness, accidental discoveries, or results of long experience, seventeenth-century science and mathematics were reaching a stage that permitted systematic and frequent improvements in the applied arts and sciences.

One should not conclude that science and mathematics became concerned exclusively with the solution of problems facing society. It is true that the scientists of the seventeenth century worked on many specific practical problems, the invention of a clock, the improvement in methods of determining longitude, the design of better lenses, and so on. And they focused even their more general theoretical efforts on those fields of pure science—astronomy, motion, and optics—in which practical problems predominated or whose investigation gave promise of solving practical problems. But the desire to understand nature's ways was by no means lost; it remained the outstanding motivation for the truly great scientists and mathematicians.

Indeed one might say that for the scientist as opposed to the engineer this motivation is the genuine one, even though he may take a practical problem as his starting point or be sufficiently sensitive to practical needs to hope that his results will also prove useful. The desire to understand not only has been but must be the driving force. Few practical problems are solved simply because they exist and people work on them. They are solved indirectly and even accidentally because the gradual expansion of scientific knowledge incidentally reveals uses which happen to be advantageous to society. Further, the mathematician is especially responsive to intellectual challenges. He may indeed start on a practical or physical problem of general import, formulate this problem mathematically, and proceed to solve it. But he will very likely introduce generalizations or modifications (see Chapter 5) and try to solve the more general or modified problem, even though the latter may have no practical significance or scientific application. Large portions of mathematics were built up in this way, and the results obtained often have had unexpected uses, but it is nonetheless true that the mathematician sought only the satisfaction of solving a problem or of obtaining a beautiful result. As Descartes put it, mathematics possesses very subtle creations which can serve to satisfy the curious as well as facilitate the arts and diminish the work of man.

Exercise

1. What were some of the factors causing extensive scientific activity in the seventeenth century?

14–3 The search for scientific method. We have tried to point out thus far that the scientific needs and interests of seventeenth-century Europe were great. But need and interest do not in themselves produce results. A need for money and an interest in money do not provide money. The question still remains, How did the European scientists go about solving scientific problems? How does one come to grips with nature either to understand or subjugate her?

One might be tempted to guess that the Europeans found the proper scientific method in the revived Greek literature. But this was not the case. We shall review a few principles of Greek science and late medieval science such as it was to appreciate the changes made in the seventeenth century. First of all most Greeks and medieval thinkers believed that the basic truths exist within the human mind. They are already implanted at birth and are called upon when desired, or they are so clearly truths that when proposed, the mind immediately recognizes them as such. Thus the axioms of Euclidean geometry were accepted by the Greeks as self-evident truths. The medievalists added revelation from God as another source of truth, but again a source communicated to man's mind. The task of science, then, was to determine the implications of these principles by reasoning.

To this source of knowledge, Aristotle and his followers added observation and induction on the basis of observations. Although Aristotle, Galen, the celebrated physician, and astronomers such as Hipparchus and Ptolemy certainly made observations, inductive conclusions did not play a great role. Also, observational results were more likely to be forced to fit a preconceived notion than allowed to suggest some new conclusion. For example, the principle that heavenly motions must somehow consist of circular paths because only circular motion is complete and perfect dominated all Greek and medieval astronomy.

Another methodological principle employed by the Greeks was classification, an approach stressed by Aristotle and taken over by his medieval followers. Thus one observed varieties of animals, flowers, fruits, and humans and classified them according to genus and species. This method is, of course, still used in biology and has some general applicability. It at least reduces the variety of organisms to a few major types and permits systematic study of whole classes in one swoop. It is relevant that Aristotle himself was a physician.

The Aristotelians pursued another scientific doctrine which is best described by the key words "qualitative study of nature." They believed that all phenomena could be explained in terms of the acquisition or loss of basic substances. Thus they and the Platonists believed that heat, coldness, wetness, and dryness were basic substances, and these substances, combined in different proportions, produced other substances. Heat and dryness produced fire; heat and wetness produced air; coldness and wetness produced water; and coldness and dryness produced earth. The hardness or softness, coarseness or fineness of various substances was accordingly determined by the relative abundance of the four basic elements in them. Solids, fluids, and gases were also distinguished by the possession of special substances. Thus a fluid such as mercury possessed some quality, fluidity, which gold did not have. To change mercury into gold meant that one had to take away the fluidity and substitute a new quality which supplied rigidity. Today we recognize that solidity, fluidity, and gaseousness are states of the same matter. However, explanation in terms of special substances was employed right up to modern times. Early chemists, for example Robert Boyle, ascribed the fact that substances such as sulphur were easily set afire to the presence of a special substance called phlogiston. Until the nineteenth century heat was considered to be a substance called caloric which bodies lost or gained as they lost or gained heat. Electricity in the eighteenth century was conceived of as a fluid which flowed through metals.

Thus the Aristotelians hypostatized the obvious sensuous qualities of objects, that is, regarded them as independent substances. Things which appeared to be different were different because they contained these different basic substances. To study changes in objects meant to study which qualities were being lost and which were being acquired.

Aristotelian and medieval science also emphasized another objective for science, namely explanation. To explain meant to give the cause of a phenomenon. However, there were four distinct types of causes, each important in its own way. Suppose that an architect builds a church. The material cause of the church is the brick, stone, and mortar of which the church is constructed. The formal cause is the design which the architect has in mind. Then there is the effective cause, that is, the actual building process. The fourth type, called final cause, is the purpose which the entire project serves. In the present example, the purpose might be to provide a house of worship or to glorify God. Of these four types the final cause was considered most important because it supplied the meaning people usually seek. Thus when we ask why some one was killed and are told that the killer sought revenge, we are satisfied. An entirely different explanation might be furnished in terms of the physical and physiological processes which took place. But such an explanation is usually not of as much interest.

In medieval thought, the final cause dominated. Rain falls to water the crops and supply drinking water. Plants grow to supply food for man. Balls fall to earth because all objects seek their natural place, and the natural place of heavy objects is the center of the universe which is the center of the earth.

The medieval world accepted Aristotle's concept of science. Whenever new problems arose, they turned to Aristotle's explanations of similar phenomena and applied or attempted to apply his explanations. This procedure led to endless disputation about the meaning of the principles and their applicability, and the results were inconclusive, vague, and certainly unsatisfactory.

By the sixteenth century many scholars realized that science could not be advanced by such means. They recognized that new principles and entirely new methods were needed, but did not have a clear conception of what these should be. Prior to Galileo's work, one idea emerged distinctly from the writings of Aristotle's critics, namely the need for systematic experimentation. Francis Bacon issued the manifesto for the experimental method. He attacked preconceived philosophical systems, barren speculations, and idle displays of learning. Scientific work, he said, should not become entangled in a search for final causes which belonged to philosophy. In his *Advancement of Learning* (1605) and in his *Novum Organum* (1620) he points out the feebleness of efforts and the paucity of results in past studies of nature. Man, he observes, has put very little thought and labor into science. Let us come to grips with nature. Let us not have desultory and haphazard experimentation, but let it be thorough and directed. He then makes the acute and most important statement that the only hope for progress lies in a change of method for science. All knowledge begins with observations. But *it must proceed by gradual and successive inductions rather than by hasty generalizations.* He contrasts the anticipation of nature with the interpretation of nature. The one skims; the other is orderly. We gain our ends only if we start with correct laws of nature. He criticizes the then current notions of substance, quality, action, being, heaviness, lightness, density, rareness, moistness, dryness, generation, corruption, attraction, and repulsion. The Aristotelian emphasis on form, he says, is fantastic and ill defined. Man masters nature by understanding her.

Bacon's lofty eloquence, wide learning, comprehensive views, and bold predictions made men listen attentively to what he was saying. His rank, gravity, and caution, his strong expressions and images, and his wise and acute maxims moved people to think about the "Great Instauration" he depicted. Actually he got more credit than he deserved. He was mistakenly hailed as the author and leader of the revolution in science which he had merely perceived sooner and understood better than most of his contemporaries.

14–4 The scientific method of Galileo. We have already noted the new trend of European science toward experimentation and observation. The biologists Vesalius, Cesalpinus, and Harvey were mentioned in Chapter 9. Famous for his systematic experimentation is William Gilbert (1540–1603), physician to Queen Elizabeth. In his *De Magnete* (1600) he presents the details and results of his clear and fruitful work on magnetism, a phenomenon about which practically nothing was known. Gilbert states explicitly that we must start from experiments. He remarks, incidentally, that he does not despise the ancients but respects them, and asserts that they would gladly embrace the new knowledge contained in his work. Kepler's regard for observational facts has already been mentioned. Galileo did not limit himself to astronomical observations, but performed some key experiments on motion; about his results in this area we shall say more later. Moreover, he and his pupils, notably Evangelista Torricelli (1608–1647), having convinced themselves that air has weight, proceeded to carry out relevant experiments. Torricelli also investigated the flow of water through nozzles. Blaise Pascal and Robert Boyle (1627–1691) worked on the pressure of fluids. Boyle and the French priest Edme Mariotte (1620–84) studied gases such as air. Otto von Guericke (1602–1686) invented the air pump and used it to demonstrate the pressure of air. René Descartes experimented in chemistry, biology, and optics. Robert Hooke was a famous experimenter, whose work on springs we shall discuss later, and Christian Huygens (1629–95) obtained distinguished results from his experiments with the pendulum. Newton's work on light was one of the greatest experimental achievements of the seventeenth century.

It is also true that the artists, engineers, and craftsmen, concerned with the practical problems of their trades or professions, did not wait for new scientific methods to gain further knowledge of nature. They investigated mechanical forces, the design of lenses, the chemistry of paints, the motion of cannon balls, and other phenomena and discovered new facts. To this class belongs the self-educated sixteenth-century mathematician Tartaglia, who worked on projectile motion and arrived at results which contradicted Aristotelian physics. The Dutch engineer Simon Stevin (1548–1620) learned about the pressure exerted by water on the walls of canals, and made precise observations of the nature of stable and unstable equilibrium of bodies. He also studied the motion of bodies on slopes. It was spectacle makers who, without discovering a single law of optics, nevertheless invented the telescope and microscope. Many of these men sought not ultimate meanings but common useful knowledge.

There is no doubt that experimentation and practical investigations by technicians and engineers did produce new facts and even opened up new lines of inquiry, but the rise of experimentation was not the reason that science suddenly blossomed in the seventeenth century. The value and import of seventeenth-century experimentation has been vastly overrated. Modern science owes its origins and present flourishing state to a new scientific method which was fashioned almost entirely by Galileo Galilei. Galileo's method is doubly important to us because, as we shall see, it assigned a major role to mathematics.

Galileo, born in Pisa in 1564, entered the university of his native city to study medicine. He also took private lessons in mathematics and was so strongly attracted to the subject that he decided to make mathematics his profession. At the age of 23 when his application for a teaching position at the University of Bologna was rejected because he did not seem worthy of an appointment, he accepted a professorship of mathematics at Pisa. Galileo was one of the men who attacked Aristotelian science, and he did not hesitate to express his views even though these criticisms alienated his colleagues. He had also begun to write important mathematical papers which aroused jealousy in the less competent. Galileo was made to feel uncomfortable, and left in 1592 to accept the position of professor of mathematics at the University of Padua. After 18 years at Padua he was invited to Florence by the Grand Duke Cosimo II. He appointed Galileo "Chief

Mathematician" of his court, gave him a home and handsome salary, and protected him from the Jesuits who had gained domination of the Papacy and who had already threatened Galileo because of his defense of Copernican theory. In Florence Galileo had leisure to pursue his studies and to write. There he spent 23 years. In gratitude Galileo named the satellites of Jupiter, which he discovered in the first year of his service under Cosimo de' Medici, *the Medicean stars.*

The details of his harassment by the Church have in part been related. After his condemnation by the Roman Inquisition in 1633 he was forbidden to publish any more. But he undertook to write up his years of thought and work on phenomena of motion and the strength of materials. The manuscript, entitled *Discourses and Mathematical Demonstrations Concerning Two New Sciences* (also referred to as *Dialogues Concerning Two New Sciences*), was secretly transported to Holland and published there in 1638. Galileo defended his actions with the words that he had never "declined in piety and reverence for the Church and my own conscience." He died in 1642.

Galileo excelled in many fields. As we know from our work on the heliocentric theory, he was an extraordinary astronomical observer. He is often called the father of modern invention. Though he did not invent the telescope, he was immediately able to construct one when he heard of the idea, and he independently invented the microscope. He designed the first pendulum clock, and his son used the design to construct one.

Galileo's writings, though concerned with scientific subjects, are still regarded as literary masterpieces. His *Sidereal Messenger* (1610) in which he announced his astronomical observations was an immediate success. His two greatest classics, *Dialogue on the Great World Systems* and *Dialogues Concerning Two New Sciences,* are considered to be the best Italian prose of the seventeenth century. The style is lucid, direct, witty, and yet profound. In both dialogues Galileo has one of his characters present the traditional views against which another argues cleverly and tenaciously to show the fallacies and weaknesses of these older views and the strengths of the new ones. The clarity of his exposition is in part due to his insight that scientific problems should not become enmeshed in and beclouded by theological and mystical arguments. Indeed one of his achievements in science is that he recognized clearly the domain of science and severed it sharply from religion.

Galileo began his investigation of the methodology of science by asking what is fundamental about the world of phenomena perceived by the senses, a question also considered by Descartes. Both agreed, as some philosophers had asserted earlier, that color, tastes, smells, sounds, and the various sensations of heat, hardness, and softness of objects are not distinct physical substances, but are effects which physically existing properties produce in human beings. What then does exist outside of man and is independent of man? The extension of objects, their shapes and sizes, and their motion are real and external to human perception. Galileo says, "If ears, tongues, and noses were removed, I am of the opinion that shape, quantity [size], and motion would remain, but there would be an end of smells, tastes, and sounds, which, abstractedly from the living creature, I take to be mere words." Descartes' famous words in this connection are, "Give me extension and motion and I will construct the universe." The idea advocated by these two men is known as the doctrine of primary and secondary qualities. The primary qualities exist in the physical world, and their effects on the sense organs of human beings produce the secondary qualities.

Thus in one sweeping blow Descartes and Galileo stripped away a thousand phenomena and qualities to concentrate on matter and motion. But this was only the first step in the new approach to nature which Galileo was fashioning. His next thought, one also voiced by Descartes and even by Aristotle, was that any branch of science should be patterned on the model of mathematics. This implies two essential steps. Mathematics starts with axioms, that is, clear, self-evident truths. From these it proceeds by deductive reasoning to establish new truths. So any branch of science should start with axioms or first principles and proceed deductively.

Galileo departs radically from the Greeks, medievalists, and even Descartes in the method of obtaining these first principles. As noted earlier, the pre-Galileans believed that the mind supplies the basic principles. That all objects in the universe should have a natural place seemed self-evident. The state of rest was "clearly" more natural than the state of motion. To people who had every reason to believe that the heavenly bodies were eternal and invariable, it followed indubitably that they were also perfect. The principle that a force was required to set and keep bodies in motion needed no proof, for how else could a body be gotten to move. To believe that the mind supplies the fundamental principles does not preclude some role for observation. Observation might evoke the correct principle just as the sight of a familiar face might call to mind facts about that person.

The Greeks and medievalists were so convinced of the truth of these fundamental principles that when observations occasionally yielded contradictory results, they invented special explanations which preserved the principles and yet accounted for the anomalies. These men, we might say, first decided how the world should function and then fitted what they saw into their preconceived principles.

Galileo decided that in physics as opposed to mathematics basic principles must come from experience and experimentation; they will be correct if attention is paid to what nature says rather than what the mind prefers. He openly criticized scientists and philosophers who accepted principles which conformed to their preconceived ideas of how nature should and must behave. He said that nature did not first make men's brains and then arrange the world so that it would be acceptable to human intellects. To the medievalists who kept repeating Aristotle and debating the meaning of his works, Galileo addressed the criticism that knowledge comes from observation and not from books. It was useless to debate about Aristotle. Those who did he called paper scientists who fancied that science was to be studied like the *Aeneid* or the *Odyssey* or by collation of texts. "When we have the decrees of nature, authority goes for nothing; . . ." Of course some Renaissance thinkers and Galileo's contemporary, Francis Bacon, had also arrived at the conclusion that experimentation was necessary. With respect to this particular plank of his new method Galileo was not ahead of all others. Yet even a modernist as great as Descartes did not grant the wisdom of Galileo's reliance upon experimentation. The facts supplied by the senses, he said, can only lead to delusion. Reason penetrates such delusions. Particular phenomena of nature can be deduced from, and understood in terms of, the innate general principles. In much of his scientific work, Descartes did experiment and require that theory fit facts, but in his philosophy he was still tied to truths of the mind.

Galileo is a transitional figure in the history of experimentation. His decision to experiment in order to derive basic principles involved both a little more and a little less than appears on the surface. He and even Isaac Newton 50 years later believed that a few key or critical experiments would yield correct fundamental principles. Many of Galileo's so-called experiments were really thought experiments; that is, he relied upon common experience to imagine what would happen if an experiment were performed. He then drew a conclusion as confidently as if he had actually performed the experiment. When in his *Dialogue on the Great World Systems* he describes the motion of a ball dropped from the mast of a moving ship, Simplicio, one of the characters, asks whether he has made an experiment. Galileo replies, "No, and I do not need it, as without any experience I can confirm that it is so, because it cannot be otherwise." He says, in fact, that he experimented rarely and then primarily to refute those who did not follow the mathematical arguments. Although Newton performed some famous and ingenious experiments, he, too, states that he used experiments to make his *results* physically intelligible and to convince the vulgar.

The truth of the matter is that Galileo had some preconceptions about nature, and these made him confident that a few experiments would suffice. He believed, for example, that nature was simple. Hence when he considered freely falling bodies, which fall with increasing velocity, he supposed that the increase in velocity is the same for each second of fall. This was the simplest "truth." He also believed in the mathematical design of nature, and hence any mathematical law that seemed to apply even on the basis of rather limited experimentation appeared to him to be correct. Both Galileo and Newton relied most heavily upon deductive reasoning from a few experimentally obtained principles. They emphasized the importance of and concentrated their efforts on extracting from these few principles as many conclusions as possible. While still at Pisa, Galileo said, "But as ever, we employ reason more than examples." Of course, both Galileo and Newton understood that one may glean an incorrect principle from experimentation and that as a consequence the conclusions deduced from it could be incorrect. Hence they proposed to and did use experiments to check the conclusions of their reasoning as well as to acquire basic principles.

The phenomena one observes are so numerous, so varied, so unlike each other that one can well despair of finding any principles at all in nature. Galileo decided that he must penetrate to the core of a phenomenon and begin there. He says in his *Two New Sciences* that it is impossible to treat the infinite variety of weights, shapes, and velocities. But he had observed that different objects fall with more nearly equal speeds in air than in water. Hence the thinner the medium, the smaller the difference in speed of fall among bodies. "Having observed this I came to the conclusion that in a medium totally devoid of resistance all bodies would fall with the same speed." What Galileo was doing here was to strip away the incidental or minor effects in the effort to get at the essential or major one. He started from observations and then imagined what would happen if all resistance were removed, that is, if bodies *fell in a vacuum,* and he obtained the principle that in a vacuum all bodies fall according to the same law. Since he suspected that friction, too, is a secondary effect, he experimented with smooth balls rolling down a smooth slope to obtain laws about frictionless motion. Having discovered that pendulum motion is little affected by air resistance, he studied pendulum motion to obtain fundamental principles. Thus Galileo did not just experiment and infer from experiments. He tried to discard the relatively unimportant and nonessential, and here he showed genius, for, as any card player knows, to recognize what to discard is wisdom. Galileo experimented and interpreted his observations; in other words, he *idealized.* He did just what the mathematician does in studying real figures. The mathematician strips away molecular structure, color, and thickness of lines, to get at some basic properties, and he concentrates on these. So did Galileo penetrate to basic physical principles.

Of course, actual bodies do fall in resisting media. What could Galileo say about such motions? His answer was ". . . hence, in order to handle this matter in a scientific way, it is necessary to cut loose from these difficulties (air resistance, friction, etc.) and having discovered and demonstrated the theorems, in the case of no resistance, to use them and apply them with such limitations as experience will teach."

Thus far Galileo had formulated a number of methodological principles, many of which were suggested by the pattern mathematics employed in algebra and in geometry. His next principle was to apply mathematics itself. Galileo proposed to seek for science axioms and theorems of a special kind. Unlike the Aristotelians and the medieval scientists, who had fastened upon the notion of fundamental qualities, studied the acquisition and loss of these qualities, or debated their meaning, Galileo proposed to seek *quantitative* axioms. The change is most important, and we shall see its full significance in several succeeding chapters. But an elementary example may help at the moment to demonstrate some of its implications. The Aristotelians said that a ball falls because it has weight and it falls to the earth because it, like every object, seeks its natural place, and the natural place of heavy bodies is the center of the earth. The natural place of a light body, such as fire, is in the heavens, and hence fire rises. These principles are qualitative. Even Kepler's first law of motion (the path of each planet is an ellipse) is a qualitative statement. By contrast let us consider the statement that the speed (in feet per second) with which a ball falls is 32 times the number of seconds it has been falling. This statement can be expressed more briefly in symbols. If we denote by v the speed of the body and by t the number of seconds it has been falling, then the above assertion amounts to $v = 32t$. This simple statement illustrates many important ideas. But the relevant one at the moment is that it is primarily quantitative. It tells us the speed that a ball will acquire in a given number of seconds. In two seconds its speed will be 64 feet per second; in 3 seconds, 96 feet per second; and so on. In the expression $v = 32t$, the letters v and t stand for many values. We can substitute for t any number we please and calculate the corresponding value of v. Technically, v and t are called variables, and the relation $v = 32t$ is called a *formula*. We shall pursue this concept in later chapters.

Galileo intended to adopt such formulas as his axioms, and he expected, by mathematical means, to deduce from them new formulas which would serve as theorems. Since formulas give quantitative knowledge, we can perhaps begin to comprehend the meaning of the statement that Galileo sought quantitative knowledge. Moreover we see that mathematics was to be the essential medium in his scientific reasoning.

A quantitative approach to the study of nature had been suggested by others. Kepler, who was Galileo's contemporary, said, "As the ear is made to perceive sound and the eye to perceive color, so the mind of man has been formed to understand not all sorts of things, but quantities. It perceives any given thing more clearly in proportion as that thing is closer to bare quantities as to its origin, but the farther a thing recedes from quantities, the more darkness and error inheres in it." To weigh, to measure, and to count became a leading seventeenth-century goal which spread to fields such as botany and physiology.

The decision to seek quantitative knowledge expressed in formulas engendered another decision which was also radical, although at first contact it hardly reveals its full significance. As pointed out earlier in this chapter, the Aristotelians believed that one of the tasks of science was to explain why things happened, and explanation meant unearthing the causes of a phenomenon. The statement that a body falls because it has weight gives the effective cause of the fall, and the statement that it seeks its natural place gives the final cause. But the quantitative statement $v = 32t$, for whatever it may be worth, does not explain why a ball falls. It tells only how speed changes with time. In other words, formulas do not explain; they describe. And the knowledge of nature Galileo sought was descriptive. He says, for example, in his *Two New Sciences* that he will investigate and demonstrate some of the properties of motion without regard to what the causes might be. Positive scientific inquiries were to be separated from questions of ultimate causation.

First reactions to this thought of Galileo are likely to be negative. Descriptions of phenomena in terms of formulas hardly seem to be more than a first step. It would appear that the Aristotelians had really grasped the true function of science, namely, to explain why phenomena happened. Even Descartes protested Galileo's decision to seek descriptive formulas. He said, "Everything that Galileo says about bodies falling in empty space is built without foundation: he ought first to have determined the nature of weight." Further, said Descartes, Galileo should reflect about ultimate reasons. But we shall see more clearly in the space of a few chapters that Galileo's decision to aim for description was the most profound and the most fruitful thought that anyone has had about scientific methodology. We merely wish to recapitulate here that the scientific knowledge which Galileo envisioned was to consist of a series of mathematical formulas deduced from a few fundamental ones.

Since the laws Galileo proposed to find were to be quantitative, they obviously had to relate measures, sizes, or amounts of some physical quantities, just as $v = 32t$ relates measures of speed and time. Here, too, Galileo made a fundamental contribution. Whereas the Aristotelians had talked in terms of qualities such as earthiness, fluidity, rigidity, essences, natural places, natural and violent motion, potentiality, actuality, and purpose, Galileo not only introduced an entirely new set of concepts but chose concepts which were measurable so that their measures could be related by formulas. Some of his concepts, such as distance, time, speed, acceleration, force, mass, and weight, are, of course, familiar to us, and so the choice does not surprise us. But to Galileo's contemporaries these choices, and in particular their adoption as *fundamental* concepts, were startling. However, these very ones did prove to be most instrumental in the task of understanding and mastering nature.

We have described the essential features of Galileo's program. Some of his ideas had been espoused by others. Some were entirely original with him. But what establishes Galileo's greatness in the invention of this methodology is that he saw clearly what was wrong or deficient in the scientific efforts of his age, completely shed the older ways, and formulated the new steps—almost in so many words. Moreover, he applied his method to problems of motion and in this work not only managed to provide a lucid example of the procedure but succeeded in obtaining brilliant results. He showed, in other words, that it worked. Galileo was fully conscious of what he had accomplished. He says toward the end of his *Two New Sciences*, "So that we may say the door is now opened, for the first time, to a new method fraught with numerous and wonderful results which in future years will command the attention of other minds." But others were also aware of Galileo's greatness. The seventeenth-century philosopher Thomas Hobbes said of Galileo, "He has been the first to open to us the door to the whole realm of Physics. . ."

Since we are interested in the role of mathematics in the modern world, it may be worth while to emphasize one point. The scholars who fashioned modern science, Descartes, Galileo, and Newton, approached the study of nature as mathematicians. They proposed to find broad, profound, but also simple and clear mathematical principles either through intuition or through crucial observations and experiments and then expected to deduce new laws from these principles, entirely in the manner in which mathematics proper had constructed its geometry and algebra. Mathematical deduction was to take up the major share of scientific activity. Galileo says he valued a scientific principle, whether or not obtained by experimentation, far more because of the abundance of theorems which he could deduce from it than because of the knowledge afforded by the principle itself.

What these great thinkers envisioned did in fact prove to be the profitable course. For the next two centuries, scientists formulated precise and sweeping mathematical laws of nature on the basis of slim, almost trivial, observations and experiments. The greatest progress in the seventeenth and eighteenth centuries occurred in mechanics and in astronomy, and in both these fields experimental results were hardly startling and certainly not decisive. The significant contribution, as we shall see, was the creation of vast branches of mathematical theory.

The expectations of these scientists, seemingly rash, can be explained. These men were convinced that nature is mathematically designed and therefore saw no reason why they could not proceed in scientific matters as mathematics had proceeded in the study of numbers and geometric figures. As Randall says in his *Making of the Modern Mind,* "Science was born of a faith in the mathematical interpretation of Nature, held long before it had been empirically verified."

Exercises

1. What properties of physical objects did Descartes and Galileo regard as fundamental and real?
2. What is the distinction between a qualitative and a quantitative study of nature?
3. Describe the essential principles in Galileo's plan of scientific activity and contrast them with those of his predecessors.
4. Contrast the Greek objectives in the study of nature with those advocated by Bacon and Descartes.
5. How does mathematics enter into Galileo's scientific method?

Topics for Further Investigation

1. Galileo's scientific work.
2. Huygens' scientific work.
3. The importance of experimental work versus that of mathematical deduction from basic principles in seventeenth-century science.
4. The scientific ideas espoused by Francis Bacon.

Recommended Reading

Bell, A. E.: *Christian Huygens and the Development of Science in the Seventeenth Century,* Edward Arnold and Co., London, 1947.

Burtt, E. A.: *The Metaphysical Foundations of Modern Physical Science,* 2nd ed., Chaps. 1 to 6, Routledge and Kegan Paul Ltd., London, 1932.

Butterfield, Herbert: *The Origins of Modern Science,* Chaps. 4 to 7, The Macmillan Co., New York, 1951.

Crombie, A. C.: *Augustine to Galileo,* Chap. 6, Falcon Press Ltd., London, 1952. Also in paperback under the title: *Medieval and Early Modern Science,* 2 vols., Doubleday and Co. Anchor Books, New York, 1959.

Dampier-Whetham, Wm. C. D.: *A History of Science and its Relations with Philosophy and Religion,* Chap. 3, Cambridge University Press, London, 1929.

Farrington, Benjamin: *Francis Bacon: Philosopher of Industrial Science,* Henry Schuman, Inc., New York, 1949.

Galilei, Galileo: *Dialogues Concerning Two New Sciences,* reprint by Dover Publications, Inc., New York, 1952.

Holton, Gerald and Duane H. D. Roller: *Foundations of Modern Physical Science,* Chaps. 13 to 15, Addison-Wesley Publishing Co., Inc., Reading, Mass., 1958.

Randall, John Herman Jr.: *Making of the Modern Mind,* rev. ed., Chaps. 9 and 10, Houghton Mifflin Co., Boston, 1940.

Smith, Preserved: *A History of Modern Culture,* Vol. I, Chap. 6, Henry Holt & Co., New York, 1930.

Strong, Edward W.: *Procedures and Metaphysics,* University of California Press, Berkeley, 1936.

Taylor, Henry Osborn: *Thought and Expression in the Sixteenth Century,* 2nd ed., Vol. II, Chaps. 30 to 35, The Macmillan Co., New York, 1930.

Whitehead, Alfred N.: *Science and the Modern World,* Chap. 3, Cambridge University Press, London, 1926.

Wolf, Abraham: *A History of Science, Technology and Philosophy in the 16th and 17th Centuries,* 2nd ed., Chap. 3, George Allen and Unwin Ltd., London, 1950. Also in paperback.

CHAPTER 15

THE SIMPLEST FORMULAS IN ACTION

When you can measure what you are talking about and express it in numbers, you know something about it; . . .

Lord Kelvin

15–1 Introduction. We intend to pursue in the next few chapters the seventeenth-century developments in science initiated by Galileo and to pay particular attention to the role of mathematics in Galileo's method. Let us recall just what Galileo set out to do. He proposed to find fundamental quantitative physical principles or laws and to apply mathematical reasoning to these quantitative statements in order to deduce new physical laws. These physical laws would then provide the answers to a variety of scientific and practical problems.

To express the physical principles in the manner he regarded as significant, Galileo introduced a new mathematical concept, the extremely important concept of a function. For the next two centuries mathematicians devoted themselves to the construction of functions and to the study of their properties. But the purely mathematical aspect of these creations is in itself rather barren. It is merely the sketch of a picture. And the picture in the present case is precisely the physical world which Galileo set out to investigate. Hence, as we study functions, we shall also study the situations which gave rise to them and the good that was accomplished with them. In fact, it is artificial to separate the physical thinking from the accompanying mathematics, for the two were developed as one. The leading mathematicians of the seventeenth and eighteenth centuries were also the leading scientists. And the accomplishments of these two centuries were a triumph of mathematics and science conjoined.

15–2 Functions and formulas. Before proceeding with Galileo's work, we shall familiarize ourselves with the notion of a function. Let us consider the situation in which a ball is dropped from some point above the ground and let us suppose that we wish to describe the distance the ball falls with increasing time. (Why we should seek such a description and what we can do with it are questions we shall answer later.) It is understood that the distance is measured downward from the point at which the ball begins to drop, and the time of fall is measured from the instant the ball begins to fall. Then the correct description, which we can accept for the moment as a fact, says that *the distance the ball falls, measured in feet,* is *16 times the square of the number of seconds it falls.* The italicized statement is an example of a *function.* As such it is important in two respects. First of all, it deals with varying quantities, or *variables.* The number of seconds that the ball falls increases from zero to larger and larger values. The distance that the ball falls also increases from zero to larger and larger values. Secondly, the statement specifies exactly the relationship between the variables time and distance. What is characteristic of functions, then, is that they are precise statements of relationships among variables.

We know that verbal statements are clumsy to work with. Our experience with algebra teaches us that we can be more effective by introducing symbols. Let us then introduce the symbol t to stand for any number of seconds that the ball has been falling and the symbol d to stand for the distance that the ball falls in t seconds. In these symbols the italicized statement above says that

$$d = 16t^2.$$

The algebraic expression of a functional relationship is called a *formula*. Several facts about formulas are important for their proper understanding and use. In the present case, for example, we must be sure to note that the letters d and t represent not just one particular value of distance and time but whole ranges of values. Thus, if the ball falls for 5 seconds, the variable t can represent any number from 0 to 5. The variable d can represent any distance which the ball may have fallen during the 5 seconds. Of course, the values of d are not independent of the values of t. In fact, the whole point of the formula is to tell us precisely what d is for a given t. Thus when t is 2, for example, d is $16 \cdot 2^2$, or 64. That is, by substituting a particular value of t in the formula, we can calculate the distance d that an object has fallen in that number of seconds chosen for t. The values of d depend upon the values of t and, for this reason, t is called the *independent variable* and d, the *dependent variable*. One also says that the formula expresses d as a function of t. Since we can calculate d for millions of values of t, the formula is indeed a compact representation of millions of bits of information.

Suppose that the dropped ball falls for just 5 seconds. It then hits the ground and remains at rest. However, the formula $d = 16t^2$ does not "stop" at the end of 5 seconds. We could substitute 6 for t and find that d is $16 \cdot 36$, or 576. Likewise, we could substitute 7, or $9\frac{1}{2}$, or even -2 for t and in each case calculate the corresponding value of d. Thus, the mathematical formula has meaning for all positive and negative values of t. However, if the ball falls for only 5 seconds, the formula represents the physical situation only for values of t from 0 to 5. In other words, the mathematical formula is more extensive than the physical situation.

We used the letters d and t to represent the variables distance and time. We could have used y and x, in which case the very same formula would read

$$y = 16x^2.$$

The letters d and t happen to be better because they suggest the physical meaning. But nothing would be altered mathematically if we used y and x.

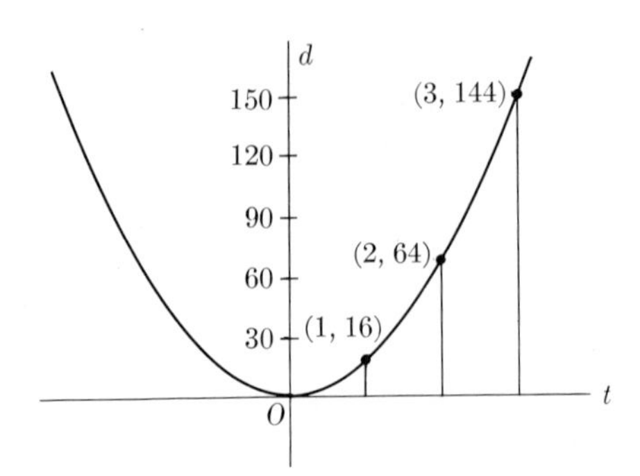

FIG. 15–1. The graph of $d = 16t^2$.

Discussion of a formula and of its physical significance is often aided by utilizing the ideas of coordinate geometry. We can think of $d = 16t^2$ as the equation of a curve. Since the choice of particular letters does not have any mathematical significance, we can introduce axes d and t (Fig. 15–1). The curve corresponding to $d = 16t^2$ consists of those points whose abscissa and ordinate or whose t- and d-coordinates satisfy the equation. Thus, since for $t = 1$, we have $d = 16$, the point whose abscissa is 1 and whose ordinate is 16 lies on the curve. (For convenience we use a smaller unit on the d-axis.) The curve need not be and is not here a picture of the physical motion. Nevertheless, it does show that compared with t, d increases very rapidly as t increases beyond the value 0. Moreover, the curve reveals that the formula has mathematical meaning for all positive and negative values of t, whereas only that part of the curve which extends from $t = 0$ to $t = 5$ represents the physical situation.

EXERCISES

1. What is a function?
2. Distinguish between a function and a formula.
3. "Cook up" some mathematical formulas of your own, whether or not they have any physical significance.
4. For the formula $v = 32t$, calculate the value of v when t is 0, 3, 7, $4\frac{1}{2}$, and -6.
5. In the formula $A = \pi r^2$, which quantity would you regard as the independent variable and which as the dependent one?
6. For the formula $d = 16t^2$, calculate the value of d when $t = 2\frac{1}{2}$, -4, and 7.
7. For the formula $v = 32t$, calculate the value of t when $v = 64$, 80, 128.
8. For the formula $d = 16t^2$, calculate the value of t when $d = 144$. Are both answers physically significant?
9. For any given temperature, the relationship between readings in the fahrenheit scale and the centigrade scale is

$$\text{F} = \tfrac{9}{5}\,\text{C} + 32.$$

What is F when $\text{C} = 0$? What is F when $\text{C} = 100$? Do these values of F and C have special physical significance?

15-3 The formulas describing the motion of dropped objects. Galileo not only formulated the general program for science described in the preceding chapter, but he put it into effect. And here, too, he showed immense wisdom. He did not try, as had scientists and philosophers before him, to tackle the whole universe or to embrace man and nature in one theory. He decided to concentrate on a few classes of phenomena, and principally on motions near the surface of the earth. Galileo possessed the restraint which proves the master.

To appreciate the significance of Galileo's contribution, which, incidentally, is the modern theory of motion if we ignore the recent modifications introduced by the theory of relativity, we must compare his approach with the essentially Aristotelian theory of motion current in his time. Ironically Aristotle's fundamental principles of motion proved to be ineffective because he hewed too closely to experience.

Aristotle distinguished sharply the motions of objects on and near the earth from motions in the heavens. The motions of heavenly bodies could be of only one kind, natural motion, and the natural path of such motion was circular. Motions on and near the earth were of two kinds, natural and violent, the latter being, for example, man-made. As for natural earthly motions, he believed that every object has a natural place. The natural place of heavy objects is the center of the earth, and so heavy objects seek that place. The natural place of light objects, smoke for example, is in the sky and so smoke rises to seek that place. In natural earthly motions objects travel along straight-line or broken-line paths.

Another of Aristotle's principles declares that rest is the natural state of things. An object does not move unless some force is being applied. In other words, force is required to set and keep a body in motion. Now if one merely releases a ball from his hand, the ball falls. What force acts in this case? Aristotle decided that in such natural motions the weight of the object is the force. But objects do not fall as readily in water as in air. The reason, according to Aristotle, is that the medium offers resistance and speed is the force divided by the resistance. In the case of a homogeneous medium such as air, the resistance is constant. Since weight also is constant, the speed, which is the weight divided by the resistance, is constant. If we consider two objects, one heavier than the other and both falling in air, the force is greater for the heavier object, whereas the resistance is the same. Hence the heavier object falls faster, though with a constant speed. This theory also led to the conclusion that a vacuum is impossible because the resistance would be zero and the speed infinite, a situation which clearly could never occur.

In violent motions, say the motion of a ball thrown up into the air, the force is provided by the thrower, and the weight of the object is the resistance. The resulting speed is again the force divided by the resistance. Thus, when a horse pulls a cart, the horse provides the force, and the weight of the cart supplies the resistance.

Let us consider some further implications of the Aristotelian concept of violent motion. Since an object thrown out into space by hand continues to move even though the hand is no longer in contact with it, the question arises, Where is the force which keeps the object in motion? Aristotle answered that the hand communicates a power of movement to the surrounding air; this volume of air transmits the power to an adjacent volume; and so on; thus the air supplies the force which keeps the object in motion. A more detailed analysis explained that the air in front of the object, disturbed by the original motion of the hand, rushes around to the back of the object and so presses it forward. This action of the air is repeated at the new location of the object, and continues to take place all along the path of the motion.

All of the above principles seem to be reasonable enough generalizations of what one observes. Actually, there were a few phenomena known to Aristotle and his successors which were not covered by these principles. The Greeks had observed that a falling object gains speed; that is, its motion is accelerated. No really satisfactory explanation of how this acceleration was caused was ever given, although many were offered from Greek through medieval times. Often the action of the air was resorted to, or, as Aristotle once put it, a falling body moves more jubilantly as it nears its home at the center of the earth.

These Aristotelian laws were criticized in the medieval period, but all comments or changes were intended to be only corrections or modifications. The medievalists sought to improve Aristotle—not to deny him. When the motion of cannon balls was seriously studied in the sixteenth century, as, for example, by the mathematician Tartaglia, it was realized that Aristotle's theory was inadequate, but the radical revision had to wait for Galileo.

Galileo approached the whole problem of motion quite differently. As we noted in the preceding chapter, Galileo thought as a mathematician, and he began his work by *idealizing* the problem he set out to solve. He, too, considered the motion of a ball, say rolling along the ground, and he asked, What if air resistance, friction between ball and ground, and any other hindering forces were not present? What would the ball do once it were set into motion? He concluded that the ball will continue indefinitely to move at a constant speed in a straight line. In more general terms, *if no force acts on a body and the body is at rest, it will remain at rest; if no force acts and the body is in motion, it will continue to move at a constant speed in a straight line.* This fundamental principle of motion or axiom of physics is now known as Newton's first law of motion. We should note that it contains two important assertions. The first, that a body in motion will continue to move in a straight line, is no innovation. It says that straight-line motion is the natural motion of bodies, the motion they will pursue unless they are forced to deviate from such a path. But the statement that the body will continue to move at a constant speed indefinitely is a radical departure from Aristotle, for Galileo was saying that no force is needed to keep the body going once it is set into motion.

But what if a force is applied to a body? Galileo answered that if the body is at rest, the force will set it in motion and change its speed from zero to some nonzero quantity. If the body is already in motion, the *force will change the speed, the direction of the motion, or both*. Thus, an object set into motion along a rough surface encounters friction. Friction is a force, and its effect is to reduce the speed of the object. Galileo's second principle, then, states that force produces *change* in speed or direction.

In other words, Galileo said that force causes acceleration. Let us consider an object which moves along a straight line, but which is being accelerated. The acceleration is a gain or loss in speed per unit time. Thus, if an object has been moving at a speed of 30 feet per second and if in one second its speed increases to 40 feet per second, then the acceleration is 10 feet per second for that one second, or 10 feet per second per second or, in scientific shorthand, 10 ft/sec^2. If the increase in speed had been 10 feet per second over two seconds, then the acceleration would have been 5 ft/sec^2.

Now, as the Greeks had observed, a body which is dropped falls with increasing speed; its motion is accelerated. Since falling bodies possess acceleration, that is, they do not move at a constant speed, it follows that some force must be causing the change in speed. By Galileo's time the concept of gravity had become more or less accepted. The earth exerts a force on any object and this force, if not offset by some other force, gives the object an acceleration. The surprising fact which Gallileo discovered is that if one neglects air resistance, then an object falls to earth with a constant acceleration, and, moreover, this constant is the same for *all* bodies, namely 32 ft/sec^2. Thus if we let a stand for acceleration, the third fundamental law of motion, an axiom of physics, states that for all bodies falling to earth*

$$a = 32. \tag{1}$$

We now have a few fundamental principles about motion. Let us see next whether, in accordance with Galileo's plan, mathematical reasoning can lead to new information. Let us consider the motion of an object which is dropped, that is, whose initial speed is zero. Galileo's third principle says that the object gains speed each second at the rate of 32 ft/sec. Hence, at the end of one second its speed is 32 ft/sec. At the end of two seconds its speed is 2 times 32 ft/sec or 64 ft/sec. At the end of t seconds its speed is t times 32 ft/sec or $32t$ ft/sec. If we let v denote the speed at the end of t seconds, then

$$v = 32t. \tag{2}$$

We now have a formula which tells us the precise speed which a dropped body acquires in t seconds. It can, of course, be used to calculate v for any given value of t. Thus at the end of 6 seconds the speed of the body is 192 ft/sec.

---

* This axiom applies only to objects near the surface of the earth. We shall say more about it in Chapter 17.

Formula (2) is of some interest but hardly a surprise. Let us see whether we can obtain more significant conclusions by the further application of mathematics. We wish to determine the distance which a dropped body falls in t seconds. To be specific, let us consider for the moment that $t = 6$. Now the speed at the end of 6 seconds is 192 ft/sec. To obtain the distance traveled in 6 seconds, one is tempted to multiply 192 by 6, that is, the speed by the time. However, the object did not travel at 192 ft/sec throughout the 6 seconds. In fact, it started with zero speed and only gradually increased its speed to 192. Which speed should we use to compute the distance traveled? Presumably, the average speed.

A reasonable guess would be that the average speed is the arithmetic average of the initial and final speeds, that is, $(0 + 192)/2$ or 96 ft/sec. Let us see whether we can establish the correctness of this guess. We note from formula (2) that when $v = 96$, then $t = 3$. Hence the velocity of 96 ft/sec is attained by the falling object at the end of 3 seconds of fall, that is, after half the time has elapsed. But now let us consider the speed one-half of a second before $t = 3$, that is, at $t = 2\frac{1}{2}$. At this instant, formula (2) tells us that the speed is 80 ft/sec. The speed one-half of a second after $t = 3$, that is at $t = 3\frac{1}{2}$, is 112 ft/sec. The average of 80 and 112 is 96. What this calculation shows is that if we take the speed at some instant before $t = 3$ and then take the speed at an instant as much after $t = 3$ as the first instant is before, the average of the two speeds will be 96, the speed at $t = 3$.

This argument can be generalized. Let h be any interval of time. Then $3 - h$ is some instant before 3 seconds, and, by formula (2), the speed of the falling body at the instant $t = 3 - h$ is $32(3 - h) = 96 - 32\,h$. At the instant $t = 3 + h$, which is h seconds after 3 seconds, the speed is $32(3 + h)$ or $96 + 32h$. We see that the speed at $t = 3 - h$ is $32h$ less than 96, and the speed at $t = 3 + h$ is $32h$ more than 96. Hence the average speed for these two instants is 96 because

$$\frac{96 - 32h + 96 + 32h}{2} = 96.$$

Since the object falls for as many instants during the interval from $t = 0$ to $t = 3$ as it does from $t = 3$ to $t = 6$, and since the pairing of instants produces the average of 96, the average speed over the entire interval of 6 seconds should be 96 ft/sec. This average speed is the speed actually attained after one-half of the time of travel has elapsed. We ought to caution that the argument used to derive the average speed holds only when the acceleration is constant. The argument depends upon formula (2) and thus is correct only when (1) is correct.

If instead of 6 seconds we had used the general value of t seconds, then our conclusion would read that the average speed is that attained after $t/2$ seconds. By formula (2) this average speed is $32(t/2)$ or $16t$. The distance d fallen in t seconds is the average speed times the time, or $16t \cdot t$, or $16t^2$.

Thus, if we let d represent the distance fallen in t seconds, we have the new result:

$$d = 16t^2. \tag{3}$$

Formula (3) says, for example, that in 3 seconds the object falls $16 \cdot 3^2$, or 144 feet. With a little mathematics we have derived an important law of falling bodies. It tells us the distance which any body that is dropped and freely falling travels in t seconds.

We can derive a few significant consequences of formulas (2) and (3) by the application of simple algebra. Dividing both sides of formula (3) by 16 and taking the square root of both sides of the resulting equation, we obtain

$$t = \pm\sqrt{\frac{d}{16}}.$$

This result tells us the time required for a dropped body to fall d feet. Of course, of the two roots (one positive and one negative), only the positive value possesses physical significance because we are dealing with a physical situation in which time is positive and measured from the instant the body begins to fall. Hence we shall forget about the negative root and consider that

$$t = \sqrt{\frac{d}{16}}. \tag{4}$$

If we now wished to calculate how long it takes an object to fall 1000 feet, we would substitute this value in formula (4) and calculate t.

From formula (4) one can draw a most significant conclusion. The formula does not tell us the name of the President of the United States, but this is not so surprising. However, it is surprising that the formula does not involve the weight or any other property of the falling body. This means that *all* bodies take the *same* time to fall a given distance, provided, of course, that air resistance is neglected. A feather and a piece of lead take the same time to fall a given distance in a vacuum. This is the lesson which Galileo is supposed to have learned by dropping various objects from the leaning tower of Pisa. Many people still hesitate to accept this conclusion because they observe bodies falling in air, and the resistance encountered by feathers is quite different from that offered to lead. Undoubtedly it was this difference gleaned from actual observations which led the Aristotelians to the conclusion that heavier bodies fall faster.

Formula (4) was derived by merely rearranging, so to speak, formula (3). But Galileo's plan envisaged also combining existing formulas to obtain new knowledge. To illustrate this process, suppose one takes the value of t given by (4) and substitutes it in the formula $v = 32t$. This yields

$$v = 32\sqrt{\frac{d}{16}}.$$

Now the square root of a fraction is equal to the square root of the numerator divided by the square root of the denominator. Hence

$$v = 32\frac{\sqrt{d}}{4} = 8\sqrt{d}. \tag{5}$$

The new formula enables us to calculate the speed which a dropped body will acquire in falling d feet. While this information is implicit in formulas (3) and (4), which yield (5), we now see clearly something we might not have appreciated before. We should note that formula (5) says that the speed increases as the square root of d. The predecessors of Galileo believed that the speed increased directly with distance.

Exercises

1. Was Aristotle wrong in asserting that, to keep an object moving at a constant speed in a real medium, a force must constantly be applied?

2. Suppose that gravity does not exist and a man steps off the roof of a building. What would his subsequent motion be? What would it be in the presence of gravity?

3. An automobile travels at the speed of 10 mi/hr for 59 min and at a speed of 50 mi/hr for 1 min. What is its average speed?

4. What is the speed of an object 4 sec after it is dropped? What is its average speed during the 4 sec? At what instant does the object actually possess this average speed?

5. Distinguish between speed and acceleration.

6. We may regard the formula $v = 32t$ as an equation in v and t, and we may therefore plot t-values as abscissas and v-values as ordinates. Draw the curve of the formula $v = 32t$.

7. Using the formula $d = 16t^2$, calculate how far a body will drop in 5 sec, $6\frac{1}{2}$ sec, 10 sec. Is the drop the same from one second to another?

8. Apply the instructions of Exercise 6 to the formula $d = 16t^2$. What is the name of the resulting curve?

9. Graph the curve of $d = 16t^2$, but let the downward direction of the vertical axis, that is the d-axis, be positive.

10. A window washer at the 50th floor of a skyscraper (500 ft above the street) steps back to observe the results of his work. Describe mathematically his subsequent behavior.

11. Using the formula $v = 8\sqrt{d}$, calculate the speed with which an object dropped from the top of the Empire State Building (about 1000 ft above street level) hits the ground.

12. If the relation between speed and distance were $v = 8d$ instead of $v = 8\sqrt{d}$, what difference would there be in the behavior of falling bodies?

13. Suppose that we are considering the motion of an object which is dropped from a point near the surface of the moon. On the moon all objects also fall to the surface with a constant acceleration, and the value of this acceleration is 5.3 ft/sec^2. What change would you make in formulas (2) and (3) to have them represent speed acquired and distance traveled for objects falling to the moon's surface? Incidentally, the moon has no atmosphere, and hence air resistance can surely be neglected.

14. Suppose that an object is dropped and falls with a constant acceleration a. What would you propose as formulas for speed and distance fallen in time t?

15. Show first that formula (5) implies $d = v^2/64$. Now suppose a dropped object acquires a speed of 88 ft/sec. What distance must it fall to acquire this speed?

16. Suppose an object is traveling along a straight line with a speed of 88 ft/sec and then starts to lose speed, that is decelerates, at the constant rate of 32 ft/sec^2. What distance must it travel for its speed to become zero? [*Suggestion:* The distance it must travel to reach zero speed is the same as the distance it would travel if it started with zero speed and accelerated at 32 ft/sec^2 to attain a speed of 88 ft/sec.]

17. As a direct generalization of the thought in Exercise 16, we may state that if an object is traveling in a straight line at a speed of v ft/sec and then loses speed at the rate of 32 ft/sec^2, the distance d it travels before attaining zero speed is $d = v^2/64$. Suppose the deceleration is 11 ft/sec^2. What formula gives the distance the object travels before attaining zero speed?

18. Using the result of Exercise 17, answer the following question. An automobile is traveling at 60 mi/hr (or 88 ft/sec), and the brakes are applied. The action of the brakes decelerates the automobile at the rate of 11 ft/sec^2. How far will the automobile travel before stopping? The answer gives the minimum distance in which one can, even under most favorable road conditions, stop a car traveling at 60 mi/hr. However, it takes about 1 sec before a person who decides to apply the brakes actually does so. What distance will the automobile travel in that time?

19. A man drops a stone into a well and listens for the sound of the splash. He finds that $6\frac{1}{2}$ sec elapse from the instant the stone is dropped until he hears the sound. How far below is the surface of the water? Assume that sound travels at 1152 ft/sec.

15–4 The formulas describing the motion of objects thrown downward. Thus far we have seen how simple formulas describe the motion of a body which is dropped. By employing slightly more complicated formulas Galileo was able to tackle further phenomena of motion. Suppose that instead of being dropped a ball is thrown downward. Now the ball does not start its motion with zero speed but with whatever speed the hand imparts to it. The problem we shall look into is, What is the subsequent motion of the ball? To be specific, suppose the hand imparts to the ball a speed of 96 ft/sec. Neglecting for the moment the action of the force of gravity, we can say that the ball will continue to travel downward in a straight line with a speed of 96 ft/sec. The basis for this assertion is, of course, the first law of motion. We know, however, that gravity will also act on the ball and give it a speed of $32t$ ft/sec in t seconds. Since both speeds will operate simultaneously to make the ball move downward, the total speed, v, of the ball is represented by the formula:

$$v = 96 + 32t. \tag{6}$$

Let us compare this formula with $v = 32t$. We see that the term 96 in formula (6) represents the speed given to the ball by the hand. Both formulas are said to be of the first degree in t because the independent variable, t, appears only to the first power. That is, the formulas contain $32t$ as opposed to $32t^2$, or $32t^3$, or some other power of t. First-degree formulas are often called *linear functions* because the curve representing each is a straight line.

We can also obtain the formula for the distance, d, which the ball will fall in t seconds. If there were no gravity, the ball would fall a distance of $96t$ feet in t seconds because it would have the constant speed of 96 ft/sec imparted by the hand. But during the same t seconds, the force of gravity will exert an additional downward pull which, according to formula (3), will cause the ball to fall $16t^2$ feet. Since both forces, the hand and gravity, cause the ball to fall downward, the total distance, d, traveled in t seconds is

$$d = 96t + 16t^2. \tag{7}$$

Comparison of formula (7) with formula (3) shows that formula (7) contains a new term, $96t$, which represents the contribution made by the action of the hand to the distance the ball falls. Formulas (3) and (7), incidentally, are of the second degree in t because the independent variable t occurs to the second power. Second-degree functions are also called *quadratic functions.*

Exercises

1. If a ball, instead of being merely dropped, is thrown downward with a speed of 128 ft/sec, will its speed be greater in t seconds? Will the distance fallen in t seconds be greater?
2. Write the formula representing the speed acquired and distance traveled in t seconds by a ball which is thrown downward with a speed of 128 ft/sec.
3. Suppose a ball is thrown downward with a speed of 96 ft/sec. What are the speed and distance traveled after 3 sec? after $4\frac{1}{2}$ sec?
4. Graph formula (7) by plotting points whose coordinates satisfy the equation. What is the name of the resulting curve?
5. Graph formula (7) by applying the method of change of coordinates presented in Chapter 13.

15–5 Formulas for the motion of bodies projected upward. A more interesting phenomenon both physically and mathematically is the motion of a ball thrown straight up into the air. Suppose, for example, that the ball is thrown upward withh a speed of 96 ft/sec, and let us again consider the questions, What are the speed and distance traveled after t seconds of motion? If gravity is neglected, then the action of the hand will cause the ball to start upward with a speed of 96 ft/sec and, according to the first law of motion, it should continue to travel upward at that speed indefinitely. However, we know that the downward pull of gravity causes the ball to acquire in t seconds a downward speed of $32t$ ft/sec. Since the hand gives the ball an upward speed of 96 and gravity gives it a downward speed of

$32t$, the net speed, v, of the ball at the end of t seconds is

$$v = 96 - 32t. \tag{8}$$

The minus sign in formula (8) takes care of the fact that the speed resulting from the action of gravity reduces the speed imparted by the hand. Formula (8) should be compared with formula (6).

Let us turn to the second question: How far does the ball travel? Since the ball travels upward to some maximum height and then falls down, we shall instead ask the more pertinent question, What height above the ground does the ball possess at any time t? If there were no gravity, the ball would move upward at the constant velocity of 96 ft/sec. Hence in t seconds it would travel upward $96t$ feet. However, we know that a ball moving above the surface of the earth for t seconds will experience a downward pull of gravity amounting to a distance of $16t^2$ feet in t seconds. Hence the net height, d, reached by the ball is

$$d = 96t - 16t^2. \tag{9}$$

As in formula (8), the minus sign here represents the fact that the action of gravity offsets the action of the hand.

We can now answer some questions about the motion of the ball. We know from experience that the ball will rise to some height and then fall back to the ground. How high will it go? We would expect the ball to continue to rise until its upward speed, which is continually decreasing, becomes zero. This fact can be put to use through formula (8). We now ask the question, What is t when $v = 0$? Suppose we denote by t_1 this particular unknown value of t. Then we may say on the basis of formula (8) that

$$0 = 96 - 32t_1.$$

To find t_1 we have but to solve this simple equation. Clearly $t_1 = 3$.

We have determined the time it takes the ball to reach its maximum height but not the height itself. However, formula (9) gives us the height at any time t. Suppose then that we substitute the value $t_1 = 3$ in (9) and calculate d_1, the height of the ball above the ground at this instant. Substitution of the quantity 3 for t in (9) yields

$$d_1 = 96 \cdot 3 - 16 \cdot 3^2 = 144.$$

Thus the maximum height above the ground to which the ball will rise is 144 feet.

We know that the ball will fall to the ground after reaching this maximum height. Will it take as much time to reach the ground from its maximum height as it did to travel from the ground to the maximum height? Those people who have lots of confidence in their intuition should answer this question before we settle it by mathematical reasoning.

To obtain an answer we must proceed somewhat indirectly. Our information about the motion is contained in formulas (8) and (9). Of these two formulas, (9) offers some prospect of being useful because it relates time and height reached by the ball. There is one bit of information which might be used in connection with formula (9), namely, that when the ball reaches the ground, the height of the ball above the ground is zero. Let us therefore find the time at which the ball reaches the ground and then see what it suggests.

We denote by t_2 the value of t at which the ball reaches the ground. Hence, by formula (9),

$$0 = 96t_2 - 16t_2^2. \tag{10}$$

Our problem now is to determine the value of t_2 which, according to (10) satisfies a second-degree equation. This equation is easily solved. We apply the distributive axiom to justify writing

$$0 = 16t_2(6 - t_2). \tag{11}$$

We are seeking the value or values of t_2 for which the right side of (11) equals the left side. Clearly when $t_2 = 0$, one factor on the right side is zero, and hence the product is zero. Likewise, when $t_2 = 6$, the product is zero. Hence, there are two values of t, namely 0 and 6, when d, the height above the ground, is zero.

Why two values? Mathematically the two values result from the fact that we are solving a quadratic or second-degree equation. Physically the two values are readily understandable. The value $t_2 = 0$ is the value of t at the instant at which the ball is about to start out; $t_2 = 6$ is the value of t at the instant at which the ball hits the ground after traveling up and down. With respect to the problem in hand, the second value is the interesting one because it tells us that 6 seconds elapse during the upward and downward travel of the ball. Since we found earlier that the ball requires 3 seconds to reach its highest position, it is evident that only 3 seconds are required for the ball to return to the ground. Hence it takes exactly the same time for the ball to go up as it does to come down.

We can now ask mathematics to answer another question for us. What speed does the ball possess when it strikes the ground? Is this speed the same, more, or less than the 96 ft/sec with which it was thrown up? The answer can be obtained at once. Formula (8) gives the speed of the ball at any instant of its flight. The ball strikes the ground at the instant $t_2 = 6$. If we substitute 6 for t in (8), we find that v_2, the speed at the instant the ball strikes the ground, is

$$v_2 = 96 - 32 \cdot 6 = -96.$$

Thus mathematics tells us that the speed is 96 ft/sec, the very same speed with which it was projected upward. Obligingly, mathematics also tells us, through the minus sign, that the speed is in the opposite direction to that of the upward throw.

Exercises

1. For a ball thrown upward with a speed of 128 ft/sec, what formula describes the relationship between the subsequent height above the ground and the time of travel?

2. If a ball is thrown upward with a speed of 160 ft/sec, then the formula relating its subsequent height above the ground and the time of travel is $d = 160t - 16t^2$, and the speed of the ball is given by the formula $v = 160 - 32t$. (a) How high is the ball after 4 sec? (b) What is its speed at the end of the fourth second? (c) How high will the ball go?

3. If the height of a ball thrown upward is given by the formula $d = 144t - 16t^2$, what is d when $t = 9$? Interpret the result physically.

4. If the height above the ground of a ball is representable by the formula $d = 192t - 16t^2$, then its height after 4 sec is 512 ft, and its height after 8 sec is also 512 ft. Verify these heights and account for the fact that the height is the same after 4 additional seconds.

5. If a gun capable of firing a bullet at the speed of 1000 ft/sec is fired straight upward, how high will the bullet go?

6. Suppose that a ball is dropped from the top of a building 100 ft high. Let d represent the *height of the ball above the ground* and t the time of travel measured from the instant the ball is dropped. Write a formula representing the motion in terms of d and t.

7. Since man seems to be preparing for experiences on the moon, it may be well to consider the following question. Suppose that a ball is thrown up from the surface of the moon with a speed of 96 ft/sec. How high will it go and how long will it take to reach its maximum height? Remember that on the moon the value of 5.3 ft/sec^2 corresponds to the acceleration of 32 ft/sec^2 which holds on the earth.

8. Suppose that a bullet shot straight up into the air returns to the ground 60 sec later. What was the initial speed? [*Suggestion:* Use formula (9). However, the initial speed, which was 96 in formula (9), is now unknown.]

9. A rocket is shot straight up into the air to a height of 50 mi at which point its velocity is 300 mi/hr. Its fuel is now exhausted, and hence the rocket receives no further acceleration from this source. Write a formula which describes the subsequent motion of the rocket. You can choose the 50-mi point as the origin for height and zero time as the instant when the rocket is at that height.

10. A ball is thrown up into the air from the roof of a building with a speed of 96 ft/sec. Write a formula for its subsequent height above the roof as a function of time.

11. Using the data of Exercise 10 and the additional fact that the roof is 112 ft above the ground, find the time when the ball reaches the ground.

15–6 Mass and weight. The formulas and applications considered so far deal with motion which is independent of the size or weight of the moving objects. There are very interesting examples of straight-line motion in which we must take the factor of weight into account. However, we must first learn a few physical facts about weight and mass.

Newton's first law of motion says that if no force is applied, bodies continue at the speed they already have. Stated otherwise, the law says that bodies have inertia; they persist in the motion they already have unless compelled to do otherwise by the application of force. This inertia or resistance of matter to change in speed is called *inertial mass* or just *mass*.

Do all objects have the same mass? Not at all. Since mass exhibits itself in an object's resistance to change in speed, we can appeal to experience to see that different objects may possess different masses. Suppose, for example, that a small and a large ball of lead are at rest on the ground and one wishes to start them moving. Experience tells us that we must exert more force to get the larger ball rolling than to get the smaller one rolling with the same speed. Since more force is required in the first case, the larger ball must possess more mass. Or we can imagine the force that might be required to stop these balls if they were rolling toward us at the same speed. Again, more force would be required to stop the larger one. Thus the masses of objects are not the same.

We shall not present physical methods of measuring mass. It suffices to know that by adopting a unit of mass, just as we adopt a unit of length, we can compare all other masses with this unit and so determine exactly how much mass there is in any individual piece of matter. Mass is measured in pounds or grams, the pound being approximately 454 grams.

Bodies falling to earth possess acceleration. Hence some force must be acting to produce this change in speed. The force, as Galileo and others realized, is the pull of the earth or the force of gravity. We feel this pull when we hold an object in our hands. This particular force applied to an object is called the *weight* of the object. Hence weight and mass are by no means the same. Mass is inertia or resistance to change in speed, and weight is a force exerted by the earth.

However, there is a remarkable, experimentally determined relationship between the mass and weight of an object, namely, near the surface of the earth the weight is always 32 times the mass; in symbols:

$$W = 32M. \tag{12}$$

The quantity 32 is precisely the acceleration which all bodies falling to earth possess. Thus equation (12) says that the force, W, which the earth exerts on a mass, M, is the acceleration with which it causes the mass to fall times the mass. When the mass M is measured in pounds, the weight W is measured in poundals. Thus a mass of one pound has a weight of 32 poundals. For convenience, 32 poundals are called one *pound of weight.* Of course, a pound of mass and a pound of weight are not the same physical quantities; and it is confusing at times to use the same unit for both mass and weight. Yet, as we shall see in a moment, this confusion is not too serious. (If the mass is measured in grams, the quantity which replaces 32 is 980, and the units are centimeters per sec^2. Then instead of (12) we have

$$W = 980M,$$

and the weight W is measured in dynes.)

Because weight and mass are so intimately related, we do not trouble in ordinary life to distinguish between the two properties. Large masses have large weights, and so, even in those instances where we are actually concerned with the mass of an object, we often tend to think of weight. For example, if one were to try to start an automobile rolling by pushing it, he would have to exert considerable force. The average person relates this fact to the great weight of the automobile. However, weight plays no role here because the force of gravity acts downward and has no effect on motion along the ground. The forceful push is required because the mass resists change in speed. Hence, it is the mass of the automobile rather than the weight which calls for the exertion of great force.

Exercises

1. Why do people usually fail to distinguish between mass and weight?
2. Let us assume that the relationship between weight and mass on the moon has the same mathematical form as on the earth; that is, the weight is a constant times the mass. The acceleration which the moon gives to all bodies falling toward its surface is 5.3 ft/sec^2. If a man weighs 160 pounds on the earth, what will he weigh on the moon?
3. If the acceleration which the sun imparts to bodies falling toward its surface is 27 times that imparted by the earth, what would a man whose weight on the earth is 160 lb weigh on the sun?

15-7 Vertical motion in water. The distinction between weight and mass becomes important, for example, when an object is placed in water. Almost two thousand years before Galileo's time Archimedes discovered a famous principle: a body placed in water is buoyed up by a force equal to the weight (in air) of the water displaced. If a body is placed in water, its weight is reduced because the buoyant force of the water offsets, or partly offsets, the force of gravity. Hence the very same mass weighs less in water than in air. Everyone has experienced this fact. A person feels lighter in water because he is buoyed up by the water; yet his mass remains the same. In fact, swimming is possible only because the buoyant force of the water offsets the force of gravity and the swimmer's net weight in water is zero.

Now suppose that an object of mass M is partially or completely immersed in water. Let m be the *mass* of the displaced water. The weight of the object in air is $32M$, and the weight of the displaced water (if the quantity of water were weighed in air) is $32m$. Then the net weight, w, of the immersed object is

$$w = 32M - 32m. \tag{13}$$

This net weight will determine whether the object sinks or swims. If a piece of lead is placed in water, it displaces very little water compared with its own mass, and hence $32m$, the weight of the displaced water, will be small compared with $32M$. Then the net weight, or force of gravity experienced by the piece of lead, will be almost as great as it is in air; the lead will be pulled downward, and its acceleration will be close to 32 ft/sec^2. All of this means no more than that lead will sink quickly in water. On the other hand, a submarine traveling under water displaces a mass of water just equal to the mass of the submarine, and so the vessel neither sinks nor rises.

We can be somewhat more precise about the motion of an object which sinks in water. The net weight of an object in water is given by formula (13). Let us apply the distributive axiom to write (13) as

$$w = \left(32 - 32\frac{m}{M}\right)M. \tag{14}$$

Now when an object of mass M is placed in water, its net weight, w, in water, which is a force, produces the acceleration with which it sinks. According to (12), the weight is acceleration times mass. Then, by (14), the quantity $32 - (32m/M)$ should be the acceleration in water, because the falling object has the same inertia or mass M everywhere. Hence the acceleration, a, of the mass in water is

$$a = 32 - 32\frac{m}{M}\cdot \tag{15}$$

Let us be specific. Suppose that we place a mass of gold in water. Since gold is a rather compact or dense substance, it will not displace much water. In fact, the mass of the water it displaces is found by measurement to be $\frac{1}{20}$ of the mass of the gold itself*. Then the ratio m/M in (15) is $\frac{1}{20}$ and $a = 32 - (32/20)$, or 30.4. To calculate the speed and distance fallen in water, we may now use the reasoning we employed to derive formulas (2) and (3) from (1). The reasoning is indeed exactly the same as before except that the number 30.4 replaces 32. Then the downward speed of a piece of gold placed in water is given by

$$v = 30.4t \tag{16}$$

and the distance traveled downward in t seconds is given by

$$d = 15.2t^2. \tag{17}$$

The principles of "motion" in water can be applied to the interesting problem of calculating the size of an iceberg and, in particular, the size of its submerged part. One might expect ice to sink in water, but ice is actually *less* dense than cold water. Stated otherwise, the mass of the displaced water is greater than the mass of the immersed ice. The ratio of the two masses, cold water and ice, is 1.08.

Let us determine how much of the iceberg's mass is in the water. Let M_1 be the mass of the iceberg above water (Fig. 15–2) and let M_2 be the

---

* We need only weigh the gold and the water displaced. If the two weights (in air) are in the ratio of 20 to 1, then their masses are in the ratio of 20 to 1 because $W = 32m$ for all objects in air (near the surface of the earth).

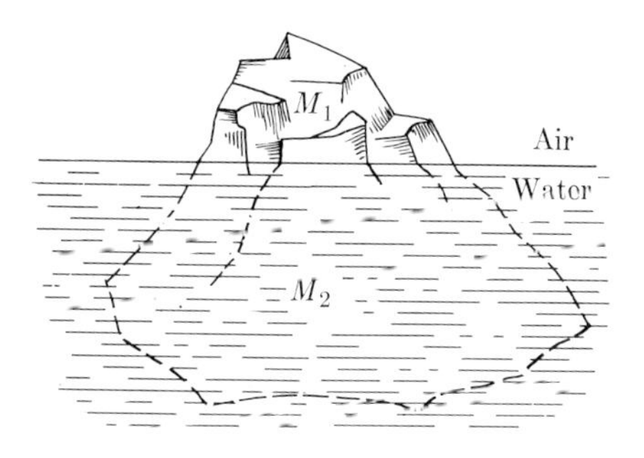

FIG. 15-2. An iceberg in water.

mass below water. The force exerted on the iceberg by the downward pull of gravity is $32(M_1 + M_2)$. However, the iceberg is buoyed upward by the weight of the water displaced. But the mass of the displaced water is $1.08M_2$, and the weight (in air) of the displaced water is therefore $32(1.08M_2)$. Since the iceberg floats, the net force acting on it must be zero. This means that the pull of gravity is just offset by the buoyant force, which is the weight of the displaced water. Hence

$$32(M_1 + M_2) = 32(1.08M_2).$$

We may divide both sides of this equation by 32 and then subtract M_2 from both sides of the resulting equation. We find that

$$M_1 = 0.08M_2.$$

Thus only about $\frac{1}{12}$ of the mass of the iceberg is above water. A ship striking what appears to be a small iceberg may suffer heavy damage from the submerged part.

EXERCISES

1. Calculate the velocity acquired and distance fallen in 10 sec by a piece of gold placed in water.
2. If a piece of lead is immersed in water, the mass of displaced water is $\frac{1}{11}$ of the mass of the lead. Write a formula for the distance which the lead will sink in t seconds.

15-8 Motion along an inclined plane. The preceding discussion of the motion of objects in water shows, among other things, that the motion of freely falling objects can be slowed down by the buoyant force of water, which counteracts the force of gravity. Since the effective, or net downward, force is reduced, the object sinks more slowly than it does in air, or does not sink at all, as, for example, in the case of the iceberg. There are other important situations in which the action of gravity is, so to speak,

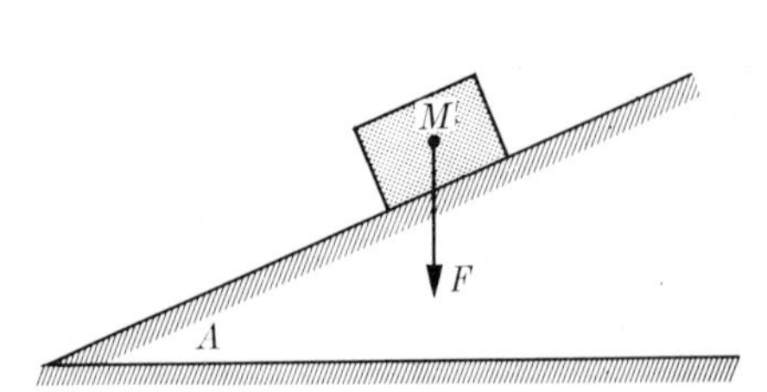

Fig. 15–3. A mass on an inclined plane.

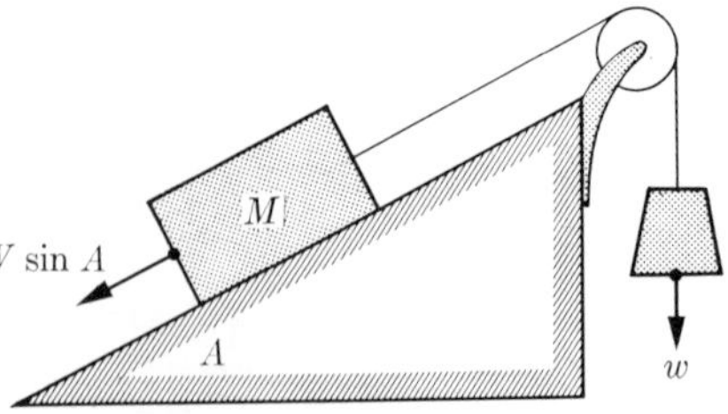

Fig. 15–4. The force down the plane exerted on a mass M of weight W is $W \sin A$.

reduced. One is motion along slopes. The uphill or downhill motion of an automobile or a train is an important practical example of motion on a slope. In the mastery of such motions the simple formulas we have just been using prove to be very effective and enlightening.

Historically there was another reason for investigating motion along slopes. In his efforts to study the motion of bodies falling to earth, Galileo encountered difficulties in obtaining numerical data such as the velocity acquired and distance fallen in various numbers of seconds. The chief source of the difficulty was that Galileo did not have a good clock at his disposal. Although very large mechanical clocks such as those seen on public buildings existed in Europe by the fourteenth century, homes or individual experimenters used water clocks which measure time by the quantity of water flowing out of a slowly emptying, graduated vessel. Such clocks could not measure minutes very accurately, let alone seconds. Galileo noticed that an object sliding downhill moves more slowly than one falling straight down. He could, therefore, expect to obtain more accurate data about sliding objects. And so Galileo decided to study motion on slopes. He hoped to be able to relate this motion to the motion of freely falling objects and thus deduce knowledge about the latter from the former type of motion. Let us see what he learned.

A hill of constant slope is idealized as an inclined plane. Let us suppose that this plane makes an angle A with the horizontal (Fig. 15–3), and that an object of mass M is placed on this plane. Of course, gravity acts on this mass, but the force of gravity pulls it straight down. However, the plane does not permit such a motion. The object is free to move only along the plane. The question we shall try to answer first is, To what extent is gravity effective in causing the object to slide downhill?

Let us study the situation. Certainly, if the inclination of the plane, angle A of Fig. 15–3, were zero, that is, if the plane were perfectly horizontal, the object would not slide. As the plane is inclined more and more, one observes that the mass slides faster, and one therefore expects the downward force acting along the plane to be stronger. When the inclination is vertical, that is, when the angle of inclination is 90°, the full force of gravity applies and the mass falls freely. Hence the force of gravity acting along the slope depends upon the angle of inclination of the plane. But precisely how great is this force for any given angle A?

A mathematician studying some variable quantity which seems to depend upon an angle would naturally think of the trigonometric ratios and try to fit one of these to the problem at hand. In our case, the force producing a downward pull along the plane depends upon angle A. When $A = 0$, the force is zero; when $A = 90°$, the force is the full force of gravity or the weight W of the object. Hence the weight W is modified by some factor involving angle A, and this factor is zero when $A = 0$, and is 1 when $A = 90°$. What can this factor be? A reasonable conjecture is that the factor is $\sin A$ and that the downward force along the plane is $W \sin A$.

This conjecture can be checked by an experiment. We place the object on an inclined plane (Fig. 15–4) and connect it by a rope passed over a pulley at the top of the plane to another object of weight w which is hanging freely in space. Suppose w is chosen so that it balances the force acting on the mass of weight W which is on the inclined plane. Then one finds that

$$w = W \sin A. \tag{18}$$

In other words, the force pulling the object down the plane, the effective force of gravity, is no longer W but $W \sin A$.

The mathematician trying to determine the downward force along the the plane would not resort to experiment but would try to reason. Reasoning in this case leads to a line of thought which we shall find helpful in other situations. Forces are physical phenomena which are slightly more complicated than, say, size. The length of an object, for example, can be represented by a mere number, but a number alone will not suffice to represent a force. Every force has direction as well as magnitude. That the direction is important can be seen immediately. We have but to imagine what our world would be like if gravity pulled all objects along the surface of the earth rather than toward the center. Hence a mathematical approach to forces must take into account direction as well as magnitude, and the most convenient way to do this is to represent a force by a line segment whose length exhibits the magnitude and whose direction exhibits the direction of the force. Directed line segments are called *vectors,* and forces are physical quantities which are represented by vectors.

Because the direction of a force is as important as its magnitude, a combination of forces acting on an object usually results in a behavior which seems peculiar to people who have not previously thought about forces. Suppose, for example, that two people pull a box along a floor. One pulls along OP (Fig. 15–5) with a force of 3 pounds, and another along OQ with a force of 4 pounds. If the box is free to slide along the floor, in what direction will it move? An axiom of physics, based, of course, on experience, states that the effect of two forces such as OP and OQ will be a force which acts along the diagonal OR of the rectangle $OPRQ$, and the magnitude of this *resultant force* is exactly the length of the diagonal OR. The direction

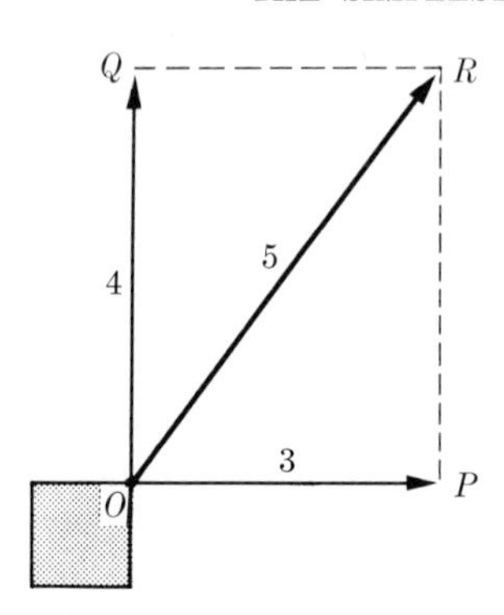

FIG. 15–5. The resultant of the forces OP and OQ is OR.

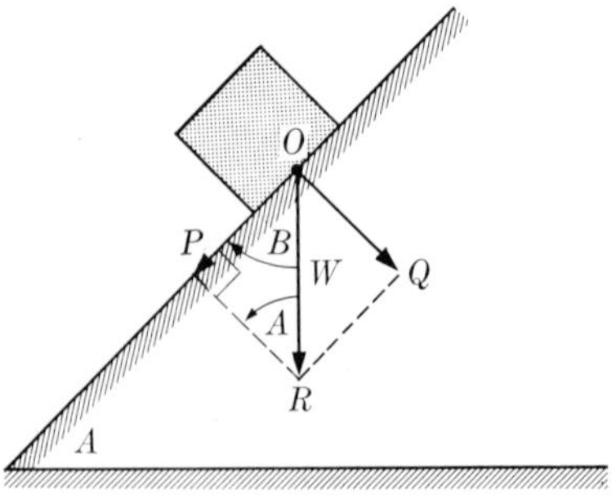

FIG. 15–6. The forces OP along the plane and OQ perpendicular to the plane exerted on an object of weight W.

of the resultant is not so surprising, but its magnitude is because the magnitude is not the sum of the 3- and the 4-pound forces. As we can see from the right triangle OPR, the magnitude is 5. In other words, when we add forces, we cannot simply add the magnitudes of the summands, but must go through the reasoning just explained. However, nature also is obliging in that it does cause forces to combine in a way which mathematically is not very complicated.

We shall now look at the above situation in reverse, so to speak. Suppose that a force OR of 5 pounds is applied to the box. We can, if we wish to, regard this force as the resultant of two forces, the force OP of 3 pounds and the force OQ of 4 pounds because the latter two forces yield the resultant OR. The forces OP and OQ are called *components* of the force OR. The point in thinking about the force OR as the resultant of two other forces, that is, as having two components, is that it permits us to answer the following question: How much does the force OR contribute to the movement of the box in the direction OP? Since the force OR is equivalent to OP and OQ, and since OQ has no effect in making the box move along the direction OP, the effectiveness of OR in the direction OP is OP itself. Hence the answer is 3 pounds. We should note that the magnitude of OP is determined by dropping a perpendicular from R onto the line OP.

Let us now return to our original problem. An object of weight W (Fig. 15–6) rests on an inclined plane making an angle A with the horizontal. The weight W is, of course, the force of gravity producing a straight downward pull. To what extent is gravity effective in forcing the object to slide downhill? To answer this question we apply results of our discussion on the action of forces. Let OR represent the magnitude and direction of the weight W. We drop a perpendicular from R to the line representing the direction of the inclined plane, thus obtaining the component OP, and OP is the magnitude of the effective force of gravity along the inclined plane. How big is OP? From Euclidean geometry we see that angle ORP equals angle A because, like angle A, it is the complement of angle B. Then, in the right triangle ORP,

$$\sin A = \frac{OP}{OR}$$

or

$$OP = OR \sin A.$$

Since the magnitude of OR is W, we have

$$OP = W \sin A. \tag{19}$$

This result is identical with the statement made in (18) above, namely, the effective force due to gravity and acting along the inclined plane is not W but $W \sin A$.

And now we are prepared to discuss the motion of the object down the plane. If the force of gravity acts on an object which is free to fall straight down, then it imparts to that object an acceleration of 32 ft/sec^2. If, however, the effective force of gravity is not W, the weight of the object, but $W \sin A$, we should expect, since the force causes the acceleration, that the acceleration of the object sliding down the plane will not be 32 but $32 \sin A$. Hence the acceleration is less than 32 and the value to which it is reduced depends upon $\sin A$. If angle A is small, then $\sin A$ is small and $32 \sin A$ will be appreciably less than 32. If the acceleration of an object is small, the object will gain speed slowly and travel less distance in a given amount of time than if it were falling straight down. This result is exactly what Galileo had hoped for and expected in turning to motion on inclined planes. Objects sliding down a plane move more slowly than in falling straight down, and he could therefore study the motion more easily.

Now that we know the acceleration of objects sliding down inclined planes, what can we deduce about the subsequent motion? For example, if a mass M starts from rest and slides down an inclined plane, what speed will it acquire and how far will it travel in a given time t? Since the acceleration, that is, the gain in speed per second, is $32 \sin A$ and the mass starts with zero speed, the speed, v, at any time, t, is

$$v = (32 \sin A)t.$$

Comparing this result with formula (2), we see that the quantity $32 \sin A$ replaces the quantity 32 in (2). It is customary and clearer to write the present result as

$$v = 32t \sin A. \tag{20}$$

To obtain the distance traveled in time t, we may employ the very same argument used to derive formula (3) from formula (2), except that the number $32 \sin A$ again replaces the quantity 32 in (2). Hence the distance d traveled by the object sliding down the plane in time t is given by

$$d = 16t^2 \sin A. \tag{21}$$

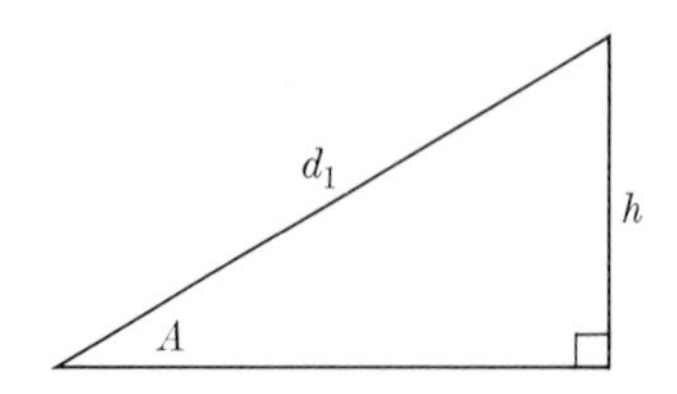

FIG. 15–7. An inclined plane of height h.

Let us apply these formulas. How long does it take an object to slide from a height h to the ground along a slope whose length is d_1 (Fig. 15–7)? The time required to travel the distance d_1 can be obtained from formula (21). What we are seeking is the value of t when $d = d_1$. Let us denote this value by t_1. Then

$$d_1 = 16t_1^2 \sin A.$$

Dividing both sides of this equation by $16 \sin A$ yields

$$t_1^2 = \frac{d_1}{16 \sin A}.$$

Taking the square root of both sides and retaining only the physically meaningful positive square root, we arrive at

$$t_1 = \sqrt{\frac{d_1}{16 \sin A}}.$$

We may write this result in a slightly simpler form. Since

$$\sqrt{\frac{d_1}{16 \sin A}} = \sqrt{\frac{1}{16} \cdot \frac{d_1}{\sin A}} = \sqrt{\frac{1}{16}}\sqrt{\frac{d_1}{\sin A}} = \frac{1}{4}\sqrt{\frac{d_1}{\sin A}},$$

we have

$$t_1 = \frac{1}{4}\sqrt{\frac{d_1}{\sin A}}. \tag{22}$$

This answer is by no means surprising. First of all, let us compare it with formula (4), which gives the time required for a freely falling object to travel a distance d_1:

$$t_1 = \tfrac{1}{4}\sqrt{d_1}. \tag{23}$$

The value of t_1 given by formula (22) is larger than that given by (23), for $\sin A$ is less than unity and therefore $d_1/\sin A$ is greater than d_1. Hence $\sqrt{d_1/\sin A}$ has a greater numerical value than $\sqrt{d_1}$. What we have accomplished so far is really much ado about nothing. We have shown merely that it takes longer for an object to slide a distance d_1 along a plane inclined at an angle A than to fall a distance d_1 straight down. Let us note, however, that formula (22) applied to the special case of $A = 90°$, that is, when $\sin A = 1$, yields formula (23). Hence from a study of motion along inclined planes one obtains facts about freely falling objects, which is just what Galileo hoped to do and did.

We should also note that the results (20) and (22) obtained for the speed and time of motion on an inclined plane are independent of the mass of the object, just as in the case of freely falling bodies. Hence a ton of lead and a one-pound piece of wood take the same time and attain the same speed in sliding down the same hill. Of course, friction is ignored in these formulas.

Exercises

1. Suppose that an object slides down a plane inclined 30° to the horizontal. How far will it slide in 3 sec? How far would a freely falling object travel in that time? Which distance would you have expected to be greater?
2. Suppose that a Cadillac and a Ford slide down a hill of height h and inclination A. Which car will reach the bottom sooner?
3. A truck weighing 5000 lb stands on a hill inclined 20° to the horizontal. What force tends to push the truck downhill?
4. A man wishes to push a box weighing 100 lb up a board which extends from the street level to the floor of a truck 4 ft above the ground. The board is 12 ft long. What force must the man exert on the box?
5. Suppose that a man pulls a cart by a rope. He pulls with a force of 100 lb, and the rope makes an angle of 40° with the horizontal. How much of the force is effective in pulling the cart?
6. Suppose that an object slides down an inclined plane from a point 16 ft above the ground. The inclination of the plane is 30°. What speed does the object have when it reaches the ground?

15–9 Motion along planes with different slopes. Before continuing our study of motion along inclined planes, let us derive another way of writing formula (22), which will be more helpful. From Fig. 15–7 we see that

$$\sin A = \frac{h}{d_1}.$$

If we multiply both sides of this equation by d_1 and then divide both sides by $\sin A$, we have

$$d_1 = \frac{h}{\sin A}. \tag{24}$$

Substituting this value of d_1 in (22), we obtain

$$t_1 = \frac{1}{4}\sqrt{\frac{h}{\sin A}\cdot\frac{1}{\sin A}} = \frac{1}{4}\sqrt{\frac{h}{\sin^2 A}},$$

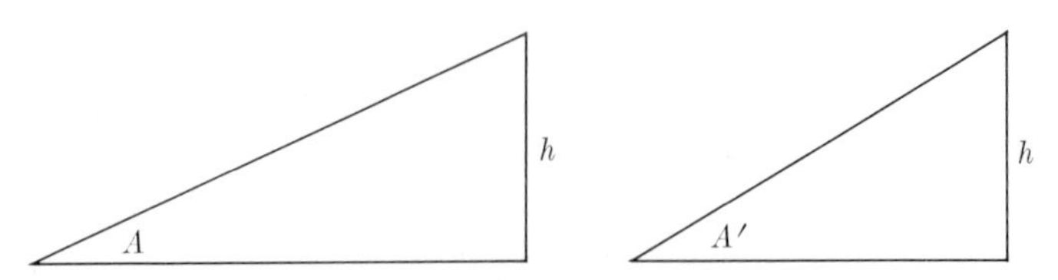

FIG. 15–8. Two inclined planes with the same height but different slopes.

where $\sin^2 A$ means $\sin A \cdot \sin A$. But the square root of a fraction is the square root of the numerator divided by the square root of the denominator. Hence

$$t_1 = \frac{1}{4 \sin A} \sqrt{h}. \tag{25}$$

This result expresses the time of descent in terms of the *vertical* distance descended and the inclination of the plane.

With formula (25) we can answer another question rather easily. Suppose that an object slides down two different hills of the *same height h,* but the slope of one hill is gradual and that of the other is steep (Fig. 15–8). What can we say about the respective times required to reach the bottom of the hills? The gradually inclined hill has a smaller inclination than the steep one. Hence angle A is smaller and therefore $\sin A$ is smaller. Since h is the same in both cases, formula (25) tells us that t_1 will be greater for the more gradual hill. Here again the conclusion does not surprise us. All we have learned is that it takes longer to slide down a gradual hill of height h than a steep hill of height h. Any child who has coasted downhill on a sled knows as much.

Let us consider next the *speed* which an object acquires in sliding down two hills of the same vertical height h but of different inclinations. Which slope will produce the greater speed, the long gradual one or the short steep one? We propose to answer this question by considering first any given hill (Fig. 15–9). The speed of the sliding object at any time, t, is given by formula (20). Let us call v_1 the speed which the object acquires in sliding down the entire distance d_1. To calculate v_1 we need the time t_1 which it takes the object to slide the distance d_1. Since t_1 is given by formula (25), we shall substitute this value of time in (20). This gives

$$v_1 = 32 \sin A \cdot \frac{1}{4 \sin A} \sqrt{h},$$

or

$$v_1 = 8\sqrt{h}. \tag{26}$$

We note that the speed attained by the object in sliding the distance d_1 does *not* depend upon d_1 or upon the inclination of the plane, but depends solely on the height descended. In other words, whether an object slides down a long gradual hill or a short steep one—as long as it descends a *vertical* distance of h feet in both cases the speed at the bottom is the same.

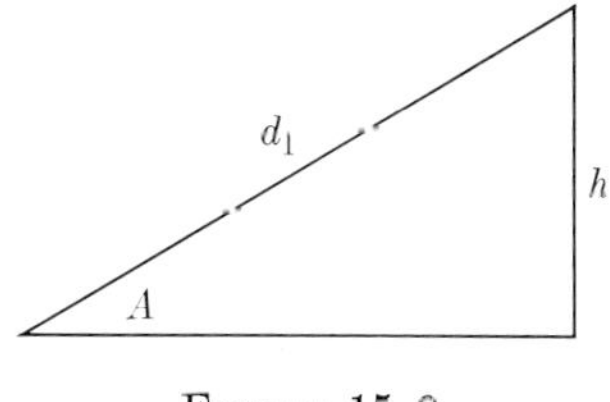

FIGURE 15–9

Since the speed at the bottom depends only upon the vertical distance descended by the object, then this speed must be the same as that of a freely falling body which falls straight down the distance of h feet. This conclusion can be immediately verified. Formula (5) says that the speed v of an object which starts from rest and falls h feet is

$$v = 8\sqrt{h}.$$

Hence the speed acquired in falling h feet is precisely the same as that given by formula (26).

Inclined planes have many practical applications; thus they are very commonly used to raise objects from one level to another. Indeed one often sees objects being slid up an inclined flat board extending from the street level to the floor of a truck. Why is the inclined plane used? We found earlier that an object of weight W resting on a plane of inclination A exerts a force $W \sin A$ down the plane. To counter this force one need apply only the force $W \sin A$ up the plane, and a slight additional force will set the object into motion up the plane. Now we can make $W \sin A$ as small as we please by reducing A which in turn reduces $\sin A$. (Physically we have but to use a long board so as to make angle A small.) Hence the force necessary to move an object up an inclined plane can be made arbitrarily small. Even a child can push a piano up a (frictionless) inclined plane if the inclination of the plane is chosen small enough.

But the age of miracles has not quite arrived. We have overlooked something. The more gradual the slope, the longer the board required to reach a given height h (Fig. 15–8). Hence the force which is applied to move the object up the plane must be exerted over a longer distance. According to (24), the distance, d_1, required to reach height h is $h/\sin A$. The force that must be exerted is $W \sin A$. A mathematician would immediately notice that the product of these two quantities, $h/\sin A$ and $W \sin A$, is simple, for

$$\text{force} \cdot \text{distance} = W \sin A \cdot \frac{h}{\sin A} = Wh. \tag{27}$$

This result says that the force required to push or pull the object up the plane multiplied by the length d_1 of the slope, which is the distance over which the force acts, is independent of the inclination A and the distance

d_1; the product depends only upon the weight W itself and the height h to which the object is raised. But the result states even more. We know that W is the full weight of the object. If this object were to be raised straight up, one would have to exert a force of magnitude W. To raise this weight a distance h, the product of force and distance would be Wh. Hence to raise a weight W to a height h, the product of force applied and the distance over which it is applied is the same whether one pulls it along a very gradual slope over the requisite long distance, a steep slope over the corresponding short distance, or straight up. To put the result in physical terms one can exert himself rather little over a long distance or strongly over a short distance, and the total exertion will be the same. Apparently, nature cares not about the force required but about the product of force and the distance over which it acts. As a consequence physicists have defined the technical concept of *work* as the product of force applied and the distance over which the force acts. The work required to do a given job is the same, no matter how it is achieved.

To do work one must expend energy. Hence the above result amounts to the statement that one cannot cheat on the energy required to perform a given task. Energy will be neither gained nor lost by doing the job in different ways. This example is a minor illustration of a major principle of science known as the conservation of energy. Energy is neither lost nor gained in the operations of nature.

Exercises

1. Suppose that a ton of lead and a pound of wood slide down a plane from a height of 100 ft; the plane has an inclination of 40°. Which has the greater speed at the bottom?

2. Suppose that a ton of lead starts from a height of 100 ft and slides down a plane inclined 20°; a pound of wood starts from the same height and slides down a plane inclined 70°. Which has the greater speed at the bottom?

3. A man pushes a 10-lb box up a slope whose inclination is 30°, to a point 5 ft above the ground. How much work does he do?

4. One man pushes a 10-lb box up a slope 12 ft long, to a point 6 ft above the ground, and another pushes a 10-lb box up a slope 24 ft long, to a point 6 ft above the ground. Which man has to exert more force? Which man does more work?

5. Suppose that an object slides down two different hills of the same height h; however, one hill has inclination A_1 and length d_1, and the other has inclination A_2 and length d_2. Show that the times of descent are proportional to the lengths; that is, $t_1/t_2 = d_1/d_2$.

15–10 Summary. From a mathematical standpoint, the purpose of this chapter is to acquaint us with various first- and second-degree functions and with some of the algebraic processes one performs on their formulas to derive new formulas or numerical results. However, we have already noted in the "Introduction," that as purely mathematical entities these functions have rather little significance. What gives them importance is the variety of physical problems to which they can be applied.

This chapter has also sought to show how Galileo's program for science operates. Physical axioms expressed as formulas are the starting points of mathematical deductions. The conclusions give us totally new information about the given physical situations. Although the examples and results considered in this chapter are by no means startling—thus far we have confined ourselves to the simplest situations—we can nonetheless see how a few laws of motion encompass a multitude of distinct physical phenomena. All objects rising and falling near the surface of the earth or sliding along slopes obey the laws we have examined. Hence knowledge of these few laws grants us understanding of thousands of observable motions and the ability to make predictions.

Recommended Reading

BONNER, FRANCIS T. and MELBA PHILLIPS: *Principles of Physical Science,* pp. 37–65, Addison-Wesley Publishing Co., Inc., Reading, Mass., 1957.

COHEN, I. BERNARD: *The Birth of a New Physics,* Chap. 5, Doubleday and Co., Anchor Books, New York, 1960.

GALILEI, GALILEO: *Dialogues Concerning Two New Sciences,* pp. 147–233, Dover Publications, Inc., New York, 1952.

HOLTON, GERALD and DUANE H. D. ROLLER: *Foundations of Modern Physical Science,* Chaps. 1 and 2, Addison-Wesley Publishing Co., Inc., Reading, Mass., 1958.

KLINE, M.: *Mathematics and the Physical World,* Chaps. 12 and 13, T. Y. Crowell Co., New York, 1959.

MOODY, ERNEST A.: "Galileo and Avempace: Dynamics of the Leaning Tower Experiment," an essay in PHILIP P. WIENER and AARON NOLAND: *Roots of Scientific Thought,* Basic Books, Inc., New York, 1957.

SAWYER, W. W.: *Mathematician's Delight,* Chaps. 8 and 9, Penguin Books Ltd., Harmondsworth, England, 1943.

TAYLOR, LLOYD WM.: *Physics, The Pioneer Science,* Chaps. 3 to 7, Dover Publications, Inc., New York, 1959.

WHITEHEAD, ALFRED N.: *Introduction to Mathematics,* Chaps. 2 to 4, Henry Holt and Co., New York, 1939.

CHAPTER 16

PARAMETRIC EQUATIONS AND CURVILINEAR MOTION

I now propose to set forth those properties which belong to a body whose motion is compounded of two other motions, namely, one uniform and one naturally accelerated; these properties, well worth knowing, I propose to demonstrate in a rigorous manner.

Galileo

16–1 Introduction. We saw in the preceding chapter that simple functions can be used to express physical principles and that by applying algebra to these functions or to the formulas which express the functions symbolically, we can obtain new physical knowledge. To some extent, then, we have come to recognize the broader significance and usefulness of functions and mathematical processes for science in general. However, we have hardly penetrated as yet the mathematical domain of functions nor have we learned enough applications to sense its real power.

In this chapter we shall slightly extend the use of functions. In Chapter 15, we represented the acceleration and speed attained and distance traveled by a falling body by using one formula for each physical quantity. We were enabled thereby to study motion along straight-line paths. We shall now examine motion along curved paths, for example, the motion of an object dropped from a moving plane, or the motion of a projectile shot out from a cannon. The best mathematical approach to such *curvilinear* motions is to treat the horizontal and vertical motions separately. This method requires two sets of formulas, one set for the acceleration, speed, and distance traveled horizontally, and another set for the same physical quantities but applying to the vertical motion. The path itself will be described by the two formulas for distance, that is, horizontal distance and vertical distance. These two formulas will furnish what we shall call the *parametric representation* of the curved path. We shall study this mathematical idea and see what we can do with it.

It was again Galileo who perceived the basic principle underlying the phenomenon of curvilinear motion. He presented the concept and its mathematical treatment to the world in the *Dialogues Concerning Two New Sciences,* the very same book in which he treated motion in a straight line. Galileo's purpose in investigating curvilinear motion was to study the behavior of cannon balls, or projectiles in general. The cannon, introduced in the fourteenth century, had undergone such improvement by Galileo's time that it could fire a projectile over several miles. However, the theory of projectile motion was not well understood before Galileo's work because mathematicians and physicists had attempted to apply Aristotle's laws of motion (see the preceding chapter), and these were not correct.

The problems that Galileo treated, e.g., the motion of cannon balls, unfortunately did not lose their importance in the succeeding centuries. In fact, they have become even more common and more complicated in our times, since such phenomena as the motion of bombs dropped from moving airplanes, the trajectories of death-dealing projectiles capable of traveling thousands of miles, and similar problems of modern "civilization," also fall within the puissance of Galileo's method. However, the value of this phase of Galileo's work is not limited to meting out death and destruction. Aside from using his results as an illustration of the power of mathematics, we shall see in the space of one chapter how an extension of Galileo's ideas on projectile motion led, in the hands of Newton, to the greatest advance in science which our civilization has achieved.

16–2 The concept of parametric equations. Let us suppose that a stone is thrown out horizontally from the top O of a cliff (Fig. 16–1). We know from physical experience that the stone will travel out and down and follow the curved path OAB. If we introduce a set of coordinate axes on which the positive direction of the Y-axis is downward, then we know from our work in coordinate geometry that this curve can be represented by an equation. Let us suppose, for definiteness, that this equation is $y = x^2$. Then it follows from equation (9) of Chapter 13 that the path is part of a parabola opening downward. We shall call $y = x^2$, which is, of course, also the formula that tells us how y changes when x changes, the *direct* relationship between x and y.

As the stone travels out and down, the horizontal distance and the downward distance which it travels from the point O keep changing with time. Thus, at the point A, the horizontal distance traveled may be 3, and since $y = x^2$, the vertical distance must be 9. At the point B, the horizontal distance may be 4, in which case the vertical distance must be 16.

The direct relationship between x and y is frequently useful, but it does not involve the time that the object is in motion. We may wish instead to utilize the equation which gives the relationship between horizontal distance traveled and time and the equation which relates vertical distance and time. Let us suppose for the moment that the stone travels *straight out* at the rate of 3 ft/sec. Then the relationship between horizontal distance and time is $x = 3t$. Since $y = x^2$, then in terms of t, $y = (3t)^2$ or $y = 9t^2$.

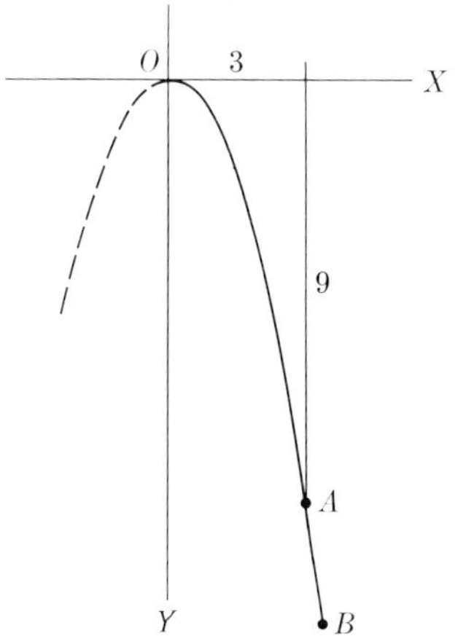

O 3 X 9 A Y B

FIG. 16–1. The path of a stone thrown out horizontally from the top O of a cliff.

The two equations:

$$x = 3t, \qquad y = 9t^2, \tag{1}$$

are called the *parametric equations* of the curve

OAB. They describe the curve OAB just as well as does the single equation $y = x^2$, provided that we understand how parametric equations are to be used. For each value of t, equations (1) yield a value of x and a value of y. These values of x and y which belong to the same value of t are the coordinates of one point on the curve OAB. Thus for $t = 1$, $x = 3$ and $y = 9$. Then $(3, 9)$ are the coordinates of a point on the curve, namely, the point A, which we discussed earlier. For $t = 4/3$, $x = 4$ and $y = 16$, and $(4, 16)$ are the coordinates of the point B.

We may also say that the two formulas $x = 3t$ and $y = 9t^2$ are equivalent to the single formula $y = x^2$. Whether we speak of equations of curves or formulas is really immaterial. The word formula emphasizes the idea of change because formulas are relationships among variables, and we often like to think of what happens to one variable as another, related variable changes. On the other hand, when a curve is given in its entirety, the concept of change may not be relevant, and then we speak of the equation of the curve.

If the two formulas in (1) are entirely equivalent to the single formula $y = x^2$, why do we bother with two formulas instead of one? There are two reasons: (1) When one argues from physical principles, it is often easier to arrive at the parametric representation of a given phenomenon, and (2) it is easier to study the phenomenon by working with parametric equations. We shall recognize the utility of parametric representations in the next few sections.

There is one more mathematical detail. Suppose that we find the parametric formulas describing a motion and we wish to determine the direct relationship between x and y. Can we do this? Yes indeed. For example, if $x = 3t$ and $y = 4t^2$ are the parametric formulas, we can solve the first one for t and obtain $t = x/3$. We substitute this value of t in $y = 4t^2$ and obtain $y = 4(x/3)^2$, or $y = 4(x^2/9)$, or $y = 4x^2/9$. This is the direct relationship between x and y.

Exercises

1. If the parametric formulas representing a phenomenon are $x = 2t$ and $y = 3t$, what is the direct relationship between x and y? What curve represents the parametric formulas or the direct relationship?

2. If the parametric formulas are $x = 4t$ and $y = 5t^2$, what is the direct relationship between x and y? What curve represents the direct relationship?

3. Suppose the parametric formulas are $x = 2t$ and $y = 10t + 4t^2$. What is the direct relationship between x and y and what curve describes it?

4. Suppose $x = 3t$ and $y = (4/3)t$ are the parametric equations of a curve. Sketch the curve by using the parametric equations only.

16–3 The motion of a projectile dropped from an airplane. Let us see now how parametric formulas arise in the study of physical phenomena and how they can be useful in deducing new information about the phenomena.

Suppose a bomb is released from an airplane which is flying horizontally at 60 miles per hour (an unrealistic figure used for computational convenience). If there were no gravity, the bomb would continue to move forward alongside the airplane at the rate of 60 miles per hour. This fact seems surprising, but it is a consequence of the first law of motion, which states that if an object is in motion and no force is applied to alter that motion, then the object will continue to move indefinitely at the speed it already has. Since the bomb has been moving with the airplane, it already possesses a horizontal speed of 60 miles per hour. We have assumed that no forces are acting on the bomb and hence it will continue to move forward at that speed. There are more familiar analogous situations which may make the truth of what was just said a little more acceptable. Suppose that a person rides in an automobile which is moving at the rate of 60 miles per hour and the driver suddenly applies the brakes. The automobile's motion is then checked, but the passenger's motion is not, and he continues to move forward at 60 miles per hour, at least until he hits the windshield.

Let us return to the motion of the bomb released from the plane. We had assumed that gravity was not acting. But it does act and it pulls the bomb downward at the same time as the bomb moves forward so that the bomb follows a curved path. Here Galileo made a discovery applying to projectile motion, namely, that one could study its horizontal and vertical motions as though they were occurring separately, and that the position of the bomb at any time could be determined by finding how far it had traveled horizontally and vertically. This idea was new and radical in Galileo's time. Aristotle had argued that one motion would interfere with the other, and that only one could operate at any given time. Thus he would have said that the violent motion imparted to the bomb by the airplane would prevail until the acting force was used up, and then the natural motion downward would take over and cause the bomb to fall straight down.

Let us apply Galileo's way of analyzing the motion. The bomb moves horizontally at the constant speed of 60 miles per hour, or 88 feet per second. If we measure time from the instant the bomb is released from the plane, and if we measure horizontal distance from the point at which it is released, then the horizontal distance x covered by the bomb in t seconds is given by the formula

$$x = 88t. \tag{2}$$

This formula describes the horizontal motion.

According to Galileo, the vertical motion downward takes place as though it were independent of the horizontal motion. But the vertical motion is due to gravity only, and we know that an object which falls straight down under the action of gravity and starts with zero speed falls $16t^2$ feet in t seconds. Hence, if we let y represent the distance downward from the point at which the bomb is released, then

$$y = 16t^2. \tag{3}$$

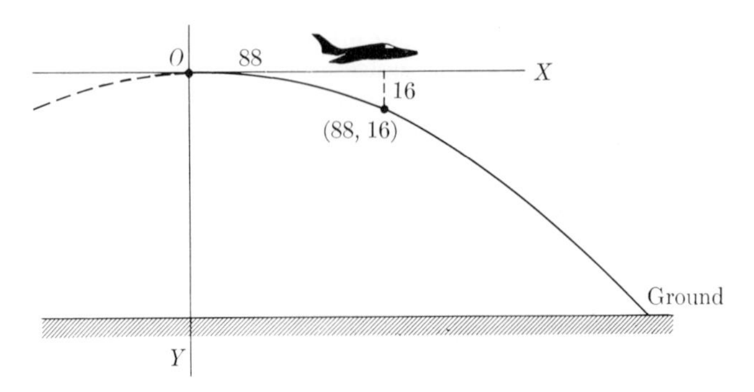

FIG. 16–2. Path of a bomb released from an airplane flying horizontally.

Formulas (2) and (3) together describe the entire motion. We observe that x and y, the horizontal and vertical distances traveled, are given in terms of a third variable, t. In fact, they are the parametric formulas for the motion in question. To draw the graph of the path described by the bomb, we may adopt either of two methods. We can choose various values of t, say $t = 0, 1, 2, 3$, and so on, and calculate the values of x and y for each value of t. Thus when $t = 1$, $x = 88$ and $y = 16$; then (88, 16) are the coordinates of one point on the curve. Calculating many such sets of coordinates will give us some idea of the shape of the curve (Fig. 16–2).

Or we may proceed by the second method, that is, determine the direct relationship between x and y. Solving equation (2) for t, we have $t = x/88$. Substituting this expression for t in (3) yields

$$y = 16\left(\frac{x}{88}\right)^2$$

or

$$y = \frac{x^2}{484}. \tag{4}$$

From formula (9) of Chapter 13 we know that the curve is a parabola. We have thus called upon our knowledge of curve and equation to determine that the curve is a parabola. If we had not been familiar with the curve of equation (4), we would have had to analyze the equation or plot points whose coordinates satisfy (4) and thereby determine the curve. In other words, we would have been faced with a problem of coordinate geometry.

We should note that only part of the parabola is of physical interest. The full parabola extends to the right and left of the Y-axis. However, only the part to the right, that is, the half corresponding to positive x-values, represents the motion of the bomb. And of this right-hand half, which mathematically extends downward indefinitely, only an arc is of physical interest, namely the arc from O to the ground.

We have learned so far that the path of the bomb released from an airplane traveling horizontally is an arc of a parabola. Let us now see whether

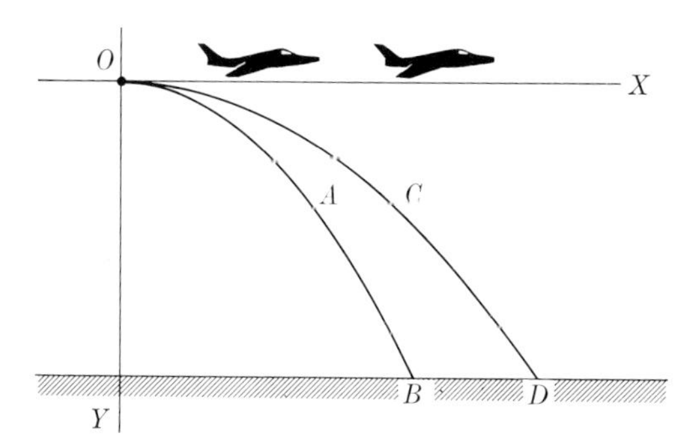

Fig. 16–3. Paths of two bombs released from two airplanes with different horizontal speeds.

we can use mathematics to derive more information about the motion of bombs or, in general, about objects which move outward and downward. Suppose that two airplanes flying horizontally at speeds of 60 and 120 miles per hour, respectively, release bombs from the same point at the same instant of time. Which of these bombs would reach the ground sooner? The reader might try to answer this question by using his intuition before resorting to mathematics.

Both bombs must fall the same vertical distance to reach the ground. The vertical motion is independent of the horizontal motion and is governed by formula (3). Hence this formula applies to both bombs. When they reach the ground, the value of y will be the same for both bombs. It follows that the value of t will also be the same for both. That is, both bombs will reach the ground at the same time.

How does the difference in the speeds of the two airplanes affect the motion? The plane flying at 60 miles per hour gives its bomb a horizontal speed of 60 miles per hour, or 88 feet per second, and the plane flying at 120 miles per hour imparts to its bomb a speed of 120 miles per hour, or 176 feet per second. Hence the bombs move with different horizontal speeds, and in the same time, t, the second one will travel farther horizontally. Thus OCD, the path of the second bomb, will be a wider parabola than OAB, the path traveled by the first bomb (Fig. 16–3).

Exercises

1. Suppose that there were no force of gravity, and that an object is released from an airplane flying horizontally at the rate of 100 mi/hr. Describe the subsequent motion of the object.

2. One object is dropped from an airplane flying horizontally at the rate of 100 mi/hr and another from a plane flying horizontally at 200 mi/hr. Both planes are at the same altitude. Compare the times required for the two objects to reach the ground. What principle is involved?

3. From a cliff 500 ft high a stone is thrown horizontally with a speed of 100 ft/sec. How long does it take the stone to reach the ground below? What horizontal distance has the stone traveled by the time it strikes the ground?

4. A gun installed in a plane which is flying in a horizontal line at a speed of 2000 ft/sec fires a bullet in the direction of the plane's motion at the initial speed of 1000 ft/sec. What is the horizontal speed of the bullet relative to the ground?

5. A bullet fired horizontally hits a point on a wall 300 ft away. The point is 1 ft below the level at which the bullet is fired. What is the horizontal speed of the bullet?

6. A plane is traveling in a horizontal line at a speed of 300 ft/sec and at an altitude of 1 mi. Where (at what horizontal distance from the target) should the gunner release a bomb to hit a given point on the ground?

7. Suppose that a plane flying in a horizontal line at the rate of 200 ft/sec releases a bomb and continues to fly horizontally at the same rate. Where is the plane in relation to the bomb when the bomb strikes the ground?

16–4 The motion of projectiles launched by cannons. A slight extension of the mathematics just introduced to treat the motion of bombs dropped from airplanes will enable us to handle the motion of projectiles shot out from cannons inclined at some angle to the ground. It was this latter problem which Galileo investigated in the seventeenth century. We shall see how neatly mathematics answers a variety of problems raised by such motions.

Suppose that a cannon inclined at an angle of 30° to the ground fires a shell with a velocity* of 1000 ft/sec (Fig. 16–4). What is the subsequent motion of the shell? We know from intuition or experience with balls thrown at a similar angle of elevation that the shell will travel out and up along some curved path and will then return to the ground. This qualitative knowledge is not, of course, sufficient to answer significant questions about the motion.

The initial velocity of the shell is in the direction which makes an angle of 30° to the ground. To treat the motion of the shell, it is mathematically simpler to consider its horizontal and vertical motions separately, that is, to obtain the parametric formulas. For this purpose we must know the horizontal and vertical velocities of the shell. Velocity is analogous to force in that direction as well as magnitude is important; in other words, velocity is another physical example of a vector. What we need, then, are the horizontal and vertical components of the vector which represents the initial velocity of the shell.

Suppose that the shell travels for one second in the direction OR in which it is fired. How far will it travel horizontally and vertically in that second? Let us drop a perpendicular from R onto the X-axis and from R onto the Y-axis. Thus we determine the lengths OP and OQ, respectively. The length OP is the horizontal distance which the shell travels in one second and the length OQ is the corresponding vertical distance. Since OP and OQ are distances traveled in one second, they also represent the horizontal and

* The terms speed and velocity are often used interchangeably. However, the word *velocity* implies that direction as well as magnitude of the speed are under discussion.

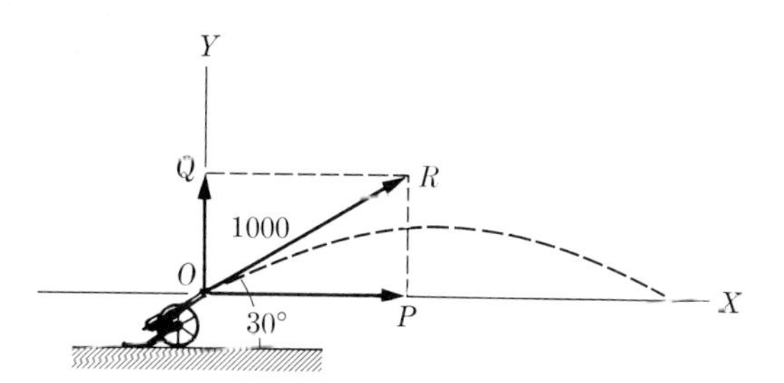

FIG. 16–4. Shell fired with an initial velocity of 1000 ft/sec from a cannon inclined at an angle of 30° to the ground.

vertical velocities. Thus the components of the velocity OR in any direction are obtained in precisely the same way that we used to determine the components of a force.

The horizontal and vertical velocities are then OP and OQ. What are their magnitudes? We see from Fig. 16–4 that

$$\cos 30° = \frac{OP}{1000}$$

or

$$OP = 1000 \cos 30° = 1000(0.8660) = 866 \text{ ft/sec.}$$

Similarly,

$$\sin 30° = \frac{PR}{1000}$$

or

$$PR = 1000 \sin 30° = 1000(0.5000) = 500 \text{ ft/sec.}$$

Since $PR = OQ$, the horizontal and vertical velocities of the shell are 866 ft/sec and 500 ft/sec, respectively.

We now utilize the physical fact that the horizontal and vertical motions can be treated independently. Let us begin with the horizontal motion. The shell has an initial horizontal velocity of 866 ft/sec, and no force acts to accelerate or decelerate the horizontal motion. Hence the shell will continue to move horizontally at a constant speed of 866 ft/sec, and the horizontal distance x traveled in time t is given by

$$x = 866t. \tag{5}$$

Next we consider the vertical motion of the shell. Gravity gives the shell a constant acceleration downward of 32 ft/sec^2. Since the upward direction has been chosen to be positive, the downward acceleration must be written

$$a = -32. \tag{6}$$

The downward velocity acquired in time t is $-32t$. However, the shell has an initial upward velocity of 500 ft/sec which, by the first law of motion, would continue indefinitely, were it not affected by gravity. The net velocity v is then (compare Section 15–5)

$$v = -32t + 500. \tag{7}$$

To obtain the distance traveled upward in any time t, we use the same reasoning as in the preceding chapter. If only the velocity of 500 ft/sec were acting, the distance traveled upward in t seconds would be $500t$. But in that time gravity pulls the shell downward a distance of $16t^2$. Hence y, the net height above the ground, is

$$y = -16t^2 + 500t. \tag{8}$$

Formulas (5) and (8) give the horizontal and vertical distances from the starting point O. We note that once more motion is represented by parametric formulas.

Several questions about the motion arise in practice. The first is, What path does the shell take? We may save ourselves the work of plotting the curve by determining the direct relationship between x and y, provided we recognize the curve of the resulting equation. Let us try this. Solving (5) for t yields $t = x/866$. We substitute this value of t in (8) and obtain

$$y = -16\left(\frac{x}{866}\right)^2 + 500\left(\frac{x}{866}\right)$$

or

$$y = -\frac{x^2}{46{,}872} + \frac{250}{433}x. \tag{9}$$

In Section 13–5 we discussed equations of the form (9)—albeit by means of numerically simpler examples. We could have proved the quite general statement that an equation of the form

$$y = -ax^2 + bx, \tag{10}$$

where a and b are any positive numbers, represents a parabola which opens downward and passes through the origin. (See Exercise 3 of Section 13–5, which is a special case.) Hence with just a little more work in coordinate geometry, we could have proved what we shall now accept without proof, namely, that equation (9) describes a parabola. Thus the parabola, as Galileo readily established, appears once more as the path of a projectile which is shot into the air at some angle to the ground.

What is the *range* of the shell? That is, how far from the starting point will the projectile strike the ground again? The answer is important because it tells us whether a given target on the ground can be reached. Unfortunately neither formula (5) nor formula (8) answers this question directly.

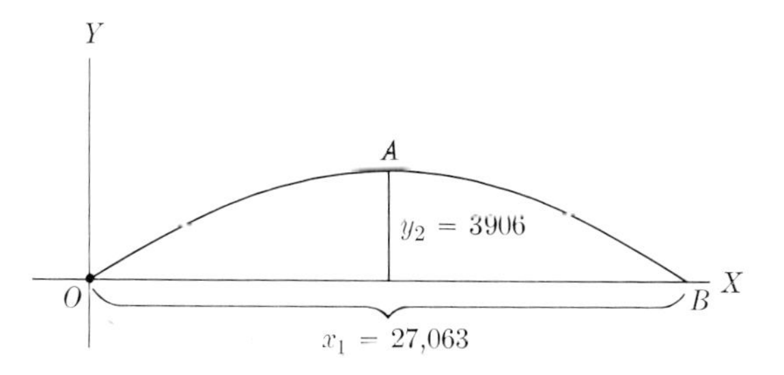

FIG. 16-5. Path of a shell shot from a cannon.

However, when the shell reaches the ground, the value of y in (8) should be zero. Let us determine, then, the value of t, say t_1, when $y = 0$. From (8)

$$0 = -16t_1^2 + 500t_1. \tag{11}$$

Equation (11), which is of the second degree in t_1, is rather easy to solve. Applying the distributive axiom, we may write

$$0 = t_1(-16t_1 + 500). \tag{12}$$

The right side of (12) equals zero when either factor is zero, that is, when $t_1 = 0$ and when

$$-16t_1 + 500 = 0.$$

The second alternative leads to

$$t_1 = \frac{125}{4}. \tag{13}$$

The first value, $t_1 = 0$, corresponds to the instant when the shell first starts its flight. Then the second value, 125/4, must be the time when the shell returns to the ground.

To determine the range, one more step is necessary. Formula (5) tells us how far the shell travels horizontally in any time t. Since the shell travels 125/4 seconds by the time it reaches the ground, we have but to substitute this value of t in (5) to get the range. If x_1 denotes the range, then (see Fig. 16-5)

$$x_1 = 866\,\frac{125}{4} = 27{,}063 \text{ feet}. \tag{14}$$

We might also like to know how high the shell will go in its flight and how long it takes to reach that height. These questions are readily answered. At the highest point in its flight, the *vertical* velocity is zero, else the shell would continue to rise. Formula (7) gives us the vertical velocity at any time t.

Let us ask for the value of t, say t_2, when $v = 0$. Then

$$0 = -32t_2 + 500$$

or

$$t_2 = \frac{500}{32} = \frac{125}{8}. \tag{15}$$

Hence it takes 125/8 seconds for the shell to reach the highest point. Now formula (8) tells us how high the shell is at any time t. Let us therefore find the height, y_2, when $t = 125/8$. We substitute 125/8 for t in (8) and obtain

$$y_2 = -16\left(\frac{125}{8}\right)^2 + 500\left(\frac{125}{8}\right)$$

or

$$y_2 = 3906 \text{ feet}. \tag{16}$$

Hence the shell reaches a maximum height of 3906 feet above the ground.

Another interesting question is whether the shell takes as long to travel from the cannon to its maximum height as it does to return from the latter position to the ground. Or, to fit the situation shown in Fig. 16–5, we may restate the question and ask, Does it take as long for the shell to travel from O to A as from A to B? We considered an analogous problem in the preceding chapter while discussing the motion of an object thrown straight up into the air and found that the two time intervals were equal. What does intuition suggest as the answer in the present case?

We can show at once that the time required to get from O *to* A is the same as the time required to travel from A to B. Equation (13) supplies the time it takes the shell to reach B, that is, to travel the path OAB. Equation (15) gives the time required to travel the path OA. We see at once that the value of t_1 is twice the value t_2. Hence the time of travel along path AB must equal the time of travel along path OA.

Exercises

1. Suppose that a shell is fired in a direction making an angle of 40° with the ground and with a velocity of 300 ft/sec. What are the horizontal and vertical velocities of the shell? What are the parametric equations describing the motion?

2. Suppose that the parametric formulas for the motion of a projectile are $x = 20t$ and $y = -16t^2 + 30t$. What is the direct relationship between x and y? What is the nature of the curve represented by the direct relationship between x and y?

3. Suppose that the parametric formulas for the motion of a projectile are $x = 3t$ and $y = -16t^2 + 5t$. Working with these formulas, plot a few points of the path.

4. Find the range of the projectile whose motion is described in Exercise 2.

5. Find the maximum height of the projectile whose motion is described in Exercise 2.

6. What velocity does the shell whose motion is treated in Section 16–4 have on striking the ground? How does this terminal velocity compare with the initial velocity?

16–5 The motion of projectiles fired at an arbitrary angle. In the preceding section we saw how we could study the motion of a shell fired from a cannon which is inclined at an angle of 30° to the ground. The initial velocity of the shell in this direction was 1000 ft/sec. Since the angle of fire and initial velocity given in our example are representative values, the example teaches a great deal about the phenomenon of projectile motion. However, suppose that we sought to answer such questions as: What is the effect of the initial velocity on the range and on the maximum height attained by the projectile? What is the effect of the angle of fire on the range and on the maximum height of the projectile? At what angle should one fire a projectile to hit a given target? One could repeat the procedures pursued in the preceding section, using different initial velocities and angles of fire, and thus perhaps obtain answers to some of these questions. But the work would be considerable and still leave us with the problem of trying to infer a general conclusion from a number of special cases. The mathematician would not proceed in this way. He would suppose that the initial velocity is an arbitrary value, V, and that the angle of fire is an arbitrary angle, A, and then study the motion with these arbitrary values V and A. He might thereby obtain conclusions about all such motions because his results would hold for *any* initial velocity and *any* angle of fire.

Let us pursue this *general* investigation of projectile motion. Suppose that a shell is fired from a cannon which is inclined at an angle A to the ground (Fig. 16–6), and that the initial velocity of the shell is V ft/sec. What is the subsequent motion of the shell?

We shall follow the procedure of the preceding section. The first major point to remember is the physical principle that the horizontal and vertical motions of the projectile can be studied as though the motions were taking place independently. Hence let us find the initial horizontal and vertical

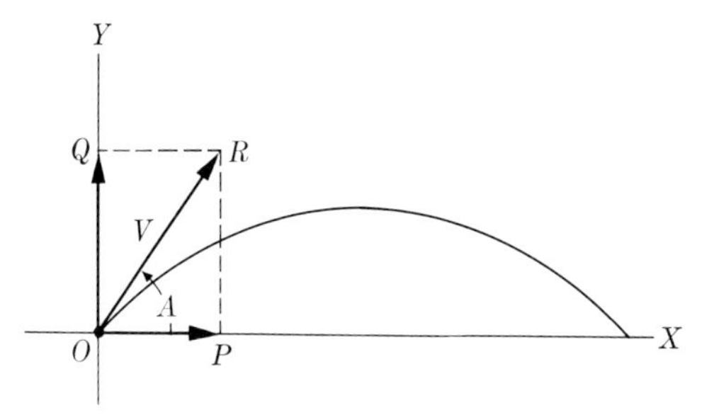

FIG. 16–6. Shell fired with an initial velocity of V ft/sec from a cannon inclined at an angle A to the ground.

velocities of the shell. By the very same argument that we used in Section 16–4, we know that the components of the velocity OR in the horizontal and vertical directions (Fig. 16–6) are obtained by dropping a perpendicular from R onto the X- and Y-axes, respectively. Thus OP is the horizontal velocity, and OQ is the vertical velocity. Now

$$\cos A = \frac{OP}{OR}.$$

Hence

$$OP = OR \cos A$$

or

$$OP = V \cos A. \tag{17}$$

Since

$$\sin A = \frac{PR}{OR},$$

and since $OQ = PR$, we have

$$OQ = OR \sin A$$

or

$$OQ = V \sin A. \tag{18}$$

Formulas (17) and (18) give us the initial horizontal and vertical velocities, respectively. We must now see what happens when the shell is in motion. The horizontal motion is uniform; i.e., no force acts to speed it up or slow it down. The shell will therefore continue to travel indefinitely in the horizontal direction at the velocity given by (17). Let us use v_x to indicate the velocity in the X-direction. Then at any time t,

$$v_x = V \cos A. \tag{19}$$

Since the velocity in the horizontal direction is constant, the distance traveled is velocity multiplied by time. Then

$$x = (V \cos A)t.$$

This expression is best written as

$$x = Vt \cos A, \tag{20}$$

so that there is no confusion about the fact that the quantity whose cosine is to be taken is A, whereas if we had written $V \cos At$, one might think that the quantity is At. Formula (20) is a generalization of formula (5), for the 866 in (5) is just $1000 \cos 30°$.

To obtain the vertical velocity of the shell we must take into account the fact that gravity does produce a vertical acceleration which affects the vertical velocity. This vertical acceleration is 32 ft/sec^2 and is downward.

Since we have chosen the upward direction as positive, we have

$$a = -32.$$

Because the acceleration is constant, the downward velocity gained by the shell in t seconds is $-32t$. However the shell has an upward initial velocity of $V \sin A$. Hence the net vertical velocity v_y is

$$v_y = -32t + V \sin A. \tag{21}$$

Next we use an old argument to determine the vertical height attained by the shell in t seconds. If only the velocity $V \sin A$ were acting, then in t seconds the shell would reach the height $(V \sin A)t$ or $Vt \sin A$. However, in these t seconds gravity pulls the shell downward a distance of $16t^2$. The net height, y, of the shell therefore is

$$y = -16t^2 + Vt \sin A. \tag{22}$$

Formulas (19) through (22) supply the general equations of projectile motion. We are now in a position to answer with respect to the arbitrary values V and A the same questions that were discussed in the preceding section for a specific numerical example. Thus, by solving (20) for t and substituting this result in (22), we could find the direct relationship between x and y. We could then see that for any fixed value of V and any fixed value of A, the path is a parabola. Similarly, we could obtain general expressions for the maximum height reached by the shell, the time required to reach that height, and, say, the time required for the shell to return to the ground. In other words, we could reproduce for any V and A the results derived in the preceding section for special values of V and A.

Let us turn instead to answering questions which we could not treat before. Let us study the effect of the initial velocity, V, and angle of fire, A, on the range of the cannon. To do this we must first determine the general expression for the range. The method is the same as the one used in the preceding section.

We begin with the physical fact that when the shell strikes the ground, the y-value of its position is zero. Hence let t_1 be the value of t when $y = 0$. Setting $y = 0$ in (22), we have

$$0 = -16t_1^2 + Vt_1 \sin A.$$

We may apply the distributive axiom to write

$$0 = t_1(-16t_1 + V \sin A).$$

Now the right side is zero when $t_1 = 0$ and when

$$-16t_1 + V \sin A = 0. \tag{23}$$

Since $t_1 = 0$ corresponds physically to the instant when the shell starts its motion, it follows that the value given by (23) is the value of t at the instant the shell strikes the ground. If we solve (23) for t_1, we find

$$t_1 = \frac{V \sin A}{16}. \tag{24}$$

To determine the shell's range, that is, its horizontal distance from the starting point at the instant when t_1 has the value just found, we use formula (20). Thus, if x_1 is the value of x when t has the value t_1, then

$$x_1 = V \cos A \, \frac{V \sin A}{16},$$

or the range is

$$x_1 = \frac{V^2}{16} \sin A \cos A. \tag{25}$$

Formula (25) answers one question immediately. If the angle A is held fixed, then the range depends upon V^2. If V is increased, then x_1 increases, and indeed x_1 increases rapidly because it depends upon V^2 rather than upon just V. Also, if we wished to attain a given range with a given angle of fire, that is, if x_1 and A were specified, we could use (25) to calculate the necessary initial velocity, V.

The more practical problem is to study the dependence of range upon angle of fire, for it is easier to change the angle of fire of a cannon than it is to change the initial velocity of the shells. Let us suppose, then, that the initial velocity V is fixed and ask the question, What is the maximum range that can be obtained by varying angle A? Since V is fixed, our question, in view of (25), amounts to: For what value of A is the product $\sin A \cos A$ a maximum? A little mathematics provides the answer.

We know from our work on the trigonometric ratios that $\sin A$ and $\cos A$ are certain ratios of sides of a right triangle. The size of the right triangle used to determine $\sin A$ and $\cos A$ does not matter because all possible right triangles containing a definite angle A are similar and therefore the ratio of two particular sides in any one of these triangles is always the same.* Hence let us choose a right triangle whose hypotenuse AB (Fig. 16–7) is the diameter of a definite circle. The vertex of the right angle of this right triangle must lie on the circle because a theorem of plane geometry states that the vertices of all right angles with vertex C and hypotenuse AB must lie on a circle with AB as diameter. Let us denote the sides of the right triangle ABC by a, b, and c. We draw the perpendicular CD and denote it by h. Then, using the right triangle ADC, we obtain

$$\sin A = \frac{h}{b}.$$

* The reader can review this point in Chapter 7, where it was first made.

From the right triangle ABC we have

$$\cos A = \frac{b}{c}.$$

So far then we have found that

$$\sin A \cos A = \frac{h}{b} \cdot \frac{b}{c} = \frac{h}{c}. \quad (26)$$

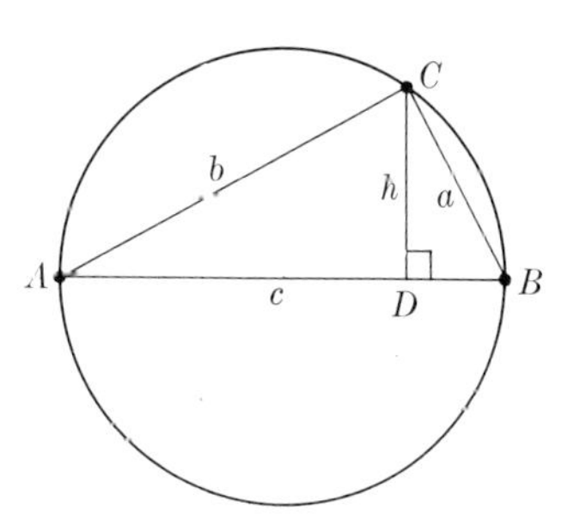

FIGURE 16–7

We may vary angle A and continue to regard it as an angle of a right triangle with hypotenuse AB because, as we noted above, the size of the right triangle is immaterial. Of course, variations in the angle A will produce changes in the position of C, but for the reason given earlier, C will continue to be a point on the circle with diameter AB. Hence we can consider all possible acute angles A by considering all triangles ABC with AB fixed and C varying on the circle. Let us now look at (26) again. Since the quantity c is fixed, $\sin A \cos A$ will be a maximum when h is a maximum, and h is greatest when it is the radius of the circle. But when h is the radius, C is directly above the center of the circle. Then $AC = BC$. In this case angle $A = 45°$ because the right triangle ABC is isosceles. Hence

$$\sin A \cos A$$

is a maximum when angle $A = 45°$.

We may now return to formula (25). We have found that when V is fixed, the maximum range, that is, the maximum possible value of x_1, is obtained when $A = 45°$. Since for $A = 45°$, $\sin A = \cos A = \sqrt{2}/2$, the maximum range is given by the formula

$$\text{maximum } x_1 = \frac{V^2}{16} \cdot \frac{\sqrt{2}}{2} \cdot \frac{\sqrt{2}}{2} = \frac{V^2}{32}.$$

This famous result was first proved by Galileo.

Let us denote the maximum range by OP (Fig. 16–8); thus for a fixed V but for all possible values of A, P is the farthest point on the ground which can be reached by a shell. Suppose that we wish to reach the point Q which is not quite so far out. At what angle A should the shell be fired? The question can be answered by substituting the desired value of x_1 in (25) and solving for angle A, but the technique of solution would lead us a bit farther into trigonometry than we shall take time for. What we shall prove is that there are two angles of fire which will yield the desired range OQ. Let A_1 be the angle shown in Fig. 16–8; that is, A_1 is one angle of fire for which the range is OQ. Then by (25),

$$OQ = \frac{V^2}{16} \sin A_1 \cos A_1. \quad (27)$$

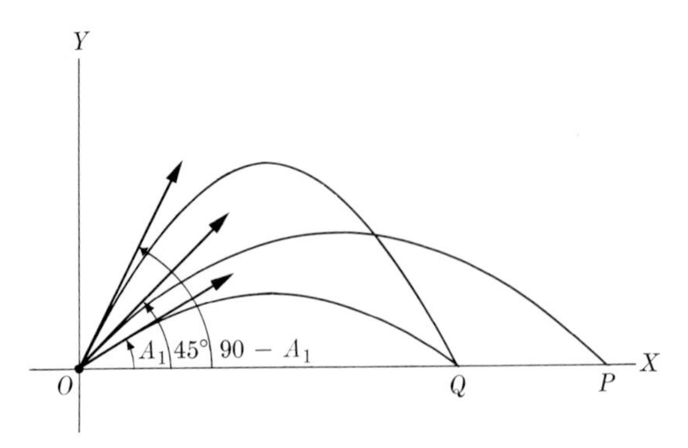

FIG. 16–8. The projectile fired at angle A and the projectile fired at angle $90 - A$ have the same range.

Now let us consider the angle $90 - A_1$. One of the elementary facts in trigonometry is that* for the acute angle A_1,

$$\sin (90 - A_1) = \cos A_1 \qquad \text{and} \qquad \cos (90 - A_1) = \sin A_1.$$

Then

$$\sin (90 - A_1) \cos (90 - A_1) = \cos A_1 \sin A_1.$$

By multiplying both sides by $V^2/16$, we obtain

$$\frac{V^2}{16} \sin (90 - A_1) \cos (90 - A_1) = \frac{V^2}{16} \cos A_1 \sin A_1. \tag{28}$$

Since the right side of (28) is the range OQ, as (27) shows, the left side must also equal OQ. But the left side states that the angle of fire is $90 - A_1$. Hence $90 - A_1$ and A_1 yield the same range. Thus there are two angles of fire which yield the same range. Moreover, one is the complement of the other.

The paths of projectiles, called trajectories, have many other interesting and important properties, but time and space do not permit us to discuss them in detail. We might, however, briefly note one or two. If we keep V fixed and vary angle A, we obtain a series of trajectories, all parabolas (Fig. 16–9). Each of these parabolas contains the points which can be reached by the shell fired at the angle A belonging to that parabola. Figure 16–9 also shows the curve which separates all those points which can be reached by the gun from all those which cannot. In mathematical language it is called the *envelope* of the *family* of trajectories. Now the interesting point, which we shall not prove, is that this envelope is itself a parabola whose focus is at the gun. This parabola is called the *parabola of surety*, the word suggesting that one is safe if he stays outside of it.

* See Exercise 5, Section 7–2.

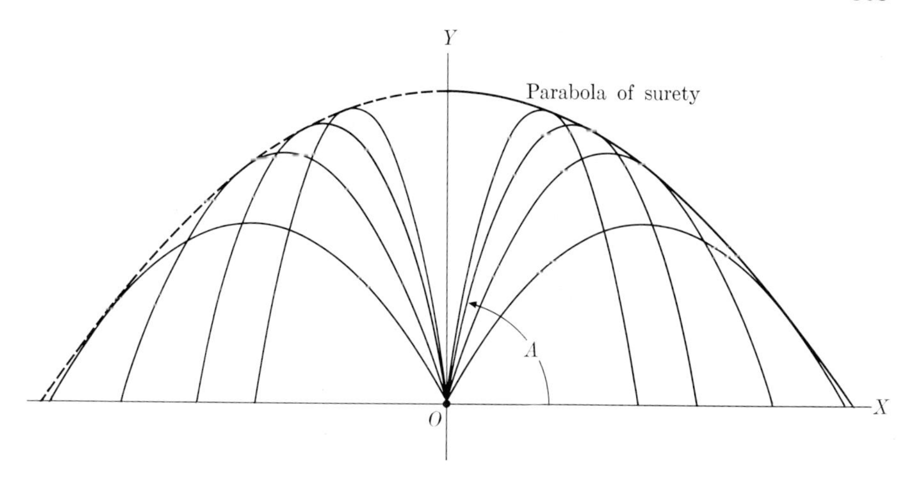

FIG. 16–9. The envelope of the parabolic paths of projectiles fired at different angles is itself a parabola.

Another interesting feature of the family of trajectories is that they all have the same directrix. This common directrix is the horizontal line tangent to the parabola of surety at its vertex.

EXERCISES

1. Formulas (20) and (22) give the parametric representation of the motion of a projectile. Find the direct relationship between x and y.
2. Derive the formula in terms of V and A for the time it takes a projectile to reach its maximum height.
3. What is the formula in terms of V and A for the maximum height reached by a projectile?
4. Show generally that it takes as long for a projectile to return from its highest position to the ground as it does to travel from the cannon to the highest position.
5. What is the range of a projectile which is fired with an initial velocity of 2000 ft/sec and at an angle of 40° to the ground?
6. What is the maximum range of a projectile fired with an initial velocity of 800 ft/sec?
7. How long does it take for a projectile fired with an initial velocity of 2000 ft/sec to reach a target located at maximum range?
8. Show that the highest point which a gun can reach for all possible angles of fire but with the same initial velocity is attained by firing straight up.

16–6 Summary. In this chapter we have seen how the simultaneous application of two simple formulas, the parametric formulas, makes it possible to represent an entire class of curves which physically happen to be the paths of projectiles. The value of the parametric formulas is, then, that they enable us to answer readily a variety of questions about projectile motion.

To appreciate how much mathematics accomplishes in this area, one might consider how he would proceed experimentally to find, for example, the dependence of the range of a projectile on the angle of fire. One would have to fire at least dozens of projectiles at different angles, making certain that other factors, such as the velocity with which the projectiles are fired, the shape of the projectiles, and the state of the atmosphere, are constant, and accurately measure the range and angle each time. With all these precautions taken and the information secured, the experimenter might obtain some limited results. He might, for example, learn that as the angle of fire increases in 1°-degree steps from 1° to 40°, the range increases steadily, but he might miss the all-important fact that above 45° the range decreases. The dependence of range upon velocity would still be entirely unknown and require further experimentation. But we have seen how a little mathematics costing only pencil and paper can supply the full story of the dependence of range upon angle of fire and initial velocity.

Thus the present chapter, too, illustrates how the combination of mathematics and simple physical axioms, such as the fact that the acceleration of bodies near the surface of the earth is 32 ft/sec^2, the first law of motion, and the independence of the horizontal and vertical motions, permits us to deduce a vast amount of knowledge about our physical world. The knowledge referred to at the moment concerns projectile motion under idealized conditions; that is, the resistance of air is neglected; the earth is assumed to be flat over the short distances which the projectiles cover; and the projectiles are limited to travel near the surface of the earth. One might regard the whole story as of minor interest because it deals with just one phenomenon and one which seems limited to bombs and guns. However, the study of this phenomenon has proved to be of immeasurable scientific importance. First of all, the deductions made from the physical axioms mentioned above can be checked experimentally If the deductions agree with experience we have some reason to believe that the axioms are correct. We must remember in this connection that physical axioms are generalizations from limited experience and that our confidence in them depends upon how well they continue to lead to new physical facts. Secondly, the study of motions near the surface of the earth, projectile motion in particular, led to the most important advance in science since 1600, namely Newtonian mechanics. The step to the broad science of mathematical mechanics will be taken in the next chapter.

Recommended Reading

Galilei, Galileo: Dialogues Concerning Two New Sciences, pp. 234 through 282, Dover Publications, Inc., New York, 1952.

Holton, Gerald and Duane H. D. Roller: *Foundations of Modern Physical Science,* Chap. 3, Addison-Wesley Publishing Co., Inc., Reading, Mass., 1958.

Kline, M.: *Mathematics and the Physical World,* Chap. 14, T. Y. Crowell Co., New York, 1959.

CHAPTER 17

THE APPLICATION OF FORMULAS TO GRAVITATION

. . . from motion's simple laws
Could trace the secret hand of Providence,
Wide-working through this universal frame.

James Thomson in his memorial
poem to Newton

17–1 Introduction. We noted in an earlier chapter that beginning with the year 1600 approximately, mathematics and science have grown through interaction with each other. Physical problems have suggested mathematical themes, and these when developed and applied to physical problems have produced new physical knowledge. The domain of functions, with which we are presently concerned, was considerably expanded as mathematicians, responding to the needs of science, developed new relationships to represent physical laws.

In this chapter, our chief mathematical tool will be new formulas, in particular, the formula which represents the action of the force of gravitation. The physical interpretations of the results obtained by applying algebraic processes to this formula and by commingling it with other simple formulas were a series of laws which described all motions in the universe. The motions of objects near the surface of the earth and the motions of the planets, the comets, and the stars were shown to be encompassed in one body of laws commonly described as Newtonian mechanics. This branch of science furnishes the most impressive evidence we have for the mathematical design of our universe.

17–2 The problem of relating earthly and heavenly motions. Before we consider the new mathematical and physical developments, let us survey the problem which attracted the attention of the great physical scientists of the seventeenth century: Robert Hooke, the famous experimental physicist, Sir Christopher Wren, physicist and later renowned architect of London, Edmond Halley, the astronomer of Halley's comet fame, Christian Huygens, mathematician and physicist whose contributions enriched many fields and who is now most often remembered as the originator of a wave theory of light, and Sir Isaac Newton, the most important mathematical physicist of our era.

By the time of Galileo's death in 1642, two monumental accomplishments had become part of modern science. Galileo, himself, had found the basic physical principles from which he deduced the behavior of objects in motion near or on the surface of the earth. Some of this work was examined in the preceding two chapters. The second achievement is due to Copernicus and

Kepler (see Chapter 12). After Copernicus had established that a heliocentric view of planetary motion was simpler than the geocentric doctrine, Kepler discovered the three famous laws which further simplified astronomical theory and at the same time improved the accuracy of the representation. One might justifiably conclude that the science of motion was complete.

But to scientists who seek the ultimate design of our universe, the two accomplishments we have just described immediately suggested more profound and greater problems. A comparison of these two classes of laws, namely Galileo's for terrestrial motions and Kepler's for heavenly motions, revealed several basic differences. In the first place, Galileo had started with clear physical principles, such as the first law of motion and the constant downward acceleration of objects moving near the surface of the earth, and had *deduced* the formulas which describe straight-line and curvilinear motions. Kepler's three laws, though they fitted observations within the limits of observational errors, did not rest on physical principles. They were merely accurate mathematical descriptions of collections of data. Moreover, the three laws were logically independent of one another. Secondly, for terrestrial motions, the parabola was found to be the basic path of curvilinear motion, whereas for planetary motion, the ellipse was the basic path.

This comparison raised several questions. Could one establish any logical relationship among the Keplerian laws or were they really independent? What physical principles determined planetary motions? The mathematical laws, accurate and succinct as they were, presented after all only a rather bleak account, without giving any insight into, or rationale for, the motions. Why should planets move in ellipses? And why should parabolic paths prevail on earth and elliptical paths in the heavens?

The overriding question, however, which bothered the leading scientists of the latter half of the seventeenth century was: Could one establish a connection between the laws of terrestrial motion and the laws of planetary motion? Perhaps the very same physical principles which Galileo had used to deduce the paths of objects moving near the earth could lead to the laws describing the motion of the planets. In this event, the two classes of laws would be united; the Keplerian laws would be related to each other by being deduced from a common basis; and the physical reasons for planetary motion would be revealed.

The thought that *all* the phenomena of motion should follow from one set of physical principles might seem grandiose and inordinate to reasonable people, but it occurred very naturally to the religious mathematicians of the seventeenth century. God had designed the universe, and it was to be expected that all phenomena of nature would follow one master plan. One mind designing a universe would almost surely have employed one set of basic principles to govern as many related phenomena as possible. Since the scientists of the seventeenth century were engaged in the quest for God's design of nature, it seemed very reasonable to them that they should seek the unity underlying the diverse earthly and heavenly motions. As phrased by Newton, this goal was "to derive two or three general principles of motion from phenomena, and afterwards to tell us how the properties and actions of all corporeal things follow from these manifest principles. . . ." A less cogent but to mathematicians nonetheless significant indication of the existence of some unity was furnished by the fact that parabola and ellipse were both conic sections. The common mathematical origin of these curves warranted some belief that parabolic and elliptical motions were but special cases of some fundamental principle of motion.

In the seventeenth century, there were other less weighty but perhaps more pressing reasons to pursue the study of motion beyond the stage reached by Galileo and Kepler. Galileo had deliberately ignored the resistance of air on the ground that it was a secondary effect which could be taken into consideration later. This was indeed true, but the problem of calculating the effect of air resistance remained unsolved. Another open question was how to relate heavenly and earthly motions in a more limited but practical connection. This was the problem described earlier of determining the longitude of a ship at sea. Although navigators had used the stars, sun, and moon to determine the locations of their ships, the positions of these celestial bodies at various times of the year had yet to be related more precisely to the longitudes of points on the earth. In the seventeenth century it seemed that the moon would be most suitable for the determination of longitude because its closeness permitted accurate observation of its position from points on the earth. Hence, more precise information about the motion of the moon around the earth was needed. This became a major scientific problem of the age.

17-3 A sketch of Newton's life. Any great advance in mathematics and science is almost always the work of many men contributed bit by bit over hundreds of years. Then one man smart enough to distinguish the worthy ideas of his predecessors from the welter of suggestions and results and imaginative and audacious enough to fit the significant ideas into a master plan makes the culminating and definitive step. In the problem of unifying all the phenomena of motion, the decisive step was made by Isaac Newton.

He was born in 1642, premature and weak. His mother was already widowed and so preoccupied with running the family farm that she could pay no attention to the boy. The elementary education Newton received in local schools of a small English town could hardly have given him much of a start, and in his youth Newton showed no promise. His family sent him to Cambridge University, where he entered Trinity College in 1661. Here, at last, Newton got the opportunity to study the works of Copernicus, Kepler, and Galileo, and here he had at least one good teacher, the distinguished mathematician Isaac Barrow. His university work was not outstanding and he had, in fact, such difficulties with geometry that he almost changed his course of study from science to law. However, Barrow did recognize that Newton had ability.

Newton finished his undergraduate work; at that point an outbreak of the plague in the area around London led to the closing of the university. He, therefore, spent the years 1665 and 1666 in the quiet of the family home at Woolsthorpe. During this period Newton initiated his great work in mechanics, mathematics, and optics. He realized that the law of gravitation, which we shall examine shortly, was the key to an embracing science of mechanics; he obtained a general method for treating the problems of the calculus (see Chapter 18); and through experiments he made the epochal discovery that white light such as sunlight is really composed of all colors from violet to red. "All this," Newton said later in life, "was in the two plague years of 1665 and 1666, for in those days I was in the prime of my age for invention, and minded mathematics and philosophy [science] more than at any other time since."

Newton returned to Cambridge in 1667 and was elected a Fellow of Trinity College. In 1669, Isaac Barrow resigned his professorship of mathematics to devote himself to theology, and Newton was appointed in Barrow's place. He apparently was not a successful teacher, for few students attended his lectures; nor did anyone comment on the originality of the material he presented.

At first Newton did not publish his discoveries. He is said to have had a fear of criticism and opposition. When in 1672 he did publish his work on light accompanied by his philosophy of science, he was seriously criticized by most contemporaries, including Robert Hooke and Huygens, who had different ideas on the nature of light. Newton was taken aback and decided that he would not publish in the future. However, in 1675 he did publish another paper on light which contained his idea that light was a stream of particles. Again he was met with criticism and claims by others that they had already discovered these ideas. This time Newton decided that he would leave his results to be published after his death. However, in 1684 his friend Edmond Halley urged him to publish his work on gravitation and even assisted him editorially and financially. Thus in 1687 the classic of science, the *Mathematical Principles of Natural Philosophy*, often briefly referred to as the *Principia* or the *Principles*, appeared. This book did receive much acclaim and, aside from three Latin editions, appeared in many languages. One popularization was entitled *Newtonianism for Ladies*. The *Principia* is written in the deductive manner of Euclid; that is, it contains definitions, axioms, and hundreds of theorems and corollaries. Its conciseness makes it difficult reading. To excuse this aspect, Newton told a friend that he had made the *Principia* difficult on purpose "to avoid being baited by little smatterers in mathematics." He thereby hoped to avoid the criticisms heaped on his earlier papers on light.

After about thirty years of creative activity which included some work in chemistry, Newton became depressed and suffered a nervous breakdown. He left Cambridge University to become Warden of the British Mint in 1696 and thereafter confined his scientific activities to the investigation of an occasional problem. He did, however, devote himself to theological studies, which he regarded as more fundamental than science and mathematics because the latter disciplines concerned only the physical world. In fact, had Newton been born two hundred years earlier he would almost surely have become a theologian. An example of his theological writing is *The Chronology of the Ancient Kings Amended,* in which he sought to determine the dates of Biblical events by utilizing astronomical facts mentioned in connection with these events.

During his last years and posthumously he was honored in many ways. He was President of the Royal Society of London from 1703 to his death; he was knighted in 1705; and he was buried in Westminster Abbey.

17–4 The law of gravitation and the second law of motion. In his philosophy and method of science, Newton followed Galileo. He, too, believed that the universe was mathematically designed by God and that mathematics and science should strive to uncover that glorious design. Like Galileo, he was convinced that fundamental physical principles should be quantitative statements about the real qualities of the world, space, time, mass, weight, and force. From these principles and with the axioms and theorems of mathematics, it should be possible to deduce the laws of nature. Newton expressed this philosophy in the preface to his *Principles:* ". . . for the whole burden of philosophy [science] seems to consist in this—from the phenomena of motion to investigate the forces of nature, and from these forces to demonstrate the other phenomena . . ." By investigating the forces of nature, he meant to arrive at the basic laws governing the operation of these forces and then, of course, to deduce the consequences.

The first problem, then, in executing such a program is to discover the fundamental principles. Like Galileo, Newton insisted on obtaining these by direct study of the physical world rather than by searching one's mind for hypotheses that seemed to be reasonable or by accepting Biblical passages. Here, too, Newton is explicit. In another of his famous books, *Opticks,* first published in 1704, he says: "Thus analysis consists in making observations and experiments and in drawing general conclusions by induction, and admitting of no objections against the conclusions, but such as are taken from experiments or other certain truths." What Newton sought to emphasize and what required emphasis in his time is that generalizations must be based on some experimental or observational grounds, and that no hypothesis can be tolerated which is contrary to a single bit of physical evidence. Further, deductions made from the basic principles must also be in accord with physical evidence, for only by continued agreement between deductively established conclusions and experimental tests can one acquire confidence that the original generalizations are correct.

With such principles of scientific method clearly in mind, Newton turned to the problem of finding the physical principles which would lead to a unifying theory of earthly and celestial motions. He was, of course, familiar with the principles unearthed by Galileo. But these were presumably not enough. It was clear from the first law of motion that the planets must be

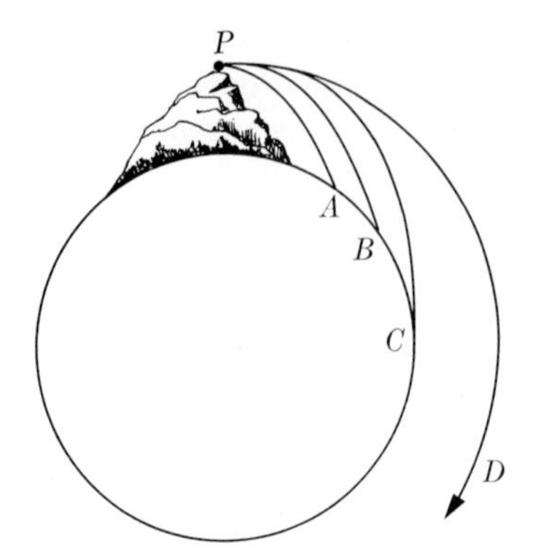

FIG. 17–1. Projectiles shot out horizontally from the top of a mountain with increasing horizontal velocities.

acted on by a force which pulls them toward the sun, for if no force were acting, each planet would move in a straight line. The idea of a force which constantly pulls each planet toward the sun had occurred to many men, Kepler, Hooke, Wren, Halley, and others, even before Newton set to work. It had also been conjectured that this force exerted on a distant planet must be weaker than that exerted on a nearer one and, in fact, that this force must decrease as the square of the distance between sun and planet increased. But, prior to Newton's work, none of these thoughts about a gravitational force advanced beyond speculation.

Newton adopted these ideas. However, in his attempt to tie in the action of the gravitational force with motions on the earth, a line of thinking occurred to him which was highly imaginative and certainly original in his time but which is now an almost daily experience. He considered the problem of what happens when a projectile is shot out horizontally from the top of a mountain. As Newton knew and as we know from our study of Galileo's work, the projectile follows a parabolic path to earth (see Chapter 16, Fig. 16–3). If the horizontal speed of the projectile is increased, then the path is wider but remains parabolic. However, Galileo had assumed that the earth was flat and that the projectiles were given moderate initial horizontal speeds such as a cannon might impart to shells. Newton then asked himself what would happen if the sphericity of the earth were taken into account, and if the horizontal speeds of the projectiles were gradually increased. If the sphericity of the earth is taken into account, then projectiles with small horizontal speeds will follow the paths PA and PB of Fig. 17–1. As the speed is increased somewhat, the projectile might take a path such as PC. Suppose now that the speed is increased still more. Would the projectile fall off into space? Not necessarily. As the projectile travels into space, it is pulled toward the earth. But the pull of a spherical earth is directed toward the center, and hence the projectile, subjected to this continuous pull toward the center, need not fall off into space. It might, in fact, continue to circle the earth indefinitely if the earth pulled it in just enough so that it would not wander out into space and yet not fall to earth.

And so Newton concluded in his *Principles:* "And after the same manner that a projectile, by the force of gravity, may be made to revolve in an orbit, and go round the whole earth, the moon also, either by the force of gravity, if it is endowed with gravity, or by any other force, that impels it toward the earth, may be continually drawn aside towards the earth, out of the rectilinear [straight-line] way which by its innate force [inertia] it would pursue; and would be made to revolve in the orbit which it now describes; nor could the moon without some such force be retained in its orbit. If this force were too small, it would not sufficiently turn the moon out of a rectilinear course; if it were too great, it would turn it too much, and draw the moon from its orbit toward the earth. It is necessary that the force be of a just quantity, and it belongs to the mathematicians to find the force that may serve exactly to retain a body in a given orbit with a given velocity; . . ." This argument which showed how the motion of the moon around the earth could be related to motions occurring on earth was immediately extended to the motions of the planets about the sun. The planets, set into motion somehow, are attracted by the sun and are presumably pulled in just enough to keep them from flying off into space or from crashing into the sun.

Thus Newton had some reason to suppose that the same force which pulled projectiles to earth caused the moon to revolve around the earth and the planets to revolve around the sun. He had now to determine precisely how strong the force of gravitation is, that is, how it depends upon the masses of the bodies involved and upon the distances between the bodies. Newton adopted the conjecture already made by his contemporaries, namely, that the force of attraction, F, between *any* two bodies of masses m and M, respectively, separated by a distance r is given by the formula

$$F = G\frac{mM}{r^2}. \tag{1}$$

In this formula, G is a contant; i.e., it is the same number, no matter what m, M, and r may be. The numerical value of this constant, which we shall determine later, depends upon the units used for mass, force, and distance.

From the mathematical standpoint, formula (1) represents a new type of functional relationship. The quantity F is the dependent variable which depends upon three independent variables, m, M, and r, the quantity G being a constant. If we give values to m, M, and r, then the value of F is determined. Such a function is, of course, more complicated than, say the formula $d = 96t - 16t^2$, which contains just one independent variable, t, and one dependent variable, d. When one is working in a situation in which m and M are fixed and only r can vary, then F is a function of just one independent variable. For example, if G were 1, m were 2, and M were 3, then the relationship between F and r would be $F = 6/r^2$. This formula expresses the dependence of the gravitational force between two fixed masses on the distance between them.

Newton had yet to show that formula (1) was the correct quantitative expression for this force. To apply formula (1) and to work with forces in general, Newton adopted a second quantitative physical principle which proved to be just as important as his law of gravitation. Galileo had found that if an object is in motion and no force is applied to it, then the object continues to travel in a straight line and with a constant speed. He had also concluded that a force acting on a body had the effect of imparting acceleration to the body. Thus the force which the earth exerts on an object traveling near the surface of the earth, called the weight of the object, gives the object an acceleration, and the relation between weight, mass, and acceleration of the object is

$$W = 32m. \tag{2}$$

That is, the force W gives the object of mass m an acceleration of 32 ft/sec². Newton generalized formula (2) to assert that the formula relating any force applied to a body, its mass, and its acceleration is

$$F = ma. \tag{3}$$

In the special case where F is the weight W, the value of a is 32 ft/sec². Formula (3) is known as Newton's second law of motion. It applies to any force, whether or not it be the force of gravity. As in the case of formula (2), if m is measured in pounds and a in feet and seconds, then F is measured in poundals. Thus a force of 32 poundals gives a mass of 1 pound an acceleration of 32 ft/sec². The unit, pound, is also used for forces with the understanding that one pound of force equals 32 poundals.

Let us see how Newton tested his law of gravitation. We shall write (1) in the slightly different form

$$F = m\frac{GM}{r^2}\cdot \tag{4}$$

If we compare (3) and (4), we observe that the quantity GM/r^2 in (4) plays the role of a in (3); that is, the law of gravitation can be viewed as stating that the gravitational force F gives a mass m the acceleration GM/r^2. In symbols,

$$a = \frac{GM}{r^2}\cdot \tag{5}$$

Now let M be the mass of of the earth and let m be the mass of a small body near the surface of the earth. Then there is the question of what r in (5) represents. It is supposed to represent the distance between the two masses. Shall we take it then to be the distance from the mass m to the surface of the earth or to some point in the interior? If the two masses were separated by millions of miles, as are the earth and sun, one might idealize each mass and regard it as concentrated at one point because the size of each mass is

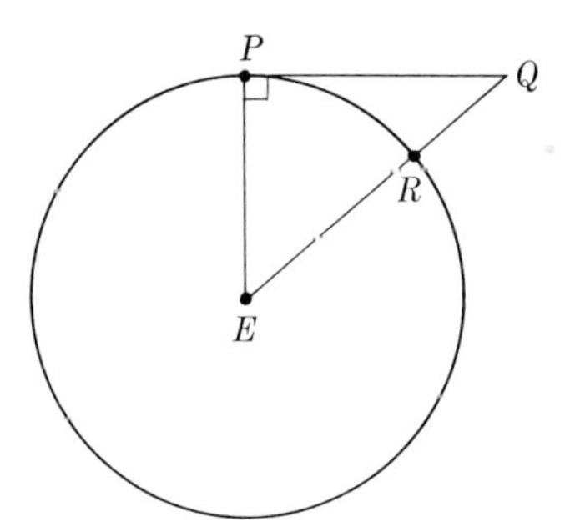

FIG. 17–2. The earth's pull on the moon causes the moon to "fall" the distance QR in one second.

small compared with the distance between them. But for objects near the surface of the earth, the value of r depends heavily upon what point in or on the surface of the earth is chosen as the position of the earth. Newton conjectured (and later proved) that for purposes of gravitational attraction the mass of the earth could be regarded as though it were concentrated at the earth's center. Hence, with respect to the earth's gravitational acceleration acting on a mass m near the surface of the earth, r in (5) can be taken to be 4000 miles or 21,120,000 feet. This value of r is, in essence, the same for all objects *near* the surface of the earth. Moreover, the mass M of the earth is constant, and so is G. Hence, for all objects near the surface of the earth, the entire right side of (5) is constant. Consequently, the acceleration which gravity imparts to all objects near the surface of the earth is constant. This is precisely what Galileo had found and, in fact, he had determined that the constant is 32 ft/sec^2. Thus Newton's law of gravitation met its first test, for it yielded as a special case a well established fact.

And now Newton sought to show that the force of gravitation given quantitatively by formula (1) was precisely the right force to account for the motion of the moon around the earth. We shall not reproduce his entire mathematical reasoning, but we can readily see how he went about the task. To simplify the discussion, let us suppose that the orbit of the moon is circular. Both moon and earth are regarded as points because their sizes are small compared with the distance between them. Suppose that the moon is at the point P of its path around the earth E (Fig. 17–2). If the gravitational attraction of the earth were not effective, then, according to the first law of motion, the moon would shoot off in the direction PQ, which is tangent to its path, at whatever speed the moon might have at the point P. Let us suppose that PQ is the distance that the moon would travel in one second. However, the earth's gravitational force causes the moon to be at the point R at the end of that second, so that we can regard QR as the distance which the moon "falls" toward the earth in one second. The calculation of QR itself is a matter of simple trigonometry. Now if the moon falls with constant acceleration, say a, then it follows from what we have learned about motion in a straight line with constant acceleration, that the distance d that the moon should fall in t seconds is $d = (1/2)at^2$. In one second it should fall $d_1 = a/2$. Thus by substituting the value of QR for d_1, we can find the acceleration of the moon. We should note that this acceleration can be obtained without resorting to the law of gravitation.

On the other hand, when M is the mass of the earth, formula (5) gives the acceleration which the earth imparts to objects located at a distance of r units. When $r = 4000$ miles or $4000 \cdot 5280$ feet, we know that $a = 32$. Then

$$32 = \frac{GM}{(4000 \cdot 5280)^2} = \frac{GM}{(4000)^2(5280)^2}. \tag{6}$$

When $r = 240{,}000$ miles or $240{,}000 \cdot 5280$ feet, which is the distance of the moon from the earth, then the acceleration which the earth exerts, a_1 say, is

$$a_1 = \frac{GM}{(240{,}000 \cdot 5280)^2} = \frac{GM}{(240{,}000)^2(5280)^2}. \tag{7}$$

If we divide equation (7) by equation (6), we obtain

$$\frac{a_1}{32} = \frac{(4000)^2}{(240{,}000)^2} = \frac{1}{(60)^2}.$$

Hence, according to the law of gravitation, a_1, the acceleration which the earth produces on the moon, is

$$a_1 = \frac{1}{(60)^2} \cdot 32 = 0.0089 \text{ ft/sec}^2. \tag{8}$$

When Newton compared this last result with the acceleration determined by the method of the preceding paragraph, he found that the two values were practically the same. Hence he had obtained a second and even more imposing piece of evidence supporting the law of gravitation. This evidence was so striking that Newton felt justified in accepting the law of gravitation as a correct physical principle. With this law, the two laws of motion, and the axioms and theorems of mathematics, he set forth to *deduce* further laws of terrestrial and celestial motions.

Exercises

1. Suppose that the gravitational force varies with the distance between two definite masses according to the formula $F = 6/r^2$. Show graphically how F varies with r.

2. Newton satisfied himself that the law of gravitation was correct by obtaining two pieces of evidence for it. What were they?

3. Is Newton's argument for the law of gravitation acceptable as a mathematical proof? Is his method good scientific procedure?

4. Knowing that the acceleration of objects near the surface of the earth is 32 ft/sec^2, use formula (5) to calculate the acceleration which the earth exerts on objects 1000 mi above the surface of the earth.

5. Suppose that an object falls to earth from a point 1000 mi above the surface. May we use the formula $d = 16t^2$ to compute the time it takes to fall this distance?

6. What is the mass of an object which weighs 150 lb? (One pound of *weight* is 32 poundals.)

7. How much force is required to give an automobile weighing 3000 lb an acceleration of 12 ft/sec^2?

8. Can you suggest a way of calculating QR in Fig. 17–2 if arc PR is the known distance the moon actually travels in one second? You need not carry out the calculations.

17–5 Further discussion of weight and mass. With reasonable support for the law of gravitation we can now, following Newton's example, adopt it as an axiom of physics and see what conclusions we may draw from this axiom and the other axioms of physics and mathematics. The law itself states that the force of gravitation F between any two masses m and M is given by the formula

$$F = G\frac{Mm}{r^2}, \tag{9}$$

where r is the distance between the masses. Formula (9) leads immediately to a better understanding of the relationship between weight and mass and to an extension of the concept of weight. Let M be the mass of the earth and let m be the mass of some other object. Since F is the force with which the earth attracts this object, we can regard F as the weight of the object, for this attractive force is what we have meant by weight. However, we now see that the force or weight depends upon the distance r between the two masses. Hence, the weight of an object is not really a fixed number but varies with the distance of the object from the earth or, more precisely, from the center of the earth (see Section 17–4). If an object of mass m is at the surface of the earth, its weight is given by

$$F_1 = G\frac{Mm}{(4000 \cdot 5280)^2}, \tag{10}$$

but the same object taken 1000 miles above the surface of the earth will have the weight:

$$F_2 = G\frac{Mm}{(5000 \cdot 5280)^2}.$$

The value F_2 is considerably less than F_1 because the denominator in the second expression is much larger. We see, then, that the farther an object of mass m is from the surface of the earth, the less is its weight. On the other hand, the mass of the object, that is its resistance to change in speed, is the same at all locations. Thus we can see more clearly that the weight and mass of an object are quite different properties.

But Galileo had expressed the relationship between weight and mass by

$$W = 32m. \tag{11}$$

How does his result square with what we have been saying? As we have already pointed out in connection with equation (5), the quantity $GM/(4000 \cdot 5280)^2$ is 32, and so in the special case of objects near the surface, equation (10) reduces to equation (11). Thus Galileo's relation between weight and mass is correct only for objects near the surface of the earth.

If the weight of an object decreases with an increase in its distance from the center of the earth, then an object should weigh less on top of a high mountain than at sea level. Robert Hooke had tried to detect such a difference in weight, but failed to do so. However, today's scientists have better instruments and can observe this difference, which represents one more confirmation of the law of gravitation.

The concept of weight can, and in the present scientific era must, be extended still further. So far we have considered the weight of an object to be the force with which the earth attracts the object. But now let us imagine that the object were taken to the moon and, for simplicity, let us suppose that no matter other than the moon and the object exist in space. May we speak of the weight of the object on the moon? The law of gravitation applies to moon and object, and so the moon will attract the object. This attractive force will be the weight of the object on the moon. To calculate this weight, we have but to let M in (9) be the mass of the moon, m, the mass of the object, and r, the radius of the moon. We know that the radius of the moon is 1080 miles or $1080 \cdot 5280$ feet. The mass of the moon can be determined by methods similar to those used later (Section 17–6) to compute the masses of the earth and sun. The result of the calculation, which we may accept for present purposes, is that the weight of an object on the moon is $\frac{1}{6}$ its weight on earth.

We can extend the notion of weight still further. Suppose that an object is in space somewhere between earth and moon. According to the law of gravitation, the earth attracts the object, and so does the moon. Since these attractions oppose each other, we may regard the weight of the object as the net attraction. If we now think of the object as moving from the earth to the moon, then the attractive force of the earth decreases while that of the moon increases. At the outset, the earth's force is stronger, but at some point in the path to the moon the two forces will be equal and oppositely directed so that the net weight of the object will be zero. This point is located at a distance of about 24,000 miles from the moon along the line from the earth to the moon. All of the above considerations about weight are now no longer purely academic flights of fancy but are important factors in the process of determining the paths of satellites which circle the moon or of rockets which are sent out to strike the moon.

Exercises

1. Suppose that a person weighs 150 lb at the surface of the earth where, of course, his distance from the center of the earth is 4000 mi. What would this person weigh at a point 4000 mi above the surface of the earth?

2. How does the law of gravitation enable us to further differentiate between the mass and weight of an object?

3. Suppose that of two objects on the earth, one has twice the mass of the other. Show that the force with which the earth attracts the first one is twice the force with which the earth attracts the second.

4. What would a man whose present weight is 150 lb weigh if the earth's mass were one-tenth of what it is?

5. Suppose that the earth's mass were twice as large as it is. What change would there be in the acceleration of falling bodies? Would a body which is dropped from a height of 1000 ft reach the ground sooner than it now does?

6. It is stated in the text that all bodies near the surface of the earth fall with the same acceleration. Suppose that an object is several thousand miles from the surface of the earth. How would the acceleration of its fall to the earth compare with the acceleration of a body near the surface?

7. Suppose that the mass of the moon were the same as the mass of the earth. The radius of the moon is about ¼ the radius of the earth. What would a man who weighs 150 pounds on the earth weigh on the moon?

8. The earth's attractive force acts quite differently on objects in the interior of the earth than on objects outside the earth. In the former case the force is given by the formula $F = GmMr/R^3$, where m is the mass of the object, M the mass of the earth, R the radius of the earth, and r the distance of the object from the center of the earth. Compare the variation of this attractive force as r varies, with the force given by formula (1).

9. Suppose that the law of gravitation were $F = GmM/r$ instead of formula (1). Compare the variation of weight with distance from the center of the earth according to this formula with the variation of weight according to (1).

10. Consider all objects at a distance of 5000 mi from the center of the earth. Is the ratio of weight to mass the same for all these objects?

17-6 Some deductions from the law of gravitation. The essence of the scientific method created by Galileo and Newton is to establish basic quantitative physical principles and to apply mathematical reasoning to these principles. The law of gravitation and the first and second laws of motion are such physical principles. We shall see now that Newton was able to make some remarkable deductions from these principles.

The law of gravitation contains the constant G. Many calculations based on the law of gravitation require that one know G. In principle, this quantity is easily measured. One has but to take two known masses, place them a measured distance apart, and measure the force with which the two masses attract each other. Then, since

$$F = G\frac{mM}{r^2}, \tag{12}$$

we see that every quantity in (12) is given except G, so that we have a simple algebraic equation for G. The actual experiments which have been made to measure G are a little more complicated because the force F is small for ordinary masses. However, the experiments have been performed, and the value of G turns out to be $1.07/10^9$. The notation 10^9 is scientific shorthand for the product in which 10 occurs as a factor 9 times, i.e., one billion. This value of G presupposes that masses are measured in pounds, distances in feet, and forces in poundals (practical English system). (In the centimer-gram-second (cgs) system of units, G is $6.67/10^8$.)

With the value of G known, it is a simple matter to calculate the mass of the earth. We may recall that formula (5), which is an immediate consequence of the law of gravitation and the second law of motion, states that the acceleration which the earth imparts to any other mass is

$$a = G\frac{M}{r^2}, \tag{13}$$

where r is the distance between the two masses. We have also learned that when r is 4000 miles or 21,120,000 feet, then $a = 32$. Let us substitute these values and the value of G in (13). Then

$$32 = \frac{1.07}{10^9} \cdot \frac{M}{(21{,}120{,}000)^2}, \tag{14}$$

and we have obtained a simple equation for the unknown M. To shorten the somewhat complicated arithmetic, let us approximate and write 21,120,000 as 21,000,000 or as $21 \cdot 10^6$. Then

$$(21 \cdot 10^6)^2 = (21)^2 \cdot (10^6)^2 = 441 \cdot 10^{12}.$$

Substituting the value just obtained in (14) yields

$$32 = \frac{1.07}{10^9} \cdot \frac{M}{441 \cdot 10^{12}}.$$

The factors 10^9 and 10^{12} can be combined, for the first factor means $10 \cdot 10 \cdot 10 \cdots$, wherein 10 occurs 9 times, and the second factor means that 10 occurs 12 times. Then in the product of these two factors 10 occurs 21 times. Hence

$$32 = \frac{1.07M}{441 \cdot 10^{21}}.$$

Multiplying both sides of this equation by $441 \cdot 10^{21}$ and dividing both sides by 1.07, we obtain

$$M = \frac{32 \cdot 441 \cdot 10^{21}}{1.07}$$

or

$$M = 13.1 \cdot 10^{24} \text{ pounds.}$$

Since there are 2000 pounds in one ton, we may divide the right side by 2000 and write

$$M = 6.5 \cdot 10^{21} \text{ tons.} \tag{15}$$

Hence some simple algebra applied to formula (13) was all that was needed to calculate the mass of the earth. Let us note clearly that this quantity is *not* the weight of the earth. Technically the earth has no weight since weight is, by definition, the force which the earth exerts on *other* masses. However, a mass of $6.5 \cdot 10^{21}$ tons would weigh the same amount of tons, and so one can get some idea of the earth's mass.

From the knowledge just obtained we can deduce some information about the interior of the earth. The earth is approximately spherical in shape, and since the volume, V, of a sphere is $4\pi r^3/3$, where r is the radius, we can compute the volume of the earth. Thus

$$V = \tfrac{4}{3}\pi(4000 \cdot 5280)^3 = \tfrac{4}{3}\pi(4 \cdot 528)^3(10^4)^3 = \tfrac{4}{3}\pi(2112)^3 \cdot 10^{12}.$$

We shall approximate 2112 by $21 \cdot 10^2$ and use the value of 3.14 for π. Then

$$V = \tfrac{4}{3}(3.14)(21 \cdot 10^2)^3 \cdot 10^{12} = \tfrac{4}{3}(3.14)(21)^3 \cdot 10^6 \cdot 10^{12}.$$

Since $(4/3)(3.14)(21)^3$ is about 39,000, we have

$$V = 39{,}000 \cdot 10^6 \cdot 10^{12} = 39 \cdot 10^{21} = 3.9 \cdot 10^{22} \text{ cubic feet.}$$

We next divide the mass of the earth, in pounds, by the volume to find the mass per cubic foot. Thus

$$\frac{M}{V} = \frac{13.1 \cdot 10^{24}}{3.9 \cdot 10^{22}} = \frac{1310 \cdot 10^{22}}{3.9 \cdot 10^{22}} = \frac{1310}{3.9} = 336. \tag{16}$$

The mass per cubic foot of water is 62.5 pounds. We see then that the mass per cubic foot of earth is about 5.5 times the mass per cubic foot of water. This figure of 5.5, incidentally, is the *density* of the earth.

Examination of the earth's surface shows that it consists mostly of water and sand. Since the quantity of rock visible on the surface does not account for the ratio 5.5, the conclusion follows that the interior of the earth must contain heavy minerals.

Only a little more work is required to compute the mass of the sun. We shall again begin with the law of gravitation and the second law of motion. The two masses involved now are the mass of the sun, S, and the mass of the earth, E. Then the law of gravitation states that the force with which the sun attracts the earth is

$$F = G\frac{SE}{r^2}, \tag{17}$$

where r is the distance from the earth to the sun. According to Newton's

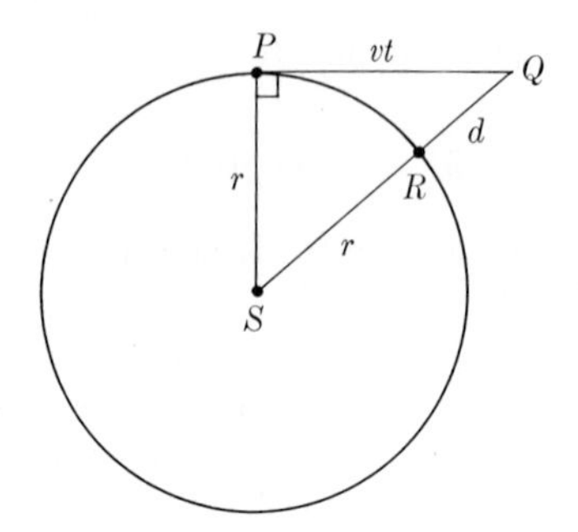

FIG. 17–3. The sun's pull on the earth causes the earth to "fall" the distance QR in t seconds.

second law of motion, the force which the sun exerts on the earth gives the earth an acceleration a such that

$$F = Ea. \tag{18}$$

Since the forces in (17) and (18) are the same, we may equate the right sides. Then

$$Ea = G\frac{SE}{r^2}\cdot \tag{19}$$

We may next divide both sides of (19) by E and obtain

$$a = \frac{GS}{r^2}\cdot \tag{20}$$

In this last equation we know G and r. If we knew a, the acceleration of the earth, we could calculate S. Let us see what we can do about calculating a.

The acceleration which the sun imparts to the earth causes the earth to depart from a straight-line path, which it might otherwise pursue, and "fall" toward the sun just enough to keep it on its elliptical path. (The acceleration which the earth imparts to the moon has the same effect on the lunar orbit.) We shall suppose, for the sake of simplicity, that the path of the earth is circular. Let us imagine that the earth is at the point P (Fig. 17–3) in its path around the sun. If there were no gravitational force, the earth would shoot straight out along the tangent at P into space in accordance with the first law of motion. Let us suppose that in time t, the earth would have reached the point Q. The distance traveled would be the velocity of the earth in its path around the sun, v say, multiplied by t. Hence $PQ = vt$. However, during that time t, the sun pulls the earth in a distance QR or d. Since SPQ is a right triangle,

$$(r + d)^2 = r^2 + (vt)^2.$$

Squaring $r + d$ and substituting the result, we obtain

$$r^2 + 2\,dr + d^2 = r^2 + v^2t^2.$$

We subtract r^2 from both sides of this equation and find that

$$2\,dr + d^2 = v^2t^2.$$

Applying the distributive axiom on the left side permits us to write

$$2\,d\left(r + \frac{d}{2}\right) = v^2t^2. \tag{21}$$

Now d is the distance that the earth falls in time t. Let us suppose that it falls with constant acceleration. (We shall soon let t become very small so that the acceleration can well be taken as constant.) If a body falls a distance d with constant acceleration a, then we know from our work in Chapter 15 that

$$d = \tfrac{1}{2}at^2$$

or

$$2d = at^2. \tag{22}$$

Let us substitute this value of $2d$ in (21). Then

$$at^2\left(r + \frac{d}{2}\right) = v^2t^2.$$

We now divide both sides of this equation by t^2 and obtain

$$a\left(r + \frac{d}{2}\right) = v^2. \tag{23}$$

Thus far t was arbitrarily chosen, and d was the distance the earth fell toward the sun in time t. Our result so far, then, is valid for any value of t. If we now let t become smaller and smaller, d will also decrease. When $t = 0$, it follows from (22) that $d = 0$. In this case, (23) becomes

$$ar = v^2$$

or

$$a = \frac{v^2}{r}. \tag{24}$$

This result states that the acceleration which the sun imparts to the earth at each point P of the earth's path is the square of the earth's velocity divided by the distance of the earth from the sun. This acceleration is called *centripetal* (i.e., center-seeking) acceleration, because it causes the earth to move toward the center of its path.

We now have the quantity a which we needed in (20). Substitution of (24) in (20) yields

$$\frac{v^2}{r} = \frac{GS}{r^2}.$$

We may multiply both sides by r and divide both sides by G to obtain

$$S = \frac{v^2 r}{G}. \tag{25}$$

Every term on the right side of this equation is known. The distance r is 93,000,000 miles or $4.9 \cdot 10^{11}$ feet. The velocity, v, of the earth is the circumference of the earth's path divided by the number of seconds in one year:

$$v = \frac{2\pi 4.9 \cdot 10^{11}}{365 \cdot 24 \cdot 60 \cdot 60} = \frac{30.8 \cdot 10^{11}}{3.15 \cdot 10^7} = 9.8 \cdot 10^4 \text{ ft/sec.}$$

Hence

$$v^2 = (9.8 \cdot 10^4)^2 = (9.8)^2 \cdot 10^8 = 96 \cdot 10^8. \tag{26}$$

In Section 17–6 we learned that $G = 1.07/10^9$. Thus, using these values of r, v^2, and G in (25), we have

$$S = \frac{96 \cdot 10^8 \cdot 4.9 \cdot 10^{11}}{1.07/10^9} = \frac{96 \cdot 10^8 \cdot 4.9 \cdot 10^{11} \cdot 10^9}{1.07}$$

or

$$S = 440 \cdot 10^{28} = 4.40 \cdot 10^{30}. \tag{27}$$

Hence the mass of the sun is $4.40 \cdot 10^{30}$ pounds. Since the earth's mass was previously found to be $1.31 \cdot 10^{25}$ pounds, we see that the mass of the sun is $3.36 \cdot 10^5$ or 336,000 times the mass of the earth.

We can determine the mass per cubic foot of the sun in the same manner as we calculated the mass per cubic foot of the earth. The mass of the sun is now known, and the radius, computed in Chapter 7, is 432,000 miles, or $2.28 \cdot 10^9$ feet. We shall not reproduce the calculations, but state the result: the mass per cubic foot proves to be 90 pounds. Since a cubic foot of water has a mass of 62.5 pounds, we see that the mass per cubic foot of the sun is about $1\frac{1}{2}$ that of water; that is, the density of the sun is about $1\frac{1}{2}$.

The examples given in this section further illustrate how mathematical reasoning can be applied to physical laws (in our case, to the second law of motion and the law of gravitation) in order to deduce fundamental knowledge about the universe. We did, of course, also use some experimentally obtained facts such as the value of G and the acceleration of bodies near the earth's surface. However, mathematics has been the main tool, and it obtains for us such remarkable information as the mass of the earth and the mass of the sun.

Exercises

1. Suppose an object moves in a circle at a constant speed. Is the motion subject to an acceleration?

2. Does the formula for the acceleration of the earth given in (24) depend upon the law of gravitation?

3. If you whirl an object on a string of radius 5 ft, at the rate of 50 ft/sec, what is the centripetal acceleration acting on the object? What force exerts this centripetal acceleration?

4. Use formula (24) with the understanding that v is the velocity of the moon and r is the distance of the moon from the earth, to calculate the acceleration of the moon. (The period of the moon's path around the earth is $27\frac{1}{3}$ days, and the distance of the moon from the earth is 240,000 mi.) Does your result agree with that given in (8)?

5. The result given in (8) does depend upon the law of gravitation. The result obtained in Exercise 4 does not. Hence, have you verified the law of gravitation?

6. Using the figures in the text for the mass and radius of the sun, calculate the ratio of the mass to the volume of the sun.

17–7 The rotation of the earth. We have repeatedly used the quantity 32 ft/sec^2 as the acceleration which the earth gives to objects near its surface. This figure is perfectly satisfactory for most purposes, but it is not strictly accurate even for motions near the earth's surface. Actually, the acceleration of falling bodies decreases from 32.257 ft/sec^2 at either pole to 32.089 ft/sec^2 at the equator.* The discovery of this decrease was at first not surprising to the seventeenth-century scientists. Newton had already proved that the earth is not strictly spherical, but has the shape of a somewhat flattened sphere (Fig. 17–4); that is, for example, the lengths OA, OB, OC, and OD are not equal, but are successively larger. Since the general formula for the acceleration due to gravity [see (5)] is GM/r^2, where G and M are fixed and r is the distance from the center of the earth, this acceleration is less at C, say, than at B because r is larger at C than at B. Hence we should expect the acceleration due to gravity to decrease as the location varies from A to D. Now the values of G and M were known. Moreover, Newton and Huygens had computed lengths such as OA, OB, and so forth, and therefore were able to determine what the acceleration should be at points such as A, B, C, and D. These calculations, based on the expression GM/r^2, call for only a small percentage of the decrease actually measured. Thus precise measurement revealed a discrepancy between the acceleration predicted by the law of gravitation and the actual acceleration of falling bodies. This discrepancy required explanation.

The problem was solved by Huygens. Objects on the surface of the earth would fly off into space if the earth did not pull them toward the center, just as an object whirled at the end of a string would fly off into

* The numerical values can, in principle, be obtained by measuring the accelerations with which bodies near the surface fall to earth. However, a more accurate method utilizes the formula for the period of a pendulum.

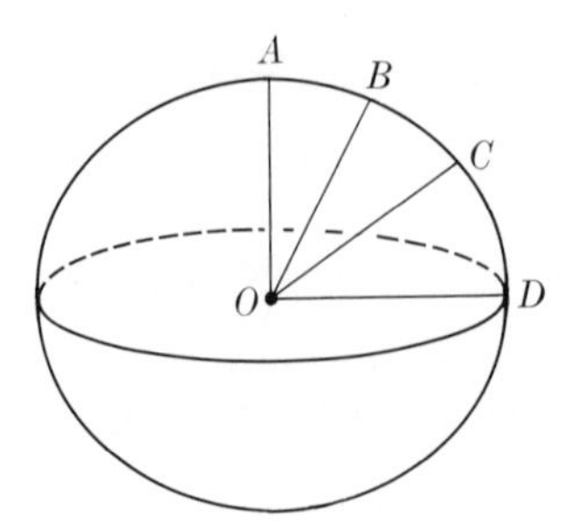

FIG. 17–4. The spheroidal shape of the earth.

space if the hand at the center did not exert an inward pull. Thus the earth's gravitational force has two effects. Even if the earth did not rotate, it would pull all objects toward the center, simply because the earth's mass attracts the object. But since the earth does rotate, it must also exert an inward pull so that objects do not fly off into space but remain on or near the surface of the earth. This latter effect is a centripetal force. In a sense, the two effects of the earth's gravitational force, that is, the force which causes objects to fall to the surface, or weight, and the centripetal force, are of the same nature. The centripetal force also pulls objects toward the earth's center, but pulls them in just enough to keep them on a circular course. The weight, on the other hand, pulls objects toward the earth from the circular course to which they are kept by the centripetal force.

Let us express quantitatively what we have just described. By Newton's second law, the centripetal force must produce an acceleration (centripetal acceleration) on the object. Now formula (24) gives the centripetal acceleration which the sun exerts on the earth. However, this formula is really quite general, that is, if we replace sun and earth in the argument which led to (24) by the earth and an object on or near the earth's surface, then the argument still holds, provided that v is the velocity of the object and r is its distance from the center of the earth. Newton's second law then tells us that the centripetal force must be the mass of the object times the centripetal acceleration, that is mv^2/r.

The force with which the earth pulls an object straight down, i.e., the weight, equals the mass of the object times the acceleration of its fall. It is this acceleration which we measure when we observe the fall of objects and which varies from pole to equator. We shall now denote it by g. Then the weight is mg.

According to Huygens the gravitational force which the earth exerts on objects supplies both the centripetal force and the weight. In symbols,

$$\frac{GMm}{r^2} = \frac{mv^2}{r} + mg. \tag{28}$$

But why does g decrease from either pole to the equator? The answer is that g decreases because the centripetal force varies. Any object on or near the surface of the earth and rotating with the earth makes a complete rotation in one day. But the velocity of the object depends upon its latitude. Thus, an object on the equator travels about 25,000 miles in one day, whereas an object at 60°- latitude will have to cover only half this distance to complete one rotation. Hence the object on the equator moves with greater velocity, and more centripetal force is required to keep it on the earth. With this situation in mind let us re-examine (28). Since, as we have already noted, the slight change in r from one point to another on the earth's surface has very little effect, we may neglect it and agree that the earth exerts the same gravitational force, namely GmM/r^2, everywhere on the earth's surface. If it must exert more centripetal force at the equator to keep an object on the surface, then mg must be smaller there. Thus g must be smaller at the equator than at, say latitude 60°.

Almost the full decrease in g, as it decreases from either pole to equator, is then due to *the rotation of the earth.* Or we can say that, to explain the measured variation of g, we must assume that the earth rotates. Thus we have obtained a new argument for the rotation of the earth.

The numerical value of g, that is the acceleration of falling bodies, has of course been of importance for centuries. Any scientific work which involves the action of gravity must be founded on good quantitative knowledge of gravitational attraction. But it has additional importance today. Let us consider a satellite which circles the earth once every hour. Although the satellite moves at a height of a few hundred miles above the surface of the earth, we shall ignore this distance and suppose that it travels very near the surface. What is significant is that the satellite covers 25,000 miles per hour. Hence the centripetal force required to keep it in its path is considerably greater than that required to keep an object which travels 25,000 miles in 24 hours from flying off into space. We see this fact from the middle term in (28) which tells us that the centripetal force increases with the square of the velocity. Thus a great deal of the earth's gravitational force must be expended in centripetal force. In fact, since the satellite does not fall to earth, the value of g, that is, the acceleration with which it should fall to earth, must be zero. In other words, the full gravitational force of the earth is expended in keeping the satellite on its circular (strictly elliptical) path around the earth, and the satellite neither flies off into space nor falls to earth.

But the weight of any object is the product of its mass and the acceleration, g, with which gravity makes it fall to earth. The weight, then, for objects near the surface is

$$W = gm. \tag{29}$$

Since for the satellite, $g = 0$, it follows that the satellite has no weight. Objects contained in the satellite would also be weightless and so would not experience any earthward pull.

In view of the importance which satellites are likely to have in future scientific investigations, it is desirable to know the velocity which a satellite must possess if it is to stay in orbit at some desired distance from the center of the earth. This velocity is readily calculated from (28). Since the satellite does not fall to earth, the value of g for it must be zero. Then

$$\frac{GMm}{r^2} = \frac{mv^2}{r}.$$

If we divide both sides by m and multiply both sides by r, we obtain

$$v^2 = \frac{GM}{r}. \tag{30}$$

We know G and M, the mass of the earth. When r, the distance of the satellite from the center of the earth, is chosen, then we know all the quantities on the right side of (30). The quantity GM can be calculated once and for all. Thus

$$GM = \frac{1.07}{10^9}(13.1)10^{24} = 14 \cdot 10^{15}.$$

The value of r must be in feet. We can now calculate v.

Exercises

1. Since the weight of an object is mg, how does a person's weight change as he travels from the North Pole to the equator?

2. Suppose that a satellite stays close to the earth's surface. How fast would it have to travel to stay on its circular path and not fall to earth? [*Suggestion:* Use formula (30).]

3. The moon is a satellite of the earth. Since the moon stays on its path and does not fall to the earth, we may conclude that the earth's entire gravitational force acts as centripetal force on the moon. Using the assumption that the moon's path is a circle and that it is 240,000 mi from the earth, calculate the velocity of the moon. [*Suggestion:* Use (30).]

4. Using the result of Exercise 3, calculate the time it takes the moon to make one complete revolution around the earth.

5. Calculate the speed required to maintain a satellite in an orbit 500 mi above the surface of the earth.

17–8 Gravitation and the Keplerian laws. Thus far in this chapter we examined the evidence which convinced Newton that the law of gravitation was correct, and we have seen how it can be applied to answer a variety of questions about objects and motions on the earth and in the heavens. We should now recall that one of the major problems challenging seventeenth-century scientists was the question whether the same physical principles could account for terrestrial and celestial motions. Since the law of gravitation when applied to bodies falling near the earth's surface reduces to fall with constant acceleration (see Section 17–4), Newton's principles certainly encompassed earthly motions. As to heavenly motions, the three famous laws of Kepler, which he had inferred from observations, were seemingly independent of the law of gravitation. The truly great triumph of Newton was his demonstration that all three Keplerian laws were mathematical consequences of the law of gravitation and the two laws of motion.

We shall illustrate what Newton did by showing how the third Keplerian law can be deduced from the basic laws just mentioned. However, we shall simplify Newton's work and suppose that the path of a planet around the sun is circular, whereas the true path, as Kepler proved, is an ellipse.

Let m be the mass of any planet, M the mass of the sun, and r, the distance between them. Then the law of gravitation says that the force F exerted by the sun on the planet is

$$F = \frac{GmM}{r^2}. \tag{31}$$

We also know that the sun's force causes any planet to depart from straight-line motion and "fall" toward the sun with some acceleration. This acceleration, a, is none other than the centripetal acceleration given by formula (24), that is v^2/r. The derivation of (24) dealt with the sun and earth, but it applies to any planet, provided that v is the velocity of the planet and r is its distance from the sun. We may also assert, by the second law of motion, that the centripetal force F with which the sun attracts that planet is

$$F = m\frac{v^2}{r}. \tag{32}$$

The velocity v of any planet is the circumference of its path divided by the time T of revolution around the sun; that is, $v = 2\pi r/T$. Hence, from (32),

$$F = \frac{m}{r}\cdot\frac{4\pi^2 r^2}{T^2} = \frac{4\pi^2 rm}{T^2}. \tag{33}$$

Now formulas (31) and (33) yield two different expressions for the force with which the sun attracts any one planet.* Hence we may equate these two expressions and obtain

$$\frac{GmM}{r^2} = \frac{4\pi^2 rm}{T^2}.$$

Dividing both sides of this equation by m eliminates that quantity. Multiplying both sides by T^2, we obtain

$$\frac{GMT^2}{r^2} = 4\pi^2 r.$$

* In the light of Section 17–7 we can say that the gravitational force equals the centripetal force because the sun does not cause any planet to fall toward the sun *from the circular path.*

If we now multiply both sides of this last equation by r^2/GM, we find

$$T^2 = \frac{4\pi^2}{GM} r^3. \tag{34}$$

The quantity $4\pi^2/GM$ is the same no matter what planet is being considered, because G is a constant, M is the mass of the sun, and $4\pi^2$ is a constant. Hence formula (34) says that T^2 is the product of some constant, say K, and r^3; in symbols,

$$T^2 = Kr^3. \tag{35}$$

Thus the square of the time of revolution of any planet is a constant (i.e., the same for all planets) times the cube of that planet's distance from the sun. Formula (35) is, then, Kepler's third law of planetary motion. We have derived it from the two laws of motion and the law of gravitation by a purely mathematical argument.

As we remarked earlier, Newton demonstrated that all three of Kepler's laws, which the latter had obtained only after years of observation and trial and error, were mathematical consequences of the laws of motion and gravitation. Hence the laws of planetary motion, which prior to Newton's work seemed to have no relationship to earthly motions, were shown to follow from the same basic principles as did the laws of earthly motions. In this sense, Newton "explained" the laws of planetary motion. These facts were as much a consequence of basic physical laws as the straight-line motion of objects falling to earth from rest or of projectiles following parabolic paths. Newton's original conjecture that the parabolic motion of projectiles should be intimately related to the elliptical motion of the planets was gloriously established. Further, since the Keplerian laws agree with observations, their derivation from the law of gravitation constituted superb evidence for the correctness of that law.

The few deductions from the laws of motion and gravitation which we have presented are just a sample of what Newton and his colleagues were able to accomplish. Newton applied the law of gravitation to explain a phenomenon which heretofore had not been understood, namely the tides in the oceans. He showed that these were due to the gravitational forces exerted by the moon and, to a lesser extent, the sun on large bodies of water. From data collected on the height of lunar tides, that is, tides due to the moon, Newton calculated the moon's mass. Newton and Huygens calculated the bulge of the earth around the equator. Newton and others showed that the paths of comets are in conformity with the law of gravitation. Hence the comets, too, were recognized as lawful members of our solar system and ceased to be viewed as accidental occurrences or visitations from God intended to wreak destruction upon us. Newton then showed that the attraction of the moon and the sun on the earth's equatorial bulge cause the axis of the earth to describe a cone over a period of 26,000 years instead of always pointing to the same star in the sky. This motion of the earth's axis causes a slight change each year in the time of the spring and fall equinoxes, a fact which had been observed by Hipparchus 1800 years earlier. Thus Newton explained the precession of the equinoxes.

Finally Newton solved a number of problems involving the motion of the moon. The plane in which the moon moves is inclined somewhat to the plane in which the earth moves. He was able to show that this phenomenon follows from the interaction of the sun, earth, and moon under the law of gravitation. As the moon travels around the earth, it cuts the plane of the earth's motion around the sun. The points in which it intersects are called the nodes. The nodes change in position, and this variation (regression of the nodes) also proved to be a consequence of the gravitational effect of the sun and earth on the moon. As the moon moves around the earth in an almost elliptical path, the point farthest from the earth, called the apogee, shifts about 2° per revolution. This effect, Newton showed, was due to the sun's attraction. Newton and his immediate successors deduced so many and such weighty consequences about the motions of the planets, the comets, the moon, and the sea, that their accomplishments were viewed as "the explication of the System of the World."

Actually Newton's work was but the beginning of a century of superb achievements. As observational astronomy improved during the eighteenth century, the departures of the planets from purely elliptical paths became noticeable and measurable. The deviation in itself was not surprising, since a planet does not move in an elliptical orbit. An elliptical path is merely an approximation which would be exact if the planet and the sun were the only bodies in the sky. There are, of course, many planets, and several of these have satellites. All of these bodies attract one another in accordance with the law of gravitation. Hence the motion of any one body is influenced by all the others. One question of particular importance to mathematicians was whether the observed deviations from elliptical paths could be predicted from the theory. In fact, the correctness of the law of gravitation was at stake, for a theory which does not lead to results in agreement with observations is at best inadequate. Mathematicians and astronomers also had other reasons to be worried. Observations showed that Jupiter was slowly losing speed and Saturn was slowly gaining speed. According to Kepler's third law, Jupiter should therefore gradually move farther away from the sun and Saturn should approach it. These facts portended bad fates for Jupiter and Saturn, but also raised the spectre of whether the same might not be happening to other planets. The most significant question was, Is the solar system stable?

We cannot undertake here a detailed account of the brilliant work performed by the leading mathematicians of the eighteenth century. The Swiss Leonhard Euler (1707–83), an amazingly prolific mathematician, the Italian Joseph Louis Lagrange (1736–1813), and the Frenchman Pierre Simon Laplace (1749–1827) showed that all observed departures of the planets from purely elliptical motions could be calculated and that the results agreed with observations. In particular, they demonstrated that the various irregularities in the motions of the planets, such as the slowing down of Jupiter and the speeding up of Saturn were periodic irregularities; that is, the irregularities would not become progressively larger and disrupt the solar system, but reach a maximum, decrease to zero, and then repeat their former behavior. Their grand conclusion was that *all* irregularities were periodic; the motions merely departed slightly from fixed paths, first in one direction and then in the other. The theory of gravitation proved, in other words, to be marvelously adequate for all the happenings and observations which demanded explanation. As Laplace said of the theory, "Such has been the fate of this brilliant discovery that each difficulty which has arisen has become for it a new subject of triumph—a circumstance which is the surest characteristic of the true system of nature."

Today we have almost daily evidence that Newton had found sound physical principles which govern the operation of the universe. By applying just those principles man can now create satellites which circle the earth. In fact, Newton's suggestion that projectiles shot out horizontally and with large velocities from the top of a mountain would circle the earth is, in essence, the one used to launch satellites. Strictly speaking, scientists do not operate from mountain tops because accessible peaks are not high enough to ensure that the satellite will clear other mountains, and because the air resistance at such altitudes is still considerable. Instead rockets project the satellite upward to a high altitude where the air resistance is negligible; there a mechanism turns the satellite to a horizontal direction and another rocket gives it a horizontal velocity. Then the satellite follows an elliptical path.

Newton went further in his speculations and conjectured that the planets must have been shot from the sun at some angle and, upon reaching their present distances, must have retained enough "horizontal" velocity to start moving in their elliptical paths around the sun. This conjecture is still the accepted theory of the origin of our solar system.

Exercises

1. What reason would there be for calling Newton's law of gravitation a universal law?
2. In what sense did Newton incorporate the Keplerian laws in his science of motion?
3. What support did the heliocentric theory receive from Newton's work on gravitation?
4. What support did Newton's principles derive from the heliocentric theory?

17–9 Implications of the theory of gravitation. The work on gravitation presented mankind with a new world order, a universe controlled throughout by a few universal mathematical laws which in turn were derived from a common set of mathematically expressible physical principles. Here was a majestic scheme which embraced the fall of a stone, the tides of the oceans, the moon, the planets, the comets which seemed to sweep defiantly through the orderly system of planets, and the most distant stars. This view of the universe came to a world seeking to secure a new approach to truth and a body of sound truths which were to replace the already discredited doctrines of medieval culture. Thus it was bound to give rise to revolutionary systems of thought in almost all intellectual spheres. And it did. We shall examine the new culture, for it was a new culture, in a series of chapters, but, for the moment, we wish to confine ourselves to the implications and consequences of the theory of gravitation for mathematics proper.

Newton's work followed and considerably broadened the plan laid down by Galileo, who proposed to find basic quantitative physical principles and to deduce from them the description of physical phenomena. Galileo had discovered and utilized such axioms as the first law of motion, the constant acceleration of bodies moving near the surface of the earth, and the independence of the horizontal and vertical motions of projectiles. His results were confined to terrestrial motions. Newton added to the axioms the second law of motion and replaced the principle of constant acceleration of falling bodies by the more general law of gravitation. He then found that the resulting set of principles enabled him to deduce the description of all motions of matter on earth and in the heavens. Thus the scientific method of Galileo and Newton involves mathematics not only in the expression of axioms and the laws which are deduced but also in the deductive process itself. Indeed, mathematics offered not merely the vehicle for scientific expression but the most powerful tool for the real work of science, that is the acquisition of knowledge about the physical world and the organization of that knowledge in coherent systems. From the time of Newton, these roles of mathematics have been unquestionably accepted and utilized. Hence, as the success of Newtonian mechanics spurred efforts in other physical domains, mathematics was confronted with new challenges and received new suggestions for the creation of concepts and methods which in turn gave greater power to science. This interaction of mathematics and science has grown immensely since its beginning in the seventeenth century and has become the outstanding feature of the intellectual life of our own century.

The work of Newton secured the triumph of the heliocentric theory and thereby enhanced the importance of mathematics. Copernicus' and Kepler's defense of heliocentrism on the ground that it gave a simpler mathematical account of the heavenly motions seemed to place too much importance on the mathematical structure. Their theory could have been regarded by everyone and had been regarded by many as a mathematical contrivance convenient for the calculations of the paths of the moon and planets but not physically true. Moreover, in addition to being isolated from the main body of scientific knowledge, the heliocentric theory created difficulties in accounting for the phenomena of motion readily observed here on earth (Chapter 12), and hence encountered legitimate objections.

But the introduction of the law of gravitation resolved all these difficulties and incorporated the theory of heavenly motions into the very same physical theory which treated terrestrial motions. As a result, the heliocentric theory acquired incontestable strength. It was now proved beyond any doubt that the other planets were *not* different from the planet earth and that the substance of the other planets could be identified with the rock and clay beneath man's feet, for this is the very essence of the law of gravitation. Earthly and heavenly motions were bound together in one theory, and one could no longer doubt the heliocentric view without doubting the entire structure. Hence the importance which Copernicus and Kepler ascribed to the mathematical element in the theory was vindicated.

However, the most surprising development of the theory of gravitation and one which established a new and unanticipated role for mathematics took place after Newton had deduced a number of conclusions about our solar system. Galileo and Newton had set about finding quantitative laws that related matter, space, time, forces, and other physical properties, but had wisely decided not to look into causal relationships; that is, they had deliberately avoided such questions as why bodies fall to earth or why planets move around the sun. In other words, they had concentrated on description. Nevertheless, they did utilize the force of gravitation, a concept which had been vaguely suggested even before Galileo's time—for example, by Copernicus and Kepler. Since the force of gravitation now assumed central importance, it was natural to ask, What is the mechanism that enables the earth to attract objects and the sun to attract planets? The heightened emphasis on this universal force could not but push such questions to the fore. The properties ascribed to the force of gravitation were indeed remarkable. It acted over distances of inches and millions of miles. It acted instantaneously and through empty space. Nor could the action of the force be suspended or blocked. Even when the moon was between the earth and the sun, the sun continued to attract the earth.

Kepler had considered this question of how the sun could exert its attractive force over so many millions of miles. Impressed by the phenomenon of magnetism which William Gilbert had made popular through a series of famous experiments, he tried to explain gravitational attraction as the action of a magnetic force. He thought that planets were huge magnets attracted by a magnetic force in the sun. But he failed to supply a quantitative expression for this force and to show that it accounted exactly for the paths of the planets. Seventy-five years later, Newton demonstrated that the magnetic force exerted by one magnet upon another did not vary inversely with the square of the distance between the magnets and thus proved that the gravitational force could not be magnetic. However, although he tried to provide some physical explanation for the action of gravity, he did not succeed, and he concluded, "I have not been able to deduce from phenomena the cause of the properties of gravity and I frame no hypotheses."

In spite of his ignorance of the workings of gravitation, Newton insisted on adopting the laws of motion and gravitation. He says, "But to derive two or three general principles of motion from phenomena, and afterwards to tell us how the properties and actions of all corporeal things follow from those manifest principles, would be a very great thing *though the causes of those principles were not yet discovered:* and therefore I scruple not to propose the principles of motion above mentioned, they being of very general extent, and leave their causes to be found out." Concerning his work in his *Principles,* he says, "But our purpose is only to trace out the quantity and properties of this force [gravitation] from the phenomena, and to apply what we discover in some simple cases as principles, by which, in a mathematical way, we may estimate the effects thereof in more involved cases; for it would be endless and impossible to bring every particular to direct and immediate observation. We said, *in a mathematical way* [italics are Newton's], to avoid all questions about the nature or quality of this force, which we would not be understood to determine by any hypothesis; . . ."

Newton was indeed troubled that he could give no explanation. In a letter to the theologian William Bentley, he says, "That gravity should be innate, inherent and essential to matter, so that one body may act upon another at a distance, through a vacuum, without the mediation of anything else by and through which their action may be conveyed from one to another, is to me so great an absurdity that I believe no man, who has in philosophical matters a competent faculty of thinking, can ever fall into it." But all he could do to justify the introduction of this force is summed up at the end of his *Principles,* "And to us it is enough that gravity does really exist, and act according to the laws we have explained, and abundantly serves to account for all the motions of the celestial bodies, and of our sea."

Newton's contemporaries were alarmed by the introduction of this occult force into physics. Huygens did not hesitate to say that Newton's principle of attraction appeared to him incredible. Gottfried Leibniz, Newton's great contemporary, called gravitation an incorporeal and inexplicable power, philosophically false and even absurd. The foremost mathematician, John Bernoulli, denounced "the two suppositions of an attractive faculty and a perfect void" as "revolting to minds unaccustomed to receiving no principle in physics save those which are incontestable and evident." Bishop George Berkeley, a famous theologian and philosopher, questioned, "Whether the mathematicians of the present age act like men of science in taking so much more pains to apply their principles than to understand them." All these men and others found the concept of a force which acted through a vacuum over millions of miles more difficult to understand than the most outlandish superstitions.

To account for the action of gravity, scientists began to use one of Newton's phrases, "action at a distance," although Newton had not used these words as an explanation. Gradually, after many repetitions, the phrase was uncritically accepted as though it were an explanation. By the nineteenth century what had previously been regarded as an "uncommon unintelligibility" became accepted as a "common unintelligibility." No one was any longer disturbed by the concept of gravitation, and it was universally regarded as a true force.

Contrary to popular belief, no one ever discovered gravitation, for the physical reality of this force has never been demonstrated. However, the mathematical deductions from the quantitative law proved so effective that the phenomenon has been accepted as an integral part of physical science. What science has done, then, in effect is to sacrifice physical intelligibility for the sake of mathematical description and mathematical prediction. This basic concept of physical science is a complete mystery, and all we know about it is a mathematical law describing the action of a force *as though it were real.* We see therefore that the best knowledge we have of a fundamental and universal phenomenon is a mathematical law and its consequences. And it has become more and more true since Newton's days that our best knowledge of the physical world is mathematical knowledge.

Recommended Reading

BELL, E. T.: *Men of Mathematics,* Chaps. 6, 9, 10, and 11, Simon and Schuster, New York, 1937.

BONNER, FRANCIS T. and MELBA PHILLIPS: *Principles of Physical Science,* Chap. 4, Addison-Wesley Publishing Co., Inc., Reading, Mass., 1957.

BURTT, E. A.: *The Metaphysical Foundations of Modern Physical Science,* rev. ed., pp. 202–262, Routledge and Kegan Paul Ltd., London, 1932.

BUTTERFIELD, HERBERT: *The Origins of Modern Science,* Chap. 8, The Macmillan Co., New York, 1951.

COHEN, I. BERNARD: *The Birth of a New Physics,* Chap. 7, Doubleday and Co., Anchor Books, New York, 1960.

DAMPIER-WHETHAM, WM. C. D.: *A History of Science and Its Relations with Philosophy and Religion,* pp. 160–195, Cambridge University Press, London, 1929.

HALL, A. R.: *The Scientific Revolution,* Chap. 9, Longmans Green and Co., Inc., New York, 1954.

HISTORY OF SCIENCE SOCIETY: *Sir Isaac Newton,* The Williams and Wilkins Co., Baltimore, 1928.

HOLTON, GERALD and DUANE H. D. ROLLER: *Foundations of Modern Physical Science,* Chaps. 4, 5, 11, and 12, Addison-Wesley Publishing Co., Inc., Reading, Mass., 1958.

JEANS, SIR JAMES: *The Growth of Physical Science,* 2nd ed., Chap. 6, Cambridge University Press, London, 1951.

JONES, SIR HAROLD SPENCER: "John Couch Adams and the Discovery of Neptune," in JAMES R. NEWMAN: *The World of Mathematics,* Vol. II, pp. 820–839, Simon and Schuster, Inc., New York, 1956.

MASON, S. F.: *A History of the Sciences,* Chaps. 17 and 25, Routledge and Kegan Paul Ltd., London, 1953.

MORE, LOUIS T.: *Isaac Newton,* Charles Scribner's Sons, Inc., New York, 1934.

Newman, James R.: *The World of Mathematics,* Vol. I, pp. 254–285, Simon and Schuster, Inc., New York, 1956.

Smith, Preserved: *A History of Modern Culture,* Vol. II, Chap. 2, Henry Holt & Co., New York, 1934.

Sullivan, John Wm. N.: *Isaac Newton,* The Macmillan Co., New York, 1938.

Taylor, Lloyd Wm.: *Physics, The Pioneer Science,* Chaps. 9, 10, and 13, Dover Publications, Inc., New York, 1959.

Wightman, Wm. P. D.: *The Growth of Scientific Ideas,* Chaps. 8, 10 and 11, Yale University Press, New Haven, 1951.

Wolf, Abraham: *A History of Science, Technology and Philosophy in the Sixteenth and Seventeenth Centuries,* 2nd ed., Chap. 7, George Allen and Unwin Ltd., London, 1950. Also in paperback.

CHAPTER 18

THE DIFFERENTIAL CALCULUS

No nature except an extraordinary one could ever easily formulate a theory.

Plato

18–1 Introduction. The mathematical ideas explored in the preceding chapters, arithmetic, algebra, Euclidean geometry, trigonometry, coordinate geometry, and the various types of functions, comprise a considerable amount of mathematics. Of course, the development of each of these ideas is far more extensive than we have indicated or than school courses usually cover. But the seventeenth century, which inspired and initiated the modern scientific movement, provided the problems and suggestions for new branches of mathematics which dwarf in extent, depth, and power the mathematics we have examined thus far. The most significant mathematical creation of that century and the one which proved to be most fruitful for the modern development of mathematics and science is the calculus. Like Euclidean geometry, it is a landmark of human thought.

It is not possible to offer, in only a few words, a clear description of the basic idea of the calculus. We can tentatively say that it is concerned with the instantaneous rate of change of one variable with respect to another as opposed to an average rate of change; for example, for a moving object, it is concerned with the rate of change of distance with respect to time at a given instant of time. This brief description of the basic idea of the calculus is, of course, vague and certainly gives no indication of its power in applications. To attain a fuller comprehension we must examine these matters more carefully and pursue concrete illustrations.

18–2 The problems leading to the calculus. The mathematicians of the seventeenth century who were gradually developing the ideas and processes which now comprise the calculus were beset by several problems. We have seen that the seventeenth century was primarily concerned with the study of motion, the motion of objects on or near the earth and the motion of heavenly bodies. In this study the problem of determining the speed and acceleration of moving bodies is, of course, quite important. Now speed, we usually say, is the rate at which distance changes with time, but if an object moves with varying speed, then to determine its speed, one must compute the rate of change of distance with time at any instant, or its instantaneous speed. The same remarks apply to acceleration. We shall see that the determination of such instantaneous rates presents a new kind of difficulty.

It is true that we did determine and work with speed and acceleration of falling bodies, but we treated simple motions and so circumvented the essential difficulty. The problem is no longer simple when, for example, one seeks the speed and acceleration of a planet moving on an elliptical path.

The converse problem is equally important. Suppose one knows the acceleration of a moving body at each instant of time. How does one find the speed and distance traveled at any instant? When the acceleration is constant, one can multiply the acceleration by the time of travel and obtain the speed acquired, but this procedure does not yield correct results when the acceleration is variable.

Another problem of motion is that of determining the direction in which an object is moving at any instant of its flight. Depending upon its direction, a projectile may make a direct hit on a target or merely strike a glancing blow. Also, the direction in which a projectile is fired determines the horizontal and vertical components of its velocity (see Chapter 16). Hence it is desirable to know the direction in which an object is moving. Generally this direction varies from one instant to another, and therein lies the difficulty.

The third major problem was that of finding the maximum and minimum values of a function. When a bullet is fired straight up, one may wish to know how high it will go. For simple motions near the surface of the earth, we were able to find the maximum height. But the methods used will not suffice to compute, for example, the maximum or minimum distance of a planet from the sun or from another planet. Nor do they suffice to discuss the motion of a rocket which travels sufficiently far up so that one must take into account the variation in the acceleration due to gravity.

The fourth major problem confronting the seventeenth century was that of determining lengths, areas, and volumes. Let us consider, for example, the volume of the earth. The true shape of the earth is that of an oblate spheroid, a sphere flattened somewhat at the top and bottom.* How does one find the volume of such a figure? Or let us consider the motion of a planet along its elliptical path. How does one find the length of the path over which the planet travels in a given period of time? This information is important if one wishes to predict the position of the planet at some future instant of time. One could also ask, What is the total distance traveled by a planet in one complete revolution; in other words, What is the length of a given ellipse?

All these questions and many others that we shall encounter in the present and later chapters bedeviled the mathematicians of the seventeenth century, and hundreds of capable men worked on them. When Newton and Leibniz made their contributions to the calculus, it became clear that all of the above problems and others too could be solved by means of one basic concept, the instantaneous rate of change of one variable with respect to another. Hence we shall begin with this concept.

---

* Recent observations made from satellites indicate that this description is not quite accurate.

18–3 The concept of instantaneous rate of change. There are three closely related ideas: change, average rate of change, and instantaneous rate of change. These three ideas should be carefully distinguished. The concept of change itself is by now a familiar one. When a ball is thrown up into the air, its height above the ground changes. The pursuit of physical problems involving functions soon obliges one to consider not just the mere fact of change but the rate of change of one variable with respect to another. In the case of a ball thrown up into the air, one might wish to know what initial speed will enable the ball to reach a height, say of 100 feet, or what speed the ball has on returning to the ground; that is, information about the speed, which is the rate of change of height with respect to time, is desirable. The statement that the earth travels around the sun in one year is a fact about rate of change rather than about mere change. Our great concern in this age for faster transportation and communication is a concern with rate of change. Circulation of the blood in one's body means quantity of blood per unit time passing through a specific artery or a collection of arteries, and here, too, it is rate of change which counts. The rate of physiological activity, that is the metabolic rate, measured in terms of the rate of consumption of oxygen per second, is a rate of change. To sum up: the rate of change of one variable with respect to another is a physically useful quantity in many situations.

The rates of change which are of interest to laymen and even to many specialists are average rates. Thus, if a motorist travels 500 miles in 10 hours, the average speed, i.e., the distance traveled divided by the time of travel, is 50 miles per hour. This average speed is what usually matters, and in most instances it is quite irrelevant that the driver may occasionally have stopped for food and thus had no speed at all during those periods of the trip. Most people like to increase their wealth and are satisfied if the rate of growth, that is the growth in wealth per month or per year, is appreciable. The increase of a country's population is usually measured per year because this average rate tells the story which is of importance for most purposes.

However, the average rate of change is not the significant quantity in many practical and scientific phenomena. If a person traveling in an automobile strikes a tree, it is not his average rate of speed for the time he has traveled from the starting point to the tree that matters. It is his speed at the instant of collision which determines whether or not he will survive the accident. Here we have an instantaneous speed or instantaneous rate of change of distance with respect to time.

There are two mathematical and physical facts involved in this event which require some elaboration. First of all there is the matter of time. As the person travels, time elapses. Mathematically this time is represented by a variable, t, say, and the values of t increase continually as the trip goes on. If time is measured from the instant at which the man starts out and if he has been traveling for 20 minutes, say, then t varies from 0 to 20. We also speak of the 20 minutes as an interval of time or an amount of time. We have, of course, referred to and used this mathematical representation of time right along. It is important now, however, to recognize that the collision of automobile and tree does not last an interval of time but occurs in what is called an instant. Many other events take place at an instant or are instantaneous. A lightning flash is instantaneous or at least happens so fast that we describe it as happening at an instant. The clock strikes at an instant. A bullet strikes a target at an instant.

The mathematical representation of an instant is simple. Mathematically, we say that when $t = 20$ or some other value, we are dealing with an instant of time; that is, an instant is merely one value of t, whereas an interval is some range of t-values, as, for example, from $t = 0$ to $t = 20$. Just as we have used the notion of an interval of time in past work, so have we used the notion of an instant. For example, we have spoken of the height of a ball at the end of the third second of flight, that is, when $t = 3$.

The second fact which must be clearly understood about the phenomenon of the automobile striking the tree is that the automobile has a speed at the instant of collision. This physical fact is apparent enough, and yet when we pursue the notion, we find that it presents difficulties. There is no difficulty in defining and calculating average speed, which is simply the distance traveled during some *interval* of time divided by that amount of time. But suppose we were to try to carry over this concept to instantaneous speed. The distance the automobile travels in one instant is zero, and the time that elapses during one instant also is zero. Hence the distance divided by the time is 0/0, and this expression is meaningless (Chapter 4). Thus, although instantaneous speed is a physical reality, there seems to be a difficulty in stating precisely what it means, and unless we can do so, we shall not be able to work with it mathematically.

To tackle some of the major problems which scientists and mathematicians have faced since the seventeenth century, it is necessary to solve the problem of calculating instantaneous rates, for some of the most important motions take place at continually changing speeds and accelerations. For example, even the simple motion of a body falling to earth occurs at a continually changing speed. Moreover, we know from our study of the law of gravitation that the acceleration of objects moving at great distances from the earth also varies from instant to instant. In view of the wide range of phenomena which are subject to the law of gravitation, it should be clear that motions with continually changing speeds and accelerations are common. Because these speeds and accelerations are continually changing, that is, are different from one instant to the other, the calculation of such quantities must be made for the instant concerning which one desires the information.

We should note, incidentally, that in treating phenomena of motion in earlier chapters, we did occasionally use instantaneous speeds and accelerations. When we calculated the speed of a falling body three seconds after it had begun to fall, we were calculating an instantaneous speed. But whenever we dealt with instantaneous rates, we relied upon physical experience to assure us that a moving object has speed at each instant, without inquiring more closely into the precise meaning of this concept, and we were able to calculate the speed because the situations under consideration involved simple formulas for speed. Specifically, we treated constant acceleration, which implies a linear or first-degree formula for speed (see Chapter 15), and so we were able to obtain the results we required.

18–4 The concept of instantaneous speed. The problem of defining and calculating instantaneous rates such as speed and acceleration attracted almost all the mathematicians of the seventeenth century. Descartes, Fermat, Newton's teacher Isaac Barrow, Newton's friend John Wallis, Huygens, and hosts of other scholars worked on this and related problems. The men who finally grasped, formulated, and applied the general ideas of the calculus, which their predecessors had only partially understood, were Newton and Gottfried Wilhelm Leibniz, about whom we shall learn more later. The fact that every major mathematician of the century took up the problem of instantaneous rates of change is in itself of interest. It illustrates how even the best minds become absorbed in the problems of their times. Genius makes its contributions to the advancement of civilization, but the substance of its thoughts is determined by its age.

To explain the concept and method of finding instantaneous speeds and accelerations, we shall begin with the problem of determining the instantaneous speed of a falling body. Let us take the simplest case, that of a body which is dropped near the surface of the earth. Our method presupposes that we know the formula relating distance and time. We know from our work in Chapter 15 that this formula is $d = 16t^2$, where d is the distance fallen and t, the time elapsed. Let us seek the speed at the end of the fourth second of fall, that is, the speed at the instant $t = 4$. We have already pointed out (Section 18–3) that we cannot obtain this speed in the same manner in which we calculate the average speed over some interval of time since it is meaningless to divide the zero distance covered at $t = 4$ by the zero time elapsed. A practical solution of the difficulty might be to calculate the average speed *during the fourth second.* Though this solution will not yield the desired result, let us see what it yields. At the beginning of the fourth second, that is, when $t = 3$, the distance covered by the falling body is obtained by substituting 3 for t in the formula $d = 16t^2$. This distance is then $16 \cdot 3^2$, or 144. The distance covered by the end of the fourth second, that is, when $t = 4$, is $16 \cdot 4^2$, or 256. Hence the ratio of distance covered during the fourth second to the time elapsed is

$$\frac{256 - 144}{1}, \quad \text{or} \quad \frac{112}{1}.$$

The average speed during the fourth second is then 112 ft/sec.

As we have already stated, the average speed during the fourth second is not the speed at $t = 4$ itself, for during the fourth second the speed of the body keeps changing. Hence the quantity 112 can be no more than an approximation to the instantaneous speed. We may, however, improve the approximation by calculating the average speed in the interval of time from 3.9 to 4 seconds, for during this interval the average speed can, on physical grounds, be expected to approximate more closely the speed actually possessed by the body at $t = 4$. We therefore repeat the procedure of the preceding paragraph, this time using the values 3.9 and 4 for t. Thus for $t = 3.9$,

$$d = 16(3.9)^2 = 16(15.21) = 243.36;$$

and for $t = 4$,

$$d = 16 \cdot 4^2 = 256.$$

The average speed during the interval $t = 3.9$ to $t = 4$ seconds is then

$$\frac{256 - 243.36}{0.1} = \frac{12.64}{0.1} = 126.4 \text{ ft/sec.}$$

We note that the average speed during this one-tenth of a second is quite different from the value 112 for the fourth second.

Of course, the average speed during the interval $t = 3.9$ to $t = 4$ is not yet the speed at $t = 4$ because even during one-tenth of a second the speed of the falling body changes and the average is not the value finally attained at $t = 4$. We can obtain a still better approximation to the speed at $t = 4$ if we calculate the average speed during the one-hundredth of a second from $t = 3.99$ to $t = 4$, because the speed during this short interval of time near $t = 4$ ought to be almost equal to that at $t = 4$. Hence we shall apply our previous procedure once more. For $t = 3.99$,

$$d = 16(3.99)^2 = 16(15.9201) = 254.7216;$$

and for $t = 4$,

$$d = 16 \cdot 4^2 = 256.$$

Thus the average speed during the interval $t = 3.99$ to $t = 4$ is

$$\frac{256 - 254.7216}{0.01} = \frac{1.2784}{0.01} = 127.84 \text{ ft/sec.}$$

We could continue the above argument and process. The speed during the interval $t = 3.99$ to $t = 4$ is not the exact speed at $t = 4$ because the speed of the falling body changes even in one-hundredth of a second. We could therefore calculate the average speed in the interval $t = 3.999$ to $t = 4$ and expect that the average would be even closer to the speed at $t = 4$ than the preceding averages. The result incidentally would be 127.989 ft/sec. Of course, no matter how small the interval over which the average speed is calculated, the result is *not* the speed at the instant $t = 4$. How far then should the process be continued? The answer to this question is the core of the new idea supplied by the seventeenth-century mathematicians. The new thought is that one should compute average speeds over smaller and smaller intervals of time and note whether these average speeds get closer and closer to one *fixed* number. If so, *this number is taken to be the instantaneous speed at* $t = 4$. Let us pursue this idea.

In our case the average speeds over the intervals of time 1, 0.1, 0.01, and 0.001 proved to be 112, 126.4, 127.84, and 127.989. These numbers seem to be approaching, or getting closer to, the fixed number 128. Hence we take 128 to be the speed of the falling body at $t = 4$. This number is called the *limit* of the set of average speeds. We should note that the instantaneous speed is *not* defined as the quotient of distance and time. Rather it is the limit approached by average speeds as the intervals over which these average speeds are computed approach zero.

Two objections to what we have done may occur. The first is, What right do we have to take the number approached by the average speeds to be the speed at $t = 4$? The answer is that mathematicians have adopted a definition which makes good physical sense. They argue that the smaller the interval of time bordering $t = 4$ over which the average speeds are computed, the closer must the behavior of the falling body be to that at $t = 4$. Hence the number approached by average speeds over the smaller and smaller intervals of time bordering $t = 4$ should be the speed at $t = 4$. Since mathematics seeks to represent physical phenomena, it quite naturally adopts definitions that seem to be in accord with physical facts. It can then expect that the results obtained by mathematical reasoning and calculations will fit the physical world.

The second possible objection to our definition of instantaneous speed is a more practical one. Apparently, one must calculate average speeds over many intervals of time and attempt to discern what number these average speeds seem to be approaching. But there appears to be no guarantee that the fixed number chosen is the correct one. Thus, if in our above calculations one had obtained only the average speeds 112, 126.4 and 127.84, he might decide that these speeds are approaching the number 127.85, and his result would then be in error by 0.15 ft/sec. The answer to this objection is that we can generalize the entire process of obtaining the instantaneous speed, so that it can be carried out more quickly and with certainty. We shall now illustrate how the new method operates.

18–5 The method of increments. Let us again calculate the instantaneous speed of a dropped body at the end of the fourth second of fall, that is, at the instant $t = 4$. The formula which relates distance fallen and time of travel is, of course,

$$d = 16t^2. \tag{1}$$

Again, as in our earlier work, we can calculate at once the distance fallen by the end of the fourth second. This distance, which we shall denote by d_4, is $16 \cdot 4^2$, or

$$d_4 = 256. \tag{2}$$

The generality of our new process consists in calculating the average speed, not over a specific interval of time such as 0.1 of a second, but over an arbitrary interval of time. That is, we introduce a quantity h which is to represent any interval of time beginning at $t = 4$ and extending before or after $t = 4$. The quantity h is called an increment in t because it is some additional interval of time before or beyond $t = 4$. If h is positive, then it represents an interval after $t = 4$; if it is negative, then it denotes an interval before $t = 4$.

We shall first calculate the average speed in the interval 4 to $4 + h$ seconds. To do this, we must find the distance traveled in this interval of time. We therefore substitute $4 + h$ for t in (1) and obtain the distance fallen by the body in $4 + h$ seconds. This distance will be denoted by $d_4 + k$. Here d_4 is the distance the body falls in four seconds, and k is the additional distance fallen, or the increment in distance, in the interval of h seconds. Thus

$$d_4 + k = 16(4 + h)^2.$$

Multiplying $4 + h$ by itself gives

$$d_4 + k = 16(16 + 8h + h^2).$$

Application of the distributive axiom of algebra yields

$$d_4 + k = 256 + 128h + 16h^2. \tag{3}$$

To obtain k, the distance traveled in the interval of h seconds, we have but to subtract equation (2) from equation (3). The result is

$$k = 128h + 16h^2. \tag{4}$$

The average speed in the interval of h seconds is the distance traveled in that time divided by the time, that is, k/h. Let us therefore divide both sides of equation (4) by h. Then

$$\frac{k}{h} = \frac{128h + 16h^2}{h}. \tag{5}$$

When h is *not* zero, it is correct to divide the numerator and denominator on the right-hand side of (5) by h. The result is

$$\frac{k}{h} = 128 + 16h. \tag{6}$$

Hence (6) is also a correct expression for the average speed in the interval h.

To obtain the instantaneous speed at $t = 4$, we must determine the number approached by the average speeds as the interval h of time over which these speeds are computed becomes smaller and smaller. From (6) we can now readily obtain what we seek. If h decreases, $16h$ must also decrease, and when h is very close to zero, $16h$ is also close to zero. In view of (6), then, the fixed number which the average speed approaches is 128. This number is the speed at $t = 4$.

The process we have just examined, called the method of increments, is basic in the calculus. It is more subtle than appears at first sight. One should not expect to note and appreciate the finer points on first contact, any more than one gets to know another person well on the basis of one meeting. As a step in the right direction, however, we shall make one or two observations. First we wish to emphasize the fact that we sought the number or limit approached by average speeds as the intervals of time during which the average speeds were computed became smaller and smaller and close to zero. The correct expression for the average speed in any time interval h is given by (5). Since h is not zero, we may divide numerator and denominator in (5) by h. The resulting expression for the average speed, namely (6), happens to be especially simple, and from (6) we can easily determine what the limit of the average speeds is; that is, we observe that as h approaches zero, so does $16h$, and thus the number approached by the average speeds is readily seen to be 128.

In the present case of the rather elementary function $d = 16t^2$, we may let h be zero in (6) and find that the result is also 128. This agreement between the value of k/h when h is zero and the number approached by k/h as h gets closer to zero will show up in a number of fairly simple functions. However, let us not lose sight of the fact that what we seek is a limit of k/h as h approaches zero rather than the value of k/h when h is zero. If the two values happen to be the same in some cases, as in (6), we are lucky, but let us not press this luck too hard.* The reader who wishes to tempt fate may substitute zero for h in simplified expressions such as (6).

The main point that emerges from this section is the possibility of finding instantaneous speed by a general process, that is, the method of increments. No tedious arithmetical calculations are necessary, nor is there any doubt about what the limit approached by the average speeds is.

To appreciate what the limit process achieves we might consider an analogy. Suppose that a marksman seeks to hit a particular spot on a target. Even if he is a good shot, he is not likely to hit the given spot squarely, but will hit all around it and indeed come close. A bystander observing the location of the hits will readily determine the exact spot at which the marksman is aiming, by noting the concentration of the hits. This process of inferring the precise location at which the marksman is aiming is analogous to determining the instantaneous speed from a knowledge of the average speeds. We note what number the average speeds are approaching by examining (5) or the simplified form (6), and this limit is taken to be the instantaneous speed.

---

* We could pursue the point further and learn just when the limit approached by k/h must agree with the value of k/h when h is zero. But to do so would involve a long digression into theory which, at the moment, is of secondary importance.

Exercises

1. Distinguish between the change in distance which results when an object moves for some interval of time and the rate of change of distance compared to time in that interval.

2. Distinguish between average speed and instantaneous speed.

3. What mathematical concept is used to define instantaneous speed?

4. If the distance d, in feet, which a body falls in t seconds is given by the formula $d = 16t^2$, calculate the average speed of the body during the first five seconds of fall and during the fifth second of fall.

5. If the distance d, in feet, which a body falls in t seconds is $d = 16t^2$, calculate the instantaneous speed of the body at the end of the fifth second of fall, that is, when $t = 5$.

6. If the formula which relates the height above the ground and the time of travel of a ball thrown up into the air is $d = 128t - 16t^2$, calculate the speed at the instant $t = 3$.

18–6 The method of increments applied to general functions. We have calculated the instantaneous speed at the end of the fourth second for an object which falls according to the law $d = 16t^2$. Obviously the process used would have limited value if it were applicable to just the fourth second and to the formula $d = 16t^2$. Let us investigate the possibility of generalizing the procedure and see whether it might apply to any instant of time and perhaps to other formulas. We begin by considering the formula

$$y = ax^2, \tag{7}$$

where a is some constant and y and x are any variables related by (7). (After all, the fact that d represented distance and t time in the formula $d = 16t^2$ played no role in the purely mathematical process of calculating the rate of change of d with respect to t at $t = 4$.) By using the letters y and x and the constant a we emphasize the fact that we are considering a strictly mathematical relationship, and we shall calculate the rate of change of y with respect to x at a given value of x. Such rates, incidentally, are also called instantaneous rates, even though x does not always represent time. The word "instantaneous" has been carried over because the original and many current applications of the calculus contain time as the independent variable.

Let x_1 denote the value of x at which we are to compute the instantaneous rate of change of y compared to x. Thus x_1 is analogous to the value 4 of t used in the preceding section. To compute the desired rate of change, we shall repeat the process employed there. We first compute the value of y when x has the value x_1. This value of y, which we shall call y_1, is obtained by substituting x_1 for x in (7). Then

$$y_1 = ax_1^2. \tag{8}$$

We now consider an increase or increment h in the value of x, so that the new value of x is $x_1 + h$. To compute the new value of y, which we shall denote by $y_1 + k$, we must substitute the new value of x in (7). Then

$$y_1 + k = a(x_1 + h)^2.$$

Since

$$(x_1 + h)^2 = x_1^2 + 2x_1h + h^2,$$

it follows that

$$y_1 + k = ax_1^2 + 2ax_1h + ah^2. \tag{9}$$

Our next step is to determine the change k in y which results from the change h in x, by subtracting equation (8) from (9). Thus

$$k = 2ax_1h + ah^2. \tag{10}$$

To arrive at the *average* rate of change of y in the interval h, we must find k/h. Accordingly, we divide both sides of (10) by h and obtain

$$\frac{k}{h} = \frac{2ax_1h + ah^2}{h}. \tag{11}$$

Equation (11), which gives the average rate of change of y with respect to x in the interval h, is the generalization of equation (5).

To secure the instantaneous rate of change of y compared to x at the value x_1 of x, we must now determine the limit of the right side of (11) as h approaches zero. We are again fortunate in that we may divide the numerator and denominator of (11) by h and obtain

$$\frac{k}{h} = 2ax_1 + ah. \tag{12}$$

As h becomes smaller and smaller, the quantity ah, which is merely a constant times h, also becomes smaller, and the quantity k/h approaches the value $2ax_1$. This last quantity is the limit approached by the average rates of change, k/h, and so is the rate of change of y with respect to x at the value x_1 of x. Just to check our result, we note that when $a = 16$ and $x_1 = 4$, the quantity $2ax_1$ is 128, and this is the limit we obtained in the special case treated earlier.

Since y and x are variables which have no physical meaning, we cannot speak of the limit $2ax_1$ as an instantaneous speed. Instead we must describe it as the instantaneous rate of change of y compared to x at the value x_1 of x. To avoid stating this lengthy phrase, the quantity is called the *derivative* of y with respect to x at the value x_1. We shall denote it by $\dot{y}$, the notation used by Newton. (Leibniz devised the notation dy/dx. However, this notation, though suggestive of what takes place, can be misleading, for the instantaneous rate of change of y with respect to x is not a quotient but rather the limit approached by the quotient k/h.) Thus we have established that at the value x_1 of x

$$\dot{y} = 2ax_1. \tag{13}$$

Actually, we have arrived at a more general result. The quantity x_1 was any value of x. Hence we may as well emphasize this fact by dropping the subscript and writing

$$\dot{y} = 2ax. \tag{14}$$

Equation (14) states that when $y = ax^2$, the instantaneous rate of change of y compared to x at any value of x is $2ax$, or the *derivative* of y with respect to x is $2ax$. Since (14) holds at any value of x, it is a function; that is, the derivative of y with respect to x is itself a function of x. The process of deriving (14) from (7) is called *differentiation.*

The result (14) holds regardless of the physical meaning of y and x. Hence in any situation in which the formula $y = ax^2$ applies, we may conclude at once that the instantaneous rate of change of y compared to x is $2ax$. The generality of this result is immensely valuable, since a general mathematical result can always be applied to many different physical situations. To illustrate this point for the derivative (14), let us first reconsider our old friend $d = 16t^2$. In this case, d plays the role of y; t plays the role of x; and 16 is the value of a. Hence

$$\dot{d} = 2 \cdot 16t = 32t. \tag{15}$$

Now the instantaneous rate of change of distance compared to time is the instantaneous speed, and since speed occurs so often in applications, it is usually denoted by a special symbol, v; that is, $\dot{d} = v$. Hence (15) says that

$$v = 32t. \tag{16}$$

Knowing the formula which relates distance and time of a dropped object, we have derived the formula for the instantaneous speed. Thus from one formula we may derive another significant formula by applying the process of determining the instantaneous rate of change, i.e., by differentiation.

Now let us apply (14) to the formula for the area of a circle, namely, $A = \pi r^2$. Here A plays the part of y; r plays the part of x; and the constant π is the value of a. Formula (14) then tells us that

$$\dot{A} = 2\pi r. \tag{17}$$

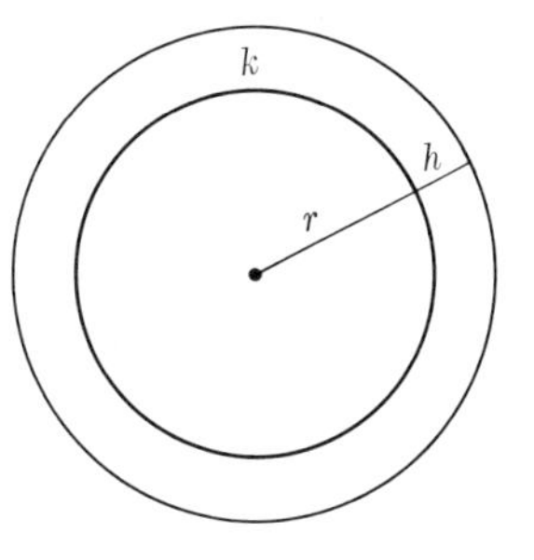

FIGURE 18–1

The result, (17), has a very simple geometrical meaning (Fig. 18–1). It says that the instantaneous rate of change of the area of a circle with respect to the radius at any given value of the radius is the circumference. More loosely stated, the rate at which the area increases when r increases is the size of the circumference. This result is very reasonable. When the radius r is increased by an amount h, the area A of the circle increases by an amount k. We may think of k as made up of a sum of circumferences and of h as the number of such circumferences. The ratio k/h is then an average circumference in the region k. As h approaches zero, this average circumference approaches the circumference for the radius r. This latter circumference is the instantaneous rate with which the area increases at the given value of r.

Of course, the process of finding the instantaneous rate of change can be applied to all functions and not just to the simple function $y = ax^2$. For example, if y represents the pressure of the atmosphere and x represents the height above the surface of the earth, then $\dot{y}$ represents the rate of change of pressure with respect to height at a given height. If y represents the price level of a commodity and x represents time, then $\dot{y}$ represents the rate of change of price with respect to time at a given instant. Various other examples will be presented in the course of our subsequent work.

To make effective use of the calculus, one must learn how to determine the instantaneous rate of change for many types of formulas, because the variety of functions occurring in applications is very great. Since our purpose is primarily to gain some idea of what the calculus has to offer, we shall limit ourselves to just the simplest ones. Thus if

$$y = bx, \tag{18}$$

where b is any constant, then by using the method of increments, we would find that the instantaneous rate of change of y with respect to x is

$$\dot{y} = b. \tag{19}$$

This result applies, for example, to a body which falls with the velocity

$$v = 32t. \tag{20}$$

Formula (20) is just a special case of (18) where y becomes v, x is replaced by t, and b is 32. Hence (19) tells us that

$$\dot{v} = 32. \tag{21}$$

Since $\dot{v}$ is the instantaneous rate of change of velocity with respect to time, it is the instantaneous acceleration. Hence (21) tells us that a body which falls with a velocity $v = 32t$ has an acceleration at each instant of 32, that is, $a = 32$.

Were we to go through the process of determining the instantaneous rate of change of y compared to x when

$$y = ax^3, \tag{22}$$

where a is any constant, we would find that

$$\dot{y} = 3ax^2. \tag{23}$$

Occasionally we shall also treat a formula consisting of a sum of two terms instead of a single term. Thus suppose that the functional relationship between the variables y and x is given by the formula

$$y = ax^2 + bx, \tag{24}$$

where a and b are constants. The method of increments can, of course, still be applied to find the instantaneous rate of change of y with respect to x. Actually the work amounts to treating simultaneously a formula such as (7) and a formula such as (18). The result can be anticipated. In view of the rate of change (14) which applies to $y = ax^2$ and the rate of change (19) which applies to $y = bx$, we should expect that

$$\dot{y} = 2ax + b. \tag{25}$$

This is the correct result.

Exercises

1. By going through the full process of finding the instantaneous rate of change, that is, the method of increments, prove that (a) if $y = bx$, then $\dot{y} = b$; (b) if $y = ax^3$, then $\dot{y} = 3ax^2$; (c) if $y = c$, where c is a constant, then $\dot{y} = 0$.

2. Apply the method of increments to find the instantaneous rate of change of $y = x^2 + 5$ and compare the result with the instantaneous rate of change of $y = x^2$. Does this example suggest a general conclusion?

3. Find the derivative, or the instantaneous rate of change of the dependent variable compared to the independent variable, for the following functions. [You may use formulas (14), (19), (23), and (25).]

(a) $y = 2x^2$ (b) $d = 2t^2$ (c) $y = (\frac{1}{2})x^2$ (d) $y = 4x^3$
(e) $y = -2x^2$ (f) $d = -16t^2$ (g) $h = -16t^2 + 128t$ (h) $h = 128t - 16t^2$

4. If an object is thrown downward with the initial velocity of 100 ft/sec, then the distance it falls in t seconds is given by the formula $d = 100t + 16t^2$. Calculate the speed of the object at the end of the fourth second of fall. [*Suggestion:* Apply formula (25).]

5. (a) In geometrical terms the instantaneous rate of change of the area of a circle compared to the radius is the circumference. What is your guess as to the geometrical interpretation of the instantaneous rate of change of the volume of a sphere compared to the radius?

(b) Now determine $\dot{V}$ mathematically by applying formula (23) of the text to the formula for the volume of a sphere, $V = (4/3)\pi r^3$, and check your answer to part (a).

6. (a) When $y = ax^2$, then $\dot{y} = 2ax$; when $y = ax^3$, then $\dot{y} = 3ax^2$. Now suppose $y = ax^4$. What would you expect $\dot{y}$ to be?

(b) Verify your conjecture in part (a) by applying the method of increments to $y = ax^4$.

7. Find the rate of change of the area of a square with respect to a side at a given value of the side. Is the result intuitively reasonable?

8. The area of a rectangle is given by the formula $A = lw$, where l and w are the length and width, respectively. Suppose that l is kept fixed. What is the rate of change of area with respect to width? Interpret the result geometrically.

18–7 The geometrical meaning of the derivative. The instantaneous rate of change of y with respect to x can be interpreted geometrically. This interpretation not only clarifies the meaning of such a rate, but at the same time points the way to new uses of the concept. Let us consider the function

$$y = x^2, \tag{26}$$

and let us interpret geometrically the instantaneous rate of change of y with respect to x at $x = 2$. To find this rate of change by the method of increments we first calculate y at $x = 2$. This value of y, denoted by y_2, is

$$y_2 = 2^2 = 4.$$

The values 2 for x and 4 for y are, of course, the coordinates (2, 4) of a point, denoted by P in Fig. 18–2, on the curve which represents $y = x^2$. The second step in the method of increments is to increase the independent variable by an amount h so that its value becomes $2 + h$. The dependent variable then changes by an amount k so that its new value is $4 + k$. Now the quantities $2 + h$ and $4 + k$ can be interpreted as the coordinates of another point on the curve which represents $y = x^2$, because when x is $2 + h$, y becomes $4 + k$. The new point is shown as the point Q in Fig. 18–2. Next we calculate the average rate k/h. As the figure shows, k is the difference in the y-values of P and Q, whereas h is the difference in the x-values of P and Q. The ratio k/h is the slope of the line PQ, which, as in plane geometry, is called a secant. Thus far, then, we see that for any value of h and the corresponding value k, the ratio k/h is the slope of the secant through two points of the curve representing $y = x^2$.

Finally we consider the limit approached by the ratio k/h as h gets closer and closer to zero. As h decreases, the point Q on the curve of Fig. 18–2 moves closer to the point P. The secant through P and Q changes position, always, of course, going through the fixed point P and the point Q, wherever

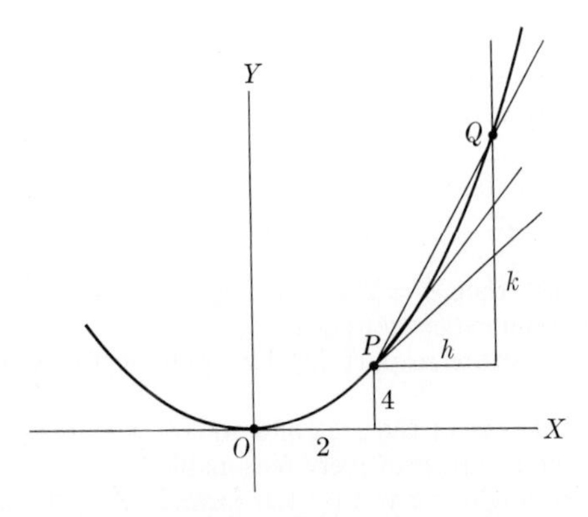

FIG. 18–2. The secant PQ approaches the tangent at P as Q approaches P along the curve.

the latter happens to be. As h approaches zero, the point Q approaches the point P, and the secant PQ comes closer and closer to the line which just touches the curve at P; that is, PQ approaches the tangent at P. Since k/h is the slope of PQ, the limit approached by k/h must be the slope of the line approached by PQ. In other words, *the instantaneous rate of change of y with respect to x at $x = 2$, is the slope of the tangent to the curve at P, the point whose coordinates are* (2, 4). Of course, the value of 2 for x has been arbitrarily chosen to present a typical, yet concrete example. We could have been more general and have carried through the entire discussion for the value a, say, of x; that is, the rate of change of y with respect to x at any given value of x is the slope of the tangent to the corresponding curve at the point having that given value of x as abscissa.

We see therefore that the derivative of a function has a precise geometrical counterpart: the slope. Since slope is the rise (or fall) of a line per unit of horizontal distance (Chapter 13), the geometrical meaning is a rather simple one. Thus, since the value of the derivative of $y = x^2$ at $x = 2$ is 4, the slope of the tangent at $x = 2$ is 4; Fig. 18–2 does not show this because the scale on the Y-axis is not the same as that on the X-axis.

From the standpoint of application, the fact that the derivative is the slope of the tangent is very significant. The slope of a curve at a point on that curve is, very reasonably, defined to be the slope of the tangent at that point. Knowing the slope of the tangent thus means knowing the slope of the curve. Just to get some idea of how useful this information is, let us consider for a moment the roadway of a bridge which is pictured as the arc AOB in Fig. 18–3. For the purpose of our illustration we can assume that this arc is part of the parabola $y = -x^2$. Now the slope of the curve at $x = -2$ is given by the derivative. Since the derivative of $y = -x^2$ at an arbitrary value of x is $-2x$, the derivative at $x = -2$ is $+4$. This then is the slope of the roadway at $x = -2$; that is, the roadway is rising at the rate of 4 feet for every foot of horizontal distance. This rate of climb is totally impractical, since no automobile or truck has the power to climb

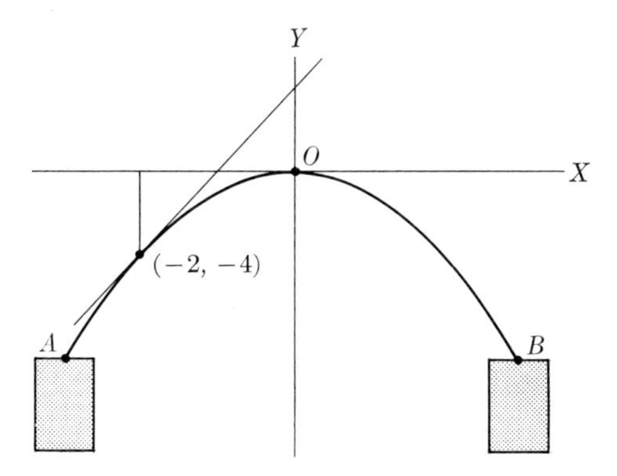

FIG. 18–3. The slope of the roadway of a bridge at $x = -2$.

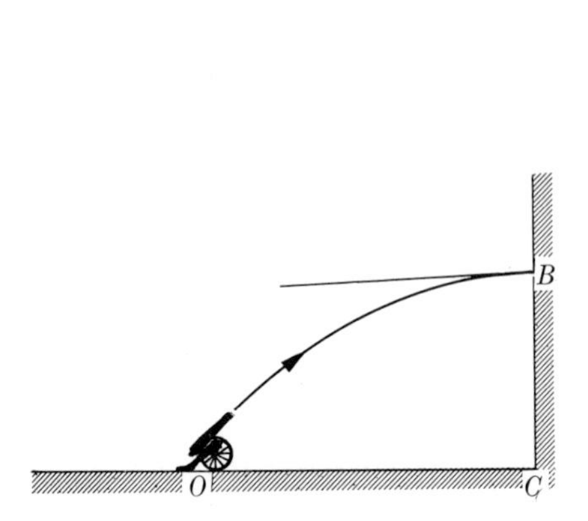

FIG. 18–4. The slope of a projectile's path when it strikes a wall at B is the slope of the tangent at B.

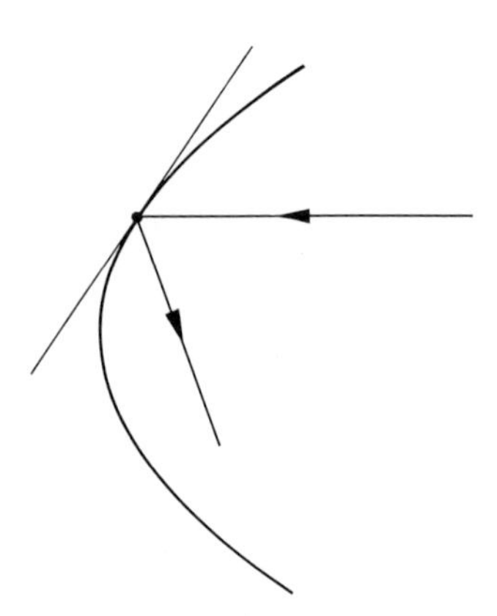

FIG. 18–5. The angles which light rays make with a curve are determined by the slope of the curve.

at such a rate. Our example thus makes the general point that the derivative enables us to calculate the slope of a curved roadway and to determine whether the slope is or is not too steep for the vehicles that are to use that route.

As another illustration suppose a projectile shot up and out from the point O (Fig. 18–4) is to strike the wall BC at the point B. Knowing the equation of the path of the projectile (Chapter 16), we can calculate the slope at the point B. This slope amounts to the direction that the projectile possesses at the point B, for the slope is the rate at which the curve is rising or falling.* One might want the direction of the projectile at B to be perpendicular to the wall because such an impact would damage the wall more effectively than a hit in a glancing direction. If necessary one could adjust the angle of fire and initial velocity to achieve the desired direction at B.

A third example illustrating the usefulness of knowing the slope is furnished by the phenomena of reflection and refraction of light. Let us consider the case of reflection. Suppose one wishes to design a mirror such that all rays of light coming from some source are reflected to one point. We know from Chapter 6 that when a light ray strikes a mirror, the angle of reflection equals the angle of incidence. Suppose that we consider a plane section of the mirror which contains the incident and the reflected rays (Fig. 18–5). This plane section is a curve. The angle which the incident ray makes with the mirror is, in fact, the angle between the incident ray and the tangent. To discuss this angle as well as the corresponding angle of reflection, we must know the direction, and hence the slope, of the tangent.

* Sometimes the word direction is taken to mean the angle which the tangent to the curve makes with the horizontal. However the slope is an equally good indication of direction.

Exercises

1. Suppose that an uphill path can be represented by the equation $y = (1/100)x^2$. (a) What is the slope of the hill at $x - 3$? (b) Is the slope steeper or more gradual at $x = 3$ or at $x = 5$? (c) Determine the slope at $x = 0$, and interpret the result geometrically.

2. Suppose that the path of a projectile is represented by the equation $y = 4x - x^2$. (a) What direction does the projectile have when $x = 1$? (b) At what value of x is the direction of the projectile horizontal?

3. The variation of y with x of a certain function is illustrated in Fig. 18–6. Describe how the derivative of y with respect to x varies as x increases from A to B.

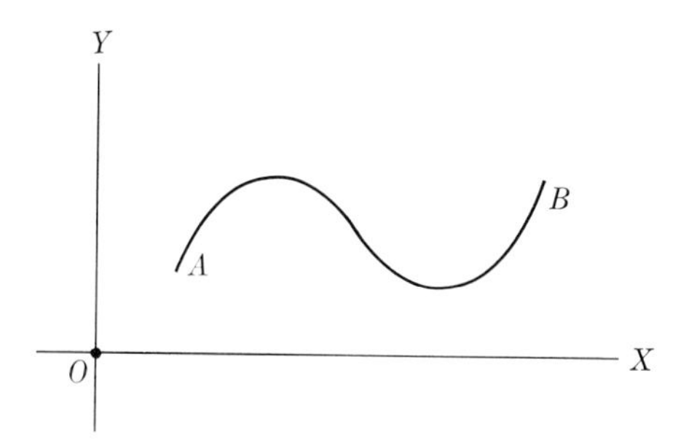

Figure 18–6

4. Can you explain geometrically why the functions $y = x^2$ and $y = x^2 + 5$ should have the same derivative at, say, $x = 2$?

18–8 The maximum and minimum values of functions. We have had occasion to apply our elementary algebra and geometry to problems in which the objective was to maximize or minimize some important physical quantity. For example, in Chapter 6 we found the dimensions of the rectangle having maximum area among all rectangles with the same perimeter. In Chapter 15 we found the maximum height attained by an object thrown or shot up into the air. The methods used to solve these problems were rather limited; they worked for the problems in question, but could hardly be applied to other types. One of the advantages of the calculus is that the concept of instantaneous rate of change of a function proves to be the key to a general method of finding the maximum or minimum values of variable quantities.

Let us reconsider the problem of determining the maximum height reached by a ball which is thrown up into the air. If the ball leaves the hand with a speed or velocity of 128 ft/sec, then, according to Chapter 15, the formula which relates d, the height of the ball, and t, the time that the ball has been in motion, is

$$d = 128t - 16t^2. \tag{30}$$

During our earlier discussion of the motion represented by formula (30), we had to resort to an independent physical argument, to prove that the speed of the ball at any instant is given by

$$v = 128 - 32t, \tag{31}$$

whereas now the purely mathematical process of differentiation immediately yields (31) as the formula for the instantaneous speed of the ball.

To determine the maximum height reached by the ball, we argued in Chapter 15 that the speed of the ball at the highest point must be zero or the ball would continue to rise. Hence, to find the instant t_1 at which $v = 0$, we set v equal to zero, that is, we set

$$128 - 32t_1 = 0, \tag{32}$$

and, by solving this equation for t_1, we found that $t_1 = 4$. We then substituted this value of t in (30) to arrive at the maximum value of d.

We can now see that, translated into the language of the calculus, our above procedure of determining the maximum value of the variable d given by formula (30) consisted in setting the instantaneous rate of change, $\dot{d}$, equal to zero and finding the value (or values) of the independent variable, t in the present case, at which the rate of change is zero. This example suggests a general procedure. If y is a function of x, and if we wish to find the maximum value of y, we set the instantaneous rate of change of y with respect to x equal to zero; find the value of x for which this rate of change, or derivative, is zero; and then substitute this value of x in the formula for y. The resulting value of y is the maximum value of y.

Of course, we really do not know that this general procedure is justified. For a ball thrown up into the air, we used the *physical* argument that the velocity must be zero at the highest point. This argument might be suitable for the motions of balls, but it certainly is not applicable to formulas which represent quite different phenomena. However, we shall now introduce a geometrical argument which proves that the procedure is indeed justified.

Let us use a specific function to illustrate the idea. The argument we shall give could be phrased in general terms. Suppose that we wish to find the maximum value of the function

$$y = 10x - x^2, \tag{33}$$

represented by the curve in Fig. 18–7. We observe that at the point on the curve where y has a maximum value, the tangent is horizontal; that is, the slope of the tangent is zero. Now the slope of the curve at any value of x is the value of the derivative, or instantaneous rate of change of y with respect to x at that value of x. Hence to determine the value x_1 of x at which the slope of the curve is zero, we find the derivative of y in (33), that is, we find $\dot{y}$, and set this derivative equal to zero. Thus for formula (33), we have

$$10 - 2x_1 = 0.$$

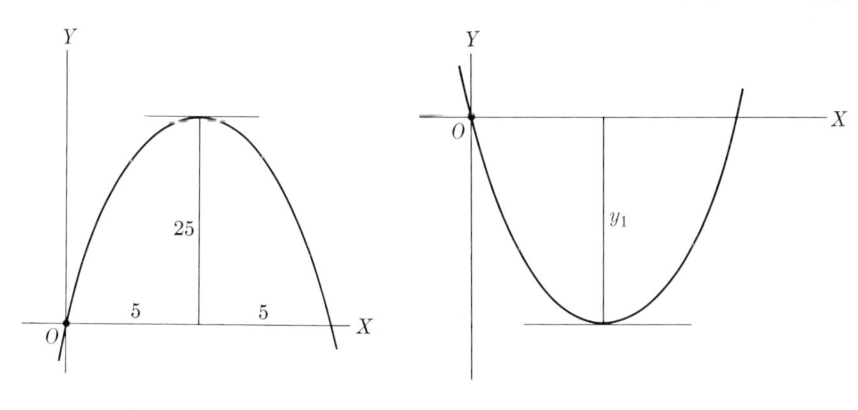

FIGURE 18–7

FIGURE 18–8

We see at once that $x_1 = 5$. To obtain the maximum y-value of (33), we substitute the value 5 for x and find that y_1, the maximum y-value, is 25.

This geometrical argument proving that the derivative of a function is zero at the function's maximum value also applies to its minimum value. The minimum value of the function $y = x^2 - 10x$ is the length y_1 of Fig. 18–8. The slope of the curve at the point where y has this minimum value is zero. Hence, as before, at this point the derivative $\dot{y}$ must be zero, and we may use the process already described in connection with maxima of functions to determine minima also.

The question arises, If the same process yields a maximum and a minimum, how do we know in a particular problem whether we are obtaining one or the other? For physical problems the answer is given by the sense of the problem. But there are also purely mathematical criteria which enable us to determine whether we have found the maximum or the minimum value of a function.

EXERCISES

1. Calculate the instantaneous speed at $t = 4$ of a body whose height d above the ground at time t is given by the formula $d = 128t - 16t^2$. Interpret the result physically and geometrically.

2. To illustrate the power of the calculus, Fermat showed how it can be used to prove that of all rectangles with the same perimeter the square has maximum area. Carry out this task. [*Suggestion:* Let p be the perimeter common to all the rectangles. If x and y are the dimensions of any one of the rectangles, then $2x + 2y = p$ or $y = (p/2) - x$. The area A of any rectangle is given by $A = xy$. Express A as a function of x only and apply the calculus.] Which method do you prefer, that of Euclidean geometry or the calculus?

3. A farmer wishes to enclose a rectangular piece of land which borders on a river, so that no fence will be required along the bank. He has 100 ft of fencing at his disposal. What dimensions should he choose to obtain maximum area? [*Suggestion:* If y is the side which parallels the river, then the amount of fencing re-

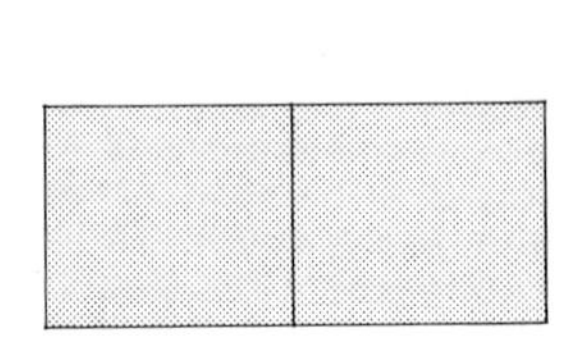

FIGURE 18–9

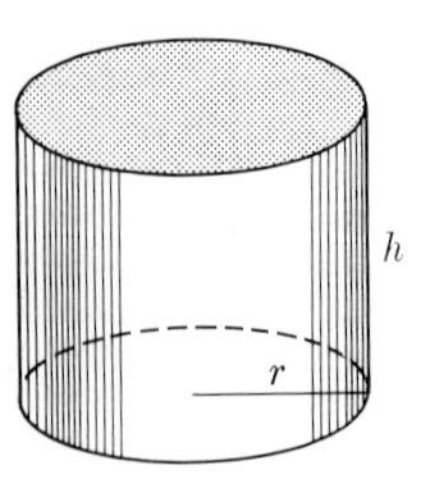

FIGURE 18–10

quired is $y + 2x$. This must equal 100. The area A of the rectangle is $A = xy$. Replace y by its value from $y + 2x = 100$ and find the maximum value of A.] Do you prefer this method or the method of Euclidean geometry?

4. A farmer wishes to use 100 ft of fencing to enclose a rectangular area and to divide the area into two rectangles by running a fence down the middle (Fig. 18–9). What dimensions should he choose to enclose the maximum total area?

5. A manufacturer wishes to construct cylindrical tin cans (Fig. 18–10) such that each can is made of a fixed amount of tin, say 100 in^2, and has maximum volume. What should the radius r of the base and the height h of the cylinder be? [*Suggestion:* The amount of tin used equals the surface area of the can, that is the sum of the area of the side, $2\pi rh$, and the area of the top and bottom, $2\pi r^2$. Hence

$$2\pi rh + 2\pi r^2 = 100. \tag{1}$$

The volume V of the can is

$$V = \pi r^2 h. \tag{2}$$

Solving (1) for h, we have

$$h = \frac{50 - \pi r^2}{\pi r}. \tag{3}$$

If we substitute this value of h in (2), we obtain

$$V = \pi r^2 \left(\frac{50 - \pi r^2}{\pi r}\right) = r(50 - \pi r^2) = 50r - \pi r^3.$$

Now apply the calculus.]

RECOMMENDED READING

BALL, W. W. ROUSE: *A Short Account of the History of Mathematics,* Chaps. 16 and 17, Dover Publications, Inc., New York, 1960.

KASNER, EDWARD and JAMES R. NEWMAN: *Mathematics and the Imagination,* Chap. 9, Simon and Schuster, Inc., New York, 1940.

SAWYER, W. W.: *Mathematician's Delight,* Chaps. 10 through 12, Penguin Books Ltd., Harmondsworth, England, 1943.

WIENER, PHILIP P. and AARON NOLAND: *Roots of Scientific Thought,* pp. 412–442, Basic Books, Inc., New York, 1957.

WIGHTMAN, WM. P. D.: *The Growth of Scientific Ideas,* Chap. 9, Yale University Press, New Haven, 1953.

CHAPTER 19

THE INTEGRAL CALCULUS

More laws are vain where less will serve.
Robert Hooke

19–1 Differential and integral calculus compared. The material examined in the preceding chapter belongs to the differential calculus. The basic process in this subject is to start with the formula relating two variables and to find the instantaneous rate of change of one variable with respect to the other. Suppose, however, that one began with the rate of change of one variable with respect to another and wished to find the formula which relates the two variables. For example, if we should happen to know that $\dot{y} = 2x$, could we find the relation between y and x? One might expect that the answer is affirmative because it would seem that among the various functions whose derivatives we have obtained, there should surely be one whose derivative is $2x$, and this function is the answer to our question. Except for a minor difficulty which we shall consider later, this expectation is correct. In this connection one might also ask whether there is any point in determining functions whose derivatives are given. The answer decidedly is yes. As we shall see, in numerous physical problems the most readily available information is an instantaneous rate of change, whereas the information sought can be best obtained from the function which relates the variables in question. Hence the process of finding the function from its derivative is immensely valuable—indeed, even more valuable than the basic process of finding derivatives from given formulas.

The major idea characterizing the integral calculus is the inverse to that underlying the differential calculus: namely, instead of finding the derivative of a function from the function, one proceeds to find the function from the derivative. Of course, all really significant ideas prove to have extensions and applications far beyond what is immediately apparent, and we shall find this to be true of the integral calculus also.

19–2 Finding the formula from the given rate of change. The key concern, then, of the integral calculus is to determine the formula which relates two variables from the given instantaneous rate of change of one variable with respect to another. Before we can see how useful this idea is, we must examine and learn a few facts about the mathematical process itself.

Suppose we happen to know that the instantaneous rate of change of some variable y with respect to another variable x is $2x$, that is $\dot{y} = 2x$. What formula relates y and x? The mathematician's method of answering this question is to survey all the rates of change of functions obtained in the past and to locate the function whose rate of change he has previously found to be $2x$. In our case, his eye will soon light on the function $y = x^2$. Hence this function is the answer to the problem of finding the relation between y and x such that $\dot{y} = 2x$. The function $y = x^2$ is called the *indefinite integral,* or antiderivative, or often just the integral of the derivative $\dot{y} = 2x$, and the process of obtaining it is called *integration* or antidifferentiation.

However, the formula $y = x^2$ is not the only integral of $\dot{y} = 2x$. We had occasion to observe in the preceding chapter that the presence of a constant term in a formula has no effect on the instantaneous rate of change. For example, $y = x^2$ and $y = x^2 + 5$ both lead to $\dot{y} = 2x$. Hence $y = x^2 + 5$ is as much an integral of $\dot{y} = 2x$ as $y = x^2$ is. In fact, $y = x^2 + C$, where C is any constant, is an integral of $\dot{y} = 2x$. If C is chosen to be zero, we obtain $y = x^2$, and if C is chosen to be 5, we obtain $y = x^2 + 5$. It may seem unfortunate that there should be more than one answer, but we shall see in a moment that the reverse is the case.

The general problem of finding the formula relating y and x when we are given $\dot{y}$ as a function of x is handled by the method illustrated in our example of $\dot{y} = 2x$; that is, we must examine the formulas whose rates of change we have previously determined and try to locate among these derivatives the rate of change we are concerned with. Since this rate of change has been previously derived from some formula relating y and x, that formula is the answer to our problem; in addition, we can add any constant to the formula and still have the correct answer. The process of searching among all formulas whose rates of change have previously been found may seem to be haphazard. But in practice mathematicians tabulate these formulas according to distinctive properties, so that a little experience with the tables usually enables one to find the desired formula. Since we are limiting the variety of formulas and their derivatives to a few cases, we shall not bother to become acquainted with a table. Instead we shall seek to recall the formulas and their derivatives which were calculated in the preceding chapter.

Exercises

For the following problems, find the formula which relates the variables whose instantaneous rate of change is given:

(a) $\dot{y} = 3x^2$ (b) $\dot{y} = 5$ (c) $\dot{y} = x$ (d) $\dot{y} = 3x$
(e) $\dot{d} = 2t$ (f) $\dot{d} = 32t$ (g) $\dot{v} = 32$ (h) $\dot{d} = 2t + 10$
(i) $\dot{d} = -32t + 128$ (j) $\dot{v} = -32$ (k) $\dot{v} = 32t$

19–3 Applications to problems of motion. We shall now present some examples of the usefulness of integration in physical problems. Galileo had found that all objects falling to earth from points near the surface of the earth possess the same acceleration, namely 32 ft/sec^2. This acceleration is constant; that is, it is the same at each instant of the fall. Now the acceleration at any one instant is the instantaneous rate of change of speed with respect to time. Hence, instead of writing $a = 32$, we can equally well write

$$\dot{v} = 32. \tag{1}$$

The physically important question is, What formula relates v and t? By reviewing the formulas for which we obtained rates of change [see formula (20) of Chapter 18] we find that $v = 32t + C$, where C is any constant.

In a particular physical problem, the quantity C can be chosen to fit the situation. Thus suppose that the object is merely dropped to earth; that is, at the instant it begins to fall its speed is zero. If time is measured from the instant the object begins to fall, then the speed v at $t = 0$ is zero. Hence, to make the formula

$$v = 32t + C \tag{2}$$

fit the physical fact that v must be zero when $t = 0$, we must have

$$0 = 32 \cdot 0 + C,$$

or $C = 0$. Hence

$$v = 32t \tag{3}$$

is the answer to this particular problem in which the object is dropped, and time is measured from the instant it begins to fall.

Physical problems often require knowledge of the distance which an object falls in time t. Since the instantaneous speed is the rate of change of distance with respect to time, then if d denotes the distance the object falls, $\dot{d} = v$. In view of (3), which applies when an object is dropped, we may state that

$$\dot{d} = 32t. \tag{4}$$

We now wish to find the formula which relates d and t. Again we appeal to our experience with formulas and their derivatives [see formula (15) of Chapter 18] and note that the formula $d = 16t^2$ has the derivative given by (4). However, the formula:

$$d = 16t^2 + C, \tag{5}$$

where C is any constant, also has the derivative (4). Since we have no reason to ignore the constant, we must accept (5) as the formula for the distance fallen in time t. However, if we agree to measure the distance fallen from the point where the object happens to be at the instant it starts to fall, and if time of fall is also measured from this instant, then it follows that $d = 0$ when $t = 0$. Substituting these values in (5) yields

$$0 = 16 \cdot 0 + C,$$

and we see that C must be zero if formula (5) is to represent our situation. Hence

$$d = 16t^2 \tag{6}$$

gives the distance the dropped object falls in time t if time is measured from the instant the object begins to fall and if distance is measured from the point where the object is at $t = 0$.

We have been able to reverse or invert the process of finding the rate of change of a function and thus proceed from a knowledge of acceleration to speed as given by (3), and from speed to distance fallen as given by (6). Before we comment further, let us consider some other situations.

Suppose that an object is *thrown* downward and leaves the hand with a speed of 100 ft/sec. The acceleration is still given by (1), and so the speed is still given by (2). However, if time is measured from the instant the object leaves the hand, then at the instant $t = 0$, $v = 100$. To make the formula

$$v = 32t + C$$

fit this new situation, we must have $v = 100$ at $t = 0$, or

$$100 = 32 \cdot 0 + C,$$

or $C = 100$. Hence

$$v = 32t + 100 \tag{7}$$

is the final formula for the speed of an object thrown downward with an initial speed of 100 ft/sec.

Now let us seek the distance covered in time t. We know that the instantaneous speed is the instantaneous rate of change of distance with respect to time. Thus, if we denote distance by d, we may write $\dot{d} = v$. In view of (7), we have

$$\dot{d} = 32t + 100. \tag{8}$$

We must now ask, What formula relates d and t? By reviewing the derivatives and the functions from which they were obtained, we find that the term $32t$ in (8) must come from the term $16t^2$ and the term 100 must come from $100t$. The formula for d therefore is presumably $d = 16t^2 + 100t$. However, we must recall that the formula

$$d = 16t^2 + 100t + C, \tag{9}$$

where C is any constant, also has the derivative (8). Hence, so far (9) is the general formula for distance fallen. If we agree to measure distance from the point at which the object happens to be when it begins to fall, and if time is measured from the instant the object begins to fall, then $d = 0$ when $t = 0$. Substituting these values in (9), we have

$$0 = 16 \cdot 0 + 100 \cdot 0 + C,$$

whence $C = 0$, and

$$d = 16t^2 + 100t \tag{10}$$

is the final formula for our situation.

We see from the examples already presented that the occurrence of the constant C in the integration is not a disadvantage but rather an advantage. It permits us to adjust the formulas for speed and distance to the specific situation we wish to describe, although the basic fact in all instances is $\dot{v} = 32$.

The applications of integration made thus far have involved proceeding from the constant acceleration of 32 ft/sec^2 to the formula for distance. But this we were also able to do in Chapter 15 without depending upon the calculus. It might seem that, thus far at least, the process of integration has not added at all to the power of mathematics. However, there are two points to be taken into consideration. The derivation of formula (10), for example, from the basic physical fact, $a = 32$, is much more readily done by integration than by the arguments given in Chapter 15. But the second and more important point is that the method displayed here for the derivation of the formula for velocity from that for acceleration and of the formula for distance from that for velocity applies to all formulas, whereas the argument given in Chapter 15 is limited to constant acceleration. Thus, if an object should move with variable acceleration, as is the case when an object falls to the earth from a great height, then the method of Chapter 15 no longer applies, whereas integration does. We shall treat such problems later.

Since the motions of objects *thrown up* into the air are very important, let us note that our present method applies to them also except for minor modifications. We shall again restrict ourselves to objects which do not rise very far from the surface of the earth, so that we can continue to use the physical fact that the acceleration is constant and equal to 32 ft/sec^2. When we studied the motion of a freely falling object, we decided, for convenience, to consider the acceleration to be positive. As a consequence, the speed at any instant of time and the distance fallen turned out to be positive. However, an object thrown up into the air will, of course, rise and then fall. Hence, if we regard the speed in the upward direction to be positive, then we must take the acceleration to be negative because it causes speed in the downward direction. We start then with the basic fact that

$$\dot{v} = -32. \tag{11}$$

By integration we obtain

$$v = -32t + C. \tag{12}$$

To fit the value of C to our situation, we shall use the physical fact that at $t = 0$, that is, at the instant at which the object is thrown upward, the hand or possibly a gun imparts to the object a speed of, say 100 ft/sec. Thus at $t = 0$, $v = 100$. If we substitute these values in (12), we have

$$100 = -32 \cdot 0 + C.$$

Hence $C = 100$, and the final formula for speed is

$$v = -32t + 100. \tag{13}$$

Since, as we know, instantaneous speed is the instantaneous rate of change of distance with respect to time, we can now apply integration to find the distance traveled. Let us use d to represent the height above the ground reached by the object in time t. Then by integrating (13) we obtain [cf. (9)]

$$d = -16t^2 + 100t + C. \tag{14}$$

We now wish to adjust the value of C to fit the physical situation. At the instant $t = 0$, the object is about to be thrown up, and at this instant, $d = 0$. If we substitute 0 for d and 0 for t in (14), we obtain

$$0 = -16 \cdot 0 + 100 \cdot 0 + C,$$

and so $C = 0$. Thus the final formula for height above ground is

$$d = -16t^2 + 100t. \tag{15}$$

Having obtained various formulas for speed and distance such as (13) and (15) or (7) and (10), we can now proceed to solve problems of the type considered in Chapter 15. We shall not repeat this work here, but instead propose to show in the following sections how integration, or the inverse of differentiation, produces useful formulas from basic physical facts.

Exercises

In all of the following problems the motions involved take place near the surface of the earth. Hence you may assume that the acceleration is constant.

1. Suppose that an object is thrown up into the air with an initial velocity of 150 ft/sec. Derive the formulas for the speed and height above the ground.

2. Given that an object is dropped and falls to earth. Suppose that distance is measured from a point 50 ft above the point at which the object is dropped, but that time is measured from the instant the object begins to fall. What formula relates distance fallen and time of fall?

3. Suppose that an object is dropped from a point 75 ft above the ground. Derive the formulas for the speed and height above the *ground*.

4. Suppose that an object is thrown up into the air from the roof of a building 50 ft high. The initial speed is 100 ft/sec. Derive the formulas for the speed and for the height above the *ground*.

5. Suppose an object is thrown downward from the roof of a building 50 ft high and that the initial speed is 100 ft/sec. Derive the formulas for the speed and height above the *ground*.

19–4 Areas obtained by integration. The derivation of formulas useful in the study of motion was one of the seventeenth-century problems which motivated the creation of the calculus. Another basic class of problems was concerned with finding the lengths of curves, the areas bounded by curves, and the volumes bounded by surfaces. In Section 18–2 of the preceding chapter, we mentioned a few problems which called for the determination of lengths, areas, and volumes. The expansion of science and technology has brought about literally thousands of new uses of curves and surfaces for which the very same quantities are required. The distance a ship travels along the spherical surface of the earth is the length of a curve. Cables and roadways of bridges are curves, and in planning the construction one must know the lengths of these cables and roadways. The weights of various objects employed in scientific and engineering projects are easily obtained once the volumes are known. For example, if a steel beam of some particular shape is to be used in the framework of a building, then, since the material is the same throughout the beam, the weight is merely the volume multiplied by the weight per cubic foot of the metal. Hence volume is the essential quantity to be determined.

But problems concerning lengths of curves, areas, and volumes had already been solved in Euclidean geometry. Why should they then have presented special difficulties to the scientists of the seventeenth century? The answer is that Euclidean geometry is adequate only to treat figures bounded by straight line segments and by circles. This limitation is inherent in the subject. Examination of the axioms of Euclidean geometry shows that they state properties of lines and circles. Naturally the theorems which can be deduced readily must also be limited to such figures. Although the Greeks managed to compute a few areas and volumes of figures bounded by other geometrical shapes, they were able to do so only with great difficulty and by introducing special methods limited to the figures in question. The variety and number of problems which arose in the seventeenth century demanded more general and more easily applicable methods.

Though it is by no means evident, the calculus proves to be the very mathematical tool which enables us to calculate the lengths of curves, the areas bounded by curves, and the volumes bounded by surfaces. We shall illustrate this fact by treating the problem of area. Let us try to determine the area $DEFG$ of Fig. 19–1. This area is bounded by the vertical line

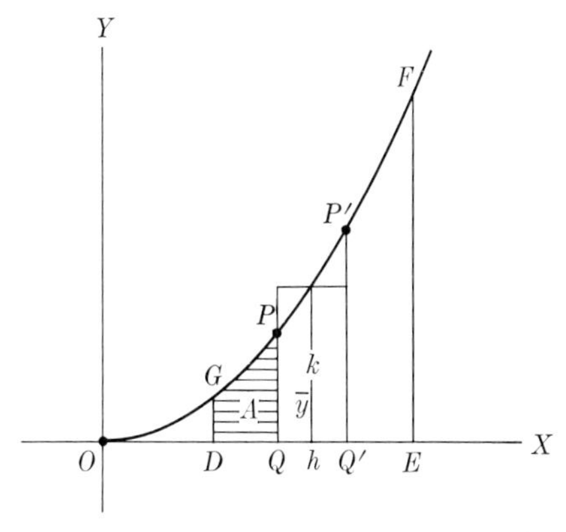

Fig. 19–1. The area under the curve is that swept out by the vertical line QP as it moves to the right.

segments DG and EF, by the segment DE and by the arc FG of the curve whose equation is, say $y = x^2$. We may think of this area as being swept out by a vertical line segment PQ which starts at the position DG and moves to the right. Naturally the length of PQ varies as it moves. Let us suppose that PQ has reached the position shown in the figure. The area swept out by this moving segment depends, of course, upon the position it has reached. This position can be specified by the x-value of the point Q. Hence the variable area, which we shall denote by A, is a function of x, the abscissa of the point Q. We now propose to find the formula which relates A and x. Our procedure is as follows: We begin by determining the rate of change of A with respect to x at any given x and integrate this derivative to arrive at the desired formula.

Our first task therefore is to find the rate of change of A with respect to x.* To do so, let us suppose that PQ has moved a little farther to the position $P'Q'$. The abscissa of Q' is, of course, somewhat larger than that of Q. Let us denote the abscissa of Q' by $x + h$, so that the increase in the abscissa from Q to Q' is h. Obviously, the variable area A also increases when PQ moves to $P'Q'$. Let us use k to denote this increase which geometrically is the area $QQ'P'P$. It is immediately evident that the increase is equal to the area of a rectangle whose base is h and whose height is an ordinate, $\bar{y}$, which is larger than PQ and smaller than $P'Q'$. (We do not know how large $\bar{y}$ is, but we shall see in a moment that this does not matter.) We have, then,

$$k = \bar{y}h. \tag{16}$$

Let us divide both sides of (16) by h. Then

$$\frac{k}{h} = \bar{y}. \tag{17}$$

Now k/h is the average rate of change of area with respect to the abscissa in the interval h. By the very definition of an instantaneous rate, the rate of change of area with respect to abscissa at the x-value of Q should be the limit of the average rate of change as h approaches zero. But as h approaches zero, $\bar{y}$ approaches the y-value of P, or the length PQ. Thus

$$\dot{A} = y. \tag{18}$$

Since the y in (18) is the ordinate of the point P and P lies on the curve OGF, it follows that $y = x^2$. Then

$$\dot{A} = x^2. \tag{19}$$

We now have the rate of change, with respect to x, of the variable area A.

* At this stage, the reader might reconsider the example in the preceding chapter dealing with the rate of change of the area of a circle with respect to the radius and try to guess the answer to our present problem of rate of change.

To find A itself, we must ask ourselves what formula has the derivative x^2. A review of previously obtained derivatives tells us that the derivative of x^3 is $3x^2$, and that therefore $A = x^3/3$. We know, however, that the integral may contain a constant term and that this will not affect the derivative, which will remain unchanged. Hence the full answer is:

$$A = \frac{x^3}{3} + C. \tag{20}$$

To determine the value of C, we make use of the fact that when PQ is at DG, the area is zero because DG was the starting position of PQ. Suppose that the x-value of D is 3. Then, by substituting 0 for A and 3 for x in (20) we have

$$0 = \frac{3^3}{3} + C,$$

or $C = -9$. Thus

$$A = \frac{x^3}{3} - 9, \tag{21}$$

and this formula gives the area between DG and the variable position of the moving line segment PQ. If we wish to determine the area from DG to EF, we may assume that PQ has reached the position EF. Let us suppose that the x-value of E is 6. If we now substitute 6 for x in (21), we obtain the area $DEFG$. Hence

$$\text{area } DEFG = \frac{6^3}{3} - 9 = 72 - 9 = 63. \tag{22}$$

Thus we have found the area bounded by a curve through the process of integration. We have, of course, used the equation of the curve, which, thanks to Descartes and Fermat, should be known to us.

In working problems, one can eliminate some writing by neglecting to introduce the constant C in (20) and using just the formula $A = x^3/3$. We then substitute 6, which is the abscissa of the point E, in this formula; next, we substitute 3, which is the abscissa of the point D; and finally we subtract the second result from the first. These steps lead to the result given by (22).

Exercises

1. Find the area bounded by the curve $y = x^2$, the X-axis, and the ordinates at $x = 2$ and $x = 6$.
2. Find the area bounded by the curve $y = x^2$, the X-axis, and the ordinates at $x = 4$ and $x = 6$.
3. Find the area bounded by the straight line $y = x$, the X-axis, and the ordinates at $x = 4$ and $x = 6$.
4. Find the area bounded by the curve $y = x^2 + 9$, the X-axis, and the ordinates at $x = 3$ and $x = 6$.

19–5 The calculation of work. An important quantity for scientific and engineering purposes is the work done in various physical operations. We use the word "work" here in its technical sense (see Chapter 15, Section 15–9), that is, in the sense of force times distance. This quantity is important, for example, in the operation of machinery. One must know how much work a machine is capable of doing, to decide whether it is suitable for a particular task. A train pulling a load over some distance and an airplane carrying freight or passengers do work, and again the capacity of these carriers and the fuel requirements must be known for proper design.

We shall now consider how the calculus can be used to calculate work. Let us suppose that we wish to compute the work required to raise a 500-pound load to a height of 100 miles. This problem arises, for example, in determining the quantity of fuel required to raise a rocket to some desired height. Now the force that will accomplish this goal must be great enough to offset the force of gravity, which pulls the object down. The force of gravity, as we know, is given by

$$F = \frac{GMm}{r^2}, \tag{23}$$

where G is the gravitational constant, M is the mass of the earth, m is the mass of the object, and r is the variable distance between the position of the object and the center of the earth. In our problem, we shall regard G, M, and m as constant, so that the only variables in (23) are F and r. (However, in actual rocket problems, the fuel itself is part of the load which must be raised, and, since the fuel is gradually burned up as the rocket rises, the mass m also is a variable.) Since the object is to travel a distance of 100 miles up, the force which must be applied varies over the distance. Hence it is not possible to calculate the work by merely multiplying the force by the distance.

Let W be the work required to raise the object from the surface to some distance r from the center of the earth. Of course, W is a function of r and is unknown. Let us suppose that the object is raised an additional distance h, that is, from r to $r + h$ (Fig. 19–2). The corresponding extra work, k, again depends upon the force which must be applied [given by (23)] and the distance h, over which it operates. However, the force varies even as r increases to $r + h$. Let $\bar{r}$ be some value of r between r and $r + h$ such that the corresponding force $GMm/\bar{r}^2$ is the average force required during the interval r to $r + h$. This last-mentioned force is entirely analogous to the quantity $\bar{y}$ introduced in the treatment of area as an average ordinate in the interval h on the X-axis. We shall see in a moment that the precise value of

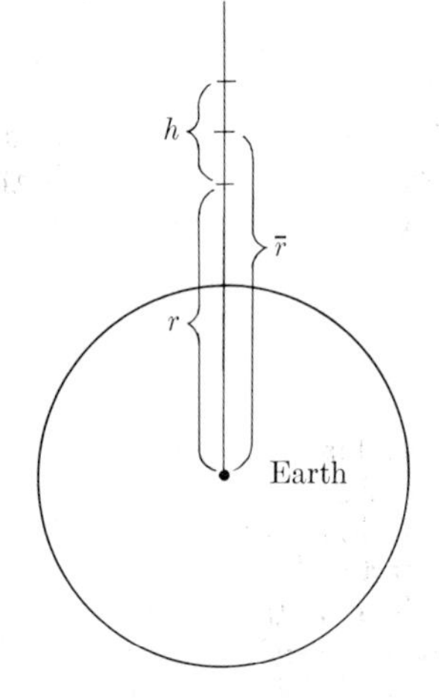

FIGURE 19–2

$\bar{r}$ plays no role. Then the work done in raising the mass m from r to $r+h$ is

$$k = \frac{GMm}{\bar{r}^2}\,h. \tag{24}$$

Equation (24) yields the additional work required to raise the object the distance h. We shall determine next the average rate of change of work with respect to distance. This quantity, k/h, is obtained from equation (24) by dividing both sides by h. Thus

$$\frac{k}{h} = \frac{GMm}{\bar{r}^2}.$$

We compute next the instantaneous rate of change of work with respect to distance. This rate is obtained by letting h approach zero and finding the limit approached by k/h. However as h approaches zero, the quantity $\bar{r}$ must approach r because $\bar{r}$ is always intermediate between r and $r+h$. Then the instantaneous rate of change of work with respect to distance is given by

$$\dot{W} = \frac{GMm}{r^2}. \tag{25}$$

Since we now know $\dot{W}$, we can find W by integration; that is, we examine the various functions we have differentiated, to find one which yields (25) as its derivative. It so happens that in our work we did not encounter the rate of change given by (25), but we can take for granted now and check later that the function corresponding to this derivative is

$$W = -\frac{GMm}{r} + C. \tag{26}$$

To determine the constant of integration, we recall that when $r = 4000$ miles or $4000 \cdot 5280$ feet, the object is on the surface of the earth and, hence $W = 0$. Since we do not wish to manipulate large numbers at the moment, we denote the radius of the earth by R. Then when $r = R$, $W = 0$. We substitute these values in (26) and obtain

$$0 = -\frac{GMm}{R} + C$$

or

$$C = \frac{GMm}{R}.$$

Hence

$$W = -\frac{GMm}{r} + \frac{GMm}{R}. \tag{27}$$

We now have the function which expresses the work done in raising an object of mass m from the surface of the earth to a height r units from the center. By substituting the numerical data at our disposal we can calculate the work done in the example proposed at the outset. We know G and the mass M of the earth. The quantity R is $4000 \cdot 5280$ feet, and the value of r is 4100 miles or $4100 \cdot 5280$ feet. The mass m, in our example, is the mass of an object which weighs 500 pounds or $500 \cdot 32$ poundals at the surface of the earth. Hence the mass m is 500 pounds.

We can simplify the arithmetic somewhat since we know that the acceleration due to the earth's gravitational attraction is [formula (5) of Chapter 17]

$$a = \frac{GM}{r^2},$$

and that when $r = R$, then $a = 32$ ft/sec^2. Thus

$$32 = \frac{GM}{R^2},$$

whence

$$GM = 32R^2.$$

If we substitute this result in (27), we obtain

$$W = -\frac{32R^2m}{r} + \frac{32R^2m}{R}$$

or

$$W = -\frac{32R^2m}{r} + 32Rm. \tag{28}$$

By applying the distributive axiom we may write

$$W = 32Rm\left(1 - \frac{R}{r}\right). \tag{29}$$

Formula (29) is the useful form for the calculation of the work, except for one detail. The unit of work is usually taken to be foot-pounds. However if the mass m is given in pounds, then the formulas we have employed here and elsewhere express the force in poundals. Hence formula (29) yields the work done in foot-poundals. To obtain an answer in foot-pounds we just ignore the factor 32 in (29).

Exercises

1. Calculate the work done in raising the 500-lb weight to a height of 100 mi.

2. Suppose that one neglected the variation in gravitational force with height and assumed that the 500-lb weight remains constant over the 100 mi that it is raised. What is the work required to raise it?

3. A cable weighing 2 lb/ft is suspended in a well 100 ft deep; a tool weighing 300 lb is attached at the cable's lower end. Find the work done to raise the tool to the surface. [*Suggestion:* Let W be the work done to raise the tool x feet. Since now only $100 - x$ ft of cable remain, the work k done in raising the tool h feet more is $k = [300 + 2(100 - \bar{x})]h$, where $\bar{x}$ is some value of x between x and $x + h$. Now find $\dot{W}$, and then W. Determine the constant and compute the work done.]

4. Use the method of increments to show that the function $W = c/r$, where c is any constant, has the derivative $\dot{W} = -c/r^2$.

19–6 The calculation of escape velocity. We can use the theory of the preceding section to answer a question which is of special interest today, namely, what velocity one must give to a rocket to ensure that it just reaches a specified height. The condition intended here is that the rocket will have zero velocity when it reaches this height, for if it still possessed some velocity, it would continue to rise.

As the rocket travels upward, it loses velocity because the acceleration of gravity, which is directed downward, continually decreases the velocity. However, if the initial velocity V is properly chosen then the rocket will have zero velocity at the required height. We wish to determine V.

In the preceding section we calculated the work done in raising an object to a height of d feet above the surface of the earth. However the result did not involve the initial velocity V. We shall therefore obtain another expression for this work. It is physically clear that the work done against gravity in the process of sending an object up with some initial velocity V to reach a height of d feet should equal the work done by gravity acting on the object when it falls d feet. Hence let us calculate the latter. We have an object which begins its fall with zero velocity and falls d feet. Since it gains acceleration on the way down in just the reverse order in which it loses velocity on the way up, it strikes the ground with a velocity of V ft/sec.

To make matters simple, let us suppose that the object falls with a constant acceleration of 32 ft/sec². This is, of course, not true when an object falls from a great height, but we shall limit our proof to cases where we may use this constant acceleration. The result is, as a matter of fact, correct even when the acceleration is variable.* According to formula (5) of Chapter 15, an object which falls a distance d with a constant acceleration of 32 ft/sec² has a terminal velocity of

$$V = 8\sqrt{d}$$

or

$$V^2 = 64d. \tag{30}$$

Since the object falls with constant acceleration, the force of gravity is $32m$, and the work W which gravity does over the distance d is $32md$ or $32dm$. Then, since by (30), $32d = V^2/2$, we have

$$W = \frac{mV^2}{2}. \tag{31}$$

Formula (31), then, is an expression for the work that gravity does in causing an object to fall a distance d and to acquire, at the end of the fall, the velocity V. As we have already noted, this is the work we must do to raise the object from the surface of the earth where it has velocity V to the point where it has velocity zero.

---

* Just a little more mathematics would allow us to demonstrate this more general fact.

Formula (29), namely

$$W = 32Rm\left(1 - \frac{R}{r}\right),$$

also gives the work required to raise an object from the surface of the earth to the distance r from the center. Let $r = R + d$ so that the object is raised to the height d above the surface. Then

$$W = 32Rm\left(1 - \frac{R}{R + d}\right). \tag{32}$$

We now have two expressions, formulas (31) and (32), for the work. We equate them and obtain

$$\frac{mV^2}{2} = 32Rm\left(1 - \frac{R}{R + d}\right).$$

Dividing both sides by m and multiplying by 2 yield

$$V^2 = 64R\left(1 - \frac{R}{R + d}\right). \tag{33}$$

This then is the expression for the initial velocity required to send an object up so that it just reaches the height of d feet above the surface. In applications, the quantities occurring in formula (33) must be expressed in feet.

Formula (33) has a very interesting consequence. If we wished to send an object farther and farther up so that d becomes indefinitely large, then, since R is fixed, the quantity $R/R + d$ approaches zero. The result is

$$V^2 = 64R$$

or

$$V = \sqrt{64R} = 8\sqrt{R}. \tag{34}$$

This velocity is often described as the velocity required to reach infinity and is called the escape velocity. Of course, infinity is not a geographical location, and what is really meant is that the object will keep going out indefinitely and never return. If the initial velocity is less than the escape velocity, then the object will attain zero velocity at some finite, though possibly large, distance from the earth and fall back to earth.

We can readily calculate the escape velocity. Since

$$R = 4000 \cdot 5280 = 21{,}120{,}000 \text{ ft},$$

we have

$$\begin{aligned} V &= 8\sqrt{21{,}120{,}000} = 8\sqrt{2112} \cdot \sqrt{10{,}000} \\ &= 8 \cdot 4600 = 36{,}800 \text{ ft/sec} \\ &= 7 \text{ mi/sec, approximately.} \end{aligned}$$

This is the velocity necessary to escape from our trouble-infested earth.

EXERCISES

1. Calculate the velocity required to send an object 240,000 miles up (this is the distance to the moon) and arrive there with zero velocity. [*Suggestion:* Use (33).]

2. Does formula (34) give the escape velocity from the moon, that is, does it give the velocity required to send an object from the moon to infinity? If not, how should it be modified?

19–7 The equation of the cable of a bridge. We shall now present another type of problem in which integration again provides the key to the answer. We shall derive the equation which represents the shape of the cable of a bridge when the load, that is, the roadway, has the same weight for each foot of *horizontal* distance. For simplicity, we may assume that the roadway, $B'O'B$ in Fig. 19–3, is horizontal, although in an actual case it may be curved. However, in the latter case, it must still be true that the portion of the roadway which stretches over any one foot of horizontal distance has the same weight. The weight of the cable is assumed to be negligible.

Let OP be the section of the cable which extends from the lowest point O to any point P on the cable. This section is in equilibrium; that is, it does not move. Hence the horizontal forces acting on it must offset each other, and the same must be true of the vertical forces. Let us isolate these forces and express mathematically the facts just stated.

The section OP is pulled to the left at O by the entire left-hand section of the cable. This pull or force at O is directed horizontally. We do not know the magnitude of this force, but it is constant, and so we shall denote it by q. At P the portion PQ of the cable above P pulls the section OP. This pull is exerted along the cable, and so at P must have *the direction of the tangent* at P. Let us denote the pull or force by T. There is a third force acting on OP, namely the downward pull exerted by the section of the roadway be-

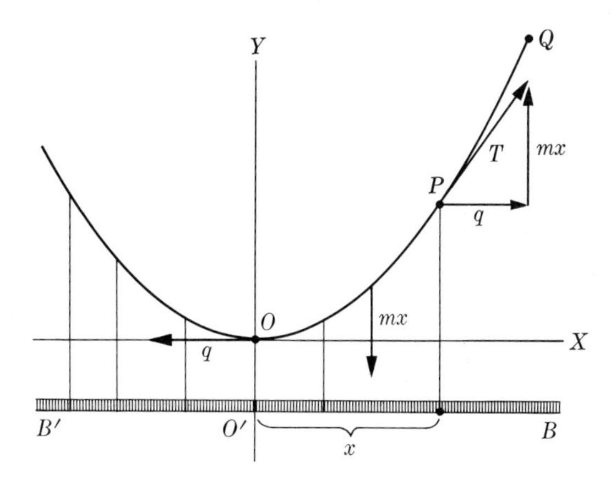

FIG. 19–3. Forces acting on a section OP of a bridge cable.

tween O and P. If the weight of the roadway is m pounds per foot of horizontal distance, and if x is the abscissa of P, then the section of the roadway between O and P weighs mx pounds, and this is the downward force on OP exerted by the roadway.

Now we know that the horizontal forces acting on OP must offset each other. There is the horizontal force q acting at O to the left. The horizontal force which offsets q can only be the horizontal component of T, which pulls to the right. Hence the horizontal component of T is q.

As to the vertical forces on OP, there is the weight, mx, of the roadway and this force is directed downward. The only vertical force which can offset mx is the vertical component of T. Hence the vertical component of T is mx.

A glance at Fig. 19–3 shows that if mx is the vertical component of T and q is the horizontal component, then the slope of the resultant force T is mx/q. But we have already pointed out that the direction in which T acts at P is along the tangent to the cable at P. Then the tangent at P has the slope mx/q. Now the slope of the tangent to a curve is the value of the derivative of the function which represents the curve. Thus, if the formula which represents the cable consists of y as some function of x, then

$$\dot{y} = \frac{m}{q} x. \tag{35}$$

To find the formula which relates y and x, we must integrate (35). Now m/q is a constant, and hence (35) is, in effect, a derivative of the form ax. The formula which has this derivative is

$$y = \frac{m}{q} \frac{x^2}{2} + C. \tag{36}$$

If we choose the coordinate axes as shown in Fig. 19–3, then $y = 0$ when $x = 0$. Thus $C = 0$. We do not know the values of m and q, nor shall we seek to determine them. We have, however, learned a very important fact. The shape of the cable is an arc of a parabola.

19–8 The concept of a differential equation. All problems of the integral calculus considered thus far have the same mathematical form in common. One begins with some physical information which mathematically amounts to knowing the derivative of some function, and one seeks the function possessing that derivative. The information with which we started, for example, $\dot{d} = 32t$, is an equation involving a derivative. An equation which gives information about the derivative of a function is called a *differential equation*. The process of finding the function itself is described as solving the differential equation. Now we managed to solve the differential equations treated in the preceding sections by integration alone, that is, by reversing differentiation. This method is possible only for the simplest differential equations.

We shall now see that the given physical information can lead to much more complicated differential equations. Let us consider the rather simple example of a body dropped in air, without, however, neglecting the air resistance. What equation expresses the physical situation? We know that the attraction of the earth imparts a constant acceleration to an object near the surface of the earth. If the motion occurred in a vacuum, we would write $\dot{v} = 32$. However we wish to take air resistance into account. We shall accept as a fact of physics that air resistance is proportional to velocity; that is, air resistance reduces acceleration by an amount which is a constant times velocity. In symbols, air resistance is cv, where c is some constant whose value does not matter for our purposes. Then the equation which describes the motion is

$$\dot{v} = 32 - cv. \tag{37}$$

This is a differential equation, but it is more complicated than those encountered earlier. If the right side contained only a constant and terms involving the *independent* variable t, we might be able to solve for v as a function of t by integration. However, the right side also contains v, and we can no longer just integrate to determine the functional relation between v and t.

This equation can be solved for v by methods we could readily learn. However, it is not our objective to acquire the techniques of solving differential equations, but merely to recognize that a natural extension of the calculus leads to a vast subject, differential equations, which is extremely important in the application of mathematics to science.

The importance of differential equations arises from the fact that the basic physical principles which man derives from the study of nature turn out to be accurately represented by such equations. The preceding example of motion in air illustrates this statement. Let us consider another. If we neglect air resistance, then for motions near the surface of the earth, we know that the fundamental physical fact is $\dot{v} = 32$. If we wish to consider motions at greater distances from the earth, we must utilize the law of gravitation, and in this case the fundamental physical fact is that the acceleration, or the instantaneous rate of change of velocity, which the earth's attraction imparts to any object is given by the differential equation [see formula (5) of Chapter 17] $\dot{v} = GM/r^2$. Were we to investigate the basic physical principles of light, sound, heat, the flow of fluids and gases, atomic theory, and a dozen other branches of physics and chemistry, we would also find that these principles are, in effect, differential equations. However, the physical laws which contain the most usable information are solutions of these differential equations. We see therefore why the subject of solving differential equations looms so large and why the calculus is really the prologue to a vast body of vital mathematics.

In Chapter 17 we deduced some consequences of the law of gravitation and we mentioned a number of other results which Newton and his contemporaries had obtained. All these were achieved by solving differential equations. We avoided this subject only by making the assumption that the paths of the moon and planets were circular, but, of course, seventeenth-century astronomy was not that crude. So many physical principles are most effectively formulated as differential equations that God has often been credited with using these equations as His starting point in designing the universe.

19–9 The integral as the limit of a sum. In our discussion of integration as a means of finding areas bounded by curves, we mentioned that from Greek times up to the seventeenth century the efforts of mathematicians to determine such areas had not been very successful. The reason, already noted, was that these men tried to use Euclidean geometry, and this geometry is limited in power. To prove theorems about the areas of figures bounded by curves, they had to overcome great difficulties by a special method known as the method of exhaustion. They approximated the area in question by figures bounded by straight lines—for such figures the area could readily be found—and then considered what happened as the approximation was improved more and more. Although, for the moment, it may seem that we are taking a step backward, we shall nevertheless reapproach the problem of area by adopting the Greek view. We shall find that our re-examination will have fruitful results for a new and wide class of problems.

Let us consider the problem [treated in Section 19–4] of finding the area $DEFG$ (Fig. 19–4) which is bounded by the arc FG of the curve whose equation is $y = x^2$, by DE, and by the vertical line segments DG and EF. We subdivide the interval DE into three equal parts, each of length h, and denote the points of subdivision by D_1, D_2, and D_3, where D_3 is the point E. Let y_1, y_2, and y_3 be the ordinates at the points of subdivision. Now y_1h, y_2h, and y_3h are the areas of three rectangles shown in Fig. 19–4, and the sum:

$$y_1h + y_2h + y_3h \tag{38}$$

is the sum of the three rectangular areas and thus an approximation to the area $DEFG$.

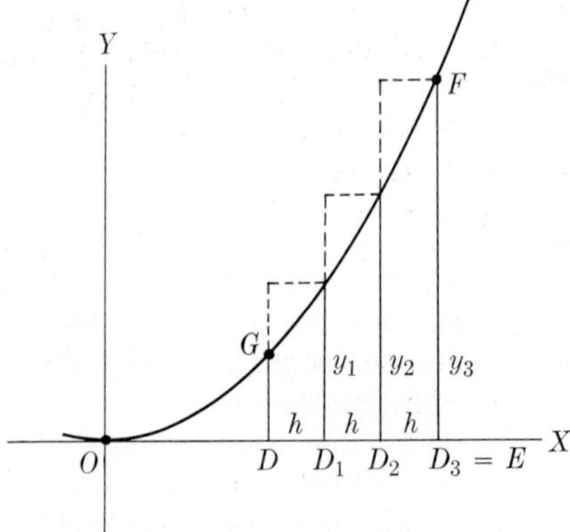

FIG. 19–4. The area under a curve approximated by a sum of rectangular areas.

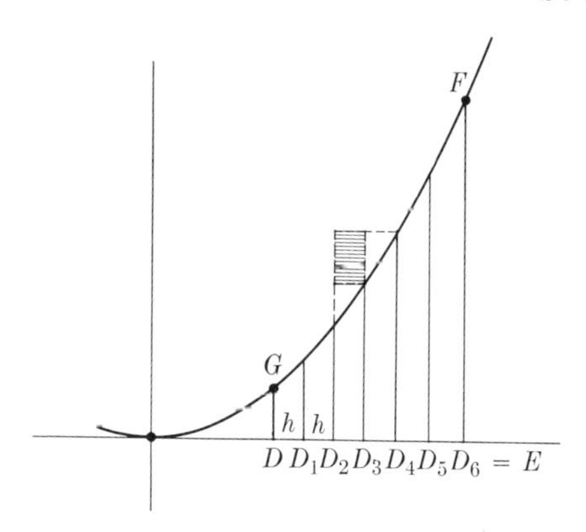

FIG. 19-5. Decreasing the widths of the rectangles improves the approximation provided by the sum of the rectangular areas.

We can obtain a better approximation to the area $DEFG$ by using smaller rectangles and more of them. To illustrate this point, suppose that we subdivide the interval DE into six parts. Figure 19-5 shows what happens to the middle rectangle of Fig. 19-4. This rectangle is replaced by two, and because we use the y-value of each point of subdivision as the height of a rectangle, the shaded area in Fig. 19-5 is no longer a part of the sum of the areas of the six rectangles which now approximates the area $DEFG$. Therefore the sum

$$y_1h + y_2h + y_3h + y_4h + y_5h + y_6h \tag{39}$$

is a better approximation to the area $DEFG$ than the sum (38).

We can make a more general statement concerning this process of approximation. Suppose that we divide the interval DE into n parts. There would then be n rectangles, each of width h. The ordinates at the points of subdivision are $y_1, y_2, \ldots, y_n$, where the dots indicate that all intervening y-values at points of subdivision are included. The sum of the areas of the n rectangles is then

$$y_1h + y_2h + \cdots + y_nh, \tag{40}$$

and the dots again indicate that all intervening rectangles are included. In view of what we said above about the effect of subdividing DE into smaller intervals, the approximation to the area $DEFG$ given by the sum (40) improves as n increases. Of course, as n gets larger, h gets smaller because $h = DE/n$.

We see so far how figures formed by line segments—rectangles in the present case—can be used to provide better and better approximations to an area bounded by a curve. Thus far we have utilized the Greek idea. We now depart from it somewhat and introduce the concept of a limit. Specifically the area $DEFG$ is the limit approached by the sum of the rectangles as the number of rectangles becomes larger and larger, or, one says, as the number of rectangles becomes infinite. Thus the number of rectangles might be successively 3, 6, 12, 24, 48, . . . , where the dots indicate that we continue to double the number indefinitely. Of course, h, the width of each rectangle approaches zero. In symbols, we write

$$\text{area } DEFG = \lim_{h \to 0} (y_1 h + y_2 h + \cdots + y_n h); \tag{41}$$

the symbol "$\lim\limits_{h \to 0}$" means that what we wish to obtain is the number approached by the sum in parentheses as h approaches zero.

What have we accomplished? We seem to have made a simple thing difficult. The innocent area $DEFG$ has been approximated by a sum of rectangles and, as the number of rectangles increases (while each rectangle becomes thinner), the sum becomes a better and better approximation of the area $DEFG$. However, we know from Section 19–4 that the area $DEFG$ may be obtained in the following way: If the equation of the curve FG is $y = x^2$, we find the formula whose derivative is x^2 or, in other words, we find the integral of x^2. This happens to be $x^3/3$. We substitute the abscissa of the point E and obtain a number. We next substitute the abscissa of the point D and obtain a number. Finally, we subtract the latter result from the former. We see, then, that limits of the kind expressed in (41) can be determined by integration and the subsequent numerical work just described.

Now insofar as obtaining areas is concerned, we do not seem to have accomplished very much. Actually, were we to pursue the subject of area a little further than we shall in this book, we would find the new point of view significant in this very connection. However, we shall go on to other applications in which the fact that a limit of the form (41) can be obtained by integration will be the key to the solution.

It is helpful to shorten the writing of an expression such as (41). The notation used in calculus books is

$$\text{area } DEFG = \int_a^b y \, dx. \tag{42}$$

This notation must not be taken too literally. The symbol $\int$ is an abbreviated S and is intended to denote that we are dealing with the limit of a sum. For areas this sum is a sum of rectangles. The y in (42) indicates that the heights of these rectangles are ordinates of some curve, and the dx indicates that the base of each rectangle is a small interval along the X-axis. The number a is the abscissa of the left-hand end point of the interval DE, and the number b is the abscissa of the right-hand end point. The entire expression on the right side of (42) is called the *definite integral* of the function represented by y. The words "definite integral" denote that we are interested in the integral regarded as the limit of a sum.

Whereas Newton had concentrated on finding the derivatives of given functions and on the inverse process, the recognition that limits of sums, such as that expressed by (41), can be obtained by reversing differentiation is due primarily to Gottfried Wilhelm Leibniz (1646–1716). Leibniz's career contrasts sharply with Newton's. Newton, as we know, had undertaken the study of mathematics and physics early in life and had pursued these two fields almost exclusively, although he did make minor contributions to chemistry and theology. His career as a professor gave him the opportunity to concentrate. Leibniz started by studying law at the University of Leipzig, the city in which he was born and lived as a youth. He secured a bachelor's degree at Leipzig and in 1666 a doctor's degree at the University of Altdorf. His first position was that of ambassador for the Elector of Mainz, and until 1672 his interest in mathematics was secondary. In 1672, during a trip to Paris on behalf of his employer, he met Huygens, who acquainted Leibniz with current scientific problems and activities. Leibniz's interests were deeply stirred and thereafter he devoted much time to mathematics. In 1676 he was appointed librarian and councillor to the Elector of Hannover and, although this position also entailed many administrative duties, he nevertheless had more leisure for academic pursuits. In 1700 he went to Berlin to work for the Elector of Brandenburg and, while there, founded the Berlin Academy of Sciences. What is amazing about the man is the vast quantity of first-rate contributions to many fields. Although his profession was jurisprudence, his work in mathematics and philosophy ranks among the best the world has produced. He also did major work in mechanics, nautical science, optics, hydrostatics, logic, philology, and geology, and was a pioneer in historical research. Throughout his life he tried to reconcile the Protestant and Catholic faiths. We may recall also his previously noted activities—his efforts to organize a society devoted to the dissemination of the new scientific knowledge and to turn the German language into a suitable vehicle for the new ideas. No subject pursued by intellectuals of his age was neglected; only Leibniz himself went unrecognized and neglected by his contemporaries.

From our present point of view, Leibniz's emphasis on area as a limit of a sum may seem to be no blessing. But the full import of what he taught, namely that such limits can be evaluated by reversing differentiation, is of vast significance because limits of sums arise naturally in physical problems. Let us consider an example. In Newton's and Leibniz's time and for one hundred years thereafter, one of the major problems was to calculate the gravitational force exerted by one mass on another. If these masses are so compact that they can be regarded as concentrated at points, then the distance between them is the definite distance between these points, and the force of attraction is given by the usual formula. If, however, one mass is the earth and the other is some small object at, say a distance of a few hundred miles from the earth, then, although the latter may in many cases be considered as concentrated at a point, the earth itself cannot be so regarded. The difficulty is that the mass of the earth is distributed over an enormous volume and hence cannot be said to be separated from the mass m by a definite distance.

We can, however, regard the volume of the earth (Fig. 19-6) as broken up into small cubes numbered from 1 to n.* Since each cube is small, a good

* Strictly speaking, there will be pieces left over since a sphere is not a sum of cubes. However, we shall see that these pieces become negligible.

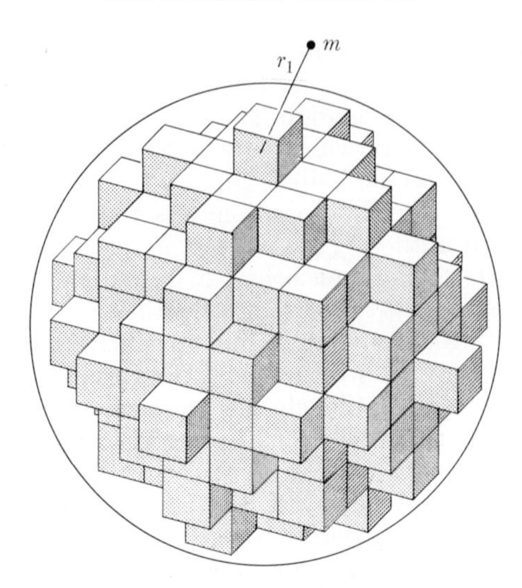

FIG. 19–6. The volume of sphere approximated by a sum of cubical volumes.

approximation of the distance of a cube from the mass m is the distance from the center of the cube to the mass m. Thus if this distance is r_1 for the first cube, then the gravitational attraction exerted by the cube on the mass m is given by

$$F_1 = \frac{Gmh}{r_1^2},$$

where h now stands for the mass of the cube. The same applies to each cube up to the nth cube. Then the total gravitational attraction F of the n cubes is

$$\frac{Gm}{r_1^2}h + \frac{Gm}{r_2^2}h + \cdots + \frac{Gm}{r_n^2}h. \tag{43}$$

Formula (43) is entirely analogous to (40). The quantity y_1 in (40) has now become $GM/r_1{}^2$; y_2 has become $Gm/r_2{}^2$; and so forth. But (43) is not the exact expression for the attraction exerted by the earth because it assumes that each cube acts as though its mass were concentrated at its center. However, if we make each cube smaller, which means that h will be smaller, and increase the number of cubes, n, so that they continue to fill as much of the sphere as possible, the sum of attractions exerted by the n cubes will be a better approximation to the force of attraction exerted by the entire sphere. The reason is that the smaller the cube, the more appropriate it is to assume that its mass can be regarded as concentrated at its center. The exact value of F is

$$F = \lim_{h \to 0} \left(\frac{Gm}{r_1^2}h + \frac{Gm}{r_2^2}h + \cdots + \frac{Gm}{r_n^2}h \right). \tag{44}$$

Formula (44) is exactly like (41). It is now clear that to determine the total gravitational attraction exerted by a distributed mass, the earth in our discussion, we must calculate the limit of a sum. We know that such limits can be calculated by reversing differentiation. The particular limit in (44) cannot be computed with the mathematics at our disposal. The reason is that h in (44) is three-dimensional (because it is mass per cubic foot times volume), whereas the h in (41) is a segment. Hence (44) is a little more complicated than (41). But our example makes the point that limits of sums arise in physical problems, and that we can evaluate them by reversing differentiation.

As a matter of history, Newton solved the very problem we have been considering and proved that the earth attracts a small mass *as if* the entire mass of the earth were concentrated at its center. In other words, although the earth's mass is distributed over a large region, it happens to be true that a spherical mass attracting a small mass can be treated as if its mass were concentrated at its center. The solution of this problem enabled Newton to make further advances in the theory of gravitation. In our discussion of the law of gravitation, we also considered the quantity r to be measured from the center of the earth; that is, we made implicit use of Newton's result.

19–10 Some relevant history of the limit concept. We could continue to study extensions and further applications of the calculus, but there are other features of this mathematical development which take precedence in view of the time we can devote to the subject. In the first place, it is most important to note that the calculus rests on a new concept, the concept of the limit of a function. We employed this concept in two essential ways. In the differential calculus we introduced the instantaneous rate of change of a function. This rate is the limit of the average rate of change of speed, that is of k/h, as h approaches zero. In the integral calculus we used the limit concept to speak of the quantity approached by the sum

$$y_1h + y_2h + \cdots + y_nh$$

as h approaches zero. This limit can represent area, the gravitational force exerted by an extended body, or other quantities, depending upon the physical or geometrical interpretation of the function relating y and x, and of h. Thus it is this new concept which distinguishes the calculus from the branches of mathematics previously studied.

The limit was defined as the number approached by some function of h as h approaches zero. This description is admittedly vague. In particular, the word "approach" is suspect. If for smaller and smaller values of h the ratio k/h should have the values $1/4$, $3/8$, $7/16$, $15/32$, . . . , are these values approaching 1? They are, in the sense of getting closer to 1, but it is also clear that they are always less than $1/2$, and so the limit might very well be $1/2$. In other words, how closely must the values of k/h approach a particular number before we can decide that that number is the limit of k/h? We shall not attempt to give a precise formulation of the limit concept. However, it may be a comfort to know that a precise definition can be given today.

The history of the efforts of mathematicians to grasp this concept properly is instructive as to how mathematics develops. The trouble started early in the seventeenth century. We have already mentioned that many mathematicians of this century made contributions to the calculus, even before Newton and Leibniz began to work on the subject. These forerunners realized that they were unable to give satisfactory expositions of their ideas and, in fact, hardly comprehended the significance of what they were creating. Despite the long tradition of rigorous proof in mathematics, the early workers in the calculus did not hesitate to advance their crude and imprecise ideas and defended themselves in ways that seem strange for mathematicians. Rigor, said Bonaventura Cavalieri, a pupil of Galileo and professor at the University of Bologna, is the concern of philosophy and not of geometry. Pascal argued that the heart intervenes to assure us of the correctness of mathematical steps. Proper finesse rather than logic is what is needed to do the correct thing, just as, he added, the appreciation of religious grace is above reason.

Although Newton and Leibniz made the most significant advances in the formulation of the ideas and methods of the calculus, neither contributed much to the rigorous establishment of the subject. They both realized that they had not presented clearly and precisely the basic ideas of instantaneous rate of change and the definite integral. Yet they were sure that their ideas were sound because they made sense physically and intuitively, and because the methods gave results which agreed with observations and experiments. Of course, mathematical results should be based on definitions, axioms, and deductive proof and not be defended by appeal to experiments. Both gave many versions of their ideas in the attempt to hit upon the precise concept, but neither was successful. In some writings Leibniz went so far as to say that the methods of the calculus were only approximate, but since its errors were smaller than observational or measurable errors, the subject was useful.

The work of Newton and Leibniz was criticized even by their contemporaries. Newton did not reply to the criticisms, but Leibniz did. In addition to defending the methods by an appeal to the agreement of the results with experience, he attacked the critics as overprecise—a strange stand for a mathematician. He also said that we should not lose the fruits of an invention by excessive scruples. Of course, such replies did not provide the missing clarity and rigor. Some writers on the calculus proceeded as though there were no difficulties. Their attitude seemed to be that what was incomprehensible needed no further explanation.

The successors of Newton and Leibniz attempted to supply better foundations for the calculus. However their efforts were blocked in two ways. First of all, the formulations presented by Newton and Leibniz were different, but both yielded correct results; hence the rigorous construction of the calculus had to reconcile the two formulations. Secondly, the whole situation became complicated by an argument between Newton and Leibniz on the question of whether Leibniz had stolen ideas from Newton. Newton's friends, and English mathematicians in general, sided with him, while continental mathematicians defended Leibniz. The quarrel between the two groups became so bitter that they stopped corresponding with each other for about one hundred years. English mathematicians continued to talk about fluents and fluxions, Newton's terms for functions and their derivatives, whereas continental scientists talked about infinitesimals, Leibniz's name for h and k or, in his notation, dx and dy.

Some of the greatest mathematicians of the eighteenth century, Leonhard Euler and Joseph Louis Lagrange, worked on the problem of clarifying the calculus, but without success. Both arrived at the conclusion that as it stood, the calculus was unsound, but that somehow errors were offsetting one another so that the results were correct. A more drastic opinion was offered by the mathematician Michel Rolle (1652–1719). He taught that the calculus was a collection of ingenious fallacies. Voltaire called the calculus "the art of numbering and measuring exactly a Thing whose existence cannot be conceived." Near the end of the eighteenth century the distinguished mathematician Jean le Rond d'Alembert (1717–83) felt obliged to advise his students that they should persist in their study of the calculus; faith would eventually come to them. All eighteenth-century attempts to supply rigorous foundations for the calculus failed.

In the first quarter of the nineteenth century, Augustin-Louis Cauchy (1789–1857), the leading French mathematician, gave the first satisfactory definitions of the derivative and definite integral. Gradually other concepts of the calculus, which we have not considered, were clarified. The differences in notation were also eliminated. In the nineteenth century, Charles Babbage, the designer of large computing machines, and some of his friends formed a society to introduce the Leibnizian notation and to banish the dotage of Cambridge, a pun, of course, on Newton's use of the dot to indicate a derivative. Actually these men were chiefly concerned to propagate the superior continental ideas in England.

This history of the development of the calculus is significant because it illustrates the way in which mathematics progresses. Ideas are first grasped intuitively and extensively explored before they become fully clarified and precisely formulated even in the minds of the best mathematicians. Gradually the ideas are refined and given the polish and rigor which one encounters in textbook presentations. In the instance of the calculus, mathematicians recognized the crudeness of their ideas and some even doubted the soundness of the concepts. Yet they not only applied them to physical problems, but used the calculus to evolve new branches of mathematics, differential equations, differential geometry, the calculus of variations, and others. They had the confidence to proceed so far along uncertain ground because their methods yielded correct physical results. Indeed, it is fortunate that mathematics and physics were so intimately related in the seventeenth and eighteenth centuries—so much so that they were hardly distinguished—

for the physical strength supported the weak logic of mathematics. Of course, mathematicians were selling their birthright, the surety of results obtained by strict deductive reasoning from sound foundations, for the sake of scientific progress, but it is understandable that the mathematicians succumbed to the lure.

It may be clear from this account of the difficulties that mathematicians experienced with the concept of limit that the two chapters devoted to the calculus do not provide a complete description of the concept and all its ramifications. The reader may justifiably feel some vagueness and uneasiness about what has been presented. Further study of the calculus would eliminate these objections. We must also point out that in illustrating the ideas of the calculus we have confined ourselves to the simplest functions. The subject is a vast one and far more powerful than we can indicate with the technique employed here. Indeed an enormous number of new branches of mathematics rest on the calculus. All of these, together with the calculus, constitute a division of mathematics, called analysis, which is considerably more extensive than algebra or geometry. But some glimpse of this most significant mathematical creation of modern times may in itself afford a rich enough reward.

Exercises

1. What essentially new idea does the calculus treat?
2. Mathematicians are logical thinkers; they reason directly and flawlessly to the desired conclusions. Discuss this assertion in the light of the history of the calculus.

Recommended Reading

Bell, Eric T.: *Men of Mathematics,* Chap. 7, Simon and Schuster, Inc., New York, 1937.

Colerus, Egmont: *From Simple Numbers to the Calculus,* Chaps. 24 to 34, Wm. Heinemann Ltd., London, 1954.

Singh, Jagit: *Great Ideas of Modern Mathematics: Their Nature and Use,* Chap. 3, Dover Publications, Inc., New York, 1959.

Smith, David Eugene: *History of Mathematics,* Vol. II, Chap. 10, Dover Publications, Inc., New York, 1958.

CHAPTER 20

THE AGE OF REASON

And the first Morning of Creation wrote
What the last Dawn of Reckoning shall read.

Omar Khayyám

20–1 Introduction. The men of the Renaissance had turned to new sources of knowledge—nature, reason, and mathematics—but were rather vague as to the specific methods by which they were to reconstruct the various branches of thought. Fortunately a series of developments and creations not only gave strong impetus to the urge to rebuild knowledge but actually supplied the method by which positive truths were to be attained. Galileo formulated this method clearly and applied it to obtain the laws of motion of objects moving on and near the surface of the earth. Newton conclusively demonstrated the efficacy of Galileo's method and supplied the doctrine which was to be the keystone in any new system of thought. All bodies in the universe were subject to one set of physical axioms, and their behavior could be deduced from these axioms. Nature was mathematically designed, and natural phenomena adhered strictly to universal mathematical laws which described the behavior of a speck of dust and of the most distant stars.

The mathematical achievements of the seventeenth century gave force and meaning to the ideas which had stirred the Renaissance. As the noted mathematician Jean le Rond d'Alembert (1717–83), chief collaborator of Denis Diderot (1713–84) in the writing of the famous *Encyclopédie,* put it, "The true system of the world has been recognized, developed, and perfected." Natural law clearly was mathematical law.

The intellectual leaders of the seventeenth and eighteenth centuries realized that they now had the tenets which vouchsafed a totally new outlook on the universe and which justified a reconstruction of all of mankind's systems of thought, institutions, and way of life. The right foundation in the form of a mathematical-mechanical explanation was available. New ideas, always far more effective in determining the course of cultures than wars or political events, began to work their influence, and the growing use of books permitted the leaders to reach large groups of people.

D'Alembert spoke for his age: ". . . a certain exaltation of ideas which the spectacle of the universe produces in us . . . [has] brought about a lively fermentation of minds. Spreading through nature in all directions like a river which has burst its dams, this fermentation has swept with a sort of violence everything along with it which stood in its way. . . . Thus from the principles of the secular sciences to the foundations of religious revelation, from metaphysics to matters of taste, from music to morals, from the scholastic disputes of theologians to matters of trade, from the laws of princes to those of people, from natural law to the arbitrary law of nations . . . everything has been discussed and analyzed, or at least mentioned."

The intellectuals were confident that reason, based on the new truths made manifest by the mathematical and scientific work of the seventeenth century and cleansed of the metaphysical and theological presuppositions of the medieval period, could rebuild philosophy, religion, literature, art, political thought, and economic life. They saw clearly that mathematics and science were offering not just a few theorems and isolated results, but a new approach to truths and a new interpretation of the universe. Thus thinking men were impelled to a sweeping reorganization of all knowledge and institutions.

The eighteenth century has been called the Age of Reason. It was not the first period of history in which reason played a dominant role, but it was the first in which the intellectual elite emboldened by some successes in physical science dared to apply reason to the reconstruction of an entire civilization. The spirit and outlook of the leaders are indicated by their reference to their own age as the Enlightenment.

In this chapter we shall discuss some of the new philosophical doctrines which the Age of Reason created. The dominant ones, materialism, mechanism, and determinism, affected profoundly man's view of his universe and of his own power and role. These revolutionary philosophical systems also furnished new points of view to the infant sciences of psychology and biology. In succeeding chapters we shall survey changes in religious thought and in the content and style of literature. In a later chapter (Chapter 28) we shall examine the influence of the new rationalism on political and economic thought.

20–2 Materialism. The first major doctrine which the Age of Reason was to propound was already foreshadowed in the philosophy of science formulated by Galileo. This sage had selected those properties or qualities of the physical world which he believed to be basic. These properties, i.e., matter, extension, solidity, and motion, he called primary qualities, while color, taste, smell, temperature, and hardness, which he called secondary qualities, were due in his opinion to the effect of the primary qualities on human beings. Thus the secondary qualities did not exist in nature but only in humans. The real qualities, those independent of man and external to him, were the primary ones.

Approaching the problem of what is real and knowable about the physical world, René Descartes, who had already seen early in the seventeenth century the necessity of rebuilding all knowledge, arrived independently at the doctrine of primary and secondary qualities. Because he was convinced that all of physics reduced to mathematics (which was largely geometry in his day) and because the primary properties are representable mathematically, Descartes affirmed that the primary properties are the essence of all those that we ordinarily attribute to the corporeal world. All real differences in objects, animal and man included, were differences in extension (shape) and motion of particles.

To comprehend the difference between primary and secondary qualities, Descartes says, take a piece of beeswax. It has sweetness, odor, color, shape, and size; it is hard, cold, and easily handled. Then place it near a fire. Its taste and smell vanish; the color changes; the shape is altered; size increases; it becomes liquid and hot; it cannot be handled; and it emits no sound when struck. Thus the beeswax has changed its properties. Why is it the same piece of beeswax? Because the mind capable of going beyond the senses recognizes the extension and motion of the beeswax. These are the basic properties. All other qualities Descartes reduces to the status of mere accidents.

It was matter that was in motion; it was matter to which forces applied; and it was matter that played a dominant role in the work of Galileo, Huygens, and Newton. Such evidence led readily to the belief that everything in the universe was reducible to matter, and thus the eighteenth century arrived at the doctrine of materialism. Matter was the sole reality. Even those phenomena which had not yet been successfully analyzed, for example, heat, light, electricity, and magnetism, were explained in terms of matter. Since the matter which conveyed light could not be detected, it was called imponderable, the word meaning a substance too small to be weighed. The matter which constituted heat was called caloric, and the matter of fire was called phlogiston. Electricity was for a long time regarded as a fluid.

The doctrine that everything is composed of matter was expressed bluntly by the philosopher Thomas Hobbes (1588–1679) in his *Leviathan* (1651): "The universe, that is, the whole mass of all things that are, is corporeal, that is to say, body, and hath the dimensions of magnitude, namely, length, breadth, and depth; also, every part of body is likewise body, and hath the like dimensions, and consequently every part of the universe is body, and that which is not body is no part of the universe; and because the universe is all, that which is no part of it is nothing; and consequently nowhere." Body, he continues, is something which occupies space, is divisible, movable, and behaves mathematically. Only matter in motion exists, and the world consists entirely of the motions and mechanical interactions of bodies. Hence the mathematical laws describing the physical world are the principles of all reality.

Materialism, then, may be said to assert that all happenings of nature, the whole physical and moral order, are reducible to matter and motion and are completely explicable in terms of these two concepts. Since man himself is part of physical nature, all of man, too, must be explainable in terms of matter and motion.

The doctrine of materialism, though more or less widely held in eighteenth-century Europe, was most heartily embraced in France. The leading French materialist, whose writings were highly influential, was Paul Heinrich Dietrich, Baron d'Holbach (1723–89), a German by birth but French in education and thought. His *System of Nature* (1770) was called the Bible of materialism. He rejected all nonmaterial conceptions, spiritual beings, God, and the soul and "explained" all things in terms of matter and motion. Thought as well as consciousness were not more than molecular motions. Mind is just brain and perishes with the body. Man, he says, has "neglected experience and stuffed himself with systems and with conjectures."

D'Holbach was strongly seconded by the French physician Julien O. de la Mettrie (1709–51), whose *Man a Machine* appeared in 1748. La Mettrie was not only an exponent of the new scientific world view, but, basing his conclusions, about man in particular, on his physiological studies, was sure that all mental processes reduced to material ones. Matter can think and thoughts are particles lodged in the brain.

20–3 Mechanism. A second dominant theme of the Age of Reason was the mechanical view of nature. All observable processes and happenings can be broken down into elementary attractions and repulsions between small particles which obey laws such as the Newtonian law of gravitational attraction and whose movements are dictated by the action of these forces. Every occurrence in nature is no more than the rearrangement of particles moving according to mathematical laws.

The founder of mechanism is Descartes. In his search for simple, clear ideas he decided that nothing better satisfied this criterion than the construction and operation of machines. Hence he sought to interpret the world as a machine and became convinced that he could prove that this was the case. Descartes gave a thoroughly mechanical view of the origin and movement of the planets in his *The World,* which he completed in 1633, but did not publish because he heard of Galileo's condemnation by the Church. [He did publish a revised version in his *Principles of Philosophy* (1644)]. The world arose as matter whirling around fixed centers. Smaller particles collected into larger ones, and thus the planets were formed. Some bodies were swept into eddies around larger ones and so the planets were drawn into the sun's eddy. Each planet was carried about the sun in a whirlpool of ether, or a vortex. The moon was similarly carried about the earth. To please theologians he added that the world was created perfect and complete, but he wished to point out the *possibility* that the world might have developed according to mechanical laws from a less perfect original condition. Descartes used the idea of vortices to explain light, heat, gravitation, and lightning. The vortex theory of planetary motion was later shown by Newton to be inconsistent with the manner in which fluids behave. However, the mechanical view of nature was not shaken, for Newton's own work, while sparse on the precise details of the mechanism, nevertheless stressed matter moving and interacting with other matter.

The largest body of evidence for the mechanical view of nature was the science of mechanics. However, as studies of light, sound, and the elasticity of bodies proceeded, it was found that matter, force, and motion were the keys to these phenomena also. Even the infant science of electricity was approached through motion, for friction was the source of the static electricity available in the seventeenth and eighteenth centuries. The notion that heat was no more than a motion of molecules was also advanced in the seventeenth century, although this idea did not prevail until the nineteenth century.

The concept of the universe as a mechanism grew so strong and popular that the reference to the universe as a clock became a commonplace. Even Newton and Huygens used this analogy. It so happens that a good deal of scientific work in the seventeenth century was directed toward the design of accurate and portable clocks and watches, and by about 1700 good clocks were available and excited admiration. Hence the likening of the universe to a clock was also an expression of respect for the marvelous mechanism of the universe. Of course, clocks must periodically be given motive power, and those who argued that the universe, too, must be supplied with motive power assigned that function to God. The real world appeared to be the totality of mathematically expressible motions of objects in space and time, and the entire universe was a great, harmonious, and mathematically designed machine.

The mechanical view of nature was not universally accepted. Leibniz sought to save mechanism while retaining a place for God, energy, and purpose. He proposed in his *Monadology* (1714) that the universe was a vast living organism composed of particles which in turn also were organisms. Each basic particle, or monad, was indivisible, dynamic, and a living center of energy. Also, each contained its past and future and mirrored the universe. These monads entered into larger congeries of organisms to make up matter, and because they cooperated with each other in a harmony of wills, a living, active, vital world resulted. The mechanical doctrine presented only the outside, the spatial and temporal aspects, but not the inner dynamism. The pre-established harmony of the functioning of the monads accounted for their response to mathematical laws. It so happened that in the seventeenth century the microscope revealed to biologists and chemists a new world of minute living organisms. Leibniz was impressed by these animalcules and referred to the discoveries by Leeuwenhoek, Malpighi, and Swammerdam to support his thesis.

But the thoroughly mechanical view of nature prevailed. As Arthur Clough put it in his poem "The New Sinai":

> Earth goes by chemic forces; Heaven's
> A Mécanique Céleste
> And heart and mind of human kind
> A watch-work as the rest!

The mechanical view of nature persisted through the nineteenth century, long after other opposing doctrines had appeared on the scene. Kant derived Newton's laws of motion from pure reason. Herbert Spencer decided that one could not conceive of motion being destroyed, and that the first law of motion therefore was a necessity of human thought. Hermann von

Helmholtz affirmed that reducing all of nature to laws of attraction and repulsion, was the only rational approach. Many people still hold mechanistic views, and, although later developments, in particular, electromagnetic theory and modern atomic theory, have shaken the doctrine, mechanism continues to be a mode of thought even in modern science.

20–4 Determinism. The most weighty implication of the mechanical view of nature is strict determinism, for a "machine" functioning according to precise mathematical laws will do just what these laws prescribe. There is no uncertainty about what the components will do. Translated into somewhat more technical terms, mechanism implies that the positions and velocities of the individual particles determine their future positions and velocities through such laws as the laws of motion and gravitation.

Since the whole universe functions according to precise laws, the course of the world is determined. For each event there is a fixed preceding and consequent event. Moreover, these laws will not change, said Descartes, because the eternal invariableness of God's will is established by his perfection. That there is an established destiny is as certain as three times three is nine, declared Leibniz. Mathematics describes this destiny, for everything in nature is determined by number, motion, and force. An omniscient mind knowing the state of the universe at any instant could, by applying mathematics, recreate the past and predict the future. As Laplace put it: "We may regard the present state of the universe as the effect of its past and the cause of its future. An intellect which at any given moment knew all the forces that animate nature and the mutual positions of the beings that compose it, if this intellect were vast enough to submit the data to analysis, could condense into a single formula the movement of the greatest bodies of the universe and that of the lightest atom: for such an intellect nothing could be uncertain; and the future just like the past would be present before its eyes." A mechanized, determined world has no ends or purposes. It just goes on existing. The ultimate goals or purposes maintained by medieval theologians are irrelevant.

The deterministic point of view was held so firmly that philosophers applied it to the actions of man as part of nature. Ideas, volitions, and actions are necessary effects of matter acting upon matter. The human will is determined by external physical and physiological causes. Hobbes explained apparent free will thus: Events from without act on our sense organs, and these press upon our brains. Motions within the brain produce what we call appetites, delights, or fears. But these feelings are no more than the presence of such motions. When appetite and aversion jostle each other, there is a physical state called deliberation. One motion prevails, and we say that we have exercised free will. But no choice is really made by the individual. We are conscious of the result, but unconscious of the process which determined it. There is no free will. It is a meaningless conjunction of words. The will is bound fast in the actions of matter.

Voltaire in his *Ignorant Philosopher* states, "It would be very singular that all nature, all the planets, should obey eternal laws, and that there should be a little animal, five feet high, who, in contempt of these laws, could act as he pleased, solely according to his caprice." Chance, also, is nothing but a word invented to express the known effect of an unknown cause.

So disturbing was this conclusion that even materialists sought to modify its severity. Some said that while human actions were determined, thoughts were not. The introduction of this dichotomy was not too comforting, for it meant that thoughts were useless in determining action and man remained an automaton. Others reinterpreted the meaning of freedom in an effort to retain some semblance of it. Voltaire hedged: "To be free means to be able to do what we like, not to be able to will what we like." Apparently we must like what is willed for us in order to be free.

20-5 Theories of knowledge. The Age of Reason accepted new truths and new sources of truths. Instead of doctrines postulating a world concerned with man and his destiny, a God who ruled the universe and could change its course at his pleasure, the spiritual life as the highest goal, and man as a free agent to mold his own destiny, the age chose materialism, mechanism, and determinism. As to the new sources of truth, clearly the physical world itself rather than innate truths or the Scriptures led man to some mathematical principles, and the application of mathematical processes to these principles revealed the less obvious and more profound principles underlying the design of the cosmos. Mathematics particularly was stressed as the foundation and key to knowledge. Typical of the appreciation of mathematics was the declaration of the philosopher Immanuel Kant that the progress of a science could be determined by noting how deeply mathematics entered into its method and contents. The only dissenting voice was that of Jonathan Swift, to whom mathematical astronomy was useful only because the prediction of a solar eclipse gave him time to provide candles.

To thinking beings these changes in the substance and source of truth suggested the question, How do we know what we know? After all, to be sure of what is accepted as truth, one must know that the means of acquiring it is sound. Of course, the Age of Reason had no doubts about its beliefs or the criteria of truth, but the philosophical problem of justifying its convictions remained open.

The question of determining what information we can trust had been tackled by Descartes even before mathematics and science had revealed their full power. We may recall that Descartes had rejected the knowledge which he had been taught in school and had decided to construct his own body of truths after determining the proper sources and methods. Sense perceptions he rejected because they are subject to sense deceptions. How then could he arrive at truths? Though he was the founder of the mechanistic view of nature, he made one large exception: man possessed a soul or mind which is not physical, but is nonmaterial and everlasting. He decided to accept as true only what was clearly and distinctly apprehended by the mind. But the very fact that he was able to think gave him his first truth, namely, that he existed, for a thinking mind meant perforce that there was a thinking being. He discovered further that the mind had clear and distinct ideas of duration, number, and a perfect Being, that is God. These ideas Descartes accepted as innate. The concept of God, for example, could not come from the reaction of the mind to sensations because such qualities as eternity, omniscience, omnipotence, and perfection were not derived from the physical world or from mortal beings. Since Descartes believed that every effect had a cause and could not be greater than the cause, he concluded that there must be a God who caused the idea of God to appear in human minds.

The mind also contained the idea of an external world. Was there one? Of course, since a perfect Being would not deceive us. The same argument led to the belief that the reality of extension, motion, and the laws of geometry was guaranteed; that is, the clear ideas man possessed in these areas and reality had to agree because both stemmed from God. Descartes also considered why the truths which man deduced by pure reasoning should agree with the physical behavior of the universe, and decided that He who made and sustained man and the universe had caused them to agree.

Although Descartes thought that his philosophical and scientific doctrines subverted Aristotelianism and scholasticism, he was at heart a scholastic or Aristotelian because he drew from his own mind propositions about the nature of being and reality. Perhaps for this very reason his writings influenced the seventeenth century more extensively than the researches of those who drew truths from observation and experimentation—sources that were wholly at variance with the traditional ones. However, because he discussed nearly all the physical problems of his day, his speculation became the ferment which induced the gradual clarification of these problems during the next century. Descartes taught men to think for themselves, and this was his greatest contribution. Pupilage to antiquity or to authority was rejected.

But Galileo and Newton did not begin with basic truths. They inferred general principles from experience, and did not presuppose an ultimate metaphysical foundation for them. The fact was original, not the principle. Thus the new method of science stressed experience as the prime source, and yet the mind obtained truths in the form of broad mathematical principles. How did the mind obtain those truths? Every philosopher of the Age of Reason offered his own theory of knowledge. We shall examine a few representative views, primarily to see how philosophers and scientists attempted to embrace the new roles of experience and mathematics and to sanction the new truths.

In his *Essay Concerning Human Understanding* (1690), John Locke (1632–1704) undertook to determine the origin, certainty, and extent of human knowledge. Unlike Descartes, he begins by asserting that there are no innate ideas in men; they are born with minds as empty as blank tablets. Experience, through the medium of the sense organs, writes on those passive tablets and produces simple ideas. Some of these are exact images of qualities actually inherent in bodies, namely, the primary qualities of solidity, extension, figure (shape), motion or rest, and number. Other ideas produced by sensations are the effects of the real properties of objects on the mind, but do not correspond to actual properties. These sensations are the secondary qualities of color, taste, smell, and sound.

God did not endow us with truths, but gave us faculties which enable us to discover all we need to know, namely, the faculties of sensation and reflection. Although the mind cannot invent or frame any simple ideas, it does have the power to reflect on, compare, distinguish, and unite simple ideas, and thus form complex ideas. Complex ideas consist of the agreement or disagreement of ideas, and relationships between ideas such as before and after, cause and effect, and the coexistence of ideas, for example, that yellow and heavy belong to gold. The mind can also form abstractions such as space and time; it can judge; and it can will. Thus, though Locke rejects innate ideas, he grants innate powers to the mind to work with the ideas arising from sensations. The mind recognizes not reality itself, but only ideas of reality and works with these. Knowledge is concerned only with the connection of ideas, such as their agreement or inconsistency.

But how do we know that there is a world of things which corresponds to our ideas? Perhaps we experience sensations which are not caused by any real objects. Locke answers that there is a manifest difference between dreaming of being in a fire and actually being in one. Something real creates the latter sensation.

To the ordinary ideas produced by sensations, Locke adds other ideas arising from intuition and demonstration. Intuition is the *perception* of self-evident truths such as our own existence. Demonstration establishes other truths by connecting ideas. Thus God's existence is established by demonstration.

Locke accepted truths arrived at by demonstration and preferred among them the mathematical truths because the ideas with which they deal were to him the clearest and most reliable. Furthermore, mathematics relates ideas by exhibiting necessary connections among them, and the mind understands such connections best. Locke not only prefers the mathematical knowledge of the physical world, but rejects all physical knowledge. He argues that many facts about the structure of matter are not clear and mentions as an example the physical forces by which objects attract or repel each other. Moreover, since we can never know the real substance of the external world but only ideas produced by sensations, physical knowledge can hardly be satisfactory. He is convinced, nevertheless, that the physical world possessing the properties described by mathematics does exist, as do God and we ourselves. Because he reflected the contents of Newtonian science, his influence on the eighteenth century, like Descartes' on the seventeenth, was enormous.

Another thinker who entered the fray on how we know, was motivated by religious zeal. Bishop George Berkeley (1685–1753), famous as a philosopher and churchman, sought to oppose the growing materialism which, of course, depended upon the primacy of an external world consisting only of matter. In his chief philosophical work, *Principles of Human Knowledge* (1710), Berkeley made a frontal assault. Locke, and Hobbes before him, had maintained that all our ideas were produced by the action upon our minds of external material things. Berkeley granted the sensations or sense impressions and the ideas derived from them, but challenged the belief that they were caused by material objects external to the perceiving mind. Since man perceived only sensations and ideas, there was no reason to believe that anything was external to him. To support his position Berkeley turned Locke's own arguments against him. The latter had distinguished ideas of primary qualities from those of secondary qualities and had maintained that the primary ones corresponded to real properties, whereas the secondary qualities existed only in the mind. Berkeley asked, Could anyone conceive of the extension and motion of a body without including other sensible qualities, such as color? Extension, figure, and motion *per se* were inconceivable. If, therefore, the secondary qualities existed only in the mind, so did the primary ones.

In other words, Berkeley argued that since men knew only sensations and ideas formed by these sensations, but did not know external objects themselves, there was no need to assume an external world at all. That world did not exist any more than the stars one sees when hit on the head. Then, continued Berkeley, "it seems no less evident that the various sensations or ideas imprinted on the sense . . . cannot exist otherwise than in a mind perceiving them." Thus he argued that the mind did exist. Berkeley's extreme views on matter and mind led to the pun, What is matter? Never mind. What is mind? Never matter.

How did Berkeley account for the organization of our sensations which scientists described by the physical laws of an external world? Berkeley said that the mind evokes ideas in the sequences described by these laws. But what caused these sequences or any ideas, for that matter, to appear in men's minds? God, the ultimate reality.

Berkeley had yet to reckon with mathematics. How was the mind able to obtain laws which not only described but *predicted* the course of the external world? What could he do to counter the strongly established eighteenth-century belief in the truths about an external world proffered by mathematics? Roused to do battle with those who opposed science to faith and incensed by Halley's preference for Newton and the *Principia* over Moses and the Ten Commandments, Berkeley decided to attack mathematics itself, the backbone of the new science. He wisely chose to criticize the calculus and picked on the concept of the rate of change of a function. As we stated earlier, this concept was not clearly understood and therefore not well presented by either Newton or Leibniz. Hence Berkeley was able to attack with justification and conviction. In *The Analyst* of 1734, addressed to an infidel mathematician (Edmond Halley), he did not mince any words. Instantaneous rates of change he condemned as "neither finite quantities, nor quantities infinitely small, nor yet nothing." These rates of change were but "the ghosts of departed quantities. Certainly . . . he who can digest a second or third fluxion (Newton's technical name for instantaneous rate of change) . . . need not, methinks, be squeamish about any point in Divinity." That the calculus nonetheless proved practical was accounted for by Berkeley, as well as by some eighteenth-century mathematicians, on the ground that somewhere errors were offsetting one another. Although the above criticism of the calculus was warranted at that time, Berkeley had not actually disposed of all the truths mathematics had produced about the physical world. Nevertheless, having given his opponents something to think about, he rested his case against mathematics at this point. By depriving materialism of its matter and by indicting mathematics, Berkeley believed he had disposed of the physical world and Newtonian science. The reader who consults his delightful *Dialogues of Hylas and Philonous* (1713) will find an extremely able and entertaining defense of his philosophy.

The sceptic Scot, David Hume (1711–76), thought Berkeley had not gone far enough. Berkeley did accept a thinking mind in which sensations and ideas existed. Hume denied the mind too. In his *Treatise of Human Nature* (1739–40), he maintained that we know neither mind nor matter. Both are fictions. We perceive sensations and simple ideas such as images, memories, and thoughts, all three of which are but faint effects of sensations. Any complex idea, for example, substance, is but a word for a collection of simple ideas. The mind is *identical* with our collection of sensations and ideas. We should not assume the existence of any substances other than those which can be tested by immediate experience; but experience yields only sensations.

Hume is equally critical of matter. Who guarantees that there is a permanently existing world of solid objects? All we *know* are our own sensations of such a world. Repeated sensations of a chair do not prove that a chair actually exists. Space and time are but a manner and order in which ideas occur to us. Similarly causality is but a customary connection of ideas. Neither space nor time nor causality is an objective reality. We are deluded by the force and firmness of our ideas into believing in such realities. The existence of an external world with fixed properties is really an unwarranted inference. The origin of our sensations is inexplicable. Whether they arise from external objects, or the mind itself, or God we cannot say. Hence there can be no scientific laws concerning a permanent, objective physical world; such laws signify merely convenient summaries of sensations. Moreover, since the idea of causality is based not on scientific proof but merely on a habit of mind resulting from the frequent occurrence of the usual order of "events," we have no way of knowing that sequences perceived in the past will recur in the future. Thus Hume stripped away the inevitability of the laws of nature, their eternality, and their inviolability.

Man himself is but an isolated collection of perceptions, that is, sensations and ideas. He exists only as such. The ego is a bundle of different perceptions. Any attempt to perceive one's self leads only to a perception. All other men and the supposed external world are just perceptions to any one man, and there is no assurance that they exist.

By destroying the doctrines of an external world following fixed mathematical laws, Hume had destroyed the value of a logical deductive structure which represented reality. But mathematics also contains theorems about numbers and geometry which follow inevitably from supposed truths about numbers and geometrical figures. Hume did not reject the axioms, but chose to deflate the value of the results obtained by deduction. The theorems of pure mathematics, he asserted, were no more than redundant statements, needless repetitions of the same fact in different ways. That $2 \cdot 2$ equals 4 is no new fact; 4 is but another way of saying or writing $2 \cdot 2$. Likewise the theorems of geometry are merely elaborate repetitions of the axioms.

Hume, then, answers the fundamental question of how man obtains truths by denying their existence: man cannot arrive at truths. Hume's work not only vitiated the efforts and results of science and mathematics, but challenged the value of reason itself. But such a conclusion, such a denial of man's highest faculty, was revolting to most eighteenth-century thinkers. Mathematics and other manifestations of human reason had accomplished too much to be cast aside as useless. The supreme philosopher Immanuel Kant (1724–1804) expressed his revulsion for Hume's unwarranted extension of Locke's and Berkeley's theories of knowledge. It appeared indubitable to Kant that man possessed ideas and truths beyond mere amalgamations of sense experiences, and mathematics supplied the evidence.

To Kant all axioms and theorems of mathematics were truths. How could one doubt the truth of the statement that the straight line is the shortest distance between two points? But why, Kant asked himself, was he willing to accept this statement as a truth? Surely experience itself did not vouchsafe it. The question could be answered if one could answer the larger question of how the very science of mathematics is possible. Kant's answer is that our minds possess the forms of space and time. Space and time are modes of perception—Kant calls them intuitions—in terms of which the mind views experience. We perceive, organize, and understand experience in accordance with these mental forms. Experience fits into them as dough into a mold. The mind imposes these modes on the sense impressions received and causes these sensations to fall into built-in patterns. Since the intuition of space has its origin in the mind, the mind automatically accepts certain properties of this space. Principles such as that the straight line is the shortest path between two points, that three points determine a plane, and the parallel axiom of Euclid are part of our mental equipment. The science of geometry merely explores the logical consequences of these principles. The very fact that the mind views experience in terms of the "spatial structure" of the mind means that experience will conform to the basic principles and the theorems. Moreover, since all experience is grasped in terms of the mental framework of space and time, mathematics must be applicable to all experience.

More generally, Kant argues that the world of science is a world of sense impressions arranged and controlled by the mind in accordance with innate categories such as space, time, cause and effect, and substance. The mind contains furniture into which the guests must fit. The sense impressions do originate in a real world, but unfortunately *this world* is unknowable. Actuality can be known only in terms of the subjective categories supplied by the perceiving mind. Hence there never would be another way to organize experience than by Euclidean geometry and Newtonian mechanics.

As experience broadens, as new sciences are formed, the mind does not formulate new principles by generalizing from these new experiences; rather unused compartments of the mind are called into use to interpret these new experiences. The mind's power of vision is lit up by experience. This accounts for the relatively late recognition of some truths, such as, for example, the laws of mechanics, compared with others known for many centuries.

Kant's philosophy, barely intimated here, reinstated reason; however, he assigned to it the role of exploring not nature but the recesses of the human mind. Experience received due recognition as a necessary element in knowledge, since sensations from the external world must supply the raw material which the mind organizes. And mathematics retained its place as the discloser of the necessary laws of the mind. Unfortunately Kant insisted that mathematics must be true and, in particular, that the geometry of the mind must be Euclidean. Nineteenth-century mathematical developments, which we shall explore later, robbed Kant's philosophy of its foundation.

20–6 Psychology. In their efforts to establish how man comes to know truths, the philosophers of the Age of Reason explored the mind. This very analysis of the processes of thought is the basic task of psychology. Thus Descartes, Locke, Berkeley, Hume, and Kant are the founders of psychology. They examined man's mind, the formation of his ideas, and memory; many considered also the nature of feelings, affections, and desires. Even the materialists, who asserted that mind is nothing but a body or a name for some function of the body, were explaining the mind, although in physical terms.

The psychologists, more properly called philosophers in the eighteenth century, also studied the perceptions obtained through our senses. An example of their work is furnished by Berkeley's analysis of the perception of space which he expounds in his *New Theory of Vision* (1709). For Berkeley, space is merely a relationship which the mind abstracts from the regular procession of visual and tactile impressions. There is no solid structure "space" outside of us in which objects have a place. Space is a derived concept, as is mass. The impressions yielded by retina, nerves, and brain are the basis one must use for inferring what, if anything, exists outside. We learn to see just as we learn to read and write. The quick judgments which we learn to make of size, distance, and positions lead us to think that we need only open our eyes to see these properties. This is an illusion.

The uniform space which we normally think underlies the entire external world of phenomena is merely a symbol for a complicated series of sense impressions. In fact, different sense organs really contribute different sensations, each of which imparts to the mind its own concept of space, for example: optical space, tactile space, and kinesthetic space.

It follows from Berkeley's theory of space that beings with different or more powerful sense organs would conceive a different world and perhaps even a different God. Logic, ethics, science, and theology disappear into anthropology. Truth is merely what human beings with particular kinds of sense organs construct from their sensations. Other thinkers of the eighteenth century shrank from this "scandal of human reason."

The philosophers had all proceeded deductively, but their basic principles, such as that all knowledge comes from experience or that the operations of the mind are or are not purely physical, were adopted on the basis of personal convictions. In the early nineteenth century, psychologists began to recognize that psychological theories derived from metaphysical presuppositions were not advancing psychology so much as controversy. They recognized the vagueness, abstractness, and inapplicability of their broad theories which did not yield a science capable of predicting human behavior, supplying objective evidence of the operations of the human mind, or suggesting therapeutic treatment of problem cases.

A new approach, described by Johann Friedrich Herbart (1775–1841) in his *Psychology as a Science Newly Founded on Experience, Metaphysics and Mathematics* (1824–25), advocated that psychology start with empirical data; that is, like the physical sciences, it should measure causes and effects. Then, by applying mathematical reasoning to the principles so obtained, it would become an exact science. Herbart himself did not throw aside metaphysical principles and his work was purely speculative. Other psychologists, especially physicians, turned to experimentation and to measuring sensation and perception of light, sound, and touch. Their objective was the derivation of laws. Ernst Heinrich Weber (1795–1878), brother of the prominent physicist Wilhelm E. Weber, approached psychology by methods taken over from physics and applied measurement to mental and physiological processes. The movement he founded is worth noting.

Everybody knows that an inch added to a man's nose is far more perceptible than an inch added to his height. By experimentation Weber obtained a quantitative statement of this relative perceptibility. He found that in any given sensation, such as weight for example, we perceive equal *relative* differences equally well. Thus we recognize a difference of one-quarter of a pound in one pound as readily as a difference of one pound in four, or two in eight. Weber's law was taken over and elaborated by Gustav Theodor Fechner (1801–87), another famous nineteenth-century physicist and philosopher. Starting from the position that body and mind were just different aspects of one reality, Fechner, like other great thinkers who lived in the shadow of the eighteenth century, at once came to suspect that mathematical relations held between the mental and physical realms. He pounced upon the possibilities in Weber's law and soon found that it could be broadened into the statement that when the stimulus of a sensation increased in geometrical progression, the sensation increased in arithmetic progression. Thus, if the intensity of a light were increased from 10 to 100 candle power, the brightness of the second light would appear to be twice as great as that of the first one; if it were increased to 1000 candle power, the brightness would appear to be three times as great, and so on.

When first expounded, the idea that a mathematical formula could describe human sensations was startling to psychologists. But the Weber-Fechner law is accepted today with minor limitations. Fechner went on to explore the philosophical implications of his contribution, the interdependence of body and soul. Other psychologists ignored the philosophical digressions but took up the more solid content of Fechner's work. The concept of measurable sensations was the first step in the development of experimental psychology, which ultimately turned to the laboratory rather than to the mind as the source of its laws. Hence the rest of this story belongs to our later work. Suffice it to say here that for many people today psychology means tests and measurements.

20–7 Biology. The Age of Reason had proclaimed that man, as part of nature, was necessarily part of the great machine. His body and its behavior were nothing but matter subject to, and acting in accordance with, the same forces and laws as the planets or a projectile shot into space. Biology was a direct beneficiary of this belief.

The doctrine of mechanism in biology was founded by the same man who first formulated the doctrine for the physical world, René Descartes. Although Descartes did reserve a place for the soul in the pineal gland and granted the soul the power to receive sensations and to direct the resulting actions of the body, in all other respects he treated the body as a machine and insisted that the functions of the body in receiving sense data and in executing the will of the soul could all be explained in terms of mechanics. Mechanical laws governed the human organism and organic life. The body was just a collection of particles subject to the laws of motion and heat; pushes and pulls determined all bodily actions.

Descartes' *Dioptrics* (1637) contains his investigations of the functioning of the eye. He removed the retina from the eye of an ox, replaced it by a thin piece of paper, and thus could see how the eye handled the light which passed through the liquid and lens and struck the retina. He also explored the role of the iris, ciliary muscles, lens, and the phenomenon of binocular vision.

After studying the functioning of the eye, he postulated that the images received by the retina were conveyed by nerves to the pineal gland. The soul, situated in this gland, inclined the gland; this motion activated nerves leading to muscles, and these caused some part of the body, a hand, say, to move. The pineal gland was for Descartes the connection between mind and matter, a thinking substance interacting with a material substance.

In his *Traité de l'homme,* he showed how the human body could be conceived as a mere machine. He tried to explain the beating of the heart in terms of the laws of motion. The phenomenon of muscular action was clarified by application of principles of the lever. He proposed models which reproduced the behavior of the various organs of the body, and attempted to verify the accuracy of these models by laboratory work. He shed much light on the reflex movements of the body and explained the process of ingesting and digesting food and the circulation of blood.

William Harvey (1578–1657), who shares with Descartes the honor of founding modern physiology, also subscribed to the mathematical-mechanical view of the functioning of the human body. His proof (1628) that blood indeed circulates within the body was regarded as a triumph for this view. Not only did he use quantitative arguments to prove that the same blood must flow throughout the body many times a minute, but he likened the heart to a pump which forces the blood around. Thus the mechanical action of the heart alone accounted for the circulation of the blood. Descartes greeted joyously Harvey's confirmation of the mechanical view.

The progress of physiology and medicine from the days of Descartes and Harvey was not rapid. Harvey worked directly from observational and experimental data, whereas Descartes had started with metaphysical presuppositions, so that a schism actually developed between the followers of the two men. There was also a school of vitalists who maintained that the actions of living creatures transcended the laws of physics. Nevertheless, the physicians of future generations continued to search for a mechanical or chemical explanation of bodily actions and disease. The conception of man as a machine placed the emphasis on the body rather than on the soul, and the conception of the entire universe as a machine united plant, animal, and human biologies.

We may pass quickly over the intervening years to the work of the confirmed materialists, d'Holbach and la Mettrie. The latter's *Man a Machine* (1748) popularized the view that man differed in no essential respect from plants and animals and that all were machines. D'Holbach said that man differed from lower animals or any inanimate matter only as Huygens' clock differed from a primitive time piece. The seeming chasm between inert matter and vital phenomena is insignificant. Thought was no more than motions in the brain. The observation that a turkey can run around for quite some time after its head is cut off convinced la Mettrie that the body was capable of functioning without the mind. Indeed la Mettrie's materialism, discussed earlier, was founded on his study of physiology.

On the basis of his studies Diderot decided that there was no distinct line between plants and animals or between animals and men. Sensation was a general property of all matter, and thought no more than sensation. The recognition that pain, pleasure, the passions, wine, narcotics, and other physical factors affected the mind as well as the body confirmed his hypothesis that mind was nothing without the body.

The doctrine of mechanism led to the idea that the higher forms of animal life might have arisen from lower ones by a natural process of development proceeding in accordance with mechanical laws. The concept of such an evolution seemed justified, for careful examination of the functioning of organs in men and animals showed that man is a more complicated form of animal life and that there are almost imperceptible gradations from man down to almost formless matter.

The theory that the human body is a mechanism has proved immensely important for biology and medicine. Our vast knowledge of the functioning of the sense organs, the nervous system, the brain, and the heart was acquired by pursuing the mechanical view of nature. Moreover, as physics and chemistry advanced and as our knowledge of light, sound, electricity, and of atomic and molecular structures increased, our knowledge of the human body, its functioning, and the causes and cures of diseases progressed in turn. Nevertheless, it is less possible to say today that human life is merely a series of mechanical actions and reactions. As we shall discover when we study wave motions, the very progress of physics has cast doubt on the mechanical view of nature.

20-8 Respite. We have seen so far how the militant rationalism engendered by the success of mathematics and science invaded the field of philosophy, created psychology, and empowered biology. In succeeding chapters, we shall consider the influences on religion, literature, art, and the social sciences. Let us note once again the leading motifs: the supremacy of human reason, the primacy of matter and motion, the mechanical view of nature including man, and by virtue of the invariability of the laws governing nature, a completely determined world. The Age of Reason began with the recognition of the importance of the individual equipped with the established power of reason. Before it had run its course it had reduced man to an insignificant bit of matter in a mechanically determined universe.

Exercises

1. What is the doctrine of materialism espoused by eighteenth-century philosophers?
2. What is the doctrine of mechanism espoused by eighteenth-century philosophers?
3. What is the doctrine of determinism espoused by eighteenth-century philosophers?
4. How does mathematics enter into the doctrines of mechanism and determinism?
5. Why did the seventeenth- and eighteenth-century philosophers feel compelled to reconsider the processes used by man to obtain truths?
6. What status does mathematical knowledge have in Locke's theory of knowledge?
7. Contrast the philosophies of Hobbes and Locke, on the one hand, and that of Berkeley, on the other, with respect to the reality of material objects.

8. Contrast the philosophies of Locke and Berkeley, on the one hand, and that of Hume, on the other, with respect to the existence of truths.

9. How does Berkeley indict mathematics?

10. What is Hume's position with respect to determinism and hence predictability of the future?

11. Why, according to Kant, do the theorems of mathematics apply to the external world?

Topics for Further Investigation

1. The philosophy of August Comte.
2. New concepts of history (see also Chapter 28).
3. The rise of the concept of progress.

Recommended Reading

Berkeley, George: *Three Dialogues Between Hylas and Philonous,* Open Court Publishing Co., Chicago, 1929.

Becker, Carl L.: *The Heavenly City of the Eighteenth-Century Philosophers,* Yale University Press, New Haven, 1932.

Boring, Edwin G.: "Gustav Theodor Fechner," in James R. Newman: *The World of Mathematics,* Vol. II, pp. 1148–1166, Simon and Schuster, Inc., New York, 1956.

Bronowski, J. and B. Mazlish: *The Western Intellectual Tradition,* Part II, Harper and Bros., New York, 1960.

Bury, J. B.: *The Idea of Progress,* Dover Publications, Inc., New York, 1955.

Cranston, Maurice: *John Locke,* Macmillan Co., New York, 1957.

Hazard, Paul: *European Thought in the Eighteenth Century,* Yale University Press, New Haven, 1954.

Hazard, Paul: *The European Mind,* Yale University Press, New Haven, 1953.

Höffding, Harald: *A History of Modern Philosophy,* 2 vols., Dover Publications, Inc., New York, 1955.

Randall, John Herman Jr.: *The Making of the Modern Mind,* Chaps. 9, 11, and 21, Houghton Mifflin Co., Boston, 1940.

Russell, Bertrand: *A History of Western Philosophy,* Chaps. 12 to 20, Simon and Schuster, Inc., New York, 1945.

Sampson, R. V.: *Progress in the Age of Reason,* Harvard University Press, Cambridge, 1956.

Smith, Preserved: *A History of Modern Culture,* Vol. I, Chap. 7, Vol. II, Chaps. 3 to 5, and 11, Henry Holt and Co., New York, 1930.

Snyder, Louis L.: *The Age of Reason,* pp. 7–31, D. Van Nostrand Co., Inc., Princeton, 1955.

Vartanian, Aram: *Diderot and Descartes,* Princeton University Press, Princeton, 1953.

Wolf, Abraham: *A History of Science, Technology and Philosophy in the Sixteenth and Seventeenth Centuries,* 2nd ed., Chaps. 24 and 26, George Allen and Unwin Ltd., London, 1950.

Wolf, Abraham: *A History of Science, Technology and Philosophy in the Eighteenth Century,* 2nd ed., Chaps. 28, 31, and 32, George Allen and Unwin Ltd., London, 1952.

CHAPTER 21

RELIGION IN THE AGE OF REASON

Enough! There is a God; of him we see some trace
Whenever steadily we look in Nature's face.
Haller

No man knows about the Gods more than another.
Herodotus

21–1 Introduction. The Age of Reason made a profound impress on religion as it did on almost all branches of our culture. Several specific features of the Enlightenment caused a re-examination of acceptable religious beliefs. First of all, prior to the seventeenth century, even those theologians who did their utmost to provide the Christian religion with rational foundations, notably Thomas Aquinas, perforce resorted to faith as the only basis for accepting some doctrines, and did not hesitate to advocate reliance upon faith. But the accomplishments of reason in the seventeenth century had exalted that faculty to the point where rational arguments were considered to be the only reliable support for truths. Secondly, the Christian religions had embraced all of man and nature and had dictated what man should believe about his universe, his own role in it, his fate, and the standards of his conduct on earth. But the new world picture constructed by Newton and rapidly developed by his eighteenth-century successors presented an entirely different account of the origin, structure, and functioning of the physical world. Moreover, the implications of this new world view concerning man, which many were quick to perceive, contradicted Christian dogmas and teachings. A resolution of the two conflicting doctrines struck the intellectual leaders of the times as imperative. Because the new approach to religion was stimulated by and rested upon the mathematical accomplishments of the age, it is worthy of our attention.

21–2 The grounds of the conflict. Let us attempt first to pin down the issues between the champions of the new mathematico-mechanical world view and the champions of the traditional religions.

We have already mentioned that one source of conflict was the superiority granted to reason over all other bases for knowledge, such as authority, revelation, and faith. The issue was stated by a man who regarded himself as a most ardent Christian. Robert Boyle affirmed, "I look upon the metaphysical and mathematical principles . . . to be truths of a transcendental kind that do not properly belong to philosophy or theology; but as universal foundations and instruments of all the knowledge we mortals can acquire." Thus Boyle was distinguishing between two kinds of knowledge and handing the palm to that derived from scientific thought. Descartes, another deeply religious scholar, did not hesitate to affirm the supremacy of reason. What he would and did accept in the sphere of religion had to be first established by reason. As the successes of science mounted, the rational spirit was more and more glorified. Reason had produced the new science, and so reason was exalted at the expense of faith, which was labeled credulity.

A basic Christian tenet proclaims the omnipotence of God. But the mathematicians and scientists of the seventeenth century limited the power of God, often without recognizing this implication of their work. Descartes had already said that God could not abolish extension or motion, and his philosophy of the invariability of the laws of nature collided with an active Providence. Newton, like Descartes, credited God with the initial act of creation. Furthermore, the unexplained irregularities in the motions of the heavenly bodies which might cause serious disturbances in the lawful functioning of the universe prompted him to credit God with correcting these irregularities and keeping the universe in order. But this was the whole extent of the power accorded to God. He was, to use a figure of speech very commonly employed in those days, just a watchmaker who had designed and set a clock into operation, but who thereafter only occasionally adjusted the mechanism so that it kept correct time. Huygens and Leibniz went further and denied that God's intervention was needed to repair the mechanism. Although they were unable to say how the irregularities in the functioning of the universe unexplained in their time were corrected, they believed that it was an insult to God to think that His handiwork might need occasional adjustment. The substance of these assertions was to reduce God, who in medieval thought was the lord of all activity and purpose in the universe, to just a means to an end. The end itself was the regular, precise functioning of all operations of the universe.

The culmination of this line of thought, which assigned to God the functions of creator and caretaker, was reached when Lagrange and Laplace showed that the supposed irregularities in the motions of the moon and planets were really periodic and would not, if unattended to, ultimately disrupt the universe. Hence even the corrective measures which Newton required of God were seen to be unnecessary. Moreover, Lagrange and Laplace believed that the universe had always existed infinitely far back in the past. There was no need even to suppose that God had created the universe. God was unnecessary, a superfluous assumption. Laplace had no hesitation in drawing this conclusion, as can be seen from the following well-known story concerning an exchange between Laplace and Napoleon. The latter once asked Laplace whether it was true that in his monumental work, the *Mécanique Céleste,* he had made no mention of God. Laplace's answer was that he had no need for this hypothesis. He was able to describe the workings of the universe by mathematical laws which sufficed in themselves.

Unwittingly, the mathematical physicists of the seventeenth century had also set up a new Bible, the independent and original truths of nature as opposed to the Scriptures. Not God's word but His work, not the testimony of the Scriptures but that of the visible world, as revealed by mathematics of course, were the new documents. The truths of nature outclassed in brightness, perspicuity, and precision the truths of the traditional Holy Writings.

In addition, the philosophies of mechanism, materialism, and determinism, which we examined briefly in the preceding chapter, presented further challenges to orthodoxy. The mechanistic doctrine stated that the universe was just a vast, impersonal machine; events took place not because God ordered them to occur, but because they were necessary elements of a fixed mathematical plan. Mechanism denied not merely the will but the intent of God. The conception of the universe as a machine did not allow for human desires, goals, and needs, and prayers to God to intercede for humans were futile. At best man could be part of the huge machine, filling roles dictated to him by the laws which governed all matter. In brief, the unvarying adherence of natural phenomena to the mathematical laws discovered by Galileo, Newton, Huygens, Laplace, and Lagrange discredited the belief in an active Providence concerned with, and able to do something for her ward, man.

Similarly devastating implications were evident in the doctrines of materialism and determinism. Materialism accepted as real only the mathematically describable properties of matter, such as size and shape, and denied the existence of mind and soul. It inevitably led to the conclusion that man was just matter and hence incapable of doing more than merely responding to the forces applying to all matter. Since there was presumably no soul, there could be no after-life. All preparation for this after-life was therefore pointless. Determinism denied free will; hence man could not be guilty of sin. Moreover, the uniformity and invariability of nature's behavior did not permit deviations and thus denied the occurrence of such events as the virgin birth, resurrection, and other miracles. As Alexander Pope put it, "Order is Heaven's first law."

Rationalism also opposed and dispelled the emotional appeal of the orthodox religions. Even the rational arguments for God, for example those based on design, presented Him as an intellectual abstraction which the mind had to appreciate. This God lacked the appeal of a mysterious superhuman force previously presented by the more imaginative, poetic, and emotional literature. The god demonstrated by logical arguments was no more than a faint vision.

21-3 The beliefs of the creators of the Age of Reason. The developments in eighteenth-century religious thought were not a consequence of a planned attack on religion. In fact, insofar as intentions were concerned, the situation was just the opposite. Let us recall that the leading mathematicians and scientists of the Renaissance were deeply religious men. They had indeed turned to the study of nature, but nature was the work of God, and so the study of nature was literally the study of God. Whereas medieval theologians had regarded the physical world as tainted and ignored it as though it were the work of Satan, the Renaissance scientists considered nature a divine creation and sought to restore this domain to God. They thought that reason would enable man to see God's wisdom, law, and order made manifest in His noblest creation. Study of the physical world was a religious quest to be pursued as piously and as devoutly as scholastic theologians had examined every syllable of the Scriptures.

The religious views of Copernicus and Kepler have already been noted. Galileo, too, though he preferred to separate religious from scientific problems, was devout and said that God did not less admirably reveal Himself in nature's ways than in the study of the sacred Scriptures. Robert Boyle, father of modern chemistry, devoted most of his time outside the laboratory to religion and regarded even his experimental work as a service to God. In his will he left funds to combat atheists, skeptics, and other infidels. Isaac Barrow (1630–77), Newton's teacher and predecessor as professor of mathematics at Cambridge, was primarily a theologian. He had found that, to be a good theologian, he had to know chronology and to know chronology meant to know astronomy, which in turn implied knowing mathematics. Hence he studied mathematics and made basic contributions to the calculus. After serving as a professor of geometry at Cambridge, he resigned in 1669 in favor of Newton and took up spiritual tasks and theological studies. We have already indicated* that Newton, too, was deeply religious. He considered the strengthening of the foundations of religion more important than his mathematical and scientific studies, for the latter were restricted merely to uncovering God's design of the natural world. He often said that the hard and at times dreary scientific work was justified only because it provided evidence of God's design of the universe.

Newton saw everywhere in the universe proof of God's majestic design and of His continuous effort to keep the universe running according to plan. In his words, "The main business of natural philosophy is to argue from phenomena without feigning hypotheses, and to deduce causes from effects, till we come to the very first cause, which certainly is not mechanical . . . What is there in places almost empty of matter, and whence is it that the sun and planets gravitate towards one another, without dense matter between them? Whence is it that nature doth nothing in vain; and whence arises all that order and beauty we see in the world? To what end are comets, and whence is it that planets move all one and the same way in orbs concentric, while comets move all manner of ways in orbs very eccentric, and what hinders the fixed stars from falling upon one another? How came the bodies of animals to be contrived with so much art, and for what ends were their several parts? Was the eye contrived without skill in optics, or the ear without knowledge of sounds? How do the motions of the body follow from the will, and whence is the instinct in animals? . . . And these things being rightly dispatched, does it not appear from phenomena that there is a being incorporeal, living, intelligent, omnipresent, who, in infinite space,

---

* See Chapter 17.

as it were in his sensory, sees the things themselves intimately, and thoroughly perceives them; and comprehends them wholly by their immediate presence to himself?"

It was clear to Newton that the universe which the science of his century had revealed must have been designed by a God who was a mathematician: "To make this [solar] system, therefore, with all its motions, required a cause which understood, and compared together the quantities of matter in the several bodies of the sun and planets, and the gravitating powers resulting from thence; the several distances of the primary planets from the sun, and of the secondary ones [i.e., moons] from Saturn, Jupiter, and the earth; and the velocities with which these planets could revolve about those quantities of matter in the central bodies; and to compare and adjust all these things together in so great a variety of bodies, argues that cause to be not blind or fortuitous, but very skilled in mechanics and geometry."

Hence religious convictions were seemingly strengthened by the work in mathematics and physics. The order of nature vouched for God. And indeed well into the eighteenth century we find the scientists thoroughly satisfied that God's existence was securely established. Leonhard Euler, the greatest of the eighteenth-century mathematicians, even believed that God's existence could be proved directly by means of principles of mathematics and physics and that one could dispense with the indirect argument of the marvelous design of the universe.

Despite the intentions of eminent mathematicians and scientists and despite their convictions as to the presence and role of God, other men began to see the disturbing implications of the new scientific doctrines and of the philosophies built upon them. Early in the seventeenth century, Blaise Pascal recognized that the study of nature was not providing clear evidence for the existence of God. Though intensely religious by training and nature, Pascal looked beyond the world so neatly described by science and was disturbed to find misery, sickness, death, and chaos in the affairs of men. In real despair, he expressed his bewilderment: "This is what I see that troubles me. I look on all sides and I find everywhere nothing but obscurity. Nature offers nothing which is not a subject of doubt and disquietude; if I saw nowhere any sign of a Deity I should decide in the negative; if I saw everywhere the signs of a Creator, I should rest in peace in my faith; but seeing too much to deny and too little confidently to affirm, I am in a pitiable state, and I have longed a hundred times that, if a God sustained nature, nature should show it without ambiguity, or that, if the signs of a God are fallacious, nature should suppress them altogether: Let her say the whole truth or nothing, so that I may see what side I ought to take." Pascal concluded that nature proved God only to those who already believed in him. God is known to the heart and not to the reason. Those who would believe in God must surrender to faith. Thus Pascal ended by defending religion, but only by separating it from science and placing it beyond science.*

---

* See Chapter 11.

Many other mathematicians and scientists who were personally quite satisfied with the evidence for the existence of God discerned the threat to religion inherent in their work and did their best to combat growing doubts and adverse interpretations of the evidence.

Newton, himself, was aware of the need to defend religion. He says in a letter written in 1692, "When I wrote my treatise [*Mathematical Principles*] about our system, I had an eye on such principles as might work with considering men for the belief of a Deity; and nothing can rejoice me more than to find it useful for that purpose."

Leibniz wrote several books in which he expounded his religious views and argued against dissenters. His *Testimony of Nature Against Atheists* tried to prove that some natural phenomena were better explained by the existence of God than by scientific description, and his very famous *Essais de Théodicée* (1710) repeated the already familiar argument for God as the intelligence who created the world. The world, he contended, was a huge mechanism designed to serve divine ends and to fulfill a pre-established harmony. Leibniz's entire philosophy was directed toward explaining how, in this "best of all possible worlds," an active, omnipotent Being could produce a mechanically functioning universe. His attempt to unite teleology and mechanism was founded on the thought that the forces which made this perfect and constant functioning possible emanated from the Divine Being.

Of course, some philosophers and theologians also rushed to the defense of religion in innumerable books and articles. The theologian William Bentley sought to use Newton's work as proof of the existence and wisdom of the Creator, and even asked Newton for books which would help him master the *Mathematical Principles*. Bishop Berkeley's attack on mathematics, which he recognized to be the chief weapon in the scientific invasion of religious grounds, was discussed in the preceding chapter. Berkeley was, in effect, saying that if we theologians speak nonsense, then so do you mathematicians.

All the efforts to defend orthodoxy, including the clearly stated beliefs of the most respected Newton, did not suffice to halt the attacks on the existing religious edifices. The results obtained in mathematics and science were used as ammunition for a crusade against orthodox religions; in fact, Newton's own name became the symbol for revolt. Many philosophers and intellectual leaders decided to re-examine the basis for religious belief in general and for the specific tenets of the Christian religions. Their goal was either to provide a creed erected on the basis of the new knowledge or to see which of the traditional doctrines could still be maintained as consistent with it.

21–4 Rational movements in religion. The problem of refounding religion to suit the intellectual temper of the eighteenth century and to harmonize with the new conceptions of the universe occupied the best minds of the Age of Reason. What characterizes all the new movements is the effort to make religion conform to reason. The title of a book published by the philosopher Immanuel Kant toward the end of the century, *Religion Within the Limits of Reason Alone,* epitomizes the position of the intellectual. Since the eighteenth century identified reason and nature, for reason merely brought forth or made explicit what was inherent in nature, the leaders expected that the application of reason to the problem of religion would produce the natural religion, i.e., a religion which was a reflection of the nature of the universe and man just as the mathematical laws of science were but an expression of the essence of nature. This natural religion would render Christianity, Judaism, and Mohammedanism superfluous. However, because the problem treated by religion is the most difficult which man faces, the intellectuals did not come forth with a unique answer.

One of the earliest prominent rational movements, often called rationalistic supernaturalism, limited its efforts to the reconstruction of Christianity. It possessed an able leader in John Locke. Approaching the problem mathematically, Locke asked in his *Reasonableness of Christianity* what axioms might be adopted as a basis for religious truths. Locke was satisfied that the principle of causality, that effects have causes, was intuitively valid. Hence, he argued, we had to accept an omnipotent God, for our own existence and the existence of a wisely contrived physical world surely must have a cause and this was God. Since there existed a God who can design and create at will, it was equally axiomatic that man should live in obedience to the will of God. From such axioms Locke drew the conclusions that man must live in accordance with the moral principles of Christianity and should seek to attain his reward in heaven. Thus reason led to some of the Christian doctrines.

However, Locke was not able to deduce from his axioms all the doctrines he was disposed to accept. And so he admitted revelation as an extension of religion based on reason. To remain somehow within the bounds of reason, he added that reason was to judge whether a revelation really came from God. Thus reason selected only those revelations which contained clear knowledge. On this basis Locke decided that resurrection of the dead was credible, but transubstantiation was not. God could cause the sun to stand still but could not change the ratio of the circumference to the diameter of a circle. To further justify his admission of revelation, Locke explained that reason was, in fact, revelation; that is, through reason God communicated truths accessible to our natural faculties. Apparently then, revelation *per se* was not any strange or uncommon process.

Locke also wished to admit certain miracles and so he added, in his *Letter on Toleration* (1685), that religion by its very nature involved a superior power who must be capable of some miracles.

The admission of revelation and miracles was objectionable to those who wished to be thoroughly rational and yet defend some of the Christian truths. Hence the philosophers who did not like Locke's justification of these two classes of phenomena preferred to argue that our judgment of what is natural as opposed to what is supernatural depended upon our capacity to understand; to superior beings all doctrines of Christianity might appear natural. Moreover, what men judged to be miracles might really be natural events beyond human experience just as snow might appear to be a miracle to natives of the tropics.

These efforts to defend revelation and miracles clearly involved transcending what human reason could encompass. At least, this is what the leading group of educated people, notably in France, contended. These men therefore founded a new religion, deism, which departed far more from Christianity than did rational supernaturalism. The deists, led by Voltaire, a champion of Newtonian mathematics and physics, aimed to erect a strictly intellectual system. They would have none of mysteries, miracles, secrets confided by God to select individuals, or the authority of the Bible. They also rejected passive conformity to the dictates of authority. Viewed from a more positive standpoint, the deists believed that there was a natural religion whose basic principles could be apprehended by studying the physical world and man. From these basic principles, others could be deduced by rational demonstration. Belief in the doctrines so obtained was the necessary outcome of reasoned judgment. The deists were explicit that Jews, Christians, Mohammedans, and heathens were heretics because these groups departed from the natural religion, which alone is authentic.

What doctrines did the deists derive from the study of nature? By nature the deists meant not sensually observed phenomena such as leaves and trees, but the inner design as portrayed by mathematical laws. From this aspect of nature the existence of God (though not necessarily an omnipotent one) was assured because the universe was unmistakably planned—witness the Newtonian laws—and such a universe could not be an accident.. Given the existence of God, one could rather readily deduce a future life in which man would obtain his just rewards. Sin was disobedience to the dictates of reason and common sense. Because the ten commandments appealed to common sense, they should be obeyed. Worship of God and repentance for sins were valuable only because they made for better living on earth. Dogma was far less important than practice. Since morality was to the deists the essence of religion, they desired more morality and less dogma.

At first glance these doctrines do not appear to deviate too sharply from Christian dogma. But since the deists refused to accept what they thought could not be defended by reasoned deductions from the facts of nature, they rejected the divinity of Christ, the virgin birth, the doctrine of original sin, miracles, special providences, and revelation. The authority of the clergy was denied, and that of the Bible limited to symbolic or figurative meaning. Voltaire mocked the Biblical account of creation and of some of the early events such as the fall of man. Geology, he remarked, breaks down the theology of creation and "Biblical physics" was his term of contempt for the cosmogony of the Scriptures. We can see why the orthodox Christians called the deists atheists and why those more sympathetic were content to characterize deism as the rationalist wraith of a departed faith.

Among the intellectuals of the eighteenth century deism was the most widely accepted doctrine. Its followers included Thomas Jefferson, Benjamin Franklin, and the first seven presidents of the United States of America. Of course, many of these referred to the Christian God in political speeches, but they were, as John Adams put it, "scanty Christians." Adams himself described the prevailing Christian religions as monsters.

Among those who sought to apply reason to the reconstruction of Christianity some concluded that it should be demolished entirely. The philosophers and leading intellectuals Hobbes, Hume, Montaigne, Diderot, d'Alembert, the historian Edward Gibbon, and the French materialists, notably d'Holbach and la Mettrie, attacked all religious beliefs. Hobbes described formal religions as accepted bodies of superstitions. "Fear of power invisible feigned by the mind or imagined from tales publicly allowed, is religion; not allowed, superstition." Religion arises as the product of human ignorance and fear. For Hobbes everything that existed consisted solely of matter. Hence there could be no soul or God. Such incorporeal entities were incomprehensible. "Reason," he says, "is not to be folded in the napkin of an implicit faith, but employed in the purchase of justice, peace, and true religion." Hobbes was, however, willing to use the authority of the Church to maintain civil government.

In his *Philosophic Thoughts* (1746), Diderot exclaimed, "Thanks to the works of great men, the world is no more God's; it is a machine which possesses wheels, ropes, pulleys, springs, and weights." He maintained that the influence of revealed religions on society had been disastrous. They cut all natural bonds between man and man. They sow dissensions and hatred among the closest friends and among blood relations. Anyone who is not satisfied with the visible and seeks the invisible causes of visible effects is no wiser than a peasant who attributes the motion of a clock whose mechanism he does not understand to a spirit concealed in it. One must choose between freedom and slavery, between clear conscience and vague emotion, between knowledge and belief.

The Encyclopedists declared open war on the doctrines and influences of the established religions. They accused these religions of having been a hindrance to intellectual progress and of lacking the capacity to build a genuine morality and a just political and social order. Nor was deism more acceptable. It was, to Diderot, a weak compromise. It cut off a dozen heads from the hydra of religion, but by leaving one it would let all the others grow again.

Instead of attempting to found rational religious doctrines, the philosopher David Hume proceeded to investigate thoroughly the actual origin of religion in human nature. He arrived at conclusions unfavorable to Christianity. The deists had maintained that human nature was stable and that man had fundamental knowledge on which he could rely. Hume assails these doctrines. Human nature is a fiction. It is a confusion of instincts, not a cosmos but a chaos. In his *Natural History of Religion* (1757), he says that man is not by nature given to abstract reasoning, but to appetites and passions, and religion is rooted in these instincts, not in logical or ethical considerations. Hope and fear, fear of supernatural powers and man's desire to propiate these imagined powers, have led men to adopt beliefs. Superstitions and the fear of demons are the origin of our conception of God. Fear is the beginning of all religion, and the various shapes and forms of religion are derived from this emotion. He concludes that no supernatural elements in any faith are to be given the slightest credence. For religious doctrines he has but contempt. He says, "If we take in hand any volume of divinity, or school metaphysics, for instance, let us ask, does it contain any abstract reasoning concerning quantity or number. No. Does it contain any experimental reasoning concerning matter of fact and existence? No. Commit it then to the flames for it can contain nothing but sophistry and illusion." Any claim of religion to convey and make accessible a higher world is illusory and untenable.

The two eighteenth-century atheists whose works were most influential were d'Holbach and la Mettrie. They attacked religion ruthlessly. D'Holbach's *System of Nature* (1770) was called the Bible of atheism, and his *Common Sense* (1772) gained immense popularity. They contended that religion was a system invented by imagination and ignorance to conciliate fictitious powers. The idea of God had no correspondence with anything real. God was but nature; soul was just body. The concept of a spiritual substance they considered a useless and self-contradictory hypothesis. This and other absurd and puerile notions were the foundations of priesthood, temples, altars, rights, authority, and dogma. The Christian dogmas were myths, no different from the myths of pagan religions.

Religion, they maintained, was useful only to priests and politicians. In their lust for power the clergy exploited man's naive assumption that spirits were the causes of unexplained events. Religion served to divert men from the evils inflicted by rulers and induced them to bear with their misery by promising happiness in another world if they agreed to be unhappy in this one. By teaching man to fear invisible tyrants his leaders made him the slave of earthly ones. Far from guaranteeing morality, religion permitted leaders to stir up war in the name of God.

Since man understood nature, there was no need for the primitive superstitious account furnished by the established religions. He who succeeds in destroying the notion of God would be mankind's greatest friend. La Mettrie said that the world would never be happy until it decided to become atheistic. Then nature, hitherto infected with a sacred poison, would resume its rights and purity.

We see from the above survey that educated eighteenth-century views on religion ranged from minor deviations from orthodoxy to outright atheism, and even to strong attacks on what some thinkers called spiritual tyranny. Of course, the broad intellectual significance of these movements does not consist of the defense of one or another belief or of atheism. Indeed, subsequent advances in mathematics and science, some of which we shall examine, show that the Enlightenment's concept of the universe was too simple; in particular, its belief in matter as the ultimate reality has been shattered by more recent developments. But the eighteenth-century attack on the blind acceptance of dogmas, on the uncritical adoption of beliefs unsupported by factual knowledge, has had immeasurable value. Dogma is the foe of knowledge and the source of the perversion of knowledge. Rational inquiry may indeed lead to belief, but belief without inquiry is sheer credulity. It is to the credit of the eighteenth-century intellectuals that they dared to initiate rational investigations.

Subsequent developments in religious thought and movements, about which we can say very little here, were in the main reactions to the extreme views of the eighteenth century. There were sharp protests that reason incited by that archvillain mathematics was not adequate to encompass the physical, emotional, and spiritual worlds. In the words of William Blake, "God is not a Mathematical Diagram . . . God forbid that Truth should be confined to Mathematical Demonstration."

Actually, by the end of the eighteenth century, most intellectuals had retreated somewhat from the position that reason was a competent judge of what men should or should not believe. The greatest philosopher of the century, Immanual Kant, believed that the ultimate truths about man and nature transcended rational understanding. The most that reason could hope to achieve was to recognize, describe, and manipulate objects and events perceived by the senses, but ultimate reality, i.e., the realm beyond the sense data, eluded the human mind. Matters which could not be apprehended through the senses, the essence of the universe, the soul, and God, fell outside of rational knowledge. On this basis man could not hope to attain certainty or proof, for example, of the existence or nonexistence of God.

21–5 The decline of superstition. Whether or not the eighteenth century went too far in its attacks on or denial of religion, the application of reason to the entire compass of religious beliefs has had immeasurably beneficial effects. Until the eighteenth century people at large and even such educated men as popes and kings, Luther, Calvin, and Cotton Mather accepted as truths doctrines that almost all of us find incredible today. They believed in the devil, witches, and evil spirits and ascribed supernatural powers to these imaginary beings. Witches could make people sin, could transform themselves into wolves and devour cattle, and consorted with the devil. Evil spirits infected people and caused them to do strange and wicked acts. Such superstitions could be dismissed as great jokes were it not for the fact that thousands of men, women, and children were cruelly tortured and put to death because they were supposed to be practitioners of the black art. A punishment milder than death was considered to be an offense against God. The most dastardly aspect of the entire history of sorcery was that anyone could be accused of witchcraft by any other, even mentally unbalanced, person, and that the accused was required to prove his innocence, a task about as simple as asking a person to prove that he is not a thief. Moreover, the judges in such trials were usually biased, warped, often sadistic specimens of humanity who could not be expected to have any concept of judicial fairness or concern for factual evidence.

The beliefs in witches and demons waned because people began to examine them critically and sought to determine how such beliefs could be tested and what evidence supported them. It is worthy of note that in 1735 Great Britain passed an act curbing witchcraft, sorcery, conjurations, and enchantment. However, the world is unfortunately not entirely rid of such menaces. In June of 1958 a Church of England commission concluded that demons could cause human illness and that exorcism of demons might be necessary in some cases. On the other hand, the commission denied the power of mediums who invoke benevolent spirits to heal because there is no evidence for the healing power of the spirit doctor.

Reason also discredited the accounts of gruesome horrors which many people accepted as their fate after death. Hell was a reality, and there hideous and unbearable tortures awaited the damned. They were condemned not to be consumed in hell, but to suffer unabating torture by eternal burning. Fear of the wrath of God and His punishment was inculcated in people from the time they could begin to understand, and thoughts of such a possible fate darkened the lives of almost all Europeans.

The Encyclopedists attacked superstitions as the major evil. Many writers pointed out that the opposite of belief is not disbelief but superstition, for the latter gnaws at the true root of faith and dries up the source from which true religious feeling springs. Voltaire also attacked the superstitions with the cry, "Eradicate the infamy." These men were effective in their battle because they could point to a universe run in accordance with clear, knowable laws, in which the mythical figures accepted by the ignorant not only played no role but could not disturb the operations of nature. As Halley put it in his 'Ode to Newton',

> " . . . . Now we know
> The sharply veering ways of comets, once
> A source of dread, no longer do we quail
> Beneath appearances of bearded stars."

The course of nature, men became convinced, could not be affected by spirits, ghosts, charms, and incantations. Superstitions about eclipses, solstices, and special conjunctions of the moon and planets, still alive in areas untouched by Western science, were discarded. The elimination of superstitions freed the human mind from the ballast of centuries and enabled it to think clearly, to view man and nature unencumbered by doctrines and fears which previously clouded its view or discouraged thought.

21–6 The rise of toleration. The atheistic outpourings of the eighteenth century sprang as much from the excesses committed by religious leaders as from the impossibility of reconstructing the accepted doctrines by rational arguments. Suppression of speech and publications and the prohibition of teachings contrary to the dogmas of the prevailing creed were among the minor crimes of the religious leaders. Individual acts or utterances displeasing to the authorities were labeled heresy, and transgressors sentenced to some of the cruelest tortures man has devised: the boot, the rack, public whippings, brandings, and burning at the stake. Rather mild was the sentence to years of imprisonment in foul dungeons. The Spanish, Roman, and Mexican Inquisitions are synonyms for cruelty. In the name of God, the members of a faith or whole nations were impelled to make war and kill those whose beliefs differed. The Crusades, the massacre of the Albigenses, the St. Bartholomew's Day massacre in France, the Thirty Years' War, and the persecution of Catholics in England, Ireland, and Scotland were perpetrated by religious zealots. During the fifteenth, sixteenth, and seventeenth centuries, vast areas of Europe were devastated by religious wars. Protestants and Catholics killed each other mercilessly or used religion as a tool to stir up opposition to political adversaries. Men who professed to speak for God and whose avowed function was to improve the lot of man, to supply spiritual values, and to teach brotherhood displayed instead arrogance, bigotry, fanaticism, and sadism.

The very concept, to say nothing of the practice, of toleration was practically unknown in Europe. When the Grand Duke Ferdinand of Tuscany (1549–1609) decreed freedom of religion in Leghorn, Italy, he did something sensational for those times. Jews, Catholics, and Protestants fled to Leghorn to avoid persecution in their native cities. Holland, having suffered deeply at the hands of the Spaniards, was another oasis of religious freedom in the seventeenth century. But such centers of toleration were rarities.

The very men who dared to question and examine the basis for religious belief and deny some or all of the Christian doctrines also attacked the power of religion and fought for understanding and tolerance. Montaigne, Montesquieu, Locke, the Encyclopedists, and Voltaire condemned bigotry and advocated the abolition of torture. D'Holbach devoted his life to spreading toleration.

Montesquieu compared oriental and occidental religions and concluded that the former were superior because they were tolerant. Locke in his *Letter on Toleration* (1685) denied the right to force on others speculative opinions and definite forms of worship. Particular ceremonies and incomprehensible dogmas were trivialities. In another essay he said, "For he that examines and, upon a fair examination, embraces an error for a truth, has done his duty more than he who embraces the profession of the truth . . . without having examined whether it be true or no." Voltaire stressed that minor points of belief had little to do with the core of religion. Condemning the excesses of fanatics, Voltaire said that if we let reason do its work, it would slowly but surely rid us of that scourge. Many pointed out that pure religion transcended differences in doctrines, customs, and rituals which were but the wrappings. Diversity concerned only the symbols.

Of course, attacks on the excesses of the major European religions tended to be suppressed by ecclesiasts, who had the power to punish by death the authors of books not licensed by the appropriate authorities. The men who dared to write as they thought were constantly threatened with confiscation of property, imprisonment, or exile. Hence attacks on Christianity were often veiled as attacks on Mohammedanism. But gradually the denunciations achieved their goal of curbing the power of the churches. Freedom of faith and freedom to worship or not to worship became recognized as human rights and necessary conditions of a civilization.

We in the United States are fortunate that the principles of our government were formulated in the latter part of the eighteenth century. There is a common belief that the colonists who settled America in the seventeenth century came here to enjoy religious freedom. With few exceptions, such as the Quakers, these people sought only the freedom to practice their own brand of intolerance. They brought with them all the narrowness, prejudices, and closed-mindedness of the European scene. Witch-hunting mars the history of New England. Quakers were hanged in Boston because they dared to visit Puritan domains. However, in the eighteenth century the leaders of our Revolution, notably Benjamin Franklin and Thomas Jefferson, influenced by the rational movement in Europe, saw the need for toleration and for restricting the influence of organized religion. Separation of church and state is the broad principle which our Founding Fathers recognized as one of the necessary guarantees of civil liberties, and it is still today one of the superior features distinguishing our form of government from European governments. Here we have, then, one of the concrete and most valuable benefits which all Americans derive from the mathematically inspired rationalism of the Enlightenment.

21–7 The problem of ethics. The rationalistic spirit of the eighteenth century was bound to invade sooner or later the domain of ethics, or the principles of human conduct, and to reopen age-old questions: What is goodness? Why should man be honest, charitable, kind to others? Why make any concessions to fellow beings which do not result in benefits to oneself? Christianity had provided answers: God required morality. Those who obeyed God would win a place in heaven; those who disobeyed would be punished by being sent to hell. The basic question, Why does evil exist, was answered by the doctrine of original sin. Because Adam sinned, all men were born sinners. Hence theology had furnished the logical basis for ethics, and acceptance of the Christian religion meant the acceptance of this basis for ethics.

But the critical attack on the religious doctrines themselves and on the very foundations of belief automatically brought into question the system of ethics which presupposed God's will and plan. Moreover, the new philosophical doctrines—mechanism, materialism, and determinism—opposed Christian ethics. Materialism, as d'Holbach put it in his books *Social System* and *Universal Morality,* maintains that in nature nothing is just or unjust, good or bad; all happenings are equal in value and validity. The philosophy of determinism destroyed the Christian basis for ethics because it denied free will. D'Holbach pointed out that all phenomena are necessary; hence no being can act otherwise than he does. The order in nature does not permit any part to deviate from the certain and necessary rules of that order. The will is bound fast in the predetermined behavior of matter. Man is not a free agent and is not responsible for his actions. Hence there is no evil or guilt.

The abandonment of all standards of morality and ethical values could not appeal to intelligent people. Since they were not satisfied with the religious justification of ethics, they sought new foundations. Morality was to be separated from religion and find its own basis. Some thought that reason alone could determine the proper standards. Others demanded that ethical principles be based on human nature which, after all, was part of the world of nature; however, reason was again called upon to determine by an analysis of human nature what system of ethics was natural. Man must know his place in the world to act rightly, and self-examination would provide the answers.

Hobbes chose to base ethics on the impulse to self-preservation. This impulse, he asserts, is a fact of human nature. But the individual's desire for power to preserve himself brings him into conflict with others. Through need and fear men war against each other. Reason moves men to seek better means of self-preservation than those they can secure when each fights for himself. Hence men on their own free will (which Hobbes denied at other times) entered into a contract to make social life possible. If he is to live in a society, each individual must renounce certain rights he possesses in the natural state. The agreements made on how to live together are like scientific principles from which other laws may be deduced. Thus it follows from the "social contract" that fidelity, gratitude, courtesy, forbearance, and justice must be practiced; pride and arrogance must be avoided; and arbitration must be accepted as a means of reducing strife among individuals. Experience must be used to check these basic principles just as the degree of agreement between observed fact and theory is used as a check on scientific principles. Then, Hobbes brings in religion. Natural moral law is also the will of God, for reason is bestowed by God. Christ, the apostles, and the prophets proclaimed the very laws which can be deduced from reasoning on the basis of the social contract.

Voltaire believed that nature's design included moral laws for man as well as physical laws. In his words, "Should nature everywhere have aimed at unity, order and complete regularity, and have missed only in the case of its highest creation, man? Should nature rule the physical world according to general and inviolable laws only to abandon the moral world completely to chaos or whim? Just as the law of gravity is not confined to this planet but extends throughout the cosmos and connects every particle of matter with every other, so the fundamental laws of morality prevail among all peoples. The moral nature of man is immutable. Man has uniform inclinations, instincts, and appetites. Nature as she realizes herself without hindrance or fetters will realize the true and the good."

The application of reason to the problem of ethics did not satisfy the thorough-going empiricist David Hume. In his *Treatise of Human Nature* (1738) and in his *Enquiry Concerning the Principles of Morals* (1751), he too sought a moral Newtonianism, a moral force of gravitation to unify the moral world. And he, too, rejecting supernatural and metaphysical sanctions for morality, found the true foundation of ethics in human nature. But whereas others had assumed that the results of reasoning were in accord with the true nature of man, Hume rejected reason in favor of nature. He said that we know only our hearts and so must be guided by these. The good is what people approve. Our feelings instigate all acts, good or evil. Hence we have but to gather statistics on what people approve. Since pleasure or utility meet with approval, acts leading to pleasant or useful results are good.

Kant said that moral truths exist *a priori,* as do the axioms of geometry, and derive from the nature of man *and society.* The moral obligation is inherent in man and does not derive from a system of religious beliefs. To Kant the purpose of moral laws was the good of society, for only in terms of society as a whole is human history intelligible and significant. Man is morally free within the limitation of acting for the good of his fellow beings. Kant therefore stated his chief ethical axiom thus: "Act only according to those maxims which thou canst at the same time wish to be universal laws." This dictum he called the categorical imperative. That act is good which is in conformity with this dictum. Kant's insistence that moral laws be universally applicable was motivated by the desire to secure for such laws the objectiveness and universality enjoyed by scientific laws.

Others said that man had a moral sense just as he had an aesthetic sense. This natural sense of right and wrong tells us how to distinguish good and evil. It is, moreover, independent of religion. In fact, to be virtuous because one expects rewards in heaven is unchristian. Another approach arrived at the conclusion that enlightened self-interest rather than obedience to God's wishes recommended morality. Virtue brings its own rewards. All these arguments sought, of course, to support the same ethical standards that Christianity had espoused, but attempted to do so by providing a basis more satisfactory to the intellectual temper of the age.

However, the problem of accounting for the presence of evil in a world regulated by nature remained unsolved. Numerous answers were suggested. One claimed that man himself through customs, institutions, and failure to exercise control had interfered with the workings of nature. The civilization man had fashioned was artificial. The true moral behavior intended by nature would be found in more primitive societies which had not perverted nature's ways. Hence the ways of the primitive societies known to Europeans, the Chinese, the Tahitians, and the American Indians, should be studied and imitated. A smug few accounted for evil by stating that the beauty and happiness of the whole was contingent upon the inferiority of the parts. Still another argument proclaimed that evil, too, was a manifestation of human nature, and hence right. To minds less limited than man's these evils may appear to be a necessary part of the workings of the whole universe. As Alexander Pope put it,

> "All discord, harmony not understood;
> All partial evil, universal good;
> And, spite of pride, in erring reason's spite
> One truth is clear, Whatever is, is right."

Curiously the attempt to find a justification for morality independent of religion led in many systems of thought to some support for the established religions. Religion became important because it favored the principles and practice of morality. In Locke's words, "The Scriptures confirm the moral laws which reason discovers." Voltaire in his *Treatise on Toleration* (1763) argued that moral consciousness was a primary phenomenon and thus the standard by which one should judge religious truth. The commands of the Bible which were in harmony with morality might be obeyed, the others rejected. The philosopher Immanuel Kant said that the Bible was valuable only insofar as it supplemented the moral code, and religion was useful in that it made acceptable to man the moral restraints within which he must live as a member of society. But morality was more important than God. One might believe that God existed because the moral obligation existed within man.

There is no one answer to the questions of what acts are moral and what is the source of moral truths. The eighteenth century did not settle these questions. They were taken up in the nineteenth century, and we shall have occasion in Chapter 28 to note another answer. What is definite is that the Age of Reason, inspired by the new view of a mathematically designed universe, launched rational inquiries into the nature and grounds for religious and ethical beliefs. Whereas the seventeenth and earlier centuries solved such problems by incorporating them in theological doctrines, the eighteenth century withdrew unqualified acceptance of the doctrines themselves and severed ethics from the domination and tutelage of religion. The latter yielded the primacy it had enjoyed and was required to submit to conditions imposed by man's reason.

Exercises

1. What was the effect of the rise of mathematical mechanics on the role of God in the creation and functioning of the universe?
2. Describe the predominant attitude of the seventeenth-century mathematicians toward religion.
3. What is the impact of the mathematical order of nature on superstitions?
4. How would you relate mathematics and toleration?
5. How does the deist position on religion derive from the mathematical order in nature?

Recommended Reading

Burtt, E. A.: *The Metaphysical Foundations of Modern Physical Science,* 2nd ed., pp. 280–299, Routledge and Kegan Paul Ltd., London, 1932.

Bury, J. B.: *A History of Freedom of Thought,* 2nd ed., Oxford University Press, New York, 1952.

Carr, Herbert W.: *Leibniz,* Dover Publications Inc., New York, 1960.

Höffding, Harald: *A History of Modern Philosophy,* Vol. I., Books III to V, Vol. II, Books VI and VII, Dover Publications, Inc., New York, 1955.

Lecky, Wm. E. H.: *History of the Rise and Influence of the Spirit of Rationalism,* 2 vols., Longmans, Green and Co., London, 1882.

Martin, Kingsley: *The Rise of French Liberal Thought,* pp. 117–132, New York University Press, New York, 1954.

Mortimer, Ernest: *Blaise Pascal, the Life and Work of a Realist,* Harper and Bros., New York, 1959.

Randall, John Herman Jr.: *The Making of the Modern Mind,* Chap. 12, rev. ed., Houghton Mifflin Co., Boston, 1940.

Smith, Preserved: *A History of Modern Culture,* Vol. I, Chaps. 12 to 15, Vol. II, Chaps. 13 to 16, Henry Holt and Co., New York, 1934.

Snyder, Louis L.: *The Age of Reason,* Chap. 3, D. Van Nostrand Co., Princeton, 1955.

Willey, Basil: *Nineteenth Century Studies,* Chaps. 1 to 4, Chatto and Windus, London, 1949.

Willey, Basil: *The Eighteenth Century Background,* Chatto and Windus, London, 1940.

Willey, Basil: *The Seventeenth Century Background,* Chatto and Windus, London, 1934.

Wilson, Arthur M.: *Diderot: The Testing Years 1713–1759,* Oxford University Press, New York, 1957.

CHAPTER 22

REASON IN LITERATURE AND AESTHETICS

Nothing is beautiful but the true.

Boileau

22–1 Introduction. What is talent? It is reason brilliantly set forth. What is spirit? It is reason well expressed. Taste is but good sense refined by reason. And genius is sublime reason. Thus the Enlightenment defined the qualities which account for great art. Like science, art must derive from reason, for this sovereign faculty knows no bounds or qualifications.

The age also glorified nature because it gave rise to the marvelous mathematical laws man had discovered; it was nature's grand design which the great minds, Descartes, Galileo, Huygens, Newton, Leibniz, and dozens of other luminaries, had unearthed. The eighteenth century ruled that the artist, too, must know the laws of nature before he can breathe real life into his creations. Nature tells the artist what to say and how to say it.

But the age was not content just to insist on reason and nature as the foundations of art. It dared to analyze beauty itself and to decree the ingredients of beauty. It produced, in other words, a philosophy of aesthetics or philosophies of aesthetics which were clearly dominated by reason, law, and the scientific approach.

In this chapter we shall examine how the Age of Reason refashioned literature and the criteria it selected for beauty.

22–2 The language of reason. Since language is the tool of literature, the reformers began with a re-examination of that element, and since the power of mathematics is often credited to the efficacy of its language, it is not surprising that the writers and critics of the Enlightenment set out to model prose as closely as possible on the language of mathematics. The art of speaking, the art of writing, and the art of thinking, these men maintained, are at bottom one and the same art. Each of these activities is a series of analytic operations, and words are the guideposts which mark and fix the steps of these analyses in our minds. Hence language should be as exact and as clear as mathematical notation.

The language of algebra was especially admired, for its symbols enable one to keep in view the identity of a quantity no matter how much it may be involved in complex transformations. It was therefore decided that language, in general, should possess this superior feature. In the ideal language, ideas would be represented by symbols. Thereby ambiguous or inadequate words and misleading metaphors would be eliminated. Furthermore, ideas would be associated by the smallest possible number of syntactical relationships just as numbers are associated by only a few operations (addition, multiplication, and so forth). It should then be possible to relate or compare two statements just as it is now possible, for example, to observe that one equation is obtained from another by multiplication by a constant. As a matter of fact, Descartes and Leibniz each began a work which advocated the use of symbols for the ideas and relationships occurring in all thought, and hoped that such a system would facilitate reasoning in all fields. However, literary critics were more concerned with attaining precision.

Under the influence of mathematics, writers began to standardize language. Arbitrary symbols intended to remain fixed were adopted for ideas just as mathematicians invariably use x to denote an unknown. Thus the words *mill* and *clock* became the accepted symbols designating the mind, girls were nymphs, and lovers became swains. Combinations such as dewy lawns, mossy streams, and limpid water were used repeatedly. Numerous scientific terms were employed in literary contexts because they rendered meaning with precision. The rainbow was refracted by a cloud; the sun was arch-chemic and gave off a magnetic beam; the telescope was presumably more accurately described as the optic tube.

Important words, men thought, exhibited common properties just as the word animal expresses a property common to dogs, horses and cats. Such general words were believed to be the essence of knowledge. Hence, in further imitation of mathematics, ordinary discourse began to use abstract concepts. A gun became a leveled tube; birds were a plumy band; fish were a scaly breed or a finny race; the ocean became a watery plain; and the sky, a vault of azure. The poets, in particular, indulged in abstract terms such as virtue, folly, joy, prosperity, and melancholy, which they personified and wrote in capital letters.

The first definite accomplishment in the standardization of language was due to the Académie Française which published the two-volume *Dictionary of the Academy* (1687–91). In this work, the members of the society tried to set standards in orthography and vocabulary. Their choices were made on the basis of usage and reason, with preference given to logical consistency.

In England, Jonathan Swift proposed in 1712 a project of language reform, aimed at correcting, enlarging, and defining the vocabulary. The language was imperfect, he said, and was daily being corrupted more than improved. The Restoration period, he complained, filled the language with affected phrases, new conceited words. Moreover, the poets abbreviated words to make them fit their verses and so added many new monosyllabic words to a language already overstocked with them. Prose had also incorporated barbarisms and unwarranted abbreviations. Swift wanted some words thrown out and some that had become obsolete restored. Primarily he wished to cut out neologisms, abbreviations, and slang and to institute simpler spelling.

Once it was improved, he wanted to see the language fixed forever. Better a language that is not wholly perfect than one which is perpetually changing.

He valued the Bible and Common Prayer Book because they had in practice proved to be a stabilizing influence on language, and were well written in the first place by reason of their simplicity of style.

In *Gulliver's Travels,* Swift employed his favorite weapon, satire, to support his ideas. During his visit to Laputa, Gulliver encountered several professors who were likewise engaged in improving the language of their country. One of their proposals was to shorten discourse by replacing polysyllabic words by monosyllabic ones and by omitting verbs and participles. These reformers also argued that all things imaginable were but nouns anyway. Carrying the idea one step further led to the suggestion of eliminating all words and having people carry objects about with them. But the women of Laputa objected to the last feature because it would not allow them to use their tongues.

The effective step toward standardization in England was made by Samuel Johnson. His *Dictionary* (1755) is one of the landmarks of the English language. Johnson proposed to legislate for language. He sought to standardize pronunciation, preserve the purity of the language, and make words serve well-defined uses. He introduced quotations from the best English writers to establish exact meanings and proper usage. It was his intent that these meanings and usages should be fixed for all time just as the word triangle has meant precisely the same thing for thousands of years. While there were many faults in Johnson's work, he did help immensely to set standards. He became, in fact, the arbiter of verbal fashions.

The eighteenth-century concept of language was, of course, unrealistic. One cannot fix forever the meanings of words. Words are like people. They are born and, by increasing usage, acquire new meaning and vitality. For a while they shine brightly, but sooner or later they become dulled by wear and die. Words, so to speak, are living stuff and not stone. In behalf of the century's writers, one can say that they endeavored to remove ambiguities and misleading metaphors and, in general, to remedy deficiencies of language. The haphazard growth of language, if uncontrolled, leads to the development of hundreds of dialects and the breakdown of communication.

22-3 The reform of style and spirit. The reform of language was but one of the steps in the movement to "improve" the literature. Writers believed—one might say, felt, but feelings they distrusted—that the style of writing could be improved by closer attention to the style of mathematical exposition. The age recognized that statements in a mathematical discussion or demonstration were concise, unambiguous, clear, and exact. Since many writers were convinced that the success of mathematics was due almost entirely to this bare style, they resolved that all good writing should adopt it.

The Fellows of the Royal Society of London, a scientific body, appointed a committee to study the problem of improving the language. The committee urged the members of the society "to reject all the amplifications, digressions and swellings of style," to seek "a return to primitive purity and shortness, when men delivered so many things in an almost equal number of words," and to use "a close, naked, natural way of speaking; positive expressions, clear senses, a native easiness; bringing all things as near the mathematical plainness as they can; . . ." Although the Royal Society's admonitions were addressed to scientists, its recommendations were identical with those which writers, in general, had approved and begun to adopt in their work.

The works of many mathematicians were held up as examples and models of perfect style. Descartes' writings possessed lucidity, precision, and readability. Hence his writings were extolled. The rationality of Pascal was hailed. His *Lettres provinciales* are still considered a model of beautiful prose. Huygens, a native of Holland, was another master of French. Galileo's dialogues, which we discussed in earlier chapters, were valued for their clear, direct, forceful, and even witty style. Galileo is, in fact, the founder of modern Italian prose. Newton's pure, austere, and exact diction provided an example for the English language. All these men were masters not only of scientific thought but of literary style.

Critics examined such works and came forth with prescriptions for clear and rational writing. The communication of facts under high standards of logical thought, "the easy intelligible intercourse of minds," was to be the goal of writing. The demand for intelligibility and clarity required that each phrase or group of words be readily grasped. Thus brief, crisp sentences became fashionable. The order of the words within the sentence was to be determined by the thought; hence inversion was not tolerated. Sentences were to be linked with one another in a manner which clearly showed the progression of thought—whence it came and where it went. Order, neatness, sobriety, conciseness to the point of epigrammatic brevity, and exactness were to control the language, while clear and sure ordering of details, formal unity of material, and strict logical concentration were to control the organization of written material. The order in the universe was to be reflected in the order in literature. Thus simplicity, plainness, perspicuity, and directness were considered great virtues.

Metaphors were rejected in favor of accurate language describing objective realities, for metaphor and symbolism, though agreeable, were not rational. The pedantic, florid style and complex Latinized constructions of medieval and Renaissance scholars were abandoned. Banished also were the impetuous flights of imagination, vigorous, emotionally charged expressions, exuberance, enthusiasm, and sonorous and highly suggestive phrases.

The style of prose became less majestic, less involved, less emotional, and less ornate. Thus Swift in his writings replaced metaphors and far-fetched words by simple, common language, though he managed nevertheless to be forcible and pungent. Lucidity and perspicuity, as exemplified by Dryden, were instated. The smooth, even style of Locke was hailed as was the plain prose of Hume. Writing did become easy to read, factual, and incisive.

The new theory of style was expressed by the naturalist George de Buffon (1707–88). "A style is good by reason of the infinite number of truths it presents. All the intellectual beauties in it are so many truths." The churchman and philosopher Étienne de Condillac (1715–80) affirmed that the whole beauty of style consisted in clarity and character, while the philosopher Claude Helvétius (1715–71) was satisfied with clarity alone. To him the worst vice of style was ambiguity.

The regulation of style was applied to poetry also. The literary critics of the period urged poets to achieve mathematical objectivity by suppressing all personal or individualistic qualities in their work. They ruled that a poet should be something of a mathematician. Dryden declared, "A man should be learned in several sciences, and should have a reasonable, philosophical, and, in some measure, a mathematical head to be a complete and excellent poet . . ." Clarity was the major virtue to be attained by correctness of grammar and justness of sentence structure. Emotion and imagination were to be curbed.

Nicholas Boileau-Despréaux (1636–1711), a French poet and critic, formulated the code which poets had already begun to follow. Love reason and moderation above all things. "Let good, sound sense be harnessed to your rhyme." Put words in the right order, confine the spirit within the rules of reason, limit yourself by the natural laws of poetry, avoid barbarisms and solecisms. Above all, follow nature for "the false is always insipid, boring, and feeble; but Nature is always true to the point of being felt as such."

Pope repeated Boileau. To follow nature, to cultivate taste, to select a pure vocabulary, to vary one's phrase, and to avoid the trite and the paradoxical are the chief maxims. The writer's job, said Pope, is

> more to guide than spur the Muse's steed;
> Restrain his fury, than to provoke his speed.

The critics believed that not only the style of poetry but the forms of the various genres of poetry should be standardized. There were natural laws for the writing of poetry just as there were for science. Verse, it was thought, could be written by rule; lyric, epic, sonnet, didactic verse, ode, and epigram could be constructed by observance of their natural laws. In his "Essay on Criticism" Pope says,

> First follow Nature and your judgment frame
> By her just standard, which is still the same.
> Unerring Nature, still divinely bright,
> One clear, unchanged and universal light,
> Life, force, and beauty, must to all impart,
> At once the source, and end, and test of Art.

In the domain of literary efforts, to follow nature did not mean precisely what it meant in the physical sciences; that is, it did not imply obedience to nature's mathematical laws; instead, by virtue of an historically justifiable association of the Greeks with nature, it meant to imitate the form of the Greek classics. Thus the rules or laws of literature evolved from an identification of nature, antiquity, and reason, with the result that to follow one was to follow all.

Curiously, the principles of form in poetry were likened to mathematical axioms because axioms determined the form as well as the content of theorems. The heroic couplet won favor because of its balance and symmetry, and, extreme as it may seem to us, because the form was analogous to a series of equal proportions. It was regarded as the essence of cadenced regularity. Form, proportion, and symmetrical structures were upheld. To the literary critics of the age, Pope, Addison, and Johnson, beauty consisted in adherence to these strict rules of versification.

Obedient to the directives laid down by these arbiters of taste, the poets tried to imitate Homer, Virgil, and Horace, and made frequent allusions to the classics. Dryden's translations of the Latin classics were considered a standard against which to measure performance. Poetry became temperate, well regulated, and intellectual, emphasizing the stylistic features of order, lucidity, moderation, elegance, and proportion. Decorum, which meant harmony of theme, matter, and form, emerged as the most cherished quality. And propriety, that is, adherence to the above standards, restraint, and coolness, became the core of beauty. The poets suppressed enthusiasm, rapture, and imagination so much that poetry degenerated to a prosaic vehicle for rational statements. It no longer offered images, symbols, or nuances. The conception of poetry as something awe-inspiring, spiritual, or divine was forgotten during the age. As Matthew Arnold said of Dryden, poets composed with their wits rather than their souls.

From about 1650 to 1750, poets accepted readily the restrictions on language, style, and form. To justify their work, many decided that poetry must be made useful by assuming didactic, polemic, or ratiocinative functions, such as upholding examples of great virtue, refining passions, or moderating fears.

Of course, poetry suffered. The belief that the art required only a narrow outlook, limited imagination, and mere obedience to rules lowered its worth. Poetry as an outlet for the inspirational synthesis of thought and feeling was destroyed, and the art sank to the level of a minor amusement, game of skill, fantasy of wishes, or, as T. H. Huxley describes it, belletristic ornament and diversion. Those few who sought to defy the times and yet be accepted resorted to irony.

The man who challenged the age without violating its rules, Jonathan Swift, did not fail to satirize the belief that great literature could be composed by obedience to mechanical and rigid rules. Gulliver reports that in his visit to Laputa he found a professor and forty pupils engaged "in a project for improving speculative knowledge by practical and mechanical operations." They were gathered about a large square frame filled with hundreds of cubes attached to one another by wires. The faces of each cube displayed the words of the native tongue in all moods, tenses, and declensions. By operating cranks installed along the edge of the square pupils could turn up different faces of the cubes. Whenever a few words which made sense appeared simultaneously, the students copied down the phrase, and these fragments were collected in volumes. The professor intended to piece these phrases together and thereby compose great works "in philosophy, poetry, politics, law, mathematics, and theology without the least assistance from genius and study."

22–4 The age of prose. In the sphere of literature the Age of Reason is known as the age of prose. It is rather easy to see why the age favored prose and employed this medium so extensively. Emphasis on the rational elements in style, factual material, and rigid forms favors rhetoric, reasoning, and narrative, while it stifles the emotions and passions that inspire great poetry. Mythology and legends were entirely alien to the times, and great writers such as Dante, Milton, and even Shakespeare were criticized or classed as inferior. The high regard for the abstractions of mathematics applauded the intellect, while it automatically deprecated the concrete images of poetry. One might feel in verse, but one must think in prose. The preference for the truths of mathematics and science, which deal with clear and distinct properties of objects, properties which can be measured and verified, produced coldness and disdain for the vague insights and expressions of poetry. The very spirit of the age imposed a sharp separation between what one thought as a man of sense and judgment and what one felt as a human being possessed of emotions. Cold reasoning opposed the images, warmth, and spontaneity of poetry.

The Age of Reason expressed itself therefore most characteristically in prose, the natural outlet for reasoned subject matter. Aside from scientific writings proper, the novel, diary, essay, satire, moral writings, newspapers, and journals commanded the field. Fiction replaced poetry as an outlet for imaginative writing, and tales of chivalry and allegories were superseded by accounts of the lives of common people. The novel became the favorite fictional form. Samuel Richardson, Laurence Sterne, Oliver Goldsmith, Henry Fielding, and Daniel Defoe were the writers who created the modern novel, which has remained a major form of literature ever since. Although the novel was fiction, its function was to make truth and virtue shine all the more brightly. For example, virtue is ennobled in Samuel Richardson's *Pamela*.

Biography was another prose form which was redefined and brought into prominence. Instead of looking upon their subjects as unalloyed examples of virtue or vice, biographers turned to factual accounts. The details of a man's life became important for their own sake. Every scrap of information was collected and exhibited as though it were a major truth. The classic example is, of course, Boswell's *Life of Samuel Johnson.*

Leading thinkers not only favored prose, but did not hesitate to condemn poetry. The members of the Royal Society insisted that one must separate the knowledge of nature from the colors of rhetoric, the devices of fancy, and the deceit of fables. Prose dealt with facts; poetry only with pleasure and fancy. Hence poetry was not knowledge, but, at best, a vehicle for delight. Locke said that poetry merely offered pleasant pictures and agreeable visions. Though it may amuse, it is not really needed by people who have seen the light of reason. Hence no thought should be expended to examine its truth. Newton was equally cold to poetry. When his opinion was asked, he cited his teacher, Barrow, who had said that poetry was a kind of ingenious nonsense. Leibniz did assign a role to poetry—to teach prudence and virtue. Hume minced no words in his attack: Poetry was the work of professional liars who sought to entertain by fiction.

Even the poets did not seem to think better of their mission. Dryden agreed in his *Apology for Heroic Poetry and Poetic License* that while readers might derive pleasure from the images of poetry, they should not allow themselves to be cozened by its fictional content. Poetry was, in other words, just agreeable fiction. Joseph Addison could say no more for poetry than that reality evoked in us a whole series of delightful *imaginary* qualities, and that poetry, by expressing these, made the material world more enjoyable. Samuel Johnson described poetry as the art of uniting pleasure with truth by letting imagination, so to speak, play with truth. Of course poetry did not lie because it affirmed nothing.

22–5 The content of the literature. While the spirit, style, and forms of literature were being restricted, writers did find new themes which they explored to the hilt. Nature, reason, mathematical law, and a designed and perfectly functioning universe replaced the themes of the birth, love, and death of man. It is, of course, impossible to reproduce here a significant portion of seventeenth- and eighteenth-century prose and poetry, but we shall give some indication of the contents.

The new views of the universe which mathematics had uncovered began to affect literature from the very onset of the revolution in astronomy. The change in the position of man was so drastic that it could not but attract the writers. One of the greatest figures of English seventeenth-century literature, John Milton, devotes a part of his *Paradise Lost* to a discussion of Ptolemaic and Copernican theory, but makes no decisive choice. He concludes this portion of his debate with himself with the words:

> Solicit not thy thoughts with matter hid,
> Leave them to God above. . . .
> Think only what concerns thee and thy being.

He then advises man to admire rather than to try to penetrate God's secret design of the universe and to seek more knowledge of God. Milton was on the whole unmoved by the new science and was more concerned to "justify the ways of God to man." But Milton's work was a "swan song of a passing world of untroubled certitude."

Even famous lyric poets, for example, John Donne, devoted some poems to the marvels of the new astronomical theory while at the same time deploring the loss of man's importance.

As the seventeenth-century mathematicians and scientists advanced the mathematical representation of natural phenomena, largely through the application of Newton's law of gravitation, and as this and similar laws were shown to account for more and more heavenly motions, the poets came to appreciate that a new and grander design was replacing and compensating for the loss of a universe centered on man. They began to express delight in the new order of nature, and became almost exclusively preoccupied with such themes as the harmony in nature, the presence of guiding and controlling laws, the mathematical plan of nature, and the mathematical art of God, the universal cause. Mathematicians were hailed as the new prophets who spoke for God and revealed His order and handiwork. The works of John Dryden, Alexander Pope, James Thomson, Joseph Addison, and many others contain numerous passages and entire poems glorifying the above themes. Pope's writings, in particular, most typically exhibit the general enthusiasm for the new world order and appreciation of the marvelous vistas revealed by science.

What is especially interesting in eighteenth-century prose is the resort to satire as a means of criticizing the customs, conventions, and institutions of the age. The worship of reason made the unreasonable, especially in political and social institutions, stand out so starkly that writers were impelled to react. To avoid the punishment which outright condemnation might have provoked, the writers had to veil their charges. Thus the supreme satirist of the age, Jonathan Swift, presented his attack on the defects of society in the disguise of stories which could, to superficial readers, appear to be fantasies, and which have, in fact, been regarded as fairy tales by young people. But each of Gulliver's reports on his travel adventures is really a satire on some phase of the European scene. The puny Lilliputians, uninformed, helpless, and rather ridiculous, are no more than the average ignorant and powerless European. Gulliver's attempts to explain the customs and ways of European life to the highly rational Houyhnhnms, the ruling members of the society of horses, succeed in exposing the irrational state of Europe. The king of Brobdingnag, who stands for nature and reason, fails to understand the strange customs of Europe, and, as a result, the irrationality or some objectionable feature of these customs becomes apparent. Swift was, of course, indignant about those who, obsessed with greed for wealth, power, and position, wreaked injustices upon the common man and trampled over him in their self-seeking.

22–6 Aesthetics. The Age of Reason sought to recognize, clarify, and fix standards in all fields. It therefore quite understandably sought to determine the nature of art itself, in other words, to construct a philosophy of aesthetics. We shall not pursue the variety of schools of thought or shades of judgment, but we shall attempt to illustrate how the generating forces of the age influenced the theory of aesthetics.

Prior to the Enlightenment, the church and the various courts of Europe were the arbiters of taste and beauty. Louis XIV of France, for example, never doubted that he was an infallible judge of art. After reason gained ascendancy, no one doubted its power to derive the laws of aesthetics.

The leading philosophy of aesthetics was founded by René Descartes, who maintained that art, like science, was to be judged by reason. Reason is a firmer and more stable foundation of art than empirical observations and haphazardly accumulated rules. This light which nature has given us must guide art too. The qualities of art must be genuine, lasting, and essential. The momentary emotions awakened in us are not a criterion of substance in art. Art is a representation of nature, and since nature is governed by principles, art must embody them. Truth and beauty, reason and nature are but different expressions for the order of existence which is revealed alike in art and science. The faculties of sense and imagination, on the other hand, are the source of all delusions. Imagination must be kept in check. It may be a stimulus to knowledge, but it does not supply knowledge.

Descartes did not give a systematic exposition of his philosophy of aesthetics. More explicit principles were presented by the followers of Cartesianism, among them Boileau. In his *Art of Poetry,* Boileau says that the true poet is born, not made. But what is true of the poet is not true of poetry. To become art, that is, to possess objective truth and perfection, an artist's work must purge itself of the subjective forces which created it; it must burn all bridges leading back to the world of fancy. The laws governing art are not derived from or produced by the imagination, but are discovered in the nature of things. Reason is the epitome of objective merit, and the artist must love reason. He is not to seek external pomp and false embellishment. If he portrays his subject in its simple truth, then he may be sure that he has satisfied the highest standards of beauty, for beauty can be approached only along the path of truth. To get at the natural object itself he must select among the phenomena constantly intruding upon him and distinguish between the changeable and the lasting, between the accidental and the necessary.

In his *Art of Poetry,* Boileau, following the spirit of mathematics, arrived at a general theory of the genres of poetry. Tragedy, comedy, elegy, epic, satire, and epigram all have their definite laws of form which no individual creation can neglect or deviate from without offending nature itself and without losing its claim to artistic truth. The artist cannot make up his own form, nor is the aesthetician the lawgiver of art, any more than the mathematician is the lawgiver of nature. Both are limited to ascertaining what is, what already exists. License in art is a crime which must never be permitted.

To be bound by the laws of art is no obstacle for genius. The laws protect genius against arbitrariness and help it to attain true artistic freedom. Though the artist is bound by form and must suit the material to the form, he has the freedom of expression, and it is in this respect that he can exhibit his real powers. But even in giving expression to his material, he should strive for faithfulness to his subject and clarity, not for novelty. The latter is restricted by the demands of simplicity, brevity, and conciseness.

The most elaborate formulation of the rational doctrine of aesthetics is due to the painter and critic Sir Joshua Reynolds, who found the essence of beauty in the expression of universal laws. Beauty is the same as truth and is apprehended by the same rational faculty. "It is the very same taste which relishes a demonstration in geometry that is pleased with the resemblance of a picture to its original and touched with the harmony of music. All these have unalterable and fixed foundations in nature." Fidelity to the object painted, subservience of color to idea, and sacrifice of detail to the general and everlasting elements are the laws of art. The painter is to address himself to the mind and not to the eye. The rules of art are so rigid, he said, that no work which breaks them will last. By selecting the universal and perfectly developed portions of a natural object or scene the artist could eliminate imperfections and correct nature.

The domination of the rationally derived and constraining system of aesthetics was broken in the same way that Descartes' approach to the natural sciences was superseded by Galileo's and Newton's method. But we shall not pursue the history of this development.

22-7 The revolt against reason. The eighteenth century had stripped language of its concrete, colorful, and picturesque words, had confined the spirit, repressed the emotions, and had forced what little expression of feeling was tolerated to follow strict rules of versification. The picture of nature insisted on by the age was that of a cold, material, colorless machine. And the mathematical and scientific pronouncements on nature crowded out all other ideas. Reaction to these bonds on thought and feeling and revolt against the domination of mathematics and science set in late in the eighteenth century and continued into the nineteenth. This revolt is known in literature as the Romantic movement.

Though Blake, Coleridge, Wordsworth, Lamb, Byron, Keats, and Shelley, the key figures in the revolt, understood what mathematics and science had accomplished, they nevertheless protested the deprecation of the senses, the suppression of feelings, the ban on the imagination, the dreary account of nature, and the denigration of emotions. They rejected the world of matter and force. The perfect order of the universe was declared an illusion, and the mechanical universe, said Coleridge, is a dead world. Reason was a form of self-deception, an unfortunate delusion; the thinking man was a sick animal. Science gave power, but not wisdom. It failed to provide happiness and moral good.

Reason, Blake said, is a devil whose high priests are Newton and Locke. Individual experience is the source of all knowledge, and nature is what the imagination says it is. Lamb and Keats said that Newton had destroyed poetry. This art should be concerned with the worth and dignity of life and man, whereas mathematics and science are indifferent to such values. The world of imagination and vision must be rebuilt. Shelley maintained that imagination was the great instrument of moral good, and that poetry ministered to the effect by acting upon the cause. Wordsworth, who in his youth had appreciated the accomplishments of science, turned against it. He contended that reason alone produced moral monsters, and he attacked scientists who pried apart nature and soul and thereby missed the grandeur and mystery. Science leaves something out of nature which is most important. Whereas the philosophy of mechanism had produced a universe of death, he claimed to seek the world of life and light. Nature as a whole, not dissected into parts and so distorted, is the source of feeling, and our feelings constitute reality. The intuitive basis for truth was reasserted also by Rousseau. His last words were, "It is true as soon as it is felt."

A new type of literature emerged from this revolt. Poetry broke the fetters of rationalism and expressed human emotions. Myth and symbol were reinstated. Imagination was set above reason, and accounts of nature reported what the imagination "saw." The poets described nature as they apprehended it and no longer refrained from expressing their joy in nature. Rousseau, Lamb, Stevenson, and Blake wrote about their personal experiences and affirmed that these were the sources of knowledge rather than some abstract formulas. The poet felt free to exercise his own genius and to express the truths of his heart. In fact, Wordsworth, for one, maintained that what nature had to say was as much dependent upon the active power within man as upon the external reality perceived by the senses. Truth is half perceived and half created by the poet. Thus the Romantic poets celebrated a new kind of nature, a nature alive and vibrant, suffused with color, light and shade, mysterious and suggestive. They spoke glowingly of individual scenes and locales, as Wordsworth did of scenery and landscape, and they gloried in their own sensations of nature.

But rebellion is not counterargument. The nineteenth-century advances in science and mathematics reinforced the mathematical, mechanical view of nature, and man was redemoted to a purely biological phenomenon. When their passionate outbursts had subsided, the poets again faced the problem of what meaning there was in the universe and in the life of man. Tennyson, notably in his "In Memoriam," alternates between courage and despair. He sees opposing visions of the world and wishes to reconcile them, but finds no solution.

> Are God and Nature then at strife
> That Nature lends such evil dreams?
> So careful of the type she seems,
> So careless of the single life. . . .

Throughout the nineteenth century, poets were torn between the account of nature furnished by mathematics and science and the account furnished by the senses, between mechanism and the aesthetic intuitions of nature. The outcome of the struggle was pessimism, nobly expressed by Matthew Arnold. In "Dover Beach" he says that faith

> Was once, too, at the full, and round earth's shore
> Lay like the folds of a bright girdle furled.
> But now I only hear
> Its melancholy, long, withdrawing roar,
> Retreating, to the breath
> Of the night-wind, down the vast edges drear
> And naked shingles of the world.

The conflict between heart and mind has not been resolved.

22–8 Retrospect. We have seen in the last three chapters how the Age of Reason attempted to extend the rule of natural law, obtained through reason and mathematics, to the domains of philosophy, psychology, biology, religion, ethics, literature, and aesthetics. In Chapter 28, we shall examine some of the influences of the age on political and economic thought. We could, if space permitted, also study the changes in painting, music, architecture, the design of furniture, landscape gardening, dress, and manners.

The age was overconfident. It attempted to bring the physical, mental, moral, and artistic worlds under the direction of reason, and it tolerated no limitations on the power of reason. It has been remarked that the medieval period was an age of faith based on reason and that the eighteenth century was an age of reason based on faith in the omnipotence of reason. But we have no assurance that reason is adequate to embrace moral and ethical values, or that reason is entitled to deny the validity of feeling and the reality of emotions. Moreover, the entire mathematical mechanical account of nature, the platform adopted by the spokesmen of the Enlightenment, rested upon the force of gravitation, a concept which, as we have already noted, is more mysterious than the mysteries condemned by the rationalists.

Nevertheless, the core of the eighteenth-century contentions and accomplishments have immeasurable value for us. This age built modern culture. The Renaissance men saw the need for a reconstruction of knowledge and values. The seventeenth century supplied the methodology and the materials for that reconstruction, and the eighteenth century carried it out. The philosophical doctrines of the age have retained their leading role. The materialistic interpretation of the physical world continues to be the rich lode which physical science is exploring, and the mechanical view of man has been the basis for biological and medical investigations which have given us new power and control over soil, plant and animal life, and human diseases. The age also supplied and impressed modern man with such ideals as confidence in man's powers to understand and master nature and the rational approach to problems. It discovered and fervently advocated the autonomy, power, and vitality of reason and established this faculty as a force in all investigations carried out by man. We do not know whether man will solve all his problems as the Age of Reason confidently expected, but we do know that he is deriving untold benefits by following the paths cleared during the Enlightenment.

Exercises

1. What changes in language did the Age of Reason enforce?
2. Why would you or would you not expect the meanings of words to remain fixed?
3. What reforms in the style of literature did the Age of Reason advocate?
4. What was the attitude of eighteenth-century writers toward emotions?
5. Why was prose the favored medium in the Age of Reason?

6. What position did the poets of the Age of Reason take toward their medium?
7. Describe one theory of aesthetics.
8. How is the Romantic movement related to Newtonian mathematics and science?
9. What is the dominating motif of the Age of Reason?

Recommended Reading

BUSH, DOUGLAS: *Science and English Poetry,* Oxford University Press, New York, 1950.

CASSIRER, E.: *The Philosophy of the Enlightenment,* Chap. 7, Princeton University Press, Princeton, 1951.

CRUM, RALPH B.: *Scientific Thought and Poetry,* Columbia University Press, New York, 1931.

EVANS, B. IFOR: *Literature and Science,* George Allen and Unwin Ltd., London, 1954.

JONES, RICHARD FOSTER: *The Seventeenth Century,* Stanford University Press, Stanford, 1951.

NICOLSON, MARJORIE HOPE: *The Breaking of the Circle,* Columbia University Press, New York, 1960.

NICOLSON, MARJORIE HOPE: *Science and Imagination,* Cornell University Press, Ithaca, 1956.

NICOLSON, MARJORIE HOPE: *Newton Demands the Muse,* Princeton University Press, Princeton, 1946.

POPE, ALEXANDER: *Essay on Man* (numerous editions).

RANDALL, JOHN HERMAN JR.: *Making of the Modern Mind,* Chaps. 16 and 21, Houghton Mifflin Co., Boston, 1940.

SMITH, PRESERVED: *A History of Modern Culture,* Vol. II, Chaps. 9, 10, and 17, Henry Holt and Co., New York, 1934.

SWIFT, JONATHAN: *Gulliver's Travels* (numerous editions).

WHITEHEAD, ALFRED NORTH: *Science and the Modern World,* Chap. 5, Cambridge University Press, Cambridge, 1926.

WILLEY, BASIL: *Nineteenth Century Studies,* Chap. 1, Chatto and Windus, London, 1949.

WILLEY, BASIL: *The Eighteenth Century Background,* Chaps. 6 and 12, Chatto and Windus, London, 1940.

WILLEY, BASIL: *The Seventeenth Century Background,* Chap. 10, Chatto and Windus, London, 1934.

CHAPTER 23

TRIGONOMETRIC FUNCTIONS AND OSCILLATORY MOTION

All the effects of nature are only mathematical results of a small number of immutable laws.

P. S. Laplace

23–1 Introduction. In the seventeenth century one of the most pressing problems of the times was time itself. The increasing scientific activity, particularly in an age which had decided to measure and to seek quantitative laws, created the need for convenient, accurate methods of measuring time. Moreover, as we have already had occasion to mention in other connections, the seventeenth and eighteenth centuries were concerned with the very practical problem of improving the method by which ships determined their longitude at sea. Here a good clock is the simplest answer. Suppose that the longitude of a given place on land is known and that a ship has on board a clock set to agree with the time prevailing at that given locality. Since the earth turns through 360° of longitude in one day, it turns through 15° in each hour. Hence for each 15° that a ship is west, say, of the fixed locale, midday occurs one hour later compared to the time at the fixed position on land. If a ship's officer notes (by means of the sun's position) when midday occurs at his position at sea and finds, for example, that his clock reads 3 o'clock whereas it should, of course, read 12 o'clock, he knows that the longitude of his position is 45° west of the given reference locality on land. We can see then why scientists decided to search for a reliable and accurate clock.

The thought which suggested itself almost at once was to look for some physical phenomenon which repeated itself regularly. The day contains 24 hours; hence when the number of repetitions per day is known, the duration of each repetition is readily calculated. Where then could one find a repetitive or periodic physical phenomenon? Two prospects attracted the attention of seventeenth-century scientists. The first of these is the motion of a mass, called a bob, attached to a spring and oscillating up and down, and the second is the motion of a pendulum, that is, a bob attached to a string and swinging to and fro. Now first reactions to the possibility of using the motion of a bob on a spring or a pendulum as a measure of time are apt to be negative. The bob on a spring, for example, does go through each cycle, that is, each complete up and down motion, in the same time so far as the eye can judge, but the motion soon dies down. The same is true for the pendulum. But, if air resistance could be minimized or perhaps compensated for, then these motions might become truly periodic and should therefore merit investigation. The scientist or mathematician who expects to see at once the solution of a problem he sets out to study will never accomplish much. The best he can hope for at the outset is an idea or a clue to pursue.

In this chapter we shall examine first the physical problem of the motion of a bob on a spring, a prime example of oscillatory motion. To study such motions mathematicians created a new class of functions, the trigonometric functions. We shall then discuss these functions and see how they are used to derive some knowledge about the physical problem which motivated their introduction. Surprisingly trigonometric functions proved to be admirably suited for the study of sound, electricity, radio, and a host of other oscillatory phenomena. Of these latter developments we shall learn more in the next two chapters.

23–2 The motion of a bob on a spring. The problem of investigating the motion of a bob on a spring was undertaken by one of the greatest experimentalists in the history of physics, the Englishman Robert Hooke (1635–1703). Hooke was professor of mathematics and mechanics at Gresham College. His claim to fame also rests upon his success as an inventor. To his credit are a telescope moved by a clock mechanism and devices for measuring the moisture in the atmosphere, the force of the wind, and the amount of rainfall. He improved the microscope, the barometer, the air pump, and the telescope. One of his findings, namely, that white light passed through thin sheets of mica breaks into many colors, parallels Newton's work on light. He also discovered the cell structure of plants. Hooke was very much interested in designing a useful clock and thought that springs would furnish the essential device. While working on the action of springs, he discovered a basic law, still known as Hooke's law, which we shall discuss later.

Let us follow Hooke in studying the motion of a bob on a spring. The upper end of the spring is attached to a fixed support, and a bob is attached to the lower end. Because gravity pulls the bob downward, the spring will be extended until the tension in the spring offsets the force of gravity. The bob then comes to rest in some position which is called the *rest* or *equilibrium position* (Fig. 23–1). If one now pulls the bob downward some definite distance below the rest position and then releases it, the bob moves up to the rest position, continues past that point to some highest position, and then moves downward. When it reaches the point to which it had been pulled down, it starts upward and repeats its former motion. Following Galileo's plan of idealizing the physical situation, let us suppose that air resistance is negligible. (Strictly speaking, energy is also lost in the expansion and contraction of

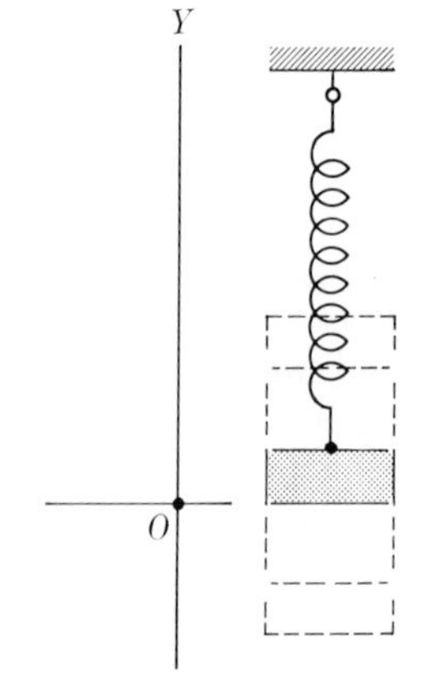

FIG. 23–1. A bob on a spring.

the spring, but this loss is negligible.) Then the bob will continue to move up and down endlessly.

To begin to get some mathematical description of this motion let us introduce a Y-axis alongside the bob (Fig. 23-1) and suppose that $y = 0$ corresponds to the rest position of the bob. When the bob is above or below the rest position, the bob is said to be displaced and the distance that it is above or below the rest position is called its *displacement*. To distinguish displacements above from those below the rest position, we shall call the former positive and the latter negative. Each displacement may then be described by a value of y. Thus $y = -\frac{1}{2}$ means that the bob is $\frac{1}{2}$ unit below the rest position.

To study the motion of the bob mathematically, it would be most helpful if we could find the formula which relates the displacement of the bob and the time it is in motion. Let us therefore seek such a formula.

23-3 The sinusoidal functions. No one of the formulas that we have considered thus far would be useful to represent the motion of the bob, for the peculiarity of the present phenomenon is that after each up and down motion, or oscillation, has been completed, the displacements go through their former sequence of values. Hence we apparently must seek a new type of formula which expresses the periodic character of the motion of the bob. We do not seem to have any clue, but a little imagination may supply one.

Suppose a point P moves around a circle of unit radius at a constant speed. Let us denote some of its positions by $P_1, P_2, \ldots$ (Fig. 23-2). We can, if we wish to, introduce a point Q on the vertical line through the center O such that Q always has the same height that P has above or below the horizontal through O. The point Q is called the *projection* of P on the vertical line. Thus to the position P_1 of P there corresponds Q_1; to P_2 there corresponds Q_2; and so on. Why should we introduce the point Q? Well, let us imagine P moving around the circle through many revolutions starting from the horizontal line at the right. What does its "shadow" Q do? It

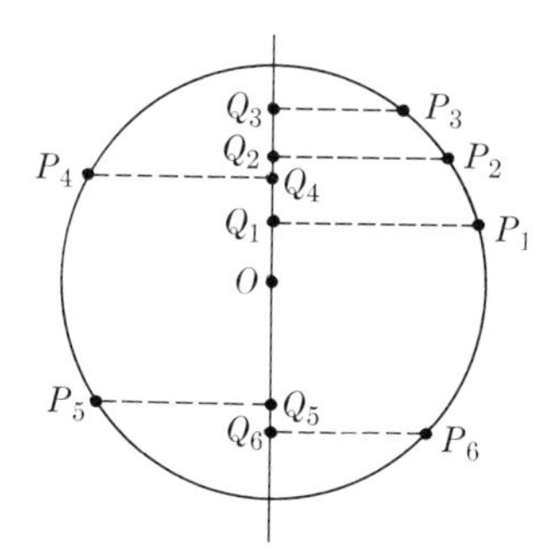

FIG. 23-2. Successive positions of a point P which moves around a circle at a constant velocity, and the corresponding positions of Q.

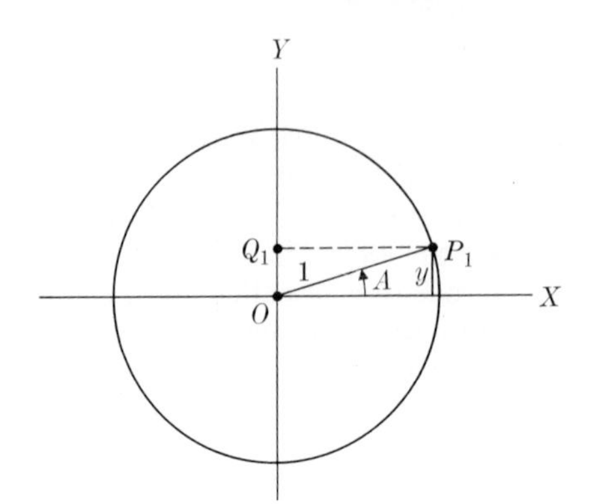

FIGURE 23–3

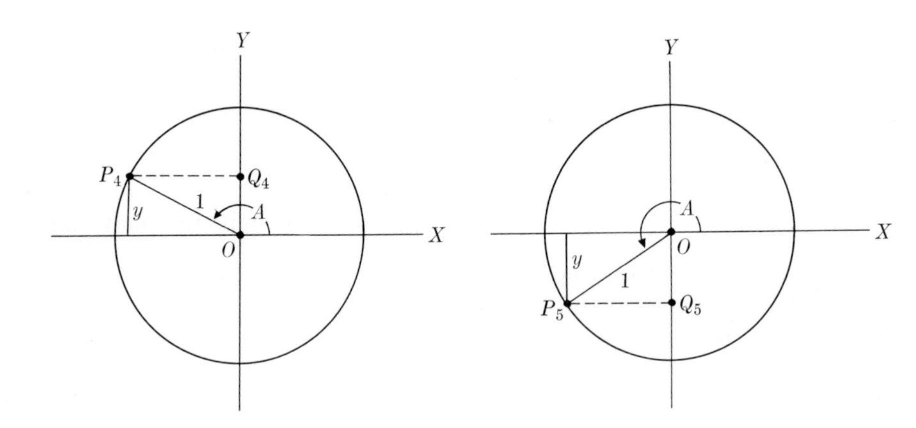

FIGURE 23–4

FIGURE 23–5

FIGURE 23–6

FIGURE 23–7

moves up from O to a highest position, moves down again to O, moves past O to a lowest position on the vertical line, moves up again to O, and then repeats this up and down motion. The motion of Q certainly seems to have the essential characteristics of the motion of the bob on the spring. Hence perhaps by pursuing further the motion of Q we may obtain the function we are seeking.

Let us introduce coordinate axes as shown in Fig. 23-3. If P starts from the X-axis and reaches, say the position P_1, then we may describe the position of P by the angle A shown in the figure. The height of Q above the X-axis *is the same* as the y-value of P. Now

$$\sin A = \frac{y}{1}.$$

Hence

$$y = \sin A. \tag{1}$$

Thus if the position of P is described by the angle A, then the position of the corresponding point Q on the vertical line is given by (1).

But now suppose P has moved to the position P_4 shown in Fig. 23-4. The angle A which describes the position of P_4 is the obtuse angle shown in the figure. This angle is no longer an acute angle of a right triangle, and we therefore have no right to speak of $\sin A$. However, let us extend the meaning of sine so that, by definition, $\sin A$ is the y-value of P_4. Since the height of Q_4 above O equals the y-value of P_4, we may continue to write $y = \sin A$ to describe the position of Q. That is, the distance of Q from O on the vertical line will be given by $y = \sin A$.

Suppose next that P occupies the position P_5 shown in Fig. 23-5. The angle A which describes how far around the circle P has moved is now the angle shown. Let us agree again that by $\sin A$ we shall mean the y-value of P_5, which is also the distance below the X-axis of the point Q_5. Then we again may write $y = \sin A$ to describe the position of Q. Note that y is now a negative quantity.

If P reaches the position P_6 shown in Fig. 23-6, then its position is represented by the angle A shown, and if we again agree to mean by $\sin A$ the y-value of P_6, we shall be able to say here too that $y = \sin A$ describes the position of Q. In this instance also, y is a negative quantity.

As P returns to the X-axis and starts to repeat its revolution, the angle A which describes the position of P will now be 360° plus some additional angle (Fig. 23-7). It is only by including 360° for each revolution of P that we can keep track of the number of revolutions. However, let us note that the y-values of P will recur in precisely the same order in which they appeared on the first revolution. Despite the fact that on the second revolution the values of A are larger than 360°, we shall continue to mean by $\sin A$ the y-value of P. Thus sin 390° will be the same as sin 30°. As P goes through its second revolution, Q repeats the motions of the first revolution. Hence it will still be true that $y = \sin A$ describes the position of Q on the vertical line. With each revolution of P, the y-values repeat, although the

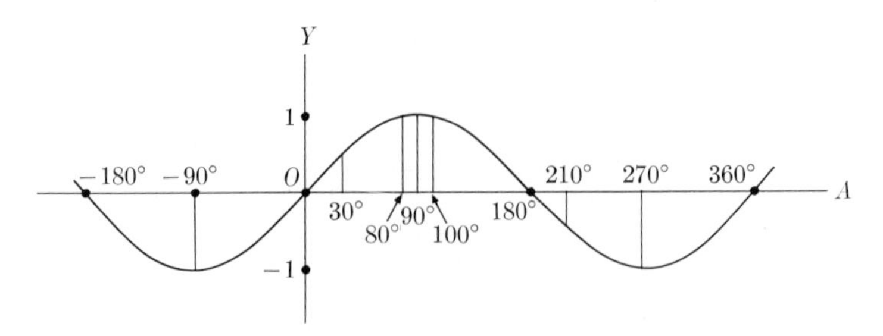

FIG. 23–8. The graph of $y = \sin A$.

angle A increases by 360°. Since the motion of Q also repeats, its position on the vertical line will continue to be represented by $y = \sin A$.

If P were to revolve in the clockwise direction, then we would make only one change, i.e., call the values of A negative. The y-value of P, wherever it is, is still, by definition, $\sin A$ and this y-value would represent the position of Q.

Let us survey what we have done. To describe mathematically the position of the point Q, we have introduced a new function: $y = \sin A$. When A is an acute angle, as in Fig. 23–3, then $\sin A$ has the old meaning; that is, it is the ratio of the side opposite angle A to the hypotenuse of the right triangle in which A lies. (In the present case the hypotenuse is 1.) But when A is larger than 90°, then the equation $y = \sin A$ is a definition of what we mean by $\sin A$. Since there is a definite y-value for each value of A, positive or negative, we do indeed have a function.

To appreciate the nature of this function let us graph it. Figure 23–8 shows the graph. The values of A are plotted along the horizontal axis, and the corresponding y-values are plotted in the usual way.

Do we know the precise numerical value of y for each value of A? We do. For values of A which are between 0° and 90°, the y-values are the ordinary sine values which we find in our trigonometric table. In the interval from 90° to 180°, the values of $\sin A$ repeat, but in *reverse* order, the values which $\sin A$ has when A varies from 0° to 90°. This statement implies that $\sin 100° = \sin 80°$, $\sin 110° = \sin 70°$, and so forth. Stated in more general terms:

$$\sin A = \sin (180° - A). \tag{2}$$

In the interval from 180° to 360°, $\sin A$ has the same numerical values as when A varies from 0° to 180°. However, now $\sin A$ is negative. Thus $\sin 210° = -\sin 30°$; $\sin 220° = -\sin 40°$; and in general:

$$\sin A = -\sin (A - 180°). \tag{3}$$

Since for each 360°-interval beyond the interval 0° to 360° $\sin A$ repeats the values that it has in the interval from 0° to 360°, $\sin 390° = \sin 30°$; sin

$400° = \sin 40°$; and so on. In symbolic form,

$$\sin A = \sin (A - 360°). \tag{4}$$

The values of $\sin A$ for negative values of A are also shown in Fig. 23–8. If we look at the figure, we see that for any negative A-value, $\sin A$ is the negative of the sine of the corresponding positive A-value. That is, $\sin(-30°) = -\sin(30°)$; $\sin(-50°) = -\sin(50°)$, and in general:

$$\sin A = -\sin(-A). \tag{5}$$

Thus we have arrived at a definition of the function $y = \sin A$ for all values of A. Since we know quantitatively what $\sin A$ is for values of A between 0° and 90°, formulas (2) through (5) enable us to calculate $\sin A$ for all other values of A. The function we have just introduced is called a *periodic function* because the y-values repeat themselves in every 360°-interval of A-values. The interval of 360° is called the period of $y = \sin A$, and the entire set of y-values in one period is called the *cycle* of y-values.

Exercises

1. Using formulas (2) through (5) or Fig. 23–8, express the following sine value as sines of angles between 0° and 90°.

(a) sin 120°	(b) sin 150°	(c) sin 210°
(d) sin 260°	(e) sin 270°	(f) sin 300°
(g) sin 350°	(h) sin 370°	(i) sin −50°
(j) sin 750°		

2. What is the largest value of $\sin A$? What is the smallest value of $\sin A$?
3. At what value of A between 0° and 360° does the function $y = \sin A$ reach a maximum?
4. Why is $y = \sin A$ called a periodic function?
5. What purpose does the function $y = \sin A$ serve with respect to the location of Q, the projection of P?
6. What is the relationship between the function $y = \sin A$ and the trigonometric ratio $\sin A$ studied in Chapter 7?
7. Describe how $\sin A$ varies as A varies from 0° to 360°; from 360° to 720°.
8. For how many values of A between 0° and 360° is $\sin A = 0.5$?
9. Distinguish between the period and the cycle of $y = \sin A$.

Thus far we have described the size of angles in degrees. There is, however, no need to stick to this unit. Let us return to the motion of the point P in Figs. 23–3 through 23–7. The size of angle A can be specified by describing the *arc length* traversed by P from its starting point on the X-axis. This arc length is as much a measure of the size of angle A as the rather arbitrary agreement that a complete revolution of one side of A should be 360°.

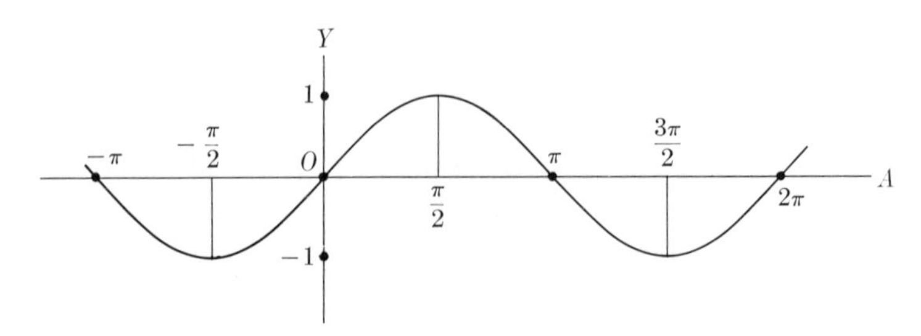

FIG. 23–9. The graph of $y = \sin A$ when A is measured in radians.

Suppose that we agree to use the arc length traversed by P as a measure of A. How do we express in this new unit an angle of 90°, for example? When A is 90°, P has traversed one-quarter of the entire circumference. But the entire circumference of a circle of unit radius is 2π. Then the size of A in the new unit is $\pi/2$, that is about 1.57. We call this new unit *radians*. Thus an angle of 90° is also one of $\pi/2$ or 1.57 radians.

The advantage of radians over degrees is simply that it is a more convenient unit. Since an angle of 90° is of the same size as an angle of 1.57 radians, we now have to deal only with 1.57 instead of 90 units. The point involved here is no different from measuring a mile in yards instead of inches. If yards are just as good on other grounds, then it is far more convenient to speak of 1760 yards than 63,360 inches.

The fact that we measure angles in radians does not disturb at all the meaning of the function $y = \sin A$. Instead of stating that $\sin 90° = 1$, we simply say that $\sin \pi/2 = 1$. The same applies to any other value of A in the sinusoidal function we have introduced. Suppose, for example, that we wished to find the value of $y = \sin A$ when $A = \pi/6$. Because our table is set up in degrees, we note first that an angle of $\pi/6$ is of the same size as 30°, for $\pi/2$ radians is the same as 90°. Now from our tables $\sin 30° = 0.5$, and so $\sin \pi/6 = 0.5$.

Since we shall be using radians a good deal, we may as well become familiar with the function $y = \sin A$ when A is expressed in radians. Figure 23–9 shows the same function as Fig. 23–8 except that the units of A are now radians.

EXERCISES

1. Express the sizes of the following angles in radians: 90°, 30°, 180°, 270°, 360°, 420°

2. The sizes of the following angles are in radians. Express the same angles in degrees.

$$\pi/2, \quad 2\pi/3, \quad 5\pi/2, \quad 3\pi, \quad -\pi/2, \quad 1.$$

3. Find the value of:

(a) $\sin \pi$ (c) $\sin \pi/3$ (e) $\sin 3\pi$
(b) $\sin \pi/2$ (d) $\sin 3\pi/2$ (f) $\sin 5\pi/2$.

4. Describe how $\sin A$ varies as A varies from 0 to 2π, as A varies from 2π to 4π.

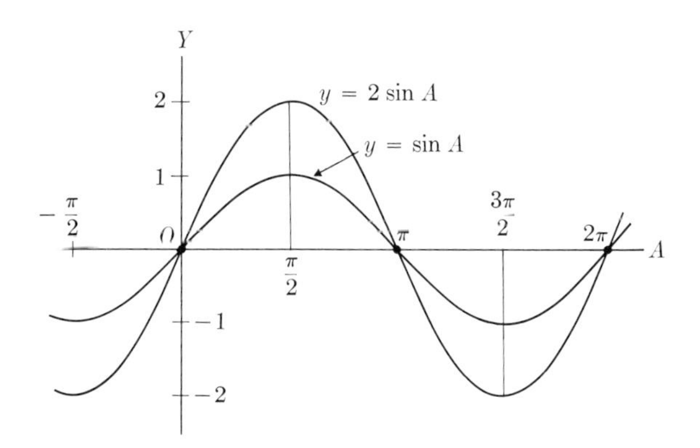

FIG. 23-10. Comparison of $y = \sin A$ and $y = 2 \sin A$.

The function $y = \sin A$ has a maximum value of $+1$ and a minimum value of -1. The maximum y-value, incidentally, is called the *amplitude* of the function. Such a function, even if it were suitable in all other respects, could not represent the motion of a bob whose maximum displacement is 2 or 3, say. This difficulty is easily obviated. Now that we have $y = \sin A$, we can readily manufacture hundreds of new functions whose amplitudes are whatever we choose to make them. Consider, for example, $y = 2 \sin A$. How does this function behave compared to $y = \sin A$? The answer is immediate. For any value of A, $y = 2 \sin A$ is twice as much as $y = \sin A$. Thus when $A = \pi/4$ or $45°$, $\sin A = 0.71$, and $2 \sin A$ is 1.42. Figure 23-10 illustrates how $y = 2 \sin A$ looks compared to $y = \sin A$. If we want a sine function with amplitudes 3, $\frac{1}{2}$, or any other number, we can write one down immediately. As is evident from the nature of the function $y = 2 \sin A$, the function

$$y = D \sin A$$

has amplitude D.

Before we can use functions such as $y = \sin A$ or $y = 3 \sin A$ to represent the motion of a bob on a spring, we must clear one more hurdle. The function we seek should represent a relationship beween displacement and time. The y-values of our functions do indeed represent the displacement of a point Q which moves up and down on a line, but our independent variable is an angle. Suppose, however, that the point P revolves around the circle f times in one second. Then, since for each revolution the angle A increases by 2π radians, the size of the angle which describes the amount of revolution of P in one second is $2\pi f$. If the point P revolves for t seconds and makes f revolutions per second, it will make ft revolutions in t seconds. The angle generated during these ft revolutions will be $2\pi ft$. Hence, the value of A in t seconds will be $2\pi ft$. Thus the function $y = \sin A$ becomes

$$y = \sin 2\pi ft. \tag{6}$$

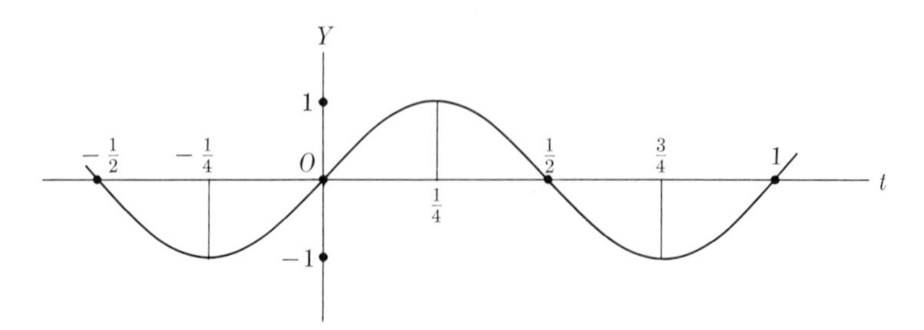

FIG. 23–11. The graph of $y = \sin 2\pi t$.

This function requires some study. Suppose the point P makes one revolution per second. Then $f = 1$. The function (6) then is $y = \sin 2\pi t$. As t increases from 0 to 1, the quantity $2\pi t$ will increase from 0 to 2π. We must now ask, How will $\sin 2\pi t$ vary as $2\pi t$ varies from 0 to 2π? Since the *angle* which $2\pi t$ describes now varies from 0 to 2π, the function will go through the entire cycle of sine values. However, if we now label our horizontal axis with time values, we obtain the graph shown in Fig. 23–11.

Next let us consider a slightly more difficult case. Suppose the point P makes 2 revolutions per second so that $f = 2$. As t increases from 0 to $1/2$, $2\pi \cdot 2t$ will increase from 0 to 2π and $\sin 2\pi \cdot 2t$ will go through the entire cycle of sine values. As t increases from $1/2$ to 1, $2\pi \cdot 2t$ increases from 2π to 4π. Then $\sin 2\pi \cdot 2t$ takes on the values corresponding to angles from 2π to 4π. But in this range the sine function takes on the same values as it does in the range from 0 to 2π. Hence in the entire interval 0 to 1 for t, the graph will be as shown in Fig. 23–12. The conclusion, which emerges clearly from the graph, is that

$$y = \sin 2\pi \cdot 2t \tag{6}$$

goes through 2 complete cycles in one second or, as one says, it has a *frequency* of 2 cycles per second.

We can now anticipate what happens for any f. The function will go

$$y = \sin 2\pi ft$$

through f cycles in one second, or it has a frequency of f cycles per second.

To increase the amplitude of any of these functions, we have but to introduce the factor D. Thus the function

$$y = D \sin 2\pi ft \tag{7}$$

will have a frequency of f cycles per second and an amplitude of D. Let us note that while f is the number of revolutions per second of P, it is also the number of oscillations per second of Q.

The y-values of formula (7) oscillate above and below the zero value as t varies. We can make the number of oscillations per second what we please

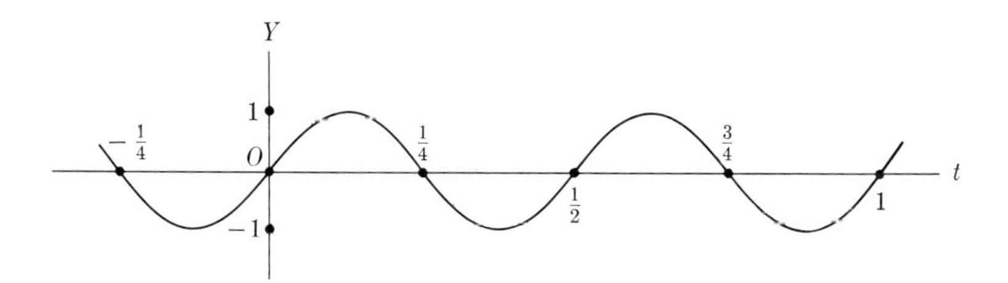

FIG. 23-12. The graph of $y = \sin 2\pi \cdot 2t$.

by merely inserting the proper value of f and we can do the same with respect to the amplitude by inserting the proper value of D. Of course, we do not know the proper values of f and D which fit the motion of a bob, but we shall see in the next section that it is not difficult to determine them.

Let us summarize what we have accomplished. We sought to represent the motion of a point which oscillates back and forth on a straight line. We were able to do so by introducing a point Q which is the projection of a point P moving around a circle at a constant velocity. Because the y-value of P equals the displacement of the oscillating point Q and because the y-value of P is expressible as a sinusoidal function, we can represent the motion of the oscillating point Q by such a function. That the approach to the oscillating point Q through the circle should be successful may be surprising, but, as Aristotle pointed out, "There is nothing strange in the circle being the origin of any and every marvel."

Exercises

1. Find the value of $2 \sin A$ when A is $30°$, $90°$, $\pi/2$, $\pi/3$.
2. What is the maximum value of $3 \sin A$? the minimum value?
3. What is the amplitude of $y = 4 \sin A$?
4. Find the value of (a) $\sin 2t$ when $t = \pi/4, \pi/2, 3\pi/4, \pi$. (b) $\sin 3t$ when $t = \pi/6, \pi/3, \pi/2, 2\pi/3$.
5. What is the shape of the graph of $y = \sin 2\pi \cdot 2t$ as t varies from 1 to 2?
6. Graph the function $y = \sin 2\pi \cdot 3t$ as t varies from 0 to 1.
7. Graph the function $y = 2 \sin 2\pi \cdot 2t$ as t varies from 0 to 1.
8. What is the frequency (in one second) of $y = \sin 2\pi \cdot 10t$?
9. Find the value of (a) $y = \sin 2\pi \cdot 2t$ when $t = 1/8, 1/4, 1/3$; (b) $y = \sin 2\pi \cdot 4t$ when $t = 1/6, 1/2, 1$; (c) $y = 2 \sin 2\pi \cdot 3t$ when $t = 1/6, 1/4, 1/12$.

23-4 The mathematical analysis of the motion of the bob. We now wish to represent mathematically the motion of the bob on the spring. We know that this motion is periodic and has a definite frequency and amplitude. However we do not know that the motion is really sinusoidal. That is, as t varies, do the displacements of the bob follow precisely the variation of y in a function of the form

$$y = D \sin 2\pi f t? \tag{8}$$

If, for example, the motion of the bob should be faster on the upper half of its path than on the lower half, it could still have the same period for each complete oscillation and perform a fixed number of oscillations per second. Yet the motion would not be of the form (8). We need a little more insight into the motion of the bob than we now have.

This insight into the action of bobs on springs was supplied by Robert Hooke. The principle he discovered, still known as Hooke's law, is very simple. We all know that if we stretch or compress a spring, the spring seeks to restore itself to its normal length; that is, when stretched or compressed the spring exerts a force. Hooke's law says that the force is a constant times the amount of compression or extension. In symbols, if L is the increase or decrease in length of the spring and F is the force exerted by the spring, then $F = kL$, where k is a constant for a given spring. The quantity k is called the spring or stiffness constant and it represents the stiffness of the spring. If k is large, the spring exerts considerable force even for small L.

We shall now see what we can deduce from Hooke's law. Suppose that a bob of mass m is attached to a spring. Then we know that gravity pulls the bob downward some distance d where the bob comes to rest (Fig. 23–13). The rest position is reached when the force of gravity acting on the bob, or the weight of the bob, just offsets the upward force exerted by the spring. Now the force of gravity is $32m$, and, according to Hooke's law, the upward force exerted by a spring which is pulled downward a distance d is kd. Since at the rest position these two forces just offset each other, we have

$$32m = kd. \tag{9}$$

Now suppose the spring is pulled downward an additional distance y. If we use the convention agreed upon in Section 23–2 that displacements above the rest position are to be positive and below the rest position negative, then the total extension of the spring is now $d - y$ because y itself is negative. The force that the spring exerts in an upward direction is, by Hooke's law,

$$k(d - y) \quad \text{or} \quad kd - ky. \tag{10}$$

However, the weight of the bob, or $32m$, exerts a constant downward force. Hence the net upward force is $kd - ky - 32m$. In view of equation (9) the *net* upward force is $-ky$. We now apply Newton's second law of motion, which says that when a force is applied to a mass, the force equals the mass times its acceleration. Thus we have

$$ma = -ky \tag{11}$$

or, by dividing both sides of this equation by m,

$$a = -\frac{k}{m}y \tag{12}$$

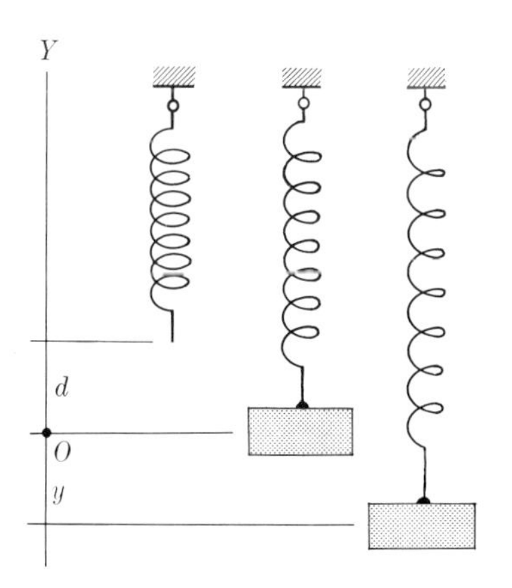

Fig. 23-13. A bob on a spring in the rest position (center) and pulled down a distance y (right).

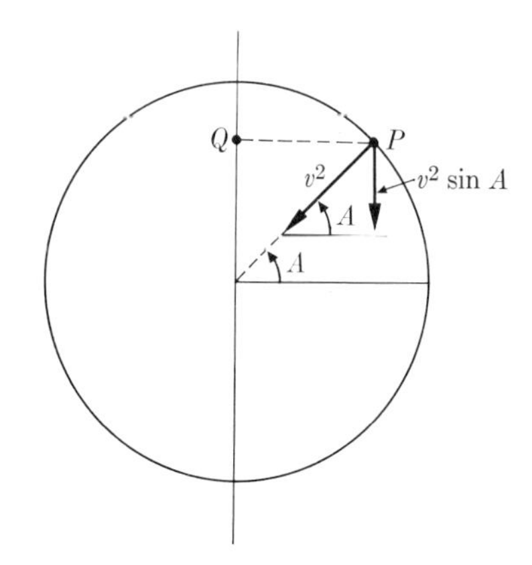

Fig. 23-14. Determination of the vertical acceleration of P.

Formula (12) is the basic law governing the motion of the bob on the spring. The reason is simply that the acceleration determines the velocity acquired by the bob and the velocity determines the distance covered in any specified interval of time. This argument is in principle the same as that followed in Chapter 15, for example, where starting with an acceleration of 32 ft/sec^2, we determined the velocity acquired and then the distance covered by the moving object. It is not so easy to proceed from formula (12) as it was to build on $a = 32$, but we shall soon utilize (12).

To relate the motion of the bob with the motion of the projection Q (Fig. 23-14) of the point P which revolves on a circle, let us determine the acceleration of Q. If an object moves along a circular (instead of a straight-line) path, there must be a centripetal acceleration acting on the object. We know from our work in Chapter 17 [see formula (24)] that this acceleration is v^2/r, where v is the velocity of the moving object and r is the radius of the circle. In our case the radius of the circle on which P moves is 1. Hence the centripetal acceleration is v^2. This acceleration, as we know, is directed toward the center of the circle. We are, however, interested in the vertical motion of P for this vertical motion is precisely the same as the motion of the point Q. Since by Galileo's principle vertical motion is independent of horizontal motion, the vertical motion of P is determined by the vertical component of the acceleration. How, then, do we compute the vertical component of the acceleration? As Fig. 23-14 shows, the vertical component, which we shall denote by a, is

$$a = v^2 \sin A.$$

We know that [see (1)]

$$\sin A = y,$$

where y is the ordinate of P. Hence

$$a = v^2 y.$$

However, since the acceleration is directed downward when y is positive, we must write

$$a = -v^2 y. \tag{13}$$

If now the moving point P makes f revolutions per second, then P covers f circumferences per second; that is, $v = 2\pi f$ and $v^2 = 4\pi^2 f^2$. We substitute this result in (13) and obtain

$$a = -4\pi^2 f^2 y \tag{14}$$

as the acceleration of the vertical motion of P or of its shadow Q on the Y-axis.

Formula (14) tells us two very important things. Since, as we pointed out earlier, the acceleration determines the motion and we know that the displacement of the point Q is given by the formula

$$y = D \sin 2\pi f t, \tag{15}$$

then an acceleration of the form (14) implies the formula (15). But the acceleration of the bob, namely

$$a = -\frac{k}{m} y, \tag{12}$$

is of exactly the same form as (14) because in both cases the acceleration is a constant times the displacement. Hence *the displacement of the bob must also be describable by a function of the form (15).*

However we do not know what f is for the bob. We note that the constant $4\pi^2 f^2$ in (14) leads to the quantity $2\pi f$ in (15). Since the constant k/m appears in (12), the formula for the bob's motion must be

$$y = D \sin \sqrt{\frac{k}{m}}\, t. \tag{16}$$

If we write this in the form

$$y = D \sin 2\pi \left(\frac{1}{2\pi}\sqrt{\frac{k}{m}}\right) t,$$

we see that the frequency per second, or the number of oscillations per second, of the bob's motion must be

$$f = \frac{1}{2\pi}\sqrt{\frac{k}{m}}\,. \tag{17}$$

We know now how the value of k, the spring's stiffness, and the value m of the mass attached to the spring determine the frequency of the bob's motion, and we can therefore compute this frequency.

We made one misleading statement in the preceding discussion. We said that the acceleration of the bob determines the motion of the bob, and so the formula of the bob's motion must be of the form (15). The acceleration does determine the essential characteristics of the motion, but the initial velocity and initial displacement do have some effect. This point may become clearer if we compare the present case with the forces affecting falling bodies. All bodies rising or falling near the surface of the earth are subject to an acceleration of 32 ft/sec^2, and this fact determines the essential nature of the motion. But if a body is thrown up into the air, the maximum height that it will reach depends upon the initial velocity and the point from which it is thrown up. Now if a bob on a spring is pulled down a distance D from the rest position, then we know from observation that in each oscillation, it will rise to a height of D above the rest position and then descend a distance D below it. That is, the amplitude of the motion, D, is determined by the initial displacement.

There is one more mathematical fact of importance. Formula (17) yields the number of complete oscillations per second of the bob. Hence the time required to make one complete oscillation, or the period T, is given by

$$T = \frac{1}{f} = \frac{2\pi}{\sqrt{k/m}} = 2\pi\sqrt{\frac{m}{k}}. \tag{18}$$

Formula (18) is in itself something of a reward for all our work, for it tells us that we can determine the period from a knowledge of m and k or that we can fix the period by fixing m and k. But Hooke observed something else in formula (18), which is at least as significant: The period is independent of the amplitude of the motion; that is, whether one pulls the bob down a great distance or a short distance and then releases it, the time required by the bob to go through each complete oscillation will be the same.

This fact is immensely useful. At the very outset of our treatment of the bob's motion we pointed out that the resistance of the air and internal energy losses in the spring will cause the motion to die down. At the time we decided to ignore this fact and to suppose that there was no loss of energy. But there is. Suppose, however, that we were to give the bob a little upward push every time it reached its lowest position, i.e., add energy to the motion and keep the bob moving. Such an action might alter the amplitude, but would *not* affect the period, and each successive oscillation of the bob would therefore continue to take the same amount of time. Hence the oscillations of the bob on the spring can be used to measure time or to regulate the motion of some hands on a dial which would show time elapsed.

Of course, the motion of a bob on a spring is not quite the practical device for a clock. The device actually used can be found in every modern pocket or wrist watch. There a spring coiled in a spiral and carrying a weight called the balance wheel expands and contracts regularly. Each second the wheel is given a little "kick" which restores the energy the spring loses on each oscillation. (The energy comes from a mainspring which is wound up by hand usually once a day.) The spiral spring regulator was invented and patented by Christian Huygens in 1675. The first chronometer which was sufficiently accurate to be used by ships to determine longitude was invented by John Harrison, who in 1772 won a prize of £20,000 offered by the British government for such a device.

Exercises

1. If a mass of 2 lb pulls a spring down 6 in., what is the spring constant? [*Suggestion:* Use (9).]

2. Suppose that one attaches a mass of 2 lb to a spring whose stiffness constant is 50. Calculate the number of oscillations per second which the mass would make if set into vibration.

3. What is the period of a mass vibrating at a rate of 100 oscillations per second?

4. Suppose a mass is set to vibrating on a spring at the rate of 50 oscillations per second. If the mass has a maximum displacement of 3 in., what formula describes the motion?

5. Suppose a mass of 3 lb is attached to a spring whose stiffness constant is 75. The mass is pulled down 3 in. below the rest position and then released. Write a formula relating displacement and time.

6. Suppose that you are given a spring with a stiffness constant of 50. Calculate the mass that you would have to place on the spring to produce a period of oscillation of one second.

7. Suppose that a mass oscillates on a spring so that the relation between displacement and time is (a) $y = 4 \sin 2\pi \cdot 5t$; (b) $y = 4 \sin 10t$. Describe the motion of the mass for (a) and (b).

8. Suppose that you wished to decrease the number of oscillations per second which a mass makes on a given spring. How would you alter the mass?

9. Suppose that a tunnel is dug through the earth and a man of mass m steps into the tunnel. Inside the earth the force of gravity on a mass m at a distance r

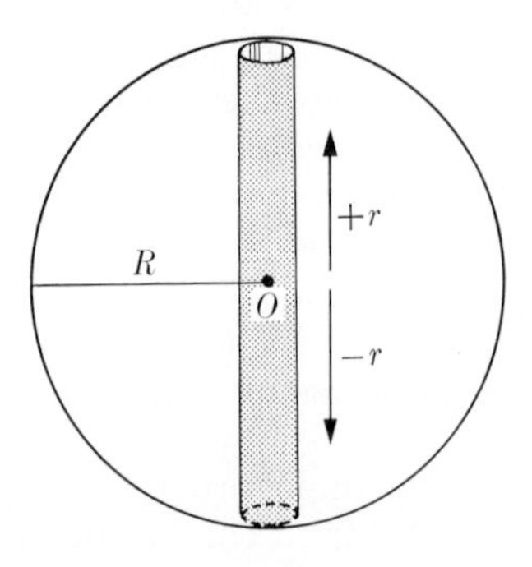

Figure 23–15

from the center is $F = GmMr/R^3$, where M is the mass and R is the radius of the earth. This force is directed toward the center. To distinguish distances above and below the center, let r be positive above and negative below the center. Then the acceleration acting on the mass m is $a = -GMr/R^3$. Discuss the subsequent motion of the man (Fig. 23–15).

23–5 Summary. The mathematical objective of this chapter was to introduce a new type of mathematical function, the sinusoidal function. There is not just one sinusoidal function, for all functions of the form $y = D \sin 2\pi ft$, no matter what D and f may be, are sinusoidal. The sinusoidal functions are also called trigonometric functions because they are obtained by extending the concept of the sine of an angle, a concept which was first created and studied in trigonometry. Other trigonometric functions can be derived from an extension of the concept of cosine and tangent of an angle and other trigonometric ratios which we did not study. All trigonometric functions are highly useful in scientific work.

The creation of trigonometric functions was motivated by the study of vibratory or oscillatory motion. We have used the motion of a bob on a spring to illustrate such a motion, and we have shown how the mathematical description of this motion can be used to deduce information about it. We have yet to see some of the major uses of sinusoidal functions.

Topics for Further Investigation

The mathematics of pendulum motion.
The trigonometric function $y = \cos A$.

Recommended Reading

Brown, Lloyd A.: "The Longitude," in James R. Newman: *The World of Mathematics,* Vol. II, pp. 780–819, Simon and Schuster, Inc., New York, 1956.

Kline, M.: *Mathematics and the Physical World,* Chap. 18, T. Y. Crowell Co., New York, 1959.

Taylor, Lloyd Wm.: *Physics, The Pioneer Science,* Chap. 15, Dover Publications, Inc., New York, 1959.

Whitehead, Alfred N.: *An Introduction to Mathematics,* Chaps. 12 and 13, Henry Holt and Co., New York, 1939.

CHAPTER 24

THE TRIGONOMETRIC ANALYSIS OF MUSICAL SOUNDS

Motion appears in many aspects—but there are two obvious kinds, one which appears in astronomy and another which is the echo of that. As the eyes are made for astronomy so are the ears made for the motion which produces harmony: and thus we have two sister sciences, as the Pythagoreans teach, and we assent.

Plato

24–1 Introduction. In this chapter we intend to show how trigonometric functions have given man his first real insight into the nature of musical sounds, and how this knowledge is utilized in the design of such devices as the telephone, the phonograph, the radio, and sound films.

The mathematical study of musical sounds did not start with the application of trigonometric functions. Indeed, it goes back to the very first emergence of any real mathematics and science, namely the beginning of the classical Greek period. We may recall the Pythagorean discovery that the lengths of two equally taut plucked strings whose sounds harmonize are related by simple numerical ratios such as 2 to 1, 4 to 3, and 3 to 2. The lower note in each case originates with the longer string. Thus the Pythagoreans discovered the cardinal principle that harmony is governed by a mathematical relationship. They also designed musical scales whose notes, as measured quantitatively by the lengths of the vibrating strings, possessed precise numerical values. From Pythagorean times onward, mathematicians and scientists were convinced that musical sounds had important mathematical properties, and music, along with arithmetic, geometry, and astronomy, became part of the quadrivium. These four subjects were studied together right through the medieval period. Although Greek, Arab, and medieval mathematicians continued to investigate musical sounds and wrote books on music, their work was essentially limited to the construction of new systems of scales for instrumental and vocal music.

It was the mathematicians and scientists of the seventeenth century who initiated other investigations and made the next series of important discoveries. Familiar names, such as Galileo, his French pupil and colleague Father Marin Mersenne (1588–1648), Hooke, Halley, Huygens, and Newton, obtained significant new results. Whereas the Pythagoreans had studied strings of different length but equal tension, Mersenne studied the effect of changing tension and mass of a string and found that an increase in mass and a decrease in tension produce lower notes in a string of given length. This discovery was very important for stringed instruments such as the violin and the piano; to secure the range of pitch which these instruments possess by variations in length only would require exceedingly long strings.

Galileo and Hooke demonstrated experimentally that each musical sound is characterized by a definite number of air vibrations per second, a statement which will mean more to us in a few moments. The determination of the velocity of sound (about 1100 feet per second in air) was another achievement. It is of interest that the clocks which some of these men designed and constructed were essential to the progress made in the study of sound because, as we can see from the results cited, the ability to measure small intervals of time was an indispensable condition for any work in this field.

The best mathematicians of the eighteenth century, Leonhard Euler, Daniel Bernoulli (1700–1782), Jean le Rond d'Alembert (1717–1783), and Joseph Louis Lagrange, studied vibrating strings, such as the violin string, and vigorously disputed whether trigonometric functions were adequate to represent the vibrations. The mathematical analysis of sound waves soon followed and proved to be the chief tool in the theoretical mastery of musical sounds. We can readily see why mathematics was invaluable in these investigations, for observation of the air, even of air in the process of propagating sound, reveals nothing.

Before we undertake to study just what the nineteenth-century mathematicians and scientists learned, we must make some distinctions. The first is a matter of terminology. We shall be interested in the analysis of musical sounds as opposed to noise. However, in the present context, the term "musical sound" is used in a technical sense and includes not only those sounds commonly understood to be music, but also the sounds of ordinary speech. As a matter of fact, the physicist's meaning might be more appropriately represented by the term *intelligible sound.* Just what is meant by either phrase will be clear in a few moments.

The second distinction one must make is between sound as a motion of air and sound as a sensation which human beings experience. The former is a physical phenomenon which takes place in space and whose physical and mathematical properties are fixed. On the other hand, the sensations which human beings may receive because moving air strikes their ears and stimulates certain nerves depend upon their auditory mechanism and may vary from one person to another. There are, for example, physical sounds which humans cannot hear at all. Though we shall have something to say about the perception of sound, our first and main concern will be to understand the physical phenomenon.

24–2 The nature of simple sounds. The variety of sounds given off by musical instruments, the human voice, phonographs, radios, and whirring machinery, for example, is so great that one cannot hope to study all of them in one swoop. Hence it would seem wise to start one's investigation with simple sounds. But which sounds are simple? If we rely upon our ears to decide this question, then the sounds given off by tuning forks seem to be simple. The ear may, indeed, be deceived here, but let us follow up this suggestion.

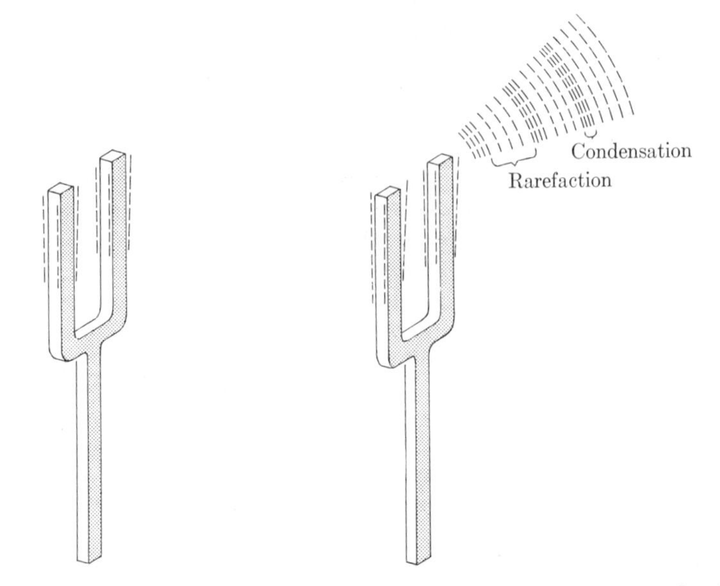

FIG. 24–1. A vibrating tuning fork.

FIG. 24–2. Motion of air molecules under pressure of a vibrating tuning fork.

If either prong of a tuning fork is struck, both prongs will move inward and then outward very rapidly and will repeat this motion for a long time. Let us consider one prong, say the right one shown in Fig. 24–1. Before the prong is struck, it occupies what we might call the rest position. After being struck, the tip is displaced some distance to the right. It then moves to the left, to a position somewhat to the left of the rest position, and then moves to the right. The sequence then repeats itself many times. The displacement of the tip varies with time, and the first question one might raise is, What is the relationship between displacement and time? There are two considerations which suggest that the formula is sinusoidal: First of all, the prong resembles a spring-and-bob arrangement. The spring is the prong itself, though the motion is a sidewise oscillation rather than an expansion and contraction. The mass which corresponds to the bob is the mass of the prong itself, though admittedly this mass is not concentrated in one place as it is in the case of a bob on a spring. The second consideration is that as the tip of the prong moves farther and farther out from the rest position, the force which the prong exerts to return to the rest position may be expected to increase with the displacement. The simplest assumption one might make in this case is that the force increases directly with the displacement. From formula (11) of the preceding chapter we can see that this is indeed the mathematical law which underlies and determines the sinusoidal motion of the bob. Hence it seems reasonable to expect that the relation between displacement of the tip of either prong and time is sinusoidal. The amplitude of this relation is the maximum displacement of the prong, and the frequency is the frequency per second with which the prong oscillates.

Of course, we are not so much interested in the motion of the tuning fork as we are in the sound it creates. Hence what matters next is, How does the air respond to the vibration of the tuning fork? The fundamental fact about the behavior of air which is of importance in this connection is that air pressure seeks to become uniform everywhere. This means that if the air pressure for any reason should become high in one place, the air will spread out from that place into neighboring regions where the pressure is lower and so try to equalize the pressure in the entire region under consideration. With this physical fact in mind, let us see what happens when the right prong of the tuning fork moves, say to the right. The prong pushes the molecules of air near it to the right and thus crowds them into a place occupied by other molecules. The pressure becomes high in this place, and since the molecules of air cannot move to the left because the prong is there, they will move off farther to the right (and in other directions) in order to equalize the pressure. But this motion means that the crowding now occurs a little farther away from the tuning fork and again, to equalize the pressure, the molecules move farther to the right. The process continues, and the crowding, or *condensation*, as it is usually called, moves off to the right.

The prong, having moved as far to the right as it can, will now move back not only to its rest position but farther to the left. This motion leaves an empty region—the place that the prong had occupied—and so the molecules of air on the right rush into this empty space. Molecules still farther to the right also move to the left because the pressure has become less to their left. Thus a state of low pressure, or *rarefaction*, as it is called, moves to the *right* as molecules move to the left to equalize the pressure in their neighborhood. With each successive vibration of the prong, a condensation and a rarefaction move off to the *right* (Fig. 24–2). The successive condensations and rarefactions also move out in other directions, but it is sufficient for our purposes to follow what happens in one direction.

The action of the air is somewhat complicated because it consists of billions of molecules, and they do not all behave in exactly the same way. However, there is an average effect. It is convenient to speak of a series of typical molecules to the right of the prong which represent the average behavior of the entire collection. If we consider the action of any one typical molecule, say one near the prong, then what it does is to move to the right when the prong moves to the right. When the prong moves to the left, the typical molecule will also move to the left because the air pressure has been lowered. Like the prong, it will move past its rest position, and continue to the left. Then, as the prong moves to the right, the molecule will be pushed to the right, will pass its rest position, go farther to the right, and, from this time on, it will continue to oscillate.

Typical molecules farther to the right will behave like the typical molecule near the prong; however, their reaction will be slightly delayed since

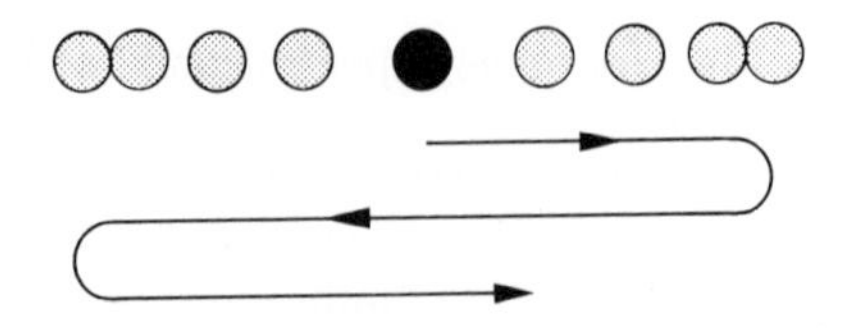

FIG. 24-3. Motion of a typical molecule.

condensations and rarefactions reach them a little later. Figure 24-3 illustrates the motion of a typical molecule reacting to a series of prong oscillations.

Two important facts emerge from the above discussion. The first is that the average, or typical, molecule follows, in effect, the motion of the prong. Any one molecule acts as if it were attached to the prong by a spring. When the prong moves to the right, it contracts the spring. The latter seeks to restore its length and so pushes the molecule to the right. While the molecule moves to the right, the prong moves to the left, and hence the spring is extended. It now seeks to contract and so pulls the molecule to the left. The molecule moves to the left, and the spring contracts. But now the prong is ready to move to the right, and consequently the motion of prong and molecule repeats itself. The action of air pressure is indeed like the action of the spring. In fact, Hooke used the phrase, "the spring of the air," to describe the effect of air pressure.

The second fact is that the sound wave which moves from the prong to some person's ear, say, consists of the series of condensations and rarefactions induced by the prong's motion. Each molecule merely oscillates about its rest position, but in doing so it produces the increase and reduction of pressure which cause the neighboring molecules to oscillate.

The nature of the sound wave may perhaps be made clearer by comparing it with a water wave. If the end of a stick is quickly moved back and forth in still water, a series of waves will spread out from the end of the stick. However, the individual water molecules do not move out. Each oscillates about its original position, but the increase and decrease in pressure which the stick creates cause the molecules farther away to duplicate the motion of the molecules near the stick.

Since the motion of any typical molecule whether near or far from the prong is the same, let us study the motion of any one of these molecules. Specifically, let us seek the relationship between the displacement from rest position and the time that the molecule is in motion. What formula relates displacement and time? We have already produced two crude physical arguments suggesting that for the prong, displacement and time are related by a sinusoidal formula. Since the motion of any typical air molecule duplicates the action of the prong, the formula which relates displacement and time for the typical air molecule should also be sinusoidal. Actually these physical arguments do not really prove that the formula is sinusoidal. However, this fact can be established either by a rather complex

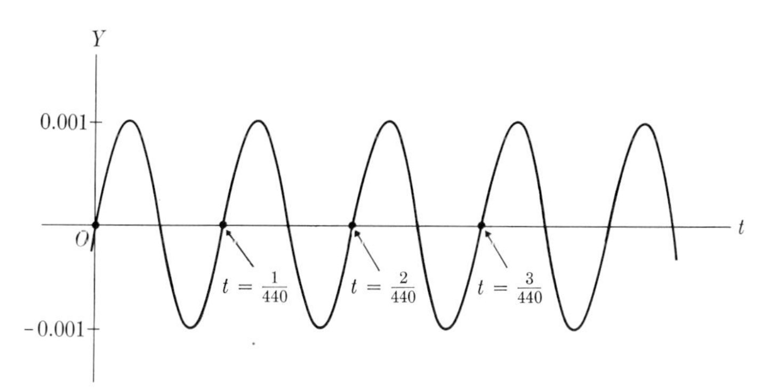

FIG. 24-4. Graph of displacement versus time of a typical molecule executing 400 oscillations per second.

mathematical analysis of air motion or, experimentally, by converting the air pressure to electric current (by means of a microphone, for example) and by then displaying the current on a cathode-ray tube (television tube).

We shall take for granted that for a typical air molecule, the formula relating displacement and time is sinusoidal. Hence, if y is the displacement and t is the time, then, in view of what we have learned in the preceding chapter, the formula is

$$y = D \sin 2\pi f t, \tag{1}$$

where D is the amplitude, or maximum displacement, and f is the number of oscillations, or the frequency of cycles per second. We wish to emphasize that the formula applies to the sounds produced by tuning forks or to what we have reason to believe are simple sounds.

To use formula (1), we must know D and f. The value of f is the frequency with which the tuning fork oscillates. A frequency commonly used to standardize the pitch of sounds is 440 per second. This then is a typical value of f. The value of D, the amplitude of the motion of a typical air molecule, is *not* the amplitude of the prong's motion, but depends upon the medium in which sound spreads out or is propagated. It depends, so to speak, on the "springiness" of the medium. In air, 0.001 inch can be considered to be a reasonable value for D. Hence a typical formula for a simple sound is

$$y = 0.001 \sin 2\pi \cdot 440t. \tag{2}$$

Thus, a typical air molecule oscillating in accordance with formula (2) shuttles back and forth about its mean, or rest, position 440 times per second or, as we say, it goes through 440 complete cycles in one second. The farthest distance from the mean position that it reaches, that is the amplitude of its motion, is 0.001 inch.

Figure 24-4 illustrates the relationship between displacement and time for a simple sound such as formula (2) represents. Of course the typical molecule shuttles backward and forward, but on the graph its displacements are plotted as ordinates and the time elapsed is shown by the corresponding abscissas.

Although formula (2) represents only simple sounds—we have yet to discuss the formulas describing more complicated sounds—it enables us to understand what we meant earlier by the phrase "intelligible sounds." We see that a simple sound has a regularity or periodicity. The motion of the air molecules repeats itself a number of times a second. When the ear receives many cycles of this motion, it can identify the sound. If, on the other hand, the motion of the air molecule is not regular but varies irregularly with time, the ear still hears sound, but sound that does not convey any meaning, i.e. noise.

Exercises

1. What is the basic mathematical formula which represents simple sounds? State the physical meaning of the various letters in the formula.
2. State the formula which describes the relationship between displacement and time for a simple sound whose frequency is 300/sec and whose amplitude is 0.0005 in.
3. If $y = 0.002 \sin 2\pi \cdot 540t$ is the mathematical description of a sound, what are the frequency and amplitude of this sound?
4. If a sound has a frequency of 400 cycles/sec, how many cycles would the ear receive in 1/20 sec?

24–3 The method of addition of ordinates. We have now a good mathematical representation of simple sounds. But interesting musical sounds, whether vocal or instrumental are, as a rule, not simple, and the really significant contribution of mathematics to the understanding of musical sounds lies in the analysis of more complex sounds. To comprehend this contribution, we must first examine a relevant mathematical idea. Instead of considering simple sinusoidal functions such as (2), let us take the function

$$y = \sin 2\pi t + \sin 4\pi t. \tag{3}$$

What sort of relationship between y and t does formula (3) represent?

A good way to investigate this question is to draw a graph of the above function. Since we wish to obtain merely some general idea of how y varies with t, we shall seek only a sketch rather than a very accurate graph. We could proceed by selecting values of t, calculating the corresponding values of y, and then plotting the points whose coordinates have thus been determined. However, there is a quicker method which is also more perspicuous. Let us consider the two functions:

$$y_1 = \sin 2\pi t \tag{4}$$

and

$$y_2 = \sin 4\pi t. \tag{5}$$

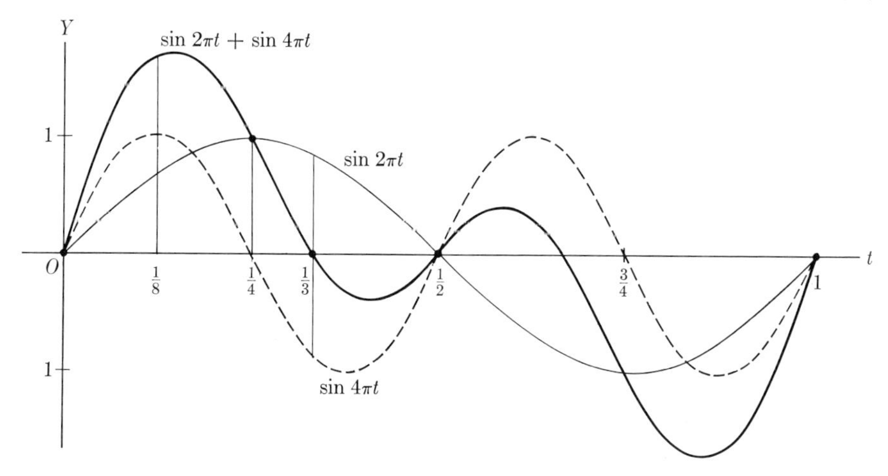

FIG. 24-5. The graph of $y = \sin 2\pi t + \sin 4\pi t$ obtained by addition of ordinates.

We have used the notation y_1 and y_2 to distinguish the dependent variables in (4) and (5) from the y in formula (3). Formulas (4) and (5) are easily graphed. Formula (4) is the ordinary sine function which goes through the regular cycle of sine values in each unit of t. Formula (5) has a frequency of 2 in each unit of t; that is, the y-values go through the complete cycle of sine values twice in each unit of t. Let us sketch both functions on the same set of axes (Fig. 24-5).

Now the y of formula (3) is clearly the sum of y_1 and y_2. Hence adding the values of y_1 and y_2 at various values of t will yield y. Since we are interested only in a sketch, let us perform the addition by using Fig. 24-5 to obtain the values of y_1 and y_2. Thus for $t = 0$, the graphs show that y_1 and y_2 are both zero. Hence y, the sum of y_1 and y_2, also is zero. At $t = 1/8$, we see from the graph that y_1 is about 0.7 and y_2 is 1. Hence $y = 1.7$ when $t = 1/8$. At $t = 1/4$, we find that $y_1 = 1$ and $y_2 = 0$. Hence $y = 1$ when $t = 1/4$. At $t = 1/3$, y_1 is about 0.85 and y_2 is about -0.85. In adding the last two values for y_1 and y_2, we must take into account that one is positive, the other negative, and their sum zero. Hence at $t = 1/3$, $y = 0$. By selecting a few more values of t and estimating the corresponding y_1- and y_2-values, we can obtain more y-values. Finally, we join the various points which belong to the graph of formula (3) by a smooth curve. The result is the heavy-lined curve shown in Fig. 24-5. The method just described for graphing y as a function of t provides a rough sketch. If one wishes to obtain a more accurate graph, he can calculate the value of y for each value of t.

How far need we carry this process of determining y-values corresponding to various t-values? We note that the function $y_1 = \sin 2\pi t$ repeats itself when t becomes larger than 1. The function $y_2 = \sin 4\pi t$ goes through two full cycles in the interval from $t = 0$ to $t = 1$ and begins its third cycle of sine values as soon as t increases beyond 1. Thus at $t = 1$ both functions begin to repeat the values which they had taken on at $t = 0$ and, in the interval from $t = 1$ to $t = 2$, both functions will repeat the behavior exhibited in the interval from $t = 0$ to $t = 1$. Since y_1 and y_2 repeat their former behavior, it follows that y, which is the sum of y_1 and y_2, will also repeat its former behavior. In other words, in the interval from $t = 1$ to $t = 2$, y will behave precisely as it did in the interval from $t = 0$ to $t = 1$. And in each succeeding unit interval of t-values, the function will repeat the behavior exhibited in the interval from $t = 0$ to $t = 1$. If we therefore determine the behavior of y in the interval from 0 to 1, we know how it behaves for all larger values of t.

There are several major facts to be learned from this example. First of all, since the function (3) repeats its behavior in every unit of t-values, it is periodic. Moreover, because the term $\sin 4\pi t$ goes through two cycles of sine values in exactly the t-interval in which $\sin 2\pi t$ goes through one cycle, the entire function repeats itself with the frequency with which $y = \sin 2\pi t$ repeats itself. Hence the frequency of formula (3) is one cycle per unit of t. Thirdly, the shape of the graph of formula (3) shows that the formula is *not* sinusoidal even though it is periodic. In other words, the sum of two sine functions can yield a function whose shape is quite different from that of a sine function, but the sum can nevertheless repeat itself.

We might expect that functions built up of three or more sine functions could have quite strange shapes and yet be periodic if the summands *all* began to repeat at some value of t, say $t = 1$, the values they had taken on at $t = 0$.

Exercises

1. By following the method described in the text sketch the graph of:
   (a) $y = \sin 2\pi t + \sin 6\pi t$;
   (b) $y = \sin 2\pi t + \frac{1}{2} \sin 4\pi t$;
   (c) $y = \sin 2\pi t + \sin 3\pi t$.
2. What is the frequency, in one unit of t, of the function:
   (a) $y = \sin 2\pi t + \sin 8\pi t$;
   (b) $y = 2 \sin 2\pi t + \sin 4\pi t$;
   (c) $y = \sin 2\pi t + \sin 4\pi t + \sin 6\pi t$;
   (d) $y = \sin 2\pi \cdot 100t + \sin 2\pi \cdot 200t + \sin 2\pi \cdot 300t$.

24–4 The analysis of complex sounds. We have already mentioned that the sounds given off by almost all musical instruments and by the human voice are not simple sounds; that is, they are not representable by functions of the form (1). Yet these sounds are intelligible, which means that they must be periodic or that the pattern of displacement versus time must repeat itself. The shapes of the curves which represent such sounds are, however, quite varied. In fact, to each sound there corresponds a characteristic shape. For example, Fig. 24–6 shows the shape corresponding to the sound of a piano note C. To obtain this graph, the sound is converted to

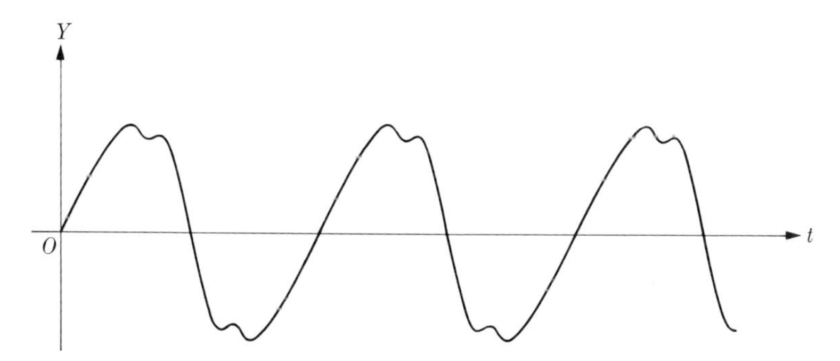

Fig. 24–6. Displacement versus time of a typical molecule for the note C on a piano.

electric current and the vibration of the current is made visible by means of a cathode-ray tube. In view of the variety of musical sounds it may seem that we have reached an impasse in our attempt to analyse all such sounds mathematically. But by a stroke of good luck mathematics provided the very theorem which gives us remarkable insight into all complex sounds. The stroke of good luck was the mathematician Joseph Fourier (1768–1830).

Fourier was the son of a French tailor. While attending a military school he became intrigued with mathematics. Since he realized that his low birth would not permit him to become an army officer, he let himself be persuaded by members of the Church to study for the priesthood. However, he abandoned the priesthood to accept a professorship of mathematics at the military school that he had attended. Later he became a professor at the École Normale and at the École Polytechnique, universities founded by Napoleon.

Fourier's main interest was mathematical physics, and his most important work in that domain concerned the conduction of heat; for example, he studied how heat travels along metals. His chief contribution, a book entitled *The Analytical Theory of Heat* (1822), is one of the great classics of mathematics. In the development of the theory of heat Fourier established a mathematical theorem whose value extends far beyond the physical application for which it was intended. Our interest in the theorem lies in what it does to analyze complex musical sounds.

Fourier's celebrated theorem says that any periodic function is a sum of simple sine functions of the form $D \sin 2\pi ft$. Moreover, the frequencies of these component functions are all integral multiples of one frequency. To illustrate the significance of this theorem, let us suppose that y is a periodic function of t. Then the formula which relates y and t must be of the form

$$y = \sin 2\pi \cdot 100t + 0.5 \sin 2\pi \cdot 200t + 0.3 \sin 2\pi \cdot 300t + \cdots \qquad (6)$$

The numbers in this formula depend, of course, on the choice of the initial periodic function, but let us suppose that they are correct and see what they stand for. The numbers 1, 0.5, 0.3 are the amplitudes of the respective sinusoidal components of the entire periodic function. The lowest frequency per second, that of the first term, is 100. The second term has frequency 200, or twice the lowest frequency. The third term has frequency 300, or three times the lowest frequency, and so on. The dots at the end of formula (6) imply that we might need additional terms like the ones shown, to represent any given periodic function. In accordance with the theorem, all frequencies occurring in such additional terms must be multiples of 100.

Before we consider the significance of Fourier's theorem for the study of musical sounds, we should satisfy ourselves that formulas such as (6) do represent periodic functions. In this connection, two results of our work in Section 24–3 should be helpful. We learned there that the sum of two sine functions can produce a rather peculiarly shaped but nevertheless periodic graph. Moreover, because the second term in formula (3) had twice the frequency of the first one, the frequency of the *entire* function was the lower of the two frequencies. The situation in (6) is very much the same. It is a sum of sine terms, and the graph of this sum may indeed have a peculiar or irregular shape. But the shape will repeat itself because during the time that the first term goes through one cycle, namely the interval $t = 0$ to $t = 1/100$, the second term will go through two cycles, and the third term through three, so that the entire function will repeat itself as soon as the first term does. Since the frequency of the first term is 100 in one unit of t, the entire function has the frequency of the first term.

And now what does Fourier's theorem have to do with the analysis of musical sounds? The application of this theorem to music was made by a German, George S. Ohm, a teacher of mathematics and physics, who lived in the first half of the nineteenth century. As pointed out earlier in this section, every musical sound, is a periodic function; that is, the relation between displacement and time of a typical air molecule oscillating under the pressure exerted originally by the source of the sound is a periodic function of t. But Fourier's theorem says that every such function is a sum of simple sine functions of the type illustrated in (6). Each simple sine function corresponds to a simple sound such as is given off by a tuning fork. Hence one arrives at the important conclusion that *every* musical sound is a sum of simple sounds. Moreover, the frequencies per second of these simple sounds are all multiples of one lowest frequency. To put the matter differently, every musical sound can be duplicated by a combination of tuning forks, each vibrating with the proper frequency and amplitude.

The musical sound, whose graph is shown in Fig. 24–6, for example, is a sum of five simple sounds. The frequencies of these sounds and their respective amplitudes are tabulated below. The amplitudes are expressed in terms of the first one entered, which is chosen to be 1.

Frequency	512	1024	1536	2048	2560
Amplitude	1	0.2	0.25	0.1	0.1

We should note that the frequencies are all multiples of the lowest one, which is 512. The formula representing this sound is then

$$y = \sin 2\pi \cdot 512t + 0.2 \sin 2\pi \cdot 1024t + 0.25 \sin 2\pi \cdot 1536t + 0.1 \sin 2\pi \cdot 2048t + 0.1 \sin 2\pi \cdot 2560t.$$

The assertion that every musical sound is no more than a combination of simple sounds is so surprising that, although it is backed by unassailable mathematics, one wishes to see it confirmed by experimental evidence. Such evidence is available. First of all, a trained ear can recognize the simple sounds present in a complex sound. Secondly, if one releases the dampers on the strings of a piano and then strikes a note, a number of other strings will also begin to vibrate, namely those whose basic frequencies are the same as the component frequencies present in the note struck. The physical explanation is that the note struck gives off several frequencies—the frequencies of its component simple sounds. Each of these frequencies sets off air vibrations which in turn force into vibration all other strings whose basic frequencies are the same as those of the simple sounds.

Perhaps the best experimental evidence is furnished by some specially designed instruments. The distinguished nineteenth-century physician, physicist, and mathematician Hermann von Helmholtz (1821–1894) gave two kinds of demonstrations. In the first one he designed special pipes, called resonators, each of which selected and rendered audible only that frequency which was suited to the dimensions of the pipe. A resonator in the neighborhood of a complex sound will pick up and render audible any component of the sound whose frequency excites the resonator. By using resonators of different sizes Helmholtz was able to show that the frequencies present in the complex sound were just those called for by Fourier's theorem. Then Helmholtz demonstrated the reverse. He set up electrically driven tuning forks of the proper frequency and amplitude such that the combination of simple sounds duplicated a given complex sound. A modern version of this latter device is the electronic music synthesizer.

There is no question, then, that any musical sound is no more than a sum of simple or sinusoidal sounds. The simple sound of lowest frequency is called the fundamental, or first partial, or first harmonic. The simple sound whose frequency is twice that of the lowest one is called the second partial or second harmonic; and so on. The frequency of the entire complex sound is the frequency of the first harmonic for the reason already given in our discussion of Fourier's theorem. The amplitudes of the individual sine terms are the amplitudes or strengths of the harmonics present.

Exercises

1. State Fourier's theorem.
2. Suppose that a complex sound is representable by the function

$$y = 0.001 \sin 2\pi \cdot 240t + 0.003 \sin 2\pi \cdot 480t + 0.01 \sin 2\pi \cdot 720t.$$

What is the frequency of the complex sound? What is the amplitude of the third harmonic?

3. Write the formula for a musical sound whose frequency is 500/sec and whose first, second, and third harmonics have amplitudes of 0.01, 0.002, and 0.005, respectively.

4. If the relationship between displacement and time for the fundamental of a musical sound is $y = 3 \sin 2\pi \cdot 720t$, what is the frequency of the third harmonic?

5. Explain why the frequency of a complex musical sound is always that of the first harmonic.

24–5 Subjective properties of musical sounds. Musical sounds as received by the ear seem to possess three essential properties; that is, the ear recognizes what are commonly called the pitch, the loudness, and the quality of a sound. One of the major values of the mathematical analysis of musical sounds is that it clarifies and makes precise just what we mean by these properties. We shall consider them in turn.

In our subjective judgment, sounds vary from low or deep tones to high or piercing ones. Verbal descriptions of the pitch of sounds are, of course, qualitative and vague. If one experiments with tuning forks of different pitch, he readily discovers that high pitch means high frequency of fork vibration and therefore high frequency of oscillation of the air molecules. Correspondingly, low pitch means that the fork and the air molecules vibrate with low frequencies. Prior to the availability of the analysis examined in the preceding section, the notion of pitch was not clear for complex sounds. But we now know that all musical sounds have a definite frequency, namely the frequency of the fundamental. Thus, although complex sounds contain other frequencies, that is, the frequencies of the higher harmonics, it is the over-all frequency of the composite sound which determines whether it appears high- or low-pitched to the ear. Hence, as one strikes the notes on a piano going from left to right, the pitch steadily rises.

The loudness of a musical sound is determined by the amplitude of the corresponding molecular motion, but the relationship between loudness and amplitude is not quite so simple as that between pitch and frequency. Let us note, first of all, that amplitude means the maximum displacement of the typical air molecule or the largest y-value of the corresponding graph. Physicists call the square of this amplitude the intensity of the sound. Thus intensity is still a physical or objective property of a musical sound. Among sounds of a given frequency, the more intense sound will seem louder to the ear. However, this is no longer true if the frequencies of the sounds differ. The average ear is most sensitive to a frequency of about 3500 per second and less so to frequencies above and below this value. Hence a very intense sound at a frequency of 1000 per second may sound softer to the ear than a less intense one at a frequency of 3500 per second. As a matter of fact, the average human ear does not hear at all sounds above about 16,000 vibrations per second, no matter how intense they are. Loudness depends not only on the intensity and the frequency of the sound but also on the shape of the graph within any one period. Two sounds may possess the same frequency and the same amplitude, but may have differently shaped graphs. Such sound will, in general, not sound equally loud to the ear.

The most interesting and from an aesthetic standpoint the most important aspect of musical sounds is their quality. It is this property which determines whether or not a sound is pleasing. The quality of a sound depends upon which harmonics are present in the sound and the amplitudes of these harmonics. Thus a sound emitted by a piano and a sound of the same frequency emitted by a violin create different effects on the ear because they differ in the harmonics present and in the amplitudes of these harmonics. Since the harmonics and their amplitudes determine the shape of the graph, it follows that, mathematically, quality is the shape of the graph within any one period.

Sounds or tones vary greatly with respect to harmonics and their amplitudes. Some sounds, for example, the sounds of tuning forks, some notes on the flute, and sounds produced by wide-stopped organ pipes, possess only a few harmonics or, in effect, merely the first. On the other hand, most instruments give off sounds containing many harmonics, but some of these may have small or almost zero amplitude. For example, the sounds of organ pipes are, in general, weak in the higher harmonics. The sounds of a violin possess a great number of harmonics and are usually strong in the first six harmonics. The relative amplitudes of the harmonics present in violin sounds are about the same for all notes; however, there are enough differences for the ear to distinguish, say the A- from the D-string, even though both are sounded at the same frequency. The uniformness of quality may explain why the sounds of a violin are so pleasing. The sounds of a piano also contain many harmonics, but the relative amplitudes of the harmonics in any one sound depend upon the velocity with which the hammer strikes the string.

The vowel sounds of the human voice are rich in harmonics. For example, the sound of "oo" as in tool, expressed at a fundamental frequency of 125 vibrations per second, has as many as 30 detectable harmonics. The relative amplitudes of the first six are 0.4, 0.7, 1, 0.2, 0.2, and 0.2, respectively. The higher harmonics, though present, have lower amplitudes. However, not only do the number of harmonics present and their relative amplitudes vary considerably from one vocal sound to another, but even the same sound issued at two different pitches will have different harmonics and amplitudes.

The physical reason for the differences in quality among the many types of musical instruments is, of course, the nature of the device itself. The piano and violin both use vibrating strings, but piano strings are struck whereas violin strings are bowed. The clarinet, oboe, and bassoon are operated by forcing air against vibrating reeds. Air is forced past the edge of an opening in the organ pipe also, but here the edge or lip is rigid. In addition, each instrument possesses a resonance device which emphasizes certain harmonics. The sounding board of the piano, the hollow box of the violin, and the pipes of an organ are resonance devices.

Although two people may not quite agree about their reactions to sounds, it is on the whole true that the qualities of sounds which we describe by such words as soft, piercing, rich, dull, braying, hollow, bright, and the like, are due to the harmonics and their relative amplitudes. Sounds which contain only the first harmonic are soft but dull. For brightness and acuteness the higher harmonics are essential. Sounds which possess the first six harmonics are grand and sonorous. If harmonics beyond the sixth or seventh are present and have appreciable amplitudes, the tones are piercing and rough. In general, the amplitudes of harmonics decrease as the frequencies increase. However, if the amplitudes of higher harmonics are too large compared with that of the fundamental, the tone is described as poor rather than rich.

Exercises

1. Suppose that two sounds are represented respectively by $y = 0.06 \sin 2\pi \cdot 200t$ and $y = 0.03 \sin 2\pi \cdot 250t$. Which one is louder? Which one is higher pitched?
2. Explain in mathematical terms the meaning of a simple sound.
3. What is the mathematical criterion of a musical sound as opposed to noise?
4. Which mathematical properties of the formula for a complex musical sound represent the pitch of the sound and the quality of the sound?
5. Discuss the assertion that music is basically just mathematics.

24–6 Some practical applications of the mathematical analysis. The contribution made by the mathematical analysis of musical sounds to the clarification of pitch, loudness, and quality is a contribution to our understanding of music. The same analysis is immensely important in the design of instruments such as the telephone, phonograph, radio, loud speaker systems, the sound track of films, and other recording and reproducing devices. To design such instruments properly one must know the effect of every component part on the sounds or on the electric currents generated by the sounds which pass through the component. It would be impossible to meet this requirement if one had to consider all sounds that are received by the ear, for the variety of such sounds is unlimited. But the Fourier analysis of sounds simplifies the problem immensely. All musical sounds are no more than simple sounds of different frequencies. Moreover, the frequencies of sounds intended for the human ear range from about 16 to about 16,000. Hence the designer's problem is reduced to learning how his components handle sounds in this range of frequencies.

The problem can be further simplified for the design of some instruments. For example, the telephone and interoffice communication systems are expected to do no more than transmit intelligible sounds. Quality on the whole can be sacrificed. Studies show that only a few harmonics need be retained to preserve the intelligibility of vocal sounds. Hence such instruments need only pass frequencies from about 100 to 1000 cycles per second, for this range will include the fundamental and a few harmonics of sounds of normal voices. The sound track of a motion picture film must reproduce voice and music with some degree of faithfulness to the original sound, and therefore requires a frequency range of 16 to about 8000 cycles per second. On the other hand, high-fidelity musical instruments must handle frequencies from about 16 to 20,000 per second to please the sensitive ear.

There are special instruments which utilize even higher frequencies. Ultrasonic sounds, that is sounds whose frequencies exceed the audible range, are used to test metals for internal flaws. These sounds are directed into the metal and are reflected by flaws. Hence by detecting the amount of reflected sound the tester can ascertain the internal state of the metal. Of course, ultrasonic radiators and detectors must be designed to handle the frequencies employed by the devices. Ultrasonic sounds, incidentally, can be heard by dogs, bats, and insects.

The technical work involved in studying the transmission of sound is considerable. For example, the telephone, phonograph, and radio contain mechanical and electrical devices, such as microphones, radio tubes, resistances, coils, condensors, and loud speakers. In studying the performance of these devices the Fourier analysis of sounds is of key importance, for all one needs to know is the performance of these components in handling only simple sounds. In the phonograph, a knowledge of the actual function which describes a sound is also useful, for the shape of the groove in a record is identical with the shape of the graph of the sound recorded by that groove. The excellence of our recording and reproducing instruments is evidence of the success with which the design of such equipment has been carried out.

24-7 Applications to physiology. It is characteristic of scientific investigations that their results embrace far more than intended and that instruments and techniques developed for one field of investigation prove to be effective in others. The study of musical sounds is primarily devoted to a physical phenomenon, the motion of air molecules. However, it has incidentally taught us much about the human voice and the ear.

Since the voice can produce musical sounds, it is a musical instrument, and the information gained in the study of man-made instruments can be applied to explain the action of the voice mechanism. The throat contains the vocal cords, which act somewhat like the vibrating reed in a clarinet. Air from the lungs is forced against the cords and causes them to vibrate. But vocal cords can be varied in shape and in tension whereas the reed is fixed in these respects. Hence in this respect the voice is superior. We have also mentioned that musical instruments contain resonating devices which accentuate the sounds given off by the basic vibrating source. For the voice, the mouth acts as a resonant chamber. By opening the mouth more or less widely and by positioning the lips and tongue, the shape of this resonant chamber can be altered in many ways so that different frequencies emanating from the vocal cords can be accentuated. The entire voice mechanism is then capable of making far more adjustments than any other musical instrument, and it is, indeed, more versatile and richer than any other source of musical sounds. We see, then, why a well-trained voice can afford so much pleasure.

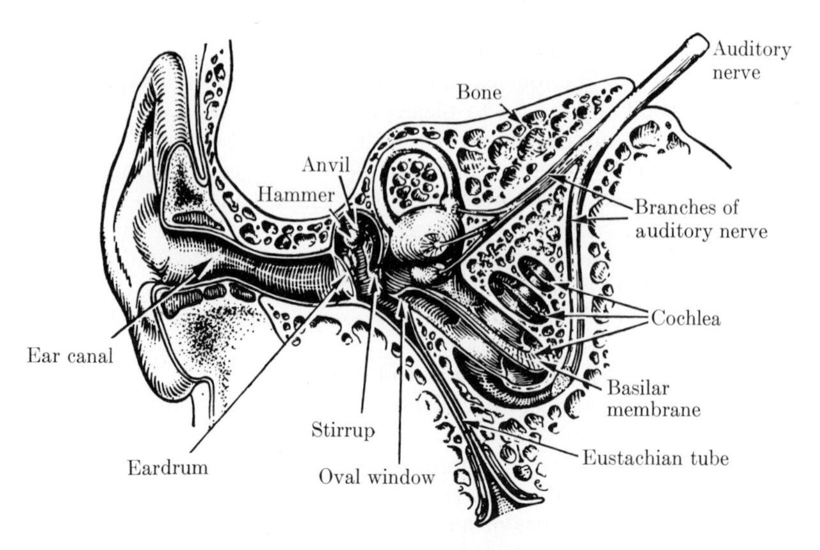

FIG. 24–7. Diagrammatic section of the right ear.

The ear, too, is a remarkable device. Within the range of frequencies from about 16 to about 16,000 cycles per second, it is exceedingly sensitive. We have already pointed out that the sensitivity, that is the ability to detect sounds, is greatest for frequencies of about 3500 per second. At this frequency the ear can hear sounds for which the amplitude of the air molecule's motion is as little as one ten-billionth of a centimeter. In fact, an amplitude of one-thousandth of a centimeter produces a loud sound. However, the sensitivity of the ear is about 1000 times poorer at 100 cycles per second than, for example, at 1000 cycles per second.

The ear may hear a sound but not comprehend it. We have mentioned that the ear identifies a musical sound by the repetition of the sound pattern. Reception of just a few cycles of a 100-cycle-per-second note will enable the ear to recognize the sound. At 8000 cycles per second, the ear must receive about 150 cycles before it can identify the note. On the average, the ear needs about $1/50$ of a second to identify a sound. In normal speech each individual sound lasts appreciably longer so that the ear has no difficulty at all in detecting and distinguishing the many sounds.

The mechanism of the ear is very complicated, and some of our knowledge about the functions of the various parts has been acquired by comparing these functions with physical devices which seem to exhibit a similar behavior (Fig. 24–7). For example, scientists have studied the vibrations of drumskins and the frequencies given off by skins of various shapes and tensions. The findings are, of course, of interest for the action of a drum itself, but they also help to explain the behavior of the vibrating eardrum. Likewise the study of resonators has made clear that the hollow chambers in the ear undoubtedly serve as resonators to accentuate the frequencies that are received. The heart of the entire hearing mechanism is a series of about 4500 nerve endings which are located in a region of the inner ear, called the cochlea. When any one or more of these nerve endings are excited, they convey a sensation to the brain. Now the ear can recognize the frequencies (ranging from about 16 to 16,000 cycles per second) of the harmonics in any musical sound. The ear can also distinguish any two frequencies whose ratio is 1000 to 1001 or less. This means roughly that the ear can distinguish between two frequencies which differ (1) by a fraction of a cycle per second at the low end of the audible range, (2) by a frequency of one cycle per second in the region around 1000 cycles, and (3) by 10 to 20 cycles at the higher frequencies. To account for the ear's ability to make so many and such sharp distinctions with the 4500 nerve endings available, Helmholtz supposed that each frequency excited several nerves and that the strength of the excitation of any one nerve varied with the frequency; that is, he hypothesized that the same nerve might respond to two frequencies, but respond more strongly to the higher one. The particular group of nerves excited, together with the strength of the excitation which that frequency induces, enables the brain to identify the frequency. However, this theory is not too well established, and the exact manner in which the ear conveys information about frequency to the brain is not fully understood.

These bits of information about the voice and ear are, of course, just a small part of the knowledge which scientists possess. It is not our intention to study these organs; we are, however, interested in noting that the mathematical analysis of musical sounds illuminates not only the physical phenomenon proper, but can be extended to shed light on the structure of man himself. Moreover, this knowledge is important for medicine. To analyze disorders of the voice or the ear and to suggest remedies, doctors must know the way in which these organs function.

24–8 Summary Having wandered so far from the main theme of this chapter, we may do well to recall what is central in the entire discussion. We have seen in other chapters that mathematics seems to be able to represent basic physical laws and that the application of mathematical reasoning to these mathematically formulated physical laws will yield new information about the physical world. We have tried to show in this chapter that the same applies to the phenomenon of sound.

Mathematically, a musical sound is a function which describes the relationship between displacement and time of the air molecules carrying that sound. Because musical sounds are periodic, scientists can apply Fourier's theorem and deduce a wealth of knowledge about musical sounds. One of the most useful results thus obtained is the recognition that all musical sounds are combinations of simple sounds.

Musical sounds are, of course, one of the great sources of beauty. We find now that this beauty can be described in the precise language of mathematics. Proclus, a Greek commentator on Euclid, once said that wherever there is mathematics, there is beauty. A modern version of this epigram states that wherever there is beauty, there is mathematics.

Topics for Further Investigation

1. The construction of musical scales. Include, in particular, the work of J. S. Bach on the equal-tempered scale.
2. The human voice as a source of musical sounds.
3. The functioning of the human ear.

Recommended Reading

FLETCHER, HARVEY: *Speech and Hearing,* D. Van Nostrand Co., Princeton, 1929.

HELMHOLTZ, HERMANN VON: *On the Sensations of Tone,* Dover Publications, Inc., New York, 1954.

JEANS, SIR JAMES H.: *Science and Music,* Cambridge University Press, London, 1937.

MILLER, DAYTON C.: *The Science of Musical Sounds,* 2nd ed., The Macmillan Co., New York, 1926.

OLSON, HARRY F.: *Musical Engineering,* McGraw-Hill Book Co., Inc., New York, 1952.

REDFIELD, JOHN: *Music, A Science and an Art,* A. A. Knopf, New York, 1926.

SEARS, FRANCIS W. and MARK W. ZEMANSKY: *University Physics,* 2nd ed., Chaps. 21 to 23, Addison-Wesley Publishing Co., Inc., Reading, Mass., 1955.

TAYLOR, LLOYD WM.: *Physics, The Pioneer Science,* Chaps. 24 to 28, Dover Publications, Inc., New York, 1959.

CHAPTER 25

TRIGONOMETRIC FUNCTIONS AND ELECTROMAGNETISM

All the pictures which science now draws of nature and which alone seem capable of according with observational fact are mathematical pictures.

Sir James Jeans

25–1 Introduction. The study of the uses of trigonometric functions leads us directly into the most significant scientific development of the nineteenth century—the theory of electromagnetism. Whereas the pursuit of mechanics dominated the seventeenth and eighteenth centuries, the investigations of electricity and magnetism dominated the nineteenth. The findings wrought changes in our civilization and culture which have been at least as profound as those inaugurated by mathematical mechanics.

Of course, mathematics was the vital instrument in the nineteenth-century developments. Various branches of mathematics were employed, of which differential equations was the most effective. However, we shall confine our discussion to a broad descriptive account of the role played by the trigonometric functions in these scientific advances. We shall see that in the domain of electromagnetic theory mathematics has made one of its most impressive and most influential contributions to the understanding and mastery of nature. Indeed the impact of mathematics became so powerful that scientists recognized explicitly what had been implicit in the work of the preceding centuries, namely that mathematical structure is the essence of scientific theory and explanation.

25–2 Historical background. Pure mathematics, that is, the developments which center about the concepts of number and geometrical figure, has rather little to say about our physical world. As we have often pointed out, to produce physical knowledge, we must conjoin basic physical principles to the axioms and theorems of mathematics. We shall be better able to appreciate how mathematics played its part in the development of electromagnetic theory if we go back a little to see what physical principles the scientists had accumulated.

Electricity and magnetism have been known since Greek times, but until the nineteenth century it had not occurred to anybody that the two phenomena might be related. Moreover, knowledge about both was fragmentary and on the whole unimpressive. The compass was the only application of significance. The first systematic study of magnetism was made by William Gilbert, physician to Queen Elizabeth of England, who published his findings in his *De Magnete* (1600). His most surprising result was that the earth itself is a huge magnet. During the seventeenth and eighteenth centuries a few more facts about electricity were gathered, but neither the understanding nor the application of the two phenomena made any noticeable progress. The chief difficulty facing scientists interested in the study of electricity was the lack of any effective means of producing a large or continuous stream of electric current. Insofar as magnetism was concerned, while natural magnets in themselves have some use, as in the compass for example, again no means of producing a strong magnet was known.

The investigations of electricity and magnetism received a great impetus in the early part of the nineteenth century. The study of electricity was immensely advanced by the development of the battery, whose modern form is found in flashlights or in the automobile, where it supplies the power to start the motor and the spark to ignite the gasoline in the cylinders. The men whose work was crucial here were the Italians Luigi Galvani (1737–98) and Alessandro Volta (1745–1827). Although the battery, too, is a limited source of electricity, primarily because it does not supply electricity for a long time (the automobile battery is constantly being recharged when the motor runs), it did give the experimenters something to work with, and discoveries of major import soon followed.

In a paper published in 1820 and covering work of the preceding eight years, the Dane Hans Christian Oersted (1777–1851) described an experiment in which he passed electric current through a wire and held a compass needle nearby and parallel to the wire. The needle was deflected; that is, it turned from the direction in which it had originally been placed. The explanation was soon forthcoming. It had been known that the opposite ends of a magnet are different. One is attracted to the north-magnetic pole of the earth, and the other to the south-magnetic pole. The north-seeking pole is called the north, or positive, pole of the magnetic, and the south-seeking is the south, or negative, pole. Like poles of two magnets repel and unlike poles attract each other so that if two like poles are near each other and the magnets are free to move, the magnets will move apart from each other. Oersted's finding suggested that the current in the wire made the wire act like a magnet; this magnet acted on the compass held nearby and turned the needle from its original orientation.

The action of magnets on each other or the effect of a wire carrying current on a magnet is described nowadays in terms of concepts with which we should become familiar. Since two magnets exert a force on each other even though they are some distance apart, the way in which the force is transmitted from one magnet to the other requires some explanation. Physicists accounted for the transmission by advancing the idea that each magnetic pole creates a field of force about it; that is, a magnet acts as though it can send out "feelers" in various directions and these attract or thrust away certain types of objects. In the case of two magnets, each field acts on the other magnet. Of course, it is natural to ask, What does the field consist of and how does it act? We shall not answer this question now, but shall have much to say about it later. What is important at the moment is the recognition that a magnet creates a special state or field in the space about it, and that this field can act on other magnets. This explanation also applies to a wire carrying a current, since such a wire, too, acts like a magnet.

Oersted had discovered an important link between the two phenomena of electricity and magnetism, which previously had been regarded as distinct from each other. A still more important link was discovered by two other scientists who worked independently of each other in places far apart. One was Michael Faraday (1791–1867) of England and the other, Joseph Henry (1799–1878) of New York. Faraday started to earn his living at the age of thirteen by serving as errand boy and apprentice to a bookbinder. In his spare time, he studied chemistry and physics and attended lectures given by well-known scientists, among them the famous chemist Sir Humphrey Davy. Faraday wrote to Davy to ask for a job and, to back his application, he enclosed notes he had taken of Davy's lectures. Davy was impressed and hired the young man. Faraday rose rapidly to become the leading experimental physicist of the nineteenth century. Hermann von Helmholtz said of him, "He smells the truth."

Henry, too, started to earn his living at an early age and in a very humble capacity. At fifteen he became a watchmaker's apprentice, although his ambition was to become an actor and writer. Like Faraday, he started to read books on science and was attracted to the field. He entered the Albany Academy in New York, completed his work there, and became professor of mathematics at that school. The experimental work begun at Albany was continued at what is now Princeton University, when Henry became professor of natural philosophy [science] there. The difficulties Henry had to surmount to find time and place for his experimental work make his success almost incredible. During the latter part of his life, from 1847 to 1878, he was director of the Smithsonian Institution in Washington.

The principle which Faraday and Henry discovered in 1832 is called electromagnetic induction. At the moment we shall describe it briefly by stating that if a wire moves in a magnetic field, the field induces an electric current in the wire. The immediate importance of the principle from the standpoint of those interested in electricity and magnetism, or electromagnetism as the combined fields are called, was that it afforded a new method of generating electricity. From the standpoint of the mathematician, it meant that knowledge about electromagnetic phenomena had arrived at the stage where fundamental, mathematically expressible principles were available and mathematics could profitably be employed.

25-3 The generation of alternating current. Our first concern will be to examine the principle discovered by Faraday and Henry. In our description we shall take advantage of the physical insight and mathematics that were supplied by dozens of mathematicians and scientists in the last hundred years.

We have already mentioned briefly that electromagnetic induction is the generation of electricity in a wire moving in a magnetic field. This statement is, of course, vague, and we must now say a little more about

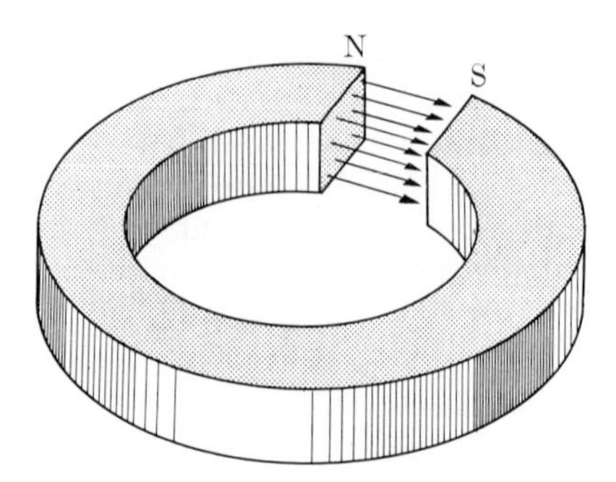

FIG. 25–1. A bar magnet bent so that opposite ends face each other.

the idea and concepts involved. To account for the attraction or repulsion which one magnetic pole exerts on another, we have introduced the idea that a magnetic pole creates a field about it and that this field acts on other magnetic poles. Let us take a long bar magnet and bend it so that the poles face each other. Now the field set up by the two poles together has the direction illustrated in Fig. 25–1; that is, if a positive or north-magnetic pole, P say, were placed between the north and south poles, N and S, of the curved magnet, the north pole, N, would repel the pole P and the south pole, S, would attract the pole P, so that P would move from left to right.

It is a physical fact that if the ends or faces of the magnet are large, the field will be constant; that is, the force which the field exerts on any magnet placed between the faces will be the same, irrespective of the magnet's greater or smaller distance from one or the other pole. Partial support for this statement is supplied by physical reasoning. If the positive pole P is placed near the south pole in Fig. 25–1, one would expect it to be attracted rather strongly, just as a mass near the earth is attracted more strongly than one farther away. On the other hand, since the positive pole P is near the south pole, it is farther from the north pole and so is only weakly repelled by the north pole of the fixed magnet. If now the positive pole P is placed near the north pole in Fig. 25–1, it will be repelled strongly, but since its distance from the south pole is now greater, it will be attracted only weakly. The total force acting on the pole P is nevertheless the same in the two positions because the repulsive force and the attractive force both act to push P from left to right.

Let us next clarify what electricity is. Modern theory states that all matter consists of atoms which normally are not electrified; that is, they are neutral. In every atom there are tiny bits of matter, the so-called electrons. These are so small that 25 trillion electrons lined up alongside one another would cover one inch. Electrons can be separated from the atoms. When metal is heated, for example, electrons are ejected from the metal and shoot off into the space around the metal. This happens in radio tubes. Each electron repels any other isolated electron or collection of electrons. We describe this property by saying that each electron, in addition to being a

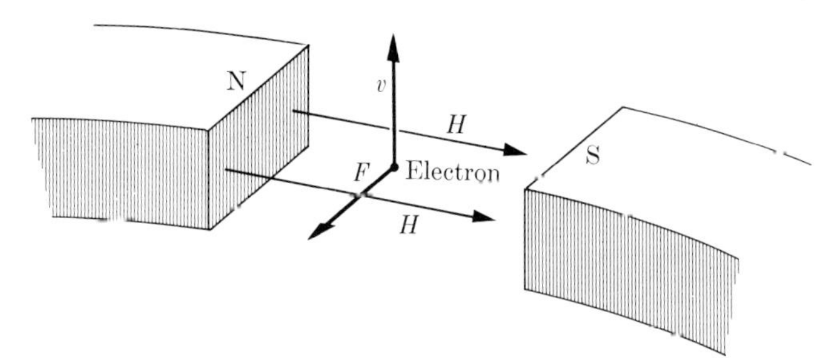

FIG. 25-2. An electron moving upward across a magnetic field is forced out of the paper.

bit of matter, possesses negative electric charge, and that the two electric charges repel each other. If extra electrons are added to an atom or to a collection of atoms, the matter is said to be negatively charged. If an atom has lost one or more electrons, the state of affairs is described by saying that the atom is now positively charged or possesses positive electricity. In reality, then, a positively charged atom does not possess something additional, but has lost electrons, and the language, a carry-over from times when electricity was not understood, is misleading.

If we wished to specify how much charge, negative or positive, a bit of matter contains, it would seem most natural to do so by indicating how many extra electrons the matter has acquired or how many of its original number it has lost. Similarly, for a collection of isolated electrons, the charge equals the number of electrons. However, in practice, the convenient unit of electricity is not the number of electrons, but the coulomb. One coulomb equals $6 \cdot 10^{18}$ electrons. (The coulomb is named after Charles Augustin Coulomb (1736–1806), who first determined that two negative charges or two positive charges repel each other, and that a negative charge attracts a positive charge.)

An electric *current* consists of electrons in motion. Now the atoms of metals, for example, contain many electrons, and it is rather easy to separate them from the atoms and set them into motion along the wire. This flow of electrons in a metal wire is the current, and it can be produced by attaching a battery to the ends of the wire.

In electromagnetic induction, an electric current is created in a wire, not by a battery, but—as we have already mentioned—by the motion of the wire itself in a magnetic field. Let us delve deeper into this phenomenon.

To understand the principle discovered by Faraday and Henry, let us first imagine that we have an isolated electron and that we can place it in the magnetic field of the magnet shown in Fig. 25-2. This magnetic field has a certain strength, or force, which it can exert on other magnets. Let us denote this strength by H. If the electron is merely placed in this field, nothing happens; that is, it remains exactly where it is placed. The magnetic field and the electron ignore each other. But if the electron *moves* in the magnetic field, then the field acts upon it in a peculiar way. Let us suppose that the electron is moving upward in the plane of the paper, so that its

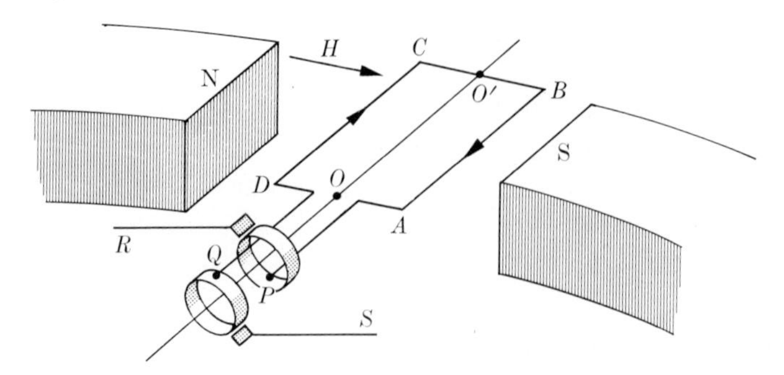

FIG. 25–3. The generation of current in a wire loop rotating in a magnetic field.

velocity is *perpendicular* to the magnetic field. Then the magnetic field will exert a force on the electron and cause it to move out of the paper. The direction of the force F is shown in Fig. 25–2. The magnitude of the force F is given by the formula

$$F = qHv, \tag{1}$$

where q is the charge on the electron, H is the strength of the magnetic field, and v is the original velocity of the electron. If the electron were moving downward, the magnitude of the force would be the same, but the force would cause the electron to move into the paper. Formula (1) is a physical fact which has been experimentally determined, and which must become part of our scientific information, along with mathematics proper, if we are to reason about electromagnetism.

Now let us see how the above information applies to the generation of electricity by electromagnetic induction. Suppose that a loop of wire as shown in Fig. 25–3 is placed in the field of a magnet. The loop of wire is rigidly attached to the rod OO'. The rod is made to rotate counterclockwise, and, since the wire is attached to it, the loop will also rotate. We must suppose, however (and this can be done), that the rod is insulated from the wire; that is, whatever electrical phenomena take place in the wire are not transmitted to the rod. Now let us suppose that the loop is turning at a constant rate and that at the moment it is horizontal. Then the electrons in the segment AB of the loop are moving upward, and we now have electrons moving in a magnetic field. Hence, according to formula (1), there is a force acting on these electrons, and the direction of the force is along the wire from B to A. Thus as the electrons move upward with the wire, there is a force acting on them which will cause them to move along the wire.

Let us consider what, at the same moment, is happening to the electrons in the segment CD. When the loop is horizontal, these electrons are moving downward. Consequently, the force acting on them is directed into the paper, that is from D to C. (The segments BC and AD play no role; the force acting on electrons in BC and AD is perpendicular to the segments

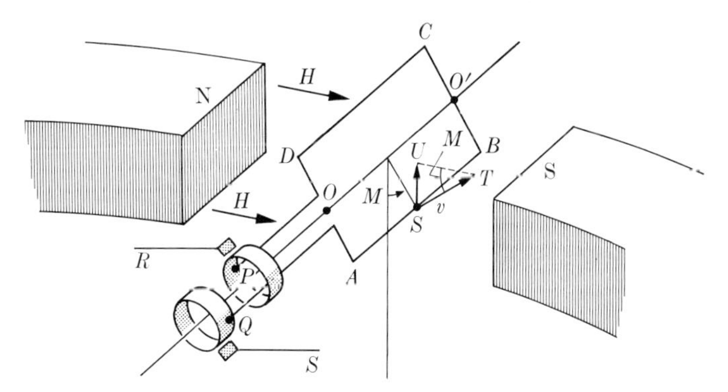

FIG. 25–4. The wire loop makes an angle M with the vertical.

instead of being directed along the segments, but the electrons will not move out of the wire.) Hence the substance of what we have found so far is that a force is acting on the electrons along the wire from D to C and from B to A.

Of course, the loop is rotating and so is only momentarily horizontal. Let us suppose that the loop was originally vertical, with AB at the lowest position, and that it is rotating counterclockwise. Then a short time later the loop might make an angle, say M, with the vertical (Fig. 25–4). Let us again consider an electron in the segment AB. This electron is moving around in a circle, and the direction of its motion at any instant is along the tangent to this circle. Figure 25–4 shows the direction of the electron's velocity, v, when the loop makes an angle M with the vertical. Now the direction of v is not perpendicular to the magnetic field. What, then, is the force acting on the electron? The answer is still given by formula (1), although, to cover the present situation, we must be a little more explicit. Formula (1) calls for the velocity of the electron to be *perpendicular* to the magnetic field. Hence we must find that component of the velocity v which is perpendicular to the magnetic field. Let us determine this component. Line ST in Fig. 25–4 shows the direction and size of the velocity v of the electron. The component of this velocity which is perpendicular to the magnetic field is the line segment SU. Now it is simple to see that angle UTS equals angle M (Fig. 25–5). Hence the component SU is $v \sin M$. If, therefore, we sub-

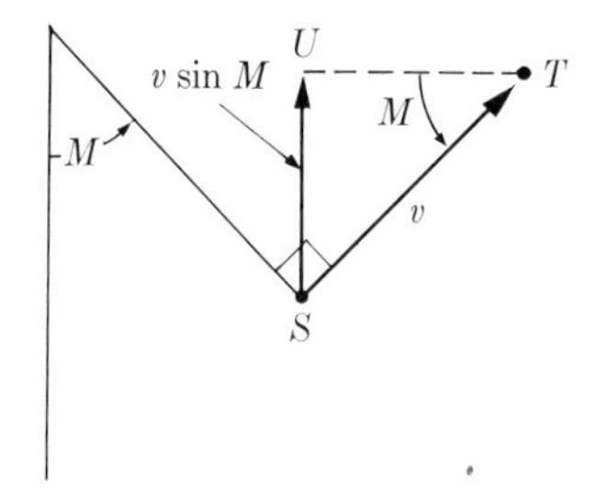

FIGURE 25–5

stitute this value of the velocity in formula (1), we have

$$F = qHv \sin M. \tag{2}$$

This, then, is the force which will be exerted on each electron. The direction of the force remains unchanged; that is, it is from B to A when AB is moving upward and from D to C when DC is moving downward.

We have learned the surprising fact that the force along the wire varies with the sine of angle M. Let us also note that when AB reaches the top and starts moving downward, it will be in the position occupied by CD during the first half of the rotation. Hence, as AB moves downward to the left, the direction of the force will be from A to B instead of from B to A. This reversal is already contained in formula (2), for when M varies from 180° to 360°, sin M is negative. Thus, when M varies from 180° to 360°, the force acting on all the electrons in AB or CD is opposite in direction to the force acting when M varies from 0° to 180°.

The force along the wire, or voltage as it is called in electricity, gives each electron an acceleration, and so the electrons acquire velocity along the wire. This flow of electrons is the electric current. However, the strength of the current depends upon the resistance which the electrons encounter in the wire, and the resistance in turn depends on the nature of the metal. Some metals are good conductors, that is, permit electrons to flow freely, while others are poor conductors. The amount of flow, then, is contingent upon the force and the material of the wire. Since the quantities q, H, and v in (2) are constants, the force varies with sin M. The resistance is constant, and so the current also varies with sin M. We cannot specify the quantity of current since that value depends upon the metal. If we call the maximum current D, then we may write for the current I:

$$I = D \sin M. \tag{3}$$

Formula (3) tells us how the current will vary in the wire as the angle M which the loop makes with the vertical direction varies.

Let us now note the physical implication of equation (3). The terminal, P, of the loop shown in Fig. 25–4 is always electrically connected to the point R, whether AB is moving upward or downward. Hence, when electrons are moving from B to A, they will move to P and then to R. When electrons are moving from A to B, then electrons will be pulled away from P and therefore from R. The electron motion along CD will at any time be opposite in direction to that along AB. Since the terminal Q of the loop is always electrically connected to S, it follows that when electrons are moving from D to C, they are being pulled into the loop at S and when electrons are moving from C to D, they will be moving out at S.

Equations (2) and (3) tell us more than we have thus far observed. The force given by (2) determines the current and depends upon q, H, and v. Now q, the charge of an electron, is fixed. The velocity v of the electron will be discussed in a moment. But H, the strength of the magnetic field, can be increased or decreased to provide a stronger or weaker force and therefore a stronger or weaker current.

We can derive a little more information from formula (3) if we introduce the time of motion as the independent variable. Let us suppose that the wire loop makes f rotations per second. In each rotation it goes through 2π radians of angle. In t seconds the loop will make ft rotations and so turn through $2\pi ft$ radians. This angle is the value of M corresponding to any given value of t. Substituting this value in (3), we obtain

$$I = D \sin 2\pi ft. \tag{4}$$

We can now see that if the loop is made to rotate f times per second, the current will have frequency f; that is, at any point on the loop the current will go through f complete sinusoidal oscillations in one second. This means that f times a second the current will increase to a maximum, decrease to zero, decrease to a minimum, and increase to zero. The current is called *alternating* because of this sinusoidal character. In the United States the value of f used is 60. Our homes, for example, are lit with 60-cycle alternating current. However, we do not notice the variations in current in our electric lights, for example, because they occur very rapidly and because the wire, which becomes white-hot, does not lose its brilliance while the current varies.

The practical significance of electromagnetic induction is that it furnishes a means of generating electricity. Of course, power must be supplied to turn the loop. This power is obtained from waterfalls, coal, oil, and atomic fission.

It is also possible to reverse the above process. We can feed electricity into the loop $ABCD$. From Oersted's discovery, we know that the current will create a magnetic field around the wire, that this field will interact with the field of the magnet already present, and that the loop will be forced to turn. This is, of course, the principle of the electric motor. Naturally, the design of a workable and efficient device involves many additional considerations.

The application of mathematics to the analysis of electromagnetic induction is of value because, starting with formula (1), we can determine the kind of current produced and the factors which affect the strength of the current. Were we to pursue the subject further, we could see how one can employ the mathematical representation to learn many new facts about the generation and control of electricity. Let us also note that an old acquaintance, the trigonometric function, has made its appearance once more.

There is one other comment which pertains to mathematical and scientific work in general, but which is especially true of the history of the early work in electricity and magnetism. The scientists pursued these phenomena for decades because they found them exciting and interesting. They did not envision electric lighting, vacuum cleaners, electric motors to run machines, and any other of the thousand and one uses which we make today of their ideas. Some interesting stories told about Faraday bear on this point. Faraday was still in the early stages of his work when he was visited by Mr. Gladstone, the famous British Prime Minister. Gladstone asked Faraday what use could be made of his methods of inducing electricity. Faraday's reply was, "What is the use of a child—it grows to be a man." On a later visit Gladstone repeated his question, and this time Faraday answered, "Why, sir, presently you will be able to tax it."

Exercises

1. What is electromagnetic induction?
2. How does current generated by electromagnetic induction vary?
3. What is the mathematical description of an alternating current as a function of time?
4. Suppose that a generator produces a current alternating at the rate of 3 cycles/sec. How fast is the loop turning in the generator?
5. Suppose that a generator produces 3-cycle alternating current with an amplitude of 1 ampere. (An ampere or "amp" is the unit of current.) Draw a graph to show the variation of current with time.
6. How many cycles of sine values does a 60-cycle alternating current complete in one second?
7. Suppose that a loop of wire in an electric generator makes 30 rotations per second and that the maximum current generated is 0.1 amp. Write a mathematical formula which relates current and time of rotation.
8. Suppose that the relation between time and the current produced by a wire loop moving in a magnetic field is $I = 0.5 \sin 2\pi \cdot 50t$. How many rotations per second does the loop make? How many cycles per second does the current go through?

25–4 Electromagnetic waves. The generation of electricity by rotating a loop of wire between the poles of a magnet or, one might say, in the field of a magnet, was but the first step in a series of scientific and mathematical developments that have had even greater impress on our world than electricity itself. Faraday's next move was to consider the following question. If current can be induced in a wire which moves in a fixed magnetic field, will current also be induced in a *fixed* wire by changes in a magnetic field located in the neighborhood of the wire?

To answer this question Faraday set up the following experiment. He started with Oersted's finding that a current flowing in a wire causes the wire to act like a magnet. This means that a magnetic field exists around the wire as it does around any magnet. If the current in the wire were alternating, the magnetic field would also alternate, that is, increase and decrease in strength. Now if another wire were held fixed in this magnetic field while the magnetic field increased and decreased around the wire, then one could test whether a current is induced in this second (fixed) wire. Faraday used as his first wire a coil such as that shown on the left of Fig. 25–6. He chose a coil instead of a simple wire because a current flowing through a coil

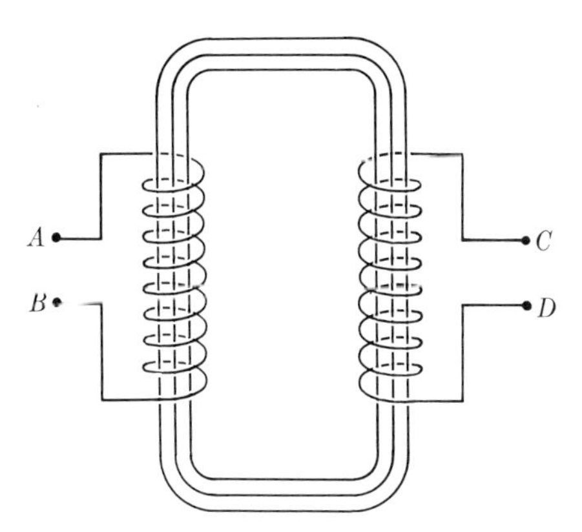

FIG. 25-6. Current is induced in a coil when a magnetic field in the neighborhood changes.

produces a strong magnetic field. Faraday fed an alternating current into the left coil by connecting the terminals A and B to a source of alternating current. He now had an alternating magnetic field around the coil. The loops shown in Fig. 25-6 illustrate how the field is spread out. Faraday now placed a second coil, the right-hand coil in Fig. 25-6, in the magnetic field of the first coil, and connected C and D, the terminals of the right-hand coil, with equipment which would detect whether a current flowed in this coil.

The upshot was that Faraday obtained a current in the right-hand coil. One import of the experiment was that current could be transferred from one coil to another even though these coils were physically some distance apart. Of course, insofar as Faraday's experiments showed, this distance could not be too great since the second coil had to be in the vicinity of the first coil around which the changing magnetic field existed. Faraday's discovery that a changing magnetic field will induce current in a wire held fixed in that field has important applications in modern electrical devices called transformers. It was, however, but a step in the direction of far bigger advances.

Faraday performed many other experiments which are classics of science, but he suffered from one limitation. His genius was confined to intuition and experimentation. He could not formulate mathematically the significance of his findings; nor was he competent to apply mathematical reasoning to such formulations in order to deduce new conclusions. We have already pointed out in other connections that although experiments may yield basic facts or serve as checks on theoretical deductions, full knowledge of a phenomenon and all its ramifications can be obtained only by good theory, and this means mathematical theory.*

Faraday's findings as well as those of the men who preceded him in the work on electricity and magnetism were formulated mathematically by

* Review the discussion of this point in Chapter 16 treating projectile motion.

James Clerk Maxwell (1831–1879), the greatest theoretical physicist of the nineteenth century. Maxwell attended the Universities of Edinburgh and Cambridge and was a brilliant student. In 1856 he became professor of physics at Marischal College in Aberdeen; later he was appointed professor at King's College in London and finally at Cambridge University. He did basic work in many fields of physics, but it was his work in electromagnetism which earned him the greatest fame and produced the most remarkable results.

Maxwell recognized the limitations of physical thinking about such elusive, invisible phenomena as electric currents, magnetic fields, and electromagnetic induction, and he proceeded at once to cast the known physical facts into mathematical form. We shall not present his mathematical equations because they involve concepts of mathematics which we have not treated, but we can give some qualitative account of his achievements. (This account will also be somewhat elusive, but it may help to show how Maxwell made his great discoveries.) First of all, in his formulation of already known facts, he gave prominence to a concept which had been introduced and, to some extent, utilized previously, namely the concept of an electric field. Just as the action of a magnetic pole in attracting or repelling another magnetic pole had been explained by assuming that there is a field of lines about each pole which extends far from the pole and acts on the other pole, so did Maxwell utilize the concept of a field surrounding a charge to explain the phenomenon of attraction and repulsion between electrical charges.

The existence of an electrical current in a wire means that free electrons are moving down the wire. Since each electron is surrounded by a field, an electric field moves along the wire when current flows in it. Of course, as Oersted had discovered, the wire is also surrounded by a magnetic field. But Maxwell now introduced the electric field with consequences yet to be related.

In electromagnetic induction, a magnetic field moves or changes around a wire, and current is *induced* in the wire, just as current is induced in the right-hand coil of Fig. 25–6. Maxwell preferred to suppose that *the magnetic field creates an electric field* and that this electric field causes the electrons in the wire to move. This assumption of an electric field may seem arbitrary, but it is not quite so. If the repulsion of one electron by another is explained by the hypothesis that every electron has a field about it which acts on the other and causes it to move, then it seems reasonable to suppose that the motion of the electrons in the coil CD, that is, the current, is caused by an electric field. Since the magnetic field is responsible for the existence of the current, it must be that the magnetic field somehow creates the electric field, which in turn causes the electrons to flow.

To describe the action of stationary electric charges, stationary magnets, charges and magnets in motion, and the fields which accompany charges and magnets, Maxwell wrote down four equations. These equations are, of course, quantitative relationships. By working with them, that is, by combining them under legitimate mathematical operations such as addition, subtraction, and differentiation, Maxwell found that these equations led to a contradiction with another physical law, namely, the law stating that charges cannot be created or destroyed. However, contradictions cannot be tolerated, and Maxwell found that he could indeed eliminate this contradiction by introducing a new term in his original equations. This term took into account the quantitative value of the rate of change with respect to time of the electric field wherever it was present. Thus, for example, Maxwell had supposed that an electric field was present about the coil AB because free electrons moved in the wire. As the electrons move, the electric field changes strength at any point around the wire, and it is the rate of change of this field which is the new term that Maxwell added. Moreover, from the way in which the revised equations read, it appeared that the current in the wire *plus* the rate of change of the electric field gave rise to the correct quantitative value of the magnetic field. In other words, the change in the electric field had as much to do with producing the magnetic field as did the current in the wire. Maxwell called the new term he had introduced the *displacement current.* He used the word "current" because the changing electric field had the same property as the current in the wire, namely, it created a magnetic field. The word "displacement" merely specified that the term represented displacement or change in the electric field around the wire.

The revised equations, now called Maxwell's electromagnetic equations, told him much more. The current in the wire and the accompanying displacement current created a magnetic field. But a magnetic field, as we have pointed out above, can create an electric field. This *new* electric field, *since it has the same properties as current in a wire, could by itself create a magnetic field.* This in turn creates a new electric field. And so the process goes on. One might guess that these successively created electric and magnetic fields do not stay in one place. They might spread out into space. And this is exactly what Maxwell's equations told him. They implied that the successively created electric and magnetic fields would move out into space.

The analogy with sound waves, though it is only an analogy, might help us here. The tuning fork creates successive condensations and rarefactions in the surrounding air. These are transmitted from one molecule to the other, and a wave moves out into space. In the case of sound there had been no question that the vibrations of the tuning fork caused the pressure, or sound waves, to move out from the fork. But prior to Maxwell, the oscillatory motion of the electrons in a wire was not believed to create waves in space. By adding the notions of electric field, rate of change of electric field, and the ideas that a changing electric field creates a magnetic field and that a changing magnetic field creates an electric field, Maxwell was led to postulate electric and magnetic fields moving out into space. In fact, he supposed that the electric and magnetic fields acted on particles of ether (which was believed to pervade all space), and that the particles became the carriers which transmitted electric and magnetic fields. This motion of electric and magnetic fields in space is called an electromagnetic wave.

Now such waves might travel a considerable distance. Thus for example, an electromagnetic wave starting from the left-hand coil in Fig. 25–6 might reach the right-hand coil although the two coils are far apart. Maxwell indeed predicted that exactly this would happen. His deduction was remarkable and, to some leading physicists of Maxwell's time, even incredible. Maxwell was asserting the existence of a new kind of physical wave, one which had hardly been suggested before. Clearly, there was need of experimental confirmation, for mathematical predictions based on physical axioms, in the present case Maxwell's equations, are not certain since physical axioms may be at fault. In 1887, about 25 years after Maxwell had predicted the existence of electromagnetic waves, Heinrich Hertz (1857–1894), another famous physicist and brilliant student of Helmholtz, generated such waves and received them in a coil some distance away from the source. These waves were called Hertzian waves for a long time and are none other than the radio waves employed today in a thousand ways.

In Maxwell's work, mathematics made possible the perhaps greatest prediction in the entire history of science. The physicist Ernst Mach said with respect to such predictions that it must sometimes seem to the mathematician that it is not he but his pencil and paper which are the real possessors of intelligence. What he meant was that once the mathematical equations are formulated, straightforward application of mathematical processes leads to new conclusions. A successful equation, he added, is a triumph of the mind.

The existence of radio waves is one thing and their use for communication, radar, and entertainment, as in modern radio and television broadcasting, is another. Here, too, mathematics plays an essential role. To generate radio waves or the changing electric and magnetic fields which spread out into space, a variable current must flow through a coil such as the left-hand coil of Fig. 25–6. The basic variable current employed is a sinusoidal current. However, to generate radio waves efficiently, the frequency of this sinusoidal current must be high. For example, for radio broadcasting, these frequencies are of the order of 1,000,000 cycles per second. For television broadcasting, the frequencies are of the order of 100,000,000 cycles per second. Such currents, incidentally, are not generated by using rotating wire loops but by special radio tubes called oscillators. The electric and magnetic fields which are created by the current and which travel out into space are also sinusoidal and have the frequency of the current source. Thus at any point in space, there will be a changing electric field whose variation in strength with time is represented by an equation of the form

$$E = D \sin 2\pi ft. \tag{5}$$

This equation does not represent all the physical properties of the electric field, but it does contain the essential feature. The quantity f, as we know from our study of trigonometric functions, is the frequency, or number of cycles per second, of the electric field.

The transmission of electric fields of the form (5)—accompanied, of course, by a magnetic field—is in itself useful. In wireless telegraphy, the generator sends out long and short "bursts" of such sinusoidal waves. A long burst may consist of, say, 5000 cycles, and a short burst of 1000 cycles. At the receiving station, these long and short bursts are picked up, and the duration of the signal received is sufficient to indicate whether it is a long or a short burst. Each combination of long and short bursts represents a letter of the alphabet, and a succession of such combinations conveys a message.

However, radio and television, which, as we know, transmit far more information than wireless telegraphy, employ more advanced ideas. If the sending station should send out a pure sinusoidal wave, this is all that would be received at the receiving end. While such a message is more meaningful than many that are currently broadcast, it will obviously not do for the communication of speech and music. We learned in Chapter 24 that to any sound there corresponds a characteristic curve of displacement versus time of the air molecules. Suppose that the amplitude D of the sinusoidal wave in (5) could itself be varied with time so that its variation represented the wave form of a genuine sound. If this variation in amplitude could be isolated at the receiving station and reconverted into sound, the receiver would be able to reproduce the very sound given out at the sending end. This scheme, called amplitude modulation, is actually employed in practice. The sounds at the sending station are converted to electric current by means of a microphone; this electric current is then used to vary the amplitude of the basic sinusoidal wave, or carrier; the amplitude-modulated wave travels through space; and the variation in amplitude is detected at the receiving station and reconverted into sound through a loudspeaker.

In another scheme for conveying vocal and musical sounds, known as frequency modulation, it is not the amplitude D of equation (5) which is varied but the frequency, and the variation of frequency with time corresponds to the variation of displacement with time of the musical sound being transmitted. Again the variation in frequency is imposed on the sinusoidal wave (5); the modulated wave travels through space; and at the receiving station the variation in frequency is detected and converted into sound. The processes of imposing modulation on the basic sinusoidal wave (5), or carrier, and then removing it involve electronic circuits, and are learned in radio engineering.

We shall not introduce the mathematical representation of amplitude-modulated or frequency-modulated radio waves. However, it is of importance and should be noted that the sinusoidal function is the basic one in these representations, and that the study of these functions enables engineers to determine the characteristics of the electronic equipment they must employ. Indeed, the entire process of transmission, whether by amplitude modulation or frequency modulation, is analyzed mathematically before the design of equipment is undertaken.

It may be of interest that radar sets, too, employ short bursts of sinusoidal waves. Each burst, which is a succession of, say 10,000 cycles, may strike a target and be reflected. The reflected burst is picked up by the same radio set, which also acts as a receiver. The fact that a signal is received indicates the presence of a reflecting obstacle, for example, an airplane or a ship.

Exercises

1. What is a displacement current?
2. What kind of mathematical argument led Maxwell to introduce a displacement current?
3. What is a radio wave?
4. Every radio station operates on a particular frequency, and when you tune in that station on your radio receiver, you adjust your set to receive that frequency. To what does the frequency pertain?

25–5 Electromagnetic waves and light. Maxwell's prediction of the existence of radio waves is in itself an unexcelled example of how mathematical reasoning can lead to discoveries which revolutionize a way of life. The uses of radio and television are now so widespread and so well known to modern man that we need not elaborate on the effect of these instruments on our civilization. But this prediction was just the first of several equally far-reaching ones. By the middle of the nineteenth century it was well agreed that light was some kind of wave motion, but the physical composition of the waves was not clear. The accepted view postulated that a medium, called ether, was spread throughout space, and that light was a motion of the particles of ether just as sound is a motion of the molecules of air. Now his equations showed Maxwell that electromagnetic waves traveled out into space. He was also able to determine the speed with which the waves move and found this to be 186,000 miles per second. The velocity of light had already been measured in several ways and was known to agree with this figure.

The coincidence of these two figures, coupled with the presence of wave motion in both phenomena, led Maxwell to conclude that *light is an electromagnetic wave motion.* Maxwell said, "If my calculations are correct, there are now two forms of energy, light and electricity, propagated through space with the same velocity. Since the only function of an ether is to provide a vehicle which will propagate energy at a specific velocity, it would be foolish to fill space with two ethers when one is sufficient." The identity of the two velocities clinched the matter in Maxwell's judgment.

Of course, Maxwell's inference that light was an electromagnetic wave remained to be verified. To some extent Hertz had already done so in the very experiments which confirmed the existence of radio waves. He showed that electromagnetic waves can be reflected from large sheets of smooth metal just as light waves are reflected from mirrors, and he demonstrated that paraboloidal reflectors will concentrate radio waves emitted by some source and send out a strong beam in one direction just as paraboloidal reflectors concentrate light coming from a source. These and subsequent experiments fully established that electromagnetic waves and light waves are the same phenomenon. Hertz's experiments also showed that the difference between light waves and radio waves produced by electrical currents is a difference of frequencies: light waves have frequencies of the order of 500 billion billion ($5 \cdot 10^{14}$) cycles per second, whereas the radio waves produced by Hertz had a frequency of 500 million ($5 \cdot 10^{8}$) cycles per second. Today we can generate radio waves with frequencies from about 20,000 cycles per second to about 30,000 million ($3 \cdot 10^{10}$) cycles per second.

The discovery that light is an electromagnetic wave unified two seemingly different phenomena and in effect doubled the knowledge available about each, for all the information on light which scientists had acquired over centuries could now be applied to radio waves, and, conversely, the theory and practice which had been developed with respect to electromagnetic fields could be applied to light. Thus the fact that light waves from a source, such as a small bulb, can be concentrated by a paraboloidal reflector into a powerful, directed beam was not only used by Hertz to verify the identity of the two phenomena, but is used today in many types of radio antennas. Large paraboloidal reflectors are used to produce powerful radio beams and this concentration enables radio waves to travel long distances—an important application in the communications field. In radar sets the sharpness of the beam enables the operator who has picked up a reflected signal to determine the direction of the reflecting obstacle quite accurately because the direction of the obstacle and the direction of the beam sent out are identical.

The electromagnetic theory of light, which considers light to consist of a succession of electric and magnetic fields, gave man his first indication of what light might be. Though prior to Maxwell's work theories of light had been put forth, none adequately explained all the phenomena. The electromagnetic theory of light proved to be satisfactory and granted new power to scientists to predict, for example, what light will do in passing through various media. In particular, the older concept which regarded light as some unknown but fixed substance traveling along rays and subject to the laws of reflection and refraction was seen to be just a good approximation, for, strictly speaking, a light wave which travels out in space is not confined to a set of lines, and its strength varies with time at any given point, and from point to point at any given time; in other words, it behaves just like water waves traveling out from a source. However, the variations are so small and so rapid that light appears to be a constant flow along each of a set of rays.

The prediction based on mathematical reasoning that light is an electromagnetic wave illustrates one of the remarkable values of mathematics. In the words of the foremost contemporary philosopher Alfred North Whitehead, "The originality of mathematics consists in the fact that in the mathematical sciences connections between things are exhibited which, apart from the agency of human reason, are extremely unobvious."

25–6 The range of electromagnetic waves. Maxwell's work uncovered vast new domains. Radio waves, as we have already mentioned, range from frequencies of about 20,000 cycles to about 30,000 million ($3 \cdot 10^{10}$) cycles per second. (We could say from 0 to about $3 \cdot 10^{10}$ cycles per second, but electromagnetic fields generated by low-frequency currents are very weak and so do not travel far from the wires in which the currents flow.) Light waves have frequencies of the order of $5 \cdot 10^{14}$ cycles per second. The obvious question is, Are there waves with frequencies between those of radio and light waves and are there waves with frequencies higher than those of light waves? In Maxwell's days physicists had already realized that sunlight contains ultraviolet rays which blacken photographic film, just as visible light waves do, and infrared rays which convey heat. It was soon established that these waves are also electromagnetic, with the frequencies of infrared rays being a little below and those of ultraviolet waves a little above the frequencies of visible light. This knowledge has enabled scientists to utilize these rays since so much was already known about radio and light waves. For example, infrared photography is now a very useful application. In 1895 Wilhelm Konrad Roentgen discovered x-rays, and the thought occurred immediately that these rays, too, were electromagnetic and of even higher frequency than ultraviolet rays. This fact was soon confirmed by experiments, though it was not until 1913 that the physicist Max von Laue determined that the frequencies of x-rays lie between 10^{16} and 10^{18} cycles per second. Radioactive elements such as uranium and radium were discovered from 1896 on, and among the radiations given off by such elements, physicists have detected the so-called gamma rays which are electromagnetic rays with frequencies of 10^{20} cycles per second.

The theory of electromagnetic waves has been carried much farther. Indeed it has led to a wave theory of all matter. One can show experimentally that the tiny electrons exhibit the properties of waves. The wave theory of matter in turn was essential to the reasoning that matter can be converted into energy, for this energy takes the form of electromagnetic waves. We are today all fully aware of the marvelous and terrifying applications which have been made of the conversion of matter into energy. But the pursuit of these thoughts leads too far into the domain of physics to be continued here. Our primary objective in this section was to see how Maxwell's work suggested that there could be electromagnetic waves of all frequencies and that this expectation was justified by subsequent physical discoveries.

25–7 Electromagnetic waves and the sense of sight. Man's exploration of the physical world cannot but produce knowledge of the most important object in the physical world; man himself. One of the apparent facts about visible light is that it transmits color. Moreover, as Newton showed, even white or colorless light is a combination of light of all colors from red to violet. The knowledge that light is an electromagnetic wave soon led to the discovery that different colors correspond to different frequencies; that is, light or visible electromagnetic waves range in frequency from $4 \cdot 10^{14}$ cycles per second for red light to $7 \cdot 10^{14}$ cycles per second for violet light. From the point of view of a scientist concerned solely with light as a physical phenomenon, all waves in this range are alike except for frequency. However, each frequency produces a different effect on the eye, and this difference is the color. This fact has led to investigations of the nerve cells in the eye to determine how different frequencies can produce different color effects. Special receptor cells in the retina, called cones, are apparently the receptors of color.

Another fact which emerges with clarity from the theory of electromagnetic waves is that the eye, like the ear, is a limited organ. It responds to electromagnetic waves in a definite range of frequencies, but does not respond to frequencies above and below this range. Moreover, the eye is most sensitive to frequencies in the middle of this range, just as the ear is most sensitive to frequencies of about 3500 cycles per second. Hence, two light waves of equal intensities but different colors will not be perceived equally clearly by the eye. The practical value of this fact for signals such as traffic or railroad light signals, for example, is obvious. The signal should be of a color most readily detected by the eye. For this effect green is an excellent choice.

The analogy between the eye in its response to light and the ear in response to sound is helpful. A combination of light waves of different frequencies, and therefore of different colors, produces a composite color sensation just as a combination of simple sounds produces a complex sound with a quality of its own. Just as some complex sounds are pleasing to the ear, so are some composite colors pleasing to the eye. However, the ear is more sensitive than the eye in that it can recognize the simple sounds in a complex sound, whereas the eye is not able to do more than detect roughly the proportions of the primary colors red, yellow, and blue which make up a given color. It certainly cannot resolve a given composite color into the component frequencies.

Our knowledge of the functioning of the eye is still in its infancy. One difficulty is that we cannot produce monochromatic light, that is, visible waves of a single frequency, and study the reaction of the eye to individual frequencies. Another is that the sensation of light is received by cells in the retina of the eye, the cones and rods, which are too small to be viewed and too inaccessible (in the living eye) for experimentation. However, we can, by means of radio tubes, produce waves of a single frequency in the radio range. Moreover, we can attempt to simulate the relevant parts of the eye. The simulated parts are larger than the corresponding parts of the eye in the ratio that the frequency of light is to the frequency of the radio wave used. The ratio is roughly 100,000 to 1. This means that the parts are easily constructed, handled, and observed. Hence we can expect that current experimentation will produce far more knowledge of the process of vision.

Scientific work in radar has given us the clue to understanding how some animals "see." It had long been a puzzling phenomenon that bats see so well in their night flights. We have already mentioned that a radar set sends out short bursts of radio waves in essentially one direction; when these bursts strike an obstacle, they are reflected and picked up by the set. Since the radio beam is highly directional, the direction of the obstacle is known. When no return signal is received, then, of course, there is no obstacle, at least within the range over which the radar set can transmit. Recent studies show that most species of bats use radar except that they send out bursts of *sound* waves with frequencies of about 50,000 cycles per second. Each burst contains about 100 cycles. When a bat receives a reflected signal, it knows that there is an obstacle in the direction in which the sound was directed, and so will avoid it.

Our knowledge of sight and sound leads to a rather strange conclusion in the domain of aesthetics. All musical sounds are no more than wave motions in the air. From a purely physical standpoint, the difference between one sound and another is just a difference in the number and strength of the component frequencies, that is, the harmonics. Similarly, from a purely physical standpoint, light is no more than a combination of electromagnetic waves of different frequencies in the visible range. Hence the real properties of sound and light, that is, the physical properties which exist outside of us, are just motions of air molecules in one case and of ether particles (supposedly) in the other. These physical motions are precisely the primary qualities of Descartes, Galileo, and Locke. On the other hand, color, the quality of musical sounds, the pleasurable effect of some colors, and consonance or dissonance of sounds are man's reaction to physical events. The sense organs, nerves, and brain contribute these secondary qualities, which exist only in man himself. But man prefers, for example, the color of a rose to the color of mud. He sees beauty in one and drabness or ugliness in the other. This beauty is man's contribution to nature and not nature's contribution to man.

25–8 Electromagnetic theory and the physical world. One of our main concerns is to see how mathematics aids us in gaining knowledge of the physical world. In electromagnetic theory we have a superb example. A number of scientists, among them Galvani, Volta, Oersted, Henry, and Faraday, discovered the fundamental physical phenomena of electromagnetism and enabled Maxwell to formulate basic equations which expressed mathematically the significance of these phenomena. His equations are the physical axioms of electromagnetic theory. By applying mathematical reasoning to these equations Maxwell made a number of fundamental discoveries and indicated the possibility of others in the frequency ranges which had not yet been explored. When the various types of waves, radio, infrared, ultraviolet, x-rays, and gamma rays, were experimentally confirmed and fully explored, it became clear that one mathematical theory, namely Maxwell's equations and their mathematical consequences, covered an immense variety of seemingly diverse phenomena.

In providing order, comprehension, unification, and the transfer of knowledge from one phenomenon to another, electromagnetic theory offers all the values which we can demand of a science. It offers not only understanding of nature, but, in view of the many uses of electromagnetic phenomena which man has made and of the new ones currently being devised, it would seem that the theory has given man a solid grip on a major segment of the physical world.

But now it is time to ask, Just what has man gripped? The entire development of electromagnetic theory rests upon some physical facts. The repulsion and attraction of magnets can be observed, and we account for this effect by stating that a magnetic pole is surrounded by a field which attracts or repels other magnets. The electric current generated by electromagnetic induction, when passed through a wire in electric light bulbs, heats the wire and makes it glow. We can feel the heat and see the glow. Hence we speak of a magnetic field inducing current in the wire. We also see objects and conclude that light streams from these objects to our eyes. We feel the heat of the sun and conclude that there are infrared rays. Photographic film, even though shielded from visible light, does become black when sun light impinges on the container. Hence we conclude that there are ultraviolet rays or waves. We introduce an amplitude-modulated sinusoidal current into a radio antenna and, at another such antenna many miles away, we receive the same amplitude-modulated current. We presume therefore that radio waves travel from one antenna to the other.

In short, we observe or feel a number of physical phenomena and we have built up a major theory of electric and magnetic fields which seeks to relate these phenomena and explain their occurrence in physical terms. But what are these electric and magnetic fields? Of what are they composed? How do they travel in space and how do they produce currents in wire? The answer is that we haven't the slightest idea.

When Maxwell first predicted the existence of radio waves and asserted that light is an electromagnetic wave, he concluded that all electromagnetic waves are motions in a medium called ether which had already been introduced in Newton's time to account for the motion of light waves. In the last quarter of the nineteenth century physicists accepted the reality of this medium. Indeed Lord Kelvin, one of the age's distinguished physicists said, "One thing we are sure of, and that is the reality and substantiality of the luminiferous ether." But subsequent thought showed that the concept of an ether was untenable. Thus, for example, electromagnetic waves travel through all space and even through many substances such as the walls of houses. Hence the ether would have to be present everywhere in space and in most substances. However, space is transparent, and so ether must also be transparent. The planets move through space with no evidence of resistance. Hence ether must be almost immaterial. To account for the high velocity of radio waves one must suppose, on the basis of experience with other wave motions, that the ether is a very dense and almost rigid medium. But this supposition conflicts with the fact that we can see through the supposed ether over great distances and that the planets experience no retarding friction. Furthermore, no such substance has been touched, smelled, weighed, seen, or detected in any way whatsoever. We must therefore conclude that there is no ether. We are left with the fact that we have no physical explanation of what an electromagnetic field is nor have we any idea of how even simpler phenomena, such as the action of magnets on each other or of electric charges on each other, are effected.

All attempts, including those of Faraday and Maxwell, to formulate a physically understandable account of the nature of electromagnetic waves and of the processes by which they produce their effects have failed. After veritable orgies of model-making the scientists had to acknowledge defeat. What, then, do we know about the electromagnetic waves which embrace so many fundamental phenomena—radio waves, light waves, ultraviolet "light," x-rays, and the like? The answer is that we only know their mathematical representation. Maxwell's equations and their mathematical consequences are the only clear and firm account. Electromagnetic theory is entirely a mathematical theory, and as such it predicts the few physically observable phenomena we described earlier. Asked for an explanation of what is the physical nature of electromagnetic waves, Heinrich Hertz replied, "To the question, What is Maxwell's theory? I know of no shorter or more definite answer than the following: Maxwell's theory is Maxwell's system of equations."

We are faced, then, with the amazing fact that one of the largest bodies of scientific theory is almost entirely mathematical. Certain formal deductions from this theory, such as the induction of current in wires or the reception of current hundreds of miles away from a source, can be confirmed by sense impressions, but the body of the theory itself is mathematical.

To some extent we should be prepared for this peculiar state of affairs. After having studied Newton's work on gravitation, we considered the question, What is gravity and how does it act? We found in that case, too, that we had no physical understanding of the action of gravitation. We have a mathematical law describing the quantitative value of this force and, by using this law and the laws of motion, we can predict effects which can be experimentally checked. But the central concept, gravitation, remains unknown.

We see, then, that the heart of our best scientific theories is mathematics or, more accurately, some formulas and their consequences. The firm, bold design of a scientific theory is mathematical. Our mental constructions have outrun our intuitive and sensuous perceptions. In both theories, gravitation and electromagnetism, we must confess our ignorance of the basic mechanisms and leave the task of representing what we know to mathematics. We may lose pride in making this confession, but we may gain understanding of the true state of affairs. We can appreciate now what Alfred North Whitehead meant when he said, "The paradox is now fully established that the utmost abstractions [of mathematics] are the true weapons with which to control our thoughts of concrete facts."

We have been discussing the relationship of mathematics to science, but the implications are deeper. Science is devoted to the study of the physical world and seeks to render an account of physical happenings. The fact that science, unlike mathematics, can give no account of light, radio waves, and the other members of the electromagnetic family or, for that matter, of gravitation means that our physical world presents us with mysteries. Electromagnetic waves may be pure fictions. Some physicists assert, as did Max Planck, "What one can measure also exists." Planck referred to the fact that radio receivers placed in an electromagnetic field can be used to measure the field. However, this dogmatic assertion is not an account of the nature of the field itself. Scientific mysteries may have the tremendous advantage that they can be translated into usable mathematical formulations, but our ignorance of the physical reality is not thereby relieved. The situation has been described in the words, Science is a classification and rationalization of mysteries; it is rationalized mythology.

Because the physical phenomena are unknown and perhaps unknowable, the modern scientist takes hold of the mathematical account of reality and seeks to bury the mysteries under a weight of mathematical symbols. Thus we find a leading physicist, the late Sir Arthur Stanley Eddington, declaring that a knowledge of mathematical relations is all that science can give us. Sir James Jeans, a contemporary of Eddington and of equal stature, has said that the mathematical description of the universe is the ultimate reality. The physical images that we use to assist us in understanding the operations of nature are a step away from reality. They are like "graven images of a spirit." We go beyond the mathematical formula at our own risk. "And from the intrinsic evidence of his creation the Great Architect of the universe now begins to appear as a pure mathematician."

Exercises

1. Name some forms of electromagnetic waves and describe how we recognize their existence.
2. What role did mathematics play in the discovery of radio waves?
3. What basis did Maxwell have for asserting that light waves are electromagnetic waves?
4. What physical difference is there among the several types of electromagnetic waves?
5. Which are the visible electromagnetic waves?
6. May we speak of the color of an infrared ray?
7. In view of our "knowledge" of electromagnetic waves, in what sense does science explain the phenomena of the physical world?

Recommended Reading

Bonner, Francis T. and Melba Phillips: *Principles of Physical Science,* Chaps. 14 to 17, Addison-Wesley Publishing Co., Inc., Reading, Mass., 1957.

Campbell, Lewis and Wm. Garnett: *The Life of James Clerk Maxwell,* Macmillan and Co. Ltd., London, 1882.

Cheronis, N. D., J. B. Parsons, and C. E. Ronneberg: *The Study of the Physical World,* 2nd ed., Chaps. 35 to 40 and 44, Houghton Mifflin, Boston, 1950.

Griffin, Donald R.: *Echoes of Bats and Men,* Doubleday and Co., Anchor Books, New York, 1959.

Holton, Gerald and Duane H. D. Roller: *Foundations of Modern Physical Science,* Chaps. 28 and 29, Addison-Wesley Publishing Co., Inc., Reading, Mass., 1958.

Jeans, Sir James: *The Growth of Physical Science,* 2nd ed., pp. 275–88, Cambridge University Press, Cambridge, 1951.

Kline, M.: *Mathematics and the Physical World,* Chaps. 20 and 21, T. Y. Crowell Co., New York, 1959.

Mason, S. F.: *A History of the Sciences,* Chaps. 37 and 38, Routledge and Kegan Paul Ltd., London, 1953.

Maxwell, James Clerk: A Commemoration Volume, Cambridge University Press, London, 1931.

Skilling, H. H.: *Exploring Electricity,* Ronald Press Co., New York, 1948.

Taylor, Lloyd Wm.: *Physics, The Pioneer Science,* Chaps. 45 to 50, Dover Publications, Inc., New York, 1959.

Turner, Dorothy Mabel: *Makers of Science, Electricity and Magnetism,* Oxford University Press, New York, 1927.

CHAPTER 26

NON-EUCLIDEAN GEOMETRIES AND THEIR SIGNIFICANCE

One must do no violence to nature, nor model it in conformity to any blindly formed chimaera; . . .

John Bolyai

26–1 Introduction. The most significant revolutions in this world are not of political nature. Political revolutions do not change the daily life of man too much or, if they do, exert a short-term effect which may even be reversed by subsequent revolutions. The significant revolutions are caused by new ideas; these far more effectively, powerfully, and lastingly alter the lives of civilized human beings. For example, the beliefs in the importance of the human being and in his right to life, liberty, and the pursuit of happiness have permanently and radically changed the lives and aspirations of hundreds of millions of human beings. Indeed, many political revolutions were inspired by the desire to realize these ideas. The two concepts which have most profoundly revolutionized our intellectual development since the nineteenth century are evolution and non-Euclidean geometry. The theory of evolution is generally well recognized as a prime influence, but non-Euclidean geometry, despite its more fundamental and more far-reaching effects, seems to escape attention.

In this chapter we shall examine the nature of non-Euclidean geometry, its value for science, its implications for the nature of mathematics, and, finally, its influence on our culture.

26–2 The historical background. Euclidean geometry, as well as developments such as arithmetic, algebra, and calculus, rests upon axioms. The Greeks, who formulated the axioms of Euclidean geometry, believed that human minds immediately recognized some truths about the geometrical properties of physical objects and of space. Thus, it seemed indubitable that two points determined one and only one line, and that equal line segments added to another pair of equal line segments gave equal sums. For two thousand years the entire intellectual world accepted the Greek doctrine that the axioms of Euclidean geometry and of mathematics, in general, were truths about the physical world, truths so clear and so evident that no one in his right mind could question them. Of course, since the axioms of geometry were truths, and since the theorems were logically necessary consequences of the axioms, the entire body of Euclidean geometry constituted a collection of indubitable truths about idealized objects and phenomena of the physical world.

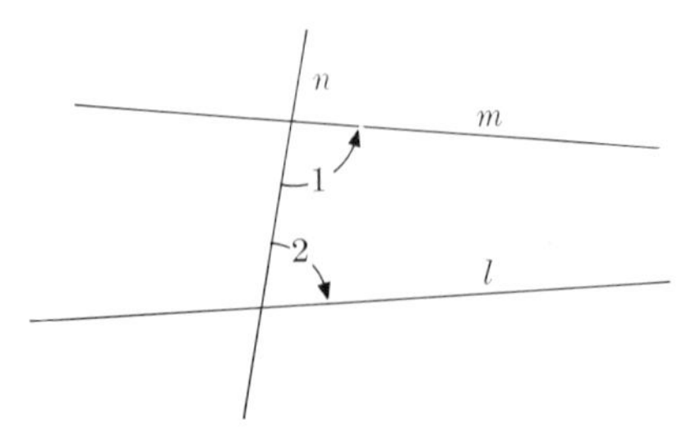

FIG. 26–1. Euclid's parallel axiom.

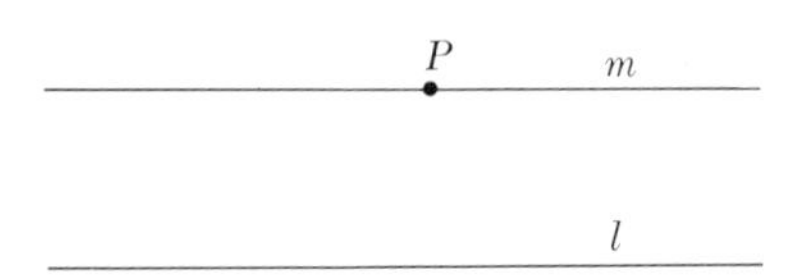

FIG. 26–2. Playfair's parallel axiom.

One slight blemish seemed to mar the collection of axioms. Euclidean geometry deals with parallel lines. By definition, two lines in the same plane are parallel if they do not meet, that is, if they do not contain any point in common. This last statement expresses what we mean by parallel lines and so is not objectionable. In itself it does not assert that there are any parallel lines. But Euclidean geometry contains an axiom which implies the *existence* of parallel lines, and our present discussion centers about the nature of this axiom. As stated by Euclid, the axiom asserts that if two lines m and l (Fig. 26–1), meet a third line, n, so as to make the sum of angles 1 and 2 less than 180°, then the lines m and l meet on that side of the line n on which the angles 1 and 2 lie. Euclid then proves, for example, that if the sum of angles 1 and 2 is 180°, then m and l are parallel. This axiom is a bit involved, and there is some reason to believe that even Euclid himself was not too happy about it. Neither he nor later mathematicians up to the nineteenth century really doubted the truth of this axiom; that is, they had no doubt that it was a correct idealization of the behavior of actual or physical lines. What bothered Euclid and his successors was that the axiom was not quite as self-evident as the axiom, say, that any two right angles are equal.

From Greek times on mathematicians sought to replace this axiom with an equivalent one. An equivalent axiom is one which, together with the other nine axioms of Euclid, will make it possible to derive the same body of theorems that Euclid deduced. Many equivalent axioms were proposed. One of these, known as Playfair's axiom [it was adopted by the mathematician John Playfair (1748–1819)], is the one we usually learn in high school. Playfair's axiom states that, given a line l (Fig. 26–2) and a point P not on that line, there is one and only one line m in the plane of P and l which passes through P and does not meet l. From Playfair's parallel axiom and the other nine axioms of Euclid one can deduce all the theorems of Euclidean geometry.

Playfair's axiom appears intuitively convincing. That is, it does seem as though straight lines in physical space possess the property asserted. However, mathematicians were not satisfied with Playfair's axiom or other proposed equivalents of Euclid's parallel axiom. They recognized that every proposed substitute involved directly or indirectly an assertion about what happened far off in space. Thus, Playfair's axiom asserts that the line m through P will not meet l even in the very distant space to which these lines extend. As a matter of fact, Euclid's axiom is superior in this respect because it does not assert that lines will not meet, but states conditions under which they will meet at some finite distance.

What is objectionable about axioms which assert what happens far out in space? The answer is that they transcend experience. The axioms of Euclidean geometry are supposed to be immediately convincing statements about the properties of space. But how can one be sure of what happens millions of miles away? How can one be sure that it is possible to extend straight lines indefinitely far out into physical space, without the lines ever being forced to meet? Thus the efforts made to find a simpler statement than Euclid's did succeed insofar as simplicity of statement was concerned, but bred doubts about the truth of any assertion regarding the existence of parallel lines.

By the eighteenth century some mathematicians decided to try a new tack. Euclid's set contained ten axioms. Perhaps nine sufficed; that is, perhaps one could *prove* an assertion about parallel lines by deduction from the other nine axioms. If this should be possible, then there would be no further problem, because the assertion about parallel lines would be a necessary consequence of the nine entirely acceptable axioms. However, all these efforts failed.

One of these efforts, though, deserves special attention. The Jesuit priest Girolamo Saccheri (1667–1733) decided to apply the indirect method of proof. Euclid's parallel axiom asserts, in effect, the existence of one and only one line through P and parallel to l. To establish the truth of this statement by contradiction, two alternatives are available: no line parallel to l through P or more than one. Saccheri's plan was to assume in turn that each alternative to Euclid's parallel axiom was true and, with this alternative and the other nine axioms of Euclid, show that deductions would lead to a contradiction. He could then proclaim that the only questionable assumption, namely the alternative to Euclid's parallel axiom, must be false. After doing this with each of the two alternatives, he could then assert that the only remaining possibility, Euclid's axiom, must be true. As a matter of history, the alternative asserting that there were no parallels to l through P produced a contradiction. However, from the second alternative (there is more than one parallel to l through P) Saccheri derived a number of strange theorems but no contradiction.

The strangeness of the theorems he obtained was enough to convince Saccheri that this second alternative could not be true, and hence that

Euclid's parallel axiom must be true because it was the only possible alternative. And so in 1733 he published a book entitled *Euclid Vindicated From All Defects*. Of course, strangeness of theorems and logical contradiction are quite different matters, and Saccheri was not justified in substituting one for the other. But he was tied to his time and, in drawing the conclusion that Euclid's assertion on parallel lines was a necessary consequence of the other nine axioms, he showed merely that if a man sets out to establish something of which he is already convinced, he will satisfy at least himself that he has proved its truth regardless of the facts.

The first to draw the conclusion which Saccheri should have drawn was Karl Friedrich Gauss (1777–1855). Gauss was one of the greatest mathematicians of all times. His father was a bricklayer who expected his son to adopt the same trade. However, Gauss showed his precocity in elementary school, and his teachers saw to it that he received a good education. At the University of Göttingen he pressed his teachers to keep up with him. In his day, one of the outstanding problems of Euclidean geometry was to construct the 17-sided regular polygon by means of straightedge and compass. One day Gauss approached his teacher, A. G. Kästner, with the proof that this polygon is constructible. Kästner was incredulous and sought to dismiss Gauss much as university teachers today dismiss angle trisectors. Rather than take the time to examine Gauss's proof and find the supposed error in it, Kästner told Gauss that the construction was unimportant because practical constructions were available. Of course, the professor knew that practical or approximate constructions were irrelevant for the theoretical problem. To interest Kästner in his proof, Gauss pointed out that he had solved a seventeenth-degree algebraic equation. Kästner replied that the solution was impossible. But Gauss rejoined that he had reduced the problem to solving an equation of lower degree. "Oh well," said Kästner, "I have already done this." Gauss was rebuffed. Later, he repaid Kästner, who thought himself a great poet, by lauding the professor as the best poet among mathematicians and the best mathematician among poets.

At 22 Gauss submitted his doctoral thesis to the University of Helmstedt. In this thesis he proved what is often called the fundamental theorem of algebra, namely, that every algebraic equation of any degree has at least one root. At the age of 30 he was appointed professor of astronomy at the University of Göttingen. He discouraged students from attending his lectures.

Gauss's scientific interests, like those of Archimedes and Newton, were unbelievably broad. He was, for example, a great inventor. He designed many instruments for use in geodesy and was one of the inventors of the electric telegraph. He devised methods of making maps, was an excellent astronomical observer, and set up systems of insurance. At the request of the Elector of Hannover he made a survey of the principality. In addition to devising instruments for the measurement of the earth's magnetic field, he himself studied the variation in the strength of this field over the earth.

The unit now used to measure the strength of magnetic fields is called the gauss.

Inspired by his scientific and practical interests, Gauss contributed to a number of major branches of mathematics. In fact, he would not think of separating mathematics from its application to science. Though his greatest achievements were in mathematics and he is therefore most often described as a mathematician, it would be more appropriate to call him a student of nature. His motto read:

> Thou, nature, art my goddess; to thy laws
> My services are bound. . . .

A famous phrase, Mathematics, Queen of the Sciences, is due to him.

Gauss was very modest. Though he had few friends, he had many admirers and achieved great fame much of which is reflected in a famous story about him. Laplace was asked one day the name of the greatest mathematician in Germany, and he replied, "Pfaff," a relatively minor man. The questioner had expected to hear Gauss's name and made a remark to this effect. "Oh," replied Laplace, "Pfaff is the greatest mathematician in Germany, but Gauss is the greatest mathematician in Europe."

Partly because he did not care much for fame and partly because he wished to put forth finished and highly polished works, Gauss did not publish much of what he created. Also, his ideas kept coming so fast that he jotted them down on papers and in a mathematical diary. Many of these results were discovered after his death. His reluctance to publish caused dismay to many of his contemporaries because, not knowing what he had done, they often proceeded to prove results only to find out later that Gauss had already arrived at the same conclusions. One of these men, Carl G. J. Jacobi, was especially frustrated because it had happened to him on several occasions that, on telling Gauss about one of his own new discoveries, Gauss pulled out from his desk drawer some papers that contained the very same results. Jacobi resolved to get even. One day he again visited Gauss to show him some new work, and the usual scene took place. Jacobi then uttered his prepared remark, "It is a pity that you did not publish this result since you have published so many poorer papers." Gauss's published papers contained no hint of motivation, intuitive meaning, or details of the proof. When criticized, he said that no architect left the scaffolding after completing the building.

One of Gauss's greatest creations and certainly the most momentous from the standpoint of its implications is his non-Euclidean geometry. Gauss started his thinking on this subject as a boy and, like others before him, he began by trying to replace Euclid's parallel axiom by a more acceptable one, that is, one which did not involve any assertion about what must happen far off in space. But he did not succeed. He also appreciated at the age of 15, as he told his friend Schumacher, that one could not hope to prove an assertion about parallel lines on the basis of the other nine axioms. That is, he realized that these nine axioms did not in themselves dictate the form of the parallel axiom. Gauss was too brilliant a man to overlook the implication of this fact. If there was some freedom in the choice of a parallel axiom, then one might choose an axiom different from Euclid's and build a new kind of geometry. Gauss did just this. He pursued the logical implications of a system of axioms which included the assumption that *more than one* parallel to a given line passed through a given point, and thus created a non-Euclidean geometry. (Gauss himself finally adopted the term non-Euclidean after having called his system anti-Euclidean geometry, and later astral geometry.)

Though Gauss realized that such a geometry could conceivably apply to physical space and hence was indeed significant, he did not publish his results. In addition to the reasons cited above for Gauss's reluctance to publish, another factor entered in the case of non-Euclidean geometry. Gauss was far ahead of his times in concluding that Euclidean geometry was not necessarily the correct description of physical space and that some non-Euclidean geometry might prove as accurate. Hence he feared that he would be laughed at. In a letter written in 1829 to his friend, the mathematician Friedrich Wilhelm Bessel, Gauss confessed that he feared the clamor of the Boeotians, a figurative reference to the dull-witted, for the Boeotians had been one of the more simple-minded Greek tribes. Gauss's work on non-Euclidean geometry was found among his papers after his death in 1855.

The men who usually receive credit for creating non-Euclidean geometry because they published their results are Nicholas I. Lobachevsky (1793–1856) and John Bolyai (1802–1860). Lobachevsky, born to a poor family and displaying brightness even as a youngster, attended the University of Kazan in Russia. One of his teachers was the German mathematician J. M. C. Bartels, a friend of Gauss, and it is very likely that the problem of the parallel axiom was called to Lobachevsky's attention by Bartels. Lobachevsky became a professor at Kazan at the age of 23 and continued to work on the problem. By 1823 he realized that there can be other geometries and that Euclidean geometry need not be the correct description of physical space. He said later that two thousand years of fruitless attempts to put the parallel axiom on an unquestionable basis had led him to suspect that it could not be done. From 1829 on Lobachevsky published books and papers in which he expounded the theorems that hold in his non-Euclidean geometry. Despite valuable services to mathematics, his university, and the Russian government, he was dismissed in 1846, but continued to work in his field until his death.

Bolyai, a Hungarian, was an Austrian army officer. He learned mathematics from his father, Wolfgang Bolyai, who set his son thinking about the parallel axiom. In 1823 John arrived at the same conclusion that had been reached by Gauss and Lobachevsky, namely, that the Euclidean parallel axiom cannot be proved and that, in fact, it was but one alternative. He proceeded to develop a non-Euclidean geometry and published his work as an appendix to his father's book on mathematics which appeared in 1833. Knowing Gauss's interest in the subject of non-Euclidean geometry,

Wolfgang sent a copy of his son's work to Gauss. The latter replied, "If I commenced by saying that I am unable to praise this work, you would certainly be surprised for a moment. But I cannot say otherwise. To praise it would be to praise myself. Indeed, the whole contents of the work, the path taken by your son, the results to which he is led, coincide almost entirely with my meditations, which have occupied my mind partly for the last thirty or thirty-five years." Though Wolfgang was pleased to learn that his son's thinking paralleled that of the great Gauss, John was not, for, like others who had found that Gauss had beaten them to a discovery, he felt cheated of the glory. The irony of the situation was compounded further. Although Lobachevsky and Bolyai are now credited with the discovery of non-Euclidean geometry because they were the first to publish on the subject, the mathematical world of the 1830's and 1840's ignored their publications until Gauss's notes on non-Euclidean geometry were found among his papers after his death. The name of Gauss attached to the idea made the mathematical world grant the proper importance to the subject.

On the surface, the history of non-Euclidean geometry seems remarkable because after two thousand years of futile work, three men suddenly saw the parallel axiom and Euclidean geometry in the proper light. However, while such coincidences of history are by no means uncommon, the present one is perhaps less remarkable than is ordinarily believed. Gauss did not hesitate to tell his mathematical friends of his radical views on the parallel axiom and of his doubts about the necessary truth of Euclidean geometry. Both Bartels and Wolfgang Bolyai were in this circle of friends, and they may have communicated Gauss's views to their respective charges. The mere derivation of theorems from new parallel axioms, though a considerable technical achievement, goes back to Saccheri, and numerous minor and major mathematicians had done similar work in the intervening one hundred-year period. But the correct evaluation of the significance of these logical developments was a new step, and this seems to have been made first by Gauss. However, Wolfgang Bolyai, perhaps to defend the originality of his son's thoughts, said of the work by the three men, ". . . because it seems to be true that many things have, as it were, an epoch in which they are discovered in several places simultaneously, just as the violets appear on all sides in the springtime."

Exercises

1. What is the definition of parallel lines?
2. What objection did Euclid's successors have to Euclid's axiom on parallel lines?
3. What objections were there to axioms such as Playfair's which replaced Euclid's parallel axiom by simpler assumptions?
4. Describe Saccheri's plan to establish the truth of Euclid's parallel axiom.
5. Did Saccheri arrive at the concept of a non-Euclidean geometry?
6. What advances did Gauss, Lobachevsky, and Bolyai make over Saccheri with respect to the nature of geometry?

26–3 The mathematical content of Gauss's non-Euclidean geometry. To appreciate the significance of what Gauss, Lobachevsky, and Bolyai created, we must look into the specific mathematical facts of their work. For brevity we shall, at present, refer to their geometry as Gauss's geometry. The main idea conceived by the three men asserted that one was logically free to adopt a parallel axiom which differs fundamentally from Euclid's, and that one could construct a new geometry which would be as valid as Euclid's and which might even be a good description of physical space.

What was the new parallel axiom? Euclid's parallel aixom, at least in the equivalent form which Playfair had given, stated that given a line l and a point P, there is one and only one line in the plane of l and P which passes through P and does not meet l (Fig. 26–2). Gauss, Lobachevsky, and Bolyai assumed that there are two lines, m and n, through P (Fig. 26–3) which are parallel to l, that is, these two lines through P do not meet l, and that any line through P falling within the angle MPN does meet l. Of course, any line such as q which passes through P and lies within the angle NPR cannot meet l because to do so, q would have to cross m or n and therefore meet m or n in a second point. But since two lines can intersect in at most one point, it follows that all lines through P which lie within the angle NPR will not meet l. Hence the assumption that there are at least two parallels to l implies that there is an infinite number of parallels.

The term *parallel lines* was reserved by Gauss, Lobachevsky, and Bolyai for m and n, whereas lines such as q were called nonintersecting lines. For this reason, their geometry is often described as a geometry with two parallel lines, although, in the Euclidean sense of the term *parallel lines* (two lines in a plane which have no point in common), it contains an infinite number of parallels. We shall use the terminology chosen by Gauss, Bolyai, and Lobachevsky.

Before proceeding to draw conclusions from this new parallel axiom and the other nine Euclidean axioms, let us dispel in advance any doubts about the common sense of the material we are about to present. We are so accustomed to Euclidean geometry (those of us who did not take our high-school geometry very seriously may be in a better position) that the idea of adopting a new parallel axiom and of proving theorems seems ridiculous. Euclid's geometry is truth to us, and to defy the truth and pursue the consequences of this folly seem a waste of time. There are two counterarguments. We can admit that from a practical standpoint the effort required to develop such a new system is pointless. Nevertheless, we should be able to comprehend that it is *logically* possible to investigate the consequences of a new set of axioms. What we are about to do is analogous to changing one of the articles in the Constitution of the United States while retaining the others, just to see what changes might follow in our laws. Thus we might decide to replace an elected president by a monarch and yet continue to elect congressmen and retain the institution of a Supreme Court. The second counterargument is based on the fact that Gauss, Bolyai, and Lobachevsky doubted the truth of Euclidean geometry and were ready to consider an

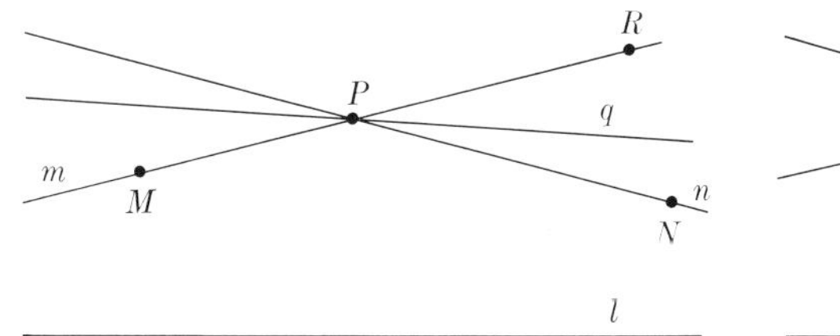

Fig. 26–3. Gauss's parallel axiom.

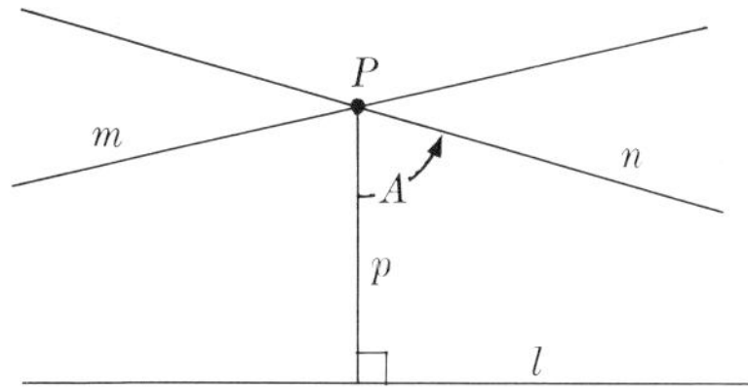

Figure 26–4

alternative geometry as a description of physical space. To what extent their doubts were justified and whether the new geometry could serve in physical applications will be clearer when we know what theorems actually resulted from the new set of axioms.

We shall describe, but not prove, the theorems of the new geometry. The methods of proofs are precisely the same as in Euclidean geometry; however, one must be careful to use only the new axioms to support any assertions.

Many theorems of the new geometry are precisely the same as those of Euclidean geometry. Indeed those theorems of Euclidean geometry which are proved with the aid of only the first nine axioms of Euclid, that is, those theorems which do not depend upon the Euclidean parallel axiom, must also be theorems of the new geometry because these nine axioms are retained in the new system. Thus the theorems which assert when two triangles are congruent, the theorem asserting that the base angles of an isosceles triangle are equal, the theorem stating that an exterior angle of a triangle is greater than either remote interior angle, and the theorem postulating that from a point off a line, there is one and only one perpendicular to that line, are all valid in the new geometry.

But now let us examine some theorems of the new geometry which are different from those of Euclidean geometry. Suppose that the point P and the line l (Fig. 26–4) are given. In the new geometry, the parallel axiom states that there are two lines, m and n, through P which are parallel to l. Let us now drop a perpendicular p from P to l. Then in this new geometry the angle A between p and n is no longer a right angle, although n and l are parallel lines. Angle A is in fact an acute angle. Moreover, the size of angle A depends upon the length of p. The shorter p is, the larger does angle A become, and as p approaches zero, angle A approaches 90° in size. Of course, in Euclidean geometry angle A is a right angle and this holds no matter how large p is. A key theorem of the new geometry asserts that the sum of the angles of a triangle is always less than 180°, whereas it is exactly 180° in Euclidean geometry. Further, in the new geometry the sum of the angles of a triangle varies with the area of the triangle. The smaller the area, the closer is the angle sum to 180°. A very surprising theorem of the new geometry states that if the three angles of one triangle equal, respectively, the three angles of another, the triangles are congruent. Of course, in Euclidean geometry two such triangles would be similar, and one could be very much larger than the other. There are many more interesting consequences of the new axioms, but we have seen enough to obtain some indication of the nature of the new geometry.

Exercises

1. Describe the parallel axiom adopted by Gauss, Bolyai, and Lobachevsky.
2. Distinguish between the Euclidean use of the term parallel lines and the usage in Gauss's geometry.
3. State three theorems which are common to Euclidean geometry and Gauss's non-Euclidean geometry.
4. State three theorems of Gauss's geometry which do not hold in Euclidean geometry.
5. Suppose that one triangle has an angle sum of 170° and another of 175°. Which has the larger area in Gauss's geometry?
6. Given two triangles, one with angles of 30°, 40°, and 100° and another with angles of 35°, 45°, and 90°. What does Gauss's geometry say about their area?
7. Suppose that the angles of one triangle are equal, respectively, to the angles of another. What may you conclude about these triangles on the basis of Gauss's geometry?

26–4 Riemann's non-Euclidean geometry. We saw in the preceding section that it is possible to investigate a new set of geometrical axioms and to deduce logical consequences. Perhaps we are not satisfied as yet that such an investigation is of value, but we shall reserve our final judgment until we have looked into one more mathematical creation in the field of non-Euclidean geometry.

Investigations of the parallel axiom had led mathematicians to question its truth because it made assertions about what must happen in physical space far beyond man's experience. Having become aware of this shortcoming, mathematicians examined the remaining axioms and soon found another which suffered from the same failing. Euclid's second axiom asserts that a straight line extends indefinitely far in either direction. This axiom attracted the attention of the nineteenth-century mathematical giant, Georg Friedrich Bernhard Riemann (1826–66). Riemann, the son of a Lutheran pastor, was born sickly and precocious. He became a student of Gauss at the University of Göttingen, and was later appointed professor at the same institution. Like Gauss, and most mathematicians for that matter, he was keenly interested in science and in the applicability of mathematics to the physical world.

Riemann observed that experience suggests not the infinite extent of the straight line but rather its endlessness. He accordingly distinguished between endlessness, or unboundedness, and infinite length. The simplest example of that distinction is furnished by the circle. One can traverse the circle endlessly, yet its length is finite. Hence Riemann proposed to replace the Euclidean axiom that a straight line extends indefinitely far by the axiom that it is unbounded.

Riemann also observed that experience does not vouch for the existence of any parallel lines. Within the limits of experience, we could equally well assume that any two lines meet. Hence Riemann proposed this axiom as an alternative to Euclid's parallel axiom. It may be recalled that Saccheri had considered the same possibility, but had found that it led to contradictions. However, Saccheri had combined this axiom with the other nine Euclidean axioms, whereas Riemann proposed an additional change, namely the unboundedness of the straight line, and had therefore reason to believe that he, unlike Saccheri, would not encounter any contradictions.

Since Riemann did retain some of Euclid's axioms, he arrived at some theorems identical to those of Euclidean geometry. Thus the theorem that two triangles are congruent when two sides and the included angle of one are equal to the corresponding parts of the other is a theorem of Riemann's geometry, as are other familiar congruence theorems.

The striking theorems of Riemann's geometry are, of course, those which differ markedly from Euclid's results. One Riemannian theorem asserts that every straight line has the same finite length. Another asserts that all perpendiculars to a line meet in one point. In Riemann's geometry the sum of the angles of a triangle is always *greater* than 180°. Moreover the sum varies with the area of the triangle and *decreases* to 180° as the area approaches zero. Two similar triangles are necessarily congruent (this is also the case in Gauss's geometry). These theorems are, of course, just a representative selection from a vast number.

Exercises

1. Why did Riemann question the Euclidean axiom that a straight line extends indefinitely far in either direction?
2. What axiom on parallel lines did Riemann adopt?
3. Some theorems of Gauss's and Riemann's non-Euclidean geometries are identical with Euclidean theorems. Why does this occur?
4. Which of the two non-Euclidean geometries would you expect to have more theorems in common with Euclidean geometry and why?
5. Compare the Euclidean, Gaussian, and Riemannian theorems about the sum of the angles of a triangle.
6. State three theorems peculiar to Riemann's geometry.
7. What is non-Euclidean geometry?
8. What happens to the Euclidean distinction between congruent and similar triangles in non-Euclidean geometry?

26–5 The applicability of non-Euclidean geometry. The very fact that there can be geometries other than Euclid's, that one can formulate axioms fundamentally different from Euclid's and prove theorems, was in itself a remarkable discovery. The concept of geometry was considerably broadened and suggested that mathematics might be something more than the study of the implications of the self-evident truths about number and geometrical figures. However, the very existence of these new geometries caused mathematicians to take up a deeper and more disturbing question, one which had already been raised by Gauss. Could any one of these new geometries be applied? Could the axioms and theorems fit physical space and perhaps even prove more accurate than Euclidean geometry? Why should one continue to believe that physical space was necessarily Euclidean?

At first blush the idea that either of these strange geometries could possibly supersede Euclidean geometry seems absurd. That Euclidean geometry is *the* geometry of physical space, that it is the truth about space is so ingrained in people's minds that any contrary thoughts are rejected. The mathematician Georg Cantor spoke of a law of conservation of ignorance. A false conclusion once arrived at is not easily dislodged. And the less it is understood, the more tenaciously is it held. In fact, for a long time non-Euclidean geometry was regarded as a logical curiosity. Its existence could not be denied, but mathematicians maintained that the real geometry, the geometry of the physical world, was Euclidean. They refused to take seriously the thought that any other geometry could be applied. However, they ultimately realized that their insistence on Euclidean geometry was merely a habit of thought and not at all a necessary belief. Those few who failed to see this were shocked into the realization when the theory of relativity actually made use of non-Euclidean geometry.

It is important to see how and why a non-Euclidean geometry can fit physical space. Let us recall, first of all, why Gauss, Bolyai, and Lobachevsky doubted the truth of the Euclidean parallel axiom. They realized that this axiom and all of the simpler forms which might serve as substitutes contained assertions about what happens in space far beyond the range of man's experience. Hence experience does not support such axioms. As a matter of fact, if the lines m and n of Fig. 26–3 and all lines, such as q, falling within the angle NPR should have almost the direction of l, then they would certainly not meet l within a short distance from the point P. Hence Gauss's axiom is as much in accord with experience as Euclid's. Our intuition is seemingly violated by such a thought, but this intuition may be conditioned by our familiarity with Euclidean geometry. In other words, insofar as experience is concerned, one can adopt Gauss's or Riemann's alternative axioms.

In view of our inability to decide *a priori* which of the several alternative axioms fits physical space, we might consider another approach to the problem. The theorems of any geometry are logical consequences of the axioms. Perhaps it would be easier to discriminate among these geometries by seeing how well their respective theorems fit physical space. This thought had already occurred to Gauss. He noted that in his geometry the sum of the angles of a triangle must be less than 180°, whereas in Euclidean geometry it is exactly 180°. Hence he had three observers stand on three mountain peaks and directed each one to measure the angle between his lines of sight to the other two observers. The sum of the angles of the triangle formed by the three peaks turned out to be 170°59′58″; that is, it was within 2″ of 180°.

This result might be interpreted as a victory for Euclidean geometry, but the situation is not so simple. In Gauss's geometry the sum of the angles of a triangle increases as the area of the triangle decreases, and the sum approaches 180° as the area approaches zero. In Riemann's geometry, the sum of the angles of a triangle is always larger than 180°, but the sum again approaches 180° as the area of the triangle decreases. Hence for *small* triangles, all three geometries call for an angle sum close to 180°. Had Gauss's measurement of the angles formed by the mountain peaks been exact, he might have been able to assert that his result was less than, equal to, or more than 180°, and so his test would have been decisive. However, every measurement contains an error because the eye and hand are not precise. We may safely suppose that the error in Gauss's measurement was more than 2″. Thus, in view of his result, the true angle sum could have been anything from a little more than 180° to a little less. Hence the result was consistent with any one of the three geometries.

For a large triangle, the angle sum should be significantly less than 180° in Gauss's geometry and significantly more in Riemann's. Hence, by using a large triangle, say one formed by three celestial bodies at some instant, one might be able to obtain a result which would fit only one of the three geometries, provided the error of measurement were kept small. Thus, a result of 175°, with an error of measurement of less than 5°, would certainly establish the physical correctness of Gauss's geometry. However, in the nineteenth century, no such result was obtained. Gauss, Lobachevsky, and Bolyai realized that, at least in their day and with the instruments at their disposal, measurement would not decide the question.

Since a test based on the sum of the angles of a triangle does not succeed in establishing which of the geometries fits physical space, one might cast about for another theorem which would serve the purpose. He might then hit upon the theorem which holds in both non-Euclidean geometries, namely the theorem stating that two similar triangles must be congruent. Here a decisive test seems likely, for it appears quite obvious that one can construct a small and a large triangle and make them similar. One only has to ensure that the angles of the small triangle equal the angles of the large one, and the similarity of the two triangles necessarily follows. Only Euclidean geometry fits this physical situation. The argument, however, has a flaw since one cannot be sure that the corresponding angles of the two triangles are really equal. After all, measurement is approximate, and at least one angle of one triangle might differ from the corresponding angle of the other triangle. If this happened, the triangles would not be similar.

The substance of the above arguments is that there is no simple test which points to one geometry rather than to another. However, if an application involved a truly large triangle, as astronomical investigations do, one might be able to determine which geometry fits better. As a matter of fact, this possibility has been realized. In the theory of relativity, Einstein employed a non-Euclidean geometry (though one more complicated than those we have studied), and the agreement between his predictions and observations was better than the result obtained by means of Euclidean geometry.

Exercises

1. Does measurement help us to show that objects in physical space are better described by the theorems of one rather than another geometry?
2. Could we use Gauss's or Riemann's non-Euclidean geometry for engineering and architecture?
3. Why did Gauss's measurement of the sum of the angles of a triangle fail to show that space is or is not Euclidean?

26–6 The applicability of non-Euclidean geometry under a new interpretation of line. Thus far we have considered the applicability of non-Euclidean geometry to our physical space, with the understanding that the physical meaning of a straight line is the stretched string or the ruler's edge. Even when we visualize a straight line so long that we could never construct it, for example, a line from the earth to the moon, we picture a long taut string or an imaginary long straight stick. However, we must now realize that the mathematical straight line is not limited to this physical or geometrical figure. Let us consider an example.

Suppose that we begin with a flat sheet of paper which can be imagined to extend indefinitely far in all directions. Now let us suppose that the paper is bent so that it curves upward to the right and to the left (Fig. 26–5) and, of course, continues to extend indefinitely far out in all directions. In mathematics, incidentally, this shape is known as a cylindrical surface. (One can form a limited cylindrical surface by merely bending upward the sides of a sheet of paper.) As a consequence of the change from a plane to a cylindrical surface, many of the straight lines of the plane become curves on the surface. Thus AC in Fig. 26–5 is an arc of a curve which, if the surface were flattened, would become a segment of a straight line. Curves which derive from straight lines we shall continue to call lines. Angles in the plane become, under the bending, angles on the surface. A triangle in the plane corresponds to a triangle formed by arcs of curves on the surface. A circle in the plane becomes a new curve on the surface which we shall continue to call a circle.

Now let us note that the bending of the plane into the cylindrical surface does not change lengths or angles, because the plane neither stretches nor contracts as one forms the surface. Hence a distance between two points

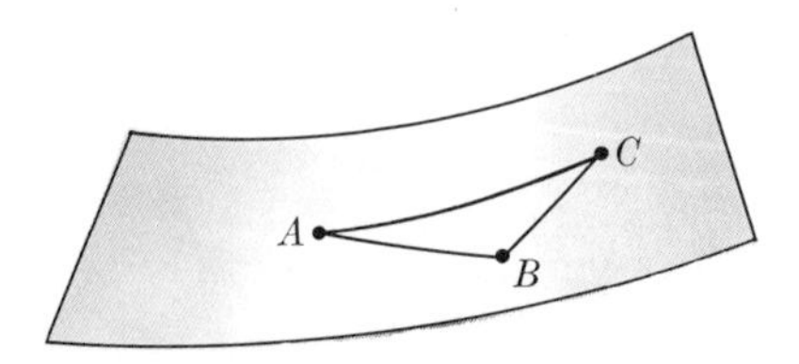

FIG. 26–5. Euclidean geometry on a cylindrical surface.

measured *along* the surface is the same as the distance between the two points when on the plane.

Since lengths and angles do not change, it follows that if the figures in the plane satisfy the axioms of Euclidean geometry, then so do the figures on the surface, provided the meaning of lines, angles, and circles on the surface is as specified above. And since theorems are logical consequences of axioms, it follows that the Euclidean theorems, too, retain their validity for figures on the surface. Thus the sum of the angles of triangle ABC in Fig. 26-5 is 180°. These last few assertions appear at first to be incredible. Our triangles are no longer formed by straight line segments, and yet the Euclidean theorems continue to hold. How is it possible to apply the word *line* of the Euclidean axioms and theorems to the curves on the surface which are no longer straight lines?

The answer to this question involves a major point about geometry and indeed about mathematics in general. In our review of Euclidean geometry in Chapter 6, we briefly mentioned that Euclid's definitions of point, line, plane, and other concepts were not quite satisfactory. The full story is that his definitions of these concepts are meaningless. He defined curve as length without breadth, but he did not say what he understood by length and breadth. He then defined the straight line as a curve which lies evenly between the ends, but he did not indicate what "lies evenly" meant. Euclid was relying on our intuitive understanding of these terms, but intuitive understanding cannot be part of a logical treatment. Mathematics does not rest logically on physical meanings. What then is the alternative? It is not possible to begin the process of developing any new branch of mathematics by defining the initial notions, for definition requires that one provide descriptions in terms of other concepts whose meaning is already established. But obviously the initial concepts cannot be defined in terms of prior ones since there are no prior ones. The point, then, is that the initial concepts cannot be defined. They must remain undefined.

This assertion raises another question. If point, line, plane, and other basic concepts are undefined, how shall we know what we mean by them and how shall we know how to treat them? The answer is that the axioms of Euclidean geometry tell us what properties these concepts possess. These properties and only these may be used in establishing the proofs. The concepts are therefore limited only to the extent that they must satisfy the axioms.

Now, as far as we can determine, physical straight lines, such as rulers' edges and stretched strings, do satisfy the axioms of Euclidean geometry. In fact, historically, the axioms were formulated by observing the properties of rulers' edges and stretched strings. What is surprising, however, is that one may encounter other "lines" which also satisfy the axioms. If they do, they are legitimate interpretations or realizations of the axioms and hence of the theorems. This is the case for the lines of our cylindrical surface. We know that they satisfy the Euclidean axioms because in deforming the plane into the cylindrical surface, we did not disturb any of the

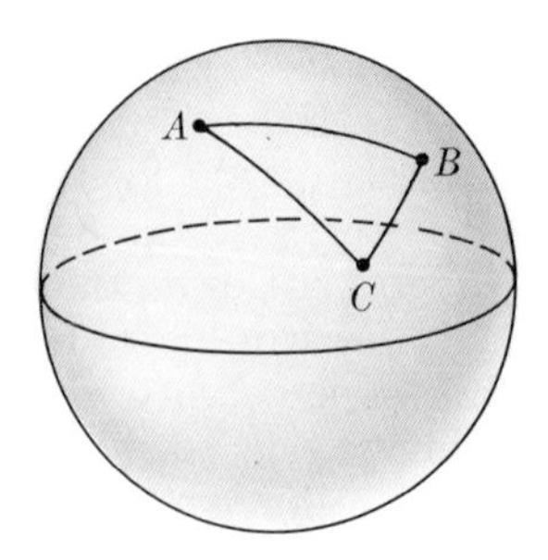

FIG. 26–6. A spherical triangle formed by arcs of great circles.

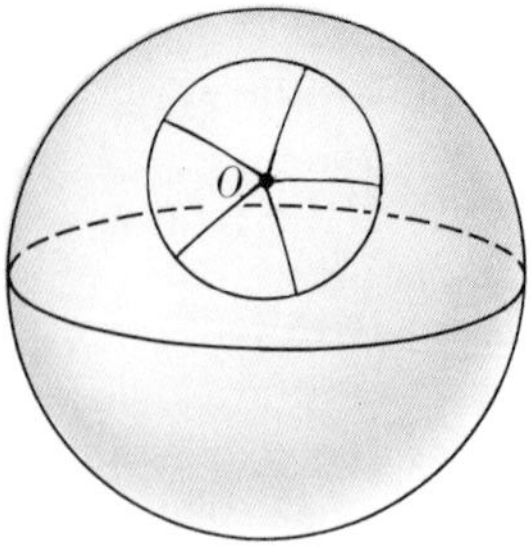

FIG. 26–7. A circle on a sphere.

properties asserted by the axioms. Consequently, Euclidean geometry as a whole applies to the points, lines, triangles, polygons, circles, and other figures on the cylindrical surface.

The significance, then, of the entire discussion about the cylindrical surface is that it is possible to associate a totally new picture with Euclidean geometry. In other words, we see that Euclidean geometry applies to more physical situations than we had previously suspected. But if one can put Euclidean geometry to new uses by adopting a new view of the straight line, perhaps one can do the same for the non-Euclidean geometries, and hence these, too, may be more valuable and more meaningful than we have hitherto suspected. We shall now show that it is possible to give a very simple and practical interpretation of Riemann's non-Euclidean geometry.

Let us consider the surface of the earth and let us suppose that we were assigned the task of developing a geometry that would fit the surface of the earth. We would, of course, choose as points the idealization of the usual physical points. We would most likely cast about next for the figure which would play the role of the straight line. Now straight lines in the usual sense do not exist on the surface of the earth because the surface is curved. Hence our line cannot be the rulers' edge or a stretched string. The most useful curve to choose would be the curve which connects two points by the shortest path. The shortest path between any two points on a sphere is the shorter arc of the great circle through these points.* Hence it seems reasonable to choose great circles on the sphere as our lines. A triangle on the sphere would be the figure formed by three arcs of three great circles (Fig. 26–6). A circle on the sphere would be the set of all points at a given distance from a fixed point (Fig. 26–7). (This distance between the fixed point and any point on the circle would, of course, be measured along the great circle through the two points.) We could, in fact, describe a variety of geometrical figures in terms of points and lines on the sphere.

The next step in the construction of our geometry would be to determine the axioms which our points, lines, and geometrical figures satisfy. Now our lines are great circles. These lines are not infinite in extent. However,

* The concept of great circle was explained in Chapter 7.

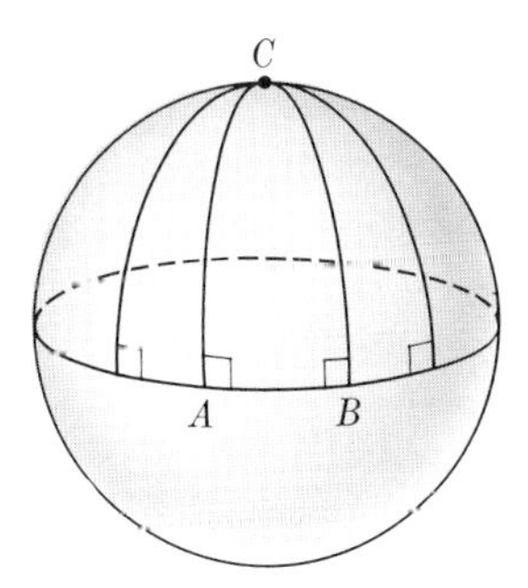

Fig. 26–8. The sum of the angles of a spherical triangle is greater than 180°.

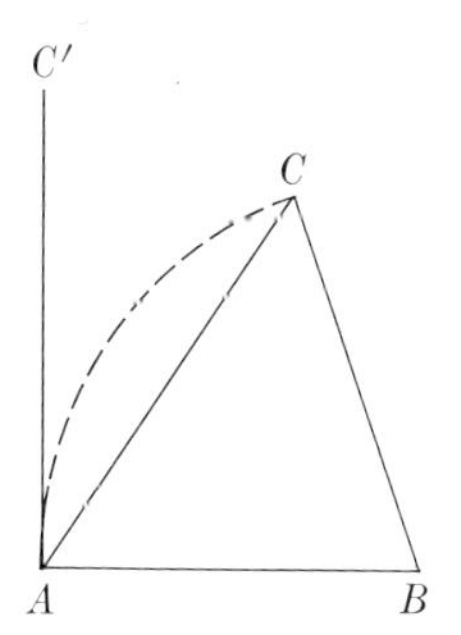

Figure 26–9

each is unbounded; that is, there is no beginning or end. Hence our axiom should assert merely that a line is unbounded. We note next that two points which are not on opposite ends of a diameter determine a unique great circle. However two points which are on opposite ends of a diameter do not determine a unique great circle. Consequently we cannot adopt the axiom that any two points determine a unique line. Let us look next into the subject of an axiom on parallel lines. Any two great circles on a sphere do meet; in fact, they meet at two points which are diametrically opposite. Hence the parallel axiom should read that any two lines meet; that is, there are no parallel lines. We shall not go further in selecting axioms for our system of geometry, for the conclusion toward which we are heading is now obvious. The axioms we would have to adopt to make our geometry fit the surface of the sphere would be exactly those which Riemann adopted for his non-Euclidean geometry.

Since the axioms of Riemann's geometry hold for the sphere, the theorems must also hold. Hence the geometry on the surface of a sphere is an application of Riemann's non-Euclidean geometry. Merely to satisfy ourselves, let us see whether a few theorems of Riemann's geometry really do apply. Let us consider the theorem that all perpendiculars to a line meet in a point. Figure 26–8 illustrates this theorem, and we see at once that it does apply. Another theorem asserts that the sum of the angles of a triangle is always greater than 180°. To verify that this theorem holds for every triangle on a sphere, we would have to call upon a theorem proved in spherical geometry which asserts that the sum of the angles of any spherical triangle is always between 180° and 540°. However, we can see at once in particular cases that Riemann's theorem does apply. Let us consider triangle ABC of Fig. 26–8. Angles A and B are each 90°. Hence, whatever the size of angle C, the sum of the three angles is larger than 180°. We could, of course, verify that all theorems of Riemann's geometry hold for the sphere.

The major point, then, which emerges from our discussion of the geometry on a sphere is that we have found a new use for Riemann's non-Euclidean geometry. It applies directly to the surface of a sphere, provided that straight line means great circle on the sphere.

Consideration of this last fact does lead to a question: Since Riemann's geometry applies so naturally to the surface of a sphere, why did not mathematicians hit upon non-Euclidean geometry in Greek times? Why did the realization that there can be non-Euclidean geometries take so long to strike home? The answer is that the Greeks, no doubt influenced by the Egyptians and Babylonians, had chosen the stretched string or ruler's edge as the physical straight line and the Euclidean axioms as the basis for their geometry. Such choices were very natural for people whose experiences were limited to a small part of the earth's surface. When they came to consider the sphere among the various surfaces, they had to approach it through concepts and axioms already adopted and hence described its properties in Euclidean terms. Thus great circles were treated as curves. As we have already noted in another connection, the Greeks and all mathematicians up to 1800 were so sure that Euclidean geometry was the true geometry of physical space that the idea of approaching the sphere directly and building a special geometry for it would have seemed nonsensical. They already possessed the true geometry and could not break what we can now see clearly was just a habit of thought.

It is possible to exhibit a surface to which the axioms and theorems of Gauss's non-Euclidean geometry apply, provided that the curves chosen as the lines on that surface are, as in the case of the sphere, those which connect two points by the shortest path. However, this surface is not widely used, and so we shall not devote any time to it. It will be more profitable to turn to the reconsideration of our familiar physical space.

We have grown accustomed to the idea that the reasonable and convenient physical interpretation of the straight line is the ruler's edge or a stretched string. Even with this interpretation we found that experience in limited regions did not enable us to exclude the non-Euclidean geometries as possible descriptions of physical space. We should now note that the ruler's edge and stretched string are, indeed, not the major physical interpretation of the mathematical straight line, and that we commonly and necessarily use another. Let us consider how surveyors determine distances. They begin by adopting a convenient base line AB (Fig. 26–9) whose length is measured by actually applying a tape measure. To determine the distance AC, a surveyor measures angle A by sighting point C in his telescope stationed at A and then swings the telescope around until he sights point B. On his theodolite he has a scale which tells him how much he has rotated his telescope, and hence he knows angle A. In a similar manner, he measures angle B. By means of trigonometry he can now calculate AC and BC. The surveyor proceeds on the assumption that the light rays which travel from C to A and from B to A follow the straight-line (stretched string) paths between those pairs of points, and, since the axioms of Euclidean geometry fit stretched strings, he applies Euclidean geometry or trigonometry to calculate AC and BC. However, the surveyor may be mistaken. The light ray from C to A may have followed the broken-line path shown in Fig. 26–9, and the surveyor at A would have to point his telescope tangentially to the light ray in order to receive the light. Hence the telescope would really be pointed to C' although the surveyor sees the point C in his telescope. Consequently, the angle he actually measures is $C'AB$ and not CAB. Thus the use of Euclidean geometry may have led to erroneous results for AC and BC.

What then should he have done? He should have applied a geometry whose axioms fit the behavior of light rays since these are the straight lines he really used. But do not light rays follow truly straight paths? We certainly know of situations in which they do not. We have had occasion to note (Chapter 1) that when light from the sun passes through the earth's atmosphere, it is bent by the refractive effect of the atmosphere. Hence there is some question of what path light rays do follow. Observation and measurement show that over short distances along or near the earth's surface light rays follow straight paths closely enough, but over larger distances this is certainly not true.

The above example may serve to suggest a major point. In astronomical measurements we necessarily depend entirely on light rays to measure angles. Since these rays travel over long distances, the paths they follow may not be truly straight, and we cannot check the true paths by laying down a tape measure or ruler. Hence we cannot be sure whether we should use Euclidean geometry.

We could try to solve this difficulty in one of two ways. We could attempt to determine by physical investigations what paths light rays follow, and, even though we may find that these are curved, treat them as if they were curves in Euclidean geometry. This is what physicists do when they study the behavior of light near the surface of the earth. The law of refraction tells us how light behaves, and all the reasoning in formulating and in applying this law is based on Euclidean geometry. Alternatively we can regard the paths of light rays as lines and construct a geometry which fits the convention that the light rays are to be the lines of that geometry. How to construct such a geometry is not obvious. One would have to determine some facts about the behavior of light rays, which would become the axioms of that geometry, and then deduce theorems. The resulting geometry might very well be non-Euclidean and might even differ from those developed by Gauss and Riemann.

All discussion of the *possibility* of applying non-Euclidean geometry is in a sense outdated. One of the basic current theories of science, the theory of relativity, does presuppose that our space is non-Euclidean and on this basis obtains better agreement between theory and experiment than the older theory of Newtonian mechanics based on Euclidean geometry was able to do. For reasons too lengthy to be discussed in this brief survey of the applicability of non-Euclidean geometry, the theory of relativity requires that the position as well as the time of events be treated together so that an object or event is described not only by the coordinates x, y, and z

which denote position, but also by the value of the time t at which the object or event occurs at position x, y, z. In other words, the relevant geometry is four-dimensional. This fact in itself does not imply that we must resort to a non-Euclidean geometry. However, in attempting to explain the phenomenon of gravitational attraction, it is necessary to regard the four-dimensional space-time as a nonhomogeneous geometry. This means that the nature of space-time varies from region to region, just as a two-dimensional mountainous surface changes its geometrical character from region to region. In the geometry of general relativity, it is also the presence of physical masses which determines the character of the geometry in any region, these masses being the earth, the moon, the sun, and other heavenly bodies. Moreover, the geometry is constructed in such a way that the "straight lines" are the shortest "paths" of that space-time. Light rays do take the shortest "paths" in the space-time system. Hence, while the straight line is not defined in terms of light rays in the non-Euclidean geometry of relativity, it is nevertheless significant that light rays do take such *paths*.

Exercises

1. Can you explain how it is that Euclidean geometry applies to a cylindrical surface even though the lines of that geometry do not have the shape of stretched strings?
2. Why must there be undefined concepts in the initial stages of any branch of mathematics?
3. Under what conditions does Riemann's non-Euclidean geometry prove to be the correct geometry on the surface of a sphere?
4. Should we distinguish between the geometries created by mathematicians and the geometry of physical space?
5. Does science really use the stretched string as its physical model of the mathematical straight line?
6. Imagine people living in a mountainous region who wish to construct a geometry for that region. They agree to consider the shortest path between two points as the line of that geometry. What kind of geometry might they arrive at?
7. Imagine people living in a mountainous region who wish to construct a geometry for that region. Since they travel by foot, using the shortest path to get from one place to another might require more time than some indirect path, because the shortest path may involve difficult mountain climbing. These people therefore agree to take as the line joining two points the path requiring least travel time. What kind of geometry might these people arrive at?

26–7 Non-Euclidean geometry and the nature of mathematics. The existence of non-Euclidean geometries which can fit physical space, to say nothing of the actual use of one of these non-Euclidean geometries in the theory of relativity, has had profound implications for mathematics itself, for science, and for some segments of our culture. In this section we shall discuss the implications for mathematics.

The most important effect of this creation has been the realization that mathematics does not offer truths. The Greeks adopted the axioms of Euclidean geometry because they believed that they were self-evident truths about our physical space. The axioms appealed to their minds as necessary truths which anyone must grant, even without experience. Since theorems are obtained by deductive reasoning and so are inescapable consequences of the axioms, the Greeks believed that the theorems, too, were truths. The observed agreement between the theorems deduced from these axioms and experience reinforced their certainty that the axioms were truths. The belief that mathematics offers truths was firmly held by every thinking being until the creation of non-Euclidean geometry. But if several geometries which contradict one another all fit physical space, then it becomes very obvious, indeed, that all of these cannot be the truth, and, worse yet, one can no longer be sure that any of these is the truth.

We see more clearly now that one must distinguish between mathematical space and physical space. Mathematicians and scientists believe that a physical world exists outside of and independently of human beings, and they seek to understand it by adopting axioms which seem to fit this physical space and then deducing theorems from these axioms. We now recognize that we have no reason to identify the mathematical construction with physical space. Indeed, several different mathematical theories may fit equally well. Mathematical theories of space are like any scientific theories; that is, the mathematical system used is the one which at the time best fits experience. If, as experience widens, it becomes clear that another geometry will fit experience better, then the older theory of space is discarded, and a new one adopted. This is precisely what happened when the theory of relativity sought to account for phenomena to an accuracy which surpassed that of the older scientific work.

The remarks we have made thus far about truth in mathematics and the provisional character of mathematical theories have been based on the developments in geometry. But the reader may have an objection. He may admit that geometry no longer offers truths, but continue to be convinced that our arithmetic, algebra, and other developments based on our number system do constitute truths. We shall devote Chapter 27 to this point and perhaps then see more clearly that with respect to truth the domain of number does not differ from geometry.

If the various branches of mathematics have only a more or less useful correspondence with physical experience, does then mathematics differ in any way from science? Mathematics had always been distinguished from science because mathematical axioms were regarded as truths, whereas the axioms of science were clearly recognized to be generalizations from limited experience or experiments. Scientists recognized that their researches produced only theories which did not provide a veridical description of what occurred in nature, but which might have to be altered to fit new facts. Is this not precisely what we now perceive to apply also to mathematics?

Insofar as the study of the physical world is concerned, mathematics has the same character as any of the sciences. It offers nothing but theories. And, as in science, new mathematical theories may replace older ones when experience or experiment shows that a new theory provides closer correspondence than an older one. As Einstein put it, "So far as the theories of mathematics are about reality, they are not certain; so far as they are certain, they are not about reality."

Yet there are basic distinctions between mathematics and science. Mathematics confines its work to numbers, geometrical figures, and relationships which obtain among numbers and among geometrical figures. All other concepts and relationships of mathematics are derived from numbers and geometry. Science deals with mass, velocity, force, energy, molecular structure, chemical processes, the structure of plants, animals, and humans, and hundreds of other concepts. That is, the subject matter of mathematics differs from that of science.

A second difference is that mathematics will always insist on deductive proof, whereas the sciences, even though they aim to be deductive, will continue to utilize any experimental or observational fact as a basis for conclusions; that is, the sciences do not insist on a thoroughly deductive structure based on a fixed number of axioms stated at the outset. There are many hypotheses of mathematics, such as, for example, that every even number is the sum of two primes, for which the inductive evidence is most conclusive. No scientist would hesitate to use an assertion so well supported by evidence. But the mathematician continues to search for a deductive proof. This difference between mathematics and science is perhaps one of degree or method of operation. Having chosen his axioms, the mathematician proceeds to derive as many conclusions as possible from them. The scientist will not hesitate to introduce new axioms if such a step seems warranted by inductive evidence.

A third difference between mathematics and science was, peculiarly, accentuated by the creation of non-Euclidean geometry. Mathematics, like science, had been devoted primarily to the exploration of nature. Yet mathematicians had always felt free to develop the implications of the axioms of number and Euclidean geometry, although there was no immediate application for any of the results pursued. Number and Euclidean geometry were regarded as so important in the study of nature that almost any information about them was welcomed. However, non-Euclidean geometry, which, at the outset and for a long time thereafter, seemed to concern axioms which could not possibly apply to the physical world, had finally proved useful in the study of the physical world. Thus history teaches us that mathematicians should feel free to investigate axioms which have no immediate or obvious bearing on the physical world. Consequently, mathematics has been given a new dimension of freedom, the freedom to explore what the mind wishes to explore, and has been released from bondage to the axioms of number and Euclidean geometry. One could say that the creation of non-Euclidean geometry had the effect of divorcing mathematics from science. By the end of the nineteenth century, Georg Cantor, one of the great mathematical minds of modern times and the creator of a strange and revolutionary theory, the theory of transfinite numbers, was able to say, "The essence of mathematics is its freedom." The enormous expansion in mathematical activity in the last century is partly the consequence of the new freedom.*

Exercises

1. Why does the existence of non-Euclidean geometry show that mathematics does not offer truths?
2. What distinction should one make between a mathematical space and physical space?
3. Is it proper to regard mathematics as one of the sciences?

26–8 The implications of non-Euclidean geometry for other branches of our culture. In view of the rôle which mathematics plays in science and the implications of scientific knowledge for all of our beliefs, revolutionary changes in man's understanding of the nature of mathematics could not but mean revolutionary changes in his understanding of science, doctrines of philosophy, religious and ethical beliefs, and, in fact, all intellectual disciplines.

Let us consider first the effect on scientific thought. Although scientists had more or less recognized that their theories in various branches of science were not the final word, in the back of their minds they continued to believe that true accounts of the various phenomena of nature were possible and that they were working toward these goals. Indeed, in the fields of astronomy and mechanics, the eighteenth-century thinkers proclaimed with certainty that they had found the true laws of nature. The influence of Newtonian mechanics on almost all thought was profound, precisely because the intellectual leaders of the eighteenth century were convinced that the mathematical account of nature's behavior was correct. The creation of non-Euclidean geometry affected scientific thought in two ways. First of all, the major facts of mathematics, i.e., the axioms and theorems about triangles, squares, circles, and other common figures are used repeatedly in scientific work and had been for centuries accepted as truths—indeed, as the most accessible truths. Since these facts could no longer be regarded as truths, every conclusion of science which depended upon strictly mathematical theorems also ceased to be truths. Or, to broaden our statement, since scientific structures were and are in large part just series of mathematical chains of reasoning, the appearance of non-Euclidean geometries raised doubts about the very framework of these structures.

* However, see Section 31–5.

Secondly, the debacle in mathematics led scientists to question whether man could ever hope to find a true scientific theory. The Greek and Newtonian views put man in the role of one who merely uncovers the design already incorporated in nature. However, scientists have been obliged to recast their goals. They now believe that the mathematical laws they seek are merely approximate descriptions and, however accurate, no more than man's way of understanding and viewing nature.

Even on the level of engineering a serious question emerged. Since bridges, buildings, dams, and other works were based on Euclidean geometry, was there not some danger that these structures would collapse? Actually there is no guarantee that they will not. But this thought did not alarm the scientists and engineers of the nineteenth century, who, despite the existence of non-Euclidean geometry, did not believe that the geometry of physical space could be other than Euclidean. The other geometries they dismissed as logical curiosities. The behavior of these scientists and even mathematicians illustrates what has been called the law of inertia in the world of ideas. Just as a body at rest or in motion exhibits inertia or unwillingness to change its velocity, so do human beings balk at changing their ideas. However, the advent of the theory of relativity drove home the point that Euclidean geometry is not necessarily the best geometry for applications. Why then do engineers continue to use Euclidean geometry for ordinary projects? They do so because on the basis of experience Euclidean geometry has been known to be reliable. This is their only assurance. For engineering involving motion with high velocities such as modern accelerators of electrons or neutrons develop, the theory of relativity is used.

In the realm of philosophy, all doctrines built on science were necessarily affected. The most majestic development of the seventeenth and eighteenth centuries, Newtonian mechanics, fostered and supported the view that the world is designed and determined in accordance with mathematical laws. The discovery of more laws in fields such as electricity and light during the early nineteenth century reinforced the belief in a highly mechanistic and deterministic universe. But once non-Euclidean geometry destroyed the belief in mathematical truth and revealed that science offered merely theories about how nature might behave, the strongest reason for belief in determinism was shattered.

Perhaps even more devastating to philosophy was the realization that man can no longer be sure of his ability to acquire truths. Through philosophy man has sought knowledge of ultimate realities, knowledge which would enable him to live wisely, and knowledge which would answer irrepressible questions about the meaning and purpose of his existence on this earth. All people, prior to non-Euclidean geometry, had shared the fundamental belief that man can obtain certainties. The solid basis for this belief had been that man had already obtained some truths—witness, mathematics. No system of thought has ever been so widely and completely accepted as Euclidean geometry. To preceding generations it was the "rock of ages" in the realm of truth. Tradition buttressed self-evidence, and experience bolstered "common sense." Men such as Plato and Descartes were convinced that mathematical truths were innate in human beings. Kant based his entire philosophy on the existence of mathematical truths. But now philosophy is haunted by the specter that the search for truths may be a search for phantoms.

The implication of non-Euclidean geometry, namely, that man may not be able to acquire truths, affects all thought. Past ages have sought absolute standards in law, ethics, government, economics, and other fields. They believed that by reasoning one could determine the perfect state, the perfect economic system, the ideals of human behavior, and the like. The standards sought were not just the most effective ones, but the unique, the correct ones. This belief in absolutes was based on the conviction that there were truths in the respective spheres. But in depriving mathematics of its claim to truth, the non-Euclidean geometries destroyed the shining knight of truth and shattered man's hope of ever attaining any truths. When the anchor of truth was lost, all bodies of knowledge were cast adrift. Apparently the intellectual process does not lead to certainties. In Henri Bergson's words, "One can always reason with reason."

Our own century is the first to feel the impact of non-Euclidean geometry because the theory of relativity brought it into prominence. It is very likely that the abandonment of absolutes has seeped into the minds of all intellectuals. We no longer search for the ideal political system or ideal code of ethics but rather for the most workable. It is almost commonplace to hear people say that one cannot expect perfection. This attitude contrasts sharply with those of the eighteenth century and the Victorian age.

Perhaps the greatest import of non-Euclidean geometry is the insight it offers into the workings of the human mind. No episode of history is more instructive. The evaluation of mathematics as a body of truths, which obtained prior to non-Euclidean geometry, was accepted at face value by every thinking being for 2000 years, in fact, practically throughout the entire existence of Western culture. This view, of course, proved to be wrong. We see therefore, on the one hand, how powerless the mind is to recognize the assumptions it makes. It would be more appropriate to say of man that he is surest of what he believes, than to claim that he believes what is sure. Apparently we should constantly re-examine our firmest convictions, for these are most likely to be suspect. They mark our limitations rather than our positive accomplishments. On the other hand, non-Euclidean geometry also shows the heights to which the human mind can rise. In pursuing the concept of a new geometry, it defied intuition, common sense, experience, and the most firmly entrenched philosophical doctrines just to see what reasoning would produce.

Exercises

1. What would you regard as the most serious implication of the creation of non-Euclidean geometry?
2. How does the existence of non-Euclidean geometry affect the goals of scientists?
3. Develop the analogy between different systems of geometry and different bodies of law.
4. Does the existence of non-Euclidean geometry augment the power of science to provide rational comprehension of natural phenomena?

Topics for Further Investigation

1. The history of attempts to find a simpler parallel postulate.
2. The work of Girolamo Saccheri.
3. The life of Carl Friedrich Gauss.
4. The use of non-Euclidean geometry in the theory of relativity.

Recommended Reading

Bell, Eric T.: *Men of Mathematics,* Chaps. 14 and 16, Simon and Schuster, Inc., New York, 1937.

Bonola, Roberto: *Non-Euclidean Geometry, A Critical and Historical Study of its Development,* Dover Publications, Inc., New York, 1955.

Carslaw, H. S.: *Non-Euclidean Plane Geometry and Trigonometry,* Chelsea Publishing Co., New York, 1959.

Durell, Clement V.: *Readable Relativity,* G. Bell and Sons Ltd., London, 1931.

Frank, Philipp: *Philosophy of Science,* Chap. 3, Prentice-Hall, Inc., Englewood Cliffs, N.J., 1957.

Gamow, George: *One Two Three . . . Infinity,* Chaps. 4 and 5, The New American Library, Mentor Books, New York, 1947.

Kline, Morris: *Mathematics in Western Culture,* Chap. 27, Oxford University Press, New York, 1953.

Poincaré, Henri: *Science and Hypothesis,* Chaps. 3 to 5, Dover Publications, Inc., New York, 1952.

Russell, Bertrand: *The ABC of Relativity,* Harper and Bros., New York, 1926.

Sommerville, D. M. Y.: *The Elements of Non-Euclidean Geometry,* Dover Publications, Inc., New York, 1958.

Wolfe, Harold E.: *Introduction to Non-Euclidean Geometry,* The Dryden Press, New York, 1945.

Young, Jacob W. A.: *Monographs on Topics of Modern Mathematics,* Chap. 3, Dover Publications, Inc., New York, 1955.

CHAPTER 27

ARITHMETICS AND THEIR ALGEBRAS

And wisely tell what hour o' the day
The clock doth strike, by Algebra
Samuel Butler

27–1 Introduction. We have seen that mathematics contains several geometries and that the very existence of these geometries has profound implications for the nature of geometry and for the relationship of geometry to our physical world. It is therefore only natural to ask whether there are also many algebras, and whether the existence of these algebras has comparable implications for mathematics and its relation to the physical world. This question is vital not only because algebra plays a most important role in physical applications but because, after non-Euclidean geometry had taught mathematicians that geometry does not offer truths, many turned to the ordinary number system and the developments built upon it and maintained that this part of mathematics still offered truths. The same thought is often expressed today by people who, wishing to give an example of an unquestionable truth, quote $2 + 2 = 4$.

Examination of the relationship between our ordinary number system and the physical situations to which it is applied will show that it does not offer truths. We shall then see that other algebras do exist and are useful, just as non-Euclidean geometries are useful.

27–2 The applicability of the real number system. Mathematicians are, of course, free to introduce the symbols $1, 2, 3, 4, \ldots$, where 2 means $1 + 1$, 3 means $2 + 1$, 4 means $3 + 1$, and so on. Moreover, as we pointed out in Chapter 4, experience suggests that for any three numbers a, b, and c, $(a + b) + c = a + (b + c)$, and so this associative property is adopted as an axiom. We may now prove readily that $2 + 2 = 4$ because, first of all, by the very meaning of 2, we have

$$2 + 2 = 2 + (1 + 1).$$

From the associative axiom it follows that

$$2 + (1 + 1) = (2 + 1) + 1.$$

According to the definition of 3,

$$(2 + 1) + 1 = 3 + 1,$$

and, according to the definition of 4,

$$3 + 1 = 4.$$

And now, by applying the axiom that things equal to the same thing are equal to one another, we can assert that $2 + 2 = 4$.

Thus by a purely logical process which employs definitions and axioms we have proved that $2 + 2 = 4$. But the question we seek to answer is not whether the mathematician can set up definitions and axioms and deduce conclusions. We grant that arithmetic is a valid deductive system. We wish to know whether this system necessarily expresses truths about the physical world.

A possible denial might be entered at once on the ground that the only justification for the associative axiom was limited experience with simple numbers, whereas the axiom asserts something about all whole numbers. There is force to this argument because a generalization on the basis of limited experience may be erroneous. However, there are many thinkers who would assert that whether or not the associative axiom was suggested by experience, it is clearly a truth. Of course, the burden of proof then rests on those who proclaim truths. We shall not insist on this point, since there are weaker links between arithmetic and the physical world.

If a farmer has two herds consisting of 10 and 25 heads of cows, respectively, he knows by adding 10 and 25 that the total number of cows is 35. That is, he need not count the cows. Suppose, however, he brings the two herds of cows to market where they are selling for $100 apiece. Will a herd of 10 cows which might bring in $1000 and a herd of 25 cows which might bring in $2500 together bring in $3500? Every businessman knows that when supply exceeds demand, the price may drop, and hence 35 cows may bring in only $3000. In some idealized world the value of the cows may continue to be $3500, but in actual situations this need not be true.

Let us consider next whether some of the slightly deeper results of arithmetic apply to the physical world. Certainly the statement that $2 \cdot \frac{1}{2} = 1$ is arithmetically correct. But do two half-sheets of paper make one whole sheet and do two half-shoes make one whole shoe? Clearly two physical halves never make one whole unless they can be joined in such a way that the halves merge into one whole. Two half-dollars, in general, equal one whole dollar in purchasing power, but in areas where silver is preferred to bills two half-dollars are worth more. To know whether the arithmetic is applicable, we must examine the physical situation.

Let us consider the addition of velocities. If a river flows at the rate of 3 miles per hour and a man capable of rowing at 5 miles per hour in still water rows downstream, his velocity relative to some fixed point in the river is the sum of 3 and 5, that is, 8 miles per hour. But if Mr. A walks along a road at the rate of 3 miles per hour and Mr. B walks along at the rate of 5 miles per hour, then B's velocity relative to some fixed point is not 8 miles per hour. Of course not, we would say. But why do we add the 3 and 5 in one case and not in the other? It is the physical situation which tells us when to add and when not to do so.

If two forces, one of 3 and the other of 4 pounds, act on an object, is the object acted on by a total force of 7 pounds? The answer, as we know from our work in an earlier chapter, depends upon the directions of these forces. If they act at right angles (Fig. 15–5), then the resultant is merely 5 pounds. Here, too, ordinary arithmetic fails to represent the combined action of the two forces.

In all of the above examples we must examine the particular physical situation to determine whether the mathematical result fits. But if we must resort to experience to decide when the results of our arithmetic apply, then it is not the truth of arithmetic on which we rely.

Let us test further the applicability of arithmetic. Suppose that we measure two boards and find them to be 3 and 4 feet long, respectively. If we place these boards end to end, will the result be 7 board feet? Probably not. All measurement is approximate, and our statement that the individual boards are 3 and 4 feet long merely means that we are unable to detect any difference between the actual lengths of the boards and the 3- and 4-foot marks on our measuring device. But the first may be 3.01 feet, and the second 4.01 feet. Together they are then 7.02 feet, and we may be able to detect a difference of 0.02 feet. One may object here and say that the trouble is due to the limitations of our senses. This is indeed true; however, can we continue to claim that $3 + 4 = 7$ applies to the physical world, insofar at least as situations involving measurement are concerned?

We learn in chemistry that when one mixes hydrogen and oxygen, he obtains water. More precisely, if one takes 2 volumes, say 2 cubic centimeters, of hydrogen and 1 volume of oxygen, one obtains 2 volumes of water vapor. Likewise 1 volume of nitrogen and 3 volumes of hydrogen yield 2 volumes of ammonia. We happen to know the physical explanation of these surprising arithmetic relationships. By Avogadro's hypothesis, equal volumes of any gas, under the same conditions of temperature and pressure, contain the same number of *particles*. If, then, a given volume of oxygen contains 10 molecules, the same volume of hydrogen will also contain 10 molecules. Then there are 20 molecules in 2 volumes of hydrogen. Now it happens that the molecules of oxygen and hydrogen are diatomic; that is, each contains two atoms. Each of these 20 diatomic hydrogen *molecules* combines with one *atom* of oxygen to form 20 molecules of water or 2 volumes of water.* The chemistry is interesting, but the main point we wish to make is that ordinary arithmetic fails to describe correctly the result of combining gases by volume.

Suppose, next, that one raindrop is added to another raindrop. Do we now have two raindrops? If one cloud is joined to another cloud do we now have two clouds? One may protest that in these examples the merged objects have lost their identity, and that the addition process of arithmetic

* This phenomenon is very clearly explained in Francis T. Bonner and Melba Phillips: *Principles of Physical Science*, Addison-Wesley Publishing Co., Inc., 1957, p. 149.

does not contemplate such loss. And precisely for this reason, arithmetic in the normal sense no longer applies.

All of the above examples lead to two general conclusions. One is that there are many physical situations where ordinary arithmetic does not apply; that is, ordinary arithmetic is unable to express proper quantitative truths about these situations. The second conclusion is that even though there are a few situations to which ordinary arithmetic does apply, such as, for example, adding herds of cattle, we must depend upon experience with those very situations to know this fact. If herds of cattle behaved like volumes of gases or like raindrops, then the arithmetic would not apply, and it is only through experience that we learn how they do behave. Hence we have no guarantee that arithmetic *per se* represents truths about the physical world.

Exercises

1. If we place the length of one 10-ft ladder on top of the length of another, do we obtain a 20-ft ladder? What is the point of the question?
2. Since measurement is approximate, can we say that pouring two 10-lb packages of flour into one bag will produce one 20-lb package of flour?
3. If we balance two objects in the pans of a scale and then add 5 lb to each pan, will the scale still balance? What axiom of arithmetic is applicable to this situation?
4. If an object is thrown downward with a velocity of 100 ft/sec and acquires a velocity due to gravity of $32t$ ft/sec, is its total downward velocity $(100 + 32t)$ ft/sec? Justify your answer.
5. If we superimpose a sinusoidal sound wave of frequency 100 cycles per second on one which has a frequency of 50 cycles per second, do we obtain a sound wave of frequency 150 cycles per second?
6. If one mixes two equal volumes of water, one having a temperature of 40°F and the other of 50°F, what is the temperature of the mixture?

27–3 Modular arithmetics and their algebras. Mathematics does not have at its disposal special arithmetics to treat all of the situations in which ordinary arithmetic fails. For example, there is no arithmetic which tells us how volumes of gases combine. Each individual combination must be analyzed on the basis of physical knowledge about the molecules involved. But there are situations which warrant the introduction of special arithmetical concepts and operations. If an arithmetic, that is the concepts and operations, accurately describes physical events and permits prediction of future behavior, just as ordinary addition predicts the result of combining two herds of cattle, then it is worth creating.

One such arithmetic is suggested by our system of recording the time of day. Six hours after 10 o'clock is not 16 o'clock but 4 o'clock; that is, in this system,

$$10 + 6 = 4.$$

Similarly, 6 hours before 3 o'clock is 9 o'clock; that is,

$$3 - 6 = 9.$$

The idea which this system of telling time suggests is that if two numbers differ by 12 or a multiple of 12, then they are equal. Thus $26 = 2$ because $26 - 2 = 2 \cdot 12$, and $9 = -3$ because $9 - (-3) = 12$. Clearly the equality sign here does not mean the same as in ordinary arithmetic, and hence we use the symbol $\equiv$ and write our new equations, which are now called *congruences,* as follows:

$$26 \equiv 2, \text{ modulo } 12;$$
$$9 \equiv -3, \text{ modulo } 12.$$

The phrase "modulo 12" after each equation repeats, in shorthand form, the condition stated above, namely, that the equation holds if and only if we neglect multiples of 12.

In this arithmetic, which is usually limited to whole numbers, any number larger than 12 is congruent to some number less than 12 because we can always subtract from the larger number some multiple of 12 to obtain a number less than 12. Thus, if we start with 35, we can subtract $2 \cdot 12$ and obtain 11. Then

$$35 \equiv 11, \text{ modulo } 12.$$

The number 12 itself is congruent to 0 because $12 - 0 = 1 \cdot 12$. Hence

$$12 \equiv 0, \text{ modulo } 12.$$

Similarly, any negative number is congruent to some positive number less than 12. For example, $-25 \equiv 11$ because $11 - (-25) = 3 \cdot 12$, or if one prefers, $-25 - 11 = -3 \cdot 12$. Thus in the arithmetic modulo 12 we need deal only with positive integers from 0 to 11. We also regard any positive integer less than 12 to be congruent to itself because the difference of the two integers is zero times 12. Thus

$$7 \equiv 7, \text{ modulo } 12.$$

Let us see what the results of simple addition and multiplication are in this modular arithmetic. For example,

$$9 + 6 \equiv 3, \text{ modulo } 12.$$

Also

$$9 + 3 \equiv 0, \text{ modulo } 12,$$

and

$$9 \times 4 \equiv 0, \text{ modulo } 12.$$

Here, then, is an arithmetic in which the sum and the product of two positive numbers can be zero, although the summands and the factors are not zero.

What we have done modulo 12 we could do modulo 4, and it would follow that $2 + 2 = 0$. Perhaps this equation will disturb some individuals, but whether or not it does, we see again that numbers are concepts invented by man to represent relations that he chooses to work with. If we design an arithmetic to fit the process of combining herds of cows, we end up with the abstract relation $2 + 2 = 4$; but for such purposes as telling time, for example, quite another arithmetic applies.

To study the properties of congruences in a systematic way, the mathematician tries to be general and to prove some fundamental theorems. It is at this point that modular *algebra* enters the scene. We shall present only two of these general theorems. The first theorem (the letters stand for integers) says that if

$$a \equiv b, \text{ modulo } m, \tag{1}$$

and

$$c \equiv d, \text{ modulo } m, \tag{2}$$

then

$$a + c \equiv b + d, \text{ modulo } m. \tag{3}$$

In other words, we may add congruences modulo m just as we may add ordinary equations.

The second theorem states that we may multiply congruences; that is, given the congruences (1) and (2) above, it follows that

$$ac \equiv bd, \text{ modulo } m. \tag{4}$$

We shall not prove these theorems. They are introduced here as examples of theorems in modular algebra.

We can use these two simple theorems on addition and multiplication of congruences to make an application to ordinary arithmetic. We know that

$$10 \equiv 1, \text{ modulo } 9. \tag{5}$$

Multiplying this congruence by itself, we obtain

$$100 \equiv 1, \text{ modulo } 9. \tag{6}$$

Multiplication of the two above congruences yields

$$1000 \equiv 1, \text{ modulo } 9. \tag{7}$$

Obviously we could continue to higher and higher powers of ten.

Now let us consider any number, say 457. This number actually is $4 \cdot 100 + 5 \cdot 10 + 7$. We may certainly state that

$$7 \equiv 7, \text{ modulo } 9. \tag{8}$$

Since

$$5 \equiv 5, \text{ modulo } 9, \tag{9}$$

we can multiply equations (9) and (5) and obtain

$$5 \cdot 10 \equiv 5, \text{ modulo } 9. \tag{10}$$

Similarly, from $4 \equiv 4$, modulo 9, and from (6), we obtain by multiplication of congruences:

$$4 \cdot 100 \equiv 4, \text{ modulo } 9. \tag{11}$$

Statement (3) says that we can add congruences, modulo the same number; hence, let us add the congruences (8), (10), and (11). The result is

$$4 \cdot 100 + 5 \cdot 10 + 7 \equiv 4 + 5 + 7, \text{ modulo } 9,$$

or

$$457 \equiv 4 + 5 + 7, \text{ modulo } 9.$$

What this result says is that a number and the sum of its digits are congruent. Hence *a number minus the sum of its digits must be a multiple of* 9. To test this statement for the number 457 itself, we note that $457 - (4 + 5 + 7)$, or 441, is 49 times 9.

This result is, of course, of interest to those who like to play with numbers, but it also provides a useful method of checking the ordinary arithmetic operations of addition, subtraction, and multiplication. For example, let us consider the product of 457 and 892. We know that

$$457 \equiv 4 + 5 + 7, \text{ modulo } 9,$$

and

$$892 \equiv 8 + 9 + 2, \text{ modulo } 9.$$

Moreover, $4 + 5 + 7$, or 16, is congruent to 7, modulo 9, and $8 + 9 + 2$, or 19, is congruent to 1, modulo 9. Hence

$$457 \equiv 7, \text{ modulo } 9,$$

and

$$892 \equiv 1, \text{ modulo } 9.$$

In view of (4), we may multiply these congruences and state that

$$457 \cdot 892 \equiv 7 \cdot 1, \text{ modulo } 9.$$

Then the product of 457 and 892 is congruent to 7, modulo 9. That is, the actual product and the product of the sums of the digits in the two factors are congruent. But the actual product is also congruent to the sum of its digits. Hence the sum of the digits in the product is congruent to the product of the sums of the digits in the factors. We have therefore a *check* on the correctness of the multiplication, which is known as the rule for casting out nines.* But we did not prove and so cannot conclude that if the congruence does hold, then the multiplication is correct.

---

* If two numbers are congruent modulo 9, then they must have the same remainders upon division by 9, for the difference must be an exact multiple of 9. Hence the rule for casting out nines is sometimes stated as follows: the product of two numbers and the product of the sums of the digits must have the same remainder upon division by 9.

The theorems concerning addition and multiplication of congruences are entirely analogous to the theorems on equalities. However, the algebra of congruences differs in many respects from the algebra of ordinary numbers. Let us consider a simple illustration. Suppose we try to solve the equation

$$3x \equiv 4, \text{ modulo } 5. \tag{12}$$

In ordinary algebra, $x = 4/3$. This answer is also correct in modular algebra if we wish to use fractions, but there are integral solutions too. For example, 3, 8, 13, and in fact all numbers obtained by adding any multiple of 5 to 3 are possible values of x. Hence we have an unending number of solutions. But if we restrict ourselves to positive integral solutions less than 5, then there is but one answer.

Let us consider next the equation

$$x^2 \equiv 1, \text{ modulo } 8. \tag{13}$$

In ordinary algebra, we have two solutions, namely $+1$ and -1. In the modular algebra, there are four positive integral solutions which are less than 8, namely 1, 3, 5, 7, to say nothing of those greater than 8.

We see therefore that over and above the arithmetic and algebra of ordinary numbers there exist the modular arithmetics and their algebras. The latter are studied in a branch of mathematics called the theory of numbers and, as we have seen, are applicable to ordinary arithmetic.

Exercises

1. Make up the addition table for the arithmetic, modulo 6.
2. Make up the multiplication table for the arithmetic, modulo 6.
3. By trials with actual numbers decide whether it is possible to subtract any whole number from another and obtain an answer in the arithmetic, modulo 6.
4. Are there any answers to the problem of dividing 4 by 2 in the arithmetic, modulo 6? of dividing 3 by 2?
5. Answer the same questions as in Exercise 4, but applied to the arithmetic, modulo 5. Does any significant conclusion suggest itself from a comparison of the answers in this exercise with those of Exercise 4?
6. Solve the equation $x + 5 \equiv 2$ in the arithmetic, modulo 12.
7. Prove that the theorem asserted in (3) of the text is correct. [*Suggestion:* Since $a \equiv b$, modulo m, $a = b + pm$.]
8. Prove that if $a \equiv b$, modulo m, and $c \equiv d$, modulo m, then $a - c \equiv b - d$, modulo m.
9. Prove that if $a \equiv b$, modulo m, and $c \equiv d$, modulo m, then $ac \equiv bd$, modulo m.
10. We know that $16 \equiv 4$, modulo 6, and $4 \equiv 4$, modulo 6. By dividing the first congruence by the second one, we obtain $4 \equiv 1$, modulo 6. What conclusion do you draw?
11. Check the addition of 578 and 642 by the "rule" of casting out nines.
12. What is the full set of positive integral solutions to $x^2 \equiv 1$, modulo 8?
13. What is the analogue for congruences of the usual axiom that things equal to the same thing are equal to each other?

27-4 The algebra of sets. We shall now examine still another algebra. Suppose that a man has inherited two different libraries of books. It would be very natural for him to merge them into one library in which, contrary to the case of inheriting dollar bills, he would not want any duplicates. He is, then, combining the two libraries and hence, mathematically speaking, he is adding one to the other. However, since he will reject any duplicates, he will not add in the usual arithmetic sense. For example, if there are 100 books in one library and 200 in the other, the combined library may contain fewer than 300 books. Indeed, if there are 50 titles in one of the collections that duplicate titles in the other, the unified library will contain only 250 books. Thus, to represent the operation of combining two libraries, we need an addition which permits $100 + 200$ to equal 250.

Mathematicians have devised an arithmetic and algebra whose addition processes represent precisely what happens in the combining of the two libraries. The system is called the algebra of sets. The arithmetic and algebra are both so simple that we may as well discuss the algebra at once.

Let A and B be any two sets of objects. Thus A and B might be the two libraries discussed above. To indicate the addition of B to A in the sense in which the libraries were to be joined, that is, an object common to A and B is to be taken only once, we write $A + B$. It must, of course, be understood that the plus sign here does not mean the same as the plus sign in $5 + 3$. We shall see in a moment, however, why we use the same sign in both cases. The addition we have just introduced implies: a book is in $A + B$ if it is in A, or in B, or in both, but if it is in both, it is counted only once.

Before proceeding, let us illustrate what the new operation of addition amounts to. Let us suppose that the books in library A are represented as points, and that the entire library consists of all points inside some curve. The set of books in library B can also be represented as the set of points inside some other curve (Fig. 27-1). Since the two libraries contain duplicate titles, these two regions will overlap. The sum $A + B$ in the sense defined above will then be represented by the collection of points inside both curves, i.e., by the entire shaded area in Fig. 27-1. Of course, the points common to the two regions, the crosshatched area in Fig. 27-1, count only once in the sum, but this just means that the sum is represented by the points in the shaded area.

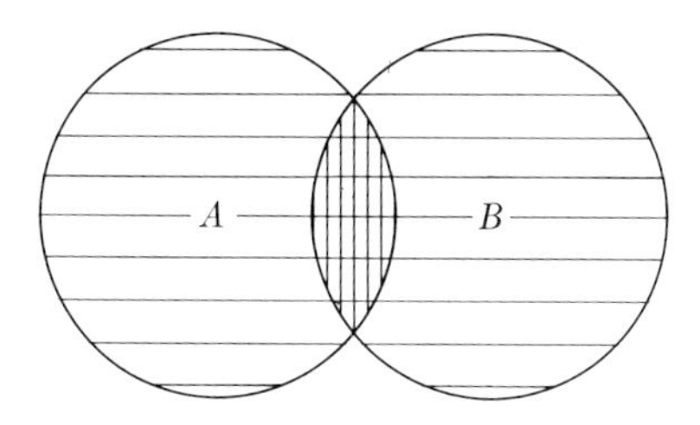

FIG. 27-1. The sum of two sets.

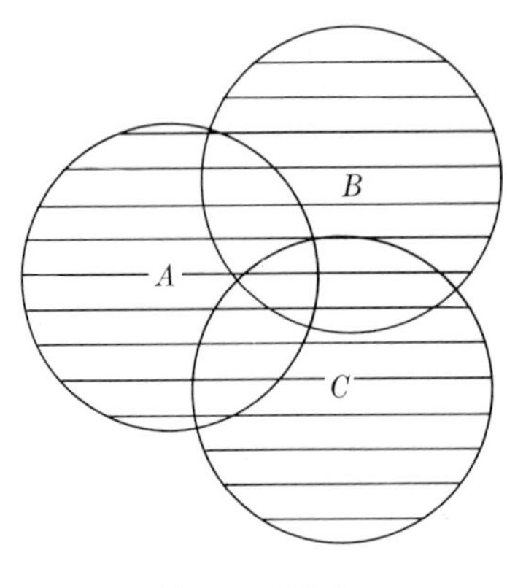

Figure 27–2

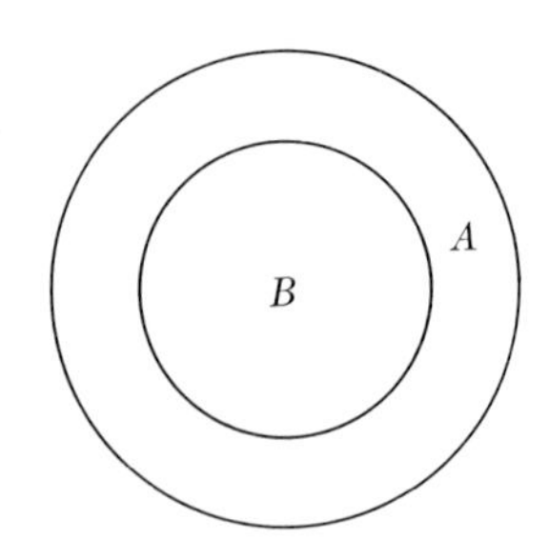

Fig. 27–3. $A + B = A$.

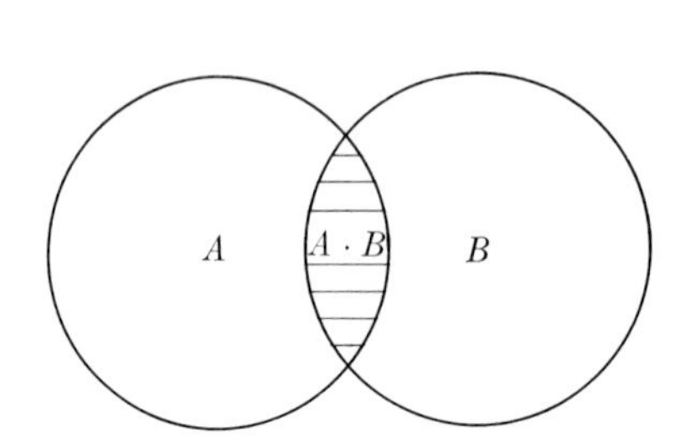

Fig. 27–4. $A \cdot B$.

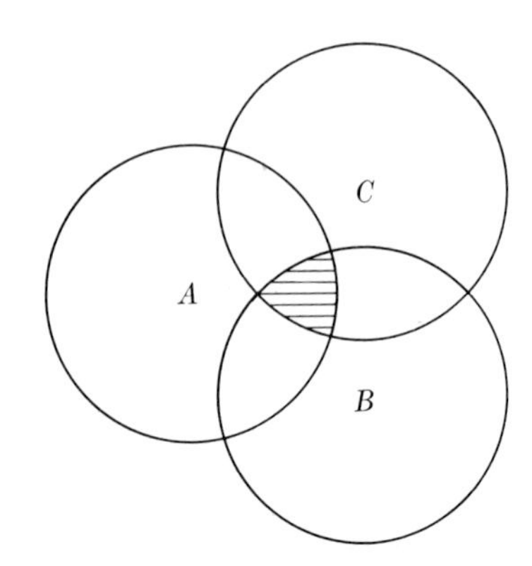

Figure 27–5

We thus have a new concept, the addition of sets. Let us now return to the matter of notation and explain why we use the plus sign of ordinary addition to denote addition in the algebra of sets. If there were no duplicate titles in the two libraries, then the sum would certainly be of the same kind as the sum in the usual sense of arithmetic. Moreover, just as in the case of ordinary addition, for any A and B,

$$A + B = B + A$$

because the same collection is formed whether we add collection B to collection A or A to B, that is, the commutative property of addition (Chapter 4) holds for this new concept of addition.

Further, the associative property holds for the addition of sets. Thus if we have three collections, A, B, and C, we may form $A + B$ and then add C to $A + B$ to obtain $(A + B) + C$, or we may begin with $B + C$ and add this sum to A to obtain $A + (B + C)$. Both procedures yield the same total collection. In other words, for this new concept of addition, we have

$$(A + B) + C = A + (B + C),$$

whether or not the sets overlap (Fig. 27–2).

It is because these familiar properties of commutativity and associativity hold for the new concept of addition that we are satisfied to use the term addition and the usual plus sign, although the operation of combining the two libraries while rejecting duplicate titles is not the same as the usual combination of collections of cows or dollars. We should, however, note that there are essential differences between the new and old concepts of addition. Suppose, for example, that all titles in library B are also in A. Then our new concept of addition requires that (Fig. 27–3)

$$A + B = A.$$

Moreover, if we add collection A to itself, we obtain A. That is

$$A + A = A.$$

The algebra of sets also has a concept of multiplication, or product. Suppose that the man combining the two libraries were interested in determining how many titles were common to the two libraries. For example, he might wish to know how many books can be sold. The set of titles common to the two libraries A and B is called the product (or sometimes intersection) of A and B and is denoted by $A \cdot B$ or simply, as in ordinary algebra, by AB. If the books in the sets A and B are again pictured as the points inside two curves (Fig. 27–4), then the product is represented by the area common to the sets of points A and B, the shaded area in the figure.

This concept of product is quite different from the ordinary meaning of the term in that multiplication in set algebra usually produces a much smaller set than is contained in either of the factors A or B. Yet we use the term product and the usual symbolism for multiplication, $A \cdot B$ or AB, because, as in the case of addition, the fundamental properties of the product for sets are the same as for ordinary multiplication. For example, it is certainly true for the product of sets that

$$AB = BA$$

because the same final set, for example, the books common to both libraries, is obtained whether we consider the objects common to A and B or to B and A. Similarly, it is true that

$$(AB)C = A(BC),$$

for, if we select the objects common to A and B (Fig. 27–5) and then those common to AB and C, we surely obtain the same set as if we selected the objects common to B and C and then the titles common to A and BC. In either case, we obtain the set of objects in all three sets, A, B, and C. Thus the commutative and associative properties of multiplication hold for this new concept of product.

There are, however, essential conceptual differences between the product for sets of objects and the product for ordinary numbers. If in our library example the set B consists of titles which are all in A, then (Fig. 27–3)

$$AB = B,$$

because the books common to A and B are those in B. Also, the product of A and A is A; that is,

$$AA = A.$$

Finally, suppose that A and B have no objects in common. What is the set of objects common to A and B? Physically, there is none. Mathematically we introduce the symbol 0 to represent an empty set of objects and write

$$AB = 0.$$

The symbol 0 possesses many of the usual properties of the number zero. Thus, by the above definitions of sum and product for sets, it is true that

$$A + 0 = A$$

and

$$A0 = 0.$$

There are other interesting concepts and operations in the algebra of sets. For example, one may be concerned with all books that exist in the world. Then this entire collection of books is called the universe of discourse, and the entire collection is denoted by 1. It follows, as it does not in ordinary algebra, that

$$A + 1 = 1.$$

On the other hand, as in ordinary algebra, we have

$$A \cdot 1 = A.$$

However, we need not explore the entire theory of the algebra of sets to appreciate that this algebra is quite distinct from the algebra of ordinary numbers.

Historically, one of the motivations for the study of the algebra of sets was provided by the study of logic. Mathematicians, notably Descartes and Leibniz, were so much impressed by the usefulness of ordinary algebra that they conceived the idea of inventing an algebra for reasoning in all fields of thought. The concepts of ethics, politics, economics, and philosophy would be the analogues of numbers, and the relationships among these concepts would be the analogues of the operations of arithmetic. They referred to this plan as a universal algebra. The work of Descartes and Leibniz was not successful because they undertook too much. (One can hardly imagine learned Republican and Democratic algebraists sitting down to calculate the solution of a vexing political problem by means of some system of algebra.) It is not likely that the ideas of disciplines such as philosophy and economics can be compactly represented by symbols and the reasoning performed by suitable algebraic operations. However, about 1850, George Boole, one of the founders of mathematical logic, showed that the reasoning processes themselves, which are studied in logic, can be formalized and carried out by an algebra of logic which is identical with the algebra of sets.

Boole's first idea was that in ordinary reasoning we deal with classes or sets of objects. The statement that all students are wise deals with the class of students and the class of wise people. Moreover, the statement itself says that the class of students is included in the class of wise people. If we let A be the class of all students and B the class of all wise people, then the statement that A is included in B can be expressed by the fact that the product of A and B is A. Thus the symbolic equivalent is

$$AB = A. \tag{14}$$

The statement that no wise people ignore mathematics can also be expressed symbolically. We let C denote the class of people who ignore mathematics. Since the statement says that there is no person common to the class B of wise people and the class C of those who ignore mathematics, then, in symbols, the statement says

$$BC = 0. \tag{15}$$

These two premises should lead to a conclusion about students and people who ignore mathematics. Hence let us derive an equation involving A and C. If we multiply the above two equations as in ordinary alegbra, we obtain

$$AB \cdot BC = A \cdot 0. \tag{16}$$

The associative property of multiplication tells us that we may arbitrarily group any two factors, just as in the product $(3 \cdot 4)(5 \cdot 6)$ we may group the 4 and 5 and write $3(4 \cdot 5)6$. Then

$$AB \cdot BC = A(B \cdot B)C.$$

However, $B \cdot B = B$. Hence

$$AB \cdot BC = ABC.$$

But by (14) $AB = A$. Therefore

$$AB \cdot BC = AC. \tag{17}$$

From (16) and (17) it now follows that

$$AC = 0. \tag{18}$$

Translated into words, this conclusion states that the class of students and the class of people who ignore mathematics have no members in common, or no student ignores mathematics. We have thus arrived at a conclusion by purely algebraic means.

This example illustrates how Boole used the symbols and algebraic operations of the algebra of sets to perform ordinary reasoning. With his algebra of logic Boole hoped not only to facilitate reasoning but to impart precision to the logical methods of reasoning. His ideas were taken up by others and became the basis of the subject now known as *symbolic logic.*

The algebra of sets and symbolic logic will not be pursued further. Neither is central in mathematics. The algebra of sets is used in a few advanced branches of mathematics, although even there it is a subsidiary notion. Symbolic logic is another very specialized field, somewhat apart from the main body of mathematics. It is used mainly by logicians who are seeking to clarify problems of logic and the relationship of mathematics to logic. The algebra of sets has been presented here only to help illustrate the variety of algebras in mathematics.

Exercises

1. If A and B are sets and $A + B = B$, what may you infer about the objects in A and B?
2. If A and B are sets and $AB = A$, what may you infer about the objects in A and B?
3. The operation of addition of sets has a different meaning from the addition of ordinary numbers. Why do we use the word addition and the symbol "+" for sets?
4. Evaluate in the algebra of sets $A + A \cdot A$.
5. Suppose A and B contain no objects in common. Evaluate $A(B + A)$.
6. Given the premises that all professors are intelligent and that no students are intelligent, translate these premises into the algebra of sets and deduce a conclusion about the relationship of students to professors.

27–5 Arithmetics and algebras as structures. We now have some familiarity not only with the arithmetic and algebra of ordinary numbers, but with those of modular number systems and of sets. Many other arithmetics and corresponding algebras have been developed in mathematics. We have, for example, briefly discussed the addition of vectors (Chapter 15). This subject could be pursued, and we would find that there is an extensive algebra of vectors, which is immensely useful for applications to science—a fact we can readily accept in view of the uses we made of vectors. There are also algebras of complex numbers, quaternions, matrices, tensors, and others. All of these would be encountered in the study of higher mathematics.

We have looked at the several algebras from the standpoint of the scientist, that is, we have tried to determine what concepts and operations man needs to represent and reason about different kinds of physical situations. From the strictly mathematical standpoint these algebras present another interesting aspect. Each algebra has its own properties, that is, axioms and consequences of these axioms. One says that each algebra has its own structure. Thus the arithmetic of ordinary numbers contains the operations of addition and multiplication. Corresponding to the operation of addition there is the operation of subtraction which is the opposite or inverse of addition. Thus, since adding 4 to 3 yields 7, the subtraction of 4 from 7 yields 3. Similarly, division is the inverse of multiplication.

In modular arithmetic the operation of division need not exist for any two members of the system. Thus the division problem $3/2$ has no solution in the arithmetic modulo 6. In the algebra of sets, there is a process of subtraction, but it is not the inverse of addition. As for division, no such operation is defined in the arithmetic of sets. The structures of the various arithmetics and their algebras may differ in other respects. We know that the associative and commutative axioms of addition and multiplication and the usual distributive axiom (Chapter 4) are adopted for the ordinary numbers. It so happens that these axioms hold in modular algebras and in the algebra of sets, but they do not hold in all algebras.

To the mathematician the different algebras are of interest precisely because they are strange and because the investigation of their properties is attractive. Indeed from the standpoint of the mathematician, algebra is the study of the properties of these various structures. We have already pointed out that these new arithmetics and their algebras have practical applications, but even if they did not, the study of such structures would be important because it makes us a little more aware of the significance of the axioms we adopt for our everyday numbers.

The traditional arithmetic and its algebra were no doubt developed first because of their immediate usefulness in daily life. The abstract structure, which was formulated much later, merely generalized the properties of numbers already verified through experience. Since the ordinary arithmetic and algebra were the only ones used for centuries, they were regarded as universal truths about the universe, just as Euclidean geometry was considered to be infallible truth. The development of other algebras has enabled mathematicians to see more clearly the relationship of ordinary arithmetic to the physical world and to appreciate that ordinary arithmetic has an empirical basis and is of limited physical applicability. Other algebras also have their role. Nature is too complex and too mysterious to be mastered with ordinary numbers alone.

27–6 Mathematics and models. Our study of non-Euclidean geometries and of exotic algebras may have prepared us to see that mathematics is a somewhat different activity from what man had presumed on the basis of the study of ordinary arithmetic and Euclidean geometry alone. The development of the latter two subjects had given rise to the belief that mathematics takes over certain truths about the physical world, adopts these as axioms, and then proceeds to study the physical world by deducing the implications of the axioms. Men did not question that the axioms were truths about the world but tried instead to account for their possession of truths by theories of knowledge or by crediting God with implanting these truths in human minds. However, as mathematicians undertook to study and encompass new classes of physical phenomena or to represent more accurately a previously studied phenomenon such as physical space, they were forced to recognize the need for new concepts and new sets of axioms.

The mathematician really creates models of reality. The concepts, axioms, and theorems of an algebra or a geometry are a model with which to think about some aspect of the physical world. Each model has a limited applicability. Moreover, one must distinguish between the mathematical model and the physical world or between mathematical theories and physical reality.

Topics for Further Investigation

1. Modular arithmetics.
2. The properties of the algebra of sets.
3. The algebra of logic.
4. The nature of symbolic logic.

Recommended Reading

Bell, Eric T.: *Men of Mathematics,* Chap. 23, Simon and Schuster, Inc., New York, 1937.

Boole, George: *An Investigation of the Laws of Thought,* Chaps. 1 to 7, Dover Publications, Inc., New York, 1951.

Courant, R. and H. Robbins: *What is Mathematics?,* pp. 31–40, 108–116, Oxford University Press, New York, 1941.

Langer, Susanne K.: *An Introduction to Symbolic Logic,* 2nd ed., Dover Publications, Inc., New York, 1953.

Newman, James R.: *The World of Mathematics,* Vol. III, pp. 1852–1900 (selections on symbolic logic), Simon and Schuster, Inc., New York, 1956.

Sawyer, W. W.: *Prelude to Mathematics,* Chaps. 7, 8, 13, and 14, Penguin Books Ltd., Harmondsworth, England, 1955.

CHAPTER 28

THE DEDUCTIVE APPROACH TO THE SOCIAL SCIENCES

Thus God and Nature fixed the general frame,
And bade self-love and social be the same.

An eighteenth-century theme

28–1 Introduction. The momentum given to scientific activity by the success of the mathematical approach to nature carried far beyond the physical sciences, philosophy, religion, and literature. It penetrated to the realms of human nature and the institutions which men establish to live together in society, notably government and economic systems. In this chapter we propose to go back and to examine briefly eighteenth-century thought in these fields. Our purpose is largely to see what the mathematically inspired notions, the belief in reason, the belief in a universe ordered by precise, harmonious laws, and the axiomatic, deductive method accomplished.

The plan of the rational approach to the social sciences is basically Descartes', but it is clearly and boldly stated by Hobbes in his *Leviathan*. "Geometry," he said, "is the only science that it hath pleased God hitherto to bestow on mankind." Of course, Hobbes did not mean that geometry proper could be applied to political and economic thought or to other social problems, but rather that the axiomatic, deductive method of geometry was the only approach capable of producing a scientific system of thought. Hobbes pointed out that mankind had previously relied only on experiment as the source of political knowledge; by this means, however, we can acquire only prudence, useful as that may be. But by means of science we acquire sapience, which is infallible.

The conviction that sciences of politics and economics could be developed had its roots in the success of the physical sciences. Man was part of the physical world. Hence there had to be universal natural laws of human behavior, a human physics so to speak, which revealed the laws of the science of human nature. From these truths about human nature and basic principles belonging to the more specialized phenomena of government and economics man should be able to deduce the laws governing the true systems of government and economics. Because such systems would be based on the correct principles of human nature and self-evident principles of the special domains, they would be in tune with nature and hence stable and proper.

28–2 The rational reconstruction of political science. Natural rights. New ideas and original writings on the subject of government date from the Renaissance. The beliefs that there are natural rights and natural laws, that kings exercise power only as representatives of the people, that the people are sovereign, and that the aim of the state is the welfare of the people can be found in the writings of the German, Johannes Althusius (1557–1638), the Dutchman, Hugo Grotius (1583–1645), and others. However, until about 1700 these ideas were not boldly applied to battle the principle of authority which was the backbone of the social order. The will and the courage to re-examine and proclaim the findings of reason in the face of tradition, authority, and special privilege were the contributions of the eighteenth century.

Though the intellectuals of that age needed no additional stimulus, reason was supplemented by other forces which also worked toward the reconsideration of political systems. During the medieval period, kings sought the backing of the Church to help control their subjects. Religion and ruler were in league. The Church confirmed the ruler, and the king willingly subjected himself to Church approval for the sake of the support he received. Thus kings became representatives of God and ruled by divine sanction. Even those monarchs who felt that they sacrificed too much by bowing to the Church and who therefore freed themselves from ecclesiastic authority still claimed direct access to God and His endorsement of their commands. But, as we have seen, the hold of religion weakened, and the claim of divine sanction no longer sufficed to convince people that they must obey the existing regime.

In the attempt to pursue a mathematical approach, the political theorists sought to identify and abstract the universal forces at work in society and to deduce from them the laws of government. Their efforts were largely dominated by the thought that men and nations act under the forces of attraction and repulsion and society's behavior is but the mechanical resultant of the action of these forces. It is quite evident that the example of mechanics was in their minds. But, whereas the mathematical physicists had found one set of basic axioms and built the unique Newtonian system of celestial mechanics, the political theorists found several.

The manner in which these men tackled the problem of finding the axiomatic truths of social organization is interesting. Some hurried rapidly from cursory observations to broad generalities, convinced that their findings were natural laws and therefore needed little verification. The type of government which they preferred undoubtedly determined what was "self-evident." Thus Hobbes says that the state is a body and hence can be studied in the same way that a physical body or object is studied. The body of the state is in fact no more than a collection of individual bodies, and the will of the state no more than a union of individual wills. The conflict of such wills can be resolved only by a covenant. This contract is not an actual agreement among people, but is the theoretical justification of what exists. It is an agreement to surrender the sovereignty of the individual to the state. The authority of the ruler establishes and maintains the whole society and he needs unlimited power to keep it together. Thus the civil state arises from the natural state. And thus Hobbes justifies an absolute, though not a divinely ordained, monarchy.

The more influential political theorists were Locke and Montesquieu. John Locke, more thorough and less biased than Hobbes and therefore less prone to prejudge his conclusions, undertook to determine the logical basis and the legitimate functions and powers of government. In agreement with his theory of knowledge, Locke maintained that all men are born equal, the differences among them being due to environment. In the natural state, the garden of Eden so to speak, political or social organization did not exist, and all men were free and guided by the laws of reason. The rights man possessed in this state were given to him by nature, and since nature and reason were equivalent in the Age of Reason, these rights were proper for man. However, to secure protection of life, liberty, and property, men made a social contract which is the basis for political and legal relations. They agreed to set up a government which would protect their rights and in return yielded to this government the right to punish offenders against society. Since differences of opinion about what was best for society might arise, the will of the majority was to decide the issue, and this will could be imposed on all. Government should interfere as little as possible with the ways of men because nature was designed in accordance with a system of divine laws and would almost automatically produce the greatest happiness possible for the people as a whole. The rulers derived their powers from the people and were intended to serve the people; hence if they transgressed their power and betrayed their constituents, revolt was clearly justified. Thus Locke spoke for the natural and inalienable rights of human beings. He is the founder of the liberal tradition in political thought.

Whereas Locke had determined the origins and powers of government by rational inquiries modeled on Descartes' approach, Baron de Montesquieu (Charles Louis de Secondat, 1689–1755), who as a youth had studied physical science and even written papers on the subject, preferred to apply the method of Galileo and Newton and obtain fundamental truths about government by generalizing from observed facts. To discover the universal laws in the domain of political institutions, he traveled all over Europe, studying the legal and political systems in existence; he also examined carefully the history of bygone political orders. He sought, in other words, the facts of experience and drew from the nature of things as he found them. His observations and inferences are striking. He thought that human nature is constant and responds in fixed ways to cultural and environmental influences. Liberty is found in cold climates because there people are energetic. On the other hand, in hot climates people are lazy, and hence despotism and slavery gain hold. Climate also determines the character of people.

He believed that a political system cannot be truly understood unless one determines the forces which produce the system. He also believed that we must know how the various types of forces act if we wish to realize a state which allows the greatest possible freedom. He found specific ones. Democracy, which he considered the ideal type of government—exemplified by the English constitutional system—rests upon patriotism, frugality, and the spirit of equality. But these forces must be balanced by counterforces. For if the spirit of equality, for example, is lost, the people grow corrupt; and if it is excessive, then the people may regard themselves as powerful as the rulers they have chosen. Within the government itself Montesquieu saw the need for checks and balances to prevent one agency of government from usurping absolute power. Hence he favored the separation of the legislative, executive, and judicial powers. To keep the actions of government within proper bounds, the people have the right to protest. Slavery, religious intolerance, and primitive penal codes he attacked because they destroyed freedom. Montesquieu's book, *The Spirit of the Laws* (1748), was highly influential in projecting a science of society and in championing democratic doctrines.

Locke, Montesquieu and others, such as Rousseau, whose ideas we cannot take up here, forthrightly condemned the absolute monarchies of the eighteenth century, enthusiastically proclaimed their new doctrines, and made every effort to spread their political philosophy. They encouraged demands for radical changes in legislation and administration, in judicial proceedings, taxation, and in trial and punishment, and they held up the ideals of democracy—rule by the people and government subject to the will and power of the governed.

Eighteenth-century thinkers, in general, were not in favor of revolution by force, but preferred gradual social reform. Nevertheless conditions became so intolerable to the people in colonial America and in France that revolutions did break out. The ideas that better forms of government than those in existence can be achieved, that the governed have the right to revolt, and that the people can set up a government subject to their will—the doctrine of democracy—were derived from the new rational approach to the study of political systems. The will and courage to act were stimulated by the spread of the new political philosophies to the people at large.

The rational approach to the problem of government and the new political philosophy of democracy are clearly expressed in the *American Declaration of Independence:*

"When in the course of human events, it becomes necessary for one people to dissolve the political bonds, which have connected them with another, and to assume among the powers of the earth, the separate and equal station to which the Laws of Nature and of Nature's God entitle them, a decent respect to the opinions of mankind requires that they should declare the causes which impel them to the separation. We hold these truths to be self-evident, that all men are created equal, that they are endowed by their Creator with certain unalienable Rights, that among these are Life, Liberty and the pursuit of Happiness. That to secure these rights, Governments are instituted among Men, deriving their just powers from the consent of the governed—That whenever any Form of Government becomes destructive of these ends, it is the Right of the People to alter or to abolish it, and to institute new Government, laying its foundations on such principles and organizing its powers in such form, as to them shall seem most likely to effect their Safety and Happiness."

The mathematical form of the *Declaration,* though a detail from the standpoint of political history, should be noted. The opening sentence is followed by the axioms of government which the colonists regarded as self-evident. The document continues with a recital of deeds by the King of England which show that the government in power has failed to provide those rights asserted in the axioms. It follows, then, from another axiom that the people are justified in abolishing the existing government and in forming a new one.

The philosophy behind the *Declaration* is more significant. It asserts that man has rights granted by nature and God. Reason, which is part of man's nature and therefore of nature, uncovers these rights; they are, in fact, self-evident to reason. Hence one can be sure of the truths so discovered. Moreover, nature and nature's God rather than the Bible or the authority of lay or religious leaders are the true source of human rights.

Jefferson, who wrote the *Declaration,* idolized Bacon, Newton, and Locke. Many of the phrases he used are Locke's. Jefferson actually wished to go further and incorporate in the *Declaration* and *Constitution* additional principles which he had derived from the political philosophers of the Enlightenment. He believed that conditions in society change and that therefore each generation should reconstruct the government to satisfy its own will and needs. He calculated that every eighteen years and eight months half of those over twenty-one years of age die. Hence to reflect properly the will of the new majority, the *Constitution* should be rewritten every nineteen years.

28-3 The philosophy of utilitarianism. The political philosophy based on the natural rights of man justified and provided the goals of the American and French revolutions. But much of the support for the revolution came from people who sought to advance their own interests. In the United States, men such as Alexander Hamilton, John Adams, and James Madison were more concerned with the protection of their own property and the power to control it than with the rights of the people. The concept of natural rights did not suit slave owners. In Europe, the rising merchant class interpreted natural rights to mean absolute power in the conduct of business and, in particular, freedom from interference by rulers. In fact, in England laborers were denied the right to education because it might open their eyes to injustices; however, the official reason given was that education might enable them to read publications against Christianity. The French Revolution, also inspired by the doctrine of natural rights, was followed by the Reign of Terror and then perverted by Napoleon, who not only re-established a monarchy but proceeded to devastate Europe. Although none of these forces and events really impugned the logic and wisdom of the philosophies of natural rights and democracy, the new doctrines were criticized, on the one hand, because in some countries they were not successful in securing the rights of men and in eliminating injustices, and, on the other, because they led to unfortunate consequences in France. There is some question as to what might have happened to the spread and practice of democracy, had not a new group of rationalists appeared on the scene with a totally new defense which also secured a strong following.

The members of the new movement were called Utilitarians, and their leader was the highly rational and courageous Jeremy Bentham (1748–1832). Though sickly as a child he soon showed a prodigious intellect. At the age of 12 he entered Oxford University and acquired some education in the sciences. Shy and retiring, he never married. Bentham devoted his life to social problems, his concern for society extending unto death. Thus he requested that his body be dissected (after death) so that mankind might reap some benefit by studying the corpse. Burial he regarded as a silly act. As to the disposal of bodies in general, Bentham had another suggestion. A good corpse, embalmed, varnished, and fully dressed, he said, could be an object of veneration. Hence instead of planting trees along the driveways of stately homes, rows of varnished and dressed forebears might be used instead. Bentham's body was embalmed at his death and now sits fully dressed in a Common Room of the University of London. It was placed there at the suggestion of Charles K. Ogden, another "eccentric" Englishman, who died in 1957.

Bentham was extreme in his rationalism. He sought logical approaches to ethics, government, and economics. In developing his theses, he critically examined every proposition and would rewrite an entire work when he found a single doubtful statement. He sought to classify all knowledge by arranging ideas in their logical relationships. His critical faculty enabled him to stand "outside of the received opinion" and analyze it objectively. He also possessed the courage to defend his ideas and attack institutions and organizations he deemed injurious to society. In an age when organized religion was still a powerful secular and spiritual factor, he stated bluntly that all dogmas were harmful and fought the alliance of church and state. He attacked the privileges of the aristocracy in his *Book of Fallacies*. His pamphlet *The Elements of the Art of Packing* charged the Crown itself with fixing juries. He dared, in fact, to urge that the monarch and the House of Lords be abolished and that universal suffrage and an annual parliament be instituted.

Bentham approached the problem of government through ethics. In his classic work of 1789, *Introduction to the Principles of Morals and Legislation,* he put forth what he regarded as a mathematical treatise. It offered primary ethical principles from which the principles of government and legislation were to be deduced. He, too, sought to establish fundamental laws of human nature and decided, as had others before him, that the desire for pleasure and the avoidance of pain are the fundamental forces determining human actions. Men continually pursue happiness and retreat from pain. The words *pleasure* and *pain* were used broadly. Malevolence may cause pain to some people but may give pleasure to the perpetrator, and so such an act has both qualities. Any realistic system of ethics must take into account the fundamental forces of pleasure and pain. Since the same act may cause pleasure to some and pain to others, its ethical value must depend upon the net effect. Thus Bentham arrived at the principle that that act is just which produces the greatest good for the greatest number. He wished, in brief, to maximize happiness The consequences of an act, and not the motive, determine its value.

Bentham proposed that one measure the effect of an act by determining the pleasure and pain it causes. First he listed the various pleasures and pains which people experience. He went on to say that any act possesses such objective properties as duration, intensity, certainty, propinquity, purity (isolation from other pleasures and pains), and fecundity (tendency to produce other pleasures and pains). Each of these, Bentham asserted, could be measured. In addition, the effect of an act depends upon the sensibilities of the people affected. A rich man who loses 100 dollars is pained less than a poor man. Thus wealth, education, sex, race, character, and other factors determine the sensibilities which also are measurable quantities. One now adds the objective measures of an act and multiplies the sum by the sensibilities of each of the various people who obtain pleasure from the act. The sum of the individual products is regarded as positive. A similar sum represents the pain caused by the act, but this number is regarded as negative. The algebraic sum of the measures of pleasure and pain is the value of the act. Thus each act has a value, and whenever two or more courses of conduct are available, one should choose the one which has the higher algebraic value.

Bentham's moral arithmetic is not readily carried out in practice. He himself gave no method of determining numerical values for acts. However, in some simple cases it can be and was applied. Thus, for example, the wisdom of vaccination against smallpox was hotly debated in his time because some people had died of the vaccination. On the other hand, of course, many had been successfuly immunized. But since the number of deaths from vaccination was far smaller than the number of deaths that would have occurred from the normal incidence of smallpox unchecked by vaccination, it was clear that the value of vaccination was positive. Though the arithmetic could not always be carried out, Bentham had given a rational criterion which could be applied, at least, qualitatively. Whether an act will produce the greatest good for the greatest number can often be decided even if a precise numerical measure cannot be obtained.

Bentham and the Utilitarian school he founded now proceeded to tackle the problem of government without invoking nature or God. Bentham, in fact, derided the theory of natural rights and called the assertion that all men were born free and equal "absurd and miserable nonsense." Everybody is born and long remains a helpless child. Human reason should be adequate to determine the best society by applying the criteria of reasonableness and social usefulness. To the principles of ethics he adds a fundamental axiom of political thought: The purpose of government is to bring about the greatest happiness for the greatest number. The validity of any form of government depends not upon any mythical contract, but on the extent to which it produces human happiness. From his ethical and political axioms Bentham deduced many conclusions. Justice is not an end in itself; it is a means to increase the total happiness. Law must consider the consequences of acts and not the motives since the purpose of government is to increase happiness. Punishment as mere retribution serves no purpose. Its only objective should be to discourage further acts which may produce pain. Moreover, since punishment itself causes pain, it is to be inflicted only when it prevents greater pain.

Most important is Bentham's argument for democracy. The rulers of a country pursue their own happiness although they should seek to achieve the greatest happiness for the greatest number. These opposing interests can be reconciled only by identifying governors and governed—the people at large. Democracy is the form of government which accomplishes this task. To support his conclusion, Bentham pointed to the example of the United States. In that country there were no useless expenditures, no corruption, and none of the other evils found in Great Britain. Clearly Bentham had never lived in the United States. But he did develop the philosophy of majority rule which in the nineteenth century superseded the doctrine of natural rights. Moreover, many of the reforms suggested and backed by the Utilitarians were adopted in England. Bentham was the leading light in nineteenth-century legal and political reform movements.

Although Bentham's illustrious disciple James Mill (1773–1836) in part duplicated the writings of his teacher, his work nonetheless contains many original ideas. The son of a shoemaker, he studied for the ministry, was ordained, but never received a parish. He later became an agnostic because he thought that the evils of this world were incompatible with the benevolence of an almighty God. Mill devoted his life to writing on his chief interest, political economy. Possessed of great physical and mental vigor, he was able to work for years from 5 a.m. to 11 p.m. He was strictly scientific in his thinking, rigorous in his reasoning, and precise in his statements. His opponents called him as dry as Euclid. They were correct in comparing him with Euclid, but hardly fair in calling Euclid dry.

Like Bentham, Mill started with laws of human nature and accepted as axiomatic that governments should secure the greatest possible happiness for the greatest number of people. Government, he says, deals with the pleasures and pains men derive from or inflict upon one another. It is a law of human nature that man, if given a chance, will not hesitate to appropriate other people's goods in order to satisfy his desires. The strong will plunder the weak. To prevent this and to secure for each man what he earns, governments are instituted, and the power is placed in the hands of a few people. But how can one prevent the guardians of the state from becoming plunderers? The hope of glutting the rulers is illusory. It is a "grand governing law of human nature" that these men, as do all human beings, covet power over others and that their desire for pleasures is infinitely great. Moreover, since pain is a more powerful instrument than pleasure for compelling compliance, rulers will seek to possess unlimited power of inflicting pain upon others. Mill adduced the example of English "gentlemen" who cruelly mistreated their West Indian slaves.

After examining monarchy, oligarchy, and democracy, he concludes that this last form of government is the solution to the problem of curbing the power of the rulers. Representative government, with representatives chosen from the ranks of the electors to ensure that rulers and ruled share the same interests, and a large body of electors to prevent coalitions of electors and elected at the expense of the general public are the answer.

The history of political thought continues, of course; brilliant, active thinkers, notably John Stuart Mill, son of James Mill, based their theories on similar premises or followed radically new lines as did Karl Marx. We do not intend to pursue this history. What is relevant to our subject, mathematics, is that the intellectuals of the eighteenth century were inspired to apply mathematical methods to social problems and to seek laws in the field of government. The *philosophes* of the Enlightenment sought a more rational, improved system of government for the masses. Their lasting achievements were the secularization of government and the promotion of democracy.

Exercises

1. What intellectual forces led eighteenth-century thinkers to believe that they could develop sciences of politics and economics?
2. What method did the social philosophers plan to use in constructing their sciences?
3. What does the phrase "the natural origins of government" mean?
4. What axioms did Locke adopt for his science of government?
5. In what respects does the *Declaration of Independence* reflect the age in which it was written?
6. What are Bentham's axioms of ethics and government?
7. Contrast Locke's argument for democracy with Bentham's.

28–4 The rational approach to economics. Just as the arbitrary exercise of power and cruelty was apparent in the political life of the eighteenth century so, in the economic sphere, were poverty, the enslavement of men, women, and children in long hours of work, and the diseases caused by malnutrition. The intellectual leaders turned to these problems with the expectation that reason would reveal the ideal economic system, and their analyses followed the rational, mathematically inspired lines of the political theories. They proceeded by deduction and they sought what they believed must exist, namely, natural laws inherent in the design of the universe. We shall illustrate the trend of thought by presenting the views of some leading schools.

The Physiocrats, led by François Quesnay (1694–1774) and Dupont de Nemours (1739–1817), believed, as the very word *physiocrat* implies, in the rule of nature. Society was fashioned by nature in accordance with eternal, immutable, and inevitable laws. Economists must discover and proclaim these laws as Newton did in the physical sciences. Neither men nor governments can make such laws. They are God's will, and men and governments must obey them. Custom and various historical accidents had given rise to various institutions, but these did not conform to nature's plan. Hence the economic ills of society were due to man's interference with the workings of nature.

What are, then, the natural laws of economic life? The Physiocrats believed that the right of property, the right of liberty in the conduct of business, and the right of security are axiomatic. The function of government is to protect these rights. A government which institutes any laws conflicting with these rights does not deserve obedience. In particular, government must not interfere with business, which must be left to businessmen whose enlightenment would ensure the successful working of the economic system. The Physiocrats thus believed that the self-interest of the individual, in general, coincided with the interests of society. Their position is often summed up in the phrase, *laissez faire,* an idiomatic expression used to describe the philosophy that government should not intervene in the affairs of business.

From these axioms, the Physiocrats deduced that nature evidently desired free competition. In fact, competition spurred humans on to perfect themselves. They also concluded that the state was the only power which could protect property and contract rights, and so must be strong enough to police such matters. Public works and education were also functions of the government. Land was the sole source of wealth, and hence a single tax on land was all that was necessary. The inequalities in income and the consequent poverty, the Physiocrats decided, must be due to inequalities in the physical powers of people.

The task of finding universal laws independent of time and place which operated in economic affairs was also undertaken by the great Scotchman Adam Smith (1723–90). Smith's chief work, *The Wealth of Nations* (1776), is generally regarded as the beginning of scientific economics. He, too, believed that God had established laws for society and that it should therefore be possible to develop a science of economics just as Newton had created the science of mechanics. In this divine order, natural law and personal freedom were basic principles. The dominant motive force of economic life is that the individual acts in his own interest. In doing so, however, he usually furthers the welfare of all. "Man's self-interest is God's providence." Like the Physiocrats, Smith believed that the rights of property and security were also axiomatic. However, Smith regarded labor rather than land as the sole source of wealth. In fact he favored labor over property owners and manufacturers, but his followers soon forgot this doctrine.

He defined such concepts as division of labor, wages, profits, interest, and rent. These concepts are basic today because he did so much to clarify them and show their importance. He maintained, for example, that the real price of an article was the amount of labor that went into it. Thus labor value was the standard of value. However, the natural price of any article consisted of three elements: wages, profits, and rent. The market price fluctuated around the natural price in accordance with the law of supply and demand, the principle which accounted for the harmonious working of society.

He then proceeded to determine the laws controlling wages, profit, and rent. Wages, for example, are determined at any given time by the number of persons seeking work. What determines this number? The abundance of the means of subsistence, that is, food, clothing, shelter, and so forth. The supply of these means does increase, but only gradually. However, unchecked population growth tends to outrun the means of subsistence. Hence wages should be low enough to keep the population down to a number that the means of subsistence will support. Any wage increase beyond this level will produce an increase in population, which in turn will drive wages down in accordance with supply and demand. Thus, in the long run, wages will always be such as to keep the population at the minimum subsistence level.

Implicit and explicit in Smith's writings was the belief that free competition was necessary, and that governments should refrain from interfering with business. The functions of government are to protect the people from external violence, maintain peace internally, and support public works and institutions which individuals would not find it to their own interest to support.

Smith was wiser than most economists in many ways. He foresaw that division of labor would increase production and raise the earnings of labor. At the same time, he warned against the stultification of mass production which deprives men of an outlet for their natural creative powers. Like Galileo and Newton, Smith attempted to draw and confirm his basic principles by observation and study of the actual workings of production and consumption rather than by pulling preconceived notions out of the recesses of his mind.

The economic theories of the Physiocrats, Adam Smith, and other eighteenth-century social thinkers were rational and even reasonable in view of the intellectual temper and economic state of Europe. The merchant and manufacturing classes had already become important factors in European life. Their needs had to be defined and defended. Industrialization had also begun to be significant, was growing rapidly, and deeply affected the lives of wage earners and the methods of production. The problems caused by this change were new and complicated. Since the laws of an ideal economic system are not to be had just for the asking, economists quite naturally sought to reason on the basis of what seemed to be truths in their times. But the economic ills of the age were not alleviated and, in fact, continued to become more severe. It was literally true that women and children worked sixteen hours a day, and the income of an entire family of workers hardly sufficed to do more than prevent starvation. Some economists felt impelled to defend the inevitability of these miserable conditions.

Thomas R. Malthus found the answer in his book *Essay on the Principles of Population,* which he first published anonymously in 1798 and later in his own name. Malthus started with axioms:

> "I think I may fairly make two postulata. First, That food is necessary to the existence of man. Second, That the passion between the sexes is necessary, and will remain nearly in its present state, . . . Assuming, then, my postulata as granted, I say, that the power of population is indefinitely greater than the power of the earth to produce subsistence for man."

Malthus then adopted a quantitative approach. Like Adam Smith, he estimated that the population would double every 25 years if people reproduced as natural desire, fecundity, and health permitted. However, the food supply increases only by a fixed amount every 25 years. Stated in more mathematical language, the population would increase in geometric progression, whereas the food supply would increase in arithmetic progression. Hence in 100 years, for example, the population would be 2^4 or 16 times as large as at the beginning of that period, whereas, if the food supply increased every 25 years by the amount available at the beginning of the 100-year period, it would be only five times the original quota at the end of the 100-year period. Clearly the population would outstrip the means of subsistence.

But the population—at least that of the European countries—did not double every 25 years. (It did in the United States at that time.) And here Malthus came up with a striking new thought. He reasoned that the population does not increase at the potential rate of growth because wars, disease, and vice kill off people. Moreover, *these evils are necessary; they are part of nature's design,* and they have the beneficial effect of keeping the population at a level which can be fed. It followed that no legislation can alleviate man's lot. In fact, interference with nature's and therefore God's plan would make matters worse. Of course, the state of affairs worked hardships on a large segment of society, but Malthus argued that no society could exist in which all members would enjoy health, happiness, and comforts.

Despite his own injunction not to interfere with nature's plan, he did urge that people should not have more children than they can support. He formulated an eleventh commandment: "Thou shall not marry until there is a fair prospect of supporting six children."

It is not clear whether Malthus believed his own explanation of the evils of society. He was accused of deliberately calculating "to lull the oppressors of mankind into a security of everlasting triumph." His writings earned him a great reputation and a professorship of history and political economy.

A somewhat different justification for the inequalities and ills of society was given by David Ricardo (1772–1823). From early youth, Ricardo was inclined to scientific pursuits; he established a laboratory, and was a member of several scientific societies. He decided to apply science to economic problems, began, as did others of his times, with the belief that God had instituted an economic order as He did the physical order, and proceeded to develop a deductive, mathematical theory of economics. He isolated and defined the factors: capital, labor, utility, rent, wages, profits, and value. These factors are variables, with labor the independent variable on which the others depend. The variables are related by inescapable laws. Thus the price of a commodity is determined by supply and demand. Scarcity means high prices, and abundance, low prices. Another law stated that man's attraction to profit-making ventures varies directly with the possible gain and inversely as the risk. The wages paid to labor are also determined by supply and demand. If wages are increased beyond this natural level, laborers will have more children and so increase the labor supply, with the result that wages will drop. Hence it is pointless to raise wages. "The natural price of labor," he says in his famous Iron Law of Wages, "is that price which enables the laborers, one with another, to subsist and perpetuate their race without either increase or diminution." That this natural price meant a near-starvation existence for a large segment of the population, was unfortunate, but nothing could be done about it. Poverty was an inevitable consequence of the workings of the natural laws of economics. It was also natural that labor, landowner, and capitalist should oppose each other. This conflict, too, was part of God's design.

The economic theories of the eighteenth and early nineteenth centuries are not satisfactory by today's standards. The "dark Satanic mills" continued to grind out the lives of men, women, and children while economists presented theories which defended the "natural" inequalities and argued against changes, against reforms, against unions, and against remedial legislation. This "dismal science" of man became the enemy of man, and the economists, consciously or unconsciously, were merely special pleaders for the privileged classes.

In the early nineteenth century, the social scientists, realizing the deficiencies of their work and perhaps less certain that God and natural rights supported their theories, instituted a new approach to economic problems. By adopting the deductive approach and initiating the search for natural laws of man and society, they had followed the mathematical method of the physical sciences. But apparently they had made the mistake of attempting to encompass entire domains such as politics and economics in one body of theory.

Galileo and Newton had proceeded quite differently. Not only had they tackled one class of phenomena at a time: earthly motions, celestial motions, sound and light, but they had formulated quantitative laws which either could be directly verified by experiments or whose implications could be tested. And so the economists decided to apply quantitative, deductive methods of analyses to limited phenomena, such as the determination of the market price under supply and demand, the principles and effects of taxation, the phenomenon of national prosperity, the factors determining profits in an industry, and so on.

This new trend in economic thought was launched by A. A. Cournot with the publication in 1838 of his *Researches into the Mathematical Principles of the Theory of Wealth.* Such investigations are being carried on extensively today, are highly technical in economic and mathematical content, and may, in the long run, furnish broad economic doctrines which will enable man to build a more satisfactory economy than the various systems now in existence. However, this new science of mathematical economics is in its infancy and does not warrant space and time in our present introductory efforts to show how far mathematics has penetrated into the social sciences.

Exercises

1. What was the evidence that led eighteenth-century thinkers to believe that there were laws of nature in the domain of economics?
2. What was the logical justification of the doctrine of *laissez-faire?*
3. How did Malthus explain the phenomenon that population did not outrun subsistence?

28–5 The reform of the philosophy of history. The study of history was also sharply influenced by the thinking of the Enlightenment. The very concept of history, the purposes it serves, and the relationship between actual events and the general conclusions drawn from them were reexamined and often drastically modified.

Although it is impossible to pinpoint any one date as the day or year in which major changes occurred in the approach to history—just as it was impossible to do so in the fields of politics, economics, and the areas surveyed in earlier chapters—one can nevertheless select a few thinkers who show clearly the influence of new times and who in turn gradually influence subsequent thought. In the domain of history, one of the innovators who responded to the Age of Reason was Pierre Bayle (1647–1706), whose major work, entitled *Historical and Critical Dictionary,* appeared in 1697. He was the first to examine critically the statements made by historians. What is factual? How can we know what to accept? These are the problems Bayle tackled in his desire to locate the solid bricks of history. He was indefatigable and insatiable in his attempts to trace down the truth or falsity of facts and was fascinated by the work of ascertaining what really happened. He painstakingly sifted all available relevant material and evaluated whatever evidence there might be in support, if not in direct corroboration, of the assertions made in the various publications. As a consequence he was able to reveal gaps, obscurities, and contradictions in the accepted histories. The Bible itself is, of course, a major historical writing, but Bayle would not accept its statements merely upon the authority of the document. And just as Galileo built science without regard to the Biblical account of the universe, so did Bayle try to build history without relying upon the Bible. Bayle is the originator of the ideal of historical accuracy.

The nature of historical studies underwent another major change when the social thinkers of the Enlightenment began to draw general conclusions from their study of history and proceeded to formulate what might be called a philosophy of history. The first such decisive attempt was made by Montesquieu in *The Spirit of the Laws.* He decided that the histories of various nations follow patterns or principles which are reflected in the form of government. Thus republics are founded on civic virtue; monarchy rests on honor; and despotism on fear. That government which comes closest to basing itself on the principle which naturally belongs to it is most nearly perfect. The patterns of education and justice, of marriage and family, and of domestic and foreign politics depend upon the fundamental principle of the state. Climate, soil, the size of a country, and its geographical location also determine the evolution of a people and the form of the government. And he makes this general observation: "As long as the principle of a form of government is as such preserved, as long as it is healthy in itself, it has nothing to fear; . . . On the other hand if the principle deteriorates, if the inner moving force weakens, then the best laws can offer no protection." Montesquieu recognized that the types which he described did not occur in pure or perfect form in any one nation, any more than a geometrical form is realized in any one physical figure. He introduced to social science a method new in his time: idealization. He grasped and expressed clearly the ideal types in history.

Voltaire, too, went beyond facts. He believed that a reduction of facts to laws should be possible in history just as it was in Newtonian science. The work of the historian is the same as that of the scientist: to seek the hidden law amid the flux and confusion of phenomena. Genuine historiography should free history from the domination of final causes, from teleological interpretation, and lead it back to the real, empirical causes. Historians, he said, should stop studying political events, the rise and fall of kingdoms, and battles. He also attacked the cult of heroes: "The more radiant their glory, the more odious they are." In one of his letters to Frederick the Great, he said that he did not like heroes who lead men to combat, sufferings, and death, and he dared to write these words after one of Old Fritz's victories. He advocated that historians should study the human race. The human spirit and the spirit of nations are what matters.

Voltaire was primarily interested in cultural history, whereas Montesquieu clung to political events. For Voltaire, the mind is the dominant feature of history. The progress of mankind can be understood only in terms of the growth of religion, art, science, and philosophy, since these are the fields in which the human spirit expresses itself. The process by which reason emerges and realizes its destiny is the fundamental meaning of history. Voltaire freed history from the weight of antiquarianism and from its confinement to the form of a chronicle.

As in all departments of philosophy, so in the philosophy of history, there is no one theory which satisfies all. And indeed the specific theories of history offered by Montesquieu and Voltaire, for example, are for us not nearly so significant as the fact that eighteenth-century thinkers applied their critical faculties to history also and sought to understand its nature and function. They endeavored to make a science out of what had been a mere recording of facts, and they succeeded at least in broadening the concept of history to encompass, in addition to military events, the developments in the intellectual. social, and political spheres.

28–6 The accomplishment in the social sciences. Most leaders in political and economic thought agreed that nature conformed to a harmonious and divinely ordered pattern. (The Utilitarians believed in nature's harmony, but disputed the divine order.) Whereas Newton had revealed the natural laws of the physical world, they expected to find and advance the natural laws of social institutions. They hoped that men would modify existing practices to make them conform to these natural laws, and that such an adjustment would cure all ills of society. They did, indeed, proclaim what they believed to be natural laws and, in the political sphere, thereby promoted the growth of democracy. In the economic domain they justified the *status quo*. However, our present scientific outlook is not in accord with the existence of natural laws instituted by God or innate in human nature. The rationalists had hastily jumped from physical to social phenomena and had concluded that what apparently held in one must hold in the other. Even today we can hardly speak of a science of politics, or a science of economics, or of a pattern in history. There are no theories or broad principles which permit us to predict human behavior or to fashion the ideal economic system. If democracy appears to be the preferred political system, it is only because experience seems to show that men can live more happily under such a government. As to economic systems, the implications of experience are less clear. However, the majority of governments have abandoned the *laissez-faire* philosophy and practice of the eighteenth century and moved far in the direction of control.

Though we still have no final answers to the problems of defining, let alone establishing, the ideal government and the ideal economic system, the eighteenth-century thinkers took the most essential step toward the realization of such goals. They dared to combat and subdue tradition, custom, and authority in order to question. They unfurled the banner of reason and rallied behind this force to invade territories cluttered with the rubble of ages and ruled by selfish and vested interests. They offered to the world the suggestion and hope that reason may produce improvements in man's social organizations even if it fails to prove that certain norms are intended by nature or are truths. The specific ideas which they espoused were hardly original with them. But by laying the foundations for a broad, ambitious program of social reform, by relying upon reason alone with a minimum of bias or inclination toward existing institutions, tradition, or vested interests, and by championing these ideas with clarity and vigor, the intellectual leaders of the Enlightenment made lasting contributions to our civilization. The very concept of social sciences is their chief positive contribution.

Kant formulated the motto of the Enlightenment: "They dared to know." They dared because the great mathematical victories in the physical sciences had engendered in all who surveyed them a spirit of rational analysis and an intoxication with the power of the human mind.

Exercises

1. Explain why the deductive approach to the social sciences failed.
2. What changes did eighteenth-century thinkers introduce in the concept of the study of history?

Topics for Further Investigation

1. The influence of mathematics and Newtonian science on the Physiocrats.
2. The influence of mathematics and Newtonian science on David Ricardo.
3. The rise and development of the concept of progress.
4. The philosophy of the Declaration of Independence.
5. The ethical and political philosophies of Jeremy Bentham.
6. David Hume's philosophy of history.

Recommended Reading

Barnes, Harry Elmer, ed.: *The History and Prospects of the Social Sciences,* Chap. 7, A. A. Knopf, Inc., New York, 1925.

Becker, Carl L.: *The Heavenly City of the Eighteenth-Century Philosophers,* Chap. 3, Yale University Press, New Haven, 1932.

Becker, Carl L.: *The Declaration of Independence,* Harcourt Brace and Co., New York, 1922.

Bronowski, J. and B. Mazlish: *The Western Intellectual Tradition,* Part III, Harper and Bros., New York, 1960.

Bury, J. B.: *The Idea of Progress,* Dover Publications, Inc., New York, 1955.

Cassirer, Ernst: *The Philosophy of the Enlightenment,* Chaps. 5 and 6, Princeton University Press, Princeton, 1951.

Davidson, William L.: *Political Thought in England, Bentham to J. S. Mill,* Oxford University Press, New York, 1915.

Hall, Everett W.: *Modern Science and Human Values,* pp. 149–184, 261–270, 354–393, D. Van Nostrand Co., Inc., Princeton, 1956.

Höffding, Harald: *A History of Modern Philosophy,* Vol. 1, pp. 441–500, Vol. II, pp. 361–433, Dover Publications, Inc., New York, 1955.

Kline, Morris: *Mathematics in Western Culture,* Chap. 21, Oxford University Press, New York, 1953.

Neill, Thomas P.: *Makers of the Modern Mind,* Bruce Publishing Co., Milwaukee, 1949.

Newman, James R.: *The World of Mathematics,* Vol. II, pp. 1200–1237 (selections from Cournot and Jevons), Simon and Schuster, Inc., New York, 1956.

Randall, John Herman, Jr.: *The Making of the Modern Mind,* rev. ed., Chaps. 13, 14, 18, and 19, Houghton Mifflin Co., Boston, 1940.

SAMPSON, R. V.: *Progress in the Age of Reason,* Harvard University Press, Cambridge, 1956.

SMITH, PRESERVED: *A History of Modern Culture,* Vol. II, Chaps. 6 to 8, Henry Holt and Co., New York, 1934.

SNYDER, LOUIS L.: *The Age of Reason,* pp. 45–68 and 125–178, D. Van Nostrand Co. (Anvil paperback) Princeton, 1955.

STEPHEN, LESLIE: *The English Utilitarians,* 3 vols., G. P. Putnam's Sons, New York, 1900.

WILLEY, BASIL: *Nineteenth Century Studies,* Chaps. 5 and 6, Chatto and Windus, London, 1949.

WOLF, ABRAHAM: *A History of Science, Technology and Philosophy in the Eighteenth Century,* 2nd ed., Chaps. 29 and 30, George Allen and Unwin Ltd., London, 1952. Also in paperback.

CHAPTER 29

THE STATISTICAL APPROACH TO THE SOCIAL AND BIOLOGICAL SCIENCES

People who don't count won't count.

Anatole France

29–1 Introduction. The success of the deductive approach employed by mathematics and the physical sciences depends upon the acquisition of correct and significant basic principles. In mathematics proper these principles are the axioms of number and geometry. In the physical sciences they are, for example, the laws of motion and gravitation. Though the social scientists sought such principles and thought they had found them when they decided that each individual acts in his own self-interest or that governments must maximize happiness, they soon realized that their deductions from these principles were not true laws of political theory or economics.

The social scientists' inability to find fundamental principles is undoubtedly due to the immense complexity of the phenomena that they wish to study. Human nature is a more complicated structure than a mass sliding down an inclined plane or a bob vibrating on a spring. A phenomenon such as national prosperity is even more complicated; not only are millions of human wills and rapacities involved, but so are natural resources, relationships with other nations, the disruptions of war, and a dozen other major factors. The difficulties which harass the social scientists are also encountered by the biologists. Although the physical sciences have provided some insight into the functioning of the eye, the ear, the heart, and muscular action, and although chemistry is making rapid advances in the study of complex molecular structures, the operation of the human body and the brain remains, on the whole, a great mystery.

If one attempts to simplify these problems by making assumptions about some of the factors involved or by neglecting what appear to be minor factors, just as Galileo, for example, neglected air resistance, one is likely to make the problem so artificial that its solution no longer has any bearing on real situations.

Very fortunately the social and biological sciences have acquired a totally new mathematical method of obtaining information about their respective phenomena—the method of statistics. By resorting to numerical data and by applying techniques which distill the essential content of those data, these sciences have made striking progress in the past one hundred years. However, the use of statistical methods has also given rise to the problem of determining the reliability of the results, and this aspect of statistics is treated by means of the mathematical theory of probability. In the present chapter we shall survey some of the concepts of statistics and in the next one we shall study the concept and applications of probability.

29–2 A brief historical review. The realization that statistics could serve as a method of attack on major social problems came first to a prosperous seventeenth-century English haberdasher, John Graunt (1620–74). Purely out of curiosity Graunt studied the death records in English cities and noticed that the percentages of deaths due to accidents, suicides, and various diseases were about the same in the localities studied and scarcely varied from year to year. Thus occurrences which superficially seemed to be a matter of chance possessed surprising regularity. Graunt also was the first to discover the excess of male over female births. On this statistic he based an argument: since men are subject to occupational hazards and war service, the number of men available for marriage approximately equals the number of women, and so monogamy has natural sanction. He also noticed the high mortality rate of children and the higher death rate in urban as compared to rural areas. In 1662 Graunt published his *Natural and Political Observations . . . upon the Bills of Mortality,* a book which might be said to have launched the trend toward scientific method in the social sciences and which certainly founded the science of statistics.

Graunt's work was followed and supported by his friend Sir William Petty (1623–85), professor of anatomy at Oxford, professor of music at Gresham College, and later an army physician. Petty wrote on medicine, mathematics, politics, and economics. His *Political Arithmetic,* written in 1676 and published in 1690, did not contain any more striking facts than Graunt's work, but is of particular significance because it calls specific attention to the method of statistics. The social sciences, he insisted, must become quantitative. He says, "The method I use is not yet very usual; for, instead of using only comparative and superlative words, and intellectual arguments, I have taken the course . . . to express myself in terms of number, weight, and measure; to use only arguments of sense, and to consider only such causes as have visible foundations in nature." To the infant science of statistics he gave the name of "Political Arithmetic," defining it as "the art of reasoning by figures upon things relating to the government." In fact, he regarded all of political economy as just a branch of statistics.

The work of Graunt and Petty was followed by studies of population and income and by extensive studies of mortality rates among which those made by the astronomer Edmond Halley are famous. Life insurance companies formed at the end of the seventeenth century and in the eighteenth century explored further data on mortality. However, though the subject of statistics did come to be known in the eighteenth century as data for statesmen, no mathematical methods for extracting significant implications from the data were developed.

Undoubtedly it was the aggravated social ills brought on by the Industrial Revolution in Europe which prompted a number of men to wade farther into important statistics, such as birth and death records, national and individual incomes, mortality, unemployment, and incidence of diseases, and to seek solutions for major problems through statistical methods. The man who revived Graunt's and Petty's basic thought that statistical methods might produce significant laws for the social sciences was a Belgian, L. A. J. Quetelet (1796–1874). Inspired by the successes of the physical sciences and conscious of the failure of the deductive approach to the social sciences, Quetelet undertook to construct and apply statistical methods suitable for social and sociological investigations. Quetelet was professor of astronomy and geodesy at the École Militaire and in 1820 became director of the Royal Belgian Observatory, which he founded. In 1835 he published his *Essay on Social Physics*. In 1848 he presented to the Royal Belgian Academy a memoir *On Moral Statistics* which contained conclusions on the science of government. Ironically, the publication of this memoir coincided with the outbreak of the revolution of 1848 in Paris. Prince Albert of Belgium remarked that the law governing the causes which led to revolutions had unfortunately come a little late.

In the latter half of the nineteenth century a number of well-known scientists, attracted by the already evident power of statistical methods, entered the field. We must content ourselves with mentioning Francis Galton (1822–1911) and Karl Pearson (1857 1936). It so happened that statistical techniques were already proving to be highly important in astronomy and in the theory of gases, and so the physical and social scientists accelerated the creation and application of statistical methods.

Before we examine the mathematics of extracting information from data, we should be clear as to how the method of statistics differs from the deductive approach. To put the matter crudely, the statistical approach to a problem is first of all a confession of ignorance. When crucial experiments, observation, or intuition fails to give us fundamental principles which can be used as premises for a significant chain of reasoning, we turn to *data* and seek to cull whatever information we can from what has happened. If we lack the knowledge which permits us to *deduce* what a new medical treatment should achieve, we apply the treatment, note results, and then attempt to draw some conclusions. Even if we come to the conclusion that the treatment is remarkably successful and should be widely applied, we still do not know what physical or chemical factors are operative. Perhaps the most important difference between the deductive approach and statistical methods is that the latter tell us what happens to large groups and do not provide definite predictions about any one given case, whereas the former predicts precisely what must happen in individual instances.

29–3 Averages. The task of the science of statistics is to summarize, digest, and extract information from large quantities of data. Our illustrations and discussions of various statistical techniques will be based on somewhat artificial and limited classes of data. Real problems usually involve large collections of data whose handling then becomes so encumbered by arithmetic that one loses sight of the essential mathematical idea.

About the simplest mathematical device for the distillation of knowledge from data is the average. A housewife who buys a 5-pound bag of potatoes once a week at prices which vary throughout a year can add the sums spent and divide by 52. She then has an average, known as the arithmetic *mean,* which represents fairly well what potatoes cost during that year.

Such an average can be meaningful in some situations and quite misleading in others. Suppose that we wish to study the wages of workers in an industry, that we have selected 1000 people as our representative sample, and that we have listed the wages of each worker. Suppose further that the mean wage turns out to be $1200. This figure gives some information about the earnings of the people, but not too much. For example, 990 people of the 1000 could be earning $1100 each; 10 people could be earning $11,100 each; and the mean of these earnings would be $1200. Hence our mean figure tells us nothing about the inequalities in the distribution of these wages.

Another type of average commonly used is called the *mode.* In a study of wages, for example, it is that wage which is earned by most people. Suppose that the distribution of wages is such that 25 people happen to receive the wage of $1150, whereas any other salary, larger or smaller, is received by fewer than 25 people. In this case, the mode is $1150. What does this figure tell us about the wage distribution if we do not know what the remaining 975 people earn? Are these others receiving wages near $100,000 a year or near $100 a year? Obviously, the mode may not be the average to represent such a situation.

A third type of average is called the *median.* Let us examine its meaning with the help of the following table:

Salary	Number of People
1,000	1
1,100	3
1,200	4
1,300	2
10,000	2
20,000	2
50,000	1

The table lists a total of 15 salaries. The median salary is the salary of the middle man, so to speak, or the eighth person; that is, there are as many who earn less than this middle person as there are who earn more. Since the eighth person occurs among the four who earn $1200, the median salary is $1200. To obtain the median of a set of data, one must arrange the data (salaries in the above example) in order of increasing magnitude and then find the datum which occurs in the middle. Of course, one must take into account how many times each datum occurs. Determining the median is a clumsy procedure. But more objectionable is the median's failure to provide any information about the level of the salaries above and below the median.

Although not one of the three averages we have discussed, or others we could discuss, is particularly informative, the mean is nevertheless the best one, for it, at least, takes into account the actual salaries earned by all the people involved. As we shall see later, the mean also proves to be the most useful concept in other statistical techniques.

Exercises

1. State the meanings of mean, mode, and median.
2. Calculate the mean and mode for the salaries listed in the above table.
3. A class of 20 students was graded as follows:

Number of Students	6	1	2	3	3	3	2
Grade	10	8	7	5	4	3	2

What are the mean, median, and mode of these data? Which average best represents the data?

4. The following weekly wages were paid to the employees of a company:

Number of Employees	4	18	10	9	13
Wages	50	40	35	30	10

Answer the same question as in Exercise 3.

5. Criticize the assertion, "Obviously there must be as many people with above-average intelligence as there are with below-average intelligence."
6. Is it safe for an adult to step into a pool whose mean depth is 4 ft?

29-4 Dispersion. We have already pointed out that no one of the averages provides detailed information about a set of data. Insofar as the mode and median are concerned, we can certainly change the salaries above and below either of these averages as much as we want to without affecting them. The mean has similar shortcomings. For example, the numbers 3, 5, and 7, each taken once, have a mean of 5, but so do the numbers 0, 5, and 10. Thus one may again change the data, though not at will, and still obtain the same mean.

To obtain further information about a set of data, statisticians seek to measure how closely the data are grouped about the mean; i.e., they seek to determine the *dispersion* of the data about the mean. Thus the data 0, 5, 10 are more widely dispersed about the mean of 5 than are the data 3, 5, 7. Various measures of dispersion might be introduced, but the one which has proved to be the most useful and also the best from the point of view of mathematical manipulation is the *standard deviation*.

Let us note first what is meant by a deviation. Suppose that the grades of six students in a class are:

Number of Students	1	1	1	1	1	1
Grade	3	4	5	6	8	10

We calculate first the mean grade; that is, we multiply each grade by the number of students earning that grade and divide by the number of students. In the present example, the mean grade is 6. The deviation of any grade from the mean is merely the difference between that grade and the mean. Thus the deviations for the above set of grades are

$$3,\ 2,\ 1,\ 0,\ 2,\ 4.$$

To obtain the *standard deviation,* one squares each deviation, calculates the mean of these squares, and then takes the square root. The squares of the deviations are

$$9,\ 4,\ 1,\ 0,\ 4,\ 16.$$

To compute the mean of these squares, we must add them *taking each with the frequency with which it occurs,* and divide by the total number of data. Thus

$$\frac{9 + 4 + 1 + 0 + 4 + 16}{6} = 5.66.$$

The standard deviation, denoted by σ (sigma), is the square root of this mean. Then

$$\sigma = \sqrt{5.66} = 2.4.$$

The standard deviation of 2.4 is merely a convenient and yet somewhat arbitrary measure of how close the various grades are to the mean. Had the grades of the six students been

Number of Students	2	2	2
Grade	2	6	10

the mean would again be 6, but the standard deviation would turn out as follows. The deviations are

$$4,\ \ 0,\ \ 4.$$

The squares of these deviations are

$$16,\ \ 0,\ \ 16.$$

The mean of these squares is

$$\frac{2 \cdot 16 + 2 \cdot 0 + 2 \cdot 16}{6} = 10.66,$$

and the standard deviation is

$$\sigma = \sqrt{10.66} = 3.3.$$

In other words, the fact that more grades in this latter distribution are farther from the mean of 6 is reflected in the change of the standard deviation from 2.4 to 3.3.

Exercises

1. The grades of eight students on a quiz were 1, 2, 4, 5, 8, 9, 9, 10. Calculate the standard deviation of this distribution of grades.
2. Calculate the standard deviation of the grades listed in Exercise 3 of Section 29–3. You may take the mean to be 6.
3. Calculate the standard deviation of the wages listed in Exercise 4 of Section 29–3.
4. Suppose that the mean height of men in a certain city is 5′7″ and the standard deviation is 2″, while in another city the mean is the same but the standard deviation is 3″. What fact is revealed by the difference in standard deviation?
5. Suppose that a student made a grade of 75 in an examination for which the mean grade was 65 and the standard deviation of the grades was 5, and another student made the same grade in an examination for which the mean was also 65, but the standard deviation was 15. Which student did better?
6. What is the significance of the standard deviation?

29–5 The graph and the normal curve. A better knowledge about some collections of data than the mean and standard deviation afford can be obtained by means of a graph. Consider, for example, the wages paid in an industry. In this case, the graph might show the wages paid as the abscissas and the numbers of people earning those wages as the ordinates. Although there may be thousands of different salaries, it is not necessary to plot that many points. One might group the salaries in $10-intervals, and enter into one interval all those earning from $50 to $60, in the next interval those earning from $60 to $70, and so on. One might then regard all those grouped in the first interval as earning $55, those in the second as earning $65, and so forth. Thus one point on the graph will have an abscissa of 55 and an ordinate equal to the number of people whose salary is somewhere between $50 and $60 or, to be precise, between $50 and $59.99. When a smooth curve is drawn through the points plotted, its shape clearly exhibits the gradual variation in the number of people earning salaries from $50 to the maximum salary paid (Fig. 29–1).

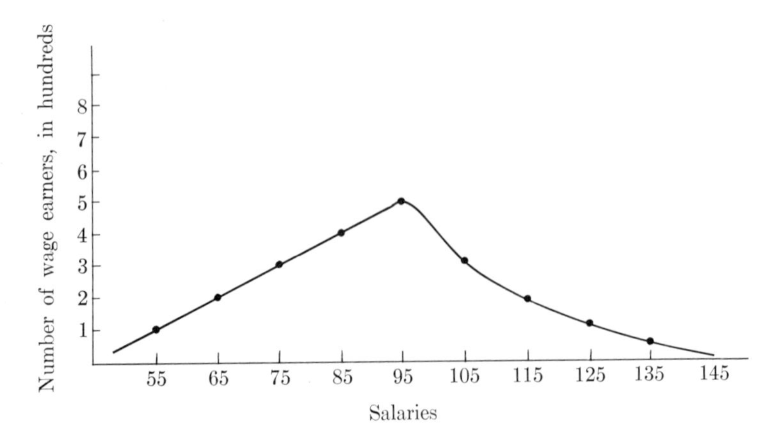

FIG. 29–1. Frequency distribution of numbers of employees earning different salaries.

Of course, the process of lumping together all those with earnings from \$50 to \$60 and using a representative figure of \$55 has introduced some element of error into the data and graph. If this error should matter for the purposes of the study, it might be necessary to use a smaller interval of, say \$5 or \$2, instead of \$10. The smooth curve may also be misleading. The graph in Fig. 29–1, for example, seems to show that to *every* possible salary in the range from \$50 to the maximum there correspond some wage earners. Actually the graph represents only the number of people earning salaries of \$55, \$65, \$75, and so on. However, the shape of the graph is reasonably accurate with respect to the *relative* frequencies. For example, the smooth curve shows more people earning \$60 than, say \$55; that is, it shows a distribution which very likely reflects the actual situation, especially if the number of wage earners is large.

The value of the graph is apparent. The mean and standard deviation of the salaries involved would not reveal the rather sharp drop in number from medium- to high-salaried employees and the small number of people earning very high salaries. Graphs, then, do provide a useful picture. A person who reads the newspapers and magazines can hardly fail to observe how commonly they are employed. Not all of these graphs are smooth curves. Bar graphs and pie charts are also frequently used.

The variables whose relationship is represented by a graph may be time and stock prices, production and consumption of coal, or hundreds of other similarly related data. The relationship which plays a central role in statistical work is called a *frequency distribution*. Thus the graph plotting wages versus the number of people earning the various wages illustrates a frequency distribution. The heights of people and the number of people possessing these various heights, intelligence quotients and the number of people possessing the various quotients, grades on an examination and the number of students earning those grades are all frequency distributions.

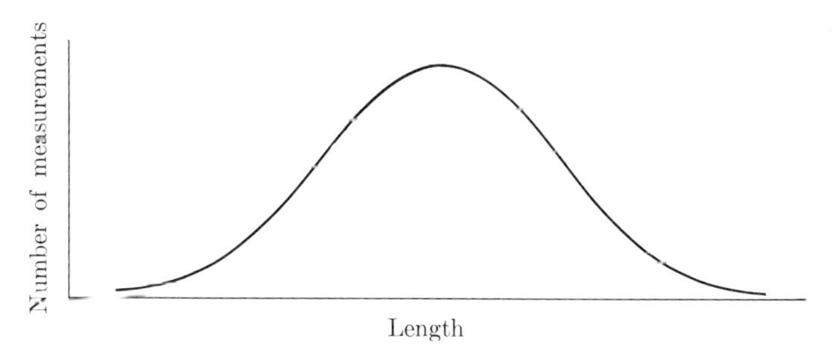

FIG. 29–2. The normal frequency curve.

Among frequency distributions one type is of particular importance. Consider the rather simple problem of measuring a length. A scientist interested in the exact length of a piece of wire, say, measures it not once but, if need be, fifty times. Partly because no measurement is exact and partly because environmental conditions such as temperature affect the length, these fifty measurements will differ from each other, sometimes perceptibly and sometimes imperceptibly. A graph plotting the results of all fifty measurements against the number of times that each measurement occurs will look like the curve in Fig. 29–2. In fact, the more measurements that are made, the more nearly will their frequency distribution follow this curve.

The graph in Fig. 29–2 has been well known to physical scientists since about 1800 because it is obtained from almost all measurements of physical quantities. The measurements cluster about one central value, just as the shots of a rifleman at a target will, if he is a marksman, cluster about the bull's-eye. The similarity of the two situations, the measurement of a length and shots at a target, suggests that the exact length should be the central value. The other lengths apparently represent random or accidental variations from the true length just as the shots near, but not on, the bull's-eye represent accidental errors in marksmanship. Because the shape of the graph in Fig. 29–2 occurs repeatedly in connection with errors of measurement, it has come to be known as the *error curve* or the *normal frequency curve*. Its very existence affirms the seemingly paradoxical but nonetheless true conclusion that accidental errors in measurements do not follow any chance pattern, but always follow the error curve. Humans may not even err at will.

The normal frequency curve is not just one curve but a class of curves possessing common mathematical properties, just as the parabola is not a single curve but a class of curves which can be defined geometrically as the loci of points which are equidistant from a fixed point and a fixed line, and which are algebraically represented by the equation $y = (1/2a)x^2$ (for the proper choice of coordinate axes). The precise definition of the class of normal frequency curves will not be stated because the formula contains a function we have not studied. But we can characterize this class of curves

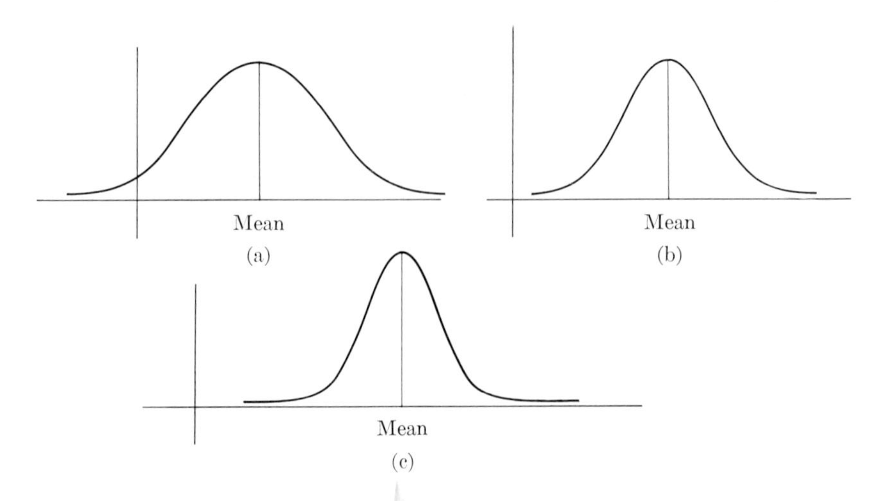

FIG. 29–3. Three different normal frequency curves.

well enough for our purposes. Figure 29–3 shows three different normal frequency curves. Their shapes resemble bells, and the curve is therefore often described as bell-shaped. Each curve is symmetric about a vertical line. The abscissa of the base point of this vertical line is the mean of the data which are plotted as abscissas, for example, lengths. The means of the three curves will generally be different values. In fact one may be a mean length; another, a mean height; and so on. It is almost apparent from the symmetry of these curves that the mode and median coincide with the mean of each distribution, and the different widths indicate that each has its own standard deviation. The left-hand curve (29–3a) evidently has the largest standard deviation of the three because the data are more widely dispersed.

Despite these differences, all normal frequency curves are characterized by their mean and standard deviation. Regardless of the value of the mean and of the standard deviation, 68.2% of the data lie within σ on either side of the mean (Fig. 29–4); 95.4% of the cases lie within 2σ of the mean; and 99.8% of the cases lie within 3σ of the mean.* Thus if the curve of Fig. 29–4 represented the heights of 100,000 men, if the mean were 67 inches, and if σ were 2, then 68,200 men would have heights falling within the range 65 to 69 inches; 95,400 men would have heights between 63 and 71 inches; and so forth. If the standard deviation were 1 instead of 2, the curve would have a sharper hump around the middle because the dispersion would be smaller. But it would still be true that 68.2% of the population would lie within σ of the mean; i.e., 68,200 men would have heights between 66 and 68 inches. Thus knowing that some frequency distribution is normal and knowing the

* The normal frequency curve, as a mathematical curve, extends indefinitely far to the right and to the left of the Y-axis although its so-called tails come closer and closer to the X-axis. However, these tails play almost no role since only 0.1% of the measured events can occur beyond 3σ to the right or to the left of the mean.

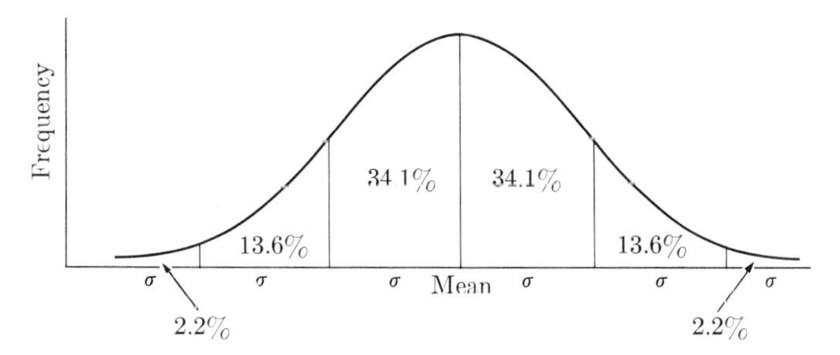

FIG. 29–4. The normal frequency curve.

mean and standard deviation of this distribution, we can draw a great number of conclusions about the data.

About 1833 Quetelet decided to study the distribution of human traits and abilities in the light of the normal frequency curve. He took many of his data, incidentally, from the thousands of anatomical measurements made by the Renaissance artists, Alberti, Leonardo, Ghiberti, Dürer, Michelangelo, and others. He found what hundreds of successors have since confirmed. All mental and physical characteristics of human beings follow the normal frequency distribution. Height, the size of any one limb, head size, body weight, brain weight, intelligence (as measured by intelligence tests), the sensitivity of the eye to the various frequencies of the visible portion of the electromagnetic spectrum, all these properties are normally distributed within one genus, which may be race or nationality. The same is true of animals and plants. The sizes and weights of grapefruits of any one variety, the lengths of the ears of corn of any one species, the weights of dogs of any one breed, and so forth, are normally distributed.

Quetelet was struck by the fact that human traits and abilities follow the same distribution curve as do errors of measurement. He concluded that all human beings, like loaves of bread, are made in one mold and differ only because of accidental variations arising in the process of creation. Nature aims at the ideal man, but misses the mark and thus creates deviations on both sides of the ideal. The differences are fortuitous, and, for this reason, the law of error applies to these distributions of physical characteristics and mental abilities. On the other hand, if there were no general type to which men conform, measuring their characteristics, height, for example, would not reveal any particular significance in the graph or any definite numerical relationships in the data.

The typical man, according to Quetelet, emerges as the result of a great number of measurements. The mean of each of the characteristics, that is, the value having the largest ordinate, belongs to this typical, or "mean," man, who is, incidentally, the center of gravity around whom society revolves. The more measurements Quetelet made, the more he noted that individual variations are effaced and that the central characteristics of mankind tend to be sharply defined. These central characteristics, he then declared, proceed from underlying forces or causes which fashion mankind. More than that, his results led him to believe that he had found decisive evidence for the existence of eternal laws of human society and of design and determinism in social phenomena.

We shall defer judgment on Quetelet's philosophical inferences to the next chapter. Let us content ourselves, for the moment, with the observation that the applicability of the error curve to social and biological problems has led to knowledge in these fields and to laws. Indeed, the conviction that the distribution of any physical or mental ability must follow the normal curve is today so firmly entrenched that any measurements on a large number of people which do not lead to this result are suspect. If, for example, a new test given to a representative group does not lead to a normal distribution of grades, it is not the conclusion about the distribution of intelligence which is challenged; the test is declared invalid. Similarly, if measurements of velocity, force, or distance failed to follow a normal distribution, the scientist would blame his measuring instruments.

Another use of the normal frequency distribution occurs in manufacture. For example, manufactured wire is continually tested for quality. Suppose that 100 samples are taken each day from the day's production and tested for tensile strength. A graph can be drawn showing strength against the number of samples having that strength. Such graphs usually are normal distributions. Now if the distribution resembled the curve in Fig. 29–3(a), it would imply a wide variety in strength of samples and hence nonuniform production. On the other hand, a distribution such as the curve in Fig. 29–3(c) shows uniform production. The two distributions differ in dispersion, and therefore in their standard deviations. If uniformity is important—and often it is more important than superior quality because a defective piece of wire in an electric circuit may do a lot of damage—then a graph such as the left-hand one indicates the need for some change in the manufacturing process.

One must not presume, however, that all, or practically all, distributions follow the normal frequency curve. The distributions of incomes of families or individuals and the number of families owning 0, 1, 2, 3 or 4 cars are not normal. The failure of incomes to follow a normal curve raises an interesting point because physical and mental abilities are normally distributed and these qualities should determine income.

Exercises

1. What is a frequency distribution?
2. Describe the normal frequency curve.
3. Given a normal frequency distribution, what percentage of the data lie within a range of 2σ on both sides of the mean?
4. Suppose that the heights of 1000 college freshmen are measured and found to follow a normal frequency distribution, with a mean of 66 in. and a standard devia-

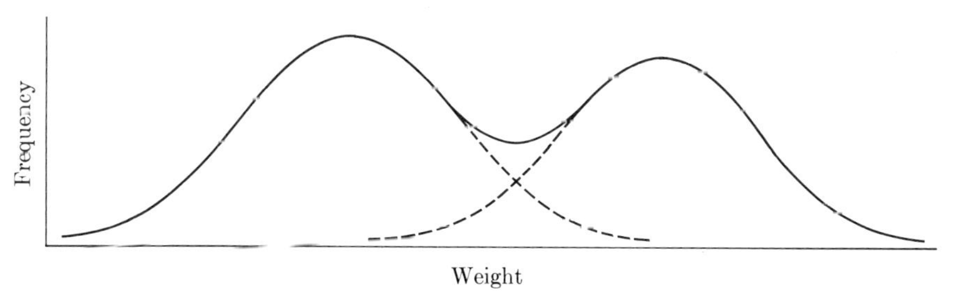

FIGURE 29–5

tion of 2 in. What percentage of the students has heights between 66 and 70 in.? between 60 and 72 in.?

5. The United States Army gives a well-designed intelligence test to all prospective soldiers and then rejects all those whose scores fall, say, below one σ to the left of the mean. Draw a graph showing the frequency distribution of intelligence of the men accepted for service.

6. Where would the mean income of the distribution of incomes shown in Fig. 29–1 lie in relation to the modal income?

7. Suppose you measured the weights of 1000 grapefruits and made a frequency distribution of the various weights. If the resulting curve followed the solid line shown in Fig. 29–5, what would you conclude about the homogeneity of the species of grapefruit?

29–6 Fitting a formula to data. We have seen that a great deal of information can be extracted from data by the application of averages, standard deviation, and graphs. When graphical methods are employed, we are particularly fortunate if the graph happens to be a normal frequency distribution. However, the major techniques of mathematics used to derive new information from given facts are designed to apply to formulas. If the data that we happen to be studying present a functional relationship, for example, the variation of population with time in some region, then it is extremely desirable to obtain a formula for this function.

Now the compression of data into formulas is usually possible, and the process is fraught with meaning which we shall examine later. For the present, we shall limit ourselves to illustrating the procedure and, for this purpose, we shall consider first a somewhat specialized and slightly oversimplified problem. By measuring the velocity of a falling body at various instants of time Galileo obtained the following data:

Time, in seconds	Velocity, in ft/sec
0	0
1	32
2	64
3	96
4	128

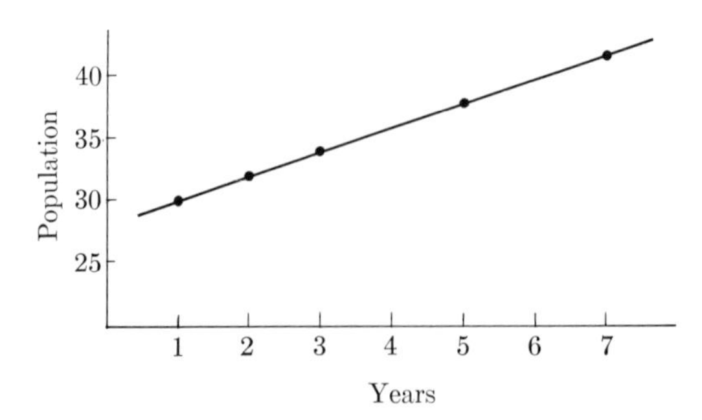

FIG. 29–6. The graph of population versus years for a particular town.

By *inspecting* this table Galileo could see that the formula relating velocity and time is

$$v = 32t. \tag{1}$$

Let us treat next the problem of obtaining a formula for data when the result is not obvious by inspection. Most towns, cities, and even the country at large are concerned with studying population changes to predict the needs for housing, water, sewers, schools, and so on. Suppose a town has the following record of growth:

Year	1951	1952	1953	1955	1957
Population	3000	3200	3400	3800	4200

Can one find a formula to fit these data?

To simplify the graphing and calculation, let us count years from 1950 and population in hundreds. Thus 1951 will be regarded as the year 1 and the population for that year as 30. The graph of the data is shown in Fig. 29–6. A straightedge placed along the plotted points shows that they lie on a straight line. We know from our work on coordinate geometry (see Exercise 13, Section 13–3) that the formula or equation of a linear graph is of the form

$$y = mx + b. \tag{2}$$

Let y represent the population of the town and x the year. Since the straight line is the curve which fits our data, let us project the straight line backward to the point where it crosses the Y-axis. At this point, we see from the graph that the value of y is 28 and, of course, $x = 0$. Since formula (2) is to fit the graph, then for $x = 0$, y must be 28. If we substitute these values in (2), we have

$$28 = m \cdot 0 + b,$$

and we see that $b = 28$. So far, then, our formula is

$$y = mx + 28. \tag{3}$$

The quantity m is unknown. However, the graph tells us that when $x = 5$, $y = 38$. Let us therefore substitute these values in (3). We obtain

$$38 = m \cdot 5 + 28$$

or

$$5m = 10$$

or

$$m = 2.$$

Hence the final equation relating y and x is

$$y = 2x + 28, \tag{4}$$

and we have found a formula which fits the data of the above table.

With this formula we can now predict the population of the town, say for the year 1970. For 1970, $x = 20$. Then, substituting 20 for x in (4), we obtain

$$y = 2 \cdot 20 + 28 = 68.$$

The formula predicts that the population in 1970 will be 68, that is, 6800 people. Of course, we are assuming that the factors which led to the increase in population from 1951 to 1957 will not only continue to operate, but will operate on the same level. Here we encounter one of the serious limitations in the use of statistics for social and economic phenomena. Since we do not know the fundamental forces which control such phenomena (although we may have some qualitative information), we cannot be at all certain that the population will continue to increase after 1957 in the same way as it did before. In fact, we can be fairly sure that it will not. Local, national, or global events slow down or accelerate the growth of a population. For example, during a financial depression young people cannot afford to marry and have children. By contrast, the study of physical phenomena reveals that the forces operating in nature are invariable.

As a matter of fact, there is a fundamental difference between the formulas obtained from data of the physical sciences and formulas derived from data of biology, psychology, the social sciences, and pedagogy. One may say that, in general, a formula developed from data of the first class *continues to hold* as added data are gathered. Three hundred years ago Kepler deduced his laws from observational data, and they are still correct. On the other hand, for problems of the second class, it does not happen very often that formulas continue to hold without corrections as additional data are gathered. We must constantly refit the formula to the enlarged collection of data. This inconstancy need not be interpreted to imply that lawlessness prevails in the social and biological sciences, for we have already en-

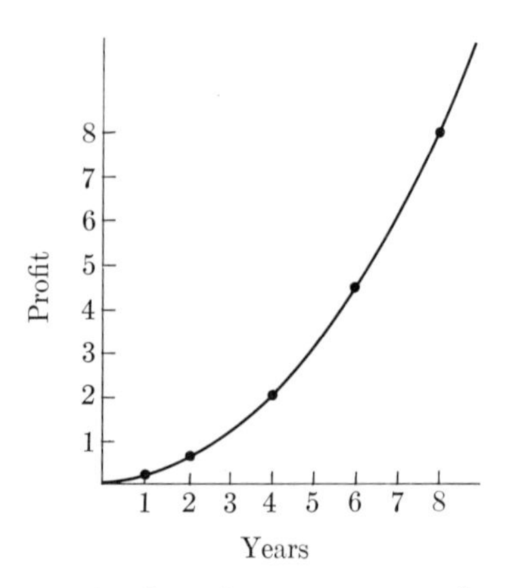

FIG. 29–7. The graph of profits versus years for a given firm.

countered instances of what appear to be well-established laws in these areas. Until we have seen more of the effectiveness of mathematics in the biological and social sciences, we shall not be in a position to discuss in detail whether or not there is design in the social order (see Chapter 30).

The above example shows data which led to a linear function. But suppose that the graph were not a straight line or sufficiently close to one to be approximated by a straight line. An example will show that we can nonetheless fit a formula to nonlinear graphs. Suppose that, beginning with the year 1951, the profits of a business concern were as follows:

Year	1951	1952	1954	1956	1958
Profits, in million dollars	0.125	0.5	2	4.5	8

To find a formula fitting these data, we first plot the data. Let us record years after 1950 as abscissas and profits as ordinates. Figure 29–7 shows the graph.

The points plotted are connected by a smooth curve whose appearance suggests a parabola. We know from our work on coordinate geometry that parabolas placed with respect to the axes as shown in the figure have equations of the form [see Chapter 13, formula (9)]

$$y = \frac{1}{2a}x^2. \tag{5}$$

Let x in (5) stand for time and let y denote the profits. To fit formula (5) to our data, we must first determine the value of a. We choose the coordinates of any one point on the curve. Thus in 1954 the profits were 2.

The corresponding point on the curve has coordinates $x = 4$, $y = 2$. Substituting these values in (5), we obtain

$$2 = \frac{1}{2a} \cdot 16.$$

Then $a = 4$, and formula (5) becomes

$$y = \frac{1}{8} x^2. \tag{6}$$

Of course, in classifying the curve as a parabola, we judged by appearance. Hence we should check whether (6) fits other data on the graph. Thus, for example, the coordinates of the point corresponding to 1956 are $x = 6$ and $y = 4.5$. If we substitute these values in (6), we see that the left side does indeed equal the right side, and so we have verified that the parabola does fit the data.

If, for example, the profits for the year 1956 had been 4.6 instead of 4.5, then the result of substituting 6 for x in (6) would not have yielded the exact figure. However, one might still accept formula (6) as a good approximation to the data. If the substitution into (6) of one or more x-values of points on the graph had not yielded even good approximations to the corresponding y-values, then our judgment about the parabolic nature of the graph would have been wrong, and we would have had to fit a different type of formula to the data. There are techniques which aid in deciding what formulas to try, but these are valuable only for the specialist.

These few examples, obviously chosen for their simplicity, show how formulas can be fitted to data. At the same time we see that the procedure presupposes a knowledge of coordinate geometry, that is, the relationship between equation and curve.

Exercises

1. The following data are given for two variables, x and y:

x	0	1	2	3	4
y	7	10	13	16	19

Find by inspection a formula of the form $y = mx + b$ relating x and y.

2. When a spring is stretched, it exerts a force which opposes the stretching force. To determine the relation between F, the force exerted by the spring, and d, the amount of stretch or displacement, the following data are available:

d, in inches	1	2	3	4
F, in pounds	4	8	11.9	16.2

Graph these data and find a formula which relates F and d. The result is Hooke's law (Chapter 23) for the particular spring.

3. Suppose that experimentation has yielded the following data on the distance fallen by a body in various time intervals:

t, in seconds	1	$1\frac{1}{2}$	2	$2\frac{1}{2}$
d, in feet	16	36	64	100

Though you undoubtedly know the answer, graph the data, and fit a formula relating d and t to the graph.

4. People are constantly concerned with the rise of prices. These are measured by an index number which represents an average cost of vital commodities and services. Suppose that the index numbers for a number of years are as follows:

Year (Y)	1951	1953	1955	1957	1959
Index number (N)	7	8.9	11.1	13	15.2

Find a formula relating the index number N and the year Y. Check the formula obtained by trying it out on those data of the table that are not used to determine the constants.

5. The profits of a concern for the years following 1950 are:

Years after 1950 (t)	1	2	3	4	5
Profits, in thousands of dollars (P)	6	12	22	36	54

Fit a formula of the form $P = at^2 + b$ to the data.

6. The volume of a given quantity of water will vary with temperature because, as we have had occasion to learn in the past, water expands or contracts with temperature. Hence suppose you had gathered the following data on the volume in cu. in. of a fixed quantity of water at various temperatures (in degrees centigrade):

T	0	2	4	6	8	10	12	14
V	2.3	1.3	1	1.3	2.2	3.7	5.8	8.3

Determine the formula relating T and V. [*Suggestion:* Try the formula $V = a + bT + cT^2$, which, of course, represents a parabola. Then, since you know V for $T = 0$, you can immediately find a. Next use two sets of data to determine b and c by solving two equations in two unknowns.]

7. What are the advantages of fitting a formula to data?

8. How would a knowledge of coordinate geometry be helpful to a scientist who is attempting to fit a formula to data?

29–7 Correlation. The process of fitting a formula to data is useful in that it summarizes data and may permit further mathematical work with the formula. However, it is not always possible to fit a formula to data. The very notion of a functional relation demands that there be a unique value of y for each value of x in the range studied. Many types of data do not meet this condition. Suppose, for example, one wished to study the relationship between height and weight of men. To any one height there correspond many weights, or if one starts with weight, to any one weight there correspond many heights. Hence one cannot ask for a formula which relates weight and height. Nevertheless there is some correspondence between these two variables, and one might wish to determine the extent and nature of this relationship.

Sir Francis Galton, a cousin of Charles Darwin and founder of the science of eugenics, faced the above problem in his study of human characteristics. Galton was a doctor who used statistics to study heredity. In particular, in his famous *Natural Inheritance,* Galton undertook to investigate the relationship between the heights of fathers and the heights of their sons. It is immediately apparent that to any given height of a father there correspond many heights of sons. Galton introduced a notion now known as correlation. This mathematical concept permits one to measure the closeness of the relationship between two sets of data which may not be functionally related.

Galton found a close relationship between the heights of fathers and their sons. Tall parents have tall children. He also discovered, incidentally, that the mean height of all the sons of tall fathers was closer to the mean of the entire population than the mean height of all the fathers. Thus, while the trait of tallness or shortness is on the average inheritable, succeeding generations regress toward a norm. He also found that the same conclusions apply to intelligence. Talent is, on the average, inherited, but the children are more mediocre than the parents. (Hence parents know better what is good for their children than do the children themselves!). After finding that the same law also holds for other human characteristics, Galton concluded, first, that human physiology is stable and, secondly, that all living organisms tend towards types. Leaving aside Galton's broad inferences, we find in his work examples of biological laws obtained solely by the use of statistics and the simplest of mathematics. Moreover, it was possible to establish these conclusions without any knowledge of the mechanism of heredity.

The precise mathematical measure of correlation which is widely used today was formulated by Karl Pearson. His formula yields a number which lies between -1 and 1. A correlation of 1 indicates that the given variables are directly related; when one variable increases or decreases, so does the other; when one of the variables assumes a high numerical value, so does the other. A correlation of -1 means that the behavior of one variable is directly opposite to that of the other; as the values of the first variable increase, those of the second decrease, and conversely. A correlation of zero means that the behavior of one variable has nothing to do with the behavior of the other; they proceed independently of each other. A correlation of three-fourths, say, indicates that the behavior of one variable is similar to, but not identical with, that of the other.

A knowledge of correlations can be extremely valuable. If stock prices correlate highly with industrial production, one can use a knowledge of the former to study and predict the behavior of the latter. This approach has a definite advantage since stock prices are much more easily compiled than data on industrial production. If general intelligence correlates highly with ability in mathematics, then a person with good intelligence can expect to do well in mathematics. If the total earnings of a nation's wage earners correlate highly with prosperity as measured by the total profits of business concerns, then industry in its own interest ought to consider the prudence of diverting a larger share of its earning to its employees. If the correlation between the frequencies of occurrence of two diseases is high, the successful analysis of one may be expected to lead to an equally favorable result for the other. Knowledge of the correlation between success in high school and success in college or between success in college and financial success in later life can be extremely valuable in predicting the future of groups of individuals.

Exercise

Suppose that a study of 1000 students reveals a very high correlation between general intelligence and ability in mathematics. A particular student is known to be very intelligent. What would you expect his ability in mathematics to be?

29–8 Cautions concerning the uses of statistics. Since statistics are now widely employed in our society to bolster arguments on both sides of controversial issues, it might be advisable on this account as well as for a general understanding of the nature of statistically established conclusions to become aware of the pitfalls in applying statistical methods and in interpreting the mathematical results. One of the first difficulties in applying statistics is to decide the meaning of the concepts involved. Suppose that one wished to make a statistical study of unemployment, say over a period of years. Who are the unemployed? Should the term include those people who do not have to work, but would like to? Or people who are employed two days a week and are looking for full-time employment? Or the well-trained engineer who cannot find a job corresponding to his qualifications and has to drive a cab? Or the man unfit for employment? Should a study of passenger cars include taxis, station wagons, and passenger cars used by salesmen for business purposes?

After one has decided what objects or groups of people are to be included in a given term, the question of whether the data are reliable arises. For example, any study of crime rates must take into account that police departments occasionally change their practices of recording and classifying crimes. A study of the incidence of mental diseases among men and women must take into account that women are less frequently hospitalized than men.

The clear delineation of the problem to be investigated is also often a difficult matter. Suppose that one wishes to compare deaths due to automobiles in the United States and Great Britain. The number of deaths is certainly larger in the United States but so are the population and the number of automobiles. Should one compute the number of deaths per inhabitant, per automobile, or per mile of automobile travel?

The largest single problem which arises in the process of using statistics is the problem of sampling. To study the incidence of tuberculosis in the United States, for example, one does not examine every person; rather a group of people *believed to be typical* of the whole population is selected for study. This group is called a *random sample.* Similarly, all physiological and mental characteristics of human beings are studied by sampling. The level of retail food prices is gauged by selecting a few important food items which are considered to be representative of all foods. The study of wages in an industry is conducted by selecting a random sample of workers. The doctor studies a person's blood by sampling a small quantity which he believes to be typical because the blood is continually circulating throughout the body. A sociologist interested in the life of families with a given income will study a selected, typical group rather than the entire class. The Gallup poll studying the country's attitude towards public questions and the astronomer studying the number and sizes of stars in a region of the sky use sampling.

Since the conclusions of a statistical study are based on the sample, it is evident that the sample should be chosen with care. If one is studying the output of a machine by sampling its products, it is essential that the products be picked at various times during the day rather than all at one time. In the morning, before there is any chance of overheating, the machine may do better work than in the afternoon.

Given a truly random sample, the next question concerns the extent to which the information derived from the sample can be trusted to be indicative of the entire population. This problem involves probabilities, and we shall discuss this subject in the next chapter.

The evaluation of statistical results presents problems of its own. Let us suppose that the meanings of terms are satisfactorily established and that representative samples have been chosen, or that the entire population has been covered so that the questions raised by sampling do not enter. Statistics do show that graduates of Harvard make more money later in life than graduates of any other university. What shall we conclude? Does Harvard's education ensure greater success for the average student there as opposed to the average student at some other university? Hardly. Many of the students at Harvard come from well-established families who take their sons into the family business or profession.

Statistics show that in 1954 among fatal accidents due to automobiles 25,930 occurred in clear weather, 370 in fog, 3640 in rain, and 860 in snow. Do these statistics show that it is safest to drive in fog? Obviously not. Fogs occur more often at night when fewer cars are on the road. When fogs occur, many people refrain from driving, and others drive more cautiously. Finally, fogs are rare.

To see more clearly the danger of drawing hasty conclusions from statistics, one might resort to some extreme and even ridiculous examples. It has been noted that among people who sit in the front rows of burlesque houses bald heads predominate. May one conclude that close observation of burlesque shows produces baldness?

These difficulties in the compilation and evaluation of statistics are real enough. They have led to false inferences and to derogatory remarks such as that there are romances, grand romances, and statistics, or to the definition of statisticians as men who draw precise lines from indefinite hypotheses to foregone conclusions. One must indeed be careful in the uses of statistics. Especially where sampling is involved, statistics do not prove anything; they tend to show. They give us guides to action. Almost always statistical results do not tell us anything certain about an individual, but they indicate what is very likely to hold among a class of individuals as, for example, the distribution of intelligence.

However, the difficulties are easily overshadowed by the effectiveness of the statistical approach in studies of population changes, stock market operations, unemployment, wage scales, cost of living, birth and death rates, extent of drunkenness and crime, distribution of physical characteristics and intelligence, and incidence of diseases. Statistics are the basis of life insurance, social security systems, medical treatments, governmental policies, educational studies, and the numbers racket. Modern business enterprises are using statistical methods to locate the best markets, test the effectiveness of advertising, gauge the interest in a new product, and so forth. Pure speculation, haphazard guesses, and the captiousness of individual judgments are being supplanted by statistical studies. Indeed statistical methods have been decisive in turning undeveloped and backward fields into sciences, and they have become a way of approaching problems and thinking in all fields.

Exercises

1. Compare the methodology of the deductive approach to a field of investigation with the statistical approach.
2. Why have economists not succeeded in finding a deductive approach to the economic system of a country?
3. Assuming a reasonable definition of an unemployed person, suppose that statistics show rising unemployment in the United States for a period of five years. Do these statistics imply that the economic condition of the country has been getting worse during those years?
4. Statistics show that every year more people die of cancer than in preceding years. Is cancer caused by factors which are becoming more common in our civilization?
5. Statistics show that cancer occurs much more frequently among men who smoke heavily than among men who smoke little or not at all. Is smoking a cause of cancer?

6. It has been shown statistically that older fathers produce more intelligent children. Do these statistics imply that men should have children later rather than earlier?

7. The average age at death of people with false teeth is higher than that of people possessing natural teeth. Do false teeth enable you to live longer?

Topics for Further Investigation

1. The work of Sir William Petty.
2. The work of John Graunt.
3. The work of Sir Francis Galton.
4. The concept of correlation.

Recommended Reading

ALDER, HENRY L. and EDWARD B. ROESSLER: *Introduction to Probability and Statistics,* Chaps. 1 to 4, W. H. Freeman & Co., San Francisco, 1960.

FREUND, JOHN E.: *Modern Elementary Statistics,* 2nd ed., Chaps. 1 to 6, Prentice-Hall, Inc., Englewood Cliffs, 1960.

HUFF, DARRELL: *How to Lie with Statistics,* W. W. Norton & Co., New York, 1954.

KLINE, MORRIS: *Mathematics in Western Culture,* Chap. 22, Oxford University Press, New York, 1953.

NEWMAN, JAMES R.: *The World of Mathematics,* Vol. III, pp. 1416–1531 (selections on statistics), Simon and Schuster, Inc., New York, 1956.

WOLF, ABRAHAM: *A History of Science, Technology and Philosophy in the Sixteenth and Seventeenth Centuries,* 2nd ed., Chap. 25, George Allen and Unwin Ltd., London, 1950. Also in paperback.

CHAPTER 30

THE THEORY OF PROBABILITY

Life is the art of drawing sufficient conclusions from insufficient premises.

Samuel Butler

30–1 Introduction. Mathematics has been created by man to help him understand the universe and utilize the resources of the physical world. But the physical world of civilized man also includes such activities as throwing dice, playing cards, betting on horse races, playing roulette, and other forms of gambling. To understand and master these very phenomena, a new branch of mathematics, the theory of probability, was created. However, the theory now has depth and significance far beyond the sphere for which it was originally intended.

The first book on the subject was written by the Renaissance roué Jerome Cardan. Cardan, being a mathematician as well as a gambler, decided that if he were to spend time on gambling, he might as well apply mathematics and make the game profitable. He thereupon proceeded to study the probabilities of winning in various games of chance and in a rare moment of altruism decided to let others profit by his thinking on the subject. He compiled his results in his *Liber De Ludo Aleae* (The Book on Games of Chance), which is a gambler's manual with advice on how to cheat and detect cheating.

In 1653 another gambler and amateur mathematician, the Chevalier de Méré, became equally interested in using mathematics to determine bets in games of chance. Since his own talent was limited, he sent some problems on dice to Pascal, and the latter in collaboration with Fermat decided to develop further the subject of probability. Whereas Cardan solved just a few problems of probability, Pascal envisioned a whole science. He aimed "to reduce to an exact art, with the rigor of mathematical demonstration, the incertitude of chance, thus creating a new science which could justly claim the stupefying title: the mathematics of chance."

Cardan, Pascal, and Fermat were attracted to probability by problems of gambling. The subject was taken up by others, notably Laplace, whose interests, equally impractical, were in the heavens. In attempting to solve major astronomical problems, Laplace found himself compelled to consider the accuracy of astronomical observations. As we shall soon see, this problem leads to the theory of probability.

The theory might have remained a minor and largely amusing branch of mathematics were it not for the fact that the use of statistical methods made recourse to probability a necessity. Perhaps the most significant statistical problems which call for probabilistic thinking arise from the process of sampling. Statistical studies must, as a practical matter, proceed by sampling, and sampling unavoidably involves the possibility of error. If a survey of wages in the steel industry is based on data collected in two or three supposedly representative mills, we cannot be sure that we shall obtain *exact* facts about the entire industry from a study of this sample. If the output of a machine is tested by sampling, the conclusion derived from the sample may not hold for the entire output. To decide upon the efficacy of a new medical treatment doctors try it on a small group of patients. Now no treatment is perfect because its effect often depends upon other factors; a good therapeutic procedure for diabetes may be disastrous to a patient with an unusually weak heart. Suppose that the new treatment cures 80% of the people on whom it is tried, whereas some older therapy effected cures in 60% of the patients. Is the new treatment really better or is the difference in percentage an accident due to the particular sample on which it was tried?

All scientific work depends upon measurement. However, all measurements are approximate. Scientists attempt to eliminate this inaccuracy by making many measurements of a given quantity and then taking the mean of the values obtained. It is true that measurements of a quantity form a normal distribution, and we have good reason to believe that the mean of the entire distribution is the true value. But a scientist cannot obtain the entire distribution of measurements in order to find the mean. He can perform twenty or even fifty measurements of a quantity and determine their mean value, but this value is not the mean of the entire distribution. How reliable is the mean computed from the actual measurements?

Although the absence of certainty in some phases of scientific work is deplorable, it is not an insuperable obstacle. As a matter of fact, very little of what we look forward to in our futures is certain. What do we do about this uncertainty? Descartes stated the course which we all consciously or unconsciously follow: "When it is not in our power to determine what is true we ought to act in accordance with what is most probable." In our daily evaluation of probabilities, we are satisfied with rough estimates; i.e., we merely wish to know whether the probability is high or low. Crossing the street involves uncertainties, but we do cross because without calculation we know that the probability of doing so safely is high. In scientific and large business ventures, however, we must do better. We can no longer accept rough estimates, but must calculate probabilities exactly, and here the mathematical theory of probability serves.

30–2 Probability for equally likely outcomes. Suppose that we wished to calculate the probability of throwing a three on one throw of a die. One could resort to experience, as many people do anyway, and throw a die 100,000 times. He would find that threes show up on about one-sixth of the throws and conclude that the probability of throwing a three is $1/6$. However, resorting to experience as a means of determining a probability is burdensome and sometimes not even possible. Pascal and Fermat suggested the following approach. In the case of throwing a die, there are six possible outcomes (if we exclude the possibility of the die's coming to rest on an edge). Each of these possible outcomes is equally likely, and of these six, one is favorable to the throw of a three. Hence the probability that a three will show is $1/6$.

If we were interested in the probability that a three or a four will turn up in the throw of a die, we would still have six possible outcomes, but now two of the six would be favorable. In this instance, Pascal's and Fermat's approach would lead to the conclusion that the probability of obtaining a three or a four is $2/6$. If the problem were to calculate the probability of *not* throwing a three, the answer would be $5/6$, because in this problem there are five favorable outcomes out of the six possible ones.

In general, the definition of a quantitative measure of probability is this: *If, of n equally likely outcomes, m are favorable to the happening of a certain event, the probability of the event happening is m/n and the probability of the event failing is $(n - m)/n$.*

From this general definition of probability it follows that if no possible outcomes were favorable, that is, if the event were impossible, the probability of the event would be $0/n$, or 0. If all n possible outcomes were favorable, that is, if the event were certain, the probability would be n/n, or 1. Hence the numerical measure of probability can range from 0 to 1, from impossibility to certainty.

As another illustration of this definition consider the probability of selecting an ace in a draw of one card from the usual deck of 52 cards. Here we have 52 equally likely outcomes, of which 4 would be favorable to the drawing of an ace. Hence the probability is $4/52$ or $1/13$.

There is often some question concerning the significance of the statement that the probability of drawing an ace from a deck of 52 cards is $1/13$. Does it mean that if one draws a card 13 times (each time replacing the card drawn), then one draw will be an ace? No, it does not. One can draw a card 30 or 40 times without obtaining an ace. However, the more times one draws, the better will the ratio of the number of aces drawn to the total number of draws approximate the ratio 1 to 13. This is a reasonable expectation because the fact that all outcomes are equally likely means that in the long run each outcome will occur its proportionate share of times.

Suppose that a coin has fallen heads five times in a row, and one now asks for the probability that the sixth throw will also be a head. Many people would argue that the probability of a head showing up on the sixth throw is no longer $1/2$, but is less. The argument generally given is that the number of heads and tails must be the same, and so a tail is more likely to show up after 5 throws of heads. But this is not so. Undoubtedly in a large number of throws, the number of heads will about equal the number of tails, but no matter how many heads have appeared already, the probability of a head on the next throw is still $1/2$. The goddess of fortune has no desire to atone for past misbehavior.

Let us consider another illustration of the definition of probability. Suppose that two coins are tossed up into the air. What are the probabilities of (a) two heads, (b) one head and one tail, and (c) two tails? To calculate these probabilities we must note first that there are *four* different, but equally likely, ways in which these coins can fall, namely: two heads, two tails, a head on the first coin and a tail on the second, and a tail on the first coin with a head on the second one. Of these four possible outcomes only one is favorable to obtaining two heads. Hence the probability of a throw of two heads is $\frac{1}{4}$. Likewise the probability of two tails is $\frac{1}{4}$. The probability that one head and one tail will show is $\frac{2}{4}$ because two of the four ways in which the coins can fall produce this result.

Let us consider next the probabilities of tossing heads and tails on a throw of three coins. The possible outcomes are:

HHH	HTH	THH	TTH
HHT	HTT	THT	TTT.

We see that there are eight possible outcomes. To calculate the probability of throwing three heads, we observe that only one possible outcome is favorable. Hence the probability of three heads is $\frac{1}{8}$. The probability of throwing two heads and one tail is, however, $\frac{3}{8}$, for of the eight possible outcomes three are favorable. Likewise, the probability of two tails and one head is $\frac{3}{8}$, and the probability of three tails is $\frac{1}{8}$.

Let us note that instead of considering the probability of throwing, say three heads on *one throw of three* coins, we could equally well consider the probability of throwing three heads on *three consecutive throws of one coin.* When three coins are tossed, each falls independently of the other two; hence the fact that they are thrown simultaneously is irrelevant; they could be thrown successively, and the result would be the same. Further, if we let the first coin take the place of the second and then of the third, the result would still be the same because one coin is just like another. Hence three throws of one coin should lead to the same probability as one throw of three coins. We know that the probability of three heads on one throw of three coins is $\frac{1}{8}$. On the other hand, the probability of a head on a throw of one coin is $\frac{1}{2}$. If we multiply the three probabilities of heads on the three throws of one coin, that is, if we form $\frac{1}{2} \cdot \frac{1}{2} \cdot \frac{1}{2}$, we can obtain the result of $\frac{1}{8}$. This example merely illustrates a general result: the probability of many separate events all happening, if the events are independent of one another, is the product of the separate probabilities.

The definition of probability we have been illustrating is remarkably simple and apparently readily applicable. Suppose that one were to argue, however, that the probability that a person will cross a street safely is $\frac{1}{2}$ because there are two possible outcomes, crossing safely and not crossing safely, and of these two, only one is favorable. If this argument were sound, people in large urban centers could not look forward to long lives. The fallacy in the argument is that the two possible outcomes, crossing and not crossing safely, are *not equally likely.* And this is the fly in the ointment. The definition given by Fermat and Pascal can be applied only if one can analyze the situation into equally likely possible outcomes.

One of the most impressive applications of the above concept of probability can now be made on the basis of some work done by Gregor Mendel (1822–84), abbot of a monastery in Moravia, who in 1865 founded the science of heredity with his beautifully precise experiments on hybrid peas. Mendel started with two pure strains of peas, yellow and green. After cross-fertilization, the peas of the second generation, despite the mixture of green and yellow, proved to be all yellow. When the peas of this second generation were cross-fertilized, three-quarters of the resulting crop of peas were yellow and one-quarter, green. Such proportions had been observed before in the breeding of two pure species, but the explanation of the rather surprising results had eluded biologists.

Mendel supplied the interpretation. He argued that the gamete, or germ cell, of the pure yellow pea contained only a yellow particle, now called a gene,* and the germ cell of the green pea contained only a green gene. When the two germ cells were mated, seed developed which contained two genes, one from each parent. Thus each seed contained a yellow gene and a green gene. Why, then, were the peas of this second generation all yellow? Because, said Mendel, yellow was the dominant color. What happens when the hybrid peas are mated? The gamete of the hybrid pea contains only one gene of the pair determining color, and hence may contain a yellow gene or a green gene. Either of these mates with the gamete of another hybrid pea which may contain either a yellow or a green gene. The seed of the offspring contains two genes, one from each parent. Hence the seed may contain one of the following combinations: yellow-yellow, yellow-green, green-yellow, and green-green. All seeds which contain a yellow gene will give rise to yellow peas because yellow is the dominant color. Hence, if all combinations are equally likely, three-fourths of the third generation should be yellow; this is precisely the proportion Mendel obtained.

Let us now look at Mendel's results from the standpoint of probability. The gamete of a hybrid pea of the second generation can be yellow or green. This is analogous to head and tail of a coin. When two such gametes mate, the combinations are analogous to the combination of heads or tails on two coins. The probability of obtaining at least one head on the two coins is three-fourths because the outcomes of head-head, head-tail, tail-head, and tail-tail are equally likely. The probability of throwing at least one head is the same as that of breeding seed with at least one yellow gene. Hence the laws of probability predict the proportion of yellow peas in the third generation.

The theory of probability may now be used to predict the proportion of yellow peas in the fourth generation or the proportions of various strains which will result when several different pairs of characteristics, such as yellow and green, tall and short, smooth and wrinkled, are simultaneously interbred. Needless to say, the theory predicts precisely what happens.

---

* The genes are contained in chromosomes, but here we wish to concentrate, in particular, on the particles which determine color.

This knowledge is now used with excellent practical results by specialists in horticulture and animal husbandry to create new fruits and flowers, breed more productive cows, improve strains of plants and animals, grow wheat free of rust, perfect the stringless string bean, and produce turkeys with plenty of white meat.

The use of the theory of probability in the study of human heredity is especially valuable. Scientists cannot control the mating of men and women, and even if they could do so, it would not be possible to obtain experimental results quickly and easily. Hence they must deduce the facts of heredity from just such considerations as were illustrated above. Moreover, because prejudices frequently enter into judgments of human characteristics, the objectiveness of the mathematical approach is far more essential in the study of human heredity than in studies of plants and animals.

Exercises

1. What is the largest value that the probability of an event can have? What does this probability mean?
2. What is the smallest value that the probability of an event can have? What does this probability mean?
3. What is the probability of throwing a two on a single throw of a die? What is the probability of throwing a three? What is the probability of throwing a three or a larger number?
4. The probability that Mr. X will live one more year is $\frac{1}{2}$ because there are two possible outcomes, namely, that he will be alive or dead at the end of the year, and only one of these possibilities is favorable. Do you accept this reasoning?
5. There are 4 caramels and 6 pure chocolate pieces in a box of candy. If a piece of candy is picked at random, what is the probability that it will be pure chocolate?
6. What is the probability of picking a diamond when one card is drawn from the usual deck of 52 cards?
7. What is the probability of throwing 3 heads and 1 tail on a throw of 4 coins?
8. The probability of throwing 4 heads and 1 tail on a throw of 5 coins is $\frac{5}{32}$. What is the probability of not throwing exactly 4 heads and 1 tail?
9. What is the probability of throwing a four or higher on a single throw of a die?
10. If the probability of an event is one in a million, is the event improbable? is it impossible?
11. What is the number of possible outcomes in throwing two dice? [*Suggestion:* A three on one die and a five on the other is not the same as a five on the first die and a three on the second.]
12. How many of the outcomes in throwing two dice yield a total of 5 on the two faces?
13. What is the probability of throwing a five on a throw of two dice?
14. In Chapter 4 we found that the number of possible variations in the genetic make-up of any one child a husband and wife may have is 2^{48}. What is the probability that two children (not identical twins) will be exactly alike? (Identical twins come from the same fertilized ovum.)
15. Since it is correct to regard a single throw of three coins as equivalent to three successive throws of one coin, present an argument proving that the probability of tossing 2 heads and 1 tail in successive throws is $\frac{3}{8}$ also.

16. A young man who dates two girls has to travel by a northbound train to see one, and by a southbound train to visit the other. He argues that since these trains run equally often in both directions, he can take the first one that comes along and, over many trips, will see both girls equally often. But suppose that at his station the train schedule is as follows.

Northbound:	8:00	8:05	8:10	8:15
Southbound:	8:04	8:09	8:14	8:19

Suppose, further, that the young man usually enters the station at any time during the 5-min intervals. Show that the probability that the "girl in the south" is favored (?) is $4/5$. The moral of this problem is not to let chance determine your dates.

17. A term frequently used in discussions of probability is "odds," which means the ratio of the probability in favor of an event to the probability against the event. What are the odds in favor of throwing a head on a single throw of a coin?

18. What are the odds in favor of throwing at least one head on a single throw of two coins?

30–3 Probability as relative frequency. The concept of probability which we have examined presupposes that we can recognize equally likely outcomes and then take into account those that are favorable to a certain event. But what is the probability that Jones, now forty years old, will live to be sixty?

One cannot say that the two possible outcomes, life or death at age sixty, are equally likely. One might then try to determine the probability that a person of age forty will die of cancer by the age of sixty, perform similar calculations to determine the probability of death due to other causes, such as heart disease, diabetes, fatal accidents, etc., and, assuming that these probabilities could be determined, somehow manage to combine these partial findings to obtain the final probability that an individual of age forty will live 20 more years. This approach would hardly be successful. Yet the probabilities that, starting from a given age, a man will live any specified number of years are of vital importance to insurance companies. Hence these companies had to find ways of determining life expectancies and mortality rates and they went about it as follows. They collected the birth and death records of 100,000 people, and found, for example, that of 100,000 people alive at age ten, 78,106 people were still alive at age forty. They then took the ratio 78,106/100,000 or 0.78 as the probability that a person of age ten will live to be forty. Of the 78,106 alive at age forty, 57,917 were alive at age sixty. The probability of living from forty to sixty was then taken to be the ratio 57,917/78,106, or about 0.74.

This approach to probability is a basic one. It is, in essence, an appeal to experience to determine the favorable outcomes out of the total number of possible outcomes. Of course, the probabilities so obtained are not exact, but a sample of 100,000 people is large enough to ensure fairly reliable probabilities. The reliance upon experience may seem quite different from the calculation of probabilities based on equally likely outcomes, but the difference is not nearly so great as it may appear to be. Why do we decide that each face on a die is equally likely to show up? Actually it is experience with dice which makes us accept the intuitively appealing argument that all faces are equally likely to appear.

Even though probabilities of life expectancies, accidents of various kinds, and the incidence of diseases are obtained from data based on past experience, once they are obtained, mathematics can be employed to calculate with these probabilities. We saw earlier that the probability of throwing two heads on one throw of two coins or two successive throws of one coin is $\frac{1}{2} \cdot \frac{1}{2}$ or $\frac{1}{4}$. If an insurance company is asked to insure the lives of a husband and wife for a twenty-year period, it is important to know the probability that both will be alive twenty years from the date on which the policy is issued. Let us suppose that both are fifty years old. Now the probability that one person of age fifty will live to be seventy is about 0.55 because of about 70,000 people alive at age fifty, 38,500 are alive at age seventy. The probability that *both* will live to be seventy can be determined in the same way as the probability of throwing two heads on two throws of one coin, namely, $0.55 \cdot 0.55$, or about 0.30. Thus, once the probability that a person of age fifty will live to be seventy has been obtained, mathematics can be employed to calculate the probability that two fifty-year old people will *both* live to be seventy. The foregoing is a fairly simple, run-of-the-mill problem. As might be expected, mathematics is employed for the solution of far more complicated probability problems arising in insurance. Thus the theory of probability, which was first developed to solve problems of gambling, takes the gamble out of the insurance business.

Exercises

1. How would you determine the probability that a person of age 40 will live to be 60?

2. Of 100,000 ten-year old children, 85,000 reach the age of 30 and 58,000 reach the age of 60. What is the probability that a person of age 10 will live to be 30? What is the probability that a person of age 30 will live to be 60?

3. Suppose that the probability that any one person of age 40 will live to be 70 is 0.5. What is the probability that three particular people of age 40 will live to be 70?

4. Suppose that it is known from long experience that 50% of the people afflicted with a certain disease die of it, say within one year. A doctor who believes that he has developed a new treatment tries it on 4 people and all 4 recover (do not die within one year). How much reliance would you place on the treatment? [*Suggestion:* What is the probability that 4 people having the disease will recover without any treatment?]

30–4 Probability in continuous variation. For the problems of probability considered so far, the number of possible outcomes was finite. Thus, for example, the number of possible outcomes in throwing three coins is

8, and the number of people who, out of a total sample of 100,000, may live 20 more years can vary from 0 to 100,000. Let us consider, however, the heights of human beings even if limited to values between 4 ft 6 in. and 8 ft 6 in. The number of possible heights in this range is infinite, since any two heights can differ by arbitrarily small amounts. Moreover we know from experience that all possible values in the range from 4 ft 6 in. to 8 ft 6 in. are not equally likely heights of human beings. What, then, is the probability that a man selected at random is between 70 and 71 in. tall?

The theory of probability can treat such problems. We shall consider the most important case, namely, where the frequencies of the various possibilities form a normal distribution (Chapter 29). Since we are now thinking in terms of probability rather than, as in the preceding chapter, of a frequency distribution of some quantity such as height or income, we should restate the normal frequency distribution in terms of probability. Let us consider the possible heights of men from the standpoint of the probability of various heights occurring. In our earlier discussion of normal frequency distributions we said that 34.1% of the cases lie within one σ, that is, one standard deviation, to the right of the mean. What this means is that if, for example, the frequency distribution of 1000 heights is normal, then 341 heights would fall within one σ to the right of the mean. If the mean should be 67 in. and σ should be 2 in., then 341 out of the 1000 men would have heights between 67 and 69 in. Then the probability that a man chosen at random out of the thousand has a height between 67 and 69 in. is 341/1000 or 0.341 because the probability is the ratio of the favorable possibilities or outcomes to the total number of possibilities (Fig. 30–1). Likewise, since in a normal frequency distribution 13.6% of all cases (or, one says, of the entire population) lie between σ and 2σ to the right of the mean, it follows that the probability of any individual's height being between σ and 2σ to the right of the mean is 0.136. In other words, every percentage which appears in the normal frequency distribution becomes a probability under the latter point of view. Thus the normal frequency curve can also be regarded as giving the probabilities of events whose possibilities are normally distributed, and it is therefore also known as the *normal probability curve.*

We can now answer questions about the probability of events when the frequencies of the various possibilities are normally distributed. For example, given that the heights of men are normally distributed with a mean of 67 in. and a standard deviation of 2 in., what is the probability of finding a man with height greater than 73 in.? Since 73 in. is 3σ to the right of the mean, a height greater than 73 in. is more than 3σ to the right and, as Fig. 30–1 shows, the probability that a height will fall more than 3σ to the right is 0.001; that is, about one man in a thousand is taller than 73 in.

Note that in dealing with infinite populations, we do not ask for the probability of a particular value, for example, the height 69 in. This probability is zero because it is one possibility in an infinite number of possibilities. But the question is not too significant. All measurements yield

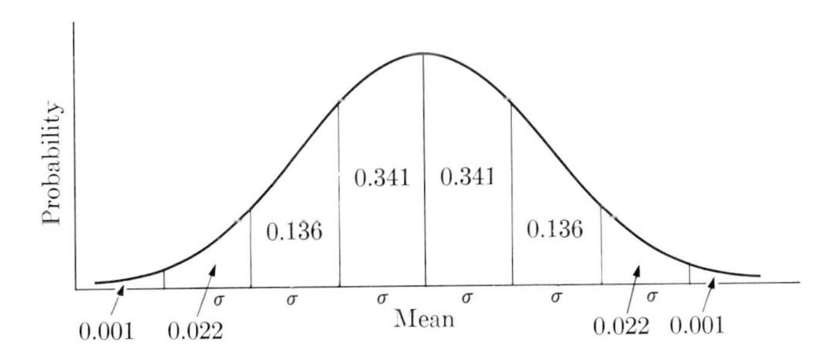

FIG. 30–1. The normal probability curve.

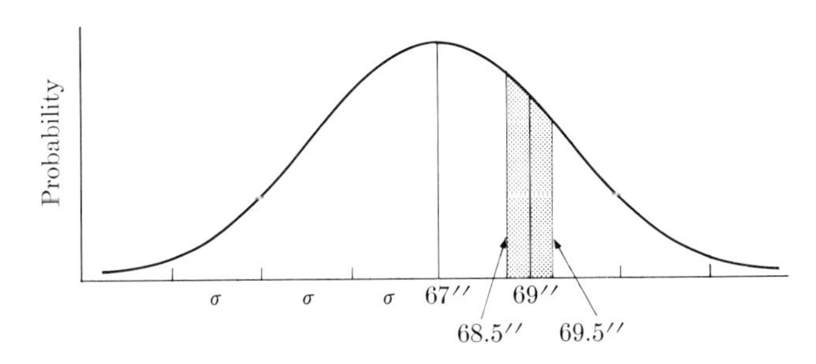

FIGURE 30–2

approximate values only. Let us suppose that the accuracy of measurement is 0.5 in. Then all we should be concerned with is a height between 68.5 in. and 69.5 in.; that is, if we are interested in men with a height of 69 in. and our accuracy of measurement is 0.5 in., then the practical problem we should set is finding the probability that a height falls between 68.5 and 69.5 in. To answer the question, What is the probability that a man chosen at random is between 68.5 and 69.5 in. tall, we would first note that this height lies between $3\sigma/4$ and $5\sigma/4$ to the right of the mean. We would next determine what percentage of the entire population in a normal frequency distribution lies between $3\sigma/4$ and $5\sigma/4$ (Fig. 30–2). We shall not bother to calculate the percentages or the corresponding probabilities which lie within fractional parts of a σ-interval because we wish to restrict our discussion to the simpler percentages or probabilities. However, the probability of a height falling within any given range can be calculated by a method of the calculus.

EXERCISES

1. Suppose that the heights of all Americans are normally distributed, that the mean height is 67 in., and that the standard deviation of this distribution is 2 in. What is the probability that any person chosen at random is between 67 and 73 in. tall? between 63 and 71 in. tall?

2. A manufacturer of electric light bulbs finds that the life of these bulbs is normally distributed and that the mean life is 1000 hr with a standard deviation of 50 hr. What is the probability that any bulb chosen at random will fail to burn at least 950 hr?

3. Given the data of the preceding problem, what is the probability that any bulb chosen at random will burn at least 1100 hr?

4. The grades obtained by a large group of students in an examination were normally distributed with a mean of 76 and a standard deviation of 3. What is the probability that a student had a grade between 76 and 79?

5. The weights of a large number of grapefruits were found to be normally distributed with a mean of 1 lb and a standard deviation of 3 oz. What is the probability that any one grapefruit has a weight between 1 lb 3 oz and 1 lb 6 oz?

30–5 Binomial distributions. Let us return for a moment to the subject of tossing coins. When one coin is tossed, there are two possible outcomes: one head and one tail. When two coins are tossed (or one coin is tossed twice), there are four possible outcomes: one yielding two heads, two yielding one head and one tail, and one yielding two tails. When three coins are tossed (or one coin is tossed three times), there are eight possible outcomes: one yielding three heads, three yielding two heads and one tail, three yielding one head and two tails, and one yielding three tails. We could now calculate the total number and distribution of outcomes in tossing four coins, five coins, and so on.

Thinking about this problem of tossing many coins led Pascal to make use of the following "triangle" (now named after him):

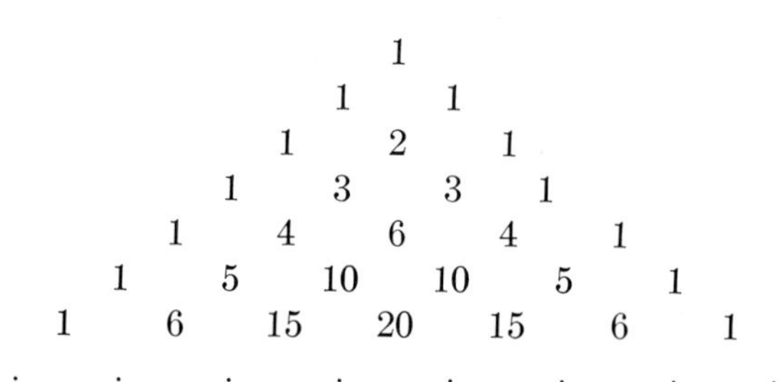

Each number in this triangle is the sum of the two numbers immediately above it (zero must be supplied where one of these two numbers is missing). Thus 4 in the fifth row down is the sum of 1 and 3, and 6 is the sum of 3 and 3. Pascal discovered how well this triangle represents the probabilities of getting heads or tails in throwing coins. Take the case of *three* coins, for example. The number of possible outcomes is 8, and this is the sum of the numbers in the *fourth* row. The probabilities involved: $\frac{1}{8}$, $\frac{3}{8}$, $\frac{3}{8}$, $\frac{1}{8}$, are obtained from the individual numbers in the fourth row, namely 1, 3, 3, 1. Similarly, the probabilities of the various alternatives which can arise in the process of flipping five coins will be found in the sixth row of the triangle, and so on. (The number 1 in the first row tells us that the probability of winning on a throw of zero coins is 1. This is the only case in which one is certain to win.)

The numbers which appear in the second row, for example, are the coefficients of a and b in $a + b$, that is, 1 and 1. The numbers which appear in the third row are the coefficients of $(a + b)^2$; for since $(a + b)^2 = a^2 + 2ab + b^2$, we see that the coefficients are 1, 2, 1. The numbers which appear in the fourth row are the coefficients of $(a + b)^3$ for $(a + b)^3 = a^3 + 3a^2b + 3ab^2 + b^3$. This relationship holds generally. The coefficients of $(a + b)^n$ are the numbers in the $(n + 1)$-row. The quantity $a + b$ is called a binomial because it consists of two terms. Hence distributions which follow any one row of Pascal's triangle are called *binomial distributions.*

If one wished to calculate the probabilities of the various outcomes in tossing 50 coins, he could employ some standard reasoning in the theory of probability or he could extend Pascal's triangle to the fifty-first row. It is, however, clear that this latter process is laborious and, as a matter of fact, calculating the various probabilities by means of the formulas of the theory would be equally laborious. There is an alternative. Let us observe the seventh row of Pascal's triangle. The numbers in this row refer to the throw of six coins, and their sum is 64. Hence they tell us, for example, that the probability of six heads is $1/64$; that of five heads and one tail, $6/64$; that of four heads and two tails, $15/64$; and so on. If we plot the number of possible heads as abscissas and the probabilities of these various numbers of heads as ordinates and draw a smooth curve through the points, we obtain the graph shown in Fig. 30–3. The shape of this graph suggests the normal probability curve. And, indeed, were we to calculate the tenth and the twentieth lines of Pascal's triangle and plot the corresponding graphs, we would find that as the number of coins increases, the graph of the probabilities of the various outcomes approaches the normal probability curve. For a

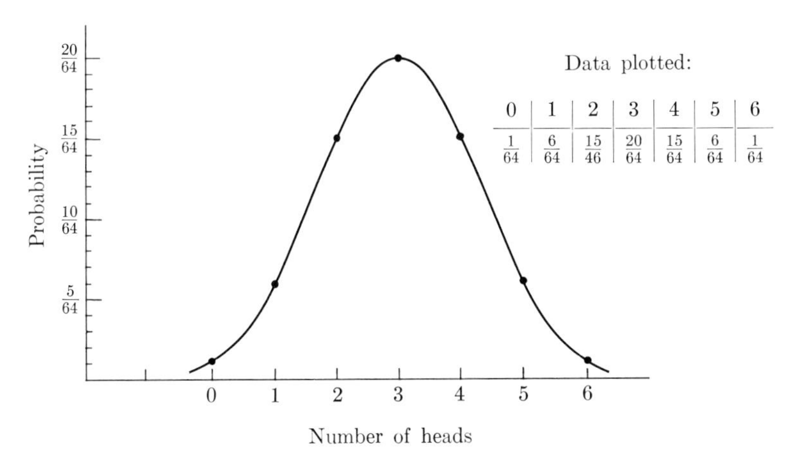

FIG. 30–3. Graph of the number of possible heads appearing on a throw of six coins versus probability.

large number of coins, 20 or more, the approximation is so good that we may as well use all the knowledge we have about the normal probability curve and abandon the calculation of the probabilities by special formulas or by extending Pascal's triangle.

To apply the normal probability curve, we must know the mean and the standard deviation. A glance at Pascal's triangle shows that the number of heads which has the highest frequency, and therefore the highest probability, is the middle number (or one of the two middle numbers if there is no one middle number). Thus in the seventh row the outcome of 3 heads has the greatest probability. Now this number 3 is the probability of throwing a head, namely, $\frac{1}{2}$, multiplied by the number of coins thrown, that is, 6. This example illustrates a general result which we shall state but not prove. *If n coins are tossed, the mean number of heads is $n/2$.**

The standard deviation could be determined by applying the procedure for finding the standard deviation of a frequency distribution to the frequencies involved in tossing coins. We would find that

$$\sigma = \sqrt{n \cdot \tfrac{1}{2} \cdot \tfrac{1}{2}} = \tfrac{1}{2}\sqrt{n}.$$

Both of these results hold for the throw of n coins and do not depend upon our approximating the frequencies of the various numbers of heads by the normal frequency distribution. However, if we do make the approximation, we can use the mean and standard deviation just indicated in connection with the normal frequency curve or with the normal probability curve.

The reader may have the impression that mathematicians have become overly absorbed in coin-tossing and accept this extensive interest as another quirk of queer minds. But, in fact, coin-tossing is merely a useful and concrete example which paves the way for more serious applications. Suppose that the probability of death from a certain disease is $\frac{1}{2}$; that is, half the people who contract the disease die within some definite period of time. A new medical treatment is given to 20 afflicted people and only 3 die within that crucial period. Is the treatment really effective or is it just chance that only 3 out of this group of 20 died? After all, to say that the probability of death from a disease is $\frac{1}{2}$ does not mean that in any one group of 20 people 10 will die. It means, as does any probability, that in the long run or in a large number half of those afflicted will die. How should we decide whether the medical treatment is effective?

The possible outcomes, that is, the possible numbers of those remaining alive without treatment among 20 afflicted people, are precisely the same as the possible outcomes in a throw of 20 coins. Just as the number of heads which may show up can vary from 0 to 20, so among the 20 people, none may live beyond the specified period, one may, and so on up to 20. Since

* In books on probability, this result is stated in more general form. One may, for example, be dealing with dice, and the probability of throwing ones on each die. This probability is 1/6 and the mean number of ones is $n/6$ or, in general, if p is the probability of the single event, the mean number in n repetitions is np.

the probability that any one person will remain alive is the same as that of tossing a head on a single throw, the probabilities of the various outcomes will be the same in the two situations. Hence the mean number of those remaining alive is $(1/2) \cdot 20$, or 10. The standard deviation of this distribution is $\sigma = (1/2)\sqrt{n}$. In our case,

$$\sigma = \tfrac{1}{2}\sqrt{20} = \tfrac{1}{2}(4.47) = 2.24.$$

Then 3σ to the right of the mean is $10 + 6.72$, or 16.72. If we now use the normal probability approximation to our distribution of probabilities (Fig. 30–1), we can say that the probability that more than 16.72 out of 20 afflicted people remain alive without treatment is less than 0.001. But actually 17 people out of the 20 treated remained alive. The probability of this happening in any group of 20 afflicted people is so small that we must credit the treatment with the remarkable record in this group of 20.

Thus the application of the theory of probability developed for coin-tossing to a serious medical problem produces a highly useful conclusion. As a matter of fact, it is precisely this theory which is used to determine the effectiveness of most medical treatments such as the Salk vaccine for poliomyelitis.

Let us consider another application. There is a common belief that boy and girl babies are equally likely or that the probability of a boy being born is $1/2$. Suppose that in 2500 births 1310 proved to be boys. Are these data consistent with the accepted probability of one boy in two births? We can answer this question. The possible numbers of boys in 2500 births are precisely the same as the possible numbers of heads in a throw of 2500 coins (or 2500 throws of one coin). Then the mean number of boys is $(1/2)2500$, or 1250. The standard deviation of the distribution of frequencies of these various numbers of boys is $\sigma = 1/2\sqrt{n}$. In our case,

$$\sigma = \tfrac{1}{2}\sqrt{2500} = \tfrac{1}{2} \cdot 50 = 25.$$

Now if we use the normal probability distribution as a good approximation to our binomial distribution, we may say that the normal probability curve has a mean of 1250 and a standard deviation of 25. Then 2σ to the right of the mean is 1300. An examination of Fig. 30–1 shows that the probability of 1300 or more boys is 0.023. This means that in only 23 cases out of 1000, on the average, will there be 1300 boys or more in 2500 births.

We are now faced with a problem rather than a conclusion, namely with the occurrence of a very improbable number of births. An event which is very improbable can, of course, occur. However, the entire reasoning is based on the assumption that boy and girl babies are equally likely or that the probability of a boy is $1/2$. It seems far more reasonable to question this hypothesis. As a matter of fact, more extensive records show that the ratio of boys to girls is 51 to 49 instead of 50 to 50. This slight difference in ratio makes a lot of difference in the result. Though our theory does not cover this case, we could show that the probability of a boy being $^{51}/_{100}$ instead of $\frac{1}{2}$ leads to the probability of 0.159 or about $\frac{1}{6}$ that 1300 or more boys will occur in 2500 births. That an event should occur whose probability is $\frac{1}{6}$ is by no means surprising.

In concluding that the medical treatment was effective and in rejecting the hypothesis of equally likely boy and girl births, we relied upon a probability. In the first case we decided that the occurrence of an event whose probability is less than 0.001 implied that the treatment was effective. In the second case we decided that the occurrence of an event whose probability is 0.023 discredited the belief that boy and girl babies are equally likely. The question of what probability to accept as evidence for or against an hypothesis must be decided by the individual concerned, and his judgment will undoubtedly be influenced by the consequences that are likely to arise from his decision.

A recent and most interesting application of the above theory has been made to "prove" the possibility of extrasensory perception, that is, the ability of some people to discern undisclosed facts by extraordinary mental power, for example, to read hidden cards. In the actual tests made by Professor J. B. Rhine and others, subjects were asked to name certain cards held face down, and these people were able to give the correct answers a far greater number of times than the mathematical probabilities of sheer guesses would predict. Thus suppose that the 4 sixes of a deck were held face down on a table. If a subject were asked to select the six of diamonds, he should be able, on the basis of sheer guess, to do so about $\frac{1}{4}$ of the times. But suppose that the subject selects the right card $\frac{1}{3}$ of the times in a large number of trials. Such an unexpectedly large ratio of correct choices is interpreted by Rhine to mean an unusual mental faculty, i.e., extrasensory perception. Of course, the argument here centers about the interpretation of the results in the light of the theory of probability. Rhine's claim is that his experiments also point to telepathy, clairvoyance, prescience, and psychokinesis (the power of the mind to control material objects).

Exercises

1. Form the eighth row in Pascal's triangle and calculate from it the probability of throwing 4 heads and 3 tails in a throw of seven coins.

2. What is the probability of 2 heads in a toss of 6 coins?

3. Suppose that 6 coins are tossed 2000 times. Approximately how often will 2 heads appear?

4. What is the probability of getting at least 5100 heads in a toss of 10,000 coins?

5. Suppose that 50% of people afflicted with a certain disease die. A medical treatment is tried on 100 people, and 65 survive. What is the probability that the treatment is effective?

6. Suppose that the conditions in Exercise 5 are changed to 1000 people of whom 650 survive. Does the probability of the treatment's effectiveness change? Justify your finding by a qualitative argument.

7. Suppose 860 boys are born in 1600 births. Do these facts support or discredit the hypothesis that boys and girls are equally likely?

8. Suppose 860 heads turn up in a toss of 1600 coins. What conclusion would you draw?

9. The probability that a person of age 40 will live to be 70 is $\frac{1}{2}$. Out of 400 people in a certain industry who were forty years old, 150 were alive at the age of 70. What may you conclude about the death rate of workers in this industry?

30-6 The problems of sampling. When one knows the distribution of the probabilities of some variable, one can calculate the probability that a particular value or a range of values will occur and draw a conclusion from this probability. Thus if one knows the distribution of the probabilities of various heights, one can calculate the probability of finding a man with a height, say, between 69 and 70 in. Likewise, if one knows the distribution of the probabilities of the various numbers of heads on a throw of, say 1000 coins, one can calculate the probability of obtaining, for example, 600 or more heads on a particular throw.

In situations of this kind one knows the probabilities of the various alternatives and calculates the probability of a particular one. Now let us consider the following problem. A manufacturer turns out millions of units of the same product each year. He sets up a standard for his product which depends upon the use to which it is put, the price at which it is to sell, the kind of machine which makes it, and other factors. Thus he might decide what, for example, the standard or mean size of his product should be. It is desirable, of course, that all articles produced by one machine be exactly alike, but even under the remarkable accuracy of modern machine performances, complete uniformity cannot be achieved. Hence he also allows for variations, and fixes the standard deviation from the mean which his machinery should hold to. However, machines deteriorate and wear out, or some part may function poorly and affect the output, and the manufacturer therefore checks the output. Since it is too expensive to test each item, he resorts to sampling. Let us say that he examines a sample of 100 units each day. But a sample may have accidental variations which may not be representative of the entire product, just as the mean height of 100 people chosen at random is not necessarily the mean height of all people, even if these people were members of an ethnically homogeneous group. By applying the theory of probability he can decide whether, on the basis of the sample, his machinery is functioning properly, that is, is producing articles with the intended mean and standard deviation.

For the problem of quality control just described, the mean and standard deviation of the entire distribution or population are known, and one judges from a sample whether the mean, say, of the population is being adhered to. There are, however, many problems of sampling in which the mean and standard deviation of the original population are not known, and one is asked to determine them by sampling. For example, a manufacturer of electric light bulbs uses new materials or a new process to make the bulbs and wishes to know how long they will burn on the average. He could test 10,000 bulbs and obtain an answer, but this is not at all necessary. He can test 100 bulbs and find the mean life of this sample. By applying methods of probability theory, that we shall not present, to the mean of his sample, he can determine the mean life of the entire output. The estimate he obtains for the mean of the entire population will not be a precise figure but will lie between certain limits with a probability of almost 1. If he wishes to obtain a better estimate, he can use a larger sample, but the surprising fact is that rather small samples give good estimates of the mean of the entire output.

30–7 Probability in the physical sciences. The development of statistical and probabilistic methods, commonly referred to as statistical methods, in the social and biological sciences was motivated by the need to gain some insight and obtain some laws. These techniques were pursued as a substitute for the deductive approach which had led to deeper and richer knowledge in the physical sciences. Had the applications of statistics remained confined to the social and biological sciences, no doubt statistical reasoning would have acquired the reputation of being a second-rate method to be employed temporarily until investigations produced some broad basic principles which could be used as axioms for a deductive approach. But statistical methods have been heavily employed in the physical sciences since the latter part of the nineteenth century, and laws have been deduced by these methods which have tremendous import. Before we discuss these implications, we should see how statistics and probability were employed in the physical sciences.

The first major application was made by Clerk Maxwell. He had set himself to studying the behavior of gases, a physical problem of considerable scientific interest because the gaseous state is one of the three states in which matter can exist, and of considerable practical importance since all steam- and gasoline-powered devices depend upon the action of gases. The molecules of a gas are particles of matter and as such attract one another in accordance with the law of gravitation. If a gas consisted of only 2 or 3 molecules, then it would be possible to deduce their motions in the same way as Newton deduced the motions of, say, the sun, earth, and moon. But a cubic centimeter of a gas contains under certain conditions $2.7 \cdot 10^{19}$ molecules. Hence, any attempt to predict the expansion, contraction, or temperature of a gas by calculating the interactions of the molecules is a hopeless undertaking.

Maxwell chose an entirely different approach. By making some assumptions about the distributions of the masses, and the separations and velocities of the many molecules in a gas, he could calculate the mean mass, the mean separation between molecules, and the mean velocity. He then introduced the concept of an ideal molecule which possesses these mean values and which can be considered to represent the properties of the actual volume of gas. These mean values are the most probable values, just as the mean height of all men is the most probable height. The most probable behavior of this ideal molecule, for example, its most probable speed and most probable direction, is then taken to be the behavior of the gas itself. By working with such ideal molecules and their most probable behaviors Maxwell was able to predict the expansion and contraction of gases under reduction or increase of external pressures, the absorption and emission of heat by gases, and a number of other actions of gases.

Now the almost unbelievable fact about Maxwell's work and the subsequent work of others who took up Maxwell's method of treating gases is that the laws obtained proved to be as useful and as accurate in predicting behavior as the Newtonian laws had proved to be for planetary motion. Most probable values and most probable behavior did and do as well for gases as the precise values and the precise laws of motion and gravitation do in ordinary mechanics.

A loose analogy may perhaps better illustrate Maxwell's achievement and the nature of his results. We know that the deaths of individuals are due to a variety of causes, and any attempt to predict the death of any one individual by deduction from these causes would be doomed to failure even if we knew far more than we do about the causes and effects of diseases. Nevertheless, by resorting to data the insurance companies can determine the probability of death in any specified number of years, and by relying on such probabilities undertake to insure millions of people. And just as the insurance companies are successful because they deal with a large number of people, so Maxwell was successful because he applied his reasoning to the behavior of large numbers of molecules.

Maxwell would have had every right to maintain that the actions of gases are governed by basic laws of nature since molecules attract each other in accordance with the law of gravitation, and that he had adopted statistical methods merely as an expedient. Hence he might have accepted as natural that he should obtain laws—irrespective of the method used to derive them—which described and predicted the behavior of gases. But Maxwell saw the deeper implications of what he had accomplished. In an essay written in 1873 he said, "I think the most important effect of molecular science on our way of thinking will be that it forces on our attention the distinction between two kinds of knowledge, what we may call for convenience, the Dynamical and the Statistical . . . Now if the molecular theory of the constitution of bodies is true, all our knowledge of matter is of the statistical kind."

Maxwell was prophetic. It was the very success of statistical methods in the physical and social sciences, and especially in the former, which raised serious philosophical problems. The essence of these problems is the conflict between two views of nature: nature determined by fixed, precise, invariable laws and nature operating in accordance with pure chance.

30–8 The statistical view of nature. Let us consider the distribution of intelligence. Even if one assumes that intelligence reduces entirely to the physical and chemical make-up of the individual, the structure of any living cell is so complex, to say nothing of the variety of such cells in the human body, that one can hardly see why the distribution of intelligence among human beings should follow any law. Yet the application of any well-designed test to a large group of people shows invariably that intelligence is normally distributed. The distribution follows a precise mathematical law. In short, we have the surprising result that a law exists where we have no reason to expect one.

Superficially, it might seem that the discovery of laws, and hence of design, where we have no reason to expect them would add strength to the doctrine of determinism. But it is necessary to look into the mouths of gift horses. Second thoughts show that the presence of such laws is disturbing. If laws can be obtained about phenomena which are basically haphazard, how much significance should be attached to the laws produced by Newtonian science? The disclosure of laws in the domain of mechanics had led men to conclude that the universe was mathematically designed and functioning according to plan. The course of the universe, presumably, was determined. Now, however, it appeared that this conclusion was perhaps not warranted.

Those who choose to read plan and determinism into the mathematical laws of mechanics account for the fact that seemingly chance phenomena, such as the fall of coins, do follow laws on the ground of underlying design. These phenomena are so complex, however, that to our limited intelligences they merely appear to be the results of chance. As Laplace put it, "Chance is but the expression of man's ignorance." Under close inspection the motions of the molecules of a gas do appear completely arbitrary; yet physicists believe that each molecule follows the same physical laws that the earth follows in its path around the sun. Similarly one could argue that the qualities which make for intelligence follow orderly physical procedures which determine the intellectual level of each individual. The same could be said of economic phenomena, the incidence of death, and other seemingly unlawful affairs. From this deterministic point of view, then, one may conclude that phenomena which appear to be haphazard are, in fact, completely determined, and that the mathematical laws obtained from statistical studies merely reflect the existence of these underlying orderly physical processes.

But let us now consider the following facts. When six coins are simultaneously flipped, any number of heads from zero to six may show. We have no way of telling what the exact number of heads will be because too many known and unknown factors determine the outcome: the strength of the wind, the force imparted to the coins by the hand, the shape of the floor on which they fall, and so on. We assume therefore that the result of tossing the coins is a matter of chance. Moreover, the greater the number of times the coins are tossed, the more is chance permitted to play a role. And yet if these six coins are tossed up a great many times, the theory of probability enables us to calculate in advance about how many times zero heads will show up, about how many times one head will show up, and so on, to the last possibility. The greater the number of throws, the closer do the results agree with the predictions of the theory. Hence, regardless of whether or not the fall of a coin is determined by some series of inviolable rules, *the assumption that chance only decides the outcome* yields mathematical laws which accurately predict the outcome. (One can say accurately because no law applies precisely to a physical situation.) Hence the assumption of chance happenings is also consistent with the presence of laws.

Similarly, chance alone may govern the motions of the planets, but we obtain seemingly precise laws because we deal with large masses whose behavior is an average for great numbers of individual molecules. In other words, the situation in the physical sciences may be really the same as that prevailing in the social sciences. The individual heights of one, two, or ten people are purely random values. Only when we consider the heights of many individuals, do we get a normal distribution which is the over-all effect of a great number of random factors.

To sum up, our observation that *some* laws cover apparently chaotic phenomena has led us to the conclusion that *all* scientific laws may, in fact, summarize random behavior. What shall we now say about the significance of mathematical laws as evidence for the existence of an orderly nature? If the earth in its motions adheres exactly to one pattern, which Kepler's laws describe so well, that pattern may still be only the average behavior; no necessity that we know of obliges the earth to repeat its behavior any more than necessity tells coins how frequently they must turn up heads. Tomorrow the earth may crash into the sun.

There seems, however, to be another essential distinction between formulas obtained from statistical procedures and the formulas of the Newtonian class. The former are based on tables of data or on probability arguments; the latter are deduced from apparently unquestionable mathematical and scientific *axioms*. For this reason one might argue that the Newtonian laws are also exact truths and therefore genuine laws which nature must follow.

The burden of this argument rests upon the truth of the axioms. Are axioms inherent in the universe or merely fitted to experience in essentially the same way as a law of population growth? Consider, for example, Newton's axiom about the force of gravity. This axiom has proved itself to be fairly accurate time after time in that deductions based upon it have led to numerical results in agreement with observations, to within the limits of accuracy of these observations. Yet the axiom may be no more than a good but approximate description of the average behavior of nature. As a matter of fact, before Newton decided upon his formula, other formulas very much like his had been tried and rejected because they did not give such accurate results. Why should Newton's axiom be the last word? Evidently man cannot be surer of the truth of such scientific axioms than he can of a law of population growth.

There is one more argument for the orderly design of the universe which rests on the experience that nature's behavior conforms to human reason. On the axioms of mathematics we build long chains of pure reasoning which are absolutely independent of experience. The conclusions we arrive at are often far removed from the axioms. For example, the Euclidean proposition which asserts that a tangent of a circle is perpendicular to the radius drawn to the point of contact is hundreds of steps removed from the axioms on which it ultimately rests. Yet the theorem is as much in accord with experience as the axiom is. Similarly, one may start with the axioms of motion and gravitation and then prove mathematically that the path of any planet is an ellipse, which it indeed is. Why should there be such a perfect agreement between the result of so many steps of *pure reasoning* and experience? Is it not because nature herself is rationally designed and lawful, and for this reason her behavior is thoroughly in accord with man's reasoning?

We may appreciate the point better if we compare the behavior of nature with the way in which humans behave. One may argue that a man who decides to follow the career of a physician, is capable of handling the school work, and has the financial means will become a physician. John Jones decides to become a physician and has the other prerequisites. One may conclude then that John Jones will become a physician. However, John Jones does not feel obliged to abide by the logic of the situation and may in the midst of his school career decide to become an artist.

From Greek times until the discovery of non-Euclidean geometry there was an almost universally accepted answer to the question of why nature conforms to man's reasoning: the universe is mathematically designed and the designer was God. Man had but to obtain facts about nature and could deduce the rest, confident that God, the architect, had followed the rational mathematical plan. Descartes had considered the question of why man's reasoning can be trusted to yield truths about nature, and his answer was that God would not deceive us by allowing our reasoning faculty to function faultily.

The more common answers today are rather pragmatic ones. One explanation of why nature conforms to reason is that scientists feel free to change the axioms if the resulting theorems do not fit. The use of a non-Euclidean geometry in the theory of relativity is such an instance. Once scientists were able to deal with large distances and high velocities, the predictions yielded by the theorems of Euclidean geometry were no longer good enough. They then sought new axioms and made sure before accepting any that the deductions from them would agree with experience.

Another explanation often given today points out that the very laws of logic are themselves derived from experience and so produce results in accord with experience. For example, one of the basic laws of reasoning, known in logic as the law of contradiction, states that an assertion cannot be true and false. An object cannot be both black and nonblack. This law or principle of logic seems, of course, indubitable; however, it may seem so, not because it is a truth, but merely because it is so invariably verified by experience that we have no reason to doubt it.

It may be difficult to see that the law of contradiction is drawn from experience, but it is far easier to recognize this origin for other principles of logic. One is right in arguing that if no mathematicians are stupid and John is stupid, then John cannot be a mathematician. A correct principle of logic has been applied to draw this conclusion. Yet we must stop and think about this reasoning to satisfy ourselves that it is correct. In fact, we often test propositions of the same form but dealing with more familiar content to see whether the conclusion agrees with experience in the more familiar context before we grant that the principle of reasoning is correct. We seem to be definitely relying upon experience to satisfy ourselves about the principles of logic.

It is relevant to recall in this connection that the codification of the laws of logic took place several centuries after man began to employ exact reasoning in geometry. The laws of reasoning were formulated by Aristotle in the fourth century B.C., and though our historical records about the classical Greeks are incomplete, it is fairly certain that the Greek experience with mathematical reasoning taught them what the correct laws of reasoning should be.

Let us recapitulate the principal tenets of the two major points of view. Determinism asserts that scientific laws state the necessary, invariable, universal behavior of natural objects. The statistical view says natural objects can behave in all sorts of ways and that the mathematical laws of science are merely convenient, usable summaries of average effects of disorderly occurrences. Nature is chaotic and unpredictable, and its laws are no more than impermanent descriptions of average effects. The determinist believes that an essential physical connection exists between objects related by law. The statistical theorist maintains that the law is merely a convenient, man-made juxtaposition, no more significant than a formula which might relate the annual exports of the United States and the annual number of poppies grown in Europe. The determinist believes that the present state of nature unalterably determines the future. If I throw a ball into the air, it must follow a parabolic path right down to earth again. The statistical view says not only that the ball may fail in any one case to follow a parabolic path, but that it may even travel directly to the sun. The essential difference between the deterministic view and the statistical view may be epitomized in the phrases "it must" and "it most likely will."

The notions that the phenomena of our universe are subject to chance, that the vast and complex operations of nature have come about accidentally and may not repeat their past behavior, and that the universe may even suddenly disintegrate have not been acceptable to many great scientists. Some, such as Sir Arthur Stanley Eddington, Sir James Jeans, Max Planck, Herman Weyl, and Albert Einstein, continued to proclaim their belief in God's design of nature. As Einstein put it, "I cannot believe that God plays dice with the universe." On another occasion he stated, "The Lord may be subtle but not capricious." The position of these men is that nature is designed and that mathematics and science will ultimately yield the true design. On the other side is one of the leading creators of quantum theory, Erwin Schrödinger. Whether or not nature is determined, Schrödinger has said, we will never know. We can decide only whether the practice of thinking that one event determines the other is more convenient for the mind than is indeterminism. Schrödinger places the burden of proof on those who would affirm a necessity underlying all sequences of events. He himself doubts such connections because they lie beyond experience and are inconsistent with the observed irregularities in the molecular world.

The question of whether determinism or the statistical view of nature is correct is not an academic one. In a designed and orderly universe life has meaning and purpose. Assurance of this design gives man courage and reason to live and build. It also reinforces his faith in a supreme being, for the strongest rational argument for the existence of God is the argument from design. A thinking, superhuman Providence or Grand Designer is almost a necessary antecedent of a mathematically guided world. The existence of a God, in turn, gives substance to vast areas of religion and ethics. On the other hand, if the statistical view of nature is correct, the physical world and man's place in it are unplanned. Occurrences obviously serve no purpose and head nowhere since they are merely accidental, chance relationships. Life offers nothing but the meaningless pleasures and pains of the moment. Man is an insignificant concomitant of accidental cosmic circumstances. Tomorrow the earth may fall apart or some unforeseen heavenly body may suddenly rush into our planetary system and alter drastically the course of every planet.

Exercises

1. What is a statistical law?
2. Distinguish between the deterministic and statistical views of nature.
3. Formulate three of the arguments for determinism.
4. Formulate three arguments for the statistical view of nature.
5. Would the proponents of the statistical view of nature deny (a) the existence of the law of gravitation? (b) the exactness of the law? (c) the unfailing operation of the law?
6. How do determinists and the advocates of the statistical view differ on the subject of prediction by means of mathematical laws of science?
7. How do the proponents of the statistical view account for the fact that (a) the laws of motion and gravitation serve to design large engineering projects which function excellently? (b) mathematical theorems deduced by pure reasoning from axioms apply to the physical world?

Topics for Further Investigation

1. Probability applied to games of chance.
2. The evidence for extrasensory perception.
3. The tests of hypotheses by sampling.
4. Probability applied to the study of heredity.
5. The statistical view of nature.

Recommended Reading

ALDER, HENRY L. and EDWARD B. ROESSLER: *Introduction to Probability and Statistics,* Chaps. 5 to 9, W. H. Freeman & Co., San Francisco, 1960.

BOHM, DAVID: *Causality and Chance in Modern Physics,* Routledge & Kegan Paul Ltd., London, 1957.

BORN, MAX: *Natural Philosophy of Cause and Chance,* Oxford University Press, New York, 1949.

COHEN, MORRIS R. and ERNEST E. NAGEL: *Introduction to Logic and Scientific Method,* Chaps. 15 and 16, Harcourt Brace and Co., New York, 1934.

FREUND, JOHN E.: *Modern Elementary Statistics,* 2nd ed., Chaps. 7 to 11, Prentice-Hall, Inc., Englewood Cliffs, 1960.

GAMOW, GEORGE: *One Two Three . . . Infinity,* Chaps. 8 and 9, The New American Library Mentor Books, New York, 1947.

KASNER, EDWARD and JAMES R. NEWMAN: *Mathematics and the Imagination,* Chap. 7, Simon and Schuster, Inc., New York, 1940.

KLINE, MORRIS: *Mathematics in Western Culture,* Chap. 24, Oxford University Press, New York, 1953.

LAPLACE, P. S.: *A Philosophical Essay in Probabilities,* Dover Publications, Inc., New York, 1951.

LEVINSON, HORACE C.: *The Science of Chance,* Rinehart & Co., New York, 1950.

MORONEY, M. J.: *Facts from Figures,* Chaps. 1 to 14, Penguin Books Ltd., Harmondsworth, England, 1951.

ORE, OYSTEIN: *Cardano, The Gambling Scholar,* pp. 143–241, Princeton University Press, Princeton, 1953.

RHINE, J. B.: *Parapsychology, Frontier Science of the Mind,* Thomas and Co., Springfield, 1957.

SCHRÖDINGER, ERWIN: *Science and the Human Temperament,* W. W. Norton & Co., New York, 1935. Reprinted under the title *Science, Theory and Man,* Dover Publications, Inc., New York, 1957.

CHAPTER 31

THE NATURE AND VALUES OF MATHEMATICS

This, therefore, is Mathematics: she reminds you of the invisible forms of the soul; she gives life to her own discoveries; she awakens the mind and purifies the intellect; she brings to light our intrinsic ideas; she abolishes oblivion and ignorance which are ours by birth . . .

Proclus Diadochus

31–1 Introduction. We have been studying the ideas, technical content, scientific applications, and cultural influences of mathematics. In our concentration on details we may have failed to note a few broader features. Also, much of what we said at the outset about the nature of mathematics was deliberately limited to ideas which would be helpful in undertaking the study of the subject. But, as we now know, both the nature and the content changed radically during the centuries. It may profit us therefore to survey the subject as mathematicians see it today, in order to gain some insights which could not be attempted at the beginning of our study.

31–2 The structure of mathematics. Mathematics, viewed as a whole, is a collection of branches. The largest branch is that which builds on the ordinary whole numbers, fractions, and irrational numbers, or what, collectively, is called the real number system. Arithmetic, algebra, the study of functions, the calculus, differential equations, and various other subjects which follow the calculus in logical order are all developments of the real number system. We shall refer to this branch as the mathematics of number. A second branch is Euclidean geometry. Projective geometry and each of the several non-Euclidean geometries are branches as are various other arithmetics and their algebras. Were we to pursue mathematics still further, we would find that it contains many more divisions.

Each branch has the same logical structure. It begins with certain concepts, such as the whole numbers in the mathematics of number, and such as point, line, and triangle in Euclidean geometry. These concepts must obey explicitly stated axioms. Some of the axioms of the mathematics of number are the associative, commutative, and distributive properties and the axioms about equalities. Some of the axioms of Euclidean geometry are that two points determine a line, all right angles are equal, and the axiom on parallel lines. The non-Euclidean geometry of Gauss, Lobatchevsky, and Bolyai contains the same axioms as Euclidean geometry does, except for the parallel axiom. From the concepts and axioms, theorems are deduced. Hence, from the standpoint of structure, the concepts, axioms, and theorems are the essential components. We shall discuss these in turn.

The basic concepts of the elementary branches of mathematics are abstractions from experience. Whole numbers and fractions were certainly suggested by obvious physical counterparts. But it is noteworthy that many more concepts are introduced which are, in essence, creations of the human mind with or without partial help from experience. Irrational numbers, such as $\sqrt{2}$, were forced upon mathematicians to represent all the lengths occurring in Euclidean geometry, for example, the length of the hypotenuse of a right triangle whose arms are both one unit long. The notion of a negative number, though perhaps suggested by the need to distinguish debits from credits, is nevertheless not wholly derived from experience, for the mind had to conceive of the notion of an entirely new type of number to which operations such as addition, multiplication, and the like can be applied. The notion of a variable to represent the quantitative values of some changing physical phenomenon, such as temperature or time, is also at least one mental step beyond the mere observation of change. The concept of a function, or a relationship between variables, is almost entirely a mental creation. The farther one proceeds with the mathematics of number, the more remote from experience are the concepts introduced and the larger is the creative role played by the mind. The derivative of a function, that is, the instantaneous rate of change of distance compared with time, is a wholly man-made and, one might also say, an ingenious construction. We do not experience instants or instantaneous velocities but, rather, small intervals of time and speeds over small intervals of time.

The gradual introduction of new concepts which more and more depart from forms of experience finds its parallel in geometry. Though point, line, triangle, circle, and a few other elementary concepts are no more than abstractions from experience, this is not true of most of the curves which geometry considers. The approach to the conic sections as sections of a cone made by a plane or by the locus definitions (Chapter 6) subsequently used to define parabola, ellipse, and hyperbola was conceived by the mind. The notions of projection and section of projective geometry and many of the specific concepts of projective geometry such as cross ratio are entirely mental creations.

This brief review of the origin of mathematical concepts may serve to emphasize several major facts. The first of these is growth. As mathematicians continue to work in any given branch, they discover new concepts which are worth introducing and developing. Secondly, as their work advances, the new concepts are less and less drawn from experience and more and more from the recesses of human minds. Moreover, the development of concepts is progressive, later concepts being built on earlier ones. These facts have unfortunate consequences. Because the more advanced ideas are purely mental creations rather than abstractions from immediate experience and because they are defined in terms of prior concepts, it is more difficult to comprehend them. At least, one cannot usually find simple and familiar physical pictures or experiences to illustrate their meanings.

Axioms constitute the second major component of any branch of mathematics. The axioms of the mathematics of number and of Euclidean geometry were suggested by experience. Up to the introduction of non-Euclidean geometry, the idea that man chose axioms suggested by experience would have been regarded as absurd. Axioms were understood to be basic, self-evident truths about the concepts involved. While minor variations in the choice and wording were tolerated, the universal belief was that man had to accept what was clearly true. These truths were supposed to be written into the universe and were considered to be inescapable. Some philosophers, such as Plato and Descartes, believed that these truths were already planted in our minds by God. We know now that this view must be discarded. The seemingly self-evident nature of the axioms of the mathematics of number and of Euclidean geometry is really the consequence of limited experience and relatively superficial observation. In the simpler uses of numbers and in the limited regions of the universe accessible to man until recent times, the well-known axioms of number and Euclidean geometry seemed inescapably true. But we can choose other axioms and produce other geometries and even other number systems. Though we can make arbitrary choices, we do not do so because we wish our new axioms to yield systems of mathematics as significant and as useful as the older systems. Were we able to continue our study of mathematics proper, we would find that there are many other systems of axioms which lead to important branches of mathematics.

The fruit of mathematical activity consists of the theorems deduced from a set of axioms. The theorems offer new knowledge by no means immediately discernible in the axioms. Whether or not the reader may prize the knowledge, he learns by pursuing the implications of the Euclidean axioms that the sum of the angles of a triangle is 180° and that the area of a circle is π times the square of the radius. The amount of information that can be deduced from some sets of axioms is almost incredible. Euclid deduced about 500 theorems from his set. The axioms of number give rise to the results of algebra, properties of functions, the theorems of the calculus, the solutions of various types of differential equations, and many other results we have not surveyed.

Mathematical theorems, as we know, must be deductively established. Since observation, measurement, induction, and a variety of other methods of obtaining knowledge are available and are used in all other scientific pursuits, the requirement that theorems be deductively established is very stringent. We have pointed out in Chapter 3 why the Greeks insisted on deductive arguments in mathematics. We are now in a position to see how much has been gained by exploring the implications of axioms. The development of precise and reliable methods of doing arithmetic, the solution of equations for unknowns, and the results of Euclidean geometry were but a first step. Simple deductive arguments about right triangles and admittedly some physical data enabled man to determine the sizes and distances of the heavenly bodies and thus obtain the first real knowledge of the solar system. By adding to mathematical axioms such physical axioms as Newton's laws of motion and gravitation, man was able to calculate the motions of projectiles, planets, the moon, and even artificial satellites. Equally valuable results were deduced about the behavior of light and sound from mathematical and physical principles. Deductive reasoning applied to the laws of electromagnetism and the axioms of number led to the conclusion that electromagnetic waves must propagate through space. The decision to explore deductively the axioms of a non-Euclidean geometry led to the construction of totally new geometries, one of which has actually been applied to the study of physical phenomena. A brief review of the contents of the preceding chapters will recall many other achievements of the deductive process. The Greeks themselves would have been astonished by the rich results which deduction produced and which the more common methods of obtaining knowledge could not have yielded. Thus much of our deepest knowledge, which we would otherwise not have had, was obtained by deductive reasoning.

The restriction of mathematics to results deduced from explicit axioms has had several other major values. It obliged human beings to apply their reasoning faculty. Though, as we pointed out in Chapter 3, imagination and invention play key roles in suggesting what to prove and how to prove it, the mathematician's goal is not any new knowledge whatever but knowledge which he has some reason to believe can be deduced from axioms. Hence it is the deductive requirement which has caused man to explore and utilize his own mental capacities to an extent which no other subject has demanded.

The third major value of the deductive process is that it enables man to predict. In a sense all reasoning enables man to predict. If one measures the angles of a dozen triangles and finds in each case that the sum of the angles is 180°, he can then predict that the sum of the angles in any other triangle is also 180°. But, as we have often emphasized, conclusions obtained by induction or analogy are not certain, whereas mathematical predictions are. Equally important is the recognition that predictions which amount to merely applying a general result to a special case are logically shallow. The predictions of mathematics are themselves general results; they are the outcome of dozens of hard-won and by no means obvious deductive arguments and are therefore profound facts which most likely could not be otherwise obtained.

The fourth value of the deductive process is its power to organize knowledge. If all the results now available in Euclidean geometry had been obtained by a multitude of observations, inductions, or measurements, they would make an unwieldy mass that could not be assimilated. The value of the information would hardly be realized. But deductive organization permits the mind to survey the whole readily, grasp what is fundamental and what is subordinate, and see the interrelationships of the many conclusions. Comprehension is vastly aided.

We have been discussing the components of mathematical branches and the values of the deductive structure. There is another feature of mathematics which a backward glance reveals, namely growth. We noted earlier that, as any branch develops, new concepts are introduced and each such concept opens up possibilities for deriving new theorems. For example, once the notion of function was conceived, it suggested hundreds of types of functions many of which have interesting or useful properties. Thus the introduction of new concepts, which takes place continually, enables mathematics to grow. New theorems are constantly being created even in such well established subjects as Euclidean geometry, a subject explored for over 2000 years. In this field it is not likely that the most recent results will prove broad or very useful though they may prove interesting. However, in fields such as the calculus, differential equations, statistics and the theory of probability, the current developments are as important as the older ones.

Growth is possible in still another way. Mathematicians learned from the history of non-Euclidean geometry that sets of axioms which seem to have no bearing on the physical world should be explored. Accordingly, mathematicians are now exploring algebras and geometries that transcend any immediate applications. Indeed, that mathematics has grown at a far more rapid rate in the last hundred years is in part due to the introduction of new systems of axioms. Of course, since mathematics receives problems and suggestions from science and science is now flourishing, the recent acceleration in the growth of mathematics owes as much to the latter factor as to the exploration of new systems of axioms. The subject is expanding at a rate inconceivable in the eighteenth century. Curiously, toward the end of that period the greatest mathematicians, Euler, d'Alembert, and Lagrange, thought that mathematics had exhausted its themes. But they were wrong. Like the mythological nine-headed monster Hydra, which grew two heads in place of each one cut off, so mathematics now spawns two problems for each one that is disposed of.

There is no doubt of the growth of mathematics, but there is some disagreement among mathematicians as to just what it is that they are adding to their structures. Succinctly phrased, the issue might be called one of discovery versus creation. Do the concepts, axioms, and theorems exist in some objective world and are merely detected by man or are they entirely human creations?

In Greek times the axioms of mathematics were regarded as necessary truths. Accordingly, mathematical theorems were also believed to be truths about the universe already incorporated in the design of the world. Hence each new theorem was regarded as a discovery, a disclosure of what already existed. That the alternate interior angles of parallel lines are equal was preordained. Mathematicians were merely uncovering theorems, but because human minds were limited, they had to labor hard and long to recognize what really lay open before them. To the mind of God all this knowledge was immediate. One might say that from this standpoint mathematics was like a mine whose riches were all there from the beginning, but had to be brought to the surface by patient digging. The existence of these riches was as independent of man as the stars and planets appear to be. This view of mathematics was undoubtedly the dominant one until well into the eighteenth century and is held by some even today.

The contrary view holds that mathematics, its concepts, axioms, and theorems, are created by man. Man distinguishes objects in the physical world and invents numbers, for example, as a way of representing one aspect which he has singled out from experience. Axioms, too, are man's generalization of how physical lines and figures seem to behave, with no guarantee given that figures actually behave this way or that the axioms really incorporate fundamental facts. Theorems may very logically follow from the axioms, but one could hardly claim more reality for them than for the axioms. Mathematics, according to this view, is a human creation in every respect. It is a consequence of what human beings are and how they think rather than what the physical world or some objective ideal world really contains.

Is then mathematics a collection of diamonds hidden in the depths of the universe and gradually unearthed one by one or is it a collection of synthetic stones manufactured by man but nevertheless so brilliant that it bedazzles those mathematicians who are already partially blinded by pride in their own creations? Several considerations incline us to the latter point of view. Historically, mathematics has not had an invariable character. To the pre-Greeks it was a practical tool. To the Greeks it was a body of pre-existing truths. It was practical and mystical knowledge in medieval Europe. The seventeenth and eighteenth centuries identified it in subject matter and method with science. Non-Euclidean geometry not only forced a separation but revealed the arbitrariness inherent in the axioms. It thus seemed to establish that mathematics is not an idealized account of the physical world but has only a correspondence with the physical world. As scientific knowledge increases, new mathematical creations are suggested and employed. It would appear as though mathematics is the creation of human, fallible minds rather than a fixed, eternally existing body of knowledge. The subject seems very much dependent upon the creator. As Alfred North Whitehead put it, "The science of pure mathematics may claim to be the most original creation of the human spirit." Only the *relatively* universal acceptance of mathematics (as opposed to the acceptance of religious, political, and ethical doctrines) may lure us into granting that subject an objective existence.*

Exercises

1. What are the fundamental components of a branch of mathematics?
2. Would you distinguish between whole numbers and irrational numbers in respect to derivation from experience?
3. Why are axioms necessary in a deductive system?

---

* See also the discussion in the preceding chapter on the issue of determinism versus statistical laws.

4. Criticize the statement: Mathematics is a fixed body of thought created in Greek times.
5. Name some branches of mathematics.
6. What are the factors that make possible the growth of mathematics?
7. What can research in mathematics mean?
8. What advantages has the requirement that man reason deductively yielded to the mathematical sciences?

31–3 The values of mathematics for the study of nature. Mathematics proper, as we have often emphasized, deals with numbers, geometrical figures, and generalizations or extensions of ideas involving numbers and geometrical figures. Mathematics proper does not deal with forces, weight, velocity, light, or the planets. The task of the so-called pure mathematician is to find and establish the implications of the axioms about mathematical concepts, that is, to prove theorems. There is much to be said for the study of mathematics itself, and we shall discuss it in a later section. However, the primary value of mathematics is not so much what the subject itself offers but what it helps man to achieve in the study of the physical world.

The greatest mathematicians from Greek times onward were interested in the physical world and in the use of mathematics to study the physical world. Although many of these men surely liked mathematics and tackled questions of mathematics proper, without thinking of immediate or even potential application, it is fairly certain that they were willing to devote time to such problems only because they were already convinced of the value of the subject for science. Actually many of the greatest mathematicians were also the greatest physicists and astronomers of their ages, and, until very recent times when the increase in knowledge forced specialization, almost all mathematicians contributed to science. One finds among the supreme mathematicians men, such as Newton, Lagrange, and Laplace, who even cared little or nothing for mathematics proper, but felt compelled to take up mathematical problems in order to solve physical problems.

We shall not review the facts of history to substantiate the above assertions. Rather we wish to summarize the ways in which mathematics works with science and the values which science derives from this collaboration.

Of course, the study of numbers and geometrical figures is to some extent physical knowledge. Quantity is an important physical fact as are the properties of geometrical figures since these are but forms of physical objects. Geometry is, moreover, a study of space. Although the belief that Euclidean geometry expresses the laws of space has been proved wrong, the various geometries that men have constructed are at least possible and, in some cases, useful descriptions of physical space.

However, the greater importance of mathematics for science lies in the fact that the physical universe is explored so effectively with mathematics. To the Greeks, who first proclaimed that the universe is mathematically designed, and to scientists up to 1600, applying mathematics to nature meant searching for geometrical patterns in nature. This quest in itself yielded Ptolemaic theory, some laws of light and mechanics, and the heliocentric theory. A more powerful mathematical approach to nature was forged by Galileo when he decided that science must seek to establish quantitative laws. Such laws as Newton's second law of motion, the law of gravitation, and Hooke's law, though they belong to science, are quantitative relationships among variables, or mathematical formulas. As we saw in earlier chapters, mathematical processes can be applied to these formulas to deduce new ones. When these new formulas are interpreted physically, new physical information is revealed.

Mathematics serves, then, to express physical laws, and the processes of mathematics are used to derive new physical information from the basic physical laws. But mathematics does far more for science. The goal of scientific efforts is not a collection of facts, whether obtained experimentally or deduced from other already established facts. The over-all goal is the formulation of theories, such as the theory of motion, the theories of light, sound, and electromagnetic waves, the theory of relativity, and quantum theory. These theories are mathematical structures. When fully developed, they are entirely analogous to the mathematics of number or to Euclidean geometry. The foundations on which any scientific theory rests are concepts and axioms, though the latter are usually called physical principles. The heart of any scientific theory is a series of results mathematically deduced from basic principles. Thus Kepler's laws are deduced from the laws of motion and gravitation and are, since Newton's work, an integral part of the theory of motion. In other words, mathematical deduction provides the structure of any scientific theory; it is the bond between one law and another. A scientific theory is a comprehensible and consistent collection of facts, and it is comprehensible and consistent because the facts are arranged in the form of a series of mathematical deductions.

In discussing earlier the value of deduction, we pointed out that, among other advantages, it permits prediction. Since science utilizes mathematical deduction, it too can predict. The prediction of the height or range of projectiles, of eclipses, and of radio waves are remarkable just because they are beyond the capabilities of any other method. The value of prediction cannot be overemphasized. On the practical side, it underlies all large-scale engineering ventures. The waste that would be entailed if inferences had to be drawn solely from models or from experiments would be enormous in even such relatively simple projects as building a bridge or a skyscraper. From the standpoint of pure science, prediction is valuable in that it confirms the scientific principles on which the predictions are based. The principles stand or fall on their predictive value, and mathematical arguments are the necessary link between the basic principles and the predictions.

The abstractness of mathematically formulated scientific principles has great value for science. Because the same abstract mathematical laws may govern two entirely different physical situations, the scientist may discover some unsuspected relationship between the two situations. An excellent example is Maxwell's discovery that electromagnetic waves and light waves are in essence identical. Maxwell based his conclusion on two findings: (1) that the same differential equation described both phenomena and (2) that both traveled with the same velocity. The important consequences which resulted from identifying these two previously dissociated phenomena have already been discussed.

The fact that the same mathematical laws govern diverse physical happenings aids the scientist in still another way. Thus, for example, the trigonometric functions apply to all wave motions, sound, radio, light, water waves, waves in gases, and many other types of wave motions. The person who understands trigonometric functions and their properties understands in one swoop all the phenomena governed by these functions. He has but to interpret the variables to suit the physical situation, and he can immediately comprehend a host of facts about the phenomenon because he knows the mathematical properties of the functions. So, too, can knowledge of the normal law of distribution be applied to heights, intelligence, the sensitivity of the eye to frequencies, and to other biological phenomena. We see, then, that the mathematics of such related areas provides an integrative value; it features the common contents. Only the names, so to speak, of the different phenomena are different. Herein lies one great value of mathematics. Abstract mathematical relationships, seemingly outside the realm of physical reality, are the kernel of our knowledge of the real world.

Another value of mathematics for science has already been treated in our presentation of gravitation and of electromagnetic waves. Quantitative physical principles, such as the law of gravitation and the laws of electromagnetism, described purportedly physical concepts, such as the force of gravitation and electric and magnetic fields. But closer examination revealed that these concepts were physical mysteries and that all we knew about them are certain quantitative laws and their mathematical consequences. The force of gravitation and electric and magnetic fields could be and undoubtedly are fictions. Hence our only precise knowledge about these supposedly real phenomena consists of a number of mathematical formulas. Mathematics, then, is the essence of our best scientific theories. Those who, admitting the paradox, deplore the fact that to achieve success, the physical sciences have to pay the price of mathematical abstractness must reconsider what it is they would look for in the ultimate scientific exposition of the nature of the physical world.

Science is indebted to mathematics in many ways and not the least of these is that mathematics provides concepts to represent physical notions. A function is a mathematical concept, but it provides the very tool for the representation of physical laws. The derivative and integral of the calculus are immensely effective in studying physical processes. The conic sections are the curves which proved to be just right for projectile motion and astronomy. Over and above concepts, mathematics provides entire theories with which to systematize and express scientific results. A non-Euclidean geometry is a complete theoretical system into which one can fit facts about space and the behavior of figures in space; such a geometry enabled Einstein to formulate his theory of relativity. This major function of mathematics is sometimes described by the statement that mathematics provides models for the scientific description of reality. The concepts and models are what really determine the thinking and theories of science, for scientists seeking to represent their ideas in precise language and to organize their findings readily adopt convenient mathematical ideas.

It is highly important that many of these models, developed on behalf of some physical problems, turned out to be just the right ones for totally new applications. The conic sections were created on behalf of investigations of light and to answer basic questions in the already significant Euclidean geometry. Given the conic sections and their properties, Kepler saw the right use for them. It seems unlikely that Kepler would have had the strength and inspiration to create the mathematics of conic sections *and* determine that they were the correct paths of the planets. The same applies to Galileo and the parabola. In recent times Einstein, in developing his theory, took full advantage of the already existing non-Euclidean geometry. The conception and development of such a geometry climaxed, as we saw, the work of hundreds of men and required, in addition, the genius of Gauss to perceive the true significance of these efforts. The value of mathematical ideas already at hand to scientific investigations was beautifully pointed out by the nineteenth-century British mathematician J. J. Sylvester: "The discovery of the conic sections attributed to Plato, first threw open the higher species of form to the contemplation of geometers. But for this discovery, which was probably regarded in Plato's time and long after him, as the unprofitable amusement of a speculative brain, the whole course of practical philosophy of the present day, of the science of astronomy, of the theory of projectiles, of the art of navigation, might have run in a different channel; and the greatest discovery that has ever been made in the history of the world, the law of universal gravitation, with its innumerable direct and indirect consequences and applications to every department of human research and industry, might never to this hour have been elicited."

Exercises

1. Describe some of the values which mathematics offers to science.
2. Why is reasoning about concepts, such as numbers, formulas, and geometric figures, likely to be more fruitful than reasoning about concrete physical phenomena?

31-4 The aesthetic and intellectual values. We have asserted that the primary value of mathematics is the assistance it renders in the study of nature, and we have summarized the ways in which mathematics supports and even molds science. This is not to deny that many of these mathematical concepts, methods, and results were suggested by physical thought. But mathematicians often carry the development of a theme or a whole branch of their subject far beyond the needs of science. The conic sections and non-Euclidean geometry are examples of such creations. If there was little or no scientific use for these extensions, why were they explored?

A partial answer is that mathematicians, already convinced of the extraordinary usefulness of some ideas, for example, the whole numbers and simple geometrical figures, were satisfied that almost any results concerning these concepts would be worth having for their potential applicability. Many men who earn enough money to satisfy their needs strive to earn more because they know that money is helpful and that uses for the surplus earnings will arise. When non-Euclidean geometry proved to be useful despite its strange, seemingly inapplicable nature, mathematicians were all the more reinforced in their proclivity to pursue their own themes. Although nature is the womb from which most basic mathematical ideas are born, mathematicians have always felt free to amplify and extend these ideas without regard for applicability to nature, confident that at some time the extensions will prove their worth.

However, immediate need in science and potential usefulness have not been the only motivations for mathematical creations. The individual may well recognize that mathematical activity is guided by physical needs, but he himself may study mathematics simply because he likes it. He may be content to know that the subject he has chosen for investigation is important for the understanding or mastery of nature and yet care little himself about the scientific bearing of the mathematics. He may even choose some problem of his own fabrication and pursue it (see Chapter 3).

In other words there are men who pursue mathematics for its own sake; they are attracted by the fact that mathematics is an art.* Among the branches of mathematics that we have examined, the subject of projective geometry and the subject of congruences (Chapter 27), a topic in the branch of mathematics called the theory of numbers, were motivated largely by aesthetic interests.

What artistic qualities do some men find in mathematics? Though experience, observation, measurement, and even guesswork suggest some results, imagination, intuition, and insight are required for major creations, and the exercise of such talents is one of the attractions offered by an art. Indeed the mathematician must be able to discern possible conclusions and methods of proof where the average person would see no hint at all. Though reason enters, on the one hand, as a guard to ward off far-fetched hypotheses and, on the other, as a guide and prompter, creation is hardly a matter of logic. If we may borrow the words of Rheticus, the man who prepared for publication the major work of Copernicus: "The mathematician . . . is surely like a blind man who, with only a staff to guide him, must make a great, endless, hazardous journey that winds through innumerable desolate

* See also the discussion in Chapter 1.

places. What will be the result? Proceeding anxiously for a while and groping his way with his staff, he will at some time, leaning upon it, cry out in despair to heaven, earth, and all the gods to aid him in his misery. God will permit him to try his strength for a period of years, that he may in the end learn that he cannot be rescued from threatening danger by his staff. Then God compassionately stretches forth His hand to the despairing man, and with His hand conducts him to the desired goal."

Since persistence is required for all creative work, the mathematician, too, must have the stamina to wrestle with a problem until he has succeeded in solving it. He must have confidence in his powers. He may be driven to creative activity, as is the poet or painter, by pride in his reasoning faculty, the spirit of exploration, and the desire to express himself, but he must persist. The greatest mathematicians have stressed the concentration and time they have devoted to problems. Gauss said, perhaps overmodestly but sincerely, "If others would but reflect on mathematical truths as deeply and as continuously as I have, they would make my discoveries."

Mathematics offers artistic outlets not only in the creation of theorems and proofs, but in the expression of its material. A painter may have a great theme but he must also present it most effectively. The same is true in mathematics. The symbolism can be employed neatly and suggestively just as words are used in poetry. The statements of theorems can be worked and reworked to be as clear and even as dramatic as possible. The theorem should establish all that can be proved from the hypothesis. Thus if the hypothesis should warrant a theorem establishing that a triangle is equilateral, then it is a defect to prove only that it is isosceles. A good proof may be intricate. Yet it must be so clearly written that it is perspicuous. Much research in mathematics is undertaken merely to secure a more appealing proof of a result already established. Some proofs are merely convincing. As a famous mathematical physicist, Lord Rayleigh, put it, they "command assent." There are others "which woo and charm the intellect. They evoke delight and an overpowering desire to say, Amen, Amen."

Despite popular impressions to the contrary, the creations of mathematicians have a highly personal character. Some men have been strongly intuitive in their thinking and have projected concepts and broad directions for activity. Others are concerned with rigorous proof. Some think best geometrically or physically, others in terms of numbers and formulas. Some have been concerned with the philosophical problems of mathematics, such as its relation to logic, and others with direct problems of science. Some prefer and obtain neat short proofs, while others are virtuosos in manipulating formulas.

Perhaps the best reason for regarding mathematics as an art is not so much that it affords an outlet for creative activity as that it provides spiritual values. It puts man in touch with the highest aspirations and loftiest goals. It offers intellectual delight and the exaltation of resolving the mysteries of the universe.

Many who accept the above described values of mathematics nevertheless insist that an art form provide emotional satisfactions. Actually a true art appeals primarily to the mind; in fact, some arts—witness modern abstract painting—can hardly have emotional import. However, mathematics meets this criterion too. There are positive and negative emotional responses to mathematics. On the negative side, there are the intense feelings of dislike which many have for the subject. On the positive side, there are pleasures ranging from the quiet satisfaction felt by many laymen who read in the subject, to the thrill of success which even young students experience when they have solved a problem, to the real delight which mathematicians who do original research derive from their work. There are satisfactions, much like those offered by great paintings, to be obtained from surveying orderly chains of reasoning. This order and harmony can be found in the development of mathematical themes, and then there is the harmony which mathematics imposes on nature and which minds such as Ptolemy, Copernicus, Kepler, Newton, and Einstein fashion. Coleridge has said that the essence of beauty is the discovery of unity within and beyond obvious variety. The mathematicians are artists who use nature as their model and provide their own orderly and unifying interpretations.

One might well ask whether there is any proof or test demonstrating that mathematics offers beauty. There are some criteria which are often applied to determine whether a piece of mathematical work is good. A general result is preferred to a special one because the former embraces many cases. A result should not be isolated, for an isolated result has no consequences or fertility. If it has no relation to other ideas, it also lacks depth and interest. A good proof must overcome a genuine difficulty. Thus the exercises in mathematics books, though they lead to new results, are too readily done. Hence they are not dignified as theorems. A good proof also offers a method which can be applied to other problems, and the method may be highly regarded for its ingenuity. But these criteria of worth are not necessarily criteria of beauty. In the final analysis beauty is subjective and what pleases some people may irritate others as is the case in modern painting and modern music. Many mathematical works enjoyed and praised by some are dismissed by others as trivial, an all too frequent comment in mathematical circles.

Somewhat distinct from the aesthetic value of mathematics is the intellectual challenge. Mathematicians respond to this challenge much as business men respond to the excitement of making money. Mathematicians enjoy the excitement of the quest, the thrill of discovery, the sense of adventure, the satisfaction of mastering difficulties, the pride of achievement or, if one wishes, the exaltation of the ego, and the intoxication of success. Mathematics is particularly attractive to people who enjoy such challenges because it offers sharp, clear problems. The fields of political theory, economics, and ethics are far more complex, and it is more difficult not only to isolate and formulate the problems but to be at all sure that one has the information which can lead to a decisive solution of these problems. By contrast an exercise in geometry, though it may contain no impressive results, offers a clear and circumscribed problem.

We have pointed out that mathematicians feel free to explore ideas which have only potential usefulness to science as well as ideas which are entirely devoid of physical implications but offer aesthetic pleasure or intellectual challenge. This freedom was enhanced by the creation of non-Euclidean geometry, which taught mathematicians that research on ideas which initially seem outlandish may nevertheless prove valuable to later generations. Mathematicians seem free to create what they will. Is there then no restriction at all on what the mathematician may pursue?

Many mathematicians have indeed drawn the conclusion that any mathematical themes of interest to themselves are worthy of study and of being thrust upon others. In fact, it is quite common to hear them speak today of pure and applied mathematics, the former having no bearing at all on the physical world and the latter being directly related to it. A flood of abstract mathematics is one of the consequences. But the pursuit of higher abstractions and broader generalities with no regard for the study of nature, potential use in this direction, or aesthetic value means an intellectual orgy which may have no more value or appeal than the study of the shapes of horses' hoofs. Felix Klein, one of the great mathematicians of the nineteenth century, hastened to point out that the privilege of freedom implies the obligation of responsibility. Good mathematics must satisfy one of the three criteria, immediate usefulness in science, potential usefulness, or beauty. Since the latter two values are not matters of fact, we find, just as in the case of literature, painting, or architecture, that only good judgment can determine what is a worthwhile undertaking. And just as one encounters excrescences in the former fields, so one does in mathematics.

Exercises

1. What are the motivations for mathematical activity?
2. Defend or attack the thesis that mathematics is an art.
3. What is commonly meant by pure and applied mathematics?
4. Are mathematicians free to create what they wish to?

31-5 Mathematics and rationalism. Among the values which mathematics offers is one which transcends the subject and its relationship to the physical sciences. Mathematics has been the advocate, essence, and embodiment of rationalism. Rationalism is not a whim. It is a spirit which stimulates, invigorates, challenges, and drives human minds to function at the highest mental level in exploring and establishing the deepest implications of knowledge already at hand. It calls for the courage to discard one's dearest and most cherished beliefs if these do not satisfy rational criteria. Mathematics also holds forth the ideal of detachment and of objective judgment. It has fostered independence of thought, adherence to the dictates of reason, careful scrutiny of arguments, and a spirit of criticism

Mathematicians have set the highest standards of reasoning by their persistence and indomitableness. We have seen that they have worked for centuries to *prove* what intuition or experience would have us believe is unquestionable. For two thousand years mathematicians sought an exact method of trisecting an angle, squaring the circle, and doubling the cube, though practical constructions of as great an accuracy as desired were theirs for the asking. In this case the long-sought methods proved to be nonexistent; that is, it is not possible to perform these constructions with straightedge and compass. But the important point at the moment is the search. Likewise for two thousand years mathematicians sought to replace the parallel axiom of Euclid by a more reliable statement. Though in this instance, too, the outcome of their search was a surprise, the important point again is the magnitude and persistence of the effort and the ultimate rejection of many substitute statements which re-examination showed would not do. To reason mathematically is to seek perfection in reasoning. The commonly used phrase "mathematical exactness" pays homage to this ideal of mathematicians.

As man's greatest and most successful intellectual experiment, mathematics demonstrates manifestly how powerful our rational faculty is. It is the finest expression of man's intellectual strength. His reason has, for example, far outstripped his imagination. He can think about stars so distant that only numbers convey any meaning, about spaces which cannot be pictured, and about electrons too small to be seen with the most powerful microscopes.

Rationalism and exact thought patterned after mathematics can be applied to many fields. We have in fact seen how mathematics has provided rational ideals for the social sciences. In this broader sense at least and often more concretely mathematics has penetrated almost all domains of inquiry and has served as the model of all intellectual enterprises. In Plato's time the word mathematics meant rational systematic knowledge, the modern equivalent of which is the German term *Wissenschaft.* Later with Aristotle it came to mean the specific subjects we study in mathematics courses. It is Plato's meaning which the subject of mathematics really encompasses today, even though the word itself still retains the more limited meaning.

31–6 The limitations of mathematics. Yet there are some limitations on what mathematics and mathematical methods seem to be able to achieve. Some aspects of the physical world and of human behavior have not yielded to either. Thus, touch, taste, and smell are sense perceptions which, unlike seeing and hearing, have defied mathematical analysis or even measurement, though the sense organs involved are physiologically simpler than the eye and the ear. Human character, desires, motivations, and emotions are more successfully studied by advertising men than by mathematicians. We have seen that neither axioms nor theorems have been found which would furnish a model of man's behavior, of the operation of his mind, and of the economic and political systems most beneficial to human society. Numbers and geometrical forms do not seem to be the applicable concepts. Only inanimate nature and, in fact, merely portions of that have yielded to mathematical analysis. It is true that statistical methods have given us some ability to predict human characteristics and behavior, but statistical conclusions wipe out all individual nuances and yield only gross effects. There is, of course, still hope that current research on mathematical models may serve other domains of inquiry and in particular solve the problems of society.

One should also question the extent to which mathematics really represents the physical world. The discipline has been effective in treating some abstractions: space, time, mass, velocity, weight, force, the frequency of light and sound, and other such concepts. It treats those physical concepts which can be represented by numbers or geometrical figures. But physical objects possess other properties as well. We do not usually think of human beings as chunks of matter moving in space and time. Nor would a poet or an artist be content to say that the mathematical laws of planetary motion represent the essence of the planets. We have become so accustomed to the analysis of the physical world in terms of space, time, form, mass, and the like, that we tend to overlook the fact that these concepts represent just some properties and narrow ones at that. They cause us to look at the world with blinders. The mathematical approach may not be the deepest possible or the most illuminating; it certainly does not answer the question of whether the solar system is designed for any special ends. Scientists may say that this question does not fall within the province of science, but it is nevertheless a question which human beings would like to see answered. The refusal of scientists to consider it does not wipe out the question, but only reveals a limitation of the mathematical approach.

The plight of man is pitiable. We are wanderers in a vast universe, helpless before the havocs of nature, dependent upon nature for food and other necessities, and uninformed as to why we were born and what to strive for. Man is alone in a cold and alien universe. He gazes upon the mysterious, rapidly changing, and endless world about him and is confused, baffled, and even frightened by his own insignificance. As Pascal put it, "For after all what is man in nature? A nothing in relation to infinity, all in relation to nothing, a central point between nothing and all and infinitely far from understanding either. The end of things and their beginnings are impregnably concealed from him in an impenetrable secret. He is equally incapable of seeing the nothingness out of which he was drawn and the infinite in which he is engulfed." Montaigne and Hobbes said the same thing in other words. The life of man is solitary, poor, nasty, brutish, and short. He is the prey of trivial happenings.

Endowed with a few limited senses and a brain, man began to pierce the mystery about him. By utilizing what the senses reveal immediately or what can be inferred from experiments man adopted axioms and applied his reasoning powers. His quest was the quest for order; his goal, to build sound knowledge as opposed to transient sensations. Amid the chaos of life and his environment, he has sought patterns of explanation and systems of knowledge that might help him to attain some mastery over his environment. The chief tool proved to be the product of man's own reason, and its accomplishments were described by Fourier. "It brings together the most diverse phenomena and discovers hidden conformities which unite them. If matter evades us, such as the air and light, because of its extreme thinness, if objects are located far from us in the immensity of space, if man wishes to understand the performance of the heavens for the successive periods which separate a large number of centuries, if the forces of gravity and of heat be at work in the interior of a solid globe at depths which will be forever inaccessible, mathematical analysis can still grasp the laws of these phenomena. It renders them present and measurable and seems to be a faculty of the human reason destined to make up for the brevity of life and for the imperfection of the senses; and what is more remarkable still, it follows the same method in the study of all phenomena; it interprets them in the same language, as if to affirm the unity and simplicity of the plan of the universe, and to make still more manifest the immutable order which presides over all natural events."

Over the centuries man has created such grand structures as Euclidean geometry, Ptolemaic theory, the heliocentric theory, Newtonian mechanics, electromagnetic theory, and in recent times the theory of relativity and quantum theory. In all of these and in other significant and powerful bodies of science, mathematics, as we now know, is the method of construction, the framework, and indeed the essence. Mathematical theories have enabled us to know something of nature, to embrace in comprehensive intelligible accounts varieties of seemingly diverse phenomena. Mathematical theories have revealed whatever order or plan man has found in nature and have given us mastery or partial mastery over vast domains.

It may be that man has introduced some limited and even artificial concepts and only in this way has managed to institute some order in nature. Man's mathematics may be no more than a workable scheme. Nature itself may be far more complex or have no inherent design. Nevertheless, mathematics remains the method *par excellence* for the investigation, representation, and mastery of nature. In those domains where it is effective it is all we have; if it is not reality itself, it is the closest to reality we can get.

Mathematics then is a formidable and bold bridge between ourselves and the external world. Though it is a purely human creation, the access it has given us to some domains of nature enables us to progress far beyond all expectations. Indeed it is paradoxical that abstractions so remote from reality should achieve so much. Artificial the mathematical account may be, a fairy tale perhaps, but one with a moral.

In the last analysis it is the picture which an age forms of its world which is its most valuable possession, for man seeks primarily to know himself, and this understanding is inseparable from his understanding of the cosmos. The knowledge so gleaned filters through philosophy, literature, religion, the arts, and social thought. It thereby fashions the whole culture and provides whatever answers man has to the major questions he raises about his own life.

RECOMMENDED READING

ELLIS, HAVELOCK: *The Dance of Life,* Chap. 3, The Modern Library, New York, 1929.

HARDY, G. H.: *A Mathematician's Apology,* Cambridge University Press, London, 1940. (Keep a copious quantity of salt on hand while reading this book.)

KLINE, M.: *Mathematics in Western Culture,* Chap. 28, Oxford University Press, New York, 1953.

NEWMAN, JAMES R.: *The World of Mathematics,* Vol. III, pp. 1756–1795, Vol. IV, pp. 2051–2063, Simon and Schuster, Inc., New York, 1956.

POINCARÉ, HENRI: *The Value of Science,* Chaps. 1 to 3, Dover Publications, Inc., New York, 1958.

POINCARÉ, HENRI: *Science and Method,* Chaps. 1 to 3, Dover Publications, Inc., New York, 1952.

RUSSELL, BERTRAND: *Our Knowledge of the External World,* George Allen & Unwin Ltd., London, 1926 (also in paperback).

RUSSELL, BERTRAND: *Mysticism and Logic,* Longmans, Green and Co. New York, 1925.

SAWYER, W. W.: *Prelude to Mathematics,* Chaps. 1 to 3, Penguin Books Ltd., Harmondsworth, England, 1955.

SPENGLER, OSWALD: *Decline of the West,* Vol. I, Chap. 2, A. A. Knopf, Inc., New York, 1926.

SULLIVAN, J. W. N.: *Aspects of Science,* Second Series, pp. 80–105, A. A. Knopf, Inc., New York, 1926.

Table of Trigonometric Ratios

Angle	Sine	Tangent	Cotangent	Cosine	
0°	0.0000	0.0000	....	1.0000	90°
1	0.0175	0.0175	57.290	0.9998	89
2	0.0349	0.0349	28.636	0.9994	88
3	0.0523	0.0524	19.081	0.9986	87
4	0.0698	0.0699	14.300	0.9976	86
5	0.0872	0.0875	11.430	0.9962	85
6	0.1045	0.1051	9.5144	0.9945	84
7	0.1219	0.1228	8.1443	0.9925	83
8	0.1392	0.1405	7.1154	0.9903	82
9	0.1564	0.1584	6.3138	0.9877	81
10	0.1736	0.1763	5.6713	0.9848	80
11	0.1908	0.1944	5.1446	0.9816	79
12	0.2079	0.2126	4.7046	0.9781	78
13	0.2250	0.2309	4.3315	0.9744	77
14	0.2419	0.2493	4.0108	0.9703	76
15	0.2588	0.2679	3.7321	0.9659	75
16	0.2756	0.2867	3.4874	0.9613	74
17	0.2924	0.3057	3.2709	0.9563	73
18	0.3090	0.3249	3.0777	0.9511	72
19	0.3256	0.3443	2.9042	0.9455	71
20	0.3420	0.3640	2.7475	0.9397	70
21	0.3584	0.3839	2.6051	0.9336	69
22	0.3746	0.4040	2.4751	0.9272	68
23	0.3907	0.4245	2.3559	0.9205	67
24	0.4067	0.4452	2.2460	0.9135	66
25	0.4226	0.4663	2.1445	0.9063	65
26	0.4384	0.4877	2.0503	0.8988	64
27	0.4540	0.5095	1.9626	0.8910	63
28	0.4695	0.5317	1.8807	0.8829	62
29	0.4848	0.5543	1.8040	0.8746	61
30	0.5000	0.5774	1.7321	0.8660	60
31	0.5150	0.6009	1.6643	0.8572	59
32	0.5299	0.6249	1.6003	0.8480	58
33	0.5446	0.6494	1.5399	0.8387	57
34	0.5592	0.6745	1.4826	0.8290	56
35	0.5736	0.7002	1.4281	0.8192	55
36	0.5878	0.7265	1.3764	0.8090	54
37	0.6018	0.7536	1.3270	0.7986	53
38	0.6157	0.7813	1.2799	0.7880	52
39	0.6293	0.8098	1.2349	0.7771	51
40	0.6428	0.8391	1.1918	0.7660	50
41	0.6561	0.8693	1.1504	0.7547	49
42	0.6691	0.9004	1.1106	0.7431	48
43	0.6820	0.9325	1.0724	0.7314	47
44	0.6947	0.9657	1.0355	0.7193	46
45°	0.7071	1.0000	1.0000	0.7071	45°
	Cosine	Cotangent	Tangent	Sine	Angle

ANSWERS TO SELECTED EXERCISES

ANSWERS TO SELECTED EXERCISES

Section 3–4

7. (a) no (b) yes (c) no (d) no (e) no (f) yes (g) yes (h) no (i) no

Section 4–3

Second set:

1. (a) $\sqrt{3}+\sqrt{5}$ (b) $\sqrt[3]{2}+\sqrt[3]{7}$ (c) $\sqrt[3]{2}+\sqrt[3]{7}$ (d) $-2a$ (e) $\sqrt{21}$ (f) $\sqrt[3]{10}$ (g) 2 (h) $a^2 - ab$ (i) $\sqrt{\frac{5}{2}}$ (j) 2 (k) $\sqrt[3]{5}$
2. (a) $5\sqrt{2}$ (b) $10\sqrt{2}$ (c) $5\sqrt{3}$

Section 4–4

1. -2
2. -13
3. $+3$
4. -5
5. 500; 500

Section 4–5

3. (a) $12a$ (b) $12a$ (c) $5a+\sqrt{2}\,a$ (d) $-2a$ (e) $6a+12b$ (f) $28a+35b$ (g) a^2+ab (h) a^2-ab (i) $16a$ (j) a^2b
4. a^2+5a+6
5. n^2+2n+1
6. yes
7. yes
8. yes
9. no

Section 4–6

First set:

1. no, $4\frac{1}{2}$ hr
2. no
3. $\dfrac{15a+10b}{6(a+b)}$
4. no
5. $\frac{5}{12}$ of a ditch per day

Second set:

1. 16
2. 12
3. 12
4. 36

Third set:

4. 13; 14; 20; 100; 120; 244
5. 5; 6; 8; 12; 36
6. 0.3
7. $\frac{1}{3}$
8. 3
9. 6

Section 5–2

5. (a) $3x+4$; (b) $3x^2+4$

Section 5–3

1. (a) $15x^2$
   (b) $x^2 + 9x + 20$
   (c) $3x^2 + 19x + 20$
   (d) $x^2 - 9$
   (e) $x^2 + 5x + \frac{25}{4}$
   (f) $x^2 - \frac{25}{4}$
2. (a) $(x + 5)(x + 4)$
   (b) $(x + 2)(x + 3)$
   (c) $(x - 2)(x - 3)$
   (d) $(x + 3)(x - 3)$
   (e) $(x + 4)(x - 4)$
   (f) $(x + 9)(x - 2)$

Section 5–4

1. 1177 ft
2. 22 mi/hr
3. 73
4. 50
5. 80
6. 1250

Section 5–5

First set:

1. (a) 6, 2
   (b) -9, 2
3. (a) $-6 \pm 3\sqrt{3}$
   (b) $6 \pm 3\sqrt{3}$

Second set:

1. (a) $4 + \sqrt{6}, 4 - \sqrt{6}$
   (b) $-4 + \sqrt{6}, -4 - \sqrt{6}$
   (c) $3 + 3\sqrt{2}, 3 - 3\sqrt{2}$
   (d) $-3, -1$
   (e) 4, 4

Section 6–2

8. 103 ft
10. 2π ft

Section 6–3

2. 418,500 mi
3. 100 by 100
4. $\frac{p}{4}$ by $\frac{p}{4}$
5. 25 by 50
7. $\sqrt{2Rh + h^2}$
8. $\sqrt{2Rh}$
9. 63 mi

Section 6–5

2. 2
4. 5

Section 7–2

3. 1 to 0
4. 0 to unlimitedly larger and larger values

Section 7–3

1. 445 ft
2. 14,265 ft
3. 11,500 ft
4. yes, by 19 ft
5. 3944 mi

Section 7–4

1. 30°
5. 139 mi
6. 263 mi
7. 18,960 mi

Section 7–5

1. 93,000,000 mi
2. 428,000 mi
3. 1065 mi
4. 36,000,000 mi

Section 7–6

1. 32°
2. 0° to 42°
4. 172,000 mi/sec

Section 9–5

3. 7.1 and 9.8
4. 137,256
5. (a) 150 (b) $2\frac{11}{12}$ hr

Section 11–2

4. $\frac{5}{4}$

Section 12–7

11. 165 years

Section 13–3

3. (a) $y = 2x$ (b) $y = \dfrac{1}{\sqrt{3}}x$ (c) $y = -4x$ (d) $y = 4x$ (e) $y = -4x$
5. Yes
8. Yes
10. m
11. (0, 7)
12. $(0, b)$
13. m; $(0, b)$

Section 13–4

3. $x = -\frac{1}{12}y^2$
4. (0, 2)
5. (a) $y = \frac{1}{16}x^2$ (b) $y = \frac{1}{24}x^2$ (c) $y = -\frac{1}{20}x^2$ (d) $x = \frac{1}{16}y^2$
6. 1, 4, 9, 16, 25
7. $1\frac{1}{10}$, $4\frac{2}{5}$, $9\frac{9}{10}$, $17\frac{3}{5}$, $27\frac{1}{2}$

Section 13–6

3. 10
5. 2, 8

Section 13–7

2. Surfaces
3. A sphere
4. A plane

Section 15–2

4. 0, 96, 224, etc.
6. 100, 256, 784
7. 2, $2\frac{1}{2}$, 4
8. +3, −3
9. 32, 212

Section 15–3

3. $10\frac{2}{3}$ mi/hr
4. 128 ft/sec; 64 ft/sec; $t = 2$
7. 400 ft; 676 ft; 1600 ft
11. 256 ft/sec, approximately
15. 121 ft
16. 121 ft
17. $d = v^2/22$
18. 352 ft; 88 ft
19. 576 ft

Section 15–4

3. 192 ft/sec, 432 ft; 240 ft/sec, 756 ft

Section 15–5

2. (a) 384 ft (b) 32 ft/sec (c) 400 ft
3. 0
5. 15,625 ft
7. 870 ft, 18.1 sec
8. 960 ft/sec
11. 7 sec

Section 15–6

2. 26.5 lb
3. 4320 lb

Section 15–7

1. 304 ft/sec; 1520 ft
2. $d = (14.5)t^2$

Section 15–8

1. 72 ft; 144 ft
3. 1710 lb
4. $33\frac{1}{3}$ lb
5. 76.616 lb
6. 32 ft/sec

Section 15–9

3. 50 ft-lb

Section 16–2

1. $y = 3x/2$
2. $y = 5x^2/16$
3. $y = 5x + x^2$

Section 16–3

3. 5.5 sec, approximately; 550 ft, approximately
4. 3000 ft/sec
5. 1200 ft/sec
6. 5450 ft, approximately

Section 16–4

1. 230 ft/sec; 193 ft/sec
2. $y = -x^2/25 + 3x/2$
4. 37.5 ft
5. 14.1 ft, approximately
6. 1000 ft/sec

Section 16–5

1. $y = -16x^2/V^2 \cos^2 A + x \sin A/\cos A$
2. $V \sin A/32$
3. $V^2 \sin^2 A/64$
5. 123,000 ft, approximately
6. 20,000 ft
7. $(62.5)\sqrt{2}$ sec

Section 17–4

4. 20.5 ft/sec^2, approximately
6. 150 lb
7. 1125 lb

Section 17–5

1. 37.5 lb
4. 15 lb
7. 2400 lb
10. Yes

Section 17–6

3. 500 ft/sec^2
4. 0.00897 ft/sec^2, approximately

Section 17–7

2. 26,000 ft/sec, approximately
3. 3300 ft/sec, approximately
4. 27.8 days, approximately
5. 24,000 ft/sec, approximately

Section 18–5

4. 80 ft/sec; 144 ft/sec
5. 160 ft/sec
6. 32 ft/sec

Section 18–6

3. (a) $4x$ (b) $4t$ (c) x
   (d) $12x^2$ (e) $-4x$ (f) $-32t$
   (g) $-32t + 128$ (h) $128 - 32t$
4. 228 ft/sec
5. (b) $\dot{V} = 4\pi r^2$
6. (a) $\dot{y} = 4ax^3$
7. $\dot{A} = 2x$
8. $\dot{A} = l$

Section 18–7

1. (a) $\frac{3}{50}$ (c) 0
2. (a) 2 (b) 2

Section 18–8

1. 0
3. 25 by 50
4. 25 by $\frac{50}{3}$
5. $r = \sqrt{50/3\pi}$, $h = 2r$

Section 19–2

1. (a) x^3 (b) $5x$ (c) $x^2/2$
   (d) $3x^2/2$ (e) t^2 (f) $16t^2$
   (g) $32t$ (h) $t^2 + 10t$ (i) $-16t^2 + 128t$
   (j) $-32t$ (k) $16t^2$

Section 19–3

1. $150 - 32t$, $150t - 16t^2$
2. $16t^2 + 50$
3. $-32t$, $75 - 16t^2$
4. $100 - 32t$, $100t - 16t^2 + 50$
5. $-100 - 32t$, $50 - 100t - 16t^2$

Section 19–4

1. $69\frac{1}{3}$
2. $50\frac{2}{3}$
3. 10
4. 90

Section 19–5

1. 257,561,000 ft-lb
2. 264,000,000 ft-lb
3. 40,000 ft-lb

Section 19–6

1. 36,500 ft/sec, approximately

Section 23–3

First set:

1. (a) $\sin 60°$ (b) $\sin 30°$ (c) $-\sin 30°$
   (d) $-\sin 80°$ (e) $-\sin 90°$ (f) $-\sin 60°$
   (g) $-\sin 10°$ (h) $\sin 10°$ (i) $-\sin 50°$
   (j) $\sin 30°$
2. 1, -1 3. 90° 8. 2

Second set:

1. $\pi/2$ or 1.57, approximately; $\pi/6$ or 0.52, approximately; π, $3\pi/2$, 2π, $8\pi/3$
2. 90°, 120°, 450°, 540°, $-90°$, 57°, approximately
3. (a) 0 (b) 1 (c) $\sqrt{3}/2$
   (d) -1 (e) 0 (f) 1

Third set:

1. 1, 2, 2, $\sqrt{3}$ 2. 3, -3 3. 4
4. (a) 1, 0, -1, 0 (b) 1, 0, -1, 0
8. 10
9. (a) 1, 0, $-\sqrt{3}/2$ (b) $-\sqrt{3}/2$, 0, 0 (c) 0, -2, 2

Section 23–4

1. 128 2. $5/2\pi$ 3. 0.01 sec
4. $y = 3 \sin 2\pi \cdot 50t$ 5. $y = 3 \sin 5t$ 6. $50/4\pi^2$

Section 24–2

2. $y = 0.0005 \sin 2\pi \cdot 300t$
3. 540, 0.002 4. 20

Section 24–3

2. (a) 1 (b) 1 (c) 1 (d) 100

Section 24–4

2. 240, 0.01
3. $y = 0.01 \sin (2\pi\ 500t) + 0.002 \sin(2\pi\ 1000t) + 0.005 \sin(2\pi\ 1500t)$
4. 2160

Section 25–3

4. 3 rps 6. 60
7. $I = 0.1 \sin 2\pi \cdot 30t$ 8. 50, 50

Section 27–3

6. 9 + any positive or negative multiple of 12
12. 1, 3, 5, and 7 plus any positive multiple of 8

Section 27–4

4. A
5. A

Section 29–3

2. 8113, 1200
3. 6.1, 5, 10
4. 31, 35, 40

Section 29–4

1. 3.24
2. 3
3. 12.8

Section 29–5

3. 95.4
4. 47.7, 99.8

Section 29–6

1. $y = 3x + 7$
2. $F = 4d$
3. $d = 16t^2$
4. $N = Y - 1944$
5. $P = 2t^2 + 4$
6. $V = 2.3 - 0.675T + 0.0875T^2$

Section 30–2

3. $\frac{1}{6}, \frac{1}{6}, \frac{4}{6}$
5. $\frac{3}{5}$
6. $\frac{1}{4}$
7. $\frac{1}{4}$
8. $\frac{27}{32}$
9. $\frac{1}{2}$
11. 36
12. 4
13. $\frac{4}{36}$
17. 1 to 1
18. 3 to 1

Section 30–3

2. 0.85, 0.68
3. $\frac{1}{8}$

Section 30–4

1. 0.499, 0.954
2. 0.159
3. 0.023
4. 0.341
5. 0.136

Section 30–5

1. $\frac{35}{128}$
2. $\frac{15}{64}$
3. $2000\ (\frac{15}{64})$
4. 0.977
5. 0.999
7. Discredit

INDEX

INDEX

CDEFGHIJKL 609876